Student Solutions Manual for
Stewart, Redlin, and Watson's

ALGEBRA AND
TRIGONOMETRY

John A. Banks
EVERGREEN VALLEY COLLEGE
SAN JOSE CITY COLLEGE

BROOKS/COLE

THOMSON LEARNING

Australia • Canada • Mexico • Singapore • Spain • United Kingdom • United States

BROOKS/COLE

THOMSON LEARNING ™

Assistant Editor: Carol Ann Benedict
Marketing Manager: Karin Sandberg
Editorial Assistant: Dan Thiem
Production Coordinator: Stephanie Andersen
Permissions Editor: Sue Ewing

Cover Design: Roy Neuhaus
Cover Illustration: Bill Ralph
Print Buyer: Micky Lawler
Printing and Binding: Webcom Limited

For more information about this or any other Brooks/Cole products, contact:
BROOKS/COLE
511 Forest Lodge Road
Pacific Grove, CA 93950 USA
www.brookscole.com
1-800-423-0563 (Thomson Learning Academic Resource Center)

ISBN: 0-534-38266-5

Table of Contents

Chapter 1   1

Chapter 2   27

Chapter 3   55

Chapter 4   102

Chapter 5   150

Chapter 6   202

Chapter 7   232

Chapter 8   253

Chapter 9   274

Chapter 10   313

Chapter 11   360

Chapter 12   406

Chapter 13   434

Chapter One
Exercises 1.1

1. $a \cdot b = b \cdot a$

3. $(a \cdot b) \cdot c = a \cdot (b \cdot c)$

5. $(a - b)(a + b) = a^2 - b^2$

7. $(a \cdot b)^2 = a^2 \cdot b^2$

9. The average of two numbers, x and y, is $a = \dfrac{x + y}{2}$.

11. The sum of a number and twice its square is $S = x + 2x^2$, where x is the number.

13. The number of days in w weeks is $d = 7w$.

15. The product of two consecutive integers is $P = n(n + 1)$, where n is the smaller integer.

17. The sum of the squares of two numbers, x and y, is $s = x^2 + y^2$.

19. The time it takes an airplane to travel d miles at r miles per hour is $t = \frac{d}{r}$.

21. The area of a square of side x is $A = x^2$.

23. The volume of a box with square base of side x and height $2x$ is $V = x \cdot x \cdot 2x = 2x^3$.

25. A box of length l, width w, and height h has two sides of area $l \cdot h$, two sides of area $l \cdot w$, and two ends with area $w \cdot h$. So the surface area is $S = 2lh + 2lw + 2wh$.

27. The race track consists of a rectangle of length x and width $2r$ and two semicircles of radius r. Each semi circle has length πr and two straight runs of length x. So the length is $L = 2x + 2\pi r$.

29. The area of a triangle is $A = \frac{1}{2}$base $\cdot$ height. Since the base is twice the height h we get
$A = (\frac{1}{2})(2h)(h) = h^2$.

31. Since the final exam counts double we must divide the total by 4 (2 exams and 2 for the final).

    (a) Average $= \dfrac{79 + 83 + 2(88)}{4} = 84.5$.

    (b) Average $= \dfrac{a + b + 2f}{4}$.

    (c) Average $= \dfrac{79 + 83 + 2f}{4} = 85 \Rightarrow 79 + 83 + 2f = 340 \Rightarrow 2f = 178 \Rightarrow f = 89$.

33. (a) The area that she can mow is $150\,\dfrac{\text{ft}^2}{\text{min}} \times 30\,\text{min} = 4{,}500\ \text{ft}^2$.

    (b) Area $= 150\,\dfrac{\text{ft}^2}{\text{min}} \times T\text{min} = 150T\ \text{ft}^2$.

    (c) The area of the lawn $= 80\,\text{ft} \times 120\,\text{ft} = 9{,}600\ \text{ft}^2$. If T is the time required, then from part (b),
    $150T = 9600$, so $T = \dfrac{9600}{150} = 64\ \text{min}$.

(d)   If R is the rate of mowing, then $60R = 9600$, so $R = \dfrac{9600}{60} = 160 \; \dfrac{\text{ft}^2}{\text{min}}$.

35.   (a) $\text{GPA} = \dfrac{4a + 3b + 2c + 1d + 0f}{a + b + c + d + f} = \dfrac{4a + 3b + 2c + d}{a + b + c + d + f}$.

(b)   Using $a = 2 \cdot 3 = 6$, $b = 4$, $c = 3 \cdot 3 = 9$, and $d = f = 0$ in the formula from part (a), we

obtain $\text{GPA} = \dfrac{4 \cdot 6 + 3 \cdot 4 + 2 \cdot 9}{6 + 4 + 9} = \dfrac{54}{19} = 2.84$.

Exercises 1.2

1. Commutative Property for addition.

3. Associative Property for addition.

5. Distributive Property.

7. $3(x + y) = 3x + 3y$

9. $4(2m) = (4 \cdot 2)m = 8m$

11. $-\dfrac{5}{2}(2x - 4y) = -\dfrac{5}{2}(2x) + \dfrac{5}{2}(4y) = -5x + 10y$

13. (a) $\dfrac{4}{13} + \dfrac{3}{13} = \dfrac{7}{13}$

    (b) $\dfrac{3}{10} + \dfrac{7}{15} = \dfrac{9}{30} + \dfrac{14}{30} = \dfrac{23}{30}$

15. (a) $\dfrac{2}{5} \div \dfrac{9}{10} = \dfrac{2}{5} \cdot \dfrac{10}{9} = \dfrac{4}{9}$

    (b) $\left(4 \div \dfrac{1}{2}\right) - \dfrac{1}{2} = \left(4 \cdot \dfrac{2}{1}\right) - \dfrac{1}{2} = 8 - \dfrac{1}{2} = 7\dfrac{1}{2}$ or $\dfrac{15}{2}$

17. (a) False

    (b) True

19. (a) False

    (b) True

21. (a) $x > 0$

    (b) $t < 4$

    (c) $a \geq \pi$

    (d) $-5 < x < \frac{1}{3}$

    (e) $|p - 3| \leq 5$

23. (a) $A \cup B = \{1, 2, 3, 4, 5, 6, 7, 8\}$

    (b) $A \cap B = \{2, 4, 6\}$

25. (a) $A \cup C = \{1, 2, 3, 4, 5, 6, 7, 8, 9, 10\}$

    (b) $A \cap C = \emptyset$

27. (a) $B \cup C = \{x \mid x \leq 5\}$

    (b) $B \cap C = \{x \mid -1 < x < 4\}$

29. $(-3, 0) = \{x \mid -3 < x < 0\}$

31. $[2, 8) = \{x \mid 2 \leq x < 8\}$

33. $[2, \infty) = \{x \mid x \geq 2\}$

35. $x \leq 1 \quad \Leftrightarrow \quad x \in (-\infty, 1]$

37. $-2 < x \leq 1 \quad \Leftrightarrow \quad x \in (-2, 1]$

39. $x > -1 \quad \Leftrightarrow \quad x \in (-1, \infty)$

41. $(-2, 0) \cup (-1, 1) = (-2, 1)$

43. $[-4, 6] \cap [0, 8) = [0, 6]$

45. $(-\infty, -4) \cup (4, \infty)$

47. (a) $|100| = 100$

    (b) $|-73| = 73$

49. (a) $\bigl||-6| - |-4|\bigr| = |6 - 4| = |2| = 2$

    (b) $\dfrac{-1}{|-1|} = \dfrac{-1}{1} = -1$

51. (a) $|(-2) \cdot 6| = |-12| = 12$

    (b) $\left|\left(-\frac{1}{3}\right)(-15)\right| = |5| = 5$

53. (a) $|17 - 2| = 15$

    (b) $|21 - (-3)| = |21 + 3| = |24| = 24$

(c) $\left|-\frac{3}{10} - \frac{11}{8}\right| = \left|-\frac{12}{40} - \frac{55}{40}\right| = \left|-\frac{67}{40}\right| = \frac{67}{40}$

55. (a) Let $x = 0.777\ldots$ So,

$$10x = 7.7777\ldots$$
$$\underline{x = 0.7777\ldots}$$
$$9x = 7$$

Thus $x = \frac{7}{9}$.

(b) Let $x = 0.2888\ldots$ So,

$$100x = 28.8888\ldots$$
$$\underline{10x = 2.8888\ldots}$$
$$90x = 26$$

Thus $x = \frac{26}{90} = \frac{13}{45}$.

(c) Let $x = 0.575757\ldots$ So,

$$100x = 57.5757\ldots$$
$$\underline{x = 0.5757\ldots}$$
$$99x = 57$$

Thus $x = \frac{57}{99} = \frac{19}{33}$.

57. (a) Is $\frac{1}{28}x + \frac{1}{34}y \le 15$ when $x = 165$ and $y = 230$? We have
$\frac{1}{28}(165) + \frac{1}{34}(230) = 5.89 + 6.76 = 12.65 \le 15$, so yes the car can travel 165 city miles and 230 highway miles without running out of gas.

(b) Here we must solve for y when $x = 280$. So $\frac{1}{28}(280) + \frac{1}{34}y = 15$ $\Leftrightarrow$ $10 + \frac{1}{34}y = 15$ $\Leftrightarrow$ $\frac{1}{34}y = 5$ $\Leftrightarrow$ $y = 170$. Thus the car can travel up to 170 highway miles without running out of gas.

59.

x	1	2	10	100	1000
$\dfrac{1}{x}$	1	$\dfrac{1}{2}$	$\dfrac{1}{10}$	$\dfrac{1}{100}$	$\dfrac{1}{1000}$

As x gets large, the fraction $\dfrac{1}{x}$ gets small. Mathematically, we say that $\dfrac{1}{x}$ goes to zero.

x	1	0.5	0.1	0.01	0.001
$\dfrac{1}{x}$	1	$\dfrac{1}{0.5} = 2$	$\dfrac{1}{0.1} = 10$	$\dfrac{1}{0.01} = 100$	$\dfrac{1}{0.001} = 1000$

As x gets small, the fraction $\dfrac{1}{x}$ gets large. Mathematically, we say that $\dfrac{1}{x}$ goes to infinity.

Exercises 1.3

1.   (a) $(-2)^4 = 16$ (b) $-2^4 = -16$

    (c) $(-2)^0 = 1$

3.   (a) $2^4 5^{-2} = \dfrac{2^4}{5^2} = \dfrac{16}{25}$ (b) $\dfrac{10^7}{10^4} = 10^{7-4} = 10^3 = 1000$

    (c) $\left(2^3 \cdot 2^2\right)^2 = \left(2^5\right)^2 = 2^{10} = 1024$

5.   (a) $\sqrt{\dfrac{4}{9}} = \dfrac{\sqrt{4}}{\sqrt{9}} = \dfrac{2}{3}$ (b) $\sqrt[4]{256} = \sqrt[4]{4^4} = 4$

    (c) $\sqrt[6]{\dfrac{1}{64}} = \sqrt[6]{\dfrac{1}{2^6}} = \dfrac{\sqrt[6]{1}}{\sqrt[6]{2^6}} = \dfrac{1}{2}$

7.   (a) $\dfrac{\sqrt{72}}{\sqrt{2}} = \sqrt{\dfrac{72}{2}} = \sqrt{36} = 6$ (b) $\dfrac{\sqrt{48}}{\sqrt{3}} = \sqrt{\dfrac{48}{3}} = \sqrt{16} = 4$

    (c) $\sqrt{\dfrac{9}{25}} = \dfrac{\sqrt{9}}{\sqrt{25}} = \dfrac{3}{5}$

9.   (a) $\left(\dfrac{4}{9}\right)^{-1/2} = \left(\dfrac{2^2}{3^2}\right)^{-1/2} = \dfrac{2^{-1}}{3^{-1}} = \dfrac{3}{2}$

    (b) $\left(-\dfrac{27}{8}\right)^{2/3} = \left(\sqrt[3]{-\dfrac{27}{8}}\right)^2 = \left(-\dfrac{3}{2}\right)^2 = \dfrac{9}{4}$

    (c) $\left(\dfrac{25}{64}\right)^{3/2} = \left(\sqrt{\dfrac{25}{64}}\right)^3 = \left(\dfrac{5}{8}\right)^3 = \dfrac{125}{512}$

11. $\sqrt[3]{108} - \sqrt[3]{32} = 3\sqrt[3]{4} - 2\sqrt[3]{4} = \sqrt[3]{4}$ 13. $\sqrt{245} - \sqrt{125} = 7\sqrt{5} - 5\sqrt{5} = 2\sqrt{5}$

15. $a^9 a^{-5} = a^{9-5} = a^4$ 17. $\left(12x^2 y^4\right)\left(\tfrac{1}{2}x^5 y\right) = \left(12 \cdot \tfrac{1}{2}\right)x^{2+5}y^{4+1} = 6x^7 y^5$

19. $\dfrac{x^9 (2x)^4}{x^3} = 2^4 \cdot x^{9+4-3} = 16x^{10}$

21. $b^4 \left(\dfrac{1}{3}b^2\right)\left(12b^{-8}\right) = \dfrac{12}{3}b^{4+2-8} = 4b^{-2} = \dfrac{4}{b^2}$

23. $(rs)^3 (2s)^{-2}(4r)^4 = r^3 s^3 2^{-2} s^{-2} 4^4 r^4 = r^3 s^3 2^{-2} s^{-2} 2^{2\cdot 4} r^4 = 2^{-2+8} r^{3+4} s^{3-2} = 2^6 r^7 s = 64r^7 s$

25. $\dfrac{(6y^3)^4}{2y^5} = \dfrac{6^4 y^{3\cdot 4}}{2y^5} = \dfrac{6^4}{2}\, y^{12-5} = 648y^7$

27. $\dfrac{(x^2 y^3)^4 (xy^4)^{-3}}{x^2 y} = \dfrac{x^8 y^{12} x^{-3} y^{-12}}{x^2 y} = x^{8-3-2} y^{12-12-1} = x^3 y^{-1} = \dfrac{x^3}{y}$

29. $\dfrac{\left(xy^2z^3\right)^4}{\left(x^3y^2z\right)^3} = \dfrac{x^4y^8z^{12}}{x^9y^6z^3} = x^{4-9}y^{8-6}z^{12-3} = x^{-5}y^2z^9 = \dfrac{y^2z^9}{x^5}$

31. $\left(\dfrac{q^{-1}rs^{-2}}{r^{-5}sq^{-8}}\right)^{-1} = \dfrac{qr^{-1}s^2}{r^5s^{-1}q^8} = q^{1-8}r^{-1-5}s^{2-(-1)} = q^{-7}r^{-6}s^3 = \dfrac{s^3}{q^7r^6}$

33. $x^{2/3}x^{1/5} = x^{(10/15+3/15)} = x^{13/15}$

35. $(4b)^{1/2}\left(8b^{2/5}\right) = \sqrt{4}\cdot 8b^{1/2}b^{2/5} = 16b^{(5/10+4/10)} = 16b^{9/10}$

37. $\left(c^2d^3\right)^{-1/3} = c^{-2/3}d^{-1} = \dfrac{1}{c^{2/3}d}$　　　　　39. $\left(y^{3/4}\right)^{2/3} = y^{(3/4)\cdot(2/3)} = y^{1/2}$

41. $\left(2x^4y^{-4/5}\right)^3\left(8y^2\right)^{2/3} = 2^3x^{12}y^{-12/5}8^{2/3}y^{4/3} = 2^{3+2}x^{12}y^{(-12/5+4/3)} = \dfrac{32x^{12}}{y^{16/15}}$ (Note that

$8^{2/3} = \left(8^{1/3}\right)^2 = 2^2$.)

43. $\left(\dfrac{x^6y}{y^4}\right)^{5/2} = \dfrac{x^{15}y^{5/2}}{y^{10}} = x^{15}y^{5/2-10} = x^{15}y^{-15/2} = \dfrac{x^{15}}{y^{15/2}}$

45. $\left(\dfrac{3a^{-2}}{4b^{-1/3}}\right)^{-1} = \dfrac{3^{-1}a^2}{4^{-1}b^{1/3}} = \dfrac{4a^2}{3b^{1/3}}$

47. $\dfrac{(9st)^{3/2}}{(27s^3t^{-4})^{2/3}} = \dfrac{27s^{3/2}t^{3/2}}{9s^2t^{-8/3}} = 3s^{3/2-2}t^{3/2+8/3} = 3s^{-1/2}t^{25/6} = \dfrac{3t^{25/6}}{s^{1/2}}$

49. $\sqrt[4]{x^4} = |x|$　　　　　　　　　　　51. $\sqrt[3]{x^3y} = \left(x^3\right)^{1/3}y^{1/3} = x\sqrt[3]{y}$

53. $\sqrt[5]{a^6b^7} = a^{6/5}b^{7/5} = a\cdot a^{1/5}b\cdot b^{2/5} = ab\sqrt[5]{ab^2}$

55. $\sqrt[3]{\sqrt{64x^6}} = (8|x^3|)^{1/3} = 2|x|$

57. (a) $\dfrac{1}{\sqrt{6}} = \dfrac{1}{\sqrt{6}}\cdot\dfrac{\sqrt{6}}{\sqrt{6}} = \dfrac{\sqrt{6}}{6}$　　　　　(b) $\sqrt{\dfrac{x}{3y}} = \dfrac{\sqrt{x}}{\sqrt{3y}} = \dfrac{\sqrt{x}}{\sqrt{3y}}\cdot\dfrac{\sqrt{3y}}{\sqrt{3y}} = \dfrac{\sqrt{3xy}}{3y}$

　　(c) $\sqrt{\dfrac{3}{20}} = \dfrac{\sqrt{3}}{2\sqrt{5}}\cdot\dfrac{\sqrt{5}}{\sqrt{5}} = \dfrac{\sqrt{3}\sqrt{5}}{10} = \dfrac{\sqrt{15}}{10}$

59. (a) $\dfrac{1}{\sqrt[3]{x}} = \dfrac{1}{\sqrt[3]{x}}\cdot\dfrac{\sqrt[3]{x^2}}{\sqrt[3]{x^2}} = \dfrac{\sqrt[3]{x^2}}{x}$　　　　　(b) $\dfrac{1}{\sqrt[5]{x^2}} = \dfrac{1}{\sqrt[5]{x^2}}\cdot\dfrac{\sqrt[5]{x^3}}{\sqrt[5]{x^3}} = \dfrac{\sqrt[5]{x^3}}{x}$

　　(c) $\dfrac{1}{\sqrt[7]{x^3}} = \dfrac{1}{\sqrt[7]{x^3}}\cdot\dfrac{\sqrt[7]{x^4}}{\sqrt[7]{x^4}} = \dfrac{\sqrt[7]{x^4}}{x}$

61. (a) $69,300,000 = 6.93\times 10^7$　　　　　(b) $0.000028536 = 2.8536\times 10^{-5}$
　　(c) $129,540,000 = 1.2954\times 10^8$

63. (a) $3.19\times 10^5 = 319,000$　　　　　　(b) $2.670\times 10^{-8} = 0.0000000267$
　　(c) $7.1\times 10^{14} = 710,000,000,000,000$

65. (a) $5,900,000,000,000$ mi $= 5.9\times 10^{12}$ mi　(b) 0.0000000000004 cm $= 4\times 10^{-13}$ cm

(c)   33 billion billion molecules $= 33 \times 10^9 \times 10^9 = 3.3 \times 10^{19}$ molecules

67. $(7.2 \times 10^{-9})(1.806 \times 10^{-12}) = 7.2 \times 1.806 \times 10^{-9} \times 10^{-12} \approx 13.0 \times 10^{-21} = 1.3 \times 10^{-20}$

69. $\dfrac{1.295643 \times 10^9}{(3.610 \times 10^{-17})(2.511 \times 10^6)} = \dfrac{1.295643}{3.610 \times 2.511} \times 10^{9+17-6} \approx 0.1429 \times 10^{19} = 1.429 \times 10^{19}$

71. $\dfrac{(0.0000162)(0.01582)}{(594621000)(0.0058)} = \dfrac{(1.62 \times 10^{-5})(1.582 \times 10^{-2})}{(5.94621 \times 10^8)(5.8 \times 10^{-3})} = \dfrac{1.62 \times 1.582}{5.94621 \times 5.8} \times 10^{-5-2-8+3}$
$0.074 \times 10^{-12} = 7.4 \times 10^{-14}$

73. $9.3 \times 10^7 \text{mi} = 186,000\dfrac{\text{mi}}{\text{sec}} \times t \text{ sec} \quad \Leftrightarrow \quad t = \dfrac{9.3 \times 10^7}{186,000} \text{ sec} = 500 \text{ sec} = 8\dfrac{1}{3} \text{ min.}$

75.   First convert 1135 feet to miles. This gives 1135 ft $= 1135 \cdot \frac{1 \text{ mile}}{5280 \text{ feet}} = 0.215$ mi. Thus the distance
you can see is given by $D = \sqrt{2rh + h^2} = \sqrt{2(3960)(.215) + (.215)^2} = \sqrt{1702.8} = 41.3$ miles.

77.   (a)

n	1	2	5	10	100
$2^{1/n}$	$2^{1/1} = 2$	$2^{1/2} = 1.414$	$2^{1/5} = 1.149$	$2^{1/10} = 1.072$	$2^{1/100} = 1.007$

So when n gets large, $2^{1/n}$ decreases to 1.

(b)

n	1	2	5	10
$\left(\frac{1}{2}\right)^{1/n}$	$\left(\frac{1}{2}\right)^{1/1} = 0.5$	$\left(\frac{1}{2}\right)^{1/2} = 0.707$	$\left(\frac{1}{2}\right)^{1/5} = 0.871$	$\left(\frac{1}{2}\right)^{1/10} = 0.933$

100
$\left(\frac{1}{2}\right)^{1/100} = 0.993$

So when n gets large, $\left(\frac{1}{2}\right)^{1/n}$ increases to 1.

79.   (a) $\dfrac{18^5}{9^5} = \left(\dfrac{18}{9}\right)^5 = 2^5 = 32$

(b) $20^6 \cdot (0.5)^6 = (20 \cdot 0.5)^6 = 10^6 = 1,000,000$

81.   Using $v = \dfrac{1}{10}c$ in the formula $m = \dfrac{m_0}{\sqrt{1 - \dfrac{v^2}{c^2}}}$, we get: $m = \dfrac{m_0}{\sqrt{1 - \dfrac{\left(\frac{1}{10}c\right)^2}{c^2}}} = \dfrac{m_0}{\sqrt{1 - \frac{1}{100}}}$. Thus

$m = \dfrac{1}{\sqrt{\frac{99}{100}}}m_0 = \dfrac{10\sqrt{11}}{33}m_0$. So the rest mass of the spaceship is multiplied by $\dfrac{10\sqrt{11}}{33} \approx 1.005$

Using $v = \frac{1}{2}c$ in the formula, we get: $m = \dfrac{m_0}{\sqrt{1 - \dfrac{\left(\frac{1}{2}c\right)^2}{c^2}}} = \dfrac{m_0}{\sqrt{1 - \frac{1}{4}}}$. Thus

$m = \dfrac{1}{\sqrt{\frac{3}{4}}}m_0 = \dfrac{2\sqrt{3}}{3}m_0$. So the rest mass of the spaceship is multiplied by $\dfrac{2\sqrt{3}}{3} \approx 1.15$.

Using $v = 0.9c$ in the formula, we get: $m = \dfrac{m_0}{\sqrt{1 - \dfrac{(.9c)^2}{c^2}}} = \dfrac{m_0}{\sqrt{1 - .81}}$. Thus

$m = \dfrac{1}{\sqrt{0.19}}m_0 \approx 2.29m_0$. So the rest mass of the spaceship is multiplied by $\dfrac{1}{\sqrt{0.19}} \approx 2.29$. As the spaceship travels very close to the speed of light, $v \to c$, so the term $\frac{v^2}{c^2}$ approaches 1. So $\sqrt{1 - \frac{v^2}{c^2}}$ approaches 0, and we obtain $m = \dfrac{m_0}{\text{very small number}}$, which means $m = (\text{very large number})m_0$.

Hence the mass of the space ship becomes arbitrarily large as its speed approaches the speed of light.

The actual value of the speed of light does not affect the calculations.

Exercises 1.4

1. $(3x^2 + x + 1) + (2x^2 - 3x - 5) = 5x^2 - 2x - 4$

3. $(x^3 + 6x^2 - 4x + 7) - (3x^2 + 2x - 4) = x^3 + 6x^2 - 4x + 7 - 3x^2 - 2x + 4 =$
   $x^3 + 3x^2 - 6x + 11$

5. $8(2x + 5) - 7(x - 9) = 16x + 40 - 7x + 63 = 9x + 103$

7. $2(2 - 5t) + t^2(t - 1) - (t^4 - 1) = 4 - 10t + t^3 - t^2 - t^4 + 1 = -t^4 + t^3 - t^2 - 10t + 5$

9. $\sqrt{x}(x - \sqrt{x}) = x^{1/2}(x - x^{1/2}) = x^{1/2}x - x^{1/2}x^{1/2} = x^{3/2} - x$

11. $\sqrt[3]{y}(y^2 - 1) = y^{1/3}(y^2 - 1) = y^{1/3}y^2 - y^{1/3} = y^{7/3} - y^{1/3}$

13. $(3t - 2)(7t - 5) = 21t^2 - 15t - 14t + 10 = 21t^2 - 29t + 10$

15. $(x + 2y)(3x - y) = 3x^2 - xy + 6xy - 2y^2 = 3x^2 + 5xy - 2y^2$

17. $(1 - 2y)^2 = 1 - 4y + 4y^2$

19. $(2x - 5)(x^2 - x + 1) = 2x^3 - 2x^2 + 2x - 5x^2 + 5x - 5 = 2x^3 - 7x^2 + 7x - 5$

21. $x(x - 1)(x + 2) = (x^2 - x)(x + 2) = x^3 + 2x^2 - x^2 - 2x = x^3 + x^2 - 2x$

23. $(2x^2 + 3y^2)^2 = (2x^2)^2 + 2(2x^2)(3y^2) + (3y^2)^2 = 4x^4 + 12x^2y^2 + 9y^4$

25. $(x^2 - a^2)(x^2 + a^2) = (x^2)^2 - (a^2)^2 = x^4 - a^4$ (the difference of squares)

27. $(1 + a^3)^3 = 1 + 3(a^3) + 3(a^3)^2 + (a^3)^3 = 1 + 3a^3 + 3a^6 + a^9$ (perfect cube)

29. $\left(\sqrt{a} - \dfrac{1}{b}\right)\left(\sqrt{a} + \dfrac{1}{b}\right) = (\sqrt{a})^2 - \left(\dfrac{1}{b}\right)^2 = a - \dfrac{1}{b^2}$ (the difference of squares)

31. $(x^2 + x - 2)(x^3 - x + 1) = x^5 - x^3 + x^2 + x^4 - x^2 + x - 2x^3 + 2x - 2 =$
    $x^5 + x^4 - 3x^3 + 3x - 2$

33. $(1 + x^{4/3})(1 - x^{2/3}) = 1 - x^{2/3} + x^{4/3} - x^{6/3} = 1 - x^{2/3} + x^{4/3} - x^2$

35. $(1 - b)^2(1 + b)^2 = \left[(1 - b)(1 + b)\right]^2 = (1 - b^2)^2 = 1 - 2b^2 + b^4$

37. $(3x^2y + 7xy^2)(x^2y^3 - 2y^2) = 3x^4y^4 - 6x^2y^3 + 7x^3y^5 - 14xy^4$
    $= 3x^4y^4 + 7x^3y^5 - 6x^2y^3 - 14xy^4$ (arranging in decreasing powers of x)

39. $2x + 12x^3 = 2x(1 + 6x^2)$

41. $6y^4 - 15y^3 = 3y^3(2y - 5)$

43. $x^2 + 7x + 6 = (x + 6)(x + 1)$

45. $x^2 - 2x - 8 = (x - 4)(x + 2)$

47. $y^2 - 8y + 15 = (y - 5)(y - 3)$

49. $2x^2 + 5x + 3 = (2x + 3)(x + 1)$

51. $9x^2 - 36 = 9(x^2 - 4) = 9(x - 2)(x + 2)$

53. $6x^2 - 5x - 6 = (3x + 2)(2x - 3)$

55. $(x - 1)(x + 2)^2 - (x - 1)^2(x + 2) = (x - 1)(x + 2)\left[(x + 2) - (x - 1)\right] = 3(x - 1)(x + 2)$

57. $y^4(y + 2)^3 + y^5(y + 2)^4 = y^4(y + 2)^3\left[(1) + y(y + 2)\right] = y^4(y + 2)^3(y^2 + 2y + 1)$
    $= y^4(y + 2)^3(y + 1)^2$

59. $(a^2 - 1)b^2 - 4(a^2 - 1) = (a^2 - 1)(b^2 - 4) = (a - 1)(a + 1)(b - 2)(b + 2)$

61. $t^3 + 1 = (t + 1)(t^2 - t + 1)$ (the sum of cubes)

63. $4t^2 - 12t + 9 = (2t - 3)^2$ (perfect square)

65. $x^3 + 2x^2 + x = x(x^2 + 2x + 1) = x(x + 1)^2$

67. $4x^2 + 4xy + y^2 = (2x + y)^2$ (perfect square)

69. $x^4 + 2x^3 - 3x^2 = x^2(x^2 + 2x - 3) = x^2(x - 1)(x + 3)$

71. $8x^3 - 125 = (2x)^3 - (5)^3 = (2x - 5)\left[(2x)^2 + (2x)(5) + (5)^2\right] = (2x - 5)(4x^2 + 10x + 25)$ (the
    difference of cubes)

73. $x^4 + x^2 - 2 = (x^2)^2 + (x^2) - 2 = (x^2 + 2)(x^2 - 1) = (x^2 + 2)(x - 1)(x + 1)$

75. $y^3 - 3y^2 - 4y + 12 = (y^3 - 3y^2) + (-4y + 12) = y^2(y - 3) + (-4)(y - 3) = (y - 3)(y^2 - 4)$
    $= (y - 3)(y - 2)(y + 2)$ (factor by grouping)

77. $2x^3 + 4x^2 + x + 2 = (2x^3 + 4x^2) + (x + 2) = 2x^2(x + 2) + (1)(x + 2) = (x + 2)(2x^2 + 1)$
    (factor by grouping)

79. $(3 + x^2)^2 - (1 + x^2)^2 = [(3 + x^2) - (1 + x^2)][(3 + x^2) + (1 + x^2)]$
    $= (3 + x^2 - 1 - x^2)(3 + x^2 + 1 + x^2) = 2(4 + 2x^2) = 4(2 + x^2)$

81. $x^{5/2} - x^{1/2} = x^{1/2}(x^2 - 1) = \sqrt{x}(x - 1)(x + 1)$

83. Start by factoring out the power of x with the smallest exponent, that is, $x^{-3/2}$. So
    $x^{-3/2} + 2x^{-1/2} + x^{1/2} = x^{-3/2}\left(1 + 2x + x^2\right) = \dfrac{(1 + x)^2}{x^{3/2}}$.

85. Start by factoring out the power of $(x^2 + 1)$ with the smallest exponent, that is, $(x^2 + 1)^{-1/2}$. So
    $(x^2 + 1)^{1/2} + 2(x^2 + 1)^{-1/2} = (x^2 + 1)^{-1/2}\left[(x^2 + 1) + 2\right] = \dfrac{x^2 + 3}{\sqrt{x^2 + 1}}$.

87. Start by factoring $y^2 - 7y + 10$, and then substitute $a^2 + 1$ for y. This gives
    $(a^2 + 1)^2 - 7(a^2 + 1) + 10 = \left[(a^2 + 1) - 2\right]\left[(a^2 + 1) - 5\right] = (a^2 - 1)(a^2 - 4) =$
    $(a - 1)(a + 1)(a - 2)(a + 2)$.

89. $3x^2(4x - 12)^2 + x^3(2)(4x - 12)(4) = x^2(4x - 12)[3(4x - 12) + x(2)(4)]$
    $= 4x^2(x - 3)(12x - 36 + 8x) = 4x^2(x - 3)(20x - 36) = 16x^2(x - 3)(5x - 9)$

91. $3(2x - 1)^2(2)(x + 3)^{1/2} + (2x - 1)^3\left(\frac{1}{2}\right)(x + 3)^{-1/2}$
    $= (2x - 1)^2(x + 3)^{-1/2}\left[6(x + 3) + (2x - 1)\left(\frac{1}{2}\right)\right] = (2x - 1)^2(x + 3)^{-1/2}\left(6x + 18 + x - \frac{1}{2}\right)$
    $= (2x - 1)^2(x + 3)^{-1/2}\left(7x + \frac{35}{2}\right)$

93. (a) $\frac{1}{2}[(a+b)^2 - (a^2+b^2)] = \frac{1}{2}[a^2 + 2ab + b^2 - a^2 - b^2] = \frac{1}{2}(2ab) = ab.$

(b) $(a^2+b^2)^2 - (a^2-b^2)^2 = [(a^2+b^2) - (a^2-b^2)][(a^2+b^2) + (a^2-b^2)]$
$= (a^2 + b^2 - a^2 + b^2)(a^2 + b^2 + a^2 - b^2) = (2b^2)(2a^2) = 4a^2b^2$

(c) LHS $= (a^2+b^2)(c^2+d^2) = a^2c^2 + a^2d^2 + b^2c^2 + b^2d^2$

RHS $= (ac+bd)^2 + (ad-bc)^2 = a^2c^2 + 2abcd + b^2d^2 + a^2d^2 - 2abcd + b^2c^2$
$= a^2c^2 + a^2d^2 + b^2c^2 + b^2d^2$

So LHS = RHS, that is, $(a^2+b^2)(c^2+d^2) = (ac+bd)^2 + (ad-bc)^2.$

(d) $4a^2c^2 - (c^2 - b^2 + a^2)^2 = (2ac)^2 - (c^2 - b^2 + a^2)$

$= \Big[(2ac) - (c^2 - b^2 + a^2)\Big]\Big[(2ac) + (c^2 - b^2 + a^2)\Big]$ (the difference of squares)

$= (2ac - c^2 + b^2 - a^2)(2ac + c^2 - b^2 + a^2)$

$= \Big[b^2 - (c^2 - 2ac + a^2)\Big]\Big[(c^2 + 2ac + a^2) - b^2\Big]$ (regrouping)

$= \Big[b^2 - (c-a)^2\Big]\Big[(c+a)^2 - b^2\Big]$ (perfect squares)

$= [b - (c-a)][b + (c-a)][(c+a) - b][(c+a) + b]$ (each factor is a difference of squares)

$= (b - c + a)(b + c - a)(c + a - b)(c + a + b)$

$= (a + b - c)(-a + b + c)(a - b + c)(a + b + c)$

(e) $x^4 + 3x^2 + 4 = (x^4 + 4x^2 + 4) - x^2 = (x^2 + 2)^2 - x^2 = \Big[(x^2 + 2) - x\Big]\Big[(x^2 + 2) + x\Big] = (x^2 - x + 2)(x^2 + x + 2)$

95. (a) $528^2 - 527^2 = (528 - 527)(528 + 527) = 1(1055) = 1055$

(b) $122^2 - 120^2 = (122 - 120)(122 + 120) = 2(242) = 484$

(c) $1020^2 - 1010^2 = (1020 - 1010)(1020 + 1010) = 10(2030) = 20{,}300$

(d) $49 \cdot 51 = (50 - 1)(50 + 1) = 50^2 - 1 = 2500 - 1 = 2499$

(e) $998 \cdot 1002 = (1000 - 2)(1000 + 2) = 1000^2 - 2^2 = 1{,}000{,}000 - 4 = 999{,}996$

97. (a) $A^4 - B^4 = (A^2 - B^2)(A^2 + B^2) = (A - B)(A + B)(A^2 + B^2)$
$A^6 - B^6 = (A^3 - B^3)(A^3 + B^3)$ (the difference of squares)
$\qquad\qquad = (A - B)(A^2 + AB + B^2)(A + B)(A^2 - AB + B^2)$ (difference and sum of cubes)

(b) $12^4 - 7^4 = 20{,}736 - 2{,}401 = 18{,}335$
$12^6 - 7^6 = 2{,}985{,}984 - 117{,}649 = 2{,}868{,}335$

(c) $18{,}335 = 12^4 - 7^4 = (12 - 7)(12 + 7)(12^2 + 7^2) = 5(19)(144 + 49) = 5(19)(193)$
$2{,}868{,}335 = 12^6 - 7^6 = (12 - 7)(12 + 7)[12^2 + 12(7) + 7^2][12^2 - 12(7) + 7^2]$
$\qquad\qquad\qquad = 5(19)(144 + 84 + 49)(144 - 84 + 49) = 5(19)(277)(109)$

Exercises 1.5

1. $\dfrac{x-2}{x^2-4} = \dfrac{x-2}{(x-2)(x+2)} = \dfrac{1}{x+2}$

3. $\dfrac{x^2+6x+8}{x^2+5x+4} = \dfrac{(x+2)(x+4)}{(x+1)(x+4)} = \dfrac{x+2}{x+1}$

5. $\dfrac{y^2+y}{y^2-1} = \dfrac{y(y+1)}{(y-1)(y+1)} = \dfrac{y}{y-1}$

7. $\dfrac{2x^3-x^2-6x}{2x^2-7x+6} = \dfrac{x(2x^2-x-6)}{(2x-3)(x-2)} = \dfrac{x(2x+3)(x-2)}{(2x-3)(x-2)} = \dfrac{x(2x+3)}{2x-3}$

9. $\dfrac{t-3}{t^2+9} \cdot \dfrac{t+3}{t^2-9} = \dfrac{(t-3)(t+3)}{(t^2+9)(t-3)(t+3)} = \dfrac{1}{t^2+9}$

11. $\dfrac{x^2+7x+12}{x^2+3x+2} \cdot \dfrac{x^2+5x+6}{x^2+6x+9} = \dfrac{(x+3)(x+4)}{(x+1)(x+2)} \cdot \dfrac{(x+2)(x+3)}{(x+3)(x+3)} = \dfrac{x+4}{x+1}$

13. $\dfrac{2x^2+3x+1}{x^2+2x-15} \div \dfrac{x^2+6x+5}{2x^2-7x+3} = \dfrac{2x^2+3x+1}{x^2+2x-15} \cdot \dfrac{2x^2-7x+3}{x^2+6x+5} =$
    $\dfrac{(2x+1)(x+1)}{(x-3)(x+5)} \cdot \dfrac{(2x-1)(x-3)}{(x+1)(x+5)} = \dfrac{(2x+1)(2x-1)}{(x+5)(x+5)} = \dfrac{(2x+1)(2x-1)}{(x+5)^2}$

15. $\dfrac{\dfrac{x^3}{x+1}}{\dfrac{x}{x^2+2x+1}} = \dfrac{x^3}{x+1} \cdot \dfrac{x^2+2x+1}{x} = \dfrac{x^3(x+1)(x+1)}{(x+1)x} = x^2(x+1)$

17. $\dfrac{x/y}{z} = \dfrac{x}{y} \cdot \dfrac{1}{z} = \dfrac{x}{yz}$

19. $\dfrac{1}{x+5} + \dfrac{2}{x-3} = \dfrac{x-3}{(x+5)(x-3)} + \dfrac{2(x+5)}{(x+5)(x-3)} = \dfrac{x-3+2x+10}{(x+5)(x-3)} = \dfrac{3x+7}{(x+5)(x-3)}$

21. $\dfrac{1}{x+1} - \dfrac{1}{x+2} = \dfrac{x+2}{(x+1)(x+2)} + \dfrac{-(x+1)}{(x+1)(x+2)} = \dfrac{x+2-x-1}{(x+1)(x+2)} = \dfrac{1}{(x+1)(x+2)}$

23. $\dfrac{x}{(x+1)^2} + \dfrac{2}{x+1} = \dfrac{x}{(x+1)^2} + \dfrac{2(x+1)}{(x+1)(x+1)} = \dfrac{x+2x+2}{(x+1)^2} = \dfrac{3x+2}{(x+1)^2}$

25. $u+1+\dfrac{u}{u+1} = \dfrac{(u+1)(u+1)}{u+1} + \dfrac{u}{u+1} = \dfrac{u^2+2u+1+u}{u+1} = \dfrac{u^2+3u+1}{u+1}$

27. $\dfrac{1}{x^2} + \dfrac{1}{x^2+x} = \dfrac{1}{x^2} + \dfrac{1}{x(x+1)} = \dfrac{x+1}{x^2(x+1)} + \dfrac{x}{x^2(x+1)} = \dfrac{2x+1}{x^2(x+1)}$

29. $\dfrac{2}{x+3} - \dfrac{1}{x^2+7x+12} = \dfrac{2}{x+3} - \dfrac{1}{(x+3)(x+4)} = \dfrac{2(x+4)}{(x+3)(x+4)} + \dfrac{-1}{(x+3)(x+4)}$
    $= \dfrac{2x+8-1}{(x+3)(x+4)} = \dfrac{2x+7}{(x+3)(x+4)}$

31. $\dfrac{1}{x+3} + \dfrac{1}{x^2-9} = \dfrac{1}{x+3} + \dfrac{1}{(x-3)(x+3)} = \dfrac{x-3}{(x-3)(x+3)} + \dfrac{1}{(x-3)(x+3)}$

$= \dfrac{x-2}{(x-3)(x+3)}$

33. $\dfrac{2}{x} + \dfrac{3}{x-1} - \dfrac{4}{x^2-x} = \dfrac{2}{x} + \dfrac{3}{x-1} - \dfrac{4}{x(x-1)} = \dfrac{2(x-1)}{x(x-1)} + \dfrac{3x}{x(x-1)} + \dfrac{-4}{x(x-1)}$

$= \dfrac{2x-2+3x-4}{x(x-1)} = \dfrac{5x-6}{x(x-1)}$

35. $\dfrac{1}{x^2+3x+2} - \dfrac{1}{x^2-2x-3} = \dfrac{1}{(x+2)(x+1)} - \dfrac{1}{(x-3)(x+1)}$

$= \dfrac{x-3}{(x-3)(x+2)(x+1)} + \dfrac{-(x+2)}{(x-3)(x+2)(x+1)}$

$= \dfrac{x-3-x-2}{(x-3)(x+2)(x+1)} = \dfrac{-5}{(x-3)(x+2)(x+1)}$

37. $\dfrac{\dfrac{x}{y} - \dfrac{y}{x}}{\dfrac{1}{x^2} - \dfrac{1}{y^2}} = \dfrac{\dfrac{x^2-y^2}{xy}}{\dfrac{y^2-x^2}{x^2y^2}} = \dfrac{x^2-y^2}{xy} \cdot \dfrac{x^2y^2}{y^2-x^2} = \dfrac{xy}{-1} = -xy.$ An alternative method is to multiply

the numerator and denominator by the common denominator of both the numerator and denominator, in this case x^2y^2.

$\dfrac{\dfrac{x}{y} - \dfrac{y}{x}}{\dfrac{1}{x^2} - \dfrac{1}{y^2}} = \dfrac{\left(\dfrac{x}{y} - \dfrac{y}{x}\right)}{\left(\dfrac{1}{x^2} - \dfrac{1}{y^2}\right)} \cdot \dfrac{x^2y^2}{x^2y^2} = \dfrac{x^3y - xy^3}{y^2-x^2} = \dfrac{xy(x^2-y^2)}{y^2-x^2} = -xy$

39. $\dfrac{1 + \dfrac{1}{c-1}}{1 - \dfrac{1}{c-1}} = \dfrac{\dfrac{c-1}{c-1} + \dfrac{1}{c-1}}{\dfrac{c-1}{c-1} + \dfrac{-1}{c-1}} = \dfrac{\dfrac{c}{c-1}}{\dfrac{c-2}{c-1}} = \dfrac{c}{c-1} \cdot \dfrac{c-1}{c-2} = \dfrac{c}{c-2}.$ Using the alternative method we

obtain:

$\dfrac{1 + \dfrac{1}{c-1}}{1 - \dfrac{1}{c-1}} = \dfrac{\left(1 + \dfrac{1}{c-1}\right)}{\left(1 - \dfrac{1}{c-1}\right)} \cdot \dfrac{c-1}{c-1} = \dfrac{c-1+1}{c-1-1} = \dfrac{c}{c-2}$

41. $\dfrac{\dfrac{5}{x-1} - \dfrac{2}{x+1}}{\dfrac{x}{x-1} + \dfrac{1}{x+1}} = \dfrac{\dfrac{5(x+1)}{(x-1)(x+1)} + \dfrac{-2(x-1)}{(x-1)(x+1)}}{\dfrac{x(x+1)}{(x-1)(x+1)} + \dfrac{x-1}{(x-1)(x+1)}} = \dfrac{\dfrac{5x+5-2x+2}{(x-1)(x+1)}}{\dfrac{x^2+x+x-1}{(x-1)(x+1)}}$

$= \dfrac{3x+7}{(x-1)(x+1)} \cdot \dfrac{(x-1)(x+1)}{x^2+2x-1} = \dfrac{3x+7}{x^2+2x-1}$

Alternatively,

$$\frac{\dfrac{5}{x-1}-\dfrac{2}{x+1}}{\dfrac{x}{x-1}+\dfrac{1}{x+1}}=\frac{\left(\dfrac{5}{x-1}-\dfrac{2}{x+1}\right)}{\left(\dfrac{x}{x-1}+\dfrac{1}{x+1}\right)}\cdot\frac{(x-1)(x+1)}{(x-1)(x+1)}=\frac{5(x+1)-2(x-1)}{x(x+1)+(x-1)}$$

$$=\frac{5x+5-2x+2}{x^2+x+x-1}=\frac{3x+7}{x^2+2x-1}$$

43. $\dfrac{x^{-2}-y^{-2}}{x^{-1}+y^{-1}}=\dfrac{\dfrac{1}{x^2}-\dfrac{1}{y^2}}{\dfrac{1}{x}+\dfrac{1}{y}}=\dfrac{\dfrac{y^2}{x^2y^2}-\dfrac{x^2}{x^2y^2}}{\dfrac{y}{xy}+\dfrac{x}{xy}}=\dfrac{y^2-x^2}{x^2y^2}\cdot\dfrac{xy}{y+x}=\dfrac{(y-x)(y+x)xy}{x^2y^2(y+x)}=\dfrac{y-x}{xy}$

Alternatively,

$$\frac{x^{-2}-y^{-2}}{x^{-1}+y^{-1}}=\frac{\left(\dfrac{1}{x^2}-\dfrac{1}{y^2}\right)}{\left(\dfrac{1}{x}+\dfrac{1}{y}\right)}\cdot\frac{x^2y^2}{x^2y^2}=\frac{y^2-x^2}{xy^2+x^2y}=\frac{(y-x)(y+x)}{xy(y+x)}=\frac{y-x}{xy}$$

45. $\dfrac{1}{1+a^n}+\dfrac{1}{1+a^{-n}}=\dfrac{1}{1+a^n}+\dfrac{1}{1+a^{-n}}\cdot\dfrac{a^n}{a^n}=\dfrac{1}{1+a^n}+\dfrac{a^n}{a^n+1}=\dfrac{1+a^n}{1+a^n}=1$

47. $\dfrac{\dfrac{1}{a+h}-\dfrac{1}{a}}{h}=\dfrac{\dfrac{a}{a(a+h)}-\dfrac{a+h}{a(a+h)}}{h}=\dfrac{\dfrac{a-a-h}{a(a+h)}}{h}=\dfrac{-h}{a(a+h)}\cdot\dfrac{1}{h}=\dfrac{-1}{a(a+h)}$

49. $\dfrac{\dfrac{1-(x+h)}{2+(x+h)}-\dfrac{1-x}{2+x}}{h}=\dfrac{\dfrac{(2+x)(1-x-h)}{(2+x)(2+x+h)}-\dfrac{(1-x)(2+x+h)}{(2+x)(2+x+h)}}{h}$

$$=\frac{\dfrac{2-x-x^2-2h-xh}{(2+x)(2+x+h)}-\dfrac{2-x-x^2+h-xh}{(2+x)(2+x+h)}}{h}=\frac{-3h}{(2+x)(2+x+h)}\cdot\frac{1}{h}$$

$$=\frac{-3}{(2+x)(2+x+h)}$$

51. $\sqrt{1+\left(\dfrac{x}{\sqrt{1-x^2}}\right)^2}=\sqrt{1+\dfrac{x^2}{1-x^2}}=\sqrt{\dfrac{1-x^2}{1-x^2}+\dfrac{x^2}{1-x^2}}=\sqrt{\dfrac{1}{1-x^2}}=\dfrac{1}{\sqrt{1-x^2}}$

53. $\dfrac{3(x+2)^2(x-3)^2-(x+2)^3(2)(x-3)}{(x-3)^4}=\dfrac{(x+2)^2(x-3)[3(x-3)-(x+2)(2)]}{(x-3)^4}$

$$=\frac{(x+2)^2(3x-9-2x-4)}{(x-3)^3}=\frac{(x+2)^2(x-13)}{(x-3)^3}$$

55. $\dfrac{2(1+x)^{1/2}-x(1+x)^{-1/2}}{1+x}=\dfrac{(1+x)^{-1/2}[2(1+x)-x]}{1+x}=\dfrac{x+2}{(1+x)^{3/2}}$

57. $\dfrac{3(1+x)^{1/3}-x(1+x)^{-2/3}}{(1+x)^{2/3}}=\dfrac{(1+x)^{-2/3}[3(1+x)-x]}{(1+x)^{2/3}}=\dfrac{2x+3}{(1+x)^{4/3}}$

59. $\dfrac{2}{3+\sqrt{5}} = \dfrac{2}{3+\sqrt{5}} \cdot \dfrac{3-\sqrt{5}}{3-\sqrt{5}} = \dfrac{2\left(3-\sqrt{5}\right)}{9-5} = \dfrac{2\left(3-\sqrt{5}\right)}{4} = \dfrac{3-\sqrt{5}}{2}$

61. $\dfrac{2}{\sqrt{2}+\sqrt{7}} = \dfrac{2}{\sqrt{2}+\sqrt{7}} \cdot \dfrac{\sqrt{2}-\sqrt{7}}{\sqrt{2}-\sqrt{7}} = \dfrac{2\left(\sqrt{2}-\sqrt{7}\right)}{2-7} = \dfrac{2\left(\sqrt{2}-\sqrt{7}\right)}{-5} = \dfrac{2\left(\sqrt{7}-\sqrt{2}\right)}{5}$

63. $\dfrac{1-\sqrt{5}}{3} = \dfrac{1-\sqrt{5}}{3} \cdot \dfrac{1+\sqrt{5}}{1+\sqrt{5}} = \dfrac{1-5}{3\left(1+\sqrt{5}\right)} = \dfrac{-4}{3\left(1+\sqrt{5}\right)}$

65. $\dfrac{\sqrt{r}+\sqrt{2}}{5} = \dfrac{\sqrt{r}+\sqrt{2}}{5} \cdot \dfrac{\sqrt{r}-\sqrt{2}}{\sqrt{r}-\sqrt{2}} = \dfrac{r-2}{5\left(\sqrt{r}-\sqrt{2}\right)}$

67. $\sqrt{x^2+1} - x = \dfrac{\sqrt{x^2+1}-x}{1} \cdot \dfrac{\sqrt{x^2+1}+x}{\sqrt{x^2+1}+x} = \dfrac{x^2+1-x^2}{\sqrt{x^2+1}+x} = \dfrac{1}{\sqrt{x^2+1}+x}$

69. $\dfrac{16+a}{16} = \dfrac{16}{16} + \dfrac{a}{16} = 1 + \dfrac{a}{16}$, so the statement is true.

71. This statement is false. For example, take $x = 2$, then LHS $= \dfrac{2}{4+x} = \dfrac{2}{4+2} = \dfrac{2}{6} = \dfrac{1}{3}$, while

     RHS $= \dfrac{1}{2} + \dfrac{2}{x} = \dfrac{1}{2} + \dfrac{2}{2} = \dfrac{3}{2}$, and $\dfrac{1}{3} \neq \dfrac{3}{2}$.

73. This statement is false. For example, take $x = 0$ and $y = 1$. Then substituting into the left side we

     obtain LHS $= \dfrac{x}{x+y} = \dfrac{0}{0+1} = 0$, while the right side yields RHS $= \dfrac{1}{1+y} = \dfrac{1}{1+1} = \dfrac{1}{2}$, and

     $0 \neq \frac{1}{2}$.

75. This statement is true: $\dfrac{-a}{b} = (-a)\left(\dfrac{1}{b}\right) = (-1)(a)\left(\dfrac{1}{b}\right) = (-1)\left(\dfrac{a}{b}\right) = -\dfrac{a}{b}$.

77. This statement is false. For example, take $x = -2$. Then LHS $= \dfrac{x^2+1}{x^2+x-1} = \dfrac{(-2)^2+1}{(-2)^2+(-2)-1}$

     $= \dfrac{4+1}{4-2-1} = \dfrac{5}{1} = 5$, while RHS $= \dfrac{1}{x-1} = \dfrac{1}{(-2)-1} = \dfrac{1}{-3} = -\dfrac{1}{3}$, and $5 \neq -\dfrac{1}{3}$.

79. (a) $R = \dfrac{1}{\dfrac{1}{R_1} + \dfrac{1}{R_2}} = \dfrac{1}{\dfrac{1}{R_1} + \dfrac{1}{R_2}} \cdot \dfrac{R_1 R_2}{R_1 R_2} = \dfrac{R_1 R_2}{R_2 + R_1}$

     (b) Substituting $R_1 = 10$ ohms and $R_2 = 20$ ohms yields $R = \dfrac{(10)(20)}{(20)+(10)} = \dfrac{200}{30} \approx 6.7$ ohms.

81. No, squaring $\dfrac{2}{\sqrt{x}}$ changes its value by a factor of $\dfrac{2}{\sqrt{x}}$.

Exercises 1.6

1. (a) $2(4) - 3 \overset{?}{=} (4) + 1$

   $8 - 3 \overset{?}{=} 5$ Yes, a solution.

   (b) $2\left(\frac{3}{2}\right) - 3 \overset{?}{=} \left(\frac{3}{2}\right) + 1$

   $3 - 3 \overset{?}{=} \frac{5}{2}$ No, not a solution.

3. (a) $\frac{1}{(-3)} - \frac{1}{(-3)+3} \overset{?}{=} \frac{1}{6}$

   The left side is undefined, not a solution.

   (b) $\frac{1}{(3)} - \frac{1}{(3)+3} \overset{?}{=} \frac{1}{6}$

   $\frac{1}{3} - \frac{1}{6} \overset{?}{=} \frac{1}{6}$ Yes, a solution.

5. (a) $a(0) - 2b \overset{?}{=} 0$

   $-2b \overset{?}{=} 0$

   Since this is true for only certain values of b, this is not a solution.

   (b) $a\left(\frac{2b}{a}\right) - 2b \overset{?}{=} 0$

   $2b - 2b \overset{?}{=} 0$ Yes, a solution.

7. $3x - 5 = 7 \quad \Leftrightarrow \quad 3x = 12 \quad \Leftrightarrow \quad x = 4$

9. $x - 3 = 2x + 6 \quad \Leftrightarrow \quad -9 = x$

11. $-7w = 15 - 2w \quad \Leftrightarrow \quad -5w = 15 \quad \Leftrightarrow \quad w = -3$

13. $\frac{1}{2}y - 2 = \frac{1}{3}y \quad \Leftrightarrow \quad 3y - 12 = 2y$ (multiply both sides by the LCD, 6) $\quad \Leftrightarrow \quad y = 12$

15. $2(1 - x) = 3(1 + 2x) + 5 \quad \Leftrightarrow \quad 2 - 2x = 3 + 6x + 5 \quad \Leftrightarrow \quad 2 - 2x = 8 + 6x \quad \Leftrightarrow$
    $-6 = 8x \quad \Leftrightarrow \quad x = -\frac{3}{4}$

17. $4\left(y - \frac{1}{2}\right) - y = 6(5 - y) \quad \Leftrightarrow \quad 4y - 2 - y = 30 - 6y \quad \Leftrightarrow \quad 3y - 2 = 30 - 6y \quad \Leftrightarrow$
    $9y = 32 \quad \Leftrightarrow \quad y = \frac{32}{9}$

19. $\frac{1}{x} = \frac{4}{3x} + 1 \quad \Rightarrow \quad 3 = 4 + 3x$ (multiply both sides by the LCD $3x$) $\quad \Leftrightarrow \quad -1 = 3x \quad \Leftrightarrow$
    $x = -\frac{1}{3}$

21. $\frac{2}{t+6} = \frac{3}{t-1} \quad \Rightarrow \quad 2(t - 1) = 3(t + 6)$ (multiply both sides by the LCD $(t - 1)(t + 6)$) $\quad \Leftrightarrow$
    $2t - 2 = 3t + 18 \quad \Leftrightarrow \quad -20 = t$

23. $r - 2[1 - 3(2r + 4)] = 61 \quad \Leftrightarrow \quad r - 2(1 - 6r - 12) = 61 \quad \Leftrightarrow \quad r - 2(-6r - 11) = 61 \quad \Leftrightarrow$
    $r + 12r + 22 = 61 \quad \Leftrightarrow \quad 13r = 39 \quad \Leftrightarrow \quad r = 3$

25. $\sqrt{3}\,x + \sqrt{12} = \frac{x + 5}{\sqrt{3}} \quad \Leftrightarrow \quad 3x + 6 = x + 5$ (multiply both sides by $\sqrt{3}$) $\quad \Leftrightarrow \quad 2x = -1$
    $\Leftrightarrow \quad x = -\frac{1}{2}$

27. $\frac{2}{x} - 5 = \frac{6}{x} + 4 \quad \Rightarrow \quad 2 - 5x = 6 + 4x \quad \Leftrightarrow \quad -4 = 9x \quad \Leftrightarrow \quad -\frac{4}{9} = x$

29. $\frac{3}{x+1} - \frac{1}{2} = \frac{1}{3x+3} \quad \Rightarrow \quad 3(6) - (3x + 3) = 2$ (multiply both sides by the LCD $6(x + 1)$)
    $\Leftrightarrow \quad 18 - 3x - 3 = 2 \quad \Leftrightarrow \quad -3x + 15 = 2 \quad \Leftrightarrow \quad -3x = -13 \quad \Leftrightarrow \quad x = \frac{13}{3}$

31. $\dfrac{2x-7}{2x+4} = \dfrac{2}{3}$ $\Rightarrow$ $(2x-7)3 = 2(2x+4)$ (cross multiply) $\Leftrightarrow$ $6x - 21 = 4x + 8$ $\Leftrightarrow$
    $2x = 29$ $\Leftrightarrow$ $x = \dfrac{29}{2}$

33. $x - \frac{1}{3}x - \frac{1}{2}x - 5 = 0$ $\Leftrightarrow$ $6x - 2x - 3x - 30 = 0$ (multiply both sides by the LCD 6) $\Leftrightarrow$
    $x = 30$

35. $\dfrac{1}{z} - \dfrac{1}{2z} - \dfrac{1}{5z} = \dfrac{10}{z+1}$ $\Rightarrow$ $10(z+1) - 5(z+1) - 2(z+1) = 10(10z)$ (multiply both sides
    by the LCD $10z(z+1)$) $\Leftrightarrow$ $3(z+1) = 100z$ $\Leftrightarrow$ $3z + 3 = 100z$ $\Leftrightarrow$ $3 = 97z$ $\Leftrightarrow$
    $\frac{3}{97} = z$

37. $\dfrac{u}{u - \frac{u+1}{2}} = 4$ $\Rightarrow$ $u = 4\left(u - \dfrac{u+1}{2}\right)$ (cross multiply) $\Leftrightarrow$ $u = 4u - 2u - 2$ $\Leftrightarrow$
    $u = 2u - 2$ $\Leftrightarrow$ $2 = u$

39. $\dfrac{x}{2x-4} - 2 = \dfrac{1}{x-2}$ $\Rightarrow$ $x - 2(2x-4) = 2$ (multiply both sides by the LCD $2(x-2)$) $\Leftrightarrow$
    $x - 4x + 8 = 2$ $\Leftrightarrow$ $-3x = -6$ $\Leftrightarrow$ $x = 2$. But substituting $x = 2$ into the original
    equation does not work, since we cannot divide by 0. Thus there is no solution.

41. $\dfrac{3}{x+4} = \dfrac{1}{x} + \dfrac{6x+12}{x^2+4x}$ $\Rightarrow$ $3(x) = (x+4) + 6x + 12$ (multiply both sides by the LCD
    $x(x+4)$) $\Leftrightarrow$ $3x = 7x + 16$ $\Leftrightarrow$ $-4x = 16$ $\Leftrightarrow$ $x = -4$. But substituting $x = -4$ into
    the original equation does not work, since we cannot divide by 0. Thus there is no solution.

43. $x^2 = 49$ $\Rightarrow$ $x = \pm 7$

45. $x^2 - 24 = 0$ $\Leftrightarrow$ $x^2 = 24$ $\Rightarrow$ $x = \pm\sqrt{24} = \pm 2\sqrt{6}$

47. $8x^2 - 64 = 0$ $\Leftrightarrow$ $x^2 - 8 = 0$ $\Leftrightarrow$ $x^2 = 8$ $\Rightarrow$ $x = \pm\sqrt{8} = \pm 2\sqrt{2}$

49. $x^2 + 16 = 0$ $\Leftrightarrow$ $x^2 = -16$ which has no real solution.

51. $(x+2)^2 = 4$ $\Leftrightarrow$ $(x+2)^2 = 4$ $\Rightarrow$ $x + 2 = \pm 2$. If $x + 2 = 2$, then $x = 0$. If $x + 2 = -2$,
    then $x = -4$. The solutions are -4 and 0.

53. $x^3 = 27$ $\Leftrightarrow$ $x = 27^{1/3} = 3$.

55. $0 = x^4 - 16 = (x^2+4)(x^2-4) = (x^2+4)(x-2)(x+2)$. $x^2 + 4 = 0$ has no real solution. If
    $x - 2 = 0$, then $x = 2$. If $x + 2 = 0$, then $x = -2$. The solutions are ± 2.

57. $x^4 + 64 = 0$ $\Leftrightarrow$ $x^4 = -64$ which has no real solution.

59. $(x+2)^4 - 81 = 0$ $\Leftrightarrow$ $(x+2)^4 = 81$ $\Leftrightarrow$ $[(x+2)^4]^{1/4} = \pm 81^{1/4}$ $\Leftrightarrow$ $x + 2 = \pm 3$. So
    $x + 2 = 3$, then $x = 1$. If $x + 2 = -3$, then $x = -5$. The solutions are: $-5, 1$.

61. $3(x-3)^3 = 375$ $\Leftrightarrow$ $(x-3)^3 = 125$ $\Leftrightarrow$ $(x-3) = 125^{1/3} = 5$ $\Leftrightarrow$ $x = 3 + 5 = 8$.

63. $\sqrt[3]{x} = 5$ $\Leftrightarrow$ $x = 5^3 = 125$

65. $2x^{5/3} + 64 = 0$ $\Leftrightarrow$ $2x^{5/3} = -64$ $\Leftrightarrow$ $x^{5/3} = -32$ $\Leftrightarrow$
    $x = (-32)^{3/5} = (-2^5)^{1/5} = (-2)^3 = -8$.

67. $2.15x - 4.63 = x + 1.19$ $\Leftrightarrow$ $1.15x = 5.82$ $\Leftrightarrow$ $x = \frac{5.82}{1.19} \approx 5.06$

69. $3.16(x + 4.63) = 4.19(x - 7.24)$ ⇔ $3.16x + 14.63 = 4.19x - 30.34$ ⇔ $44.97 = 1.03x$
     ⇔ $x = \frac{44.97}{1.03} \approx 43.66$

71. $PV = nRT$ ⇔ $R = \dfrac{PV}{nT}$

73. $\dfrac{1}{R} = \dfrac{1}{R_1} + \dfrac{1}{R_2}$ ⇔ $R_1 R_2 = R R_2 + R R_1$ (multiply both sides by the LCD, $R R_1 R_2$). Thus

     $R_1 R_2 - R R_1 = R R_2$ ⇔ $R_1(R_2 - R) = R R_2$ ⇔ $R_1 = \dfrac{R R_2}{R_2 - R}$

75. $\dfrac{ax + b}{cx + d} = 2$ ⇔ $ax + b = 2(cx + d)$ ⇔ $ax + b = 2cx + 2d$ ⇔ $ax - 2cx = 2d - b$

     ⇔ $(a - 2c)x = 2d - b$ ⇔ $x = \dfrac{2d - b}{a - 2c}$

77. $a^2 x + (a - 1) = (a + 1)x$ ⇔ $a^2 x - (a + 1)x = -(a - 1)$ ⇔ $(a^2 - (a + 1))x = -a + 1$
     ⇔ $(a^2 - a - 1)x = -a + 1$ ⇔ $x = \dfrac{-a + 1}{a^2 - a - 1}$

79. $V = \dfrac{1}{3}\pi r^2 h$ ⇔ $r^2 = \dfrac{3V}{\pi h}$ ⇒ $r = \pm\sqrt{\dfrac{3V}{\pi h}}$

81. $a^2 + b^2 = c^2$ ⇔ $b^2 = c^2 - a^2$ ⇒ $b = \pm\sqrt{c^2 - a^2}$

83. $V = \frac{4}{3}\pi r^3$ ⇔ $r^3 = \dfrac{3V}{4\pi}$ ⇔ $r = \sqrt[3]{\dfrac{3V}{4\pi}}$

85. (a) Substituting we get $0.30(60) + 0.38(3400)^{1/2} - 3(650)^{1/3} \approx 18 + 0.38(58.31) - 3(8.66)$
          $\approx 18 + 22.16 - 25.98 \approx 14.18$. Since this value is less than 16, the sail boat qualifies for the
          race.

     (b) Solve for A when $L = 65$ and $V = 650$. Substituting we get:
          $0.30(65) + 0.38A^{1/2} - 3(650)^{1/3} \le 16$ ⇔ $19.5 + 0.38A^{1/2} - 25.99 \le 16$ ⇔
          $0.38A^{1/2} - 6.49 \le 16$ ⇔ $0.38A^{1/2} \le 22.49$ ⇔ $A^{1/2} \le 59.18$ ⇔ $A \le 3502.8$.
          Thus the largest possible sail is just over of 3500 ft^2.

87. When we multiplied by x, we introduced $x = 0$ as a solution. When we divided by $x - 1$, we are
     really dividing by 0, since $x = 1$ ⇔ $x - 1 = 0$.

Review Exercises for Chapter 1

1. Commutative Property for addition.

3. Distributive Property.

5. $(-1,3] = \{x \mid -1 < x \le 3\}$

7. $x > 2 \iff x \in (2,\infty)$

9. $\left|3 - |-9|\right| = |3 - 9| = |-6| = 6$

11. $2^{-3} - 3^{-2} = \dfrac{1}{8} - \dfrac{1}{9} = \dfrac{9}{72} - \dfrac{8}{72} = \dfrac{1}{72}$

13. $216^{-1/3} = \dfrac{1}{216^{1/3}} = \dfrac{1}{\sqrt[3]{216}} = \dfrac{1}{6}$

15. $\dfrac{\sqrt{242}}{\sqrt{2}} = \sqrt{\dfrac{242}{2}} = \sqrt{121} = 11$

17. $2^{1/2}8^{1/2} = \sqrt{2} \cdot \sqrt{8} = \sqrt{16} = 4$

19. $\dfrac{1}{x^2} = x^{-2}$

21. $x^2 x^m (x^3)^m = x^{2+m+3m} = x^{4m+2}$

23. $x^a x^b x^c = x^{a+b+c}$

25. $x^{c+1}(x^{2c-1})^2 = x^{(c+1)+2(2c-1)} = x^{c+1+4c-2} = x^{5c-1}$

27. $(2x^3y)^2(3x^{-1}y^2) = 4x^6y^2 \cdot 3x^{-1}y^2 = 4 \cdot 3x^{6-1}y^{2+2} = 12x^5y^4$

29. $\dfrac{x^4(3x)^2}{x^3} = \dfrac{x^4 \cdot 9x^2}{x^3} = 9x^{4+2-3} = 9x^3$

31. $\sqrt[3]{(x^3y)^2y^4} = \sqrt[3]{x^6y^4y^2} = \sqrt[3]{x^6y^6} = x^2y^2$

33. $\dfrac{x}{2 + \sqrt{x}} = \dfrac{x}{2 + \sqrt{x}} \cdot \dfrac{2 - \sqrt{x}}{2 - \sqrt{x}} = \dfrac{x(2 - \sqrt{x})}{4 - x}$. Here simplify means to rationalize the denominator.

35. $\dfrac{8r^{1/2}s^{-3}}{2r^{-2}s^4} = 4r^{(1/2)-(-2)}\,s^{-3-4} = 4r^{5/2}s^{-7} = \dfrac{4r^{5/2}}{s^7}$

37. $78{,}250{,}000{,}000 = 7.825 \times 10^{10}$

39. $\dfrac{ab}{c} \approx \dfrac{(0.00000293)(1.582 \times 10^{-14})}{2.8064 \times 10^{12}} = \dfrac{(2.93 \times 10^{-6})(1.582 \times 10^{-14})}{2.8064 \times 10^{12}}$
$= \dfrac{2.93 \cdot 1.582}{2.8064} \times 10^{-6-14-12} \approx 1.65 \times 10^{-32}$

41. $12x^2y^4 - 3xy^5 + 9x^3y^2 = 3xy^2(4xy^2 - y^3 + 3x^2)$

43. $x^2 + 3x - 10 = (x + 5)(x - 2)$

45. $4t^2 - 13t - 12 = (4t + 3)(t - 4)$

47. $25 - 16t^2 = (5 - 4t)(5 + 4t)$

49. $x^6 - 1 = (x^3 - 1)(x^3 + 1) = (x - 1)(x^2 + x + 1)(x + 1)(x^2 - x + 1)$

51. $x^{-1/2} - 2x^{1/2} + x^{3/2} = x^{-1/2}(1 - 2x + x^2) = x^{-1/2}(1 - x)^2$

53. $4x^3 - 8x^2 + 3x - 6 = 4x^2(x - 2) + 3(x - 2) = (4x^2 + 3)(x - 2)$

55. $(x^2+2)^{5/2} + 2x(x^2+2)^{3/2} + x^2\sqrt{x^2+2} = (x^2+2)^{1/2}((x^2+2)^2 + 2x(x^2+2) + x^2)$
    $= \sqrt{x^2+2}(x^4 + 4x^2 + 4 + 2x^3 + 4x + x^2) = \sqrt{x^2+2}(x^4 + 2x^3 + 5x^2 + 4x + 4)$
    $= \sqrt{x^2+2}(x^2 + x + 2)^2$

57. $a^2y - b^2y = y(a^2 - b^2) = y(a-b)(a+b)$

59. $(x+1)^2 - 2(x+1) + 1 = [(x+1) - 1]^2 = x^2$. You can also obtain this result by expanding each
    term and then simplifying.

61. $(2x+1)(3x-2) - 5(4x-1) = 6x^2 - 4x + 3x - 2 - 20x + 5 = 6x^2 - 21x + 3$

63. $(2a^2 - b)^2 = (2a^2)^2 - 2(2a^2)(b) + (b)^2 = 4a^4 - 4a^2b + b^2$

65. $(x-1)(x-2)(x-3) = (x-1)(x^2 - 5x + 6) = x^3 - 5x^2 + 6x - x^2 + 5x - 6$
    $= x^3 - 6x^2 + 11x - 6$

67. $\sqrt{x}(\sqrt{x}+1)(2\sqrt{x}-1) = (x+\sqrt{x})(2\sqrt{x}-1) = 2x\sqrt{x} - x + 2x - \sqrt{x} = 2x^{3/2} + x - x^{1/2}$

69. $x^2(x-2) + x(x-2)^2 = x^3 - 2x^2 + x(x^2 - 4x + 4) = x^3 - 2x^2 + x^3 - 4x^2 + 4x$
    $= 2x^3 - 6x^2 + 4x$

71. $\dfrac{x^2 - 2x - 3}{2x^2 + 5x + 3} = \dfrac{(x-3)(x+1)}{(2x+3)(x+1)} = \dfrac{x-3}{2x+3}$

73. $\dfrac{x^2 + 2x - 3}{x^2 + 8x + 16} \cdot \dfrac{3x + 12}{x - 1} = \dfrac{(x+3)(x-1)}{(x+4)(x+4)} \cdot \dfrac{3(x+4)}{(x-1)} = \dfrac{3(x+3)}{x+4}$

75. $\dfrac{x^2 - 2x - 15}{x^2 - 6x + 5} \div \dfrac{x^2 - x - 12}{x^2 - 1} = \dfrac{(x-5)(x+3)}{(x-5)(x-1)} \cdot \dfrac{(x-1)(x+1)}{(x-4)(x+3)} = \dfrac{x+1}{x-4}$

77. $\dfrac{1}{x-1} - \dfrac{x}{x^2+1} = \dfrac{x^2+1}{(x-1)(x^2+1)} - \dfrac{x(x-1)}{(x-1)(x^2+1)} = \dfrac{x^2+1-x^2+x}{(x-1)(x^2+1)} = \dfrac{x+1}{(x-1)(x^2+1)}$

79. $\dfrac{1}{x-1} - \dfrac{2}{x^2-1} = \dfrac{1}{x-1} - \dfrac{2}{(x-1)(x+1)} = \dfrac{x+1}{(x-1)(x+1)} - \dfrac{2}{(x-1)(x+1)}$
    $= \dfrac{x+1-2}{(x-1)(x+1)} = \dfrac{x-1}{(x-1)(x+1)} = \dfrac{1}{x+1}$

81. $\dfrac{\dfrac{1}{x} - \dfrac{1}{2}}{x-2} = \dfrac{\dfrac{2}{2x} - \dfrac{x}{2x}}{x-2} = \dfrac{2-x}{2x} \cdot \dfrac{1}{x-2} = \dfrac{-1(x-2)}{2x} \cdot \dfrac{1}{x-2} = \dfrac{-1}{2x}$

83. $\dfrac{3(x+h)^2 - 5(x+h) - (3x^2 - 5x)}{h} = \dfrac{3x^2 + 6xh + 3h^2 - 5x - 5h - 3x^2 + 5x}{h}$
    $= \dfrac{6xh + 3h^2 - 5h}{h} = \dfrac{h(6x + 3h - 5)}{h} = 6x + 3h - 5$

85. $3x + 12 = 24 \iff 3x = 12 \iff x = 4$.

87. $7x - 6 = 4x + 9 \iff 3x = 15 \iff x = 5$.

89. $\frac{1}{3}x - \frac{1}{2} = 2 \iff 2x - 3 = 12 \iff 2x = 15 \iff x = \frac{15}{2}$.

91. $2(x+3) - 4(x-5) = 8 - 5x \iff 2x + 6 - 4x + 20 = 8 - 5x \iff -2x + 26 = 8 - 5x$
    $\iff 3x = -18 \iff x = -6$.

93. $\dfrac{x+1}{x-1} = \dfrac{2x-1}{2x+1}$ $\Leftrightarrow$ $(x+1)(2x+1) = (2x-1)(x-1)$ $\Leftrightarrow$ $2x^2 + 3x + 1 = 2x^2 - 3x + 1$
$\Leftrightarrow$ $6x = 0$ $\Leftrightarrow$ $x = 0$.

95. $x^2 = 144$ $\Rightarrow$ $x = \pm 12$.

97. $5x^3 - 15 = 0$ $\Leftrightarrow$ $5x^3 = 15$ $\Leftrightarrow$ $x^3 = 3$ $\Leftrightarrow$ $x = \sqrt[3]{3}$.

99. $(x+1)^3 = -64$ $\Leftrightarrow$ $x + 1 = -4$ $\Leftrightarrow$ $x = -1 - 4 = -5$.

101. $\sqrt[3]{x} = -3$ $\Leftrightarrow$ $x = (-3)^3 = -27$.

103. $4x^{3/4} - 500 = 0$ $\Leftrightarrow$ $4x^{3/4} = 500$ $\Leftrightarrow$ $x^{3/4} = 125$ $\Leftrightarrow$ $x = 125^{4/3} = 5^4 = 625$.

105. This statement is false. For example, take $x = 1$ and $y = 1$; then LHS $= (x+y)^3 = (1+1)^3 = 2^3$ $= 8$, while RHS $= x^3 + y^3 = 1^3 + 1^3 = 1 + 1 = 2$, and $8 \neq 2$.

107. This statement is true: $\dfrac{12 + y}{y} = \dfrac{12}{y} + \dfrac{y}{y} = \dfrac{12}{y} + 1$.

109. This statement is false. For example, take $a = -1$; then LHS $= \sqrt{a^2} = \sqrt{(-1)^2} = \sqrt{1} = 1$, which does not equal $a = -1$. The true statement is $\sqrt{a^2} = |a|$.

111. This statement is false. For example, take $x = 1$ and $y = 1$, then LHS $= x^3 + y^3 = 1^3 + 1^3$ $= 1 + 1 = 2$, while RHS $= (x+y)(x^2 + xy + y^2) = (1+1)\left[1^2 + (1)(1) + 1^2\right] = 2(3) = 6$.

113. Substituting we obtain $\sqrt{1 + t^2} = \sqrt{1 + \left[\dfrac{1}{2}\left(x^3 - \dfrac{1}{x^3}\right)\right]^2} = \sqrt{1 + \dfrac{1}{4}\left(x^6 - 2\dfrac{x^3}{x^3} + \dfrac{1}{x^6}\right)}$

$= \sqrt{1 + \dfrac{x^6}{4} - \dfrac{1}{2} + \dfrac{1}{4x^6}} = \sqrt{\dfrac{x^6}{4} + \dfrac{1}{2} + \dfrac{1}{4x^6}} = \sqrt{\left(\dfrac{x^3}{2} + \dfrac{1}{2x^3}\right)^2}$. Since $x > 0$, $\sqrt{\left(\dfrac{x^3}{2} + \dfrac{1}{2x^3}\right)^2}$

$= \dfrac{x^3}{2} + \dfrac{1}{2x^3} = \dfrac{1}{2}\left(x^3 + \dfrac{1}{x^3}\right)$.

Chapter 1 Test

1. (a) $H = 24n$ (b) $P = (n-1) \cdot n \cdot (n+1) = n^3 - n$

2. (a) $[-3, 2]$ $(4, \infty)$

    (b) $x < 5 \Leftrightarrow x \in (-\infty, 5)$ $-2 \le x \le 1 \Leftrightarrow x \in [-2, 1]$

    (c) Distance $= |-22 - 31| = |-53| = 53$

3. (a) $(-3)^4 = 81$ (b) $2^{-4} = \dfrac{1}{2^4} = \dfrac{1}{16}$

    (c) $\dfrac{5^{18}}{5^{12}} = 5^{18-12} = 5^6 = 15{,}625$

4. (a) $\left(\dfrac{2}{3}\right)^{-1} = \dfrac{3}{2}$ (b) $\dfrac{\sqrt{32}}{\sqrt{8}} = \sqrt{\dfrac{32}{8}} = \sqrt{4} = 2$

    (c) $16^{-3/4} = \left(2^4\right)^{-3/4} = 2^{-3} = \dfrac{1}{8}$

5. $\dfrac{\left(x^2\right)^a \left(\sqrt{x}\right)^b}{x^{a+b}\, x^{a-b}} = \dfrac{x^{2a} \cdot \left(x^{1/2}\right)^b}{x^{a+b+a-b}} = \dfrac{x^{2a} \cdot x^{b/2}}{x^{2a}} = x^{2a+(b/2)-2a} = x^{b/2}$

6. (a) $\sqrt{200} - \sqrt{8} = 10\sqrt{2} - 2\sqrt{2} = 8\sqrt{2}$

    (b) $(2a^3 b^2)(3ab^4)^3 = 2a^3 b^2 \cdot 3^3 a^3 b^{12} = 54 a^6 b^{14}$

    (c) $\left(\dfrac{x^2 y^{-3}}{y^5}\right)^{-4} = \left(x^2 y^{-3-5}\right)^{-4} = \left(x^2 y^{-8}\right)^{-4} = x^{2(-4)} y^{-8(-4)} = x^{-8} y^{32} = \dfrac{y^{32}}{x^8}$

    (d) $\left(\dfrac{2x^{1/4}}{y^{1/3} x^{1/6}}\right)^3 = \dfrac{2^3 x^{3/4}}{y^{3/3} x^{3/6}} = \dfrac{8x^{3/4}}{yx^{1/2}} = \dfrac{8x^{(3/4)-(1/2)}}{y} = \dfrac{8x^{1/4}}{y} = \dfrac{8\sqrt[4]{x}}{y}$

7. (a) $\dfrac{x^2 + 3x + 2}{x^2 - x - 2} = \dfrac{(x+1)(x+2)}{(x-2)(x+1)} = \dfrac{x+2}{x-2}$

    (b) $\dfrac{x^2}{x^2 - 4} - \dfrac{x+1}{x+2} = \dfrac{x^2}{(x-2)(x+2)} - \dfrac{x+1}{x+2} = \dfrac{x^2}{(x-2)(x+2)} + \dfrac{-(x+1)(x-2)}{(x-2)(x+2)}$

    $= \dfrac{x^2 - (x^2 - x - 2)}{(x-2)(x+2)} = \dfrac{x+2}{(x-2)(x+2)} = \dfrac{1}{x-2}$

    (c) $\dfrac{\dfrac{y}{x} - \dfrac{x}{y}}{\dfrac{1}{y} - \dfrac{1}{x}} = \dfrac{\dfrac{y}{x} - \dfrac{x}{y}}{\dfrac{1}{y} - \dfrac{1}{x}} \cdot \dfrac{xy}{xy} = \dfrac{y^2 - x^2}{x - y} = \dfrac{(y-x)(y+x)}{x-y} = \dfrac{-(x-y)(y+x)}{x-y} = -(y+x)$

8. (a) $325{,}000{,}000{,}000 = 3.25 \times 10^{11}$ (b) $0.000008931 = 8.931 \times 10^{-6}$

9. (a) $4(3 - x) - 3(x + 5) = 12 - 4x - 3x - 15 = -7x - 3$

    (b) $(x - 5)(2x + 3) = 2x^2 + 3x - 10x - 15 = 2x^2 - 7x - 15$

(c) $\left(\sqrt{x} + \sqrt{y}\right)\left(\sqrt{x} - \sqrt{y}\right) = \left(\sqrt{x}\right)^2 - \left(\sqrt{y}\right)^2 = x - y$

(d) $(3t + 4)^2 = (3t)^2 + 2(3t)(4) + (4)^2 = 9t^2 + 24t + 16$

(e) $(2 - x^2)^3 = (2)^3 - 3(2)^2(x^2) + 3(2)(x^2)^2 - (x^2)^3 = 8 - 12x^2 + 6x^4 - x^6$

10. (a) $9x^2 - 25 = (3x - 5)(3x + 5)$

   (b) $6x^2 + 7x - 5 = (2x - 1)(3x + 5)$

   (c) $x^3 - 4x^2 - 3x + 12 = x^2(x - 4) - 3(x - 4) = (x^2 - 3)(x - 4)$

   (d) $x^4 + 27x = x(x^3 + 27) = x(x + 3)(x^2 - 3x + 9)$

   (e) $3x^{3/2} - 9x^{1/2} + 6x^{-1/2} = 3x^{-1/2}(x^2 - 3x + 2) = 3x^{-1/2}(x - 2)(x - 1)$

   (f) $x^3y - 4xy = xy(x^2 - 4) = xy(x - 2)(x + 2)$

11. $\dfrac{\sqrt{10}}{\sqrt{5} - 2} = \dfrac{\sqrt{10}}{\sqrt{5} - 2} \cdot \dfrac{\sqrt{5} + 2}{\sqrt{5} + 2} = \dfrac{\sqrt{50} + 2\sqrt{10}}{5 - 4} = \dfrac{5\sqrt{2} + 2\sqrt{10}}{1} = 5\sqrt{2} + 2\sqrt{10}$

12. (a) $2x + 7 = 12 + \frac{5}{2}x \quad \Leftrightarrow \quad 4x + 14 = 24 + 5x \quad \Leftrightarrow \quad x = -10$

   (b) $\dfrac{2x}{x + 1} = \dfrac{2x - 1}{x} \quad \Leftrightarrow \quad (2x)(x) = (2x - 1)(x + 1) \ (x \neq -1, x \neq 0) \quad \Leftrightarrow$
   $2x^2 = 2x^2 + x - 1 \quad \Leftrightarrow \quad 0 = x - 1 \quad \Leftrightarrow \quad x = 1$

   (c) $3(x - 2)^2 - 16 = 0 \quad \Leftrightarrow \quad 3(x - 2)^2 = 16 \quad \Rightarrow \quad (x - 2)^2 = \frac{16}{3} \quad \Leftrightarrow \quad x - 2 = \pm\frac{4}{\sqrt{3}}$
   $\Leftrightarrow \quad x = 2 \pm \frac{4\sqrt{3}}{3}$

   (d) $x^5 = -243 \quad \Leftrightarrow \quad x = (-243)^{1/5} = -3$

   (e) $3x^{2/3} - 12 = 0 \quad \Leftrightarrow \quad 3x^{2/3} = 12 \quad \Leftrightarrow \quad x^{2/3} = 4 \quad \Leftrightarrow \quad x^2 = 4^3 = 64 \quad \Leftrightarrow \quad x = \pm 8$

13. $V = 4wh^2 + 2wh \quad \Leftrightarrow \quad V = (4h^2 + 2h)w \quad \Leftrightarrow \quad w = \dfrac{V}{4h^2 + 2h}$

Principles of Problem Solving

1. Let d be the distance traveled to and from work. Let t_1 and t_2 be the times for the trip from home to work and the trip from work to home, respectively. Using $time = \dfrac{distance}{rate}$, we get $t_1 = \dfrac{d}{50}$ and $t_2 = \dfrac{d}{30}$. Since $average\ speed = \dfrac{distance\ traveled}{total\ time}$, we have $average\ speed = \dfrac{2d}{t_1 + t_2} = \dfrac{2d}{\frac{d}{50} + \frac{d}{30}}$

    $\Leftrightarrow\quad average\ speed = \dfrac{150(2d)}{150\left(\frac{d}{50} + \frac{d}{30}\right)} = \dfrac{300d}{3d + 5d} = \dfrac{300}{8} = 37.5$ mi/h.

3. We use the formula $d = rt$ (distance = rate $\times$ time). Since the car and the van each travel at a speed of 40 mi/h, they approach each other at a combined speed of 80 mi/h. (The distance between them decreases at a rate of 80 mi/h.) So the time spent driving till they meet is $t = \dfrac{d}{r} = \dfrac{120}{80} = 1.5$ hours. Thus, the fly flies at a speed of 100 mi/h for 1.5 hours, and therefore travels a distance of $d = rt = (100)(1.5) = 150$ miles.

5. $\left(\sqrt{3 + 2\sqrt{2}} - \sqrt{3 - 2\sqrt{2}}\right)^2 = 3 + 2\sqrt{2} - 2 \cdot \sqrt{\left(3 + 2\sqrt{2}\right)\left(3 - 2\sqrt{2}\right)} + 3 - 2\sqrt{2}$

    $= 6 - 2 \cdot \sqrt{9 - 8} = 6 - 2 = 4$. Therefore, $\sqrt{3 + 2\sqrt{2}} - \sqrt{3 - 2\sqrt{2}} = \sqrt{4} = 2$.

7. By placing two amoebas into the vessel, we skip the first simple division which took 3 minutes. Thus when we place two amoebas into the vessel, it will take $60 - 3 = 57$ minutes for the vessel to be full of amoebas.

9. The statement is false. Here is one particular counterexample:

	Player A	Player B
First half	1 hit in 99 at-bats: average $= \frac{1}{99}$	0 hit in 1 at-bat: average $= \frac{0}{1}$
Second half	1 hit in 1 at-bat: average $= \frac{1}{1}$	98 hits in 99 at-bats: average $= \frac{98}{99}$
Entire season	2 hits in 100 at-bats: average $= \frac{2}{100}$	99 hits in 100 at-bats: average $= \frac{99}{100}$

11. **Method 1:**
    After the exchanges, the volume of liquid in the pitcher and in the cup is the same as it was to begin with. Thus, any coffee in the pitcher of cream must be replacing an equal amount of cream that has ended up in the coffee cup.

    Method 2:
    Alternatively, look at the drawing of the spoonful of coffee and cream mixture being returned to the pitcher of cream. Suppose it is possible to separate the cream and the coffee, as shown. Then you can see that the coffee going into the cream occupies the same volume as the cream that was left in the coffee.

    Method 3 (an algebraic approach):
    Suppose the cup of coffee has y spoonfuls of coffee. When one spoonful of cream is added to the coffee cup, the resulting mixture has the following ratios:

    $$\frac{cream}{mixture} = \frac{1}{y + 1} \qquad \frac{coffee}{mixture} = \frac{y}{y + 1}$$

So, when we remove a spoonful of the mixture and put it into the pitcher of cream, we are really removing $\dfrac{1}{y+1}$ of a spoonful of cream and $\dfrac{y}{y+1}$ spoonful of coffee. Thus the <u>amount</u> <u>of</u> <u>cream</u> left in the mixture (cream in the coffee) is $1 - \dfrac{1}{y+1} = \dfrac{y}{y+1}$ of a spoonful. This is the same as the <u>amount</u> <u>of</u> <u>coffee</u> we added to the cream.

13. (a) $(\text{Baby} + \text{Uncle}) + \text{Mother} = \text{Uncle} + \text{Mother} = \text{Uncle}$

(b) $\text{Father} \cdot (\text{Grandpa} + \text{Aunt}) = \text{Father} \cdot \text{Grandpa} = \text{Father}$

(c) <u>Commutative:</u> $x + y = (\text{older of } x \text{ and } y) = (\text{older of } y \text{ and } x) = y + x$
$x \cdot y = (\text{younger of } x \text{ and } y) = (\text{younger of } y \text{ and } x) = y \cdot x$
<u>Associative:</u> $(x + y) + z = (\text{older of } x \text{ and } y) + z = \text{older of (older of } x \text{ and } y) \text{ and } z = \text{older}$
of x and y and z; $x + (y + z) = x + (\text{older of } y \text{ and } z) = \text{older of } x \text{ and (older of } y \text{ and } z)$
$= \text{older of } x$ and y and z. So $(x + y) + z = x + (y + z)$
$(x \cdot y) \cdot z = (\text{younger of } x \text{ and } y) \cdot z = \text{younger of (younger of } x \text{ and } y) \text{ and } z = \text{younger of } x$
and y and z; $x \cdot (y \cdot z) = x \cdot (\text{younger of } y \text{ and } z) = \text{younger of } x \text{ and (younger of } y \text{ and } z)$
$= \text{younger of } x$ and y and z. So $(x \cdot y) \cdot z = x \cdot (y \cdot z)$.
<u>Distributive:</u> We consider three cases:
Case (i), x is oldest of x, y, and z. Then $x \cdot (y + z) = y + z$ and $x \cdot y + x \cdot z = y + z$, so
$x \cdot (y + z) = x \cdot y + x \cdot z$.
Case (ii), y is oldest. Then $x \cdot (y + z) = x \cdot y = x$ and $x \cdot y + x \cdot z = x + x \cdot z = x$ (since
$x \cdot z$ is no older than x by definition). Thus, $x \cdot (y + z) = x \cdot y + x \cdot z$.
Case (iii), z is oldest: then $x \cdot (y + z) = x \cdot z = x$ and $x \cdot y + x \cdot z = x \cdot y + x = x$ (since
$x \cdot y$ is no older than x by definition). Thus, $x \cdot (y + z) = x \cdot y + x \cdot z$.

(d) Again we consider three cases.
Case (i), x is oldest of x, y, and z. Then $x + (y \cdot z) = x$. Also, $x + y = x$ and $x + z = x$, so
$(x + y) \cdot (x + z) = x \cdot x = x$. Thus, $x + (y \cdot z) = (x + y) \cdot (x + z)$.
Case (ii), y is oldest of x, y, and z. Then $y \cdot z = z$, and so $x + (y \cdot z) = x + z$. Also,
$x + y = y$, so $(x + y) \cdot (x + z) = y \cdot (x + z) = x + z$. Thus, $x + (y \cdot z) = (x + y) \cdot (x + z)$.
The proof in the third case, when z is the oldest, is similar to the case when y is the oldest.
No, the property does not hold for real numbers. For example, if $x = y = z = 1$, then
$1 + (1 \cdot 1) \neq (1 + 1) \cdot (1 + 1)$.

15. Let $r_1 = 8$.
$r_2 = \frac{1}{2}\left(8 + \frac{72}{8}\right) = \frac{1}{2}(8 + 9) = 8.5$
$r_3 = \frac{1}{2}\left(8.5 + \frac{72}{8.5}\right) = \frac{1}{2}(8.5 + 8.471) = 8.485$
$r_4 = \frac{1}{2}\left(8.485 + \frac{72}{8.485}\right) = \frac{1}{2}(8.485 + 8.486) = 8.485$
Thus $\sqrt{72} \approx 8.49$.

17. We continue the pattern. Three parallel cuts produce 10 pieces. Thus, each new cut produces an additional 3 pieces. Since the first cut produces 4 pieces, we get the formula $f(n) = 4 + 3(n - 1)$, $n \geq 1$. Since $f(142) = 4 + 3(141) = 427$, we see that 142 parallel cuts produce 427 pieces.

19. $8^{15} \cdot 5^{37} = (2^3)^{15} \cdot 5^{37} = 2^{45} \cdot 5^{37} = 2^8 \cdot 2^{37} \cdot 5^{37} = 2^8 \cdot 10^{37} = 256 \times 10^{37}$. Therefore, the number has $3 + 37 = 40$ digits.

21. Since $x^3 + y^3 = (x+y)(x^2 - xy + y^2)$, we need to determine the value of xy.

$$x + y = 1 \quad \Leftrightarrow \quad x^2 + xy \qquad = x$$
$$x + y = 1 \quad \Leftrightarrow \qquad \underline{xy + y^2 = y}$$
$$x^2 + 2xy + y^2 = x + y$$

So $4 + 2xy = 1 \quad \Leftrightarrow \quad xy = -\frac{3}{2}$. Thus $x^3 + y^3 = (x+y)(x^2 - xy + y^2) = (1)(4 + \frac{3}{2}) = \frac{11}{2}$.

Chapter Two
Exercises 2.1

1.

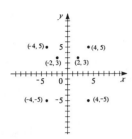

3. (a)

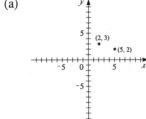

(b) $d = \sqrt{(2-5)^2 + (3-2)^2}$
$= \sqrt{(-3)^2 + (1)^2} = \sqrt{10}$

(c) midpoint: $\left(\frac{2+5}{2}, \frac{3+2}{2}\right) = \left(\frac{7}{2}, \frac{5}{2}\right)$

5. (a)

(b) $d = \sqrt{(6-(-1))^2 + (-2-3)^2}$
$= \sqrt{7^2 + (-5)^2} = \sqrt{49 + 25} = \sqrt{74}$

(c) midpoint: $\left(\frac{6-1}{2}, \frac{-2+3}{2}\right) = \left(\frac{5}{2}, \frac{1}{2}\right)$

7. (a)

(b) $d = \sqrt{(3-(-3))^2 + (4-(-4))^2}$
$= \sqrt{6^2 + 8^2} = \sqrt{36 + 64} = \sqrt{100}$
$= 10$

(c) midpoint: $\left(\frac{3+(-3)}{2}, \frac{4+(-4)}{2}\right) = (0,0)$

9. $d(A, B) = \sqrt{(1-5)^2 + (3-3)^2} = \sqrt{(-4)^2} = 4.$

$d(A, C) = \sqrt{(1-1)^2 + (3-(-3))^2} = \sqrt{(6)^2} = 6.$

$d(C, D) = \sqrt{(1-5)^2 + (-3-(-3))^2} = \sqrt{(-4)^2} = 4.$

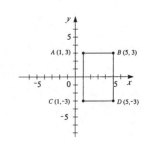

11. From the graph, the quadrilateral $ABCD$ has a pair of parallel sides,

so $ABCD$ is a trapezoid. The area is $\left(\dfrac{b_1 + b_2}{2}\right)h$. From the graph

we see that $b_1 = d(A, B) = \sqrt{(1 - 5)^2 + (0 - 0)^2} = \sqrt{4^2} = 4$;

$b_2 = d(C, D) = \sqrt{(4 - 2)^2 + (3 - 3)^2} = \sqrt{2^2} = 2$; and h is

the difference in y-coordinates $= |3 - 0| = 3$. Thus the area of the

trapezoid is $\left(\dfrac{4 + 2}{2}\right)3 = 9$.

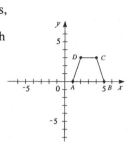

13.   15.   17.

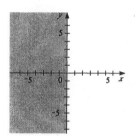   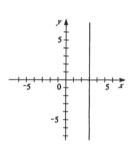   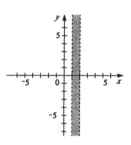

19. $x < 0$ and $y > 0$ or $x > 0$ 21.   23.
     and $y < 0$.

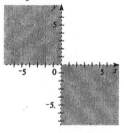   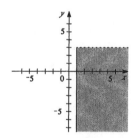   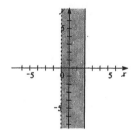

25. $d(0, A) = \sqrt{(6 - 0)^2 + (7 - 0)^2} = \sqrt{6^2 + 7^2} = \sqrt{36 + 49} = \sqrt{85}$.

$d(0, B) = \sqrt{(-5 - 0)^2 + (8 - 0)^2} = \sqrt{(-5)^2 + 8^2} = \sqrt{25 + 64} = \sqrt{89}$.

Thus point A is closer to the origin.

27. $d(P, R) = \sqrt{(-1 - 3)^2 + (-1 - 1)^2} = \sqrt{(-4)^2 + (-2)^2} = \sqrt{16 + 4} = \sqrt{20} = 2\sqrt{5}$.

$d(Q, R) = \sqrt{(-1 - (-1))^2 + (-1 - 3)^2} = \sqrt{0 + (-4)^2} = \sqrt{16} = 4$. Thus point Q is closer to
point R.

29. Since we do not know which pair are isosceles, we find the length of all three sides.

$d(A, B) = \sqrt{(-3 - 0)^2 + (-1 - 2)^2} = \sqrt{(-3)^2 + (-3)^2} = \sqrt{9 + 9} = \sqrt{18} = 3\sqrt{2}$.

$d(C, B) = \sqrt{(-3 - (-4))^2 + (-1 - 3)^2} = \sqrt{1^2 + (-4)^2} = \sqrt{1 + 16} = \sqrt{17}$.

$d(A, C) = \sqrt{(0 - (-4))^2 + (2 - 3)^2} = \sqrt{4^2 + (-1)^2} = \sqrt{16 + 1} = \sqrt{17}$. So sides AC and CB
have the same length.

31. (a) Here we have $A = (2, 2)$, $B = (3, -1)$ and $C = (-3, -3)$. So

$$d(A, B) = \sqrt{(3 - 2)^2 + (-1 - 2)^2} = \sqrt{1^2 + (-3)^2} = \sqrt{1 + 9} = \sqrt{10};$$
$$d(C, B) = \sqrt{(3 - (-3))^2 + (-1 - (-3))^2} = \sqrt{6^2 + 2^2} = \sqrt{36 + 4} = \sqrt{40} = 2\sqrt{10};$$
$$d(A, C) = \sqrt{(-3 - 2)^2 + (-3 - 2)^2} = \sqrt{(-5)^2 + (-5)^2} = \sqrt{25 + 25} = \sqrt{50} = 5\sqrt{2}.$$

Since $[d(A, B)]^2 + [d(C, B)]^2 = [d(A, C)]^2$, we conclude that the triangle is a right triangle.

(b) The area of the triangle is $\frac{1}{2} \cdot d(C, B) \cdot d(A, B) = \frac{1}{2} \cdot \sqrt{10} \cdot 2\sqrt{10} = 10$.

33. We show that all sides are the same length (its a rhombus) and then show that the diagonals are equal. So

$$d(A, B) = \sqrt{(4 - (-2))^2 + (6 - 9)^2} = \sqrt{6^2 + (-3)^2} = \sqrt{36 + 9} = \sqrt{45};$$
$$d(B, C) = \sqrt{(1 - 4)^2 + (0 - 6)^2} = \sqrt{(-3)^2 + (-6)^2} = \sqrt{9 + 36} = \sqrt{45};$$
$$d(C, D) = \sqrt{(-5 - 1)^2 + (3 - 0)^2} = \sqrt{(-6)^2 + (-3)^2} = \sqrt{36 + 9} = \sqrt{45};$$
$$d(D, A) = \sqrt{(-2 - (-5))^2 + (9 - 3)^2} = \sqrt{3^2 + 6^2} = \sqrt{9 + 36} = \sqrt{45}. \text{ So the points form a}$$

rhombus. Also

$$d(A, C) = \sqrt{(1 - (-2))^2 + (0 - 9)^2} = \sqrt{3^2 + (-9)^2} = \sqrt{9 + 81} = \sqrt{90} = 3\sqrt{10},$$

and $d(B, D) = \sqrt{(-5 - 4)^2 + (3 - 6)^2} = \sqrt{(-9)^2 + (-3)^2} = \sqrt{81 + 9} = \sqrt{90} = 3\sqrt{10}$. Since the diagonals are equal, the rhombus is a square.

35. Let $P = (0, y)$ be such a point. Setting the distances equal we get

$$\sqrt{(0 - 5)^2 + (y - (-5))^2} = \sqrt{(0 - 1)^2 + (y - 1)^2} \quad \Leftrightarrow$$
$$\sqrt{25 + y^2 + 10y + 25} = \sqrt{1 + y^2 - 2y + 1} \quad \Rightarrow \quad y^2 + 10y + 50 = y^2 - 2y + 2 \quad \Leftrightarrow$$
$12y = -48 \quad \Leftrightarrow \quad y = -4$. Thus, the point is $P = (0, -4)$. Check:

$$\sqrt{(0 - 5)^2 + (-4 - (-5))^2} = \sqrt{(-5)^2 + 1^2} = \sqrt{25 + 1} = \sqrt{26};$$
$$\sqrt{(0 - 1)^2 + (-4 - 1)^2} = \sqrt{(-1)^2 + (-5)^2} = \sqrt{25 + 1} = \sqrt{26}.$$

37. We find the midpoint, M, of PQ and then the midpoint of PM. So $M = \left(\frac{-1+7}{2}, \frac{3+5}{2}\right) = (3, 4)$, and the midpoint of PM is $\left(\frac{-1+3}{2}, \frac{3+4}{2}\right) = \left(1, \frac{7}{2}\right)$.

39. As indicated by Example 5, we must find a point $S(x_1, y_1)$ such that he midpoints of PR and of QS are the same. Thus

$$\left(\frac{4 + (-1)}{2}, \frac{2 + (-4)}{2}\right) = \left(\frac{x_1 + 1}{2}, \frac{y_1 + 1}{2}\right). \text{ Setting the}$$

x-coordinates equal, we get $\dfrac{4 + (-1)}{2} = \dfrac{x_1 + 1}{2} \quad \Leftrightarrow$

$4 - 1 = x_1 + 1 \quad \Leftrightarrow \quad x_1 = 2$. Setting the y-coordinates equal, we get $\dfrac{2 + (-4)}{2} = \dfrac{y_1 + 1}{2} \quad \Leftrightarrow \quad 2 - 4 = y_1 + 1 \quad \Leftrightarrow$

$y_1 = -3$. Thus $S = (2, -3)$.

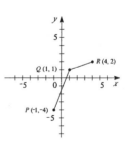

41. (a)

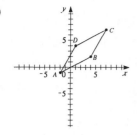

(b) The midpoint of AC is $\left(\frac{-2+7}{2}, \frac{-1+7}{2}\right) = \left(\frac{5}{2}, 3\right)$, and the midpoint of BD is $\left(\frac{4+1}{2}, \frac{2+4}{2}\right) = \left(\frac{5}{2}, 3\right)$.

(c) Since the they have the same midpoint, we conclude the diagonals bisect each other.

43. (a) The point $(5, 3)$ is shifted to $(5 + 3, 3 + 2) = (8, 5)$.

(b) The point (a, b) is shifted to $(a + 3, b + 2)$.

(c) Let (x, y) be the point that is shifted to $(3, 4)$. Then $(x + 3, y + 2) = (3, 4)$. Setting the x-coordinates equal, we get $x + 3 = 3 \quad \Leftrightarrow \quad x = 0$. Setting the y-coordinates equal, we get $y + 2 = 4 \quad \Leftrightarrow \quad y = 2$. So the point is $(0, 2)$.

(d) $A = (-5, -1)$ so $A' = (-5 + 3, -1 + 2) = (-2, 1)$;
$B = (-3, 2)$ so $B' = (-3 + 3, 2 + 2) = (0, 4)$;
$C = (2, 1)$ so $C' = (2 + 3, 1 + 2) = (5, 3)$.

45. We solve the equation $6 = \dfrac{2 + x}{2}$ to find the x coordinate of B: $6 = \dfrac{2 + x}{2} \quad \Leftrightarrow \quad 12 = 2 + x$ $\Leftrightarrow \quad x = 10$. Likewise, for the y coordinate of B, we have $8 = \dfrac{3 + y}{2} \quad \Leftrightarrow \quad 16 = 3 + y \quad \Leftrightarrow$ $y = 13$. Thus $B = (10, 13)$.

Exercises 2.2

1. $(0,0)$: $0 \stackrel{?}{=} 2(0) + 3$ $\Leftrightarrow$ $0 \stackrel{?}{=} 3$ No. $\left(\frac{1}{2}, 4\right)$: $4 \stackrel{?}{=} 2\left(\frac{1}{2}\right) + 3$ $\Leftrightarrow$ $4 \stackrel{?}{=} 1 + 3$ Yes.

   $(1,4)$: $4 \stackrel{?}{=} 2(1) + 3$ $\Leftrightarrow$ $4 \stackrel{?}{=} 2 + 3$ No.

   So $\left(\frac{1}{2}, 4\right)$ is on the graph of this equation.

3. $(0,0)$: $2(0) - 0 + 1 \stackrel{?}{=} 0$ $\Leftrightarrow$ $1 \stackrel{?}{=} 0$ No. $(1,0)$: $2(0) - 1 + 1 \stackrel{?}{=} 0$ $\Leftrightarrow$ $-1 + 1 \stackrel{?}{=} 0$ Yes.

   $(-1,-1)$: $2(-1) - (-1) + 1 \stackrel{?}{=} 0$ $\Leftrightarrow$ $-2 + 1 + 1 \stackrel{?}{=} 0$ Yes.

   So $(1,0)$ and $(-1,-1)$ are points on the graph of this equation.

5. $(0,-2)$: $(0)^2 + (0)(-2) + (-2)^2 \stackrel{?}{=} 4$ $\Leftrightarrow$ $0 + 0 + 4 \stackrel{?}{=} 4$ Yes.

   $(1,-2)$: $(1)^2 + (1)(-2) + (-2)^2 \stackrel{?}{=} 4$ $\Leftrightarrow$ $1 - 2 + 4 \stackrel{?}{=} 4$ No.

   $(2,-2)$: $(2)^2 + (2)(-2) + (-2)^2 \stackrel{?}{=} 4$ $\Leftrightarrow$ $4 - 4 + 4 \stackrel{?}{=} 4$ Yes.

   So $(0,-2)$ and $(2,-2)$ are points on the graph of this equation.

7. To find x-intercepts, set $y = 0$. This gives $0 = x - 3$ $\Leftrightarrow$ $x = 3$, so the x-intercept is 3.

   To find y-intercepts, set $x = 0$. This gives $y = 0 - 3$ $\Leftrightarrow$ $y = -3$, so the y-intercept is -3.

9. To find x-intercepts, set $y = 0$. This gives $0 = x^2 - 9$ $\Leftrightarrow$ $x^2 = 9$ $\Rightarrow$ $x = \pm 3$, so the x-intercept are ± 3. To find y-intercepts, set $x = 0$. This gives $y = (0)^2 - 9$ $\Leftrightarrow$ $y = -9$, so the y-intercept is -9.

11. To find x-intercepts, set $y = 0$. This gives $x^2 + (0)^2 = 4$ $\Leftrightarrow$ $x^2 = 4$ $\Rightarrow$ $x = \pm 2$, so the x-intercept are ± 2. To find y-intercepts, set $x = 0$. This gives $(0)^2 + y^2 = 4$ $\Leftrightarrow$ $y^2 = 4$ $\Rightarrow$ $y = \pm 2$, so the y-intercept are ± 2.

13. To find x-intercepts, set $y = 0$. This gives $x(0) = 5$ $\Leftrightarrow$ $0 = 5$, which is impossible, so there is no x-intercept. Likewise, to find y-intercepts, set $x = 0$. This gives $(0)y = 5$ $\Leftrightarrow$ $0 = 5$, which is again impossible, so there is no y-intercept.

15.

x	y
-4	-4
-2	-2
0	0
1	1
2	2
3	3
4	4

$y = 0$ $\Rightarrow$ $x = 0$. So the x-intercept is 0, and the y-intercept is also 0.

x-axis symmetry: $(-y) = x$ $\Leftrightarrow$ $-y = x$, which is not the same as $y = x$, so not symmetric with respect to the x-axis.

y-axis symmetry: $y = (-x)$ $\Leftrightarrow$ $y = -x$, which is not the same as $y = x$, so not symmetric with respect to the y-axis.

Origin symmetry: $(-y) = (-x)$ is the same as $y = x$, so it is symmetric with respect to the origin.

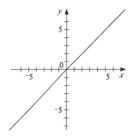

17.

x	y
-3	-4
-2	-3
-1	-2
0	-1
1	0
2	1
3	2

$y = 0$ $\Rightarrow$ $0 = x - 1$ $\Leftrightarrow$ $x = 1$, so the x-intercept is 1, and $x = 0$ $\Rightarrow$ $y = 0 - 1 = -1$, so the y-intercept is -1.

x-axis symmetry: $(-y) = x - 1$ $\Leftrightarrow$ $-y = x - 1$, which is not the same as $y = x - 1$, so not symmetric with respect to the x-axis.

y-axis symmetry: $y = (-x) - 1$ $\Leftrightarrow$ $y = -x - 1$, which is not the same as $y = x - 1$, so not symmetric with respect to the y-axis.

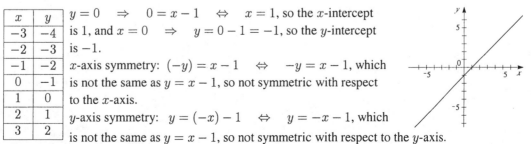

Origin symmetry: $(-y) = (-x) - 1 \quad \Leftrightarrow \quad -y = -x - 1 \quad \Leftrightarrow \quad y = x + 1$, which is not the same as $y = x - 1$, so not symmetric with respect to the origin.

19. Solve for y: $-y = -3x + 5 \quad \Leftrightarrow \quad y = 3x - 5$.

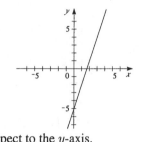

x	y
-2	-11
-1	-8
0	-5
1	-2
2	1
3	4
4	7

$y = 0 \quad \Rightarrow \quad 0 = 3x - 5 \quad \Leftrightarrow \quad 3x = 5 \quad \Leftrightarrow \quad x = \frac{5}{3}$,
so x-intercept is $\frac{5}{3}$, and $x = 0 \quad \Rightarrow \quad y = 3(0) - 5 = -5$,
so y-intercept is -5.
x-axis symmetry: $3x - (-y) = 5 \quad \Leftrightarrow \quad 3x + y = 5$, which
is not the same as $3x - y = 5$, so not symmetric with respect
to the x-axis.
y-axis symmetry: $3(-x) - y = 5 \quad \Leftrightarrow \quad -3x - y = 5$,
which is not the same as $3x - y = 5$, so not symmetric with respect to the y-axis.

Origin symmetry: $3(-x) - (-y) = 5 \quad \Leftrightarrow \quad -3x + y = 5$ which is not the same as $3x - y = 5$,
so not symmetric with respect to the origin.

21.

x	y
-3	-8
-2	-3
-1	0
0	1
1	0
2	-3
3	-8

$y = 0 \quad \Rightarrow 0 = 1 - x^2 \quad \Leftrightarrow \quad x^2 = 1 \quad \Rightarrow$
$x = \pm 1$, so the x-intercepts are 1 and -1, and $x = 0$
$\Rightarrow \quad y = 1 - (0)^2 = 1$, so the y-intercept is 1.
x-axis symmetry: $(-y) = 1 - x^2 \quad \Leftrightarrow \quad -y = 1 - x^2$,
which is not the same as $y = 1 - x^2$, so not symmetric with
respect to the x-axis.
y-axis symmetry: $y = 1 - (-x)^2 \quad \Leftrightarrow \quad y = 1 - x^2$, so it is
symmetric with respect to the y-axis.

Origin symmetry: $(-y) = 1 - (-x)^2 \quad \Leftrightarrow \quad -y = 1 - x^2$ which is not the same as $y = 1 - x^2$.
Not symmetric with respect to the origin.

23. $4y = x^2 \quad \Leftrightarrow \quad y = \frac{1}{4}x^2$

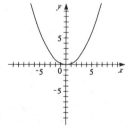

x	y
-6	9
-4	4
-2	1
0	0
2	1
4	4
3	2

$y = 0 \quad \Rightarrow \quad 0 = \frac{1}{4}x^2 \quad \Leftrightarrow \quad x^2 = 0 \quad \Rightarrow \quad x = 0$, so
the x-intercept is 0, and $x = 0 \quad \Rightarrow \quad y = \frac{1}{4}(0)^2 = 0$, so
the y-intercept is 0.
x-axis symmetry: $(-y) = \frac{1}{4}x^2$, which is not the same as
$y = \frac{1}{4}x^2$, so not symmetric with respect to the x-axis.
y-axis symmetry: $y = \frac{1}{4}(-x)^2 \quad \Leftrightarrow \quad y = \frac{1}{4}x^2$, so it is
symmetric with respect to the y-axis.

Origin symmetry: $(-y) = \frac{1}{4}(-x)^2 \quad \Leftrightarrow \quad -y = \frac{1}{4}x^2$, which is not the same as $y = \frac{1}{4}x^2$, so not
symmetric with respect to the origin.

25.

x	y
-4	7
-3	0
-2	-5
-1	-8
0	-9
1	-8
2	-5
3	0
4	7

$y = 0 \quad \Rightarrow 0 = x^2 - 9 \quad \Leftrightarrow \quad x^2 = 9 \quad \Rightarrow$
$x = \pm 3$, so the x-intercepts are 3 and -3, and $x = 0$
$\Rightarrow \quad y = (0)^2 - 9 = -9$, so the y-intercept is -9.
x-axis symmetry: $(-y) = x^2 - 9$, which is not the same as
$y = x^2 - 9$, so not symmetric with respect to the x-axis.
y-axis symmetry: $y = (-x)^2 - 9 \quad \Leftrightarrow \quad y = x^2 - 9$, so it
is symmetric with respect to the y-axis.

Origin symmetry: $(-y) = (-x)^2 - 9 \quad \Leftrightarrow \quad -y = x^2 - 9$, which is not the same as
$y = x^2 - 9$, so not symmetric with respect to the origin.

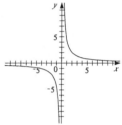

27. $xy = 2 \quad \Leftrightarrow \quad y = \dfrac{2}{x}$

x	y
-4	$-\frac{1}{2}$
-2	-1
-1	-2
$-\frac{1}{2}$	-4
$-\frac{1}{4}$	-8

x	y
$\frac{1}{4}$	8
$\frac{1}{2}$	4
1	2
2	1
4	$\frac{1}{2}$

$y = 0$ or $x = 0 \quad \Rightarrow \quad 0 = 2$, which is
impossible, so this equation has no x-intercept
and no y-intercept.
x-axis symmetry: $x(-y) = 2 \quad \Leftrightarrow \quad -xy = 2$,
whixh is not the same as $xy = 2$, so not symmetric
with respect to the x-axis.
y-axis symmetry: $(-x)y = 2 \quad \Leftrightarrow \quad -xy = 2$,

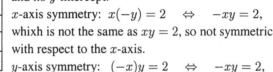

which is not the same as $xy = 2$, so not symmetric with respect to the y-axis.
Origin symmetry: $(-x)(-y) = 2 \quad \Leftrightarrow \quad xy = 2$, so it is symmetric with respect to the origin.

29.

x	y
0	0
$\frac{1}{4}$	$\frac{1}{2}$
1	1
2	$\sqrt{2}$
4	2
9	3
16	4

$y = 0 \quad \Rightarrow \quad 0 = \sqrt{x} \quad \Leftrightarrow \quad x = 0$. so the x-intercept is
0, and $x = 0 \quad \Rightarrow \quad y = \sqrt{0} = 0$, so the y-intercept is 0.
Since we are graphing real numbers and $\sqrt{x}$ is defined to be
a non-negative number, the equation is not symmetric with
respect to the x-axis nor with respect to the y-axis. Also,
the equation is not symmetric with respect to the origin.

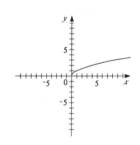

31.   Since the radicand (the inside of the square root) cannot be negative, we must have $4 - x^2 \geq 0 \quad \Leftrightarrow$
$x^2 \leq 4 \quad \Leftrightarrow \quad |x| \leq 2$.

x	y
-2	0
-1	$\sqrt{3}$
0	4
1	$\sqrt{3}$
2	0

$y = 0 \quad \Rightarrow \quad 0 = \sqrt{4 - x^2} \quad \Leftrightarrow \quad 4 - x^2 = 0 \quad \Leftrightarrow$
$x^2 = 4 \quad \Rightarrow \quad x = \pm 2$, so the x-intercept are -2 and 2,
and $x = 0 \quad \Rightarrow \quad y = \sqrt{4 - (0)^2} = \sqrt{4} = 2$, so the
y-intercept is 2.
Since $y \geq 0$, the graph is not symmetric with respect to the
x-axis.

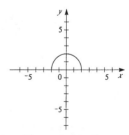

y-axis symmetry: $y = \sqrt{4 - (-x)^2} = \sqrt{4 - x^2}$, so the graph is symmetric with respect to the
y-axis. Also, since $y \geq 0$ the graph is not symmetric with respect to the origin.

33.

x	y
-3	3
-2	2
-1	1
0	0
1	1
2	2
3	3

$y = 0 \quad \Rightarrow \quad 0 = |x| \quad \Leftrightarrow \quad x = 0$, so the x-intercept is 0, and $x = 0 \quad \Rightarrow \quad y = |0| = 0$, so the y-intercept is 0.

Since $y \geq 0$, the graph is not symmetric with respect to the x-axis.

y-axis symmetry: $y = |-x| = |x|$, so the graph is symmetric with respect to the y-axis.

Since $y \geq 0$, the graph is not symmetric with respect to the origin.

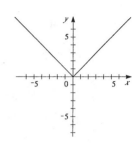

35.

x	y
-6	-2
-4	0
-2	2
0	4
2	2
4	0
6	-2

$y = 0 \quad \Rightarrow \quad 0 = 4 - |x| \quad \Leftrightarrow \quad |x| = 4 \quad \Rightarrow x = \pm 4$, so the x-intercepts are -4 and 4, and $x = 0 \quad \Rightarrow$ $y = 4 - |0| = 4$, so the y-intercept is 4.

x-axis symmetry: $(-y) = 4 - |x| \quad \Leftrightarrow \quad y = -4 + |x|$, which is not the same as $y = 4 - |x|$, so not symmetric with respect to the x-axis.

y-axis symmetry: $y = 4 - |-x| = 4 - |x|$, so it is symmetric with respect to the y-axis.

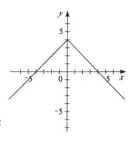

Origin symmetry: $(-y) = 4 - |-x| \quad \Leftrightarrow \quad y = -4 + |x|$, which is not the same as $y = 4 - |x|$, so not symmetric with respect to the origin.

37. Since $x = y^3$ is solved for x in terms of y, we insert values for y and find the corresponding values of x.

x	y
-27	-3
-8	-2
-1	-1
0	0
1	1
8	2
27	3

$y = 0 \quad \Rightarrow \quad x = (0)^3 = 0$, so the x-intercept is 0, and $x = 0 \quad \Rightarrow \quad 0 = y^3 \quad \Rightarrow \quad y = 0$, so the y-intercept is 0.

x-axis symmetry: $x = (-y)^3 = -y^3$, which is not the same as $x = y^3$, so not symmetric with respect to the x-axis.

y-axis symmetry: $(-x) = y^3 \quad \Leftrightarrow \quad x = -y^3$, which is not the same as $x = y^3$, so not symmetric with respect to the y-axis.

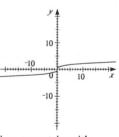

Origin symmetry: $(-x) = (-y)^3 \quad \Leftrightarrow \quad -x = -y^3 \quad \Leftrightarrow \quad x = y^3$, so it is symmetric with respect to the origin.

39.

x	y
-3	81
-2	16
-1	1
0	0
1	1
2	16
3	81

$y = 0 \quad \Rightarrow \quad 0 = x^4 \quad \Rightarrow \quad x = 0$, so the x-intercept is 0, and $x = 0 \quad \Rightarrow \quad y = 0^4 = 0$, so the y-intercept is 0.

x-axis symmetry: $(-y) = x^4 \quad \Leftrightarrow \quad y = -x^4$, which is not the same as $y = x^4$, so not symmetric with respect to the x-axis.

y-axis symmetry: $y = (-x)^4 = x^4$, so it is symmetric with respect to the y-axis.

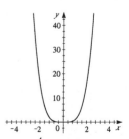

Origin symmetry: $(-y) = (-x)^4 \quad \Leftrightarrow \quad -y = x^4$, which is not the same as $y = x^4$, so not symmetric with respect to the origin.

41. x-axis symmetry: $(-y) = x^4 + x^2 \quad \Leftrightarrow \quad y = -x^4 - x^2$, which is not the same as $y = x^4 + x^2$, so not symmetric with respect to the x-axis.

y-axis symmetry: $y = (-x)^4 + (-x)^2 = x^4 + x^2$, so it is symmetric with respect to the y-axis.
Origin symmetry: $(-y) = (-x)^4 + (-x)^2$ $\Leftrightarrow$ $-y = x^4 + x^2$, which is not the same as $y = x^4 + x^2$, so not symmetric with respect to the origin.

43. x-axis symmetry: $x^2(-y)^2 + x(-y) = 1$ $\Leftrightarrow$ $x^2y^2 - xy = 1$, which is not the same as $x^2y^2 + xy = 1$, so not symmetric with respect to the x-axis.
y-axis symmetry: $(-x)^2y^2 + (-x)y = 1$ $\Leftrightarrow$ $x^2y^2 - xy = 1$, which is not the same as $x^2y^2 + xy = 1$, so not symmetric with respect to the y-axis.
Origin symmetry: $(-x)^2(-y)^2 + (-x)(-y) = 1$ $\Leftrightarrow$ $x^2y^2 + xy = 1$, so it is symmetric with respect to the origin.

45. x-axis symmetry: $(-y) = x^3 + 10x$ $\Leftrightarrow$ $y = -x^3 - 10x$, which is not the same as $y = x^3 + 10x$, so not symmetric with respect to the x-axis.
y-axis symmetry: $y = (-x)^3 + 10(-x)$ $\Leftrightarrow$ $y = -x^3 - 10x$, which is not the same as $y = x^3 + 10x$, so not symmetric with respect to the y-axis.
Origin symmetry: $(-y) = (-x)^3 + 10(-x)$ $\Leftrightarrow$ $-y = -x^3 - 10x$ $\Leftrightarrow$ $y = x^3 + 10x$, so it is symmetric with respect to the origin.

47. Symmetric with respect to the y-axis.

49. Symmetric with respect to the origin.

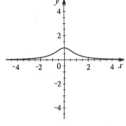

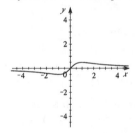

51. Using $h = 2$, $k = -1$, and $r = 3$, we get $(x - 2)^2 + (y - (-1))^2 = 3^2$ $\Leftrightarrow$ $(x - 2)^2 + (y + 1)^2 = 9$.

53. The equation of a circle centered at the origin is $x^2 + y^2 = r^2$. Using the point $(4, 7)$ we solve for r^2. This gives $(4)^2 + (7)^2 = r^2$ $\Leftrightarrow$ $16 + 49 = 65 = r^2$. Thus, the equation of the circle is $x^2 + y^2 = 65$.

55. The center is at the midpoint of the line segment, which is $\left(\frac{-1+5}{2}, \frac{1+5}{2}\right) = (2, 3)$. The radius is one half the diameter, so $r = \frac{1}{2}\sqrt{(-1 - 5)^2 + (1 - 5)^2} = \frac{1}{2}\sqrt{36 + 16} = \frac{1}{2}\sqrt{52} = \sqrt{13}$. Thus, the equation of the circle is $(x - 2)^2 + (y - 3)^2 = \left(\sqrt{13}\right)^2$ or $(x - 2)^2 + (y - 3)^2 = 13$.

57. Since the circle is tangent to the x-axis, it must contain the point $(7, 0)$, so the radius is the change in the y-coordinates. That is, $r = |-3 - 0| = 3$. So the equation of the circle is $(x - 7)^2 + (y - (-3))^2 = 3^2$, which is $(x - 7)^2 + (y + 3)^2 = 9$.

59. From the figure, the center of the circle is at $(-2, 2)$. The radius is the change in the y-coordinates, so $r = |2 - 0| = 2$. Thus the equation of the circle is $(x - (-2))^2 + (y - 2)^2 = 2^2$, which is $(x + 2)^2 + (y - 2)^2 = 4$.

61. Completing the square gives $x^2 + y^2 - 2x + 4y + 1 = 0$ $\Leftrightarrow$ $x^2 - 2x + \_\_ + y^2 + 4y + \_\_ = -1$ $\Leftrightarrow$

$x^2 - 2x + \left(\frac{-2}{2}\right)^2 + y^2 + 4y + \left(\frac{4}{2}\right)^2 = -1 + \left(\frac{-2}{2}\right)^2 + \left(\frac{4}{2}\right)^2 \Leftrightarrow$

$x^2 - 2x + 1 + y^2 + 4y + 4 = -1 + 1 + 4 \quad \Leftrightarrow \quad (x-1)^2 + (y+2)^2 = 4.$

Thus, the center is $(1, -2)$, and the radius is 2.

63. Completing the square gives $x^2 + y^2 - 4x + 10y + 13 = 0 \quad \Leftrightarrow$

$x^2 - 4x + \underline{} + y^2 + 10y + \underline{} = -13 \quad \Leftrightarrow$

$x^2 - 4x + \left(\frac{-4}{2}\right)^2 + y^2 + 10y + \left(\frac{10}{2}\right)^2 = -13 + \left(\frac{4}{2}\right)^2 + \left(\frac{10}{2}\right)^2 \quad \Leftrightarrow$

$x^2 - 4x + 4 + y^2 + 10y + 25 = -13 + 4 + 25 \quad \Leftrightarrow \quad (x-2)^2 + (y+5)^2 = 16.$

Thus, the center is $(2, -5)$, and the radius is 4.

65. Completing the square gives $x^2 + y^2 + x = 0 \quad \Leftrightarrow \quad x^2 + x + \underline{} + y^2 = 0 \quad \Leftrightarrow$

$x^2 + x + \left(\frac{1}{2}\right)^2 + y^2 = \left(\frac{1}{2}\right)^2 \quad \Leftrightarrow \quad x^2 + x + \frac{1}{4} + y^2 = \frac{1}{4} \quad \Leftrightarrow \quad \left(x + \frac{1}{2}\right)^2 + y^2 = \frac{1}{4}.$ Thus, the

center: $\left(-\frac{1}{2}, 0\right)$, and the radius is $\frac{1}{2}$.

67. Completing the square gives $x^2 + y^2 - \frac{1}{2}x + \frac{1}{2}y = \frac{1}{8} \quad \Leftrightarrow \quad x^2 - \frac{1}{2}x + \underline{} + y^2 + \frac{1}{2}y + \underline{} = \frac{1}{8}$

$\Leftrightarrow \quad x^2 - \frac{1}{2}x + \left(\frac{-1/2}{2}\right)^2 + y^2 + \frac{1}{2}y + \left(\frac{1/2}{2}\right)^2 = \frac{1}{8} + \left(\frac{-1/2}{2}\right)^2 + \left(\frac{1/2}{2}\right)^2 \quad \Leftrightarrow$

$x^2 - \frac{1}{2}x + \frac{1}{16} + y^2 + \frac{1}{2}y + \frac{1}{16} = \frac{1}{8} + \frac{1}{16} + \frac{1}{16} = \frac{2}{8} = \frac{1}{4} \quad \Leftrightarrow \quad \left(x - \frac{1}{4}\right)^2 + \left(y + \frac{1}{4}\right)^2 = \frac{1}{4}.$ Thus, the

center is $\left(\frac{1}{4}, -\frac{1}{4}\right)$, and the radius is $\frac{1}{2}$.

69. Completing the square gives $x^2 + y^2 + 4x - 10y = 21 \quad \Leftrightarrow$

$x^2 + 4x + \underline{} + y^2 - 10y + \underline{} = 21 \quad \Leftrightarrow$

$x^2 + 4x + \left(\frac{4}{2}\right)^2 + y^2 - 10y + \left(\frac{-10}{2}\right)^2 = 21 + \left(\frac{4}{2}\right)^2 + \left(\frac{-10}{2}\right)^2 \quad \Leftrightarrow$

$(x+2)^2 + (y-5)^2 = 21 + 4 + 25 = 50.$

Thus, the center is $(-2, 5)$, and the radius is $\sqrt{50} = 5\sqrt{2}$.

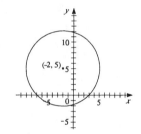

71. Completing the square gives $x^2 + y^2 + 6x - 12y + 45 = 0 \quad \Leftrightarrow$

$x^2 + 6x + \underline{} + y^2 - 12y + \underline{} = -45 \quad \Leftrightarrow$

$x^2 + 6x + \left(\frac{6}{2}\right)^2 + y^2 - 12y + \left(\frac{-12}{2}\right)^2 = -45 + \left(\frac{6}{2}\right)^2 + \left(\frac{-12}{2}\right)^2 \quad \Leftrightarrow$

$(x+3)^2 + (y-6)^2 = -45 + 9 + 36 = 0.$ Thus, the center is $(-3, 6)$,

and the radius is 0. This is a degenerate circle whose graph consists only

of the point $(-3, 6)$.

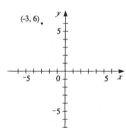

73. $\{(x, y)\mid x^2 + y^2 \le 1\}.$ This is the set of points inside (and on) the circle

$x^2 + y^2 = 1.$

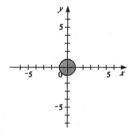

75. $\{(x, y)|\ 1 \le x^2 + y^2 < 9\}$. This is the set of points outside the circle
     $x^2 + y^2 = 1$ and inside (but not on) the circle $x^2 + y^2 = 9$.

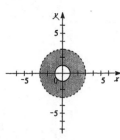

77. Completing the square gives $x^2 + y^2 - 4y - 12 = 0 \quad \Leftrightarrow \quad x^2 + y^2 - 4y + \underline{\quad} = 12 \quad \Leftrightarrow$
     $x^2 + y^2 - 4y + \left(\frac{-4}{2}\right)^2 = 12 + \left(\frac{-4}{2}\right)^2 \quad \Leftrightarrow \quad x^2 + (y - 2)^2 = 16$. Thus, the center is $(0, 2)$, and the
     radius is 4. So the circle $x^2 + y^2 = 4$, with center $(0, 0)$ and radius 2, sits completely inside the
     larger circle. Thus, the area is $\pi 4^2 - \pi 2^2 = 16\pi - 4\pi = 12\pi$.

79. Completing the square gives $x^2 + y^2 + ax + by + c = 0 \quad \Leftrightarrow$
     $x^2 + ax + \underline{\quad} + y^2 + by + \underline{\quad} = -c \quad \Leftrightarrow$
     $x^2 + ax + \left(\frac{a}{2}\right)^2 + y^2 + by + \left(\frac{b}{2}\right)^2 = -c + \left(\frac{a}{2}\right)^2 + \left(\frac{b}{2}\right)^2 \quad \Leftrightarrow$
     $\left(x + \frac{a}{2}\right)^2 + \left(y + \frac{b}{2}\right)^2 = -c + \frac{a^2 + b^2}{4}.$

     This equation represents a circle only when $-c + \dfrac{a^2 + b^2}{4} > 0$. This equation represents a point

     when $-c + \dfrac{a^2 + b^2}{4} = 0$, and this equation represents the empty set when $-c + \dfrac{a^2 + b^2}{4} < 0$.

     When the equation represents a circle, the center is $\left(-\dfrac{a}{2}, -\dfrac{b}{2}\right)$, and the radius is

     $\sqrt{-c + \dfrac{a^2 + b^2}{4}} = \dfrac{1}{2}\sqrt{a^2 + b^2 - 4ac}.$

81. (a) Symmetric about the x-axis.   (b) Symmetric about the y-axis.

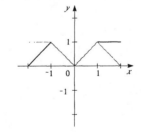

     (c) Symmetric about the origin.

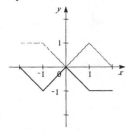

Exercises 2.3

1. $y = x^2 - 16$

   (a) $[-10, 10]$ by $[-10, 10]$

   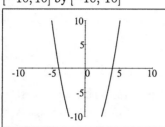

   (b) $[-5, 5]$ by $[-20, 10]$

   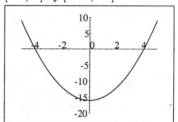

   (c) $[-2, 2]$ by $[-2, 2]$

   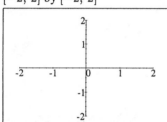

   (d) $[0, 4]$ by $[-16, 0]$

   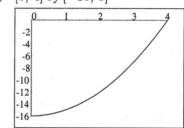

   The viewing rectangle in part (b) produces the most appropriate graph of the equation.

3. $y = x^4 + 2$

   (a) $[-2, 2]$ by $[-2, 2]$

   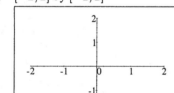

   (b) $[0, 4]$ by $[0, 4]$

   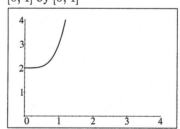

   (c) $[-8, 8]$ by $[-4, 40]$

   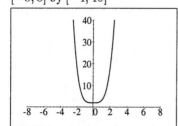

   (d) $[-40, 40]$ by $[-80, 800]$

   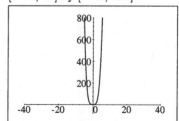

   The viewing rectangle in part (c) produces the most appropriate graph of the equation.

5. $y = 100 - x^2$

  (a) $[-4, 4]$ by $[-4, 4]$ (b) $[-10, 10]$ by $[-10, 10]$

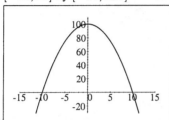   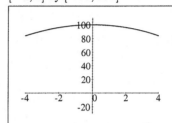

  (c) $[-15, 15]$ by $[-30, 110]$ (d) $[-4, 4]$ by $[-30, 110]$

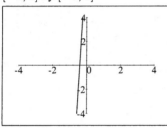   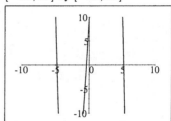

The viewing rectangle in part (c) produces the most appropriate graph of the equation.

7. $y = 10 + 25x - x^3$

  (a) $[-4, 4]$ by $[-4, 4]$ (b) $[-10, 10]$ by $[-10, 10]$

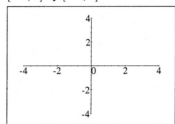   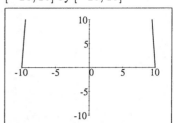

  (c) $[-20, 20]$ by $[-100, 100]$ (d) $[-100, 100]$ by $[-200, 200]$

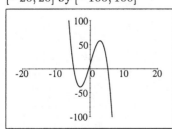   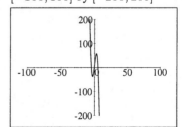

The viewing rectangle in part (c) produces the most appropriate graph of the equation.

9. $y = 100x^2$
    $[-2, 2]$ by $[-10, 400]$

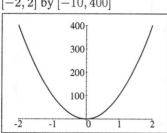

11. $y = 4 + 6x - x^2$
    $[-4, 10]$ by $[-10, 20]$

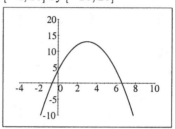

13. $y = \sqrt[4]{256 - x^2}$. We require that
    $256 - x^2 \geq 0 \quad \Rightarrow \quad -16 \leq x \leq 16$,
    so we graph $y = \sqrt[4]{256 - x^2}$ in the
    viewing rectangle $[-20, 20]$ by $[-1, 5]$.

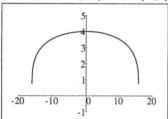

15. $y = 0.01x^3 - x^2 + 5$
    $[-50, 150]$ by $[-2000, 2000]$

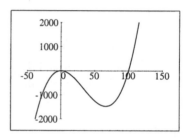

17. $y = \dfrac{1}{x^2 + 25}$. Since $x^2 + 25 \geq 25$
    $0 < \dfrac{1}{x^2 + 25} \leq \dfrac{1}{25}$. We use the
    viewing rectangle $[-10, 10]$ by $[-0.1, 0.1]$.

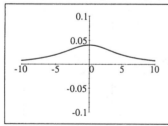

19. $y = x^4 - 4x^3$
    $[-4, 6]$ by $[-50, 100]$

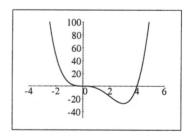

21. $y = 1 + |x - 1|$; $[-3, 5]$ by $[-1, 5]$

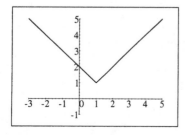

23. $y = \dfrac{|x + 1|}{x + 1}$; $[-5, 5]$ by $[-5, 5]$

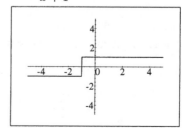

25. $x^2 + y^2 = 9 \Leftrightarrow y^2 = 9 - x^2 \Rightarrow$
$y = \pm\sqrt{9 - x^2}$. So we graph the functions
$y_1 = \sqrt{9 - x^2}$ and $y_2 = -\sqrt{9 - x^2}$ in the
viewing rectangle $[-6, 6]$ by $[-4, 4]$.

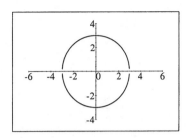

27. $4x^2 + 2y^2 = 1 \Leftrightarrow 2y^2 = 1 - 4x^2 \Leftrightarrow$
$y^2 = \dfrac{1 - 4x^2}{2} \Rightarrow y = \pm\sqrt{\dfrac{1 - 4x^2}{2}}$. So we graph
the functions $y_1 = \sqrt{\dfrac{1 - 4x^2}{2}}$ and $y_2 = -\sqrt{\dfrac{1 - 4x^2}{2}}$ in
the viewing rectangle $[-1.2, 1.2]$ by $[-0.8, 0.8]$.

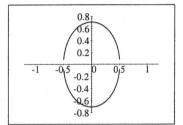

29. Although the graphs of $y = -3x^2 + 6x - \frac{1}{2}$ and
$y = \sqrt{7 - 7x^2/12}$ appear to intersect in the viewing
rectangle $[-4, 4]$ by $[-1, 3]$, there are no points of
intersection. You can verify that this is not an
intersection by zooming in.

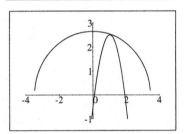

31. The graphs of $y = 6 - 4x - x^2$ and
$y = 3x + 18$ appear to have two points of
intersection in the viewing rectangle $[-6, 2]$ by
$[-5, 20]$. You can verify that $x = -4$ and
$x = -3$ are exact solutions.

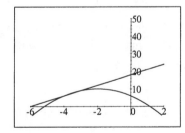

33. Answers may vary. Possible answers are given.

(a) Absolute value must be chosen from a menu or written as "abs()".

(b) Some calculators or computer graphing programs will not take the odd root of a negative
number, even though it is defined. Since $\sqrt[5]{-x} = -\sqrt[5]{x}$, you may have to enter two functions:
$Y_1 = x\char`\^(1/5)$ and $Y_2 = -(-x)\char`\^(1/5)$.

(c) Here you must remember to use parentheses and enter $x/(x - 1)$.

(d) As in part b, some graphing devices my not give the entire graph. When this is the case, you
will need to determine where the radicand (inside of the radical) is negative. Here you might
need to define 2 functions: $Y_1 = x\char`\^3 + (x + 2)\char`\^(1/3)$ and $Y_2 = x\char`\^3 + -(-x - 2)\char`\^(1/3)$.

Exercises 2.4

1. $m = \dfrac{y_2 - y_1}{x_2 - x_1} = \dfrac{4 - 0}{2 - 0} = \dfrac{4}{2} = 2.$

3. $m = \dfrac{y_2 - y_1}{x_2 - x_1} = \dfrac{0 - 2}{10 - 2} = \dfrac{-2}{8} = -\dfrac{1}{4}.$

5. $m = \dfrac{y_2 - y_1}{x_2 - x_1} = \dfrac{4 - 12}{2 - 4} = \dfrac{-8}{-2} = 4.$

7. $m = \dfrac{y_2 - y_1}{x_2 - x_1} = \dfrac{6 - (-3)}{-1 - 1} = \dfrac{9}{-2} = -\dfrac{9}{2}.$

9. For ℓ_1, we find two points, $(-1, 2)$ and $(0, 0)$ that lie on the line. Thus the slope of ℓ_1 is
   $m = \dfrac{y_2 - y_1}{x_2 - x_1} = \dfrac{2 - 0}{-1 - 0} = -2$. For ℓ_2, we find two points $(0, 2)$ and $(2, 3)$. Thus, the slope of ℓ_2
   is $m = \dfrac{y_2 - y_1}{x_2 - x_1} = \dfrac{3 - 2}{2 - 0} = \dfrac{1}{2}$. For ℓ_3 we find the points $(2, -2)$ and $(3, 1)$. Thus, the slope of ℓ_3
   is $m = \dfrac{y_2 - y_1}{x_2 - x_1} = \dfrac{1 - (-2)}{3 - 2} = 3$. For ℓ_4, we find the points $(-2, -1)$ and $(2, -2)$. Thus, the
   slope of ℓ_4 is $m = \dfrac{y_2 - y_1}{x_2 - x_1} = \dfrac{-2 - (-1)}{2 - (-2)} = \dfrac{-1}{4} = -\dfrac{1}{4}.$

11. First we find two points, $(0, 4)$ and $(4, 0)$ that lie on the line. So the slope is $m = \frac{0-4}{4-0} = -1$. Since
    the y-intercept is 4, the equation of the line is $y = mx + b = -1x + 4$. So $y = -x + 4$, or
    $x + y - 4 = 0$.

13. We choose the two intercepts as points, $(0, -3)$ and $(2, 0)$. So the slope is $m = \frac{0-(-3)}{2-0} = \frac{3}{2}$. Since
    the y-intercept is -3, the equation of the line is $y = mx + b = \frac{3}{2}x - 3$, or $3x - 2y - 6 = 0$.

15. Using the equation $y - y_1 = m(x - x_1)$, we get $y - 3 = 1(x - 2) \quad \Leftrightarrow \quad -x + y = 1 \quad \Leftrightarrow$
    $x - y + 1 = 0$.

17. Using the equation $y - y_1 = m(x - x_1)$, we get $y - 7 = \frac{2}{3}(x - 1) \quad \Leftrightarrow \quad 3y - 21 = 2x - 2 \quad \Leftrightarrow$
    $-2x + 3y = 19 \quad \Leftrightarrow \quad 2x - 3y + 19 = 0$.

19. First we find the slope, which is $m = \dfrac{y_2 - y_1}{x_2 - x_1} = \dfrac{6 - 1}{1 - 2} = \dfrac{5}{-1} = -5$. Substituting into
    $y - y_1 = m(x - x_1)$, we get $y - 6 = -5(x - 1) \quad \Leftrightarrow \quad y - 6 = -5x + 5 \quad \Leftrightarrow$
    $5x + y - 11 = 0$.

21. Using $y = mx + b$, we have $y = 3x + (-2)$ or $3x - y - 2 = 0$.

23. We are given two points, $(1, 0)$ and $(0, -3)$. Thus, the slope is
    $m = \dfrac{y_2 - y_1}{x_2 - x_1} = \dfrac{-3 - 0}{0 - 1} = \dfrac{-3}{-1} = 3$. Using the y-intercept, we have $y = 3x + (-3)$ or $y = 3x - 3$
    or $3x - y - 3 = 0$.

25. Since the equation of a horizontal line passing through (a, b) is $y = b$, the equation of the horizontal
    line passing through $(4, 5)$ is $y = 5$.

27. Since $x + 2y = 6$ $\Leftrightarrow$ $2y = -x + 6$ $\Leftrightarrow$ $y = -\frac{1}{2}x + 3$, the slope of this line is $-\frac{1}{2}$. Thus, the line we seek is given by $y - (-6) = -\frac{1}{2}(x - 1)$ $\Leftrightarrow$ $2y + 12 = -x + 1$ $\Leftrightarrow$ $x + 2y + 11 = 0$.

29. Any line parallel to $x = 5$ will have undefined slope and be of the form $x = a$. Thus the equation of the line is $x = -1$.

31. First find the slope of $2x + 5y + 8 = 0$. This gives $2x + 5y + 8 = 0$ $\Leftrightarrow$ $5y = -2x - 8$ $\Leftrightarrow$ $y = -\frac{2}{5}x - \frac{8}{5}$. So the slope of the line that is perpendicular to $2x + 5y + 8 = 0$ is $m = -\frac{1}{-2/5} = \frac{5}{2}$. The equation of the line we seek is $y - (-2) = \frac{5}{2}(x - (-1))$ $\Leftrightarrow$ $2y + 4 = 5x + 5$ $\Leftrightarrow$ $5x - 2y + 1 = 0$.

33. First find the slope of the line passing through $(2, 5)$ and $(-2, 1)$. This gives $m = \frac{1-5}{-2-2} = \frac{-4}{-4} = 1$, and so the equation of the line we seek is $y - 7 = 1(x - 1)$ $\Leftrightarrow$ $x - y + 6 = 0$.

35. (a)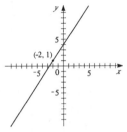

    (b) $y - 1 = \frac{3}{2}(x - (-2))$ $\Leftrightarrow$ $2y - 2 = 3(x + 2)$ $\Leftrightarrow$ $2y - 2 = 3x + 6$ $\Leftrightarrow$ $3x - 2y + 8 = 0$.

37.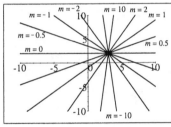

    Each of the lines contains the point $(3, 2)$ because the point $(3, 2)$ satisfies each equation $y = 2 + m(x - 3)$.

39. $x + y = 3$ $\Leftrightarrow$ $y = -x + 3$. So the slope is -1, and the y-intercept is 3.

    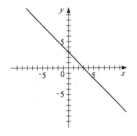

41. $x + 3y = 0$ $\Leftrightarrow$ $3y = -x$ $\Leftrightarrow$ $y = -\frac{1}{3}x$. So the slope is $-\frac{1}{3}$, and the y-intercept is 0.

    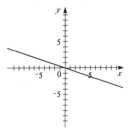

43. $\frac{1}{2}x - \frac{1}{3}y + 1 = 0$ $\Leftrightarrow$ $-\frac{1}{3}y = -\frac{1}{2}x - 1$
    $\Leftrightarrow$ $y = \frac{3}{2}x + 3$. So the slope is $\frac{3}{2}$, and the
    y-intercept is 3.

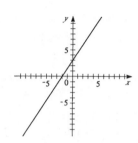

45. $y = 4$ can also be expressed as $y = 0x + 4$.
    So the slope is 0, and the y-intercept is 4.

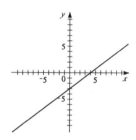

47. $3x - 4y = 12$ $\Leftrightarrow$ $-4y = -3x + 12$
    $\Leftrightarrow$ $y = \frac{3}{4}x - 3$. So the slope is $\frac{3}{4}$, and the
    y-intercept is -3.

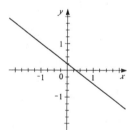

49. $3x + 4y - 1 = 0$ $\Leftrightarrow$ $4y = -3x + 1$
    $\Leftrightarrow$ $y = -\frac{3}{4}x + \frac{1}{4}$. So the slope is $-\frac{3}{4}$, and
    the y-intercept is $\frac{1}{4}$.

51. We first plot the points to find the pairs of points that determine each
    side. Next we find the slopes of opposite sides.

    The slope of AB is $\frac{4-1}{7-1} = \frac{3}{6} = \frac{1}{2}$, and the slope of DC is
    $\frac{10-7}{5-(-1)} = \frac{3}{6} = \frac{1}{2}$. Since these slope are equal, these two sides are parallel.

    The slope of AD is $\frac{7-1}{-1-1} = \frac{6}{-2} = -3$, and the slope of BC is
    $\frac{10-4}{5-7} = \frac{6}{-2} = -3$. Since these slope are equal, these two sides are
    parallel.

    Hence $ABCD$ is a parallelogram.

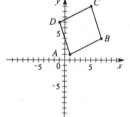

53. We first plot the points to find the pairs of points that determine each side. Next we find the slopes of opposite sides.

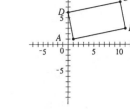

The slope of AB is $\frac{3-1}{11-1} = \frac{2}{10} = \frac{1}{5}$, and the slope of DC is $\frac{6-8}{0-10} = \frac{-2}{-10} = \frac{1}{5}$. Since these slope are equal, these two sides are parallel.

Slope of AD is $\frac{6-1}{0-1} = \frac{5}{-1} = -5$, and the slope of BC is $\frac{3-8}{11-10} = \frac{-5}{1} = -5$. Since these slope are equal, these two sides are parallel.

Since (slope of AB) × (slope of AD) $= \frac{1}{5} \times (-5) = -1$, the first two sides are each perpendicular to the second two sides. So the sides form a rectangle.

55. We need the slope and the midpoint of the line AB. The midpoint of AB is $\left(\frac{1+7}{2}, \frac{4-2}{2}\right) = (4, 1)$, and the slope of AB is $m = \frac{-2-4}{7-1} = \frac{-6}{6} = -1$. The slope of the perpendicular bisector will have slope $\frac{-1}{m} = \frac{-1}{-1} = 1$. Using the point-slope form, the equation of the perpendicular bisector is $y - 1 = 1(x - 4)$ or $x - y - 3 = 0$.

57. (a) We start with the two points $(a, 0)$ and $(0, b)$. The slope of the line that contains them is $\frac{b-0}{0-a} = -\frac{b}{a}$. So the equation of the line containing them is $y = -\frac{b}{a}x + b$ (using the slope-intercept form). Dividing by b (since $b \neq 0$) gives $\frac{y}{b} = -\frac{x}{a} + 1$ $\Leftrightarrow$ $\frac{x}{a} + \frac{y}{b} = 1$.

   (b) Setting $a = 6$ and $b = -8$, we get $\frac{x}{6} + \frac{y}{-8} = 1$ $\Leftrightarrow$ $4x - 3y = 24$ $\Leftrightarrow$ $4x - 3y - 24 = 0$.

59. Let h be the change in your horizontal distance, in feet. Then $-\frac{6}{100} = \frac{-1000}{h}$ $\Leftrightarrow$ $h = \frac{100,000}{6} \approx 16,667$. So the change in your horizontal distance is about $16,667$ feet.

61. (a) The slope is $0.0417D = 0.0417(200) = 8.34$. It represents the increase in dosage for each one-year increase in the child's age.

   (b) When $a = 0$, $c = 8.34(0 + 1) = 8.34$ mg.

63. (a)

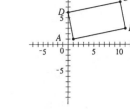

   (b) The slope is the cost per toaster oven, $6. The y-intercept, $3000, is the monthly fixed cost, the cost that is incurred no matter how many toaster ovens are produced.

65. (a) Using n in place of x and t in place of y, we find that the slope is $\frac{t_2 - t_1}{n_2 - n_1} = \frac{80 - 70}{168 - 120} = \frac{10}{48} = \frac{5}{24}$. So the linear equation is $t - 80 = \frac{5}{24}(n - 168)$ $\Leftrightarrow$ $t - 80 = \frac{5}{24}n - 35$ $\Leftrightarrow$ $t = \frac{5}{24}n + 45$.

   (b) When $n = 150$, the temperature is approximately given by $t = \frac{5}{24}(150) + 45 = 76.25°F \approx 76°F$.

67. (a) We are given $\dfrac{\text{change in pressure}}{10 \text{ feet change in depth}} = \dfrac{4.34}{10} = 0.434$. Using P for pressure and d for depth, and using the point $P = 15$ when $d = 0$, we have $P - 15 = 0.434(d - 0)$ $\Leftrightarrow$ $P = 0.434d + 15$.

   (b) When $P = 100$, then $100 = 0.434d + 15$ $\Leftrightarrow$ $0.434d = 85$ $\Leftrightarrow$ $d = 195.9$ feet. Thus the pressure is 100 lb/in^3 at a depth of approximately 196 feet.

69. (a) Using d in place of x and C in place of y, we find the slope $= \dfrac{C_2 - C_1}{d_2 - d_1} = \dfrac{460 - 380}{800 - 480} = \dfrac{80}{320}$ $= \frac{1}{4}$. So the linear equations is $C - 460 = \frac{1}{4}(d - 800)$ $\Leftrightarrow$ $C - 460 = \frac{1}{4}d - 200$ $\Leftrightarrow$ $C = \frac{1}{4}d + 260$.

   (b) Substituting $d = 1500$ we get $C = \frac{1}{4}(1500) + 260 = 635$. Thus, the cost of driving 1500 miles is $\$635$.

   (c)

   The slope of the line represents the cost per mile, $\$0.25$.

   (d) The y-intercept represents the fixed cost, $\$260$.

   (e) It is a suitable model because you have fixed monthly costs such as insurance and car payments, as well as costs that occur as you drive, such as gasoline, oil, tires, etc., and the cost of these for each additional mile driven is a constant.

71. (a) Viewing rectangle is $[0, 50]$ by $[0, 30]$.

   (b) $(p, y) = (21.82, 13.82)$

   (c) Setting the two equations equal to each other we have $-0.65p + 28 = 0.45p + 4$ $\Leftrightarrow$ $24 = 1.1p$ $\Leftrightarrow$ $p = 21.8181...$ So the price is $\$21.82$. Substituting we find $y = 0.45(21.82) + 4 = 13.82$.

73. Slope is the rate of change of one variable per unit change in another variable. So if the slope is positive, then the temperature is rising. Likewise, if the slope is negative then the temperature is decreasing. If the slope is 0, then the temperature is not changing.

Review Exercises for Chapter 2

1. (a)

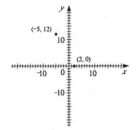

(b) The distance from P to Q is
$$d(P, Q) = \sqrt{(-5 - 2)^2 + (12 - 0)^2}$$
$$= \sqrt{49 + 144} = \sqrt{193}.$$

(c) The midpoint is $\left(\frac{-5+2}{2}, \frac{12+0}{2}\right) = \left(-\frac{3}{2}, 6\right)$.

(d) The line has slope $m = \frac{12-0}{-5-2} = -\frac{12}{7}$, and has equation $y - 0 = -\frac{12}{7}(x - 2)$ $\Leftrightarrow$
$y = -\frac{12}{7}x + \frac{24}{7}$ $\Leftrightarrow$ $12x + 7y - 24 = 0$.

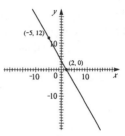

(e) The radius of this circle was found in part (b). It is $r = d(P, Q) = \sqrt{193}$. So the equation is
$$(x - 2)^2 + (y - 0)^2 = \left(\sqrt{193}\right)^2 \quad \Leftrightarrow$$
$$(x - 2)^2 + y^2 = 193.$$

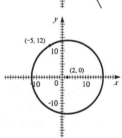

3.

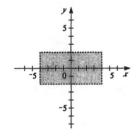

5. $d(A, C) = \sqrt{(4 - (-1))^2 + (4 - (-3))^2} = \sqrt{(4 + 1)^2 + (4 + 3)^2} = \sqrt{74}$ and
$d(B, C) = \sqrt{(5 - (-1))^2 + (3 - (-3))^2} = \sqrt{(5 + 1)^2 + (3 + 3)^2} = \sqrt{72}$. Therefore, B is closer to C.

7. The center is $C = (-5, -1)$, and the point $P = (0, 0)$ is on the circle. The radius of the circle is
$r = d(P, C) = \sqrt{(0 - (-5))^2 + (0 - (-1))^2} = \sqrt{(0 + 5)^2 + (0 + 1)^2} = \sqrt{26}$. Thus, the equation
of the circle is $(x + 5)^2 + (y + 1)^2 = 26$.

9. $x^2 + y^2 + 2x - 6y + 9 = 0$ $\Leftrightarrow$ $(x^2 + 2x) + (y^2 - 6y) = -9$ $\Leftrightarrow$
    $(x^2 + 2x + 1) + (y^2 - 6y + 9) = -9 + 1 + 9$ $\Leftrightarrow$ $(x + 1)^2 + (y - 3)^2 = 1$. This equation
    represents a circle with center at $(-1, 3)$ and radius 1.

11. $x^2 + y^2 + 72 = 12x$ $\Leftrightarrow$ $(x^2 - 12x) + y^2 = -72$ $\Leftrightarrow$ $(x^2 - 12x + 36) + y^2 = -72 + 36$
    $\Leftrightarrow$ $(x - 6)^2 + y^2 = -36$. Since the left side of this equation must be greater than or equal to
    zero, this equation has no graph.

13. $y = 2x - 3$
    x-axis symmetry: $(-y) = 2 - 3x$ $\Leftrightarrow$ $y = -2 + 3x$, which is not the same as the original
    equation, so not symmetric with respect to the x-axis.
    y-axis symmetry: $y = 2 - 3(-x)$ $\Leftrightarrow$ $y = 2 + 3x$, which is not the same as the original
    equation, so not symmetric with respect to the y-axis.
    Origin symmetry: $(-y) = 2 - 3(-x)$ $\Leftrightarrow$ $-y = 2 + 3x$ $\Leftrightarrow$ $y = -2 - 3x$, which is not the
    same as the original equation, so not symmetric with respect
    to the origin.
    Hence the graph has no symmetry.

    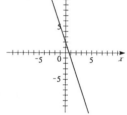

x	y
-2	8
0	2
$\frac{2}{3}$	0

15. $x + 3y = 21$ $\Leftrightarrow$ $y = -\frac{1}{3}x + 7$
    x-axis symmetry: $x + 3(-y) = 21$ $\Leftrightarrow$ $x - 3y = 21$, which is not the same as the original
    equation, so not symmetric with respect to the x-axis.
    y-axis symmetry: $(-x) + 3y = 21$ $\Leftrightarrow$ $x - 3y = -21$, which is not the same as the original
    equation, so not symmetric with respect to the y-axis.
    Origin symmetry: $(-x) + 3(-y) = 21$ $\Leftrightarrow$ $x + 3y = -21$,
    which is not the same as the original equation, so not symmetric
    with respect to the origin.
    Hence the graph has no symmetry.

    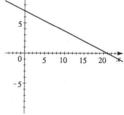

x	y
-3	8
0	7
21	0

17. $\dfrac{x}{2} - \dfrac{y}{7} = 1$ $\Leftrightarrow$ $y = \dfrac{7}{2}x - 7$
    x-axis symmetry: $\dfrac{x}{2} - \dfrac{(-y)}{7} = 1$ $\Leftrightarrow$ $\dfrac{x}{2} + \dfrac{y}{7} = 1$, which is not the same as the original
    equation, so not symmetric with respect to the x-axis.
    y-axis symmetry: $\dfrac{(-x)}{2} - \dfrac{y}{7} = 1$ $\Leftrightarrow$ $\dfrac{x}{2} + \dfrac{y}{7} = -1$, which is not the same as the original
    equation, so not symmetric with respect to the y-axis.

Origin symmetry: $\dfrac{(-x)}{2} - \dfrac{(-y)}{7} = 1 \quad \Leftrightarrow \quad \dfrac{x}{2} - \dfrac{y}{7} = -1,$

which is not the same as the original equation, so not
symmetric with respect to the origin.

Hence, the graph has no symmetry.

x	y
-2	-14
0	-7
2	0

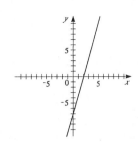

19. $y = 16 - x^2$

x-axis symmetry: $(-y) = 16 - x^2 \quad \Leftrightarrow \quad y = -16 + x^2$, which is not the same as the original
equation, so not symmetric with respect to the x-axis.

y-axis symmetry: $y = 16 - (-x)^2 \quad \Leftrightarrow \quad y = 16 - x^2$, which is the same as the original equation,
so its is symmetric with respect to the y-axis.

Origin symmetry: $(-y) = 16 - (-x)^2 \quad \Leftrightarrow \quad y = -16 + x^2$, which is not the same as the original
equation, so not symmetric with respect to the origin.

Hence, the graph is symmetric with respect to the y-axis.

x	y
-3	7
-1	15
0	16
1	15
3	7

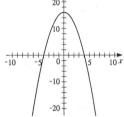

21. $x = \sqrt{y}$

x-axis symmetry: $x = \sqrt{-y}$, which is not the same as the original equation, so not symmetric with
respect to the x-axis.

y-axis symmetry: $(-x) = \sqrt{y} \quad \Leftrightarrow \quad x = -\sqrt{y}$, which is not the same as the original equation,
so not symmetric with respect to the y-axis.

Origin symmetry: $(-x) = \sqrt{-y}$, which is not the same as the original equation, so not symmetric
with respect to the origin.

Hence, the graph has no symmetry.

x	y
0	0
1	1
2	4
3	9

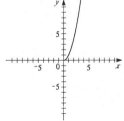

23. $y = x^2 - 6x$
Viewing rectangle $[-10, 10]$ by $[-10, 10]$.

25. $y = x^3 - 4x^2 - 5$
Viewing rectangle $[-4, 10]$ by $[-30, 20]$.

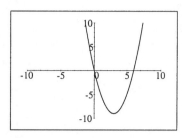   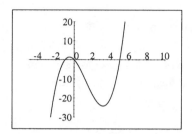

27. The line has slope $m = \frac{-4+6}{2+1} = \frac{2}{3}$, and so, by the point-slope formula, the equation is
     $y + 4 = \frac{2}{3}(x - 2) \quad \Leftrightarrow \quad y = \frac{2}{3}x - \frac{16}{3} \quad \Leftrightarrow \quad 2x - 3y - 16 = 0.$

29. The x-intercept is 4, and the y-intercept is 12, so the slope is $m = \frac{12-0}{0-4} = -3$. Therefore, by the
     slope-intercept formula, the equation of the line is $y = -3x + 12 \quad \Leftrightarrow \quad 3x + y - 12 = 0.$

31. We first find the slope of the line $3x + 15y = 22$. This gives $3x + 15y = 22 \quad \Leftrightarrow$
     $15y = -3x + 22 \quad \Leftrightarrow \quad y = -\frac{1}{5}x + \frac{22}{15}$. So this line has slope $m = -\frac{1}{5}$, as does any line parallel
     to it. Then the parallel line passing through the origin has equation $y - 0 = -\frac{1}{5}(x - 0) \quad \Leftrightarrow$
     $x + 5y = 0.$

33. (a) The slope, 0.3, represents the increase in length of the spring for each unit increase in weight w.
         The s-intercept is the resting or natural length of the spring.

     (b) When $w = 5$, $s = 0.3(5) + 2.5 = 1.5 + 2.5 = 4.0$ inches.

35. Here the center is at $(0, 0)$, and the circle passes through the point $(-5, 12)$, so the radius is
     $r = \sqrt{(-5 - 0)^2 + (12 - 0)^2} = \sqrt{25 + 144} = \sqrt{169} = 13$. The equation of the circle is
     $x^2 + y^2 = 13^2 \quad \Leftrightarrow \quad x^2 + y^2 = 169$. The line shown is the tangent that passes through the point
     $(-5, 12)$, so it is perpendicular to the line through the points $(0, 0)$ and $(-5, 12)$. This line has slope
     $m_1 = \frac{12-0}{-5-0} = -\frac{12}{5}$. The slope of the line we seek is $m_2 = -\frac{1}{m_1} = -\frac{1}{-12/5} = \frac{5}{12}$. Thus, the
     equation of the tangent line is $y - 12 = \frac{5}{12}(x + 5) \quad \Leftrightarrow \quad y - 12 = \frac{5}{12}x + \frac{25}{12} \quad \Leftrightarrow$
     $y = \frac{5}{12}x + \frac{169}{12} \quad \Leftrightarrow \quad 5x - 12y + 169 = 0.$

Chapter 2 Test

1. (a) There are several ways to determine the coordinates of S. The diagonals of a square have equal length and are are perpendicular. The diagonal PR is horizontal and has length is 6 units. So the diagonal QS is vertical and also has length 6. Thus the coordinates of S are $(3, 6)$.

   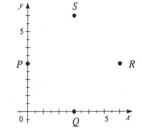

   (b) The length of PQ is $\sqrt{(0-3)^2 + (3-0)^2} = \sqrt{18} = 3\sqrt{2}$. So the area of $PQRS$ is $(3\sqrt{2})^2 = 18$.

2. (a)

   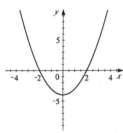

   (b) The x-intercept occurs when $y = 0$, so $0 = x^2 - 4$ $\Leftrightarrow$ $x^2 = 4$ $\Rightarrow$ $x = \pm 2$. The y-intercept occurs when $x = 0$, so $y = -4$.

   (c) x-axis symmetry: $(-y) = x^2 - 4$ $\Leftrightarrow$ $y = -x^2 + 4$, which is not the same as the original equation, so not symmetric with respect to the x-axis.

   y-axis symmetry: $y = (-x)^2 - 4$ $\Leftrightarrow$ $y = x^2 - 4$, which is the same as the original equation, so it is symmetric with respect to the y-axis.

   Origin symmetry: $(-y) = (-x)^2 - 4$ $\Leftrightarrow$ $-y = x^2 - 4$, which is not the same as the original equation, so not symmetric with respect to the origin.

3. (a) The distance from A to B is
   $$d(AB) = \sqrt{(-7-5)^2 + (4+12)^2} = \sqrt{144 + 256} = \sqrt{400} = 20.$$

   (b) The midpoint is $\left(\frac{-7+5}{2}, \frac{4-12}{2}\right) = (-1, -4)$.

   (c) The line has slope $m = \frac{4+12}{-7-5} = \frac{16}{-12} = -\frac{4}{3}$, and its equation is $y - 4 = -\frac{4}{3}(x + 7)$ $\Leftrightarrow$ $y = -\frac{4}{3}x - \frac{28}{3} + 4 = -\frac{4}{3}x - \frac{16}{3}$. Thus the equation is $y = -\frac{4}{3}x - \frac{16}{3}$ $\Leftrightarrow$ $4x + 3y + 16 = 0$.

   (d) The perpendicular bisector has slope $m = -\dfrac{1}{-(4/3)} = \dfrac{3}{4}$ and contains the midpoint $(-1, -4)$.

   Thus, from the point-slope formula, the equation is $y + 4 = \frac{3}{4}(x + 1)$ $\Leftrightarrow$ $y + 4 = \frac{3}{4}x + \frac{3}{4}$ $\Leftrightarrow$ $y = \frac{3}{4}x + \frac{3}{4} - 4 = \frac{3}{4}x - \frac{13}{4}$. Therefore, the equation is $y = \frac{3}{4}x - \frac{13}{4}$ $\Leftrightarrow$ $3x - 4y - 13 = 0$.

   (e) The length of a diameter of this circle was found in part (a), so the radius is $r = \frac{20}{2} = 10$. The center is the midpoint found in part (b), so the equation is $(x + 1)^2 + (y + 4)^2 = 10^2$ $\Leftrightarrow$ $(x + 1)^2 + (y + 4)^2 = 100$.

4. (a) $x^2 + (y - 4)^2 = 9$ $\Leftrightarrow$ $x^2 + (y - 4)^2 = 3^2$
        The center is $(0, 4)$, and the radius is 3.

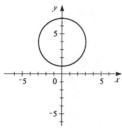

    (b) $x^2 + y^2 - 6x + 10y + 9 = 0$ $\Leftrightarrow$
        $(x^2 - 6x) + (y^2 + 10y) = -9$ $\Leftrightarrow$
        $(x^2 - 6x + 9) + (y^2 + 10y + 25) = -9 + 9 + 25$
        $\Leftrightarrow$ $(x - 3)^2 + (y + 5)^2 = 5^2$
        The center is $(3, -5)$, and the radius is 5.

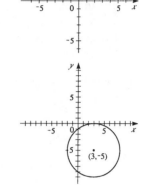

5. $3x - 4y + 24 = 0$ $\Leftrightarrow$ $-4y = -3x - 24$
    $\Leftrightarrow$ $y = \frac{3}{4}x + 6$.
    The slope is $\frac{3}{4}$, and the y-intercept is 6.

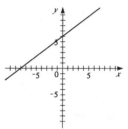

6. (a) We first find the slope of the line $2x + 6y = 17$. This gives $2x + 6y = 17$ $\Leftrightarrow$
        $6y = -2x + 17$ $\Leftrightarrow$ $y = -\frac{1}{3}x + \frac{17}{6}$. Thus, this line has slope $m = -\frac{1}{3}$, as does any line
        parallel to it. Then the parallel line passing through the point $(-2, 3)$ has the equation
        $y - 3 = -\frac{1}{3}(x + 2)$ $\Leftrightarrow$ $3y - 9 = -x - 2$ $\Leftrightarrow$ $x + 3y - 7 = 0$. (You can also find
        this equation by substituting the point $(-2, 3)$ into the equation $2x + 6y = c$ and solving for c.)

    (b) This line contains the points $(-3, 0)$ and $(0, 12)$. Thus, the slope of the line containing these
        points is $m = \frac{12 - 0}{0 + 3} = 4$, and so the equation is $y = 4x + 12$ $\Leftrightarrow$ $4x - y + 12 = 0$. (You
        can also use the two-intercept form of a line: $\dfrac{x}{a} + \dfrac{y}{b} = 1$ $\Rightarrow$ $\dfrac{x}{-3} + \dfrac{y}{12} = 1$.)

7. (a) When $x = 2{,}500$, $C = 7(2{,}500) + 17{,}500 = 17{,}500 + 17{,}500 = \$35{,}000$.

    (b) The slope represents the cost of producing one more blender, $7. The C-intercept, $17,500,
        represents the fixed cost.

Principles of Modeling

1. (a)

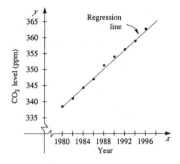

(b) Using a graphing calculator, we obtain the regression line $y = 1.5125x - 2656.4$.

(c) Using $x = 1998$ in the equation $y = 1.5125x - 2656.4$, we get $y \approx 365.6$ ppm CO_2. This value is close to the observed value of 366.7 ppm CO_2.

3. (a)

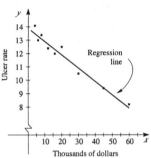

(b) Using a graphing calculator, we obtain the regression line $y = -0.0995x + 13.9$, where x is measured in thousands of dollars.

(c) Using the regression line equation, $y = -0.0995x + 13.9$, we get an estimated $y = 11.4$ ulcers per 100 population when $x = \$25,000$.

(d) Again using $y = -0.0995x + 13.9$, we get an estimated $y = 5.8$ ulcers per 100 population when $x = \$80,000$.

5. (a)

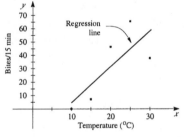

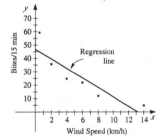

(b) Using a graphing calculator, we obtain the regression line for the temperature-bite data $y = 2.6768x - 22.19$, and for the wind speed-bite data $y = -3.5336x + 46.73$.

(c) The correlation coefficient for the temperature-bite data is $r = 0.77$, so a linear model is not appropriate. The correlation coefficient for the wind-bite data is $r = -0.89$, so a linear model is more suitable in this case.

7. (a)

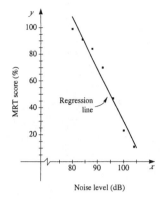

(b) Using a graphing calculator, we obtain
$y = -3.9018x + 419.7$.

(c) The correlation coefficient is $r = -0.98$,
so linear model is appropriate for x
between 80 dB and 104 dB.

(d) Substituting $x = 94$ into the regression
equation, we get
$y = -3.9018(94) + 419.7 \approx 53$. So
the intelligibility is about 53%.

9. (a)

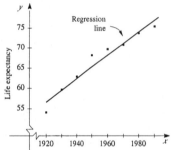

(b) Using a graphing calculator, we obtain
$y = 0.29083x - 501.8$.

(c) We use $x = 2000$ in the equation
$y = 0.29083x - 501.8$, to get $y = 79.9$,
that is, a life expectancy of 79.9 years.

11. (a)

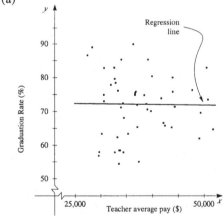

(b) Using a graphing calculator, we obtain
$y = -0.000012x + 72.0$

(c) The correlation coefficient is
$r = -0.00812$, so there is very little
correlation. It appears that teachers' salaries
do not affect graduation rates.

13. The students should find a fairly strong correlation between shoe size and height.

Chapter Three
Exercises 3.1

1. Algebraically: $2x - 9 = 5 \Leftrightarrow 2x = 14 \Leftrightarrow x = 7$.
   Graphically: We graph the two equations $y_1 = 2x - 9$ and
   $y_2 = 5$ in the viewing rectangle $[5, 10]$ by $[-1, 6]$.
   Zooming in we see that solution is $x = 7$.

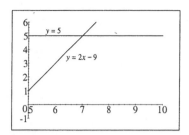

3. Algebraically: $x - 4 = 5x + 12 \Leftrightarrow -16 = 4x \Leftrightarrow$
   $x = -4$.
   Graphically: We graph the two equations $y_1 = x - 4$ and
   $y_2 = 5x + 12$ in the viewing rectangle $[-6, 4]$ by $[-10, 2]$.
   Zooming in we see that solution is $x = -4$.

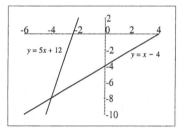

5. Algebraically: $\frac{1}{2}x - 3 = 6 + 2x \Leftrightarrow -9 = \frac{3}{2}x \Leftrightarrow$
   $x = -6$.
   Graphically: We graph the two equations $y_1 = \frac{1}{2}x - 3$ and
   $y_2 = 6 + 2x$ in the viewing rectangle $[-10, 5]$ by $[-10, 5]$.
   Zooming in we see that solution is $x = -6$.

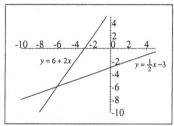

7. Algebraically: $\dfrac{2}{x} + \dfrac{1}{2x} = 7 \Leftrightarrow 2x\left(\dfrac{2}{x} + \dfrac{1}{2x}\right) = 2x(7)$
   $\Leftrightarrow 4 + 1 = 14x \Leftrightarrow x = \frac{5}{14}$.
   Graphically: We graph the two equations $y_1 = \dfrac{2}{x} + \dfrac{1}{2x}$ and
   $y_2 = 7$ in the viewing rectangle $[-2, 2]$ by $[-2, 28]$.
   Zooming in we see that solution is $x \approx 0.36$.

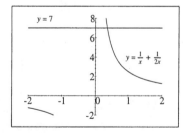

9. Algebraically: $x^2 - 32 = 0 \Leftrightarrow x^2 = 32 \Rightarrow x = \pm\sqrt{32} = \pm 4\sqrt{2}$.
   Graphically: We graph the equation $y_1 = x^2 - 32$ and
   determine where this curve intersects the x-axis. We use
   the viewing rectangle $[-10, 10]$ by $[-5, 5]$.
   Zooming in, we see that solutions are $x \approx 5.66$
   and $x \approx -5.66$.

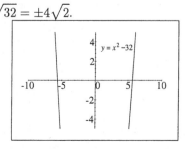

11. Algebraically: $16x^4 = 625 \quad \Leftrightarrow \quad x^4 = \frac{625}{16} \quad \Rightarrow \quad x = \pm\frac{5}{2} = \pm 2.5$.

Graphically: We graph the two equations $y_1 = 16x^4$ and $y_2 = 625$ in the viewing rectangle $[-5, 5]$ by $[610, 640]$. Zooming in we see that solutions are $x = \pm 2.5$.

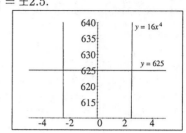

13. Algebraically: $(x - 5)^4 - 80 = 0 \quad \Leftrightarrow \quad (x - 5)^4 = 80 \quad \Rightarrow \quad x - 5 = \pm\sqrt[4]{80} = \pm 2\sqrt[4]{5} \quad \Leftrightarrow$ $x = 5 \pm 2\sqrt[4]{5}$.

Graphically: We graph the equation $y_1 = (x - 5)^4 - 80$ and determine where this curve intersects the x-axis. We use the viewing rectangle $[-1, 9]$ by $[-5, 5]$. Zooming in, we see that solutions are $x \approx 2.01$ and $x \approx 7.99$.

15. We graph $y = x^2 - 7x + 12$ in the viewing rectangle $[0, 6]$ by $[-0.1, 0.1]$. The solutions are $x = 3.00$ and 4.00.

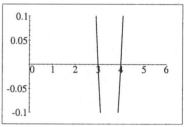

17. We graph $y = x^3 - 6x^2 + 11x - 6$ in the viewing rectangle $[-1, 4]$ by $[-0.1, 0.1]$. The solutions are $x = 1.00$, $x = 2.00$, and $x = 3.00$.

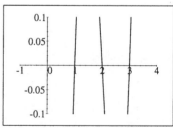

19. We first graph

$y = x - \sqrt{x + 1}$ in the viewing rectangle $[-1, 5]$ by $[-0.1, 0.1]$ and find that the solution is near 1.6. Zooming in, we see that solutions is $x \approx 1.62$.

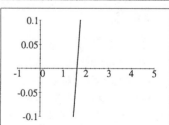

21. We graph $y = x^{1/3} - x$ in the viewing rectangle $[-3, 3]$ by
     $[-1, 1]$. The solutions are $x = -1.00$, $x = 0.00$, and
     $x = 1.00$.

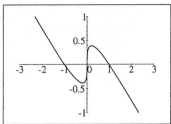

23. $x^3 - 2x^2 - x - 1 = 0$, so we start by graphing the function
     $y = x^3 - 2x^2 - x - 1$ in the viewing rectangle $[-10, 10]$ by
     $[-100, 100]$. There appear to be two solutions, one near
     $x = 0$ and another one between $x = 2$ and $x = 3$.

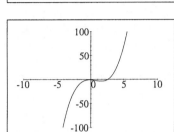

     We then use the viewing rectangle $[-1, 5]$ by $[-1, 1]$
     and zoom into the only solution at $x \approx 2.55$.

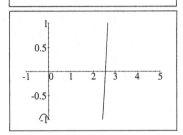

25. $x(x - 1)(x + 2) = \frac{1}{6}x$ $\quad \Leftrightarrow \quad$ $x(x - 1)(x + 2) - \frac{1}{6}x = 0$.
     We start by graphing the function $y = x(x - 1)(x + 2) - \frac{1}{6}x$
     in the viewing rectangle $[-5, 5]$ by $[-10, 10]$. There
     appear to be three solutions.

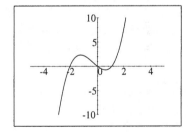

     We then use the viewing rectangle $[-2.5, 2.5]$ by $[-1, 1]$
     and zoom into the solutions at $x \approx -2.05$, $x = 0.00$, and
     $x \approx 1.05$.

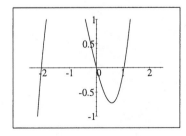

27. As in Example 4, we graph the equation
     $y = x^3 - 6x^2 + 9x - \sqrt{x}$ in the viewing rectangle $[0, 10]$ by
     $[-2, 15]$. We see the two solutions found in Example 4
     and what appears to be an additional solution near $x = 0$.

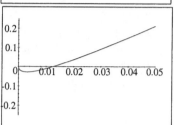

     In the viewing rectangle $[0, 0.05]$ by $[-0.25, 0.25]$, we
     find two more solutions at $x \approx 0.01$ and $x = 0$. We can
     verify that $x = 0$ is an exact solution by substitution.

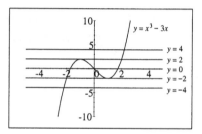

29. (a) We graph $y_1 = x^3 - 3x$ and $y_2 = k$ for $k = -4, -2, 0,$
          2, and 4 in the viewing rectangle $[-5, 5]$ by $[-10, 10]$.
          The number of solutions and the solutions are shown in
          the table below.

k	number of solutions	Solutions
-4	1	$x \approx -2.20$
-2	2	$x = -2, x = 1$
0	3	$x \approx \pm 1.73, x = 0$
2	2	$x = -1, x = 2$
4	1	$x \approx 2.20$

     (b) The equation $x^3 - 3x = k$ will have one solution for all $k < -2$ or $k > 2$, it will have exactly
          two solutions when $k = \pm 2$, and it will have three solutions for $-2 < k < 2$.

Exercises 3.2

1. If n is the first integer, then $n + 1$ is the middle integer, and $n + 2$ is the third integer. So the sum of the three consecutive integers is $n + (n + 1) + (n + 2) = 3n + 3$.

3. If n is the first integer, then $n + 1$ is the second integer. So the sum of the squares of the two consecutive integers is $n^2 + (n + 1)^2 = 2n^2 + 2n + 1$.

5. If s is the third test score, then since the other test scores are 78 and 82, the average of the three test scores is $\dfrac{78 + 82 + s}{3}$.

7. If w is the width of the rectangle, then the area of the rectangle is $50w$ in^2.

9. If s is the initial speed, then $s + 15$ is the speed during the third hour. Thus the distance traveled is $2s + s + 15 = 3s + 15$.

11. If a is the age of the firstborn, then $a - 3$ is the age of the second child, and $(a - 3) - 2 = a - 5$ is the age of the third child. Thus the average age of the three children is
$$\dfrac{a + (a - 3) + (a - 5)}{3} = \dfrac{3a - 8}{3}.$$

13. Let x be her monthly salary. Since her *annual salary* $= 12 \times$ (*monthly salary*) $+$ (*Christmas bonus*) we have $97{,}300 = 12x + 8{,}500 \quad\Leftrightarrow\quad 88{,}800 = 12x \quad\Leftrightarrow\quad x \approx 7{,}400$. Her monthly salary is \$7,400.

15. Let h be the amount that Craig inherits. So $(x + 22{,}000)$ is the amount that he invests and doubles. Thus $2(x + 22{,}000) = 134{,}000 \quad\Leftrightarrow\quad 2x + 44{,}000 = 134{,}000 \quad\Leftrightarrow\quad 2x = 90{,}000 \quad\Leftrightarrow\quad x = 45{,}000$. So Craig inherits \$45,000.

17. Let y be the age of the youngest child. Then $2y$ is the age of the oldest child. Since the *average age* $= \dfrac{\text{sum of the ages}}{4}$, we obtain $10.5 = \dfrac{y + 10 + 11 + 2y}{4} \quad\Leftrightarrow\quad 42 = 3y + 21 \quad\Leftrightarrow\quad 3y = 21 \quad\Leftrightarrow\quad y = 7$. So the youngest child is 7 years old.

19. Let x be the first integer. Then $x + 1$ and $x + 2$ are the next consecutive integers. So $x + (x + 1) + (x + 2) = 336 \quad\Leftrightarrow\quad 3x + 3 = 336 \quad\Leftrightarrow\quad 3x = 333 \quad\Leftrightarrow\quad x = 111$. Thus the consecutive integers are 111, 112, and 113.

21. Let w be width of the rectangle. Then the length is $1.5w$. The perimeter is $2 \times length + 2 \times width$. So $80 = 2(1.5w) + 2(w) = 3w + 2w \quad\Leftrightarrow\quad 80 = 5w \quad\Leftrightarrow\quad w = 16$. Thus the room is 16 feet wide.

23. Let l be the length of the garden. Since $area = width \cdot length$, we obtain the equation $1125 = 25l \quad\Leftrightarrow\quad l = \frac{1125}{25} = 45$ ft. So the garden is 45 feet long.

25. Let m be the amount invested at $4\frac{1}{2}\%$. Then $12{,}000 - m$ is the amount invested at 4%.
Since *total interest* $=$ (*interest earned at* $4\frac{1}{2}\%$) $+$ (*interest earned* at 4%), we have
$525 = 0.045m + 0.04(12{,}000 - m) \quad\Leftrightarrow\quad 525 = 0.045m + 480 - 0.04m \quad\Leftrightarrow\quad 45 = 0.005m$
$\Leftrightarrow\quad m = \frac{45}{0.005} = 9{,}000$. Thus \$9,000 is invested at $4\frac{1}{2}\%$, and $\$12{,}000 - 9{,}000 = \$3{,}000$ is invested at 4%.

27. Let x be the hours the assistant work. Then $2x$ is the hours the plumber worked. Since the
labor charge = plumber's labor + assistant's labor, we have
$4025 = 45(2x) + 25x$ ⇔ $4025 = 90x + 25x$ ⇔ $4025 = 115x$ ⇔ $x = \frac{4025}{115} = 35$
Thus the assistant works for 35 hours, and the plumber works for $2 \times 35 = 70$ hours.

29. Let q be the number of quarters. Then $2q$ is the number of dimes, and $2q + 5$ is the number of
nickels. Thus $3.00 = $ *value of the nickels + value of the dimes + value of the quarters*. So
$3.00 = 0.05(2q + 5) + 0.10(2q) + 0.25q$ ⇔ $3.00 = 0.10q + 0.25 + 0.20q + 0.25q$ ⇔
$2.75 = 0.55q$ ⇔ $q = \dfrac{2.75}{0.55} = 5$. Thus Mary has 5 quarters, $2(5) = 10$ dimes, and
$2(5) + 5 = 15$ nickels.

31. Let x be the width of the strip. Then the length of the mat is $20 + 2x$, and the width of the mat is
$15 + 2x$. Thus *perimeter = 2 × length + 2 × width*   ⇔ $102 = 2(20 + 2x) + 2(15 + 2x)$ ⇔
$102 = 40 + 4x + 30 + 4x$ ⇔ $102 = 70 + 8x$ ⇔ $32 = 8x$ ⇔ $x = 4$. Thus the strip
of mat is 4 inches wide.

33. Let x be the gross receipts for Friday. Then $2x$ is the gross receipts for Saturday. Using the
equation *average receipts* $= \dfrac{\text{sum of receipts}}{\text{number of days}}$, we get $1200 = \dfrac{650 + 550 + 300 + x + 2x}{5}$ ⇔
$6000 = 1500 + 3x$ ⇔ $4500 = 3x$ ⇔ $1500 = x$. Thus Saturday's receipts were
$2x = \$3{,}000$.

35. Let t be the time (in seconds) for the sun's light to reach earth. Using the formula
distance = rate × time, we get $1.5 \times 10^{11} = (3.0 \times 10^8)t$ ⇔ $t = \dfrac{1.5 \times 10^{11}}{3.0 \times 10^8} = 0.5 \times 10^{11-8}$
$= 0.5 \times 10^3 = 500$ seconds or 8 minutes and 20 seconds.

37. Let x be the distance from the fulcrum to where the mother sits. Then substituting the known values
into the formula given, we have $100(8) = 125x$ ⇔ $800 = 125x$ ⇔ $x = 6.4$. So the
mother should sit 6.4 feet from the fulcrum.

39. Let w be the width of the pasture. Then the length of the pasture is $2w$. Since
area = length × width we have $115{,}200 = w(2w) = 2w^2$ ⇔ $w^2 = 57{,}600$ ⇒ $w = \pm 240$.
Thus the width of the pasture is 240 feet.

41. The figure can be broken up into two rectangles.
Since the total area equals the sum of the areas
of the rectangles, we have $144 = 10x + 6x$ ⇔
$144 = 16x$ ⇔ $9 = x$. Thus the length of
x is 9 cm.

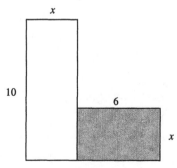

43. Let d be the distance from the lamppost, in meters. Using similar triangles, $\dfrac{d+5}{6} = \dfrac{5}{2}$ ⇔
$d + 5 = 15$ ⇔ $d = 10$. Thus the man is 10 meters from the lamppost.

45. Let x be the amount in mL of 60% acid solution to be used. Then $(300 - x)$ mL of 30% solution would have to be used to yield a total of 300 mL of solution.

	60% acid	30% acid	mixture
mL	x	$300 - x$	300
rate (% acid)	0.60	0.30	0.50
value	$0.60x$	$0.30(300 - x)$	$0.50(300)$

Thus the total amount of pure acid used is $0.60x + 0.30(300 - x) = 0.50(300)$ $\Leftrightarrow$ $0.3x + 90 = 150$ $\Leftrightarrow$ $x = \frac{60}{0.3} = 200$. So 200 mL of 60% acid solution must be mixed with 100 mL of 30% solution to get 300 mL of 50% acid solution.

47. Let x be the grams of silver added. The weight of the rings is $5 \times 18\,\text{g} = 90\,\text{g}$.

	5 rings	Pure silver	mixture
grams	90	x	$90 + x$
rate (% gold)	0.90	0	0.75
value	$0.90(90)$	$0x$	$0.75(90 + x)$

So $0.90(90) + 0x = 0.75(90 + x)$ $\Leftrightarrow$ $81 = 67.5 + 0.75x$ $\Leftrightarrow$ $0.75x = 13.5$ $\Leftrightarrow$ $x = \frac{13.5}{0.75} = 18$. Thus 18 grams of silver must be added to get the required mixture.

49. Let x be the liters of coolant removed and replaced by water.

	60% antifreeze	60% antifreeze (removed)	water	mixture
liters	3.6	x	x	3.6
rate (% antifreeze)	0.60	0.60	0	0.50
value	$0.60(3.6)$	$-0.60x$	$0x$	$0.50(3.6)$

so $0.60(3.6) - 0.60x + 0x = 0.50(3.6)$ $\Leftrightarrow$ $2.16 - 0.6x = 1.8$ $\Leftrightarrow$ $-0.6x = -0.36$ $\Leftrightarrow$ $x = \frac{-0.36}{-0.6} = 0.6$. Thus 0.6 liters must be removed and replaced by water.

51. Let t be the time in minutes it would take Candy and Tim if they work together. Candy delivers the papers at a rate of $\frac{1}{70}$ of the job per minute, while Tim delivers the paper at a rate of $\frac{1}{80}$ of the job per minute. The sum of the fractions of the job that each can do individually in one minute equals the fraction of the job they can do working together. So we have $\frac{1}{t} = \frac{1}{70} + \frac{1}{80}$ $\Leftrightarrow$ $560 = 8t + 7t$ $\Leftrightarrow$ $560 = 15t$ $\Leftrightarrow$ $t = 37\frac{1}{3}$ minutes. Thus it would take them $37\frac{1}{3}$ minutes if they worked together.

53. Let t be the time, in hours, it takes Karen to paint a house alone. Then working together, Karen and Betty can paint a house in $\frac{2}{3}t$ hours. The sum of their individual rates equals their rate working together, so $\frac{1}{t} + \frac{1}{6} = \frac{1}{\frac{2}{3}t}$ $\Leftrightarrow$ $\frac{1}{t} + \frac{1}{6} = \frac{3}{2t}$ $\Leftrightarrow$ $6 + t = 9$ $\Leftrightarrow$ $t = 3$. Thus it would take Karen 3 hours to paint a house alone.

55. Let t be the time in hours after takeoff until the planes pass each other. When the planes pass each other, the total distance they have traveled will equal 2550 km.

	Rate	Time	Distance
K.C. $\rightarrow$ S.F.	800	t	$800t$
S.F. $\rightarrow$ K.C.	900	t	$900t$

So $800t + 900t = 2550$ $\Leftrightarrow$ $1700t = 2550$ $\Leftrightarrow$ $t = 1.5$. Therefore, they will pass each other 1.5 hours after takeoff.

57. Let d be the distance in miles from Boston to Buffalo.

	Rate	Time	Distance
Boston to Buffalo	50	$\frac{d}{50}$	d
Buffalo to Boston	45	$\frac{d}{45}$	d

Using the equation $total\ time = \left(\begin{smallmatrix} time\ traveling \\ to\ Buffalo \end{smallmatrix}\right) + time\ in\ Buffalo + \left(\begin{smallmatrix} time\ traveling \\ to\ Boston \end{smallmatrix}\right)$, we have

$29 = \dfrac{d}{50} + 10 + \dfrac{d}{45}$ $\Leftrightarrow$ $19 = \dfrac{d}{50} + \dfrac{d}{45}$ $\Leftrightarrow$ $19 \cdot 45 \cdot 50 = 45d + 50d$ $\Leftrightarrow$
$42750 = 95d$ $\Leftrightarrow$ $d = \frac{42750}{95} = 450$. Thus the distance from Boston to Buffalo is 450 miles.

59. Let t be the time in hours that Wendy spent on the train. Then $\frac{11}{2} - t$ is the time in hours that Wendy spent on the bus. Using the equation $total\ distance = \left(\begin{smallmatrix} distance\ traveled \\ by\ bus \end{smallmatrix}\right) + \left(\begin{smallmatrix} distance\ traveled \\ by\ train \end{smallmatrix}\right)$, and the table

	Rate	Time	Distance
By train	40	t	$40t$
By bus	60	$\frac{11}{2} - t$	$60\left(\frac{11}{2} - t\right)$

we get the equation $300 = 40t + 60\left(\frac{11}{2} - t\right)$ $\Leftrightarrow$ $300 = 40t + 330 - 60t$ $\Leftrightarrow$ $-30 = -20t$
$\Leftrightarrow$ $t = \frac{-30}{-20} = 1.5$ hours. So the time spent on the train is $5.5 - 1.5 = 4$ hours.

61. Let r be the speed of the plane from Montreal to Los Angeles. Then $r + 0.20r = 1.20r$ is the speed of the plane from Los Angeles to Montreal.

	Rate	Time	Distance
Montreal to L.A.	r	$\dfrac{2500}{r}$	2500
L.A. to Montreal	$1.2r$	$\dfrac{2500}{1.2r}$	2500

Using the equation $total\ time = \left(\begin{smallmatrix} time\ traveling \\ Montreal\ to\ LA \end{smallmatrix}\right) + \left(\begin{smallmatrix} time\ traveling \\ LA\ to\ Montreal \end{smallmatrix}\right)$, we have

$9\frac{1}{6} = \dfrac{2500}{r} + \dfrac{2500}{1.2r}$ $\Leftrightarrow$ $\dfrac{55}{6} = \dfrac{2500}{r} + \dfrac{2500}{1.2r}$ $\Leftrightarrow$ $55 \cdot 1.2r = 2500 \cdot 6 \cdot 1.2 + 2500 \cdot 6$
$\Leftrightarrow$ $66r = 18000 + 15000$ $\Leftrightarrow$ $66r = 33000$ $\Leftrightarrow$ $r = \dfrac{33000}{66} = 500$. Thus the plane flew at a speed of 500 mph on the trip from Montreal to Los Angeles.

63. Let l be the length of the lot in feet. Then the length of the diagonal is $l + 10$. We apply the Pythagorean Theorem with the hypotenuse as the diagonal. So $l^2 + 50^2 = (l + 10)^2$ $\Leftrightarrow$ $l^2 + 2500 = l^2 + 20l + 100$ $\Leftrightarrow$ $20l = 2400$ $\Leftrightarrow$ $l = 120$. Thus the length of the lot is 120 feet.

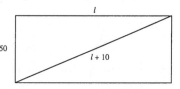

65. Let r be the radius of the running track. The running track consists of 2 semicircles and 2 straight sections 110 yards long, so we get the equation $2\pi r + 220 = 440$ $\Leftrightarrow$ $2\pi r = 220$ $\Leftrightarrow$
$r = \dfrac{110}{\pi} = 35.03$. Thus the radius of the semicircle is about 35 yards.

67. Let h be the height of the break in feet. Then the portion of the bamboo above the break is $10 - h$. Applying the Pythagorean Theorem, we obtain $h^2 + 3^2 = (10 - h)^2 \quad \Leftrightarrow$
$h^2 + 9 = 100 - 20h + h^2 \quad \Leftrightarrow \quad -91 = -20h \quad \Leftrightarrow$
$h = \frac{91}{20} = 4.55$. Thus the break is 4.55 ft above the ground.

Exercises 3.3

1. $x^2 - x - 6 = 0$ $\Leftrightarrow$ $(x - 3)(x + 2) = 0$ $\Leftrightarrow$ $x - 3 = 0$ or $x + 2 = 0$. Thus $x = 3$ or $x = -2$.

3. $x^2 - 4x + 4 = 0$ $\Leftrightarrow$ $(x - 2)^2 = 0$ $\Leftrightarrow$ $x - 2 = 0$ $\Leftrightarrow$ $x = 2$.

5. $2y^2 + 7y + 3 = 0$ $\Leftrightarrow$ $(2y + 1)(y + 3) = 0$ $\Leftrightarrow$ $2y + 1 = 0$ or $y + 3 = 0$. Thus $y = -\frac{1}{2}$ or $y = -3$.

7. $6x^2 + 5x = 4$ $\Leftrightarrow$ $6x^2 + 5x - 4 = 0$ $\Leftrightarrow$ $(2x - 1)(3x + 4) = 0$ $\Leftrightarrow$ $2x - 1 = 0$ or $3x + 4 = 0$. If $2x - 1 = 0$, then $x = \frac{1}{2}$; if $3x + 4 = 0$, then $x = -\frac{4}{3}$.

9. $x^2 = 5(x + 100)$ $\Leftrightarrow$ $x^2 = 5x + 500$ $\Leftrightarrow$ $x^2 - 5x - 500 = 0$ $\Leftrightarrow$ $(x - 25)(x + 20) = 0$ $\Leftrightarrow$ $x - 25 = 0$ or $x + 20 = 0$. Thus $x = 25$ or $x = -20$.

11. $x^2 + 2x - 2 = 0$ $\Leftrightarrow$ $x^2 + 2x = 2$ $\Leftrightarrow$ $x^2 + 2x + 1 = 2 + 1$ $\Leftrightarrow$ $(x + 1)^2 = 3$ $\Rightarrow$ $x + 1 = \pm\sqrt{3}$ $\Leftrightarrow$ $x = -1 \pm \sqrt{3}$.

13. $x^2 - 6x - 9 = 0$ $\Leftrightarrow$ $x^2 - 6x = 9$ $\Leftrightarrow$ $x^2 - 6x + 9 = 9 + 9$ $\Leftrightarrow$ $(x - 3)^2 = 18$ $\Rightarrow$ $x - 3 = \pm 3\sqrt{2}$ $\Leftrightarrow$ $x = 3 \pm 3\sqrt{2}$.

15. $x^2 + 22x + 21 = 0$ $\Leftrightarrow$ $x^2 + 22x = -21$ $\Leftrightarrow$ $x^2 + 22x + 11^2 = -21 + 11^2 = -21 + 121$ $\Leftrightarrow$ $(x + 11)^2 = 100$ $\Rightarrow$ $x + 11 = \pm 10$ $\Leftrightarrow$ $x = -11 \pm 10$. Thus $x = -1$ or $x = -21$.

17. $2x^2 + 8x + 1 = 0$ $\Leftrightarrow$ $x^2 + 4x + \frac{1}{2} = 0$ $\Leftrightarrow$ $x^2 + 4x = -\frac{1}{2}$ $\Leftrightarrow$ $x^2 + 4x + 4 = -\frac{1}{2} + 4$ $\Leftrightarrow$ $(x + 2)^2 = \frac{7}{2}$ $\Rightarrow$ $x + 2 = \pm\sqrt{\frac{7}{2}}$ $\Leftrightarrow$ $x = -2 \pm \frac{\sqrt{14}}{2}$.

19. $4x^2 - x = 0$ $\Leftrightarrow$ $x^2 - \frac{1}{4}x = 0$ $\Leftrightarrow$ $x^2 - \frac{1}{4}x + \frac{1}{64} = \frac{1}{64}$ $\Leftrightarrow$ $\left(x - \frac{1}{8}\right)^2 = \frac{1}{64}$ $\Rightarrow$ $x - \frac{1}{8} = \pm\frac{1}{8}$ $\Leftrightarrow$ $x = \frac{1}{8} \pm \frac{1}{8}$, so $x = \frac{1}{8} - \frac{1}{8} = 0$ or $x = \frac{1}{8} + \frac{1}{8} = \frac{1}{4}$.

21. $x^2 - 2x - 8 = 0$ $\Leftrightarrow$ $(x - 4)(x + 2) = 0$ $\Leftrightarrow$ $x - 4 = 0$ or $x + 2 = 0$. Thus $x = 4$ or $x = -2$.

23. $x^2 + 12x - 27 = 0$ $\Leftrightarrow$ $x^2 + 12x = 27$ $\Leftrightarrow$ $x^2 + 12x + 36 = 27 + 36$ $\Leftrightarrow$ $(x + 6)^2 = 63$ $\Rightarrow$ $x + 6 = \pm 3\sqrt{7}$ $\Leftrightarrow$ $x = -6 \pm 3\sqrt{7}$.

25. $3x^2 + 6x - 5 = 0$ $\Leftrightarrow$ $x^2 + 2x - \frac{5}{3} = 0$ $\Leftrightarrow$ $x^2 + 2x = \frac{5}{3}$ $\Leftrightarrow$ $x^2 + 2x + 1 = \frac{5}{3} + 1$ $\Leftrightarrow$ $(x + 1)^2 = \frac{8}{3}$ $\Rightarrow$ $x + 1 = \pm\sqrt{\frac{8}{3}}$ $\Leftrightarrow$ $x = -1 \pm \frac{2\sqrt{6}}{3}$.

27. $2y^2 - y - \frac{1}{2} = 0$ $\Rightarrow$ $y = \dfrac{-b \pm \sqrt{b^2 - 4ac}}{2a} = \dfrac{-(-1) \pm \sqrt{(-1)^2 - 4(2)\left(-\frac{1}{2}\right)}}{2(2)} = \dfrac{1 \pm \sqrt{1 + 4}}{4}$
$= \dfrac{1 \pm \sqrt{5}}{4}$.

29. $4x^2 + 16x - 9 = 0$ $\Leftrightarrow$ $(2x - 1)(2x + 9) = 0$ $\Leftrightarrow$ $2x - 1 = 0$ or $2x + 9 = 0$. If $2x - 1 = 0$, then $x = \frac{1}{2}$; if $2x + 9 = 0$, then $x = -\frac{9}{2}$.

31. $3 + 5z + z^2 = 0 \quad \Rightarrow \quad z = \dfrac{-b \pm \sqrt{b^2 - 4ac}}{2a} = \dfrac{-(5) \pm \sqrt{(5)^2 - 4(1)(3)}}{2(1)} = \dfrac{-5 \pm \sqrt{25 - 12}}{2}$

$= \dfrac{-5 \pm \sqrt{13}}{2}.$

33. $x^2 - \sqrt{5}x + 1 = 0 \quad \Rightarrow \quad x = \dfrac{-b \pm \sqrt{b^2 - 4ac}}{2a} = \dfrac{-\left(-\sqrt{5}\right) \pm \sqrt{\left(-\sqrt{5}\right)^2 - 4(1)(1)}}{2(1)} =$

$\dfrac{\sqrt{5} \pm \sqrt{5 - 4}}{2} = \dfrac{\sqrt{5} \pm 1}{2}.$

35. $10y^2 - 16y + 5 = 0 \quad \Rightarrow \quad x = \dfrac{-b \pm \sqrt{b^2 - 4ac}}{2a} = \dfrac{-(-16) \pm \sqrt{(-16)^2 - 4(10)(5)}}{2(10)} =$

$\dfrac{16 \pm \sqrt{256 - 200}}{20} = \dfrac{16 \pm \sqrt{56}}{20} = \dfrac{16 \pm 2\sqrt{14}}{20} = \dfrac{8 \pm \sqrt{14}}{10}.$

37. $3x^2 + 2x + 2 = 0 \quad \Rightarrow \quad x = \dfrac{-b \pm \sqrt{b^2 - 4ac}}{2a} = \dfrac{-(2) \pm \sqrt{(2)^2 - 4(3)(2)}}{2(3)}$

$= \dfrac{-2 \pm \sqrt{4 - 24}}{6} = = \dfrac{-2 \pm \sqrt{-20}}{6}.$ Since the discriminant is less than 0, the equation has no

real solutions.

39. $ax^2 - (2a + 1)x + (a + 1) = 0 \quad \Leftrightarrow \quad [ax - (a + 1)](x - 1) = 0 \quad \Leftrightarrow \quad ax - (a + 1) = 0$ or

$x - 1 = 0.$ If $ax - (a + 1) = 0$, then $x = \dfrac{a + 1}{a}$; if $x - 1 = 0$, then $x = 1$.

41. $x^2 - 0.011x - 0.064 = 0 \quad \Rightarrow \quad x = \dfrac{-(-0.011) \pm \sqrt{(-0.011)^2 - 4(1)(-0.064)}}{2(1)}$

$= \dfrac{0.011 \pm \sqrt{0.000121 + 0.256}}{2} = \dfrac{0.011 \pm \sqrt{0.256121}}{2} \approx \dfrac{0.011 \pm 0.506}{2}$

Thus $x \approx \dfrac{0.011 + 0.506}{2} = 0.259$ or $x \approx \dfrac{0.011 - 0.506}{2} = -0.248.$

43. $x^2 - 2.450x + 1.500 = 0 \quad \Rightarrow \quad x = \dfrac{-(-2.450) \pm \sqrt{(-2.450)^2 - 4(1)(1.500)}}{2(1)} =$

$\dfrac{2.450 \pm \sqrt{6.0025 - 6}}{2} = \dfrac{2.450 \pm \sqrt{0.0025}}{2} = \dfrac{2.450 \pm 0.050}{2}.$

Thus $x = \dfrac{2.450 + 0.050}{2} = 1.250$ or $x = \dfrac{2.450 - 0.050}{2} = 1.200.$

45. $h = \frac{1}{2}gt^2 + v_0 t \quad \Leftrightarrow \quad \frac{1}{2}gt^2 + v_0 t - h = 0.$ Using the quadratic formula,

$t = \dfrac{-(v_0) \pm \sqrt{(v_0)^2 - 4\left(\frac{1}{2}g\right)(-h)}}{2\left(\frac{1}{2}g\right)} = \dfrac{-v_0 \pm \sqrt{v_0^2 + 2gh}}{g}.$

47. $A = 2\pi r^2 + 2\pi rh \quad \Leftrightarrow \quad 2\pi r^2 + 2\pi rh - A = 0.$ Using the quadratic formula,

$r = \dfrac{-(2\pi h) \pm \sqrt{(2\pi h)^2 - 4(2\pi)(-A)}}{2(2\pi)} = \dfrac{-2\pi h \pm \sqrt{4\pi^2 h^2 + 8\pi A}}{4\pi} = \dfrac{-\pi h \pm \sqrt{\pi^2 h^2 + 2\pi A}}{2\pi}.$

49. We graph the equation $y = 2x^2 + 19x - 11$ in the viewing rectangle $[-15, 5]$ by $[-15, 15]$. Using Zoom and/or Trace, we get the solutions $x \approx -10.05$ and $x \approx 0.55$.

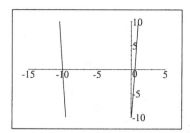

51. We graph the equation $y = 1.33x^2 + 0.26x - 0.65$ in the viewing rectangle $[-5, 5]$ by $[-5, 5]$. Using Zoom and/or Trace, we get the solutions $x \approx 0.61$ and $x \approx -0.80$.

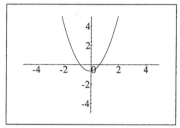

53. $D = b^2 - 4ac = (-6)^2 - 4(1)(1) = 32$. Since D is positive, this equation has two real solutions.

55. $D = b^2 - 4ac = (2.20)^2 - 4(1)(1.21) = 4.84 - 4.84 = 0$. Since $D = 0$, this equation has one real solution.

57. $D = b^2 - 4ac = (r)^2 - 4(1)(-s) = r^2 + 4s$. Since D is positive, this equation has two real solutions.

59. We want to find the values of k that make the discriminant 0. Thus $k^2 - 4(4)(25) = 0 \quad \Leftrightarrow \quad k^2 = 400 \quad \Leftrightarrow \quad k = \pm 20$.

61. Let n be one number. Then the other number must be $55 - n$, since $n + (55 - n) = 55$. Because the product is 684, we have $(n)(55 - n) = 684 \quad \Leftrightarrow \quad 55n - n^2 = 684 \quad \Leftrightarrow \quad n^2 - 55n + 684 = 0 \quad \Rightarrow$

$$n = \frac{-(-55) \pm \sqrt{(-55)^2 - 4(1)(684)}}{2(1)} = \frac{55 \pm \sqrt{3025 - 2736}}{2} = \frac{55 \pm \sqrt{289}}{2} = \frac{55 \pm 17}{2}. \text{ So}$$

$n = \frac{55+17}{2} = \frac{72}{2} = 36$ or $n = \frac{55-17}{2} = \frac{38}{2} = 19$. In either case, the two numbers are 19 and 36.

63. Let w be the width of the garden in feet. Then the length is $w + 10$. Thus $875 = w(w + 10) \quad \Leftrightarrow \quad w^2 + 10w - 875 = 0 \quad \Leftrightarrow \quad (w + 35)(w - 25) = 0$. So $w + 35 = 0$ in which case $w = -35$, which is not possible, or $w - 25 = 0$ and so $w = 25$. Thus the width is 25 feet and the length is 35 feet.

65. Let w be the width of the garden in feet. We use the perimeter to express the length of the garden in terms of width. Since the *perimeter* $= 2 \times width + 2 \times length$, we have $200 = 2w + 2length \quad \Leftrightarrow \quad 2length = 200 - 2w \quad \Leftrightarrow \quad length = 100 - w$. Using the formula for area, we have $2400 = w(100 - w) = 100w - w^2 \quad \Leftrightarrow \quad w^2 - 100w + 2400 = 0 \quad \Leftrightarrow \quad (w - 40)(w - 60) = 0$. So $w - 40 = 0 \quad \Leftrightarrow \quad w = 40$, or $w - 60 = 0 \quad \Leftrightarrow \quad w = 60$. If $w = 40$, then *length* $= 100 - 40 = 60$. And if $w = 60$, then *length* $= 100 - 60 = 40$. So the length is 60 feet and the width is 40 feet.

67. This shape is a trapezoid whose area is given by the formula $area = \dfrac{base_1 + base_2}{2} \times height$.

Substituting 1 for the top base, $x + 1$ for the bottom base, and x for height, we have

$1200 = \dfrac{1 + (1 + x)}{2} x = \dfrac{1}{2}(x + 2)x \quad \Leftrightarrow \quad 2400 = x^2 + 2x \quad \Leftrightarrow \quad x^2 + 2x - 2400 = 0 \quad \Leftrightarrow$

$(x + 50)(x - 48) = 0$. So $x = -50$ or $x = 48$. Since x represents a length, x is positive, so $x = 48$ cm.

69. Let x be the length of one side of the cardboard, so we start with a piece of cardboard x by x. When 4 inches are removed from each side, the base of the box is $(x - 8)$ by $(x - 8)$. Since the volume is 100 in^3, we get $4(x - 8)^2 = 100 \quad \Leftrightarrow \quad x^2 - 16x + 64 = 25 \quad \Leftrightarrow \quad x^2 - 16x + 39 = 0 \quad \Leftrightarrow$ $(x - 3)(x - 13) = 0$. So $x = 3$ or $x = 13$. But $x = 3$ is not possible, since then the length of the base would be $3 - 8 = -5$, and all lengths must be positive. Thus $x = 13$, and the piece of cardboard is 13 inches by 13 inches.

71. Let w be the width of the lot in feet. Then the length is $w + 6$. Using the Pythagorean Theorem, we have $w^2 + (w + 6)^2 = (174)^2 \quad \Leftrightarrow \quad w^2 + w^2 + 12w + 36 = 30276 \quad \Leftrightarrow$ $2w^2 + 12w - 30240 = 0 \quad \Leftrightarrow \quad w^2 + 6w - 15120 = 0 \quad \Leftrightarrow \quad (w + 126)(w - 120) = 0$. So either $w + 126 = 0$ in which case $w = -126$, which is not possible, or $w - 120 = 0$ in which case $w = 120$. Thus the width is 120 feet and the length is 126 feet.

73. Using $h_0 = 288$, we solve $0 = -16t^2 + 288$, for $t \geq 0$. So $0 = -16t^2 + 288 \quad \Leftrightarrow \quad 16t^2 = 288$ $\Leftrightarrow \quad t^2 = 18 \quad \Rightarrow \quad t = \pm\sqrt{18} = \pm 3\sqrt{2}$. Thus it takes $3\sqrt{2} \approx 4.24$ seconds for the ball the hit the ground.

75. We are given $v_o = 40$ ft/s
   (a) Setting $h = 24$, we have $24 = -16t^2 + 40t \quad \Leftrightarrow \quad 16t^2 - 40t + 24 = 0 \quad \Leftrightarrow$ $8(2t - 3)(t - 1) = 0 \quad \Leftrightarrow \quad t = 1$ or $t = 1\frac{1}{2}$. Therefore, the ball reaches 24 feet in 1 second (on the ascent) and again after $1\frac{1}{2}$ seconds (on its descent).
   (b) Setting $h = 48$, we have $48 = -16t^2 + 40t \quad \Leftrightarrow \quad 16t^2 - 40t + 48 = 0 \quad \Leftrightarrow$ $2t^2 - 5t + 6 = 0 \quad \Leftrightarrow \quad t = \frac{5 \pm \sqrt{25 - 48}}{4} = \frac{5 \pm \sqrt{-23}}{4}$. However, since the discriminant $D < 0$, there are no real solutions, and hence the ball never reaches a height of 48 feet.
   (c) The greatest height h is reached only once. So $h = -16t^2 + 40t \quad \Leftrightarrow \quad 16t^2 - 40t + h = 0$ has only one solution. Thus $D = (-40)^2 - 4(16)(h) = 0 \quad \Leftrightarrow \quad 1600 - 64h = 0 \quad \Leftrightarrow$ $h = 25$. So the greatest height reached by the ball is 25 feet.
   (d) Setting $h = 25$, we have $25 = -16t^2 + 40t \quad \Leftrightarrow \quad 16t^2 - 40t + 25 = 0 \quad \Leftrightarrow$ $(4t - 5)^2 = 0 \quad \Leftrightarrow \quad t = 1\frac{1}{4}$. Thus the ball reaches the highest point of its path after $1\frac{1}{4}$ seconds.
   (e) Setting $h = 0$ (ground level), we have $0 = -16t^2 + 40t \quad \Leftrightarrow \quad 2t^2 - 5t = 0 \quad \Leftrightarrow$ $t(2t - 5) = 0 \quad \Leftrightarrow \quad t = 0$ (start) or $t = 2\frac{1}{2}$. So the ball hits the ground in $2\frac{1}{2}$ seconds.

77. (a) The fish population on January 1, 1992 corresponds to $t = 0$, so $F = 1000(30 + 17(0) - (0)^2) = 30,000$. To find when the population will again reach this value, we set $F = 30,000$, giving $30000 = 1000(30 + 17t - t^2) = 30000 + 17000t - 1000t^2$ $\Leftrightarrow \quad 0 = 17000t - 1000t^2 = 1000t(17 - t) \quad \Leftrightarrow \quad t = 0$ or $t = 17$. Thus the fish population will again be the same 17 years later, that is, on January 1, 2009.
   (b) Setting $F = 0$, we have $0 = 1000(30 + 17t - t^2) \quad \Leftrightarrow \quad t^2 - 17t - 30 = 0 \quad \Leftrightarrow$ $t = \frac{17 \pm \sqrt{289 + 120}}{-2} = \frac{17 \pm \sqrt{409}}{-2} = \frac{17 \pm 20.22}{2}$. Thus $t \approx -1.612$ or $t \approx 18.612$. Since $t < 0$ is

inadmissible, it follows that the fish in the lake will have died out 18.612 years after January 1, 1992, that is on Aug. 12, 2010.

79. Let x be the rate, in mi/h, at which the salesman drove between Ajax and Barrington.

Cities	Distance	Rate	Time
Ajax $\rightarrow$ Barrington	120	x	$\dfrac{120}{x}$
Barrington $\rightarrow$ Collins	150	$x + 10$	$\dfrac{150}{x + 10}$

We have used the equation $Time = \frac{Distance}{Rate}$ to fill in the "Time" column of the table. Since the second part of the trip took 6 min. (or $1/10$ hours) more than the first, we can use the *Time* column to get the equation $\dfrac{120}{x} + \dfrac{1}{10} = \dfrac{150}{x + 10}$ $\Rightarrow$ $120(10)(x + 10) + x(x + 10) = 150(10x)$

$\Leftrightarrow$ $1200x + 12000 + x^2 + 10x = 1500x$ $\Leftrightarrow$ $x^2 - 290x + 12000 = 0$ $\Leftrightarrow$

$x = \dfrac{-(-290) \pm \sqrt{(-290)^2 - 4(1)(12000)}}{2} = \dfrac{290 \pm \sqrt{84100 - 48000}}{2} = \dfrac{290 \pm \sqrt{36100}}{2} =$

$\frac{290 \pm 190}{2} = 145 \pm 95$. Hence, the salesman drove either 50 mi/h or 240 mi/h between Ajax and Barrington. (The first choice seems more likely!)

81. Let r be the rowing rate in km/h of the crew in still water. Then their rate upstream was $r - 3$ km/h, and their rate downstream was $r + 3$ km/h.

	Distance	Rate	Time
Upstream	6	$r - 3$	$\dfrac{6}{r - 3}$
Downstream	6	$r + 3$	$\dfrac{6}{r + 3}$

Since the time to row upstream plus the time to row downstream was 2 hours 40 minutes $= \frac{8}{3}$ hour, we get the equation $\dfrac{6}{r - 3} + \dfrac{6}{r + 3} = \dfrac{8}{3}$ $\Leftrightarrow$ $6(3)(r + 3) + 6(3)(r - 3) = 8(r - 3)(r + 3)$

$\Leftrightarrow$ $18r + 54 + 18r - 54 = 8r^2 - 72$ $\Leftrightarrow$ $0 = 8r^2 - 36r - 72 = 4(2r^2 - 9r - 18)$

$= 4(2r + 3)(r - 6)$. Since $2r + 3 = 0$ $\Leftrightarrow$ $r = -\frac{3}{2}$ is impossible, the solution is $r - 6 = 0$

$\Leftrightarrow$ $r = 6$. So the rate of the rowing crew in still water is 6 km/h.

83. Let t be the time, in hours it takes Irene to wash all the windows. Then it takes Henry $t + \frac{3}{2}$ hours to wash all the windows, and the sum of the fraction of the job per hour they can do individually equals the fraction of the job they can do together. Since 1 hour 48 minutes $= 1 + \frac{48}{60} = 1 + \frac{4}{5} = \frac{9}{5}$, we

have $\dfrac{1}{t} + \dfrac{1}{t + \frac{3}{2}} = \dfrac{1}{\frac{9}{5}}$ $\Leftrightarrow$ $\dfrac{1}{t} + \dfrac{2}{2t + 3} = \dfrac{5}{9}$ $\Rightarrow$ $9(2t + 3) + 2(9t) = 5t(2t + 3)$ $\Leftrightarrow$

$18t + 27 + 18t = 10t^2 + 15t$ $\Leftrightarrow$ $10t^2 - 21t - 27 = 0$ $\Leftrightarrow$

$t = \dfrac{-(-21) \pm \sqrt{(-21)^2 - 4(10)(-27)}}{2(10)} = \dfrac{21 \pm \sqrt{441 + 1080}}{20} = \dfrac{21 \pm 39}{20}$. So $t = \dfrac{21 - 39}{20} =$

$-\frac{9}{10}$ or $t = \frac{21 + 39}{20} = 3$. Since $t < 0$ is impossible, all the windows are washed by Irene alone in 3 hours and by Henry alone in $3 + \frac{3}{2} = 4\frac{1}{2}$ hours.

85. Let x be the distance from the center of the earth to the dead spot (in thousand of mile). Now setting $F = 0$, we have $0 = \dfrac{-K}{x^2} + \dfrac{0.012K}{(239 - x)^2}$ $\Leftrightarrow$ $\dfrac{K}{x^2} = \dfrac{0.012K}{(239 - x)^2}$ $\Leftrightarrow$

$K(239 - x)^2 = 0.012Kx^2 \quad \Leftrightarrow \quad 57121 - 478x + x^2 = 0.012x^2 \quad \Leftrightarrow$
$0.988x^2 - 478x + 57121 = 0$. Using the quadratic formula, we obtain

$$x = \frac{-(-478) \pm \sqrt{(-478)^2 - 4(0.988)(57121)}}{2(0.988)} = \frac{478 \pm \sqrt{228484 - 225742.192}}{1.976} =$$

$\dfrac{478 \pm \sqrt{2741.808}}{1.976} \approx \dfrac{478 \pm 52.362}{1.976} \approx 241.903 \pm 26.499$. So either $x \approx 241.903 + 26.499 \approx$
268 or $x \approx 241.903 - 26.499 \approx 215$. Since 268 is greater than the distance from the earth to the moon, we reject it; thus $x \approx 215$ thousand miles $= 215{,}000$ miles.

87. Let x equal the original length of the reed in cubits. Then $x - 1$ is the piece that fits 60 times along the length of the field, that is, the length is $60(x - 1)$. The width is $30x$. Then converting cubits to ninda, we have $375 = 60(x - 1) \cdot 30x \cdot \frac{1}{12^2} = \frac{25}{2}x(x - 1) \quad \Leftrightarrow \quad 30 = x^2 - x \quad \Leftrightarrow$
$x^2 - x - 30 = 0 \quad \Leftrightarrow \quad (x - 6)(x + 5) = 0$. So $x = 6$ or $x = -5$. Since x must be positive, the original length of the reed is 6 cubits.

Exercises 3.4

1. $3 - 5i$: real part 3, imaginary part -5.

3. $6i$: real part 0, imaginary part 6.

5. $\sqrt{2} + \sqrt{-3} = \sqrt{2} + i\sqrt{3}$: real part $\sqrt{2}$, imaginary part $\sqrt{3}$.

7. $(4 + 3i) + (5 - 2i) = (4 + 5) + (3 - 2)i = 9 + i$.

9. $(7 - \frac{1}{2}i) + (5 + \frac{3}{2}i) = (7 + 5) + (-\frac{1}{2} + \frac{3}{2})i = 12 + i$.

11. $(-12 + 8i) - (7 + 4i) = -12 + 8i - 7 - 4i = (-12 - 7) + (8 - 4)i = -19 + 4i$.

13. $4(-1 + 2i) = -4 + 8i$.

15. $(7 - i)(4 + 2i) = 28 + 14i - 4i - 2i^2 = (28 + 2) + (14 - 4)i = 30 + 10i$.

17. $(3 - 4i)(5 - 12i) = 15 - 36i - 20i + 48i^2 = (15 - 48) + (-36 - 20)i = -33 - 56i$.

19. $(6 + 5i)(2 - 3i) = 12 - 18i + 10i - 15i^2 = (12 + 15) + (-18 + 10)i = 27 - 8i$.

21. $\dfrac{1}{i} = \dfrac{1}{i} \cdot \dfrac{i}{i} = \dfrac{i}{i^2} = \dfrac{i}{-1} = -i$.

23. $\dfrac{2 - 3i}{1 - 2i} = \dfrac{2 - 3i}{1 - 2i} \cdot \dfrac{1 + 2i}{1 + 2i} = \dfrac{2 + 4i - 3i - 6i^2}{1 - 4i^2} = \dfrac{(2 + 6) + (4 - 3)i}{1 + 4} = \dfrac{8 + i}{5}$ or $\dfrac{8}{5} + \dfrac{1}{5}i$.

25. $\dfrac{26 + 39i}{2 - 3i} = \dfrac{26 + 39i}{2 - 3i} \cdot \dfrac{2 + 3i}{2 + 3i} = \dfrac{52 + 78i + 78i + 117i^2}{4 - 9i^2} = \dfrac{(52 - 117) + (78 + 78)i}{4 + 9} =$
     $\dfrac{-65 + 156i}{13} = \dfrac{13(-5 + 12i)}{13} = -5 + 12i$.

27. $\dfrac{10i}{1 - 2i} = \dfrac{10i}{1 - 2i} \cdot \dfrac{1 + 2i}{1 + 2i} = \dfrac{10i + 20i^2}{1 - 4i^2} = \dfrac{-20 + 10i}{1 + 4} = \dfrac{5(-4 + 2i)}{5} = -4 + 2i$.

29. $\dfrac{4 + 6i}{3i} = \dfrac{4 + 6i}{3i} \cdot \dfrac{i}{i} = \dfrac{4i + 6i^2}{3i^2} = \dfrac{-6 + 4i}{-3} = \dfrac{-6}{-3} + \dfrac{4}{-3}i = 2 - \dfrac{4}{3}i$.

31. $\dfrac{1}{1 + i} - \dfrac{1}{1 - i} = \dfrac{1}{1 + i} \cdot \dfrac{1 - i}{1 - i} - \dfrac{1}{1 - i} \cdot \dfrac{1 + i}{1 + i} = \dfrac{1 - i}{1 - i^2} - \dfrac{1 + i}{1 - i^2} = \dfrac{1 - i}{2} + \dfrac{-1 - i}{2} = -i$.

33. $i^3 = i^2 \cdot i = -1 \cdot i = -i$.

35. $i^{100} = (i^4)^{25} = (1)^{25} = 1$.

37. $\sqrt{-25} = 5i$.

39. $\sqrt{-3}\sqrt{-12} = i\sqrt{3} \cdot 2i\sqrt{3} = 6i^2 = -6$.

41. $\left(3 - \sqrt{-5}\right)\left(1 + \sqrt{-1}\right) = \left(3 - i\sqrt{5}\right)(1 + i) = 3 + 3i - i\sqrt{5} - i^2\sqrt{5} =$
     $\left(3 + \sqrt{5}\right) + \left(3 - \sqrt{5}\right)i$.

43. $\dfrac{2+\sqrt{-8}}{1+\sqrt{-2}} = \dfrac{2+2i\sqrt{2}}{1+i\sqrt{2}} = \dfrac{2+2i\sqrt{2}}{1+i\sqrt{2}} \cdot \dfrac{1-i\sqrt{2}}{1-i\sqrt{2}} = \dfrac{2-2i\sqrt{2}+2i\sqrt{2}-4\,i^2}{1-2\,i^2} =$

$\dfrac{(2+4)+(-2\sqrt{2}+2\sqrt{2})\,i}{1+2} = \dfrac{6}{3} = 2.$

45. $\dfrac{\sqrt{-36}}{\sqrt{-2}\sqrt{-9}} = \dfrac{6\,i}{i\sqrt{2}\cdot 3\,i} = \dfrac{2}{i\sqrt{2}}\cdot\dfrac{i\sqrt{2}}{i\sqrt{2}} = \dfrac{2i\sqrt{2}}{2\,i^2} = \dfrac{i\sqrt{2}}{-1} = -i\sqrt{2}.$

47. $x^2+9=0 \quad\Leftrightarrow\quad x^2=-9 \quad\Rightarrow\quad x=\pm 3\,i.$

49. $x^2-4x+5=0 \quad\Rightarrow\quad x = \dfrac{-(-4)\pm\sqrt{(-4)^2-4(1)(5)}}{2(1)} = \dfrac{4\pm\sqrt{16-20}}{2} = \dfrac{4\pm\sqrt{-4}}{2}$

$= \dfrac{4\pm 2\,i}{2} = 2\pm i.$

51. $x^2+x+1=0 \quad\Rightarrow\quad x = \dfrac{-(1)\pm\sqrt{(1)^2-4(1)(1)}}{2(1)} = \dfrac{-1\pm\sqrt{1-4}}{2} = \dfrac{-1\pm\sqrt{-3}}{2} =$

$\dfrac{-1\pm i\sqrt{3}}{2} = -\dfrac{1}{2}\pm\dfrac{i\sqrt{3}}{2}.$

53. $2x^2-2x+1=0 \quad\Rightarrow\quad x = \dfrac{-(-2)\pm\sqrt{(-2)^2-4(2)(1)}}{2(2)} = \dfrac{2\pm\sqrt{4-8}}{4} = \dfrac{2\pm\sqrt{-4}}{4} =$

$\dfrac{2\pm 2\,i}{4} = \dfrac{1}{2}\pm\dfrac{1}{2}\,i.$

55. $t+3+\dfrac{3}{t}=0 \quad\Leftrightarrow\quad t^2+3t+3=0 \quad\Rightarrow\quad t = \dfrac{-(3)\pm\sqrt{(3)^2-4(1)(3)}}{2(1)} = \dfrac{-3\pm\sqrt{9-12}}{2}$

$= \dfrac{-3\pm\sqrt{-3}}{2} = \dfrac{-3\pm i\sqrt{3}}{2} = -\dfrac{3}{2}\pm\dfrac{i\sqrt{3}}{2}.$

57. $6x^2+12x+7=0 \quad\Rightarrow\quad x = \dfrac{-(12)\pm\sqrt{(12)^2-4(6)(7)}}{2(6)} = \dfrac{-12\pm\sqrt{144-168}}{12} =$

$\dfrac{-12\pm\sqrt{-24}}{12} = \dfrac{-12\pm 2i\sqrt{6}}{12} = \dfrac{-12}{12}\pm\dfrac{2i\sqrt{6}}{12} = -1\pm\dfrac{i\sqrt{6}}{6}.$

59. $\frac{1}{2}x^2-x+5=0 \quad\Rightarrow\quad x = \dfrac{-(-1)\pm\sqrt{(-1)^2-4(\frac{1}{2})(5)}}{2(\frac{1}{2})} = \dfrac{1\pm\sqrt{1-10}}{1} = 1\pm\sqrt{-9} =$

$1\pm 3i.$

61. LHS $= \overline{z}+\overline{w} = \overline{(a+b\,i)}+\overline{(c+d\,i)} = a-b\,i+c-d\,i = (a+c)+(-b-d)i = (a+c)-(b+d)i.$

RHS $= \overline{z+w} = \overline{(a+b\,i)+(c+d\,i)} = \overline{(a+c)+(b+d)\,i} = (a+c)-(b+d)i.$

Since LHS $=$ RHS, this proves the statement.

63. LHS $= (\overline{z})^2 = \left(\overline{(a+b\,i)}\right)^2 = (a-bi)^2 = a^2-2ab\,i+b^2\,i^2 = (a^2-b^2)-2ab\,i.$

RHS $= \overline{z^2} = \overline{(a+b\,i)^2} = \overline{a^2+2ab\,i+b^2\,i^2} = \overline{(a^2-b^2)+2ab\,i} = (a^2-b^2)-2ab\,i.$

Since LHS $=$ RHS, this proves the statement.

65. $z + \bar{z} = (a + b\,i) + \overline{(a + b\,i)} = a + b\,i + a - b\,i = 2a$, which is a real number.

67. $z \cdot \bar{z} = (a + b\,i) \cdot \overline{(a + b\,i)} = (a + b\,i) \cdot (a - b\,i) = a^2 - b^2 i^2 = a^2 + b^2$, which is a real number.

69. Using the quadratic formula, the solutions to the equation are $x = \dfrac{-b \pm \sqrt{b^2 - 4ac}}{2a}$. Since both solutions are imaginary, we have $b^2 - 4ac < 0 \quad \Leftrightarrow \quad 4ac - b^2 > 0$, so the solutions are

$x = \dfrac{-b}{2a} \pm \dfrac{\sqrt{4ac - b^2}}{2a}\, i$, where $\sqrt{4ac - b^2}$ is a real number. Thus the solutions are complex conjugates of each other.

Exercises 3.5

1. $x^4 = 64x^2 \quad \Leftrightarrow \quad 0 = x^4 - 64x^2 = x^2(x^2 - 64) = x^2(x - 8)(x + 8)$. So $x^2 = 0 \quad \Rightarrow \quad x = 0$, $x - 8 = 0 \quad \Leftrightarrow \quad x = 8$, or $x + 8 = 0 \quad \Leftrightarrow \quad x = -8$. The solutions are 0 and ± 8.

3. $0 = x^3 - 6x = x(x^2 - 6)$. So $x = 0$, or $x^2 - 6 = 0 \quad \Leftrightarrow \quad x^2 = 6 \quad \Rightarrow \quad x = \pm\sqrt{6}$. The solutions are 0 and $\pm\sqrt{6}$.

5. $0 = x^3 - 3x^2 + 2x = x(x^2 - 3x + 2) = x(x - 2)(x - 1) \quad \Leftrightarrow \quad x = 0, x - 2 = 0$, or $x - 1 = 0$. Thus $x = 0$, $x = 2$, or $x = 1$. The solutions are $0, 2$, and 1.

7. $0 = x^4 + 4x^3 + 2x^2 = x^2(x^2 + 4x + 2)$. So either $x^2 = 0 \quad \Leftrightarrow \quad x = 0$, or using the quadratic formula on $x^2 + 4x + 2 = 0$, we have $x = \dfrac{-4 \pm \sqrt{4^2 - 4(1)(2)}}{2(1)} = \dfrac{-4 \pm \sqrt{16 - 8}}{2} = \dfrac{-4 \pm \sqrt{8}}{2}$
$= \dfrac{-4 \pm 2\sqrt{2}}{2} = -2 \pm \sqrt{2}$. The solutions are $0, -2 - \sqrt{2}$, and $-2 + \sqrt{2}$.

9. $0 = x^3 - 5x^2 - 2x + 10 = x^2(x - 5) - 2(x - 5) = (x - 5)(x^2 - 2)$. If $x - 5 = 0$, then $x = 5$. If $x^2 - 2 = 0$, then $x^2 = 2 \quad \Leftrightarrow \quad x = \pm\sqrt{2}$. The solutions are 5 and $\pm\sqrt{2}$.

11. $x^3 - x^2 + x - 1 = x^2 + 1 \quad \Leftrightarrow$
$0 = x^3 - 2x^2 + x - 2 = x^2(x - 2) + (x - 2) = (x - 2)(x^2 + 1)$. Since $x^2 + 1 = 0$ has no real solution, the only solution comes from $x - 2 = 0 \quad \Leftrightarrow \quad x = 2$.

13. $\dfrac{1}{x - 1} + \dfrac{1}{x + 2} = \dfrac{5}{4} \quad \Leftrightarrow \quad 4(x - 1)(x + 2)\left(\dfrac{1}{x - 1} + \dfrac{1}{x + 2}\right) = 4(x - 1)(x + 2)\left(\dfrac{5}{4}\right) \quad \Leftrightarrow$
$4(x + 2) + 4(x - 1) = 5(x - 1)(x + 2) \quad \Leftrightarrow \quad 4x + 8 + 4x - 4 = 5x^2 + 5x - 10 \quad \Leftrightarrow$
$5x^2 - 3x - 14 = 0 \quad \Leftrightarrow \quad (5x + 7)(x - 2) = 0$. If $5x + 7 = 0$, then $x = -\frac{7}{5}$; if $x - 2 = 0$, then $x = 2$. The solutions are $-\frac{7}{5}$ and 2.

15. $\dfrac{x^2}{x + 100} = 50 \quad \Rightarrow \quad x^2 = 50(x + 100) = 50x + 5000 \quad \Leftrightarrow \quad x^2 - 50x - 5000 = 0 \quad \Leftrightarrow$
$(x - 100)(x + 50) = 0 \quad \Leftrightarrow \quad x - 100 = 0$ or $x + 50 = 0$. Thus $x = 100$ or $x = -50$. The solutions are 100 and -50.

17. $\dfrac{x + 5}{x - 2} = \dfrac{5}{x + 2} + \dfrac{28}{x^2 - 4} \quad \Rightarrow \quad (x + 2)(x + 5) = 5(x - 2) + 28 \quad \Leftrightarrow$
$x^2 + 7x + 10 = 5x - 10 + 28 \quad \Leftrightarrow \quad x^2 + 2x - 8 = 0 \quad \Leftrightarrow \quad (x - 2)(x + 4) = 0 \quad \Leftrightarrow$
$x - 2 = 0$ or $x + 4 = 0 \quad \Leftrightarrow \quad x = 2$ or $x = -4$. However, $x = 2$ is inadmissible since we can't divide by 0 in the original equation, so the only solution is -4.

19. $\dfrac{1}{x - 1} - \dfrac{2}{x^2} = 0 \quad \Leftrightarrow \quad x^2 - 2(x - 1) = 0 \quad \Leftrightarrow \quad x^2 - 2x + 2 = 0 \quad \Rightarrow$
$x = \dfrac{-(-2) \pm \sqrt{(-2)^2 - 4(1)(2)}}{2(1)} = \dfrac{2 \pm \sqrt{4 - 8}}{2} = \dfrac{2 \pm \sqrt{-4}}{2}$. Since the radicand is negative, there are no real solutions.

21. $0 = (x + 5)^2 - 3(x + 5) - 10 = [(x + 5) - 5][(x + 5) + 2] = x(x + 7) \quad \Leftrightarrow \quad x = 0$ or $x = -7$. The solutions are 0 and -7.

23. Let $w = \dfrac{1}{x+1}$. Then $\left(\dfrac{1}{x+1}\right)^2 - 2\left(\dfrac{1}{x+1}\right) - 8 = 0$ becomes $w^2 - 2w - 8 = 0$ $\Leftrightarrow$
   $(w-4)(w+2) = 0$. So $w - 4 = 0$ $\Leftrightarrow$ $w = 4$, and $w + 2 = 0$ $\Leftrightarrow$ $w = -2$. When $w = 4$,
   we have $\dfrac{1}{x+1} = 4$ $\Leftrightarrow$ $1 = 4x + 4$ $\Leftrightarrow$ $-3 = 4x$ $\Leftrightarrow$ $x = -\frac{3}{4}$. When $w = -2$, we
   have $\dfrac{1}{x+1} = -2$ $\Leftrightarrow$ $1 = -2x - 2$ $\Leftrightarrow$ $3 = -2x$ $\Leftrightarrow$ $x = -\frac{3}{2}$. Solutions are $-\frac{3}{4}$ and
   $-\frac{3}{2}$.

25. Let $w = x^2$. Then $x^4 - 13x^2 + 40 = (x^2)^2 - 13x^2 + 40 = 0$ becomes $w^2 - 13w + 40 = 0$ $\Leftrightarrow$
   $(w-5)(w-8) = 0$. So $w - 5 = 0$ $\Leftrightarrow$ $w = 5$, and $w - 8 = 0$ $\Leftrightarrow$ $w = 8$. When $w = 5$,
   we have $x^2 = 5$ $\Rightarrow$ $x = \pm\sqrt{5}$. When $w = 8$, we have $x^2 = 8$ $\Rightarrow$ $x = \pm\sqrt{8} = \pm 2\sqrt{2}$.
   The solutions are $\pm\sqrt{5}$ and $\pm 2\sqrt{2}$.

27. $2x^4 + 4x^2 + 1 = 0$. The LHS is the sum of two nonnegative numbers and a positive number, so
   $2x^4 + 4x^2 + 1 \geq 1 \neq 0$. This equation has no real solutions.

29. Let $u = x^{2/3}$. Then $0 = x^{4/3} - 5x^{2/3} + 6$ becomes $u^2 - 5u + 6 = 0$ $\Leftrightarrow$ $(u-3)(u-2) = 0$
   $\Leftrightarrow$ $u - 3 = 0$ or $u - 2 = 0$. If $u - 3 = 0$, then $x^{2/3} - 3 = 0$ $\Leftrightarrow$ $x^{2/3} = 3$ $\Rightarrow$
   $x = \pm 3^{3/2}$. If $u - 2 = 0$, then $x^{2/3} - 2 = 0$ $\Leftrightarrow$ $x^{2/3} = 2$ $\Rightarrow$ $x = \pm 2^{3/2}$. The solutions
   are $\pm 3^{3/2}$ and $\pm 2^{3/2}$.

31. $4(x+1)^{1/2} - 5(x+1)^{3/2} + (x+1)^{5/2} = 0$ $\Leftrightarrow$ $\sqrt{x+1}[4 - 5(x+1) + (x+1)^2] = 0$ $\Leftrightarrow$
   $\sqrt{x+1}(4 - 5x - 5 + x^2 + 2x + 1) = 0$ $\Leftrightarrow$ $\sqrt{x+1}(x^2 - 3x) = 0$ $\Leftrightarrow$
   $\sqrt{x+1} \cdot x(x-3) = 0$ $\Leftrightarrow$ $x = -1$ or $x = 0$ or $x = 3$. The solutions are -1, 0, and 3.

33. Let $u = x^{1/6}$. (We choose the exponent $\frac{1}{6}$ because the LCD of 2, 3, and 6 is 6.) Then
   $x^{1/2} - 3x^{1/3} = 3x^{1/6} - 9$ $\Leftrightarrow$ $x^{3/6} - 3x^{2/6} = 3x^{1/6} - 9$ $\Leftrightarrow$ $u^3 - 3u^2 = 3u - 9$ $\Leftrightarrow$
   $0 = u^3 - 3u^2 - 3u + 9 = u^2(u-3) - 3(u-3) = (u-3)(u^2-3)$. So $u - 3 = 0$ or $u^2 - 3 = 0$.
   If $u - 3 = 0$, then $x^{1/6} - 3 = 0$ $\Leftrightarrow$ $x^{1/6} = 3$ $\Leftrightarrow$ $x = 3^6 = 729$. If $u^2 - 3 = 0$, then
   $x^{1/3} - 3 = 0$ $\Leftrightarrow$ $x^{1/3} = 3$ $\Leftrightarrow$ $x = 3^3 = 27$. The solutions are 729 and 27.

35. $\dfrac{1}{x^3} + \dfrac{4}{x^2} + \dfrac{4}{x} = 0$ $\Rightarrow$ $1 + 4x + 4x^2 = 0$ $\Leftrightarrow$ $(1 + 2x)^2 = 0$ $\Leftrightarrow$ $1 + 2x = 0$ $\Leftrightarrow$
   $2x = -1$ $\Leftrightarrow$ $x = -\frac{1}{2}$. The solution is $-\frac{1}{2}$.

37. $\sqrt{2x+1} + 1 = x$ $\Leftrightarrow$ $\sqrt{2x+1} = x - 1$ $\Rightarrow$ $2x + 1 = (x-1)^2$ $\Leftrightarrow$
   $2x + 1 = x^2 - 2x + 1$ $\Leftrightarrow$ $0 = x^2 - 4x = x(x-4)$. Potential solutions are $x = 0$ and $x - 4$
   $\Leftrightarrow$ $x = 4$. These are only potential solutions since *squaring* is not a reversible operation. We
   must check each potential solution in the original equation.
   Checking $x = 0$: $\sqrt{2(0)+1} + 1 \overset{?}{=} (0)$, $\sqrt{1} + 1 \overset{?}{=} 0$, NO!
   Checking $x = 4$: $\sqrt{2(4)+1} + 1 \overset{?}{=} (4)$, $\sqrt{9} + 1 \overset{?}{=} 4$, $3 + 1 \overset{?}{=} 4$, Yes. The only solution is $x = 4$.

39. $\sqrt{5-x} + 1 = x - 2$ $\Leftrightarrow$ $\sqrt{5-x} = x - 3$ $\Rightarrow$ $5 - x = (x-3)^2$ $\Leftrightarrow$
   $5 - x = x^2 - 6x + 9$ $\Leftrightarrow$ $0 = x^2 - 5x + 4 = (x-4)(x-1)$. Potential solutions are $x = 4$ and
   $x = 1$. We must check each potential solution in the original equation.
   Checking $x = 4$: $\sqrt{5-(4)} + 1 \overset{?}{=} (4) - 2$, $\sqrt{1} + 1 \overset{?}{=} 4 - 2$, $1 + 1 \overset{?}{=} 2$, Yes.
   Checking $x = 1$: $\sqrt{5-(1)} + 1 \overset{?}{=} (1) - 2$, $\sqrt{4} + 1 \overset{?}{=} -1$, $2 + 1 \overset{?}{=} -1$, NO! The only solution is
   $x = 4$.

41. $\sqrt{\sqrt{x-5}+x}=5$. Squaring both sides, we get $\sqrt{x-5}+x=25$ $\Leftrightarrow$ $\sqrt{x-5}=25-x$. Squaring both sides again, we get $x-5=(25-x)^2$ $\Leftrightarrow$ $x-5=625-50x+x^2$ $\Leftrightarrow$ $0=x^2-51x+630=(x-30)(x-21)$. Potential solutions are $x=30$ and $x=21$. We must check each potential solution in the original equation.

Checking $x=30$: $\sqrt{\sqrt{(30)-5}+(30)}\overset{?}{=}5$, $\sqrt{\sqrt{(30)-5}+(30)}=\sqrt{\sqrt{25}+30}=\sqrt{35}>5$, hence not a solution.

Checking $x=21$: $\sqrt{\sqrt{(21)-5}+21}\overset{?}{=}5$, $\sqrt{\sqrt{(21)-5}+21}=\sqrt{\sqrt{16}+21}=\sqrt{25}=5$, hence a solution. The only solution is $x=21$.

43. $x^2\sqrt{x+3}=(x+3)^{3/2}$ $\Leftrightarrow$ $0=x^2\sqrt{x+3}-(x+3)^{3/2}$ $\Leftrightarrow$ $0=\sqrt{x+3}[(x^2)-(x+3)]$ $\Leftrightarrow$ $0=\sqrt{x+3}(x^2-x-3)$. If $(x+3)^{1/2}=0$, then $x+3=0$ $\Leftrightarrow$ $x=-3$. If $x^2-x-3=0$, then using the quadratic formula $x=\frac{1\pm\sqrt{13}}{2}$. The solutions are -3 and $\frac{1\pm\sqrt{13}}{2}$.

45. $\sqrt{x+\sqrt{x+2}}=2$. Squaring both sides, we get $x+\sqrt{x+2}=4$ $\Leftrightarrow$ $\sqrt{x+2}=4-x$. Squaring both sides again, we get $x+2=(4-x)^2=16-8x+x^2$ $\Leftrightarrow$ $0=x^2-9x+14$ $\Leftrightarrow$ $0=(x-7)(x-2)$. If $x-7=0$, then $x=7$. If $x-2=0$, then $x=2$. So $x=2$ is a solution but $x=7$ is not, since it does not satisfy the original equation.

47. $x^3=1$ $\Leftrightarrow$ $x^3-1=0$ $\Leftrightarrow$ $(x-1)(x^2+x+1)=0$ $\Leftrightarrow$ $x-1=0$ or $x^2+x+1=0$. If $x-1=0$, then $x=1$. If $x^2+x+1=0$, then using the quadratic formula $x=\frac{-1\pm i\sqrt{3}}{2}$. The solutions are 1 and $\frac{-1\pm i\sqrt{3}}{2}$.

49. $x^3+x^2+x=0$ $\Leftrightarrow$ $x(x^2+x+1)=0$ $\Leftrightarrow$ $x=0$ or $x=\frac{-1\pm i\sqrt{3}}{2}$. The solutions are 1 and $\frac{-1\pm i\sqrt{3}}{2}$.

51. $x^4-6x^2+8=0$ $\Leftrightarrow$ $(x^2-4)(x^2-2)=0$ $\Leftrightarrow$ $x=\pm2$ or $x=\pm\sqrt{2}$. The solutions are ±2 and $\pm\sqrt{2}$.

53. $x^6-9x^3+8=0$ $\Leftrightarrow$ $(x^3-8)(x^3-1)=0$ $\Leftrightarrow$ $(x-2)(x^2+2x+4)(x-1)(x^2+x+1)=0$ $\Leftrightarrow$ $x=2$ or $x=\frac{-2\pm2i\sqrt{3}}{2}=-1\pm i\sqrt{3}$ or $x=1$ or $x=\frac{-1\pm i\sqrt{3}}{2}$. The solutions are 2, $-1\pm i\sqrt{3}$, 1, and $\frac{-1\pm i\sqrt{3}}{2}$.

55. $\sqrt{x^2+1}+\dfrac{8}{\sqrt{x^2+1}}=\sqrt{x^2+9}$. Squaring both sides, we have $(x^2+1)+16+\dfrac{64}{x^2+1}=x^2+9$ $\Leftrightarrow$ $\dfrac{64}{x^2+1}=-8$ $\Leftrightarrow$ $\dfrac{8}{x^2+1}=-1$ $\Leftrightarrow$ $x^2+1=-8$ $\Leftrightarrow$ $x^2=-9$ $\Leftrightarrow$ $x=\pm3$ i. We must check each potential solution in the original equation.

Checking $x=\pm3\,i$: $\sqrt{(\pm3\,i)^2+1}+\dfrac{8}{\sqrt{(\pm3\,i)^2+1}}\overset{?}{=}\sqrt{(\pm3\,i)^2+9}$,

LHS $=\sqrt{-8}+\dfrac{8}{\sqrt{-8}}=2\sqrt{2}\,i+\dfrac{8}{2\sqrt{2}\,i}=2\sqrt{2}\,i-2\sqrt{2}\,i=0$.

RHS $=\sqrt{(\pm3\,i)^2+9}=\sqrt{-9+9}=0$. Since LHS $=$ RHS, $-3\,i$ and $3\,i$ are solutions.

57. Let x be the number of people originally intended to take the trip. Then originally, the cost of the trip is $\dfrac{900}{x}$. After 5 people cancel, there are now $x-5$ people, each paying $\dfrac{900}{x}+2$. Thus

$$900 = (x - 5)\left(\frac{900}{x} + 2\right) \quad\Leftrightarrow\quad 900 = 900 + 2x - \frac{4500}{x} - 10 \quad\Leftrightarrow\quad 0 = 2x - 10 - \frac{4500}{x}$$
$\Leftrightarrow \quad 0 = 2x^2 - 10x - 4500 = (2x - 100)(x + 45)$. Thus either $2x - 100 = 0$, so $x = 50$, or $x + 45 = 0$, $x = -45$. Since the number of people on the trip must be positive, originally 50 people intended to take the trip.

59. We want to solve for t when $P = 500$. Letting $u = \sqrt{t}$ and substituting, we have
$500 = 3t + 10\sqrt{t} + 140 \quad\Leftrightarrow\quad 500 = 3u^2 + 10u + 140 \quad\Leftrightarrow\quad 0 = 3u^2 + 10u - 360 \quad\Rightarrow$
$u = \frac{-5 \pm \sqrt{1105}}{3}$. Since $u = \sqrt{t}$, we must have $u \geq 0$. So $\sqrt{t} = u = \frac{-5 + \sqrt{1105}}{3} \approx 9.414 \quad\Rightarrow$
$t = \approx 88.62$. So it will take 89 days for the fish population to reach 500.

61. We have that the volume is 180 ft^3, so $x(x - 4)(x + 9) = 180 \quad\Leftrightarrow\quad x^3 + 5x^2 - 36x = 180$
$\Leftrightarrow \quad x^3 + 5x^2 - 36x - 180 = 0 \quad\Leftrightarrow\quad x^2(x + 5) - 36(x + 5) = 0 \quad\Leftrightarrow$
$(x + 5)(x^2 - 36) = 0 \quad\Leftrightarrow\quad (x + 5)(x + 6)(x - 6) = 0 \quad\Rightarrow\quad x = 6$ is the only positive solution. So the box is 2 feet by 6 feet by 15 feet.

63. Let x be the length, in miles, of the abandoned road to be used. Then the length of the abandoned road not used is $40 - x$, and the length of the new road is $\sqrt{10^2 + (40 - x)^2}$ miles, by the Pythagorean Theorem. Since the cost of the road is *cost per mile* × *number of miles*, we have
$100{,}000x + 200{,}000\sqrt{x^2 - 80x + 1700} = 6{,}800{,}000 \quad\Leftrightarrow\quad 2\sqrt{x^2 - 80x + 1700} = 68 - x$.
Squaring both sides, we get $4x^2 - 320x + 6800 = 4624 - 136x + x^2 \quad\Leftrightarrow$
$3x^2 - 184x + 2176 = 0 \quad\Leftrightarrow\quad x = \frac{184 \pm \sqrt{33856 - 26112}}{6} = \frac{184 \pm 88}{6} \quad\Leftrightarrow\quad x = \frac{136}{3}$ or $x = 16$. Since $45\frac{1}{3}$ is longer than the existing road, 16 miles of the abandoned road should be used. A completely new road would have length $\sqrt{10^2 + 40^2}$, (let $x = 0$), and would cost $\sqrt{1700} \times 200{,}000 \approx 8.3$ million dollars. So no, it would not be cheaper.

65. Let x be the height of the pile in feet. Then the diameter is $3x$ and the radius is $\dfrac{3x}{2}$ feet. Since the
volume of the cone is 1000 ft^3, we have $\dfrac{\pi}{3}\left(\dfrac{3x}{2}\right)^2 x = 1000 \quad\Leftrightarrow\quad \dfrac{3\pi x^3}{4} = 1000 \quad\Leftrightarrow$
$x^3 = \dfrac{4000}{3\pi} \quad\Leftrightarrow\quad x = \sqrt[3]{\dfrac{4000}{3\pi}} \approx 7.52$ feet.

67. Let x be the length of the hypotenuse of the triangle, in feet. Then one of the other sides has length $x - 7$ feet, and since the perimeter is 392 feet, the remaining side must have length
$392 - x - (x - 7) = 399 - 2x$. From the Pythagorean
Theorem, we get
$(x - 7)^2 + (399 - 2x)^2 = x^2 \quad\Leftrightarrow$
$4x^2 - 1610x + 159250 = 0$.
Using the quadratic formula, we get

$x = \dfrac{1610 \pm \sqrt{1610^2 - 4(4)(159250)}}{2(4)} = \dfrac{1610 \pm \sqrt{44100}}{8}$
$= \dfrac{1610 \pm 210}{8}$, and so $x = 227.5$ or $x = 175$.
But if $x = 227.5$, then the side of length $x - 7$ combined with the hypotenuse already exceeds the perimeter of 392 feet, and so we must have $x = 175$. Thus the other sides have length
$175 - 7 = 168$ and $399 - 2(175) = 49$. The lot has sides of length 49 feet, 168 feet, and 175 feet.

69. Since the total time is 3 s, we have $3 = \dfrac{\sqrt{d}}{4} + \dfrac{d}{1090}$. Letting $w = \sqrt{d}$, we have $3 = \frac{1}{4}w + \frac{1}{1090}w^2$

$\Leftrightarrow \quad \frac{1}{1090}w^2 + \frac{1}{4}w - 3 = 0 \quad \Leftrightarrow \quad 2w^2 + 545w - 6540 = 0 \quad \Rightarrow \quad w = \dfrac{-545 \pm 591.054}{4}$.

Since $w \geq 0$, we have $\sqrt{d} = w \approx 11.51$, so $d = 132.56$. The well is 132.6 ft deep.

71. $0 = a^3x^3 + b^3 = (ax + b)(a^2x^2 - abx + b^2)$. So $ax + b = 0 \quad \Leftrightarrow \quad ax = -b \quad \Leftrightarrow \quad x = -\dfrac{b}{a}$ or

$x = \dfrac{-(-ab) \pm \sqrt{(-ab)^2 - 4(a^2)(b^2)}}{2(a^2)} = \dfrac{ab \pm \sqrt{-3a^2b^2}}{2a^2} = \dfrac{ab \pm \sqrt{3}ab\,i}{2a^2} = \dfrac{b \pm \sqrt{3}bi}{2a}$. Thus the

three solutions are $-\dfrac{b}{a}$ and $\dfrac{b \pm \sqrt{3}bi}{2a}$.

73. Let $w = x^{1/6}$. Then $x^{1/3} = w^2$ and $x^{1/2} = w^3$, and so $0 = w^3 + aw^2 + bw + ab$

$= w^2(w + a) + b(w + a) = (w + a)(w^2 + b) \quad \Leftrightarrow \quad (\sqrt[6]{x} + a)(\sqrt[3]{x} + b)$. So $\sqrt[6]{x} + a = 0$

$\Leftrightarrow \quad \sqrt[6]{x} = -a$; however, since $\sqrt[6]{x}$ is positive by definition is positive and $-a$ is negative, this is

impossible. Setting the other factor equal to zero, we have $\sqrt[3]{x} + b = 0 \quad \Leftrightarrow \quad \sqrt[3]{x} = -b \quad \Rightarrow$

$x = -b^3$. Checking $x = -b^3$, we have $\sqrt{-b^3} + a\sqrt[3]{-b^3} + b\sqrt[6]{-b^3} + ab$

$= b\sqrt{b}i + a(-b) + b\left(\sqrt{-b}\right) + ab = 2b\sqrt{b}\,i \neq 0$, so this is not a solution either. Therefore, there

are no solutions to this equation.

Exercises 3.6

1. $x = -1$: $(-1) + 1 \overset{?}{\geq} 0$. Yes, $0 \geq 0$. $x = 0$: $(0) + 1 \overset{?}{\geq} 0$. Yes.

   $x = \frac{1}{2}$: $(\frac{1}{2}) + 1 \overset{?}{\geq} 0$. Yes. $x = \sqrt{2}$: $(\sqrt{2}) + 1 \overset{?}{\geq} 0$. Yes.

   $x = 2$: $(2) + 1 \overset{?}{\geq} 0$. Yes.   The solutions are $-1, 0, \frac{1}{2}, \sqrt{2}$, and 2.

3. $x = -1$: $2(-1) + 10 \overset{?}{>} 8$. No, $8 \not> 8$. $x = 0$: $2(0) + 10 \overset{?}{>} 8$. Yes.

   $x = \frac{1}{2}$: $2(\frac{1}{2}) + 10 \overset{?}{>} 8$. Yes. $x = \sqrt{2}$: $2(\sqrt{2}) + 10 \overset{?}{>} 8$. Yes.

   $x = 2$: $2(2) + 10 \overset{?}{>} 8$. Yes.   The solutions are $0, \frac{1}{2}, \sqrt{2}$, and 2.

5. $x = -1$: $\dfrac{1}{(-1)} \overset{?}{\leq} \dfrac{1}{2}$. Yes, $-1 \leq \dfrac{1}{2}$. $x = 0$: $\dfrac{1}{(0)} \overset{?}{\leq} \dfrac{1}{2}$. No, the LHS is undefined.

   $x = \frac{1}{2}$: $\dfrac{1}{(\frac{1}{2})} \overset{?}{\leq} \dfrac{1}{2}$. No. $x = \sqrt{2}$: $\dfrac{1}{(\sqrt{2})} \overset{?}{\leq} \dfrac{1}{2}$. No.

   $x = 2$: $\dfrac{1}{(2)} \overset{?}{\leq} \dfrac{1}{2}$. Yes.   The solutions are -1 and 2.

7. $3x \leq 12 \quad \Leftrightarrow \quad x \leq 4.$ Interval: $(-\infty, 4]$.   Graph:

9. $20 < -4x \quad \Leftrightarrow \quad -5 > x.$ Interval: $(-\infty, -5)$.   Graph:

11. $2x - 5 > 3 \quad \Leftrightarrow \quad 2x > 8 \quad \Leftrightarrow \quad x > 4.$ Interval: $(4, \infty)$.

   Graph:

13. $7 - x \geq 5 \quad \Leftrightarrow \quad -x \geq -2 \quad \Leftrightarrow \quad x \leq 2.$ Interval: $(-\infty, 2]$.

   Graph:

15. $2x + 1 < 0 \quad \Leftrightarrow \quad 2x < -1 \quad \Leftrightarrow \quad x < -\frac{1}{2}.$ Interval: $\left(-\infty, -\frac{1}{2}\right)$.

   Graph:

17. $3x + 11 \leq 6x + 8 \quad \Leftrightarrow \quad 3 \leq 3x \quad \Leftrightarrow \quad 1 \leq x.$ Interval: $[1, \infty)$.

   Graph:

19. $1 - x \leq 2 \quad \Leftrightarrow \quad -x \leq 1 \quad \Leftrightarrow \quad x \geq -1.$ Interval: $[-1, \infty)$.

   Graph:

21. $\frac{1}{2}x - \frac{2}{3} > 2 \quad \Leftrightarrow \quad \frac{1}{2}x > \frac{8}{3} \quad \Leftrightarrow \quad x > \frac{16}{3}.$ Interval: $\left(\frac{16}{3}, \infty\right)$.

   Graph:

23. $\frac{3}{2} - \frac{1}{2}x \leq 1 + \frac{1}{4}x \quad \Leftrightarrow \quad \frac{1}{2} \leq \frac{3}{4}x \quad \Leftrightarrow \quad \frac{2}{3} \leq x.$ Interval: $\left[\frac{2}{3}, \infty\right)$.

   Graph:

25. $4 - 3x \leq -(1 + 8x)$ $\Leftrightarrow$ $4 - 3x \leq -1 - 8x$ $\Leftrightarrow$ $5x \leq -5$ $\Leftrightarrow$ $x \leq -1$.

Interval: $(-\infty, -1]$. Graph:

27. $2 \leq x + 5 < 4$ $\Leftrightarrow$ $-3 \leq x < -1$.   Interval: $[-3, -1)$.
Graph:

29. $-1 < 2x - 5 < 7$ $\Leftrightarrow$ (add 5 to each expression) $4 < 2x < 12$ $\Leftrightarrow$ (divide each expression
by 2) $2 < x < 6$. Interval: $(2, 6)$.   Graph:

31. $0 \leq 1 - x < 1$ $\Leftrightarrow$ (add 5 to each expression) $-1 \leq -x < 0$ $\Leftrightarrow$ (multiply each expression
by -1, reverse the direction of the inequality) $1 \geq x > 0$ $\Leftrightarrow$ (expressing in standard form)
$0 < x \leq 1$.

Interval: $(0, 1]$. Graph:

33. $-2 < 8 - 2x \leq -1$ $\Leftrightarrow$ $-10 < -2x \leq -9$ $\Leftrightarrow$ $5 > x \geq \frac{9}{2}$ $\Leftrightarrow$ $\frac{9}{2} \leq x < 5$.

Interval: $\left[\frac{9}{2}, 5\right)$. Graph:

35. $\dfrac{2}{3} \geq \dfrac{2x - 3}{12} > \dfrac{1}{6}$ $\Leftrightarrow$ (multiply each expression by 12) $8 \geq 2x - 3 > 2$ $\Leftrightarrow$ $11 \geq 2x > 5$

$\Leftrightarrow$ $\frac{11}{2} \geq x > \frac{5}{2}$ $\Leftrightarrow$ (expressing in standard form) $\frac{5}{2} < x \leq \frac{11}{2}$. Interval: $\left(\frac{5}{2}, \frac{11}{2}\right]$.

Graph:

37. Let x be the *average miles driven a day*. Each day the cost of Plan A is $30 + 0.10x$, and the cost of
Plan B is 50. Plan B saves money when $50 < 30 + 0.10x$ $\Leftrightarrow$ $20 < 0.1x$ $\Leftrightarrow$ $200 < x$. So
Plan B saves money when you average more than 200 miles a day.

39. Inserting the relationship $C = \frac{5}{9}(F - 32)$, we have $20 \leq C \leq 30$ $\Leftrightarrow$ $20 \leq \frac{5}{9}(F - 32) \leq 30$
$\Leftrightarrow$ $36 \leq F - 32 \leq 54$ $\Leftrightarrow$ $68 \leq F \leq 86$.

41. (a) Let x be the number of \$3 increases. Then the number of seats sold is $120 - x$. So
$P = 200 + 3x$ $\Leftrightarrow$ $3x = P - 200$ $\Leftrightarrow$ $x = \frac{1}{3}(P - 200)$. Substituting for x we have
that the number of seats sold is $120 - x = 120 - \frac{1}{3}(P - 200) = -\frac{1}{3}P + \frac{560}{3}$.

(b) $90 \leq -\frac{1}{3}P + \frac{560}{3} \leq 115$ $\Leftrightarrow$ $270 \leq 360 - P + 200 \leq 345$ $\Leftrightarrow$
$270 \leq -P + 560 \leq 345$ $\Leftrightarrow$ $-290 \leq -P \leq -215$ $\Leftrightarrow$ $290 \geq P \geq 215$. Putting this
into standard order, we have $215 \leq P \leq 290$. So the ticket prices are between \$215 and \$290.

43. We need to solve $6400 \leq 0.35m + 2200 \leq 7100$ for m. So $6400 \leq 0.35m + 2200 \leq 7100$ $\Leftrightarrow$
$4200 \leq 0.35m \leq 4900$ $\Leftrightarrow$ $12000 \leq m \leq 14000$. She plans on driving between 12,000 and
14,000 miles.

45. $a(bx - c) \geq bc$ (where $a, b, c > 0$) $\Leftrightarrow$ $bx - c \geq \dfrac{bc}{a}$ $\Leftrightarrow$ $bx \geq \dfrac{bc}{a} + c$ $\Leftrightarrow$

$x \geq \dfrac{1}{b}\left(\dfrac{bc}{a} + c\right) = \dfrac{c}{a} + \dfrac{c}{b}$ $\Leftrightarrow$ $x \geq \dfrac{c}{a} + \dfrac{c}{b}$.

47. $ax + b < c$ (where $a, b, c < 0$) $\Leftrightarrow$ $ax < c - b$ $\Leftrightarrow$ $x > \dfrac{c - b}{a}$.

49. If $a < b$, then $a + a < a + b$ (adding a to both sides) and $a + b < b + b$ (adding b to both sides). So

$$a + a < a + b < b + b \quad \Leftrightarrow \quad \text{(dividing by 2)} \quad a < \frac{a+b}{2} < b$$

51. We are given $\dfrac{a}{b} < \dfrac{c}{d}$, where a, b, c, and d are all positive. So multiplying by d, we get $\dfrac{ad}{b} < c$ and

so by Rule 1 for Inequalities, $a + \dfrac{ad}{b} < a + c \quad \Leftrightarrow \quad \dfrac{ab + ad}{b} < a + c \quad \Leftrightarrow \quad \dfrac{a}{b}(b + d) < a + c$

$\Leftrightarrow \quad \dfrac{a}{b} < \dfrac{a+c}{b+d}$, since $b + d$ is positive. Similarly, $\dfrac{a}{b} < \dfrac{c}{d} \quad \Leftrightarrow \quad a < \dfrac{bc}{d} \quad \Leftrightarrow$

$a + c < \dfrac{bc}{d} + c \quad \Leftrightarrow \quad a + c < \dfrac{bc}{d} + \dfrac{dc}{d} \quad \Leftrightarrow \quad a + c < \dfrac{c}{d}(b + d) \quad \Leftrightarrow \quad \dfrac{a+c}{b+d} < \dfrac{c}{d}$, since

$b + d > 0$. Combining the two inequalities, $\dfrac{a}{b} < \dfrac{a+c}{b+d} < \dfrac{c}{d}$.

Exercises 3.7

1. $(x-2)(x-5) > 0$. The expression on the left of the inequality changes sign where $x = 2$ and where $x = 5$. Thus we must check the intervals in the following table.

Interval	$(-\infty, 2)$	$(2, 5)$	$(5, \infty)$
Sign of $x - 2$	$-$	$+$	$+$
Sign of $x - 5$	$-$	$-$	$+$
Sign of $(x-2)(x-5)$	$+$	$-$	$+$

From the table, the solution set is $\{x\mid x < 2 \text{ or } 5 < x\}$.
Interval: $(-\infty, 2) \cup (5, \infty)$.   Graph:

3. $x^2 - 3x - 18 \le 0 \iff (x+3)(x-6) \le 0$. The expression on the left of the inequality changes sign where $x = 6$ and where $x = -3$. Thus we must check the intervals in the following table.

Interval	$(-\infty, -3)$	$(-3, 6)$	$(6, \infty)$
Sign of $x + 3$	$-$	$+$	$+$
Sign of $x - 6$	$-$	$-$	$+$
Sign of $(x+3)(x-6)$	$+$	$-$	$+$

From the table, the solution set is $\{x\mid -3 \le x \le 6\}$.
Interval: $[-3, 6]$.   Graph:

5. $2x^2 + x \ge 1 \iff 2x^2 + x - 1 \ge 0 \iff (x+1)(2x-1) \ge 0$. The expression on the left of the inequality changes sign where $x = -1$ and where $x = \frac{1}{2}$. Thus we must check the intervals in the following table.

Interval	$(-\infty, -1)$	$(-1, \frac{1}{2})$	$(\frac{1}{2}, \infty)$
Sign of $x + 1$	$-$	$+$	$+$
Sign of $2x - 1$	$-$	$-$	$+$
Sign of $(x+1)(2x-1)$	$+$	$-$	$+$

From the table, the solution set is $\left\{x\mid x \le -1 \text{ or } \frac{1}{2} \le x\right\}$.
Interval: $(-\infty, -1] \cup \left[\frac{1}{2}, \infty\right)$.   Graph:

7. $3x^2 - 3x < 2x^2 + 4 \iff x^2 - 3x - 4 < 0 \iff (x+1)(x-4) < 0$. The expression on the left of the inequality changes sign where $x = -1$ and where $x = 4$. Thus we must check the intervals in the following table.

Interval	$(-\infty, -1)$	$(-1, 4)$	$(4, \infty)$
Sign of $x + 1$	$-$	$+$	$+$
Sign of $x - 4$	$-$	$-$	$+$
Sign of $(x+1)(x-4)$	$+$	$-$	$+$

From the table, the solution set is $\{x\mid -1 < x < 4\}$.
Interval: $(-1, 4)$.   Graph:

9. $x^2 > 3(x+6) \iff x^2 - 3x - 18 > 0 \iff (x+3)(x-6) > 0$. The expression on the left of the inequality changes sign where $x = 6$ and where $x = -3$. Thus we must check the intervals in the following table.

Interval	$(-\infty, -3)$	$(-3, 6)$	$(6, \infty)$
Sign of $x + 3$	$-$	$+$	$+$
Sign of $x - 6$	$-$	$-$	$+$
Sign of $(x + 3)(x - 6)$	$+$	$-$	$+$

From the table, the solution set is $\{x \mid x < -3 \text{ or } 6 < x\}$.
Interval: $(-\infty, -3) \cup (6, \infty)$.   Graph:

11. $x^2 < 4 \quad \Leftrightarrow \quad x^2 - 4 < 0 \quad \Leftrightarrow \quad (x + 2)(x - 2) < 0$. The expression on the left of the inequality changes sign where $x = -2$ and where $x = 2$. Thus we must check the intervals in the following table.

Interval	$(-\infty, -2)$	$(-2, 2)$	$(2, \infty)$
Sign of $x + 2$	$-$	$+$	$+$
Sign of $x - 2$	$-$	$-$	$+$
Sign of $(x + 2)(x - 2)$	$+$	$-$	$+$

From the table, the solution set is $\{x \mid -2 < x < 2\}$.
Interval: $(-2, 2)$.   Graph:

13. $-2x^2 \leq 4 \quad \Leftrightarrow \quad -2x^2 - 4 \leq 0 \quad \Leftrightarrow \quad -2(x^2 + 1) \leq 0$. Since $x^2 + 1 > 0$, $-2(x^2 + 1) \leq 0$ for all x.   Interval: $(-\infty, \infty)$ Graph:

15. $x(x^2 - 4) \geq 0 \quad \Leftrightarrow \quad x(x + 2)(x - 2) \geq 0$. The expression on the left of the inequality changes sign where $x = 0$, where $x = -2$ and where $x = 2$. Thus we must check the intervals in the following table.

Interval	$(-\infty, -2)$	$(-2, 0)$	$(0, 2)$	$(2, \infty)$
Sign of $x + 2$	$-$	$+$	$+$	$+$
Sign of x	$-$	$-$	$+$	$+$
Sign of $x - 2$	$-$	$-$	$-$	$+$
Sign of $(x + 2)(x - 2)$	$-$	$+$	$-$	$+$

From the table, the solution set is $\{x \mid -2 \leq x \leq 0 \text{ or } 2 \leq x\}$.
Interval: $[-2, 0] \cup [2, \infty)$.   Graph:

17. $\dfrac{x - 3}{x + 1} \geq 0$. The expression on the left of the inequality changes sign where $x = -1$ and where $x = 3$. Thus we must check the intervals in the following table.

Interval	$(-\infty, -1)$	$(-1, 3)$	$(3, \infty)$
Sign of $x + 1$	$-$	$+$	$+$
Sign of $x - 3$	$-$	$-$	$+$
Sign of $\dfrac{x - 3}{x + 1}$	$+$	$-$	$+$

From the table, the solution set is $\{x \mid x < -1 \text{ or } x \leq 3\}$. Since the denominator cannot equal 0 we must have $x \neq -1$.
Interval: $(-\infty, -1) \cup [3, \infty)$.   Graph:

19. $\dfrac{4x}{2x + 3} > 2 \quad \Leftrightarrow \quad \dfrac{4x}{2x + 3} - 2 > 0 \quad \Leftrightarrow \quad \dfrac{4x}{2x + 3} - \dfrac{2(2x + 3)}{2x + 3} > 0 \quad \Leftrightarrow \quad \dfrac{-6}{2x + 3} > 0$. The expression on the left of the inequality changes sign where $x = -\dfrac{3}{2}$. Thus we must check the intervals in the following table.

Interval	$\left(-\infty, -\frac{3}{2}\right)$	$\left(-\frac{3}{2}, \infty\right)$
Sign of -6	$-$	$-$
Sign of $2x + 3$	$-$	$+$
Sign of $\dfrac{-6}{2x+3}$	$+$	$-$

From the table, the solution set is $\left\{ x \mid x < -\dfrac{3}{2} \right\}$.

Interval: $\left(-\infty, -\dfrac{3}{2}\right)$.   Graph:

21. $\dfrac{2x+1}{x-5} \le 3 \quad \Leftrightarrow \quad \dfrac{2x+1}{x-5} - 3 \le 0 \quad \Leftrightarrow \quad \dfrac{2x+1}{x-5} - \dfrac{3(x-5)}{x-5} \le 0 \quad \Leftrightarrow \quad \dfrac{-x+16}{x-5} \le 0.$ The
expression on the left of the inequality changes sign where $x = 16$ and where $x = 5$. Thus we must
check the intervals in the following table.

Interval	$(-\infty, 5)$	$(5, 16)$	$(16, \infty)$
Sign of $-x + 16$	$+$	$+$	$-$
Sign of $x - 5$	$-$	$+$	$+$
Sign of $\dfrac{-x+16}{x-5}$	$-$	$+$	$-$

From the table, the solution set is $\{x \mid x < 5 \text{ or } x \ge 16\}$. Since the denominator cannot equal 0, we
must have $x \ne 5$.

Interval: $(-\infty, 5) \cup [16, \infty)$.   Graph:

23. $\dfrac{4}{x} < x \quad \Leftrightarrow \quad \dfrac{4}{x} - x < 0 \quad \Leftrightarrow \quad \dfrac{4}{x} - \dfrac{x \cdot x}{x} < 0 \quad \Leftrightarrow \quad \dfrac{4 - x^2}{x} < 0 \quad \Leftrightarrow$

$\dfrac{(2-x)(2+x)}{x} < 0.$ The expression on the left of the inequality changes sign where $x = 0$, where
$x = -2$, and where $x = 2$. Thus we must check the intervals in the following table.

Interval	$(-\infty, -2)$	$(-2, 0)$	$(0, 2)$	$(2, \infty)$
Sign of $2 + x$	$-$	$+$	$+$	$+$
Sign of x	$-$	$-$	$+$	$+$
Sign of $2 - x$	$+$	$+$	$+$	$-$
Sign of $\dfrac{(2-x)(2+x)}{x}$	$+$	$-$	$+$	$-$

From the table, the solution set is $\{x \mid -2 < x < 0 \text{ or } 2 < x\}$.

Interval: $(-2, 0) \cup (2, \infty)$.   Graph:

25. $\dfrac{x^2 - 4}{x^2 + 4} \ge 0 \quad \Leftrightarrow \quad \dfrac{(x+2)(x-2)}{x^2+4} \ge 0.$ Since the denominator is always positive, a sum of
squares, we need only consider the numerator. The expression on the left of the inequality changes
sign where $x = -2$ and where $x = 2$. Thus we must check the intervals in the following table.

Interval	$(-\infty, -2)$	$(-2, 2)$	$(2, \infty)$
Sign of $x + 2$	$-$	$+$	$+$
Sign of $x - 2$	$-$	$-$	$+$
Sign of $x^2 + 4$	$+$	$+$	$+$
Sign of $\dfrac{(x+2)(x-2)}{x^2+4}$	$+$	$-$	$+$

From the table, the solution set is $\{x|\ x \le -2 \text{ or } 2 \le x\}$.

Interval: $(-\infty, -2] \cup [2, \infty)$.   Graph:

27. $1 + \dfrac{2}{x+1} \le \dfrac{2}{x} \quad\Leftrightarrow\quad 1 + \dfrac{2}{x+1} - \dfrac{2}{x} \le 0 \quad\Leftrightarrow\quad \dfrac{x(x+1)}{x(x+1)} + \dfrac{2x}{x(x+1)} - \dfrac{2(x+1)}{x(x+1)} \le 0 \quad\Leftrightarrow$

$\dfrac{x^2 + x + 2x - 2x - 2}{x(x+1)} \le 0 \quad\Leftrightarrow\quad \dfrac{x^2 + x - 2}{x(x+1)} \le 0 \quad\Leftrightarrow\quad \dfrac{(x+2)(x-1)}{x(x+1)} \le 0.$ The expression

on the left of the inequality changes sign where $x = -2$, where $x = -1$, where $x = 0$, and where $x = 1$. Thus we must check the intervals in the following table.

Interval	$(-\infty, -2)$	$(-2, -1)$	$(-1, 0)$	$(0, 1)$	$(1, \infty)$
Sign of $x + 2$	$-$	$+$	$+$	$+$	$+$
Sign of $x - 1$	$-$	$-$	$-$	$-$	$+$
Sign of x	$-$	$-$	$-$	$+$	$+$
Sign of $x + 1$	$-$	$-$	$+$	$+$	$+$
Sign of $\dfrac{(x+2)(x-1)}{x(x+1)}$	$+$	$-$	$+$	$-$	$+$

Since $x = -1$ and $x = 0$ yield undefined expressions, we cannot include them in the solution. From the table, the solution set is $\{x|\ -2 \le x < -1 \text{ or } 0 < x \le 1\}$.

Interval: $[-2, -1) \cup (0, 1]$.   Graph:

29. $\dfrac{1}{1-x} \le \dfrac{3}{x} \quad\Leftrightarrow\quad \dfrac{1}{1-x} - \dfrac{3}{x} \le 0 \quad\Leftrightarrow\quad \dfrac{x}{x(1-x)} - \dfrac{3(1-x)}{x(1-x)} \le 0 \quad\Leftrightarrow\quad \dfrac{x - 3 + 3x}{x(1-x)} \le 0$

$\Leftrightarrow\quad \dfrac{4x - 3}{x(1-x)} \le 0.$ The expression on the left of the inequality changes sign where $x = 0$, where

$x = \frac{3}{4}$, and where $x = 1$. Thus we must check the intervals in the following table.

Interval	$(-\infty, 0)$	$(0, \frac{3}{4})$	$(\frac{3}{4}, 1)$	$(1, \infty)$
Sign of x	$-$	$+$	$+$	$+$
Sign of $4x - 3$	$-$	$-$	$+$	$+$
Sign of $1 - x$	$+$	$+$	$+$	$-$
Sign of $\dfrac{4x - 3}{x(1-x)}$	$+$	$-$	$+$	$-$

From the table, the solution set is $\{x|\ 0 < x \le \frac{3}{4} \text{ or } 1 < x\}$.

Interval: $\left(0, \frac{3}{4}\right] \cup (1, \infty)$.   Graph:

31. $\dfrac{x^2 + 2x - 3}{x^2 - 7x + 6} > 0 \quad\Leftrightarrow\quad \dfrac{(x-1)(x+3)}{(x-1)(x-6)} > 0 \quad\Rightarrow\quad \dfrac{x+3}{x-6} > 0$ (The only exception is at $x = 1$

where $\dfrac{x+3}{x-6}$ is defined and $\dfrac{(x-1)(x+3)}{(x-1)(x-6)}$ is not.) The expression on the left of the inequality

changes sign where $x = -3$ and where $x = 6$. Thus we must check the intervals in the following table.

Interval	$(-\infty, -3)$	$(-3, 6)$	$(6, \infty)$
Sign of $x + 3$	$-$	$+$	$+$
Sign of $x - 6$	$-$	$-$	$+$
Sign of $\dfrac{x+3}{x-6}$	$+$	$-$	$+$

From the table, the solution set is $\{x \mid x < -3 \text{ or } 6 < x\}$.

Interval: $(-\infty, -3) \cup (6, \infty)$. Graph:

33. $\dfrac{x-3}{2x+5} \geq 1 \Leftrightarrow \dfrac{x-3}{2x+5} - 1 \geq 0 \Leftrightarrow \dfrac{x-3}{2x+5} - \dfrac{2x+5}{2x+5} \geq 0 \Leftrightarrow \dfrac{-x-8}{2x+5} \geq 0$. The

expression on the left of the inequality changes sign where $x = -8$ and where $x = -\frac{5}{2}$. Thus we must check the intervals in the following table.

Interval	$(-\infty, -8)$	$\left(-8, -\frac{5}{2}\right)$	$\left(-\frac{5}{2}, \infty\right)$
Sign of $-x - 8$	$+$	$-$	$-$
Sign of $2x + 5$	$-$	$-$	$+$
Sign of $\dfrac{-x-8}{2x+5}$	$-$	$+$	$-$

From the table, the solution set is $\left\{x \mid -8 \leq x < -\frac{5}{2}\right\}$. Since the denominator is zero when $x = -\frac{5}{2}$, we must exclude this value.

Interval: $\left[-8, -\frac{5}{2}\right)$. Graph:

35. $\dfrac{6}{x-1} - \dfrac{6}{x} \geq 1 \Leftrightarrow \dfrac{6}{x-1} - \dfrac{6}{x} - 1 \geq 0 \Leftrightarrow \dfrac{6x}{x(x-1)} - \dfrac{6(x-1)}{x(x-1)} - \dfrac{x(x-1)}{x(x-1)} \geq 0 \Leftrightarrow$

$\dfrac{6x - 6x + 6 - x^2 + x}{x(x-1)} \geq 0 \Leftrightarrow \dfrac{-x^2 + x + 6}{x(x-1)} \geq 0 \Leftrightarrow \dfrac{(-x+3)(x+2)}{x(x-1)} \geq 0$. The

expression on the left of the inequality changes sign where $x = 3$, where $x = -2$, where $x = 0$, and where $x = 1$. Thus we must check the intervals in the following table.

Interval	$(-\infty, -2)$	$(-2, 0)$	$(0, 1)$	$(1, 3)$	$(3, \infty)$
Sign of $-x + 3$	$+$	$+$	$+$	$+$	$-$
Sign of $x + 2$	$-$	$+$	$+$	$+$	$+$
Sign of x	$-$	$-$	$+$	$+$	$+$
Sign of $x - 1$	$-$	$-$	$-$	$+$	$+$
Sign of $\dfrac{(-x+3)(x+2)}{x(x-1)}$	$-$	$+$	$-$	$+$	$-$

From the table, the solution set is $\{x \mid -2 \leq x < 0 \text{ or } 1 < x \leq 3\}$. The points $x = 0$ and $x = 1$ are excluded from the solution set because they make the denominator zero.

Interval: $[-2, 0) \cup (1, 3]$ Graph:

37. $\dfrac{x+2}{x+3} < \dfrac{x-1}{x-2} \Leftrightarrow \dfrac{x+2}{x+3} - \dfrac{x-1}{x-2} < 0 \Leftrightarrow \dfrac{(x+2)(x-2)}{(x+3)(x-2)} - \dfrac{(x-1)(x+3)}{(x-2)(x+3)} < 0 \Leftrightarrow$

$\dfrac{x^2 - 4 - x^2 - 2x + 3}{(x+3)(x-2)} < 0 \Leftrightarrow \dfrac{-2x - 1}{(x+3)(x-2)} < 0$. The expression on the left of the inequality changes sign where $x = -\frac{1}{2}$, where $x = -3$, and where $x = 2$. Thus we must check the intervals in the following table.

Interval	$(-\infty, -3)$	$(-3, -\frac{1}{2})$	$(-\frac{1}{2}, 2)$	$(2, \infty)$
Sign of $-2x - 1$	$+$	$+$	$-$	$-$
Sign of $x + 3$	$-$	$+$	$+$	$+$
Sign of $x - 2$	$-$	$-$	$-$	$+$
Sign of $\dfrac{-2x - 1}{(x + 3)(x - 2)}$	$+$	$-$	$+$	$-$

From the table, the solution set is $\left\{ x \mid -3 < x < -\frac{1}{2} \text{ or } 2 < x \right\}$.

Interval: $\left(-3, -\frac{1}{2}\right) \cup (2, \infty)$.   Graph:

39. $x^4 > x^2 \quad \Leftrightarrow \quad x^4 - x^2 > 0 \quad \Leftrightarrow \quad x^2(x^2 - 1) > 0 \quad \Leftrightarrow \quad x^2(x - 1)(x + 1) > 0$. The expression on the left of the inequality changes sign where $x = 0$, where $x = 1$, and where $x = -1$. Thus we must check the intervals in the following table.

Interval	$(-\infty, -1)$	$(-1, 0)$	$(0, 1)$	$(1, \infty)$
Sign of x^2	$+$	$+$	$+$	$+$
Sign of $x - 1$	$-$	$-$	$-$	$+$
Sign of $x + 1$	$-$	$+$	$+$	$+$
Sign of $x^2(x - 1)(x + 1)$	$+$	$-$	$-$	$+$

From the table, the solution set is $\{ x \mid x < -1 \text{ or } 1 < x \}$

Interval: $(-\infty, -1) \cup (1, \infty)$.   Graph:

41. We graph $y = x^2 - 3x - 10$ in the viewing rectangle $[-5, 6]$ by $[-12, 2]$. Thus the solution to the inequality is: $[-2.0, 5.0]$.

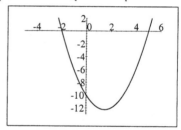

43. Since $x^3 + 11x \le 6x^2 + 6 \quad \Leftrightarrow$
$x^3 - 6x^2 + 11x - 6 \le 0$, we graph
$y = x^3 - 6x^2 + 11x - 6$ in the viewing
rectangle $[0, 5]$ by $[-5, 5]$. The solution set is
$(-\infty, 1.0] \cup [2.0, 3.0]$.

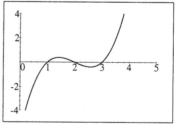

45. Since $x^{1/3} \le x \quad \Leftrightarrow \quad x^{1/3} - x < 0$, we
graph $y = x^{1/3} - x$ in the viewing rectangle
$[-3, 3]$ by $[-1, 1]$. From this, we find that the
solution set is $(-1.0, 0) \cup (1.0, \infty)$.

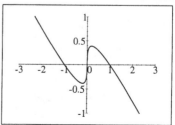

47. Since $(x+1)^2 < (x-1)^2 \quad \Leftrightarrow$
     $(x+1)^2 - (x-1)^2 < 0$, we graph
     $y = (x+1)^2 - (x-1)^2$ in the viewing
     rectangle $[-2, 2]$ by $[-5, 5]$. The solution set
     is $(-\infty, 0)$.

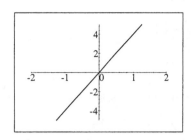

49. $128 + 16t - 16t^2 \geq 32 \quad \Leftrightarrow \quad -16t^2 + 16t + 96 \geq 0 \quad \Leftrightarrow \quad -16(t^2 - t - 6) \geq 0 \quad \Leftrightarrow$
     $-16(t-3)(t+2) \geq 0$. The expression on the left of the inequality changes sign at $x = -2$, at
     $t = 3$, and at $t = -2$. However, $t \geq 0$, so the only endpoint is $t = 3$. Thus we check the intervals in
     the following table.

Interval	$(0, 3)$	$(3, \infty)$
Sign of -16	$-$	$-$
Sign of $t - 3$	$-$	$+$
Sign of $t + 2$	$+$	$+$
Sign of $-16(t-3)(t+2)$	$+$	$-$

     So $0 \leq t \leq 3$.

51. $\dfrac{600,000}{x^2 + 300} < 500 \quad \Leftrightarrow \quad 600,000 < 500(x^2 + 300)$ (Note that $x^2 + 300 \geq 300 > 0$, so we can
     multiply both sides by the denominator and not worry that we might be multiplying both sides by a
     negative number or by zero.) $1200 < x^2 + 300 \quad \Leftrightarrow \quad 0 < x^2 - 900 \quad \Leftrightarrow$
     $0 < (x-30)(x+30)$. The expression in the inequality changes sign at $x = 30$ and $x = -30$.
     However, since x represents distance, we must have $x > 0$.

Interval	$(0, 30)$	$(30, \infty)$
Sign of $x - 30$	$-$	$+$
Sign of $x + 30$	$+$	$+$
Sign of $(x-30)(x+30)$	$-$	$+$

     So $x > 30$ and you must stand at least 30 meters from the center of the fire.

53. $240 > v + \dfrac{v^2}{20} \quad \Leftrightarrow \quad \frac{1}{20}v^2 + v - 240 < 0 \quad \Leftrightarrow \quad \left(\frac{1}{20}v - 3\right)(v + 80) < 0$. The expression in the
     inequality changes sign at $v = 60$ and $v = -80$. However, since v represents the speed, we must
     have $v \geq 0$.

Interval	$(0, 60)$	$(60, \infty)$
Sign of $\frac{1}{20}v - 3$	$-$	$+$
Sign of $v + 80$	$+$	$+$
Sign of $\left(\frac{1}{20}v - 3\right)(v + 80)$	$-$	$+$

     So Kerry must drive less than 60 mph.

55. Let x be the length of the garden. Using the fact that the perimeter is 120 ft, we must have
     $2x + 2width = 120 \quad \Leftrightarrow \quad width = 60 - x$. Now since the area must be at least 800 ft^2, we have
     $800 < x(60 - x) \quad \Leftrightarrow \quad 800 < 60x - x^2 \quad \Leftrightarrow \quad x^2 - 60x + 800 < 0 \quad \Leftrightarrow$
     $(x - 20)(x - 40) < 0$. The expression in the inequality changes sign at $x = 20$ and $x = 40$.
     However, since x represents length, we must have $x > 0$.

Interval	$(0, 20)$	$(20, 40)$	$(40, \infty)$
Sign of $x - 20$	$-$	$+$	$+$
Sign of $x - 40$	$-$	$-$	$+$
Sign of $(x - 20)(x - 40)$	$+$	$-$	$+$

The length of the garden should be between 20 and 40 feet.

57. For $\sqrt{16 - 9x^2}$ to be defined as a real number we must have $16 - 9x^2 \geq 0$ $\Leftrightarrow$
$(4 - 3x)(4 + 3x) \geq 0$. The expression in the inequality changes sign at $x = \frac{4}{3}$ and $x = -\frac{4}{3}$.

Interval	$\left(-\infty, -\frac{4}{3}\right)$	$\left(-\frac{4}{3}, \frac{4}{3}\right)$	$\left(\frac{4}{3}, \infty\right)$
Sign of $4 - 3x$	$+$	$+$	$-$
Sign of $4 + 3x$	$-$	$+$	$+$
Sign of $(4 - 3x)(4 + 3x)$	$-$	$+$	$-$

Thus $-\frac{4}{3} \leq x \leq \frac{4}{3}$.

59. For $\left(\dfrac{1}{x^2 - 5x - 14}\right)^{1/2}$ to be defined as a real number we must have $x^2 - 5x - 14 > 0$ $\Leftrightarrow$
$(x - 7)(x + 2) > 0$. The expression in the inequality changes sign at $x = 7$ and $x = -2$.

Interval	$(-\infty, -2)$	$(-2, 7)$	$(7, \infty)$
Sign of $x - 7$	$-$	$-$	$+$
Sign of $x + 2$	$-$	$+$	$+$
Sign of $(x - 7)(x + 2)$	$+$	$-$	$+$

Thus $x < -2$ or $7 < x$, and the solution set is $(-\infty, -2) \cup (7, \infty)$.

61. $\dfrac{x^2 + (a - b)x - ab}{x + c} \leq 0$ (where $0 < a < b < c$) $\Leftrightarrow$ $\dfrac{(x + a)(x - b)}{x + c} \leq 0$. The expression in
the inequality changes sign at $x = -a$, $x = b$, and $x = -c$. Endpoints: $-c, -a, b$.

Interval	$(-\infty, -c)$	$(-c, -a)$	$(-a, b)$	(b, ∞)
Sign of $x + a$	$-$	$-$	$+$	$+$
Sign of $x - b$	$-$	$-$	$-$	$+$
Sign of $x + c$	$-$	$+$	$+$	$+$
Sign of $\dfrac{(x + a)(x - b)}{x + c}$	$-$	$+$	$-$	$+$

From the table, the solution set is $\{x \mid -\infty < x < -c \text{ or } -a \leq x \leq b\}$. Note, since
$0 < a < b < c$, we must have $-c < -b < -a < 0$. Also, we exclude the point $x = -c$, because the
LHS is undefined at this point. The solution set is $(-\infty, -c) \cup [-a, b]$.

63. The rule we want to apply here is "$a < b \Rightarrow ac < bc$ if $c > 0$ and $a < b \Rightarrow ac > bc$ if $c < 0$". Thus
we cannot simply multiply by x, since we don't yet know if x is positive or negative, so in solving
$1 < \dfrac{3}{x}$, we must consider two cases

Case 1: $x > 0$ Then multiplying all sides by x, we have $x < 3$. Together with our initial condition,
we have $0 < x < 3$.

Case 2: $x < 0$ Then multiplying all sides by x, we have $x > 3$. But $x < 0$ and $x > 3$ have no
elements in common, so this gives no additional solutions.

Hence, the only solutions are $0 < x < 3$.

Exercises 3.8

1. $|3x| = 15$ $\Leftrightarrow$ $3x = \pm 15$ $\Leftrightarrow$ $x = \pm 5$.

3. $|x - 3| = 2$ is equivalent to $x - 3 = \pm 2$ $\Leftrightarrow$ $x = 3 \pm 2$ $\Leftrightarrow$ $x = 1$ or $x = 5$.

5. $|x + 4| = 0.5$ is equivalent to $x + 4 = \pm 0.5$ $\Leftrightarrow$ $x = -4 \pm 0.5$ $\Leftrightarrow$ $x = -4.5$ or $x = -3.5$.

7. $|4x + 7| = 9$ is equivalent to either $4x + 7 = 9$ $\Leftrightarrow$ $4x = 2$ $\Leftrightarrow$ $x = \frac{1}{2}$; or $4x + 7 = -9$ $\Leftrightarrow$ $4x = -16$ $\Leftrightarrow$ $x = -4$. The two solutions are $x = \frac{1}{2}$ and $x = -4$.

9. $4 - |3x + 6| = 1$ $\Leftrightarrow$ $-|3x + 6| = -3$ $\Leftrightarrow$ $|3x + 6| = 3$, which is equivalent to either $3x + 6 = 3$ $\Leftrightarrow$ $3x = -3$ $\Leftrightarrow$ $x = -1$; or $3x + 6 = -3$ $\Leftrightarrow$ $3x = -9$ $\Leftrightarrow$ $x = -3$. The two solutions are $x = -1$ and $x = -3$.

11. $3|x + 5| + 6 = 15$ $\Leftrightarrow$ $3|x + 5| = 9$ $\Leftrightarrow$ $|x + 5| = 3$, which is equivalent to either $x + 5 = 3$ $\Leftrightarrow$ $x = -2$; or $x + 5 = -3$ $\Leftrightarrow$ $x = -8$. The two solutions are $x = -2$ and $x = -8$.

13. $8 + 5|\frac{1}{3}x - \frac{5}{6}| = 33$ $\Leftrightarrow$ $5|\frac{1}{3}x - \frac{5}{6}| = 25$ $\Leftrightarrow$ $|\frac{1}{3}x - \frac{5}{6}| = 5$, which is equivalent to either $\frac{1}{3}x - \frac{5}{6} = 5$ $\Leftrightarrow$ $\frac{1}{3}x = \frac{35}{6}$ $\Leftrightarrow$ $x = \frac{35}{2}$; or $\frac{1}{3}x - \frac{5}{6} = -5$ $\Leftrightarrow$ $\frac{1}{3}x = -\frac{25}{6}$ $\Leftrightarrow$ $x = -\frac{25}{2}$. The two solutions are $x = -\frac{25}{2}$ and $x = \frac{35}{2}$.

15. $|x - 1| = |3x + 2|$, which is equivalent to either $x - 1 = 3x + 2$ $\Leftrightarrow$ $-2x = 3$ $\Leftrightarrow$ $x = -\frac{3}{2}$; or $x - 1 = -(3x + 2)$ $\Leftrightarrow$ $x - 1 = -3x - 2$ $\Leftrightarrow$ $4x = -1$ $\Leftrightarrow$ $x = -\frac{1}{4}$. The two solutions are $x = -\frac{3}{2}$ and $x = -\frac{1}{4}$.

17. $|x| < 2$ $\Leftrightarrow$ $-2 < x < 2$. Interval: $(-2, 2)$.

19. $|x - 5| \leq 3$ $\Leftrightarrow$ $-3 \leq x - 5 \leq 3$ $\Leftrightarrow$ $2 \leq x \leq 8$. Interval: $[2, 8]$.

21. $|x + 1| \geq 1$ is equivalent to $x + 1 \geq 1$ $\Leftrightarrow$ $x \geq 0$; or $x + 1 \leq -1$ $\Leftrightarrow$ $x \leq -2$. Interval: $(-\infty, -2] \cup [0, \infty)$.

23. $|x + 5| \geq 2$ is equivalent to $x + 5 \geq 2$ $\Leftrightarrow$ $x \geq -3$; or $x + 5 \leq -2$ $\Leftrightarrow$ $x \leq -7$. Interval: $(-\infty, -7] \cup [-3, \infty)$.

25. $|2x - 3| \leq 0.4$ $\Leftrightarrow$ $-0.4 \leq 2x - 3 \leq 0.4$ $\Leftrightarrow$ $2.6 \leq 2x \leq 3.4$ $\Leftrightarrow$ $1.3 \leq x \leq 1.7$. Interval: $[1.3, 1.7]$.

27. $\left|\dfrac{x - 2}{3}\right| < 2$ $\Leftrightarrow$ $-2 < \dfrac{x - 2}{3} < 2$ $\Leftrightarrow$ $-6 < x - 2 < 6$ $\Leftrightarrow$ $-4 < x < 8$. Interval: $(-4, 8)$.

29. $|x + 6| < 0.001$ $\Leftrightarrow$ $-0.001 < x + 6 < 0.001$ $\Leftrightarrow$ $-6.001 < x < -5.999$. Interval: $(-6.001, -5.999)$.

31. $4|x + 2| - 3 < 13$ $\Leftrightarrow$ $4|x + 2| < 16$ $\Leftrightarrow$ $|x + 2| < 4$ $\Leftrightarrow$ $-4 < x + 2 < 4$ $\Leftrightarrow$ $-6 < x < 2$. Interval: $(-6, 2)$.

33. $8 - |2x - 1| \geq 6$ $\Leftrightarrow$ $-|2x - 1| \geq -2$ $\Leftrightarrow$ $|2x - 1| \leq 2$ $\Leftrightarrow$ $-2 \leq 2x - 1 \leq 2$ $\Leftrightarrow$ $-1 \leq 2x \leq 3$ $\Leftrightarrow$ $-\frac{1}{2} \leq x \leq \frac{3}{2}$. Interval: $\left[-\frac{1}{2}, \frac{3}{2}\right]$.

35. $\frac{1}{2}\left|4x + \frac{1}{3}\right| > \frac{5}{6}$ $\Leftrightarrow$ $\left|4x + \frac{1}{3}\right| > \frac{5}{3}$, which is equivalent to $4x + \frac{1}{3} > \frac{5}{3}$ $\Leftrightarrow$ $4x > \frac{4}{3}$ $\Leftrightarrow$
      $x > \frac{1}{3}$, or $4x + \frac{1}{3} < -\frac{5}{3}$ $\Leftrightarrow$ $4x < -2$ $\Leftrightarrow$ $x < -\frac{1}{2}$. Interval: $\left(-\infty, -\frac{1}{2}\right) \cup \left(\frac{1}{3}, \infty\right)$.

37. $1 \le |x| \le 4$. If $x \ge 0$, then this is equivalent to $1 \le x \le 4$. If $x < 0$, then this is equivalent to
      $1 \le -x \le 4$ $\Leftrightarrow$ $-1 \ge x \ge -4$ $\Leftrightarrow$ $-4 \le x \le -1$. Interval: $[-4, -1] \cup [1, 4]$.

39. $\dfrac{1}{|x + 7|} > 2$ $\Leftrightarrow$ $1 > 2|x + 7|$ $(x \ne -7)$ $\Leftrightarrow$ $|x + 7| < \frac{1}{2}$ $\Leftrightarrow$ $-\frac{1}{2} < x + 7 < \frac{1}{2}$ $\Leftrightarrow$
      $-\frac{15}{2} < x < -\frac{13}{2}$ and $x \ne -7$. Interval: $\left(-\frac{15}{2}, -7\right) \cup \left(-7, -\frac{13}{2}\right)$.

41.   We graph the equations $y_1 = |3x| + x$ and $y_2 = 16$ in
      the viewing rectangle $[-10, 10]$ by $[0, 20]$. Using
      Zoom and/or Trace, we get the solutions $x = -8$ and $x = 4$.

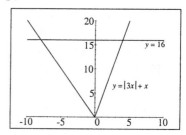

43.   We graph the equations $y_1 = |2x - 1|$ and $y_2 = \frac{1}{2}x^2$ in
      the viewing rectangle $[-5, 5]$ by $[0, 15]$. Using
      Zoom and/or Trace, we get the solutions $x \approx -4.45$,
      $x \approx 0.45$, $x \approx 0.59$, and $x \approx 3.41$.

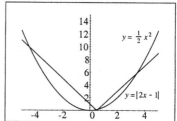

45.   We graph the equations $y_1 = |2x - 4|$ and $y_2 = |3 - x|$
      in the viewing rectangle $[-10, 10]$ by $[0, 10]$. Using
      Zoom and/or Trace, we find that the points of intersection
      are at $x = 1$ and $x \approx 2.3$. Since we want
      $|2x - 4| \le > |3 - x|$, the solution is the interval $[1, 2.3]$.

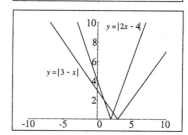

47.   We graph the equations $y_1 = x^2 - 2x + \frac{5}{4}$ and
      $y_2 = |x - 1|$ in the viewing rectangle $[-10, 10]$ by $[0, 10]$.
      Using Zoom and/or Trace, we find that the points of
      intersection are at $x \approx 0.5$ and $x \approx 1.5$. At these points, the
      equations intersect but do not cross. So
      $x^2 - 2x + \frac{5}{4} > |x - 1|$ everywhere except at $x \approx 0.5$ and
      $x \approx 1.5$. In interval notation this corresponds to
      $(-\infty, 0.5) \cup (0.5, 1.5) \cup (1.5, \infty)$.

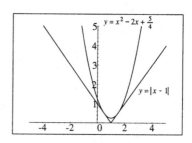

49. $|x - 1|$ is the distance between x and 1;
      $|x - 3|$ is the distance between x and 3.
      So $|x - 1| < |x - 3|$ represents those points closer to 1 than to 3, and the solution is $x < 2$, since 2
      is the point half way between 1 and 3. If $a < b$, then the solution to $|x - a| < |x - b|$ is $x < \frac{a+b}{2}$.

Review Exercises for Chapter 3

1. (a) Algebraically: $5x + 11 = 36$ $\Leftrightarrow$ $5x = 25$ $\Leftrightarrow$ $x = 5$.
   (b) Graphically: We graph the two equations $y_1 = 5x + 11$
       and $y_2 = 6$ in the viewing rectangle $[-10, 10]$ by
       $[20, 50]$. Zooming in, we see that solution is $x = 5$.

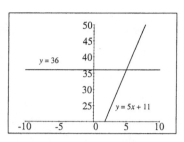

3. (a) Algebraically: $4x^2 = 49$ $\Leftrightarrow$ $x^2 = \frac{49}{4}$ $\Rightarrow$ $x = \pm\frac{7}{2}$.
   (b) Graphically: We graph the two equations $y_1 = 5x + 11$
       and $y_2 = 6$ in the viewing rectangle $[-5, 5]$ by
       $[40, 60]$. Zooming in we see that solutions are $x = \pm 3.5$.

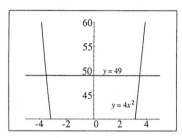

5. $\dfrac{x+1}{x-1} = \dfrac{3x}{3x-6} = \dfrac{3x}{3(x-2)} = \dfrac{x}{x-2}$ $\Leftrightarrow$ $(x+1)(x-2) = x(x-1)$ $\Leftrightarrow$
   $x^2 - x - 2 = x^2 - x$ $\Leftrightarrow$ $-2 = 0$. Since this last equation is never true, there are no real
   solutions to the original equation.

7. $x^2 - 9x + 14 = 0$ $\Leftrightarrow$ $(x-7)(x-2) = 0$ $\Leftrightarrow$ $x = 7$ or $x = 2$.

9. $2x^2 + x = 1$ $\Leftrightarrow$ $2x^2 + x - 1 = 0$ $\Leftrightarrow$ $(2x-1)(x+1) = 0$. So either $2x - 1 = 0$ $\Leftrightarrow$
   $2x = 1$ $\Leftrightarrow$ $x = \frac{1}{2}$; or $x + 1 = 0$ $\Leftrightarrow$ $x = -1$.

11. $0 = 4x^3 - 25x = x(4x^2 - 25) = x(2x-5)(2x+5) = 0$. So either $x = 0$; or $2x - 5 = 0$ $\Leftrightarrow$
    $2x = 5$ $\Leftrightarrow$ $x = \frac{5}{2}$; or $2x + 5 = 0$ $\Leftrightarrow$ $2x = -5$ $\Leftrightarrow$ $x = -\frac{5}{2}$.

13. $3x^2 + 4x - 1 = 0$ $\Rightarrow$ $x = \dfrac{-b \pm \sqrt{b^2 - 4ac}}{2a} = \dfrac{-(4) \pm \sqrt{(4)^2 - 4(3)(-1)}}{2(-3)}$
    $= \dfrac{-4 \pm \sqrt{16 + 12}}{-6} = \dfrac{-4 \pm \sqrt{28}}{-6} = \dfrac{-4 \pm 2\sqrt{7}}{6} = \dfrac{2(-2 \pm \sqrt{7})}{-6} = \dfrac{-2 \pm \sqrt{7}}{3}$.

15. $\dfrac{1}{x} + \dfrac{2}{x-1} = 3$ $\Leftrightarrow$ $(x-1) + 2(x) = 3(x)(x-1)$ $\Leftrightarrow$ $x - 1 + 2x = 3x^2 - 3x$ $\Leftrightarrow$
    $0 = 3x^2 - 6x + 1$ $\Rightarrow$ $x = \dfrac{-b \pm \sqrt{b^2 - 4ac}}{2a} = \dfrac{-(-6) \pm \sqrt{(-6)^2 - 4(3)(1)}}{2(3)}$
    $= \dfrac{6 \pm \sqrt{36 - 12}}{6} = \dfrac{6 \pm \sqrt{24}}{6} = \dfrac{6 \pm 2\sqrt{6}}{6} = \dfrac{2(3 \pm \sqrt{6})}{6} = \dfrac{3 \pm \sqrt{6}}{3}$.

17. $x^4 - 8x^2 - 9 = 0$ $\Leftrightarrow$ $(x^2 - 9)(x^2 + 1) = 0$ $\Leftrightarrow$ $(x - 3)(x + 3)(x^2 + 1) = 0$ $\Rightarrow$
    $x - 3 = 0$ $\Leftrightarrow$ $x = 3$, or $x + 3 = 0$ $\Leftrightarrow$ $x = -3$, however $x^2 + 1 = 0$ has no real
    solution. The solutions are $x = \pm 3$.

19. $x^{-1/2} - 2x^{1/2} + x^{3/2} = 0$ $\Leftrightarrow$ $x^{-1/2}(1 - 2x + x^2) = 0$ $\Leftrightarrow$ $x^{1/2}(1 - x)^2 = 0$. Since
    $x^{-1/2} \neq 0$, the only solution comes from $(1 - x)^2 = 0$ $\Leftrightarrow$ $1 - x = 0$ $\Leftrightarrow$ $x = 1$.

21. $|x - 7| = 4$ $\Leftrightarrow$ $x - 7 = \pm 4$ $\Leftrightarrow$ $x = 7 \pm 4$, so $x = 11$ or $x = 3$.

23. $|2x - 5| = 9$ is equivalent to $2x - 5 = \pm 9$ $\Leftrightarrow$ $2x = 5 \pm 9$ $\Leftrightarrow$ $x = \frac{5 \pm 9}{2}$. So $x = -2$ or
    $x = 7$.

25. $x^2 - 4x = 2x + 7$. We graph the equations $y_1 = x^2 - 4x$
    and $y_2 = 2x + 7$ in the viewing rectangle
    rectangle $[-10, 10]$ by $[-5, 25]$. Using Zoom and/or
    Trace, we get the solutions $x = -1$ and $x = 7$.

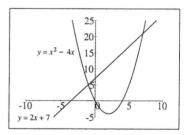

27. $x^4 - 9x^2 = x - 9$. We graph the equations $y_1 = x^4 - 9x^2$
    and $y_2 = x - 9$ in the viewing rectangle $[-5, 5]$ by
    $[-25, 10]$. Using Zoom and/or Trace, we get the solutions
    $x \approx -2.72$, $x \approx -1.15$, $x = 1$, and $x \approx 2.87$.

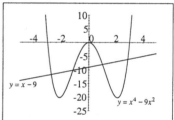

29. Let x be the number of pounds of raisins. Then the number of pounds of nuts is $50 - x$.

	raisins	nuts	mixture
pounds	x	$50 - x$	50
rate (cost per pound)	3.20	2.40	2.72
value	$3.20x$	$2.40(50 - x)$	$2.72(50)$

So $3.20x + 2.40(50 - x) = 2.72(50)$ $\Leftrightarrow$ $3.20x + 120 - 2.40x = 136$ $\Leftrightarrow$ $0.8x = 16$
$\Leftrightarrow$ $x = 20$. Thus the mixture uses 20 pounds raisins and $50 - 20 = 30$ pounds of nuts.

31. Let r be the rate the woman runs in mi/h. Then she cycles at $r + 8$ mi/h.

	Rate	Time	Distance
Cycle	$r + 8$	$\dfrac{4}{r + 8}$	4
Run	r	$\dfrac{2.5}{r}$	2.5

Since the total time of the workout is 1 hour, we have $\dfrac{4}{r + 8} + \dfrac{2.5}{r} = 1$. Multiplying by $2r(r + 8)$,
we get $4(2r) + 2.5(2)(r + 8) = 2r(r + 8)$ $\Leftrightarrow$ $8r + 5r + 40 = 2r^2 + 16r$ $\Leftrightarrow$

$$0 = 2r^2 + 3r - 40 \quad \Rightarrow \quad r = \frac{-3 \pm \sqrt{(3)^2 - 4(2)(-40)}}{2(2)} = \frac{-3 \pm \sqrt{9 + 320}}{4} = \frac{-3 \pm \sqrt{329}}{4}.$$

Since $r \geq 0$, we reject the negative value. She runs at $r = \dfrac{-3 + \sqrt{329}}{4} \approx 3.78 \, \text{mi/h}$.

33. Let x be the length of one side in cm. Then $28 - x$ is the length of the other side. Using the Pythagorean theorem, we have $x^2 + (28 - x)^2 = 20^2 \quad \Leftrightarrow \quad x^2 + 784 - 56x + x^2 = 400 \quad \Leftrightarrow \quad 2x^2 - 56x + 384 = 0 \quad \Leftrightarrow \quad 2(x^2 - 28x + 192) = 0 \quad \Leftrightarrow \quad 2(x - 12)(x - 16) = 0$. So $x = 12$ or $x = 16$. If $x = 12$, then the other side is $28 - 12 = 16$. Similarly, if $x = 16$, then the other side is 12. The sides are 12 cm and 16 cm.

35. Let w be width of the pool. Then the length of the pool is $2w$. Thus the volume of the pool is $8(w)(2w) = 8464 \quad \Leftrightarrow \quad 16w^2 = 8464 \quad \Leftrightarrow \quad w^2 = 529 \quad \Rightarrow \quad w = \pm 23$. Since $w > 0$, we reject the negative value. The pool is 23 feet wide, $2(23) = 46$ feet long, and 8 feet deep.

37. $(3 - 5i) - (6 + 4i) = 3 - 5i - 6 - 4i = -3 - 9i$.

39. $(2 + 7i)(6 - i) = 12 - 2i + 42i - 7i^2 = 12 + 40i + 7 = 19 + 40i$.

41. $\dfrac{2 - 3i}{2 + 3i} = \dfrac{2 - 3i}{2 + 3i} \cdot \dfrac{2 - 3i}{2 - 3i} = \dfrac{4 - 12i + 9i^2}{4 - 9i^2} = \dfrac{4 - 12i - 9}{4 + 9} = \dfrac{-5 - 12i}{13} = -\dfrac{5}{13} - \dfrac{12}{13}i$.

43. $i^{45} = i^{44}i = (i^4)^{11}i = (1)^{11}i = i$.

45. $(1 - \sqrt{-3})(2 + \sqrt{-4}) = (1 - \sqrt{3}i)(2 + 2i) = 2 + 2i - 2\sqrt{3}i - 2\sqrt{3}i^2$
$= 2 + (2 - 2\sqrt{3})i + 2\sqrt{3} = (2 + 2\sqrt{3}) + (2 - 2\sqrt{3})i$.

47. $x^2 + 16 = 0 \quad \Leftrightarrow \quad x^2 = -16 \quad \Rightarrow \quad x = \pm\sqrt{-16} = \pm 4i$.

49. $x^2 + 6x + 10 = 0 \quad \Rightarrow \quad x = \dfrac{-6 \pm \sqrt{6^2 - 4(1)(10)}}{2(1)} = \dfrac{-6 \pm \sqrt{36 - 40}}{2} = \dfrac{-6 \pm \sqrt{-4}}{2}$
$= \dfrac{-6 \pm 2i}{2} = -3 \pm i$.

51. $x^4 - 256 = 0 \quad \Leftrightarrow \quad (x^2 + 16)(x^2 - 16) = 0$. Thus either $x^2 + 16 = 0 \quad \Leftrightarrow \quad x^2 = -16 \quad \Rightarrow$
$x = \pm\sqrt{-16} \quad \Leftrightarrow \quad x = \pm 4i$; or $x^2 - 16 = 0 \quad \Leftrightarrow \quad (x - 4)(x + 4) = 0 \quad \Rightarrow \quad x - 4 = 0$
$\Leftrightarrow \quad x = 4$; or $x + 4 = 0 \quad \Leftrightarrow \quad x = -4$. The solutions are $\pm 4i$ and ± 4.

53. $x^2 + 4x = (2x + 1)^2 \quad \Leftrightarrow \quad x^2 + 4x = 4x^2 + 4x + 1 \quad \Leftrightarrow \quad 0 = 3x^2 + 1 \quad \Leftrightarrow \quad 3x^2 = -1$
$\Leftrightarrow \quad x^2 = -\frac{1}{3} \quad \Rightarrow \quad x = \pm\sqrt{-\frac{1}{3}} = \pm\frac{\sqrt{3}}{3}i$.

55. $3x - 2 > -11 \quad \Leftrightarrow \quad 3x > -9 \quad \Leftrightarrow \quad x > -3$.   Interval: $(-3, \infty)$.

Graph:

57. $-1 < 2x + 5 \leq 3 \quad \Leftrightarrow \quad -6 < 2x \leq -2 \quad \Leftrightarrow \quad -3 < x \leq -1$ Interval: $(-3, -1]$.

Graph:

59. $x^2 + 4x - 12 > 0 \quad \Leftrightarrow \quad (x - 2)(x + 6) > 0$. The expression on the left of the inequality changes sign where $x = 2$ and where $x = -6$. Thus we must check the intervals in the following table.

Interval	$(-\infty, -6)$	$(-6, 2)$	$(2, \infty)$
Sign of $x - 2$	$-$	$-$	$+$
Sign of $x + 6$	$-$	$+$	$+$
Sign of $(x - 2)(x + 6)$	$+$	$-$	$+$

Interval: $(-\infty, -6) \cup (2, \infty)$.   Graph:

61. $\dfrac{2x + 5}{x + 1} \le 1 \quad\Leftrightarrow\quad \dfrac{2x + 5}{x + 1} - 1 \le 0 \quad\Leftrightarrow\quad \dfrac{2x + 5}{x + 1} - \dfrac{x + 1}{x + 1} \le 0 \quad\Leftrightarrow\quad \dfrac{x + 4}{x + 1} \le 0.$ The

expression on the left of the inequality changes sign where $x = -1$ and where $x = -4$. Thus we must check the intervals in the following table.

Interval	$(-\infty, -4)$	$(-4, -1)$	$(-1, \infty)$
Sign of $x + 4$	$-$	$+$	$+$
Sign of $x + 1$	$-$	$-$	$+$
Sign of $\dfrac{x + 4}{x + 1}$	$+$	$-$	$+$

We exclude $x = -1$, since the expression is not defined at this value. Thus the solution is $[-4, -1)$.
Graph:

63. $\dfrac{x - 4}{x^2 - 4} \le 0 \quad\Leftrightarrow\quad \dfrac{x - 4}{(x - 2)(x + 2)} \le 0.$ The expression on the left of the inequality changes sign

where $x = -2$, where $x = 2$, and where $x = 4$. Thus we must check the intervals in the following table.

Interval	$(-\infty, -2)$	$(-2, 2)$	$(2, 4)$	$(4, \infty)$
Sign of $x - 4$	$-$	$-$	$-$	$+$
Sign of $x - 2$	$-$	$-$	$+$	$+$
Sign of $x + 2$	$-$	$+$	$+$	$+$
Sign of $\dfrac{x - 4}{(x - 2)(x + 2)}$	$-$	$+$	$-$	$+$

Since the expression is not defined when $x = \pm 2$, we exclude these values and the solution is
$(-\infty, -2) \cup (2, 4]$.   Graph:

65. $|x - 5| \le 3 \quad\Leftrightarrow\quad -3 \le x - 5 \le 3 \quad\Leftrightarrow\quad 2 \le x \le 8.$ Interval: $[2, 8]$.
Graph:

67. $|2x + 1| \ge 1$ is equivalent to $2x + 1 \ge 1$ or $2x + 1 \le -1$.
     <u>Case 1:</u> $2x + 1 \ge 1 \quad\Leftrightarrow\quad 2x \ge 0 \quad\Leftrightarrow\quad x \ge 0.$
     <u>Case 2:</u> $2x + 1 \le -1 \quad\Leftrightarrow\quad 2x \le -2 \quad\Leftrightarrow\quad x \le -1.$
     Interval: $(-\infty, -1] \cup [0, \infty)$.   Graph:

69. $4x - 3 \ge x^2$. We graph the equations $y_1 = 4x - 3$ and
$y_2 = x^2$ in the viewing rectangle $[-5, 5]$ by $[0, 15]$. Using
Zoom and/or Trace, we find the points of intersection are at
$x = 1$ and $x = 3$. Since we want $4x - 3 \ge x^2$, the solution
is the interval $[1, 3]$.

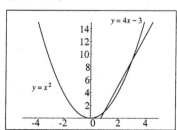

71. $x^4 - 4x^2 < \frac{1}{2}x - 1$. We graph the equations
$y_1 = x^4 - 4x^2$ and $y_2 = \frac{1}{2}x - 1$ in the viewing rectangle
$[-5, 5]$ by $[-5, 5]$. We find the points of intersection are
at $x \approx -1.85$, $x \approx -0.60$, $x \approx 0.45$, and $x = 2$. Since
we want $x^4 - 4x^2 < \frac{1}{2}x - 1$, the solution is
$(-1.85, -0.60) \cup (0.45, 2)$.

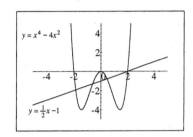

73.   (a)   For $\sqrt{24 - x - 3x^2}$ to define a real number, we must have $24 - x - 3x^2 \geq 0$ $\Leftrightarrow$
$(8 - 3x)(3 + x) \geq 0$. The expression on the left of the inequality changes sign where
$8 - 3x = 0$ $\Leftrightarrow$ $-3x = -8$ $\Leftrightarrow$ $x = \frac{8}{3}$; or where $x = -3$. Thus we must check the
intervals in the following table.

Interval	$(-\infty, -3)$	$\left(-3, \frac{8}{3}\right)$	$\left(\frac{8}{3}, \infty\right)$
Sign of $8 - 3x$	$+$	$+$	$-$
Sign of $3 + x$	$-$	$+$	$+$
Sign of $(8 - 3x)(3 + x)$	$-$	$+$	$-$

Interval: $\left[-3, \frac{8}{3}\right]$.   Graph:

(b)   For $\dfrac{1}{\sqrt[4]{x - x^4}}$ to define a real number we must have $x - x^4 > 0$ $\Leftrightarrow$ $x(1 - x^3) > 0$ $\Leftrightarrow$
$x(1 - x)(1 + x + x^2) > 0$. The expression on the left of the inequality changes sign where
$x = 0$; or where $x = 1$; or where $1 + x + x^2 = 0$ $\Rightarrow$ $x = \dfrac{-1 \pm \sqrt{1^2 - 4(1)(1)}}{2(1)}$

$= \frac{1 \pm \sqrt{1-4}}{2}$ which has no real solution. Thus we must check the intervals in the following
table.

Interval	$(-\infty, 0)$	$(0, 1)$	$(1, \infty)$
Sign of x	$-$	$+$	$+$
Sign of $1 - x$	$+$	$+$	$-$
Sign of $1 + x + x^2$	$+$	$+$	$+$
Sign of $x(1 - x)(1 + x + x^2)$	$-$	$+$	$-$

Interval: $(0, 1)$.   Graph:

Chapter 3 Test

1. (a) Algebraically: $3x - 7 = \frac{1}{2}x$ $\Leftrightarrow$ $-7 = -\frac{5}{2}x$ $\Leftrightarrow$ $x = \frac{14}{5}$.

    (b) Graphically: We graph the two equations $y_1 = 3x - 7$
         and $y_2 = \frac{1}{2}x$ in the viewing rectangle $[-5, 5]$ by
         $[-5, 5]$. Zooming in, we see that solution is $x = 2.80$.

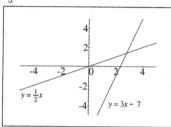

2. Let $d =$ the distance between Ajax and Bixby. Then we have the following table.

	Rate	Time	Distance
Ajax to Bixby	50	$\dfrac{d}{50}$	d
Bixby to Ajax	60	$\dfrac{d}{60}$	d

    We use the fact that the total time is $4\frac{2}{5}$ hours to get the equation $\dfrac{d}{50} + \dfrac{d}{60} = \dfrac{22}{5}$ $\Leftrightarrow$
    $6d + 5d = 1320$ $\Leftrightarrow$ $11d = 1320$ $\Leftrightarrow$ $d = 120$. Thus Ajax and Bixby are 120 miles apart.

3. (a) $(6 - 2i) - (7 - \frac{1}{2}i) = 6 - 2i - 7 + \frac{1}{2}i = -1 - \frac{3}{2}i$.

    (b) $(1 + i)(3 - 2i) = 3 - 2i + 3i - 2i^2 = 3 + i + 2 = 5 + i$.

    (c) $\dfrac{5 + 10i}{3 - 4i} = \dfrac{5 + 10i}{3 - 4i} \cdot \dfrac{3 + 4i}{3 + 4i} = \dfrac{15 + 20i + 30i + 40i^2}{9 - 16i^2} = \dfrac{15 + 50i - 40}{9 + 16} = \dfrac{-25 + 50i}{25}$
         $= -1 + 2i$.

    (d) $i^{50} = i^{48} \cdot i^2 = (i^4)^{12} \cdot i^2 = (1)^{12} \cdot (-1) = -1$.

    (e) $(2 - \sqrt{-2})(\sqrt{8} + \sqrt{-4}) = (2 - \sqrt{2}i)(2\sqrt{2} + 2i) = 4\sqrt{2} + 4i - 4i - 2\sqrt{2}i^2 =$
         $4\sqrt{2} + 2\sqrt{2} = 6\sqrt{2}$.

4. (a) $x^2 - x - 12 = 0$ $\Leftrightarrow$ $(x - 4)(x + 3) = 0$. So $x = 4$ or $x = -3$.

    (b) $2x^2 + 4x + 3 = 0$ $\Rightarrow$ $x = \dfrac{-4 \pm \sqrt{4^2 - 4(2)(3)}}{2(2)} = \dfrac{-4 \pm \sqrt{16 - 24}}{4} = \dfrac{-4 \pm \sqrt{-8}}{4}$
         $= \dfrac{-4 \pm 2\sqrt{2}i}{4} = \dfrac{-2 \pm \sqrt{2}i}{2}$.

    (c) $\sqrt{3 - \sqrt{x + 5}} = 2$ $\Rightarrow$ $3 - \sqrt{x + 5} = 4$ $\Leftrightarrow$ $-1 = \sqrt{x + 5}$. Squaring both sides
         again, we get $1 = x + 5$ $\Leftrightarrow$ $x = -4$. But this does not satisfy the original equation, so
         there is no solution. (You must always check your final answers if you have squared both sides
         when solving an equation, since extraneous answers may be introduced.)
         *Some students might stop at the statement $-1 = \sqrt{x + 5}$ and recognize that this is
         impossible, so there can be no solution.

(d) $x^{1/2} - 3x^{1/4} + 2 = 0$. Let $u = x^{1/4}$, then we have $u^2 - 3u + 2 = 0$ $\Leftrightarrow$ $(u-2)(u-1) = 0$. So $u - 2 = x^{1/4} - 2 = 0$ $\Leftrightarrow$ $x^{1/4} = 2$ $\Rightarrow$ $x = 2^4 = 16$; or $u - 1 = x^{1/4} - 1 = 0$ $\Leftrightarrow$ $x^{1/4} = 1$ $\Rightarrow$ $x = 1^4 = 1$. Thus the solutions are $x = 16$ and $x = 1$.

(e) $x^4 - 16x^2 = 0$ $\Leftrightarrow$ $x^2(x^2 - 16) = 0$ $\Leftrightarrow$ $x^2(x-4)(x+4) = 0$. So $x = 0$, or $x = -4$, or $x = 4$.

(f) $3|x-4| - 10 = 0$ $\Leftrightarrow$ $3|x-4| = 10$ $\Leftrightarrow$ $|x-4| = \frac{10}{3}$ $\Leftrightarrow$ $x - 4 = \pm\frac{10}{3}$ $\Leftrightarrow$ $x = 4 \pm \frac{10}{3}$. So $x = 4 - \frac{10}{3} = \frac{2}{3}$ or $x = 4 + \frac{10}{3} = \frac{22}{3}$. Thus the solutions are $x = \frac{2}{3}$ and $x = \frac{22}{3}$.

5. $x^3 - 9x - 1 = 0$. We graph the equation $y = x^3 - 9x - 1$ in the viewing rectangle $[-10, 10]$ by $[-10, 10]$. We find that the solutions are at $x \approx -2.94$, $x \approx -0.11$, and $x \approx 3.05$.

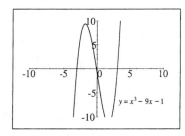

6. Let w be the width of the building lot. Then the length of the lot is $w + 70$. So $w^2 + (w+70)^2 = 130^2$ $\Leftrightarrow$ $w^2 + w^2 + 140w + 4900 = 16900$ $\Leftrightarrow$ $2w^2 + 140w - 12000 = 0$ $\Leftrightarrow$ $2(w^2 + 70w - 6000) = 0$ $\Leftrightarrow$ $2(w + 120)(w - 50) = 0$. So $w = -120$ (which we reject because $w > 0$) or $w = 50$. Thus the lot is 50 by 120.

7. (a) $-1 \le 5 - 2x < 10$ $\Leftrightarrow$ $-6 \le -2x < 5$ $\Leftrightarrow$ $3 \ge x > -\frac{5}{2}$. Expressing in standard form we have: $-\frac{5}{2} < x \le 3$. Interval: $\left(-\frac{5}{2}, 3\right]$ Graph:

(b) $x(x-1)(x-2) > 0$. The expression on the left of the inequality changes sign when $x = 0$, $x = 1$, and $x = 2$. Thus we must check the intervals in the following table.

Interval	$(-\infty, 0)$	$(0, 1)$	$(1, 2)$	$(2, \infty)$
Sign of x	$-$	$+$	$+$	$+$
Sign of $x - 1$	$-$	$-$	$+$	$+$
Sign of $x - 2$	$-$	$-$	$-$	$+$
Sign of $x(x-1)(x-2)$	$-$	$+$	$-$	$+$

From the table, the solution set is $\{x|\ 0 < x < 1 \text{ or } 2 < x\}$
Interval: $(0, 1) \cup (2, \infty)$. Graph:

(c) $|x - 3| < 2$ is equivalent to $-2 < x - 3 < 2$ $\Leftrightarrow$ $1 < x < 5$.
Interval: $(1, 5)$. Graph:

(d) $\dfrac{2x+5}{x+1} \le 1$ $\Leftrightarrow$ $\dfrac{2x+5}{x+1} - 1 \le 0$ $\Leftrightarrow$ $\dfrac{2x+5}{x+1} - \dfrac{x+1}{x+1} \le 0$ $\Leftrightarrow$ $\dfrac{x+4}{x+1} \le 0$. The expression on the left of the inequality changes sign where $x = -4$ and where $x = -1$. Thus we must check the intervals in the following table.

Interval	$(-\infty, -4)$	$(-4, -1)$	$(-1, \infty)$
Sign of $x + 4$	$-$	$+$	$+$
Sign of $x + 1$	$-$	$-$	$+$
Sign of $\dfrac{x + 4}{x + 1}$	$+$	$-$	$+$

Since $x = -1$ makes the expression in the inequality undefined, we exclude this value.

Interval: $[-4, -1)$ Graph:

8. $x^2 - 2 \le |x + 1|$. We graph the equations
    $y_1 = x^2 - 2$ and $y_2 = |x + 1|$ in the viewing rectangle
    $[-5, 5]$ by $[-5, 10]$. We find that the points of
    intersection at $x \approx -1.62$ and $x \approx 2.30$. Since we want
    $x^2 - 2 \le |x + 1|$, the solution is the interval
    $[-1.62, 2.30]$.

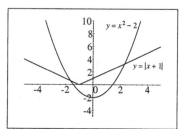

9. $5 \le \frac{5}{9}(F - 32) \le 10 \quad \Leftrightarrow \quad 9 \le F - 32 \le 18 \quad \Leftrightarrow \quad 41 \le F \le 50$. Thus the medicine is to be
    stored at a temperature between $41°F$ to $50°F$.

10. For $\sqrt{4x - x^2}$ to be defined as a real number $4x - x^2 \ge 0 \quad \Leftrightarrow \quad x(4 - x) \ge 0$. The expression on
    the left of the inequality changes sign when $x = 0$ and $x = 4$. 'Thus we must check the intervals in
    the following table.

Interval	$(-\infty, 0)$	$(0, 4)$	$(4, \infty)$
Sign of x	$-$	$+$	$+$
Sign of $4 - x$	$+$	$+$	$-$
Sign of $x(4 - x)$	$-$	$+$	$-$

From the table, we see that $\sqrt{4x - x^2}$ is defined when $0 \le x \le 4$.

Focus on Problem Solving

1. (a)   We mimic the proof in the text that $\sqrt{2}$ is irrational. Suppose that $\sqrt{6}$ is rational, so that $\sqrt{6} = \dfrac{a}{b}$, where a and b are natural numbers with no common factor. Then $a = \sqrt{6}\,b \;\Rightarrow\; a^2 = 6b^2 \;\Rightarrow\; a^2$ is a multiple of 6 (since a and b have no common factors). Thus a is a multiple of 6 $\;\Rightarrow\; a = 6m$, for some integer m. Then $a^2 = 6b^2 \;\Rightarrow\; (6m)^2 = 6b^2 \;\Rightarrow\; 36m^2 = 6b^2 \;\Rightarrow\; b^2 = 6m^2 \;\Rightarrow\; b^2$ is a multiple of 6 $\;\Rightarrow\; b$ is also a multiple of 6. Thus a and b have 6 as a common factor. This contradicts our assumption that a and b have no common factor. So $\sqrt{6}$ must be irrational.

    (b)   We prove this by contradiction. Suppose that $\sqrt{2} + \sqrt{3}$ is rational, so that $\sqrt{2} + \sqrt{3} = \dfrac{a}{b}$, where a and b are natural numbers. Then $\sqrt{2} + \sqrt{3} = \dfrac{a}{b} \;\Rightarrow\; \left(\sqrt{2} + \sqrt{3}\right)^2 = \left(\dfrac{a}{b}\right)^2$ $\;\Leftrightarrow\; 2 + 2\sqrt{6} + 3 = \dfrac{a^2}{b^2} \;\Leftrightarrow\; \sqrt{6} = \dfrac{1}{2}\left(\dfrac{a^2}{b^2} - 5\right)$. Since $\dfrac{1}{2}\left(\dfrac{a^2}{b^2} - 5\right)$ is the product of rational numbers, it is a rational number. However, this implies that $\sqrt{6}$ is a rational number, which contradicts part (a). So $\sqrt{2} + \sqrt{3}$ must be irrational.

3. Let x and y be the lengths of the sides of such a rectangle. Since $Area = xy$ and $Perimeter = 2x + 2y$, we equate these expressions and get $xy = 2x + 2y \;\Leftrightarrow\; xy - 2x - 2y = 0 \;\Leftrightarrow\; xy - 2x - 2y + 4 = 4 \;\Leftrightarrow\; (x-2)(y-2) = 4$. Since the only factorizations of 4 as a product of two positive integers are $4 = 1 \cdot 4 = 2 \cdot 2 = 4 \cdot 1$, the only integer solutions of the last equation are
    $$\begin{cases} x - 2 = 1 \\ y - 2 = 4 \end{cases} ; \qquad \begin{cases} x - 2 = 2 \\ y - 2 = 2 \end{cases} ; \qquad \begin{cases} x - 2 = 4 \\ y - 2 = 1 \end{cases} .$$
    Thus the only solutions are $x = 3$, $y = 6$; $x = y = 4$; and $x = 6$, $y = 3$.

5. We prove the assertion by contradiction. Assume there are n people at the party. Suppose the assertion is false, that is, everybody knows a different number of people. Then we can order the people so that P_1 knows 1 person (self), P_2 knows 2 people, $\ldots$, and P_n knows n people. But then P_n knows everyone. So P_n knows P_1 and P_1 knows P_n, and therefore, P_1 knows 2 people. This is a contradiction so the given assertion must be true.

7. For convenience, we label the coins $1, 2, 3, \cdots , 12$. First we weigh $\{1, 2, 3, 4\}$ against $\{5, 6, 7, 8\}$.

    <u>Case 1:</u> If the first weighing is equal, then for the second weighing, we weigh $\{1, 2, 3\}$ against $\{9, 10, 11\}$. If the second weighing is equal, then 12 must be counterfeit, and a third weighing of 12 against 1 will determine whether 12 or 1 is light or heavy. If the second weighing is unequal, suppose that $\{1, 2, 3\} < \{9, 10, 11\}$, where the symbol $<$ means "is lighter than." Then one of 9, 10, 11 is heavy. A third weighing of 9 against 10 will determine which one of the three is heavy.

    <u>Case 2:</u> If the first weighing is unequal, let us suppose that $\{1, 2, 3, 4\} < \{5, 6, 7, 8\}$. For the second weighing, we weigh $\{1, 2, 5\}$ against $\{3, 4, 6\}$. If the second weighing is equal, then either 7 or 8 is heavy, and a third weighing of 7 against 8 will determine which one. If the second weighing is unequal, suppose that $\{1, 2, 5\} < \{3, 4, 6\}$. Then either 1 or 2 is light or 6 is heavy. So if 1 is weighed against 2 on the third weighing, it will be obvious which of these three possibilities holds true.

The diagram on the next page shows the weighing and the result based on each sequence of weighings.

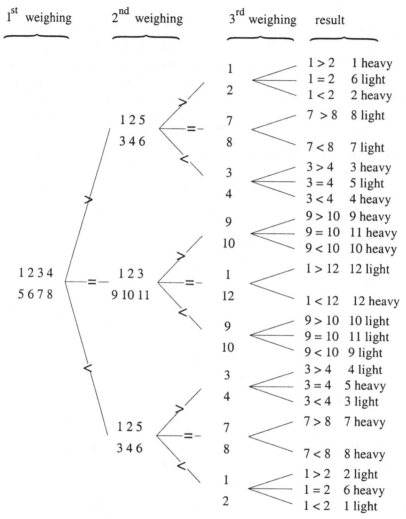

1st weighing	2nd weighing	3rd weighing	result

9. Consider any three points that are exactly one unit apart, as shown in the figure. Since only two colors are available, it must be that at least two of these points have the same color, and this proves that there are at least two points exactly one unit apart with the same color.

11. Label the points P_1, P_2, ..., P_{1000}, and consider all possible pairs of points. Each such pair (P_i, P_j) determines a slope $m_{i,j}$ (which may be infinite). There are only a finite number of slopes $m_{i,j}$, so let m be a number not equal to any of these. Move a line of slope m through the 1000 points (from bottom to top, say). Since $m \neq m_{i,j}$, the line will cross one point at a time. Stop after it has crossed 500 points. Then 500 points will be below the line and 500 above.

13. $x^2y - y^3 - 5x^2 + 5y^2 = 0 \quad \Leftrightarrow$
$y(x^2 - y^2) - 5(x^2 - y^2) = 0 \quad \Leftrightarrow$
$(y - 5)(x^2 - y^2) = 0 \quad \Leftrightarrow \quad (y - 5)(x - y)(x + y) = 0.$
So $y = 5$, $x = y$ or $x = -y$. The graph consists of 3
lines.

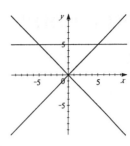

15. (a) Powers of 2, since at each step there will be an even number of winners.

   (b) Since there is exactly one winner, there are $n - 1$ losers. Since each loser loses exactly one match, there must be $n - 1$ matches.

17. By the Pythagorean Theorem, $a^2 + b^2 = c^2$. Multiplying both sides by c, we have $a^2c + b^2c = c^3$. Since the sides are in increasing order, we have $a < c \quad \Leftrightarrow \quad a^3 < a^2c$, and we have $b < c \quad \Leftrightarrow \quad b^3 < b^2c$. Thus $a^3 + b^3 < a^2c + b^2c = c^3$, so $a^3 + b^3 < c^3$.

Chapter Four
Exercises 4.1

1. $f(x) = 3x + 1$

3. $f(x) = (x + 2)^2$

5. Divide by 3, then subtract 5.

7. Square, multiply by 2, then subtract 3.

9. Machine diagram for $f(x) = \sqrt{x}$.

11. $f(x) = 2x^2 + 1$

x	$f(x)$
-1	$2(-1)^2 + 1 = 3$
0	$2(0)^2 + 1 = 1$
1	$2(1)^2 + 1 = 3$
2	$2(2)^2 + 1 = 9$
3	$2(3)^2 + 1 = 19$

13. $f(1) = 2(1) + 1 = 3$; $\quad f(-2) = 2(-2) + 1 = -3$; $\quad f(\frac{1}{2}) = 2(\frac{1}{2}) + 1 = 2$;
$f(a) = 2(a) + 1 = 2a + 1$; $\quad f(-a) = 2(-a) + 1 = -2a + 1$;
$f(a + b) = 2(a + b) + 1 = 2a + 2b + 1$.

15. $g(2) = \dfrac{1 - (2)}{1 + (2)} = \dfrac{-1}{3} = -\dfrac{1}{3}$; $\quad g(-2) = \dfrac{1 - (-2)}{1 + (-2)} = \dfrac{3}{-1} = -3$; $\quad g(\frac{1}{2}) = \dfrac{1 - (\frac{1}{2})}{1 + (\frac{1}{2})} = \dfrac{\frac{1}{2}}{\frac{3}{2}} = \dfrac{1}{3}$;
$g(a) = \dfrac{1 - (a)}{1 + (a)} = \dfrac{1 - a}{1 + a}$; $\quad g(a - 1) = \dfrac{1 - (a - 1)}{1 + (a - 1)} = \dfrac{1 - a + 1}{1 + a - 1} = \dfrac{2 - a}{a}$;
$g(-1) = \dfrac{1 - (-1)}{1 + (-1)} = \dfrac{2}{0}$, so $g(-1)$ is not defined.

17. $f(0) = 2(0)^2 + 3(0) - 4 = -4$; $\quad f(2) = 2(2)^2 + 3(2) - 4 = 8 + 6 - 4 = 10$;
$f(-2) = 2(-2)^2 + 3(-2) - 4 = 8 - 6 - 4 = -2$;
$f(\sqrt{2}) = 2(\sqrt{2})^2 + 3(\sqrt{2}) - 4 = 4 + 3\sqrt{2} - 4 = 3\sqrt{2}$;
$f(x + 1) = 2(x + 1)^2 + 3(x + 1) - 4 = 2x^2 + 4x + 2 + 3x + 3 - 4 = 2x^2 + 7x + 1$;
$f(-x) = 2(-x)^2 + 3(-x) - 4 = 2x^2 - 3x - 4$.

19. $f(-2) = 2|-2 - 1| = 2(3) = 6;$ $f(0) = 2|0 - 1| = 2(1) = 2;$ $f(\frac{1}{2}) = 2|\frac{1}{2} - 1| = 2(\frac{1}{2}) = 1;$
    $f(2) = 2|2 - 1| = 2(1) = 2;$ $f(x + 1) = 2|(x + 1) - 1| = 2|x|;$
    $f(x^2 + 2) = 2|(x^2 + 2) - 1| = 2|x^2 + 1| = 2x^2 + 2$ (since $x^2 + 1 > 0$).

21. Since $-2 < 0$, we have $f(-2) = (-2)^2 = 4$. Since $-1 < 0$, we have $f(-1) = (-1)^2 = 1$. Since
    $0 \geq 0$, we have $f(0) = 0 + 1 = 1$. Since $1 \geq 0$, we have $f(1) = 1 + 1 = 2$. Since $2 \geq 0$, we have
    $f(2) = 2 + 1 = 3$.

23. Since $-4 \leq -1$, we have $f(-4) = (-4)^2 + 2(-4) = 16 - 8 = 8$. Since $-\frac{3}{2} \leq -1$, we have
    $f(-\frac{3}{2}) = (-\frac{3}{2})^2 + 2(-\frac{3}{2}) = \frac{9}{4} - 3 = -\frac{3}{4}$. Since $-1 \leq -1$, we have
    $f(-1) = (-1)^2 + 2(-1) = 1 - 2 = -1$. Since $0 > -1$, we have $f(0) = 0$. Since $1 > -1$, we
    have $f(1) = 1$.

25. $f(x + 2) = (x + 2)^2 + 1 = x^2 + 4x + 4 + 1 = x^2 + 4x + 5;$
    $f(x) + f(2) = x^2 + 1 + (2)^2 + 1 = x^2 + 1 + 4 + 1 = x^2 + 6.$

27. $f(x^2) = x^2 + 4;$ $[f(x)]^2 = [x + 4]^2 = x^2 + 8x + 16.$

29. $f(a) = 3(a) + 2 = 3a + 2;$ $f(a + h) = 3(a + h) + 2 = 3a + 3h + 2;$
    $\dfrac{f(a + h) - f(a)}{h} = \dfrac{(3a + 3h + 2) - (3a + 2)}{h} = \dfrac{3a + 3h + 2 - 3a - 2}{h} = \dfrac{3h}{h} = 3.$

31. $f(a) = 5;$ $f(a + h) = 5;$ $\dfrac{f(a + h) - f(a)}{h} = \dfrac{5 - 5}{h} = 0.$

33. $f(a) = 3 - 5a + 4a^2;$ $f(a + h) = 3 - 5(a + h) + 4(a + h)^2$
    $= 3 - 5a - 5h + 4(a^2 + 2ah + h^2) = 3 - 5a - 5h + 4a^2 + 8ah + 4h^2;$
    $\dfrac{f(a + h) - f(a)}{h} = \dfrac{(3 - 5a - 5h + 4a^2 + 8ah + 4h^2) - (3 - 5a + 4a^2)}{h}$
    $= \dfrac{3 - 5a - 5h + 4a^2 + 8ah + 4h^2 - 3 + 5a - 4a^2}{h} = \dfrac{-5h + 8ah + 4h^2}{h} = \dfrac{h(-5 + 8a + 4h)}{h}$
    $= -5 + 8a + 4h.$

35. $f(x) = 2x$. Since there is no restrictions, the domain is the set of real numbers.

37. $f(x) = 2x$. The domain is restricted by the exercise to $[-1, 5]$.

39. $f(x) = \dfrac{1}{x - 3}$. Since the denominator cannot equal 0 we have $x - 3 \neq 0$ $\Leftrightarrow$ $x \neq 3$. Thus the
    domain is $\{x | x \neq 3\}$. In interval notation, the domain is $(-\infty, 3) \cup (3, \infty)$.

41. $f(x) = \dfrac{x + 2}{x^2 - 1}$. Since the denominator cannot equal 0 we have $x^2 - 1 \neq 0$ $\Leftrightarrow$ $x^2 \neq 1$ $\Rightarrow$
    $x \neq \pm 1$. Thus the domain is $\{x | x \neq \pm 1\}$. In interval notation, the domain is
    $(-\infty, -1) \cup (-1, 1) \cup (1, \infty)$.

43. $f(x) = \sqrt{x - 5}$. We require $x - 5 \geq 0$ $\Leftrightarrow$ $x \geq 5$. Thus the domain is $\{x | x \geq 5\}$. The
    domain can also be expressed in interval notation as $[5, \infty)$.

45. $f(t) = \sqrt[3]{t-1}$. Since the odd root is defined for all real numbers, the domain is the set of real numbers.

47. $h(x) = \sqrt{2x-5}$. Since the square root is defined as a real number only for non-negative numbers, we require that $2x - 5 \geq 0$ $\Leftrightarrow$ $2x \geq 5$ $\Leftrightarrow$ $x \geq \frac{5}{2}$. So the domain is $\{x | x \geq \frac{5}{2}\}$. In interval notation, the domain is $[\frac{5}{2}, \infty)$.

49. $g(x) = \dfrac{\sqrt{2+x}}{3-x}$. We require $2 + x \geq 0$, and the denominator cannot equal 0. Now $2 + x \geq 0$ $\Leftrightarrow$ $x \geq -2$, and $3 - x \neq 0$ $\Leftrightarrow$ $x \neq 3$. Thus the domain is $\{x | x \geq -2 \text{ and } x \neq 3\}$, which can be expressed in interval notation as $[-2, 3) \cup (3, \infty)$.

51. $g(x) = \sqrt[4]{x^2 - 6x}$. Since the input to an even root must be non-negative, we have $x^2 - 6x \geq 0$ $\Leftrightarrow$ $x(x - 6) \geq 0$. Using the method of Section 3.7, we have

	$(-\infty, 0)$	$(0, 6)$	$(6, \infty)$
Sign of x	$-$	$+$	$+$
Sign of $x - 6$	$-$	$-$	$+$
Sign of $x(x - 6)$	$+$	$-$	$+$

Thus the domain is $(-\infty, 0] \cup [6, \infty)$.

53.   This person appears to gain weight steadily until the age of 21 when this person's weight gain slows down. At age 30, this person experiences a sudden weight loss, but recovers, and by about age 40, this person's weight seems to level off at around 200 pounds. It then appears that the person dies at about age 68. The sudden weight loss could be due to a number of reasons, among them major illness, weight loss program, etc.

55.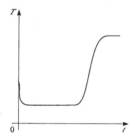

57.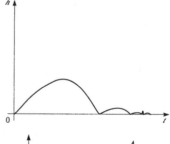

59.   The sales of Christmas cards will increase sharply approximately from October to December and decrease sharply from January to September.

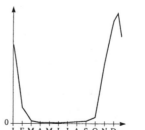

61.   (a)   At 6 A.M. the graph shows that the power consumption is about 500 megawatts. Since $t = 18$ represents 6 P.M., the graph shows that the power consumption at 6 P.M. is about 725 megawatts.

      (b)   The power consumption is lowest between 3 A.M. and 4 A.M..

      (c)   The power consumption is highest just before 12 noon.

63.
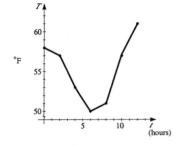

65.   Answers will vary.

Exercises 4.2

1. (a) $h(-2) = 1$; $h(0) = -1$; $h(2) = 3$; $h(3) = 4$.

   (b) Domain: $[-3, 4]$. Range: $[-1, 4]$.

3. (a) $f(0) = 3 > \frac{1}{2} = g(0)$. So $f(0)$ is larger.

   (b) $f(-3) \approx -\frac{3}{2} < 2 = g(-3)$. So $g(-3)$ is larger.

   (c) For $x = -2$ and $x = 2$.

5. The curves in parts (a) and (c) are graphs of a function of x, by the vertical line test.

7. The given curve is the graph of a function of x. Domain: $[-3, 2]$. Range: $[-2, 2]$.

9. No, the given curve is not the graph of a function of x, by the vertical line test.

11. (a) $f(x) = 1 - x$ (b) Domain: All real numbers.

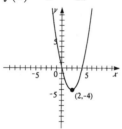

13. (a) $f(x) = x^2 - 4x$ (b) Domain: All real numbers.

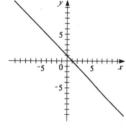

15. (a) $f(x) = \sqrt{9 - x}$ (b) We require $9 - x \geq 0$ $\Leftrightarrow$ $9 \geq x$. Domain: $(-\infty, 9]$.

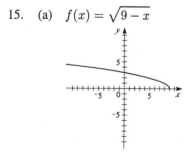

17. (a) $f(x) = \sqrt{16 - x^2}$

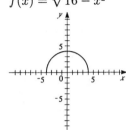

(b) We require $16 - x^2 \geq 0$ $\Leftrightarrow$ $x^2 \leq 16$ $\Leftrightarrow$ $|x| \leq 4$ $\Leftrightarrow$ $-4 \leq x \leq 4$. Therefore, the domain is $[-4, 4]$.

19. $f(x) = 3$

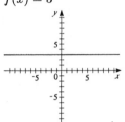

21. $f(x) = 2x + 3$

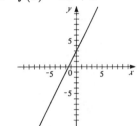

23. $f(x) = -x + 4$, $-1 \leq x \leq 4$

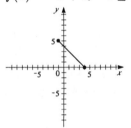

25. $f(x) = -x^2$

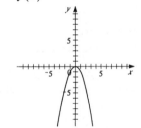

27. $g(x) = x^3 - 8$

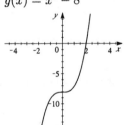

29. $g(x) = \sqrt{-x}$

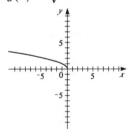

31. $F(x) = \frac{1}{x}$

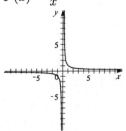

33. $H(x) = |2x|$

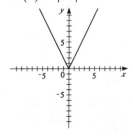

35. $G(x) = |x| + x$

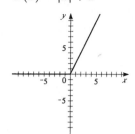

37. $f(x) = |2x - 2|$

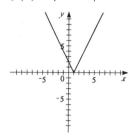

39. $g(x) = \dfrac{2}{x^2}$

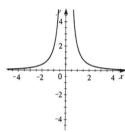

41. Solving for y in terms of x gives: $x^2 + 2y = 4 \quad \Leftrightarrow \quad 2y = 4 - x^2 \quad \Leftrightarrow \quad y = 2 - \frac{1}{2}x^2$. This defines y as a function of x.

43. Solving for y in terms of x gives: $x = y^2 \quad \Leftrightarrow \quad y = \pm\sqrt{x}$. The last equation gives two values of y for a given value of x. Thus, this equation does not define y as a function of x.

45. Solving for y in terms of x gives: $x + y^2 = 9 \quad \Leftrightarrow \quad y^2 = 9 - x \quad \Leftrightarrow \quad y = \pm\sqrt{9 - x}$. The last equation gives two values of y for a given value of x. Thus, this equation does not define y as a function of x.

47. Solving for y in terms of x gives: $x^2y + y = 1 \quad \Leftrightarrow \quad y(x^2 + 1) = 1 \quad \Leftrightarrow \quad y = \dfrac{1}{x^2 + 1}$. This defines y as a function of x.

49. Solving for y in terms of x gives: $2|x| + y = 0 \quad \Leftrightarrow \quad y = -2|x|$. This defines y as a function of x.

51. Solving for y in terms of x gives: $x = y^3 \quad \Leftrightarrow \quad y = \sqrt[3]{x}$. This defines y as a function of x.

53. (a) $f(x) = x^2 + c$, for $c = 0, 2, 4$, and 6.   (b) $f(x) = x^2 + c$, for $c = 0, -2, -4$, and -6.

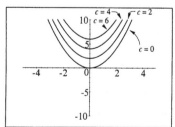
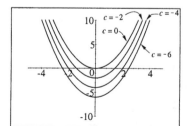

(c) The graphs in part (a) are obtained by shifting the graph of $f(x) = x^2$ upward c units, $c > 0$. The graphs in part (b) are obtained by shifting the graph of $f(x) = x^2$ downward c units.

55. (a) $f(x) = (x - c)^3$, for $c = 0, 2, 4,$ and 6. (b) $f(x) = (x - c)^3$, for $c = 0, -2, -4,$ and -6.

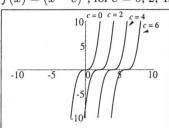   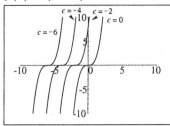

(c) The graphs in part (a) are obtained by shifting the graph of $f(x) = x^3$ to the right c units, $c > 0$. The graphs in part (b) are obtained by shifting the graph of $f(x) = x^3$ to the left $|c|$ units, $c < 0$.

57. (a) $f(x) = x^c$, for $c = \frac{1}{2}, \frac{1}{4},$ and $\frac{1}{6}$.   (b) $f(x) = x^c$, for $c = 1, \frac{1}{3},$ and $\frac{1}{5}$.

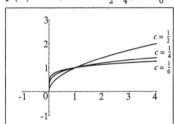   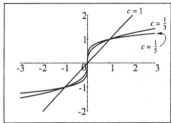

(c) Graphs of even roots are similar to $y = \sqrt{x}$, graphs of odd roots are similar to $y = \sqrt[3]{x}$. As c increases, the graph of $y = \sqrt[c]{x}$ becomes steeper near $x = 0$ and flatter when $x > 1$.

59. $f(x) = \begin{cases} 0 & \text{if } x < 2 \\ 1 & \text{if } x \geq 2 \end{cases}$ 61. $f(x) = \begin{cases} 3 & \text{if } x < 2 \\ x - 1 & \text{if } x \geq 2 \end{cases}$

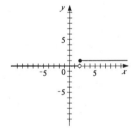   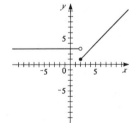

63. $f(x) = \begin{cases} x & \text{if } x \leq 0 \\ x + 1 & \text{if } x > 0 \end{cases}$ 65. $f(x) = \begin{cases} -1 & \text{if } x < -1 \\ 1 & \text{if } -1 \leq x \leq 1 \\ -1 & \text{if } x > 1 \end{cases}$

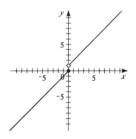   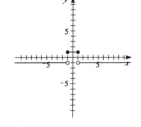

67. $f(x) = \begin{cases} 2 & \text{if } x \leq -1 \\ x^2 & \text{if } x > -1 \end{cases}$

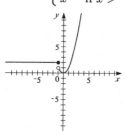

69. $f(x) = \begin{cases} 0 & \text{if } |x| \leq 2 \\ 3 & \text{if } |x| > 2 \end{cases}$

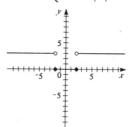

71. $f(x) = \begin{cases} 4 & \text{if } x < -2 \\ x^2 & \text{if } -2 \leq x \leq 2 \\ -x + 6 & \text{if } x > 2 \end{cases}$

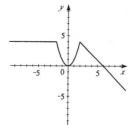

73. $f(x) = \begin{cases} x + 2 & \text{if } x \leq -1 \\ x^2 & \text{if } x > -1 \end{cases}$

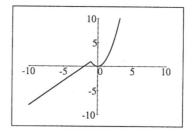

75. $C(x) = \begin{cases} 2.00 & \text{if } 0 < x \leq 1 \\ 2.20 & \text{if } 1 < x \leq 1.1 \\ 2.40 & \text{if } 1.1 < x \leq 1.2 \\ \quad\vdots \\ 4.00 & \text{if } 1.9 < x < 2 \end{cases}$

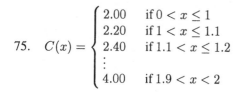

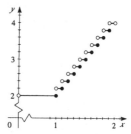

77. The slope of the line segment joining the points $(-2, 1)$ and $(4, -6)$ is $m = \frac{-6-1}{4-(-2)} = -\frac{7}{6}$. Using the point-slope form, we have $y - 1 = -\frac{7}{6}(x + 2)$ $\Leftrightarrow$ $y = -\frac{7}{6}x - \frac{7}{3} + 1$ $\Leftrightarrow$ $y = -\frac{7}{6}x - \frac{4}{3}$. Thus the function is $f(x) = -\frac{7}{6}x - \frac{4}{3}$ for $-2 \leq x \leq 4$.

79. The parabola has equation $x + (y - 1)^2 = 0$ $\Leftrightarrow$ $(y - 1)^2 = -x$ $\Rightarrow$ $y - 1 = \pm\sqrt{-x}$. Since we seek the bottom half of the parabola, we choose $y - 1 = -\sqrt{-x}$ $\Leftrightarrow y = 1 - \sqrt{-x}$. So the function is $f(x) = 1 - \sqrt{-x}$, $x \leq 0$.

81. The graph of $x = y^2$ is not the graph of a function because both $(1, 1)$ and $(-1, 1)$ satisfy the equation $x = y^2$. The graph of $x = y^3$ is the graph of a function because $x = y^3$ $\Leftrightarrow$ $x^{1/3} = y$. If n is even, then both $(1, 1)$ and $(-1, 1)$ satisfies the equation $x = y^n$, so the graph of $x = y^n$ is not the graph of a function. When n is odd, $y = x^{1/n}$ is defined for all real numbers, and since $y = x^{1/n}$ $\Leftrightarrow$ $x = y^n$, the graph of $x = y^n$ is the graph of a function.

83. $f(x) = [\![x]\!]$ $g(x) = [\![2x]\!]$ $h(x) = [\![3x]\!]$

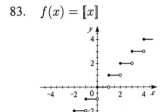   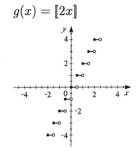   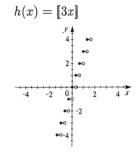

The graph of $k(x) = [\![nx]\!]$ is a step function whose steps are each $\dfrac{1}{n}$ wide.

Exercises 4.3

1. $R = kt$, where k is constant.

3. $v = \dfrac{k}{z}$, where k is constant.

5. $y = \dfrac{ks}{t}$, where k is constant.

7. $z = k\sqrt{y}$, where k is constant.

9. $V = klwh$, where k is constant.

11. $R = \dfrac{ki}{Pt}$, where k is constant.

13. Since y is directly proportional to x, $y = kx$. Since $y = 72$ when $x = 4$, we have $72 = k(4)$ $\Leftrightarrow$ $k = 18$. So $y = 18x$.

15. Since M varies directly as x and inversely as y, $M = \dfrac{kx}{y}$. Since $M = 5$ when $x = 2$ and $y = 6$, we have $5 = \dfrac{k(2)}{6}$ $\Leftrightarrow$ $k = 15$. Therefore $M = \dfrac{15x}{y}$.

17. Since W is inversely proportional to the square of r, $W = \dfrac{k}{r^2}$. Since $W = 10$ when $r = 6$, we have $10 = \dfrac{k}{(6)^2}$ $\Leftrightarrow$ $k = 360$. So $W = \dfrac{360}{r^2}$.

19. Since C is jointly proportional to l, w, and h, we have $C = klwh$. Since $C = 128$ when $l = w = h = 2$, we have $128 = k(2)(2)(2)$ $\Leftrightarrow$ $128 = 8k$ $\Leftrightarrow$ $k = 16$. Therefore, $C = 16lwh$.

21. Since s is inversely proportional to the square root of t, we have $s = \dfrac{k}{\sqrt{t}}$. Since $s = 100$ when $t = 25$, we have $100 = \dfrac{k}{\sqrt{25}}$ $\Leftrightarrow$ $100 = \dfrac{k}{5}$ $\Leftrightarrow$ $k = 500$. So $s = \dfrac{500}{\sqrt{t}}$.

23. (a) The force F needed is $F = kx$.
    (b) Since $F = 40$ when $x = 5$, we have $40 = k(5)$ $\Leftrightarrow$ $k = 8$.
    (c) From part (b), we have $F = 8x$. Substituting $x = 4$ into $F = 8x$ gives $F = 8(4) = 32$ N.

25. (a) $C = kpm$.
    (b) Since $C = 60,000$ when $p = 120$ and $m = 4000$, we get $60000 = k(120)(4000)$ $\Leftrightarrow$ $k = \frac{1}{8}$. So $C = \frac{1}{8}pm$.
    (c) Substituting $p = 92$ and $m = 5000$, we get $C = \frac{1}{8}(92)(5000) = \$57,500$.

27. (a) $R = \dfrac{kL}{d^2}$.
    (b) Since $R = 140$ when $L = 1.2$ and $d = 0.005$, we get $140 = \dfrac{k(1.2)}{(0.005)^2}$ $\Leftrightarrow$ $k = \frac{7}{2400} = 0.0029 1\overline{6}$.
    (c) Substituting $L = 3$ and $d = 0.008$, we have $R = \dfrac{7}{2400} \cdot \dfrac{3}{(0.008)^2} = \dfrac{4375}{32} \approx 137 \ \Omega$.

29. Let C be the cost and A be the area of the sheet of gold. Since C is proportional to A, $C = kA$. When $C = 75$, $A = 15 \cdot 20 = 300$, so $75 = k300$ $\Leftrightarrow$ $k = \frac{1}{4}$. So $C = \frac{1}{4}A$. When $A = 3 \cdot 5 = 15$, we have $C = \frac{1}{4}(15) = \$3.75$.

31. Let V be the value of a building lot on Galiano Island, A the area of the lot, and q the quantity of the water produced. Since V is jointly proportional to the area and water quantity, we have $V = kAq$. When $A = 200 \cdot 300 = 60{,}000$ and $q = 10$, we have $V = \$48{,}000$, so $48{,}000 = k(60{,}000)(10)$ $\Leftrightarrow$ $k = 0.08$. Thus $V = 0.08Aq$. Now when $A = 400 \cdot 400 = 160{,}000$ and $q = 4$, the value is $V = 0.08(160{,}000)(4) = \$51{,}200$.

33. Examples include radioactive decay and exponential growth in biology.

Exercises 4.4

1. Average rate of change $= \dfrac{f(3) - f(2)}{3 - 2} = \dfrac{[3(3) - 2] - [3(2) - 2]}{1} = 7 - 4 = 3.$

3. Average rate of change $= \dfrac{h(4) - h(-1)}{4 - (-1)} = \dfrac{[4^2 + 2(4)] - [(-1)^2 + 2(-1)]}{5} = \dfrac{24 - (-1)}{5} = 5.$

5. Average rate of change $= \dfrac{f(10) - f(0)}{10 - 0} = \dfrac{[10^3 - 4(10^2)] - [0^3 - 4(0^2)]}{10 - 0} = \dfrac{600 - 0}{10} = 60.$

7. Average rate of change $= \dfrac{f(2 + h) - f(2)}{(2 + h) - 2} = \dfrac{[3(2 + h)^2] - [3(2^2)]}{h} = \dfrac{12 + 12h + 3h^2 - 12}{h}$

   $= \dfrac{12h + 3h^2}{h} = \dfrac{h(12 + 3h)}{h} = 12 + 3h.$

9. Average rate of change $= \dfrac{g(1) - g(a)}{1 - a} = \dfrac{\frac{1}{1} - \frac{1}{a}}{1 - a} \cdot \dfrac{a}{a} = \dfrac{a - 1}{a(1 - a)} = \dfrac{-1(1 - a)}{a(1 - a)} = \dfrac{-1}{a}.$

11. Average rate of change $= \dfrac{f(a + h) - f(a)}{(a + h) - a} = \dfrac{\frac{2}{a + h} - \frac{2}{a}}{h} \cdot \dfrac{a(a + h)}{a(a + h)} = \dfrac{2a - 2(a + h)}{ah(a + h)}$

    $= \dfrac{-2h}{ah(a + h)} = \dfrac{-2}{a(a + h)}.$

13. (a) Average rate of change $= \dfrac{f(a + h) - f(a)}{(a + h) - a} = \dfrac{[\frac{1}{2}(a + h) + 3] - [\frac{1}{2}a + 3]}{h}$

    $= \dfrac{\frac{1}{2}a + \frac{1}{2}h + 3 - \frac{1}{2}a - 3}{h} = \dfrac{\frac{1}{2}h}{h} = \dfrac{1}{2}.$

    (b) The slope of the line $f(x) = \frac{1}{2}x + 3$ is $\frac{1}{2}$, which is also the average rate of change.

15. We use the points $(1, 3)$ and $(4, 5)$, so the average rate of change $= \frac{5-3}{4-1} = \frac{2}{3}.$

17. We use the points $(0, 6)$ and $(5, 2)$, so the average rate of change $= \frac{2-6}{5-0} = \frac{-4}{5}.$

19. (a) Average rate of change of population $= \frac{1{,}591 - 856}{1994 - 1991} = \frac{735}{3} = 245.$

    (b) Average rate of change of population $= \frac{826 - 1{,}483}{1997 - 1995} = \frac{-657}{2} = -328.5.$

    (c) The population was increasing from 1990 to 1994.

    (d) The population was decreasing from 1994 to 1999.

21. (a) Average rate of change of sales $= \frac{584 - 512}{1999 - 1989} = \frac{72}{10} = 7.2.$

    (b) Average rate of change of sales $= \frac{520 - 512}{1990 - 1989} = \frac{8}{1} = 8.$

    (c) Average rate of change of sales $= \frac{410 - 520}{1992 - 1990} = \frac{-110}{2} = -55.$

(d)

Year	CD players sold	Change in sales from previous year
1989	512	—
1990	520	8
1991	413	−107
1992	410	−3
1993	468	58
1994	510	42
1995	590	80
1996	607	17
1997	732	125
1998	612	−120
1999	584	−28

Sales *increased* most quickly between 1996 and 1997. Sales *decreased* most quickly between 1997 and 1998.

23. The function is increasing on $[-1, 1]$ and $[2, 4]$. It is decreasing on $[1, 2]$.

25. The function is increasing on $[-2, -1]$ and $[1, 2]$. It is decreasing on $[-3, -2]$, $[-1, 1]$, and $[2, 3]$.

27. (a) $f(x) = x^{2/5}$ is graphed in the viewing rectangle $[-10, 10]$ by $[-5, 5]$.

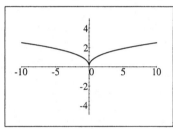

(b) The function is increasing on $[0, \infty)$. It is decreasing on $(-\infty, 0]$.

29. (a) $f(x) = x^2 - 5x$ is graphed in the viewing rectangle $[-2, 7]$ by $[-10, 10]$.

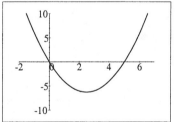

(b) The function is increasing on $[2.5, \infty)$. It is decreasing on $(-\infty, 2.5]$.

31. (a) $f(x) = 2x^3 - 3x^2 - 12x$ is graphed in the viewing rectangle $[-3, 5]$ by $[-25, 20]$.

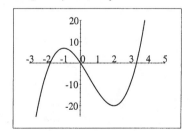

(b) The function is increasing on $(-\infty, -1]$ and $[2, \infty)$. It is decreasing on $[-1, 2]$.

33. (a) $f(x) = x^3 + 2x^2 - x - 2$ is graphed in
          the viewing rectangle $[-5, 5]$ by $[-3, 3]$.

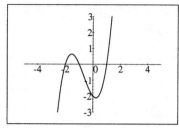

    (b) The function is increasing on $(-\infty, -1.55]$ and $[0.22, \infty)$. It is decreasing on $[-1.55, 0.22]$.

35. (a)

Time (s)	Average speed (ft/s)
0-2	17
2-4	18
4-6	63
6-8	147
8-10	237

The average speed is increasing, so the car is accelerating. From the shape of the graph, we see that the slope continues to get steeper.

    (b)

Time (s)	Average speed (ft/s)
30-32	263
32-34	144
34-36	91
36-38	74
38-40	48

The average speed is decreasing, so the car is decelerating. From the shape of the graph, we see that the slope continues to get flatter.

Exercises 4.5

1. (a) Shift the graph of $y = f(x)$ downward 4 units.

   (b) Shift the graph of $y = f(x)$ to the right 4 units.

3. (a) Stretch the graph of $y = f(x)$ vertically by a factor of 3.

   (b) Shrink the graph of $y = f(x)$ vertically by a factor of $\frac{1}{3}$.

5. (a) Reflect the graph of $y = f(x)$ about the x-axis, and then shift upward 5 units.

   (b) Reflect the graph of $y = f(x)$ about the y-axis, and then shift upward 5 units.

7. (a) Shift the graph of $y = f(x)$ to the right 2 units, and downward 3 units.

   (b) Shift the graph of $y = f(x)$ to the right 3 units, then stretch vertically by a factor of 2.

9. (a) $y = f(x - 2)$ 　　　　(b) $y = f(x) - 2$ 　　　　(c) $y = 2f(x)$

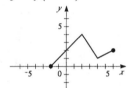　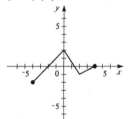　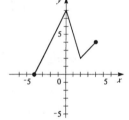

   (d) $y = -f(x) + 3$ 　　　　(e) $y = f(-x)$ 　　　　(f) $y = \frac{1}{2}f(x - 1)$

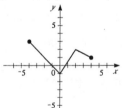　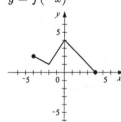　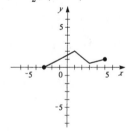

11. (a) $f(x) = \dfrac{1}{x}$

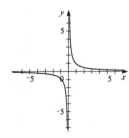

(b) (i) $y = -\dfrac{1}{x}$. Reflect the graph of f about the x-axis.

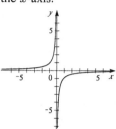

(ii) $y = \dfrac{1}{x-1}$. Shift the graph of f to the right 1 unit.

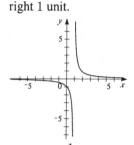

(iii) $y = \dfrac{2}{x+2}$. Shift the graph of f to the left 2 units and stretch vertically by a factor of 2.

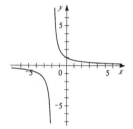

(iv) $y = 1 + \dfrac{1}{x-3}$. Shift the graph of f to the right 3 units and upward 1 unit.

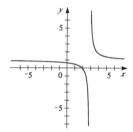

13. (a) The graph of $g(x) = (x+2)^2$ is obtained by shifting the graph of $f(x)$ to the left 2 units.

(b) The graph of $g(x) = x^2 + 2$ is obtained by shifting the graph of $f(x)$ upward 2 units.

15. (a) The graph of $g(x) = 2\sqrt{x}$ is obtained by stretching the graph of $f(x)$ vertically by a factor of 2.

(b) The graph of $g(x) = \frac{1}{2}\sqrt{x-2}$ is obtained by shifting the graph of $f(x)$ to the right 2 units, and then shrinking the graph vertically by a factor of $\frac{1}{2}$.

17. $f(x) = (x-2)^2$. Shift the graph of $y = x^2$ to the right 2 units.

19. $f(x) = -(x+1)^2$. Shift the graph of $y = x^2$ to the left 1 unit, then reflect about the x-axis.

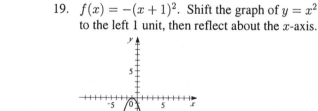

21. $f(x) = x^3 + 2$. Shift the graph of
     $y = x^3$ upward 2 units.

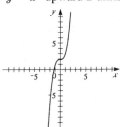

23. $y = 1 + \sqrt{x}$. Shift the graph of $y = \sqrt{x}$
     upward 1 unit.

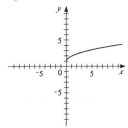

25. $y = \frac{1}{2}\sqrt{x+4} - 3$. Shift the graph of
     $y = \sqrt{x}$ to the left 4 units, shrink
     vertically by a factor of $\frac{1}{2}$, and then
     shift it downward 3 units.

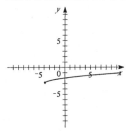

25. $y = 5 + (x+3)^2$. Shift the graph of
     $y = x^2$ to the left 3 units, then upward
     5 units.

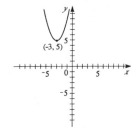

29. $y = |x| - 1$. Shift the graph of $y = |x|$
     downward 1 unit.

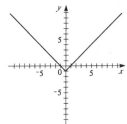

31. $y = |x+2| + 2$. Shift the graph of $y = |x|$
     to the left 2 units and upward 2 units.

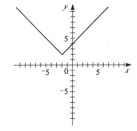

33.

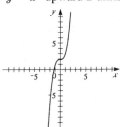

For part (b), shift the graph in (a) to the left 5
units; for part (c), shift the graph in (a) to the
left 5 units, and stretch it vertically by a factor
of 2; for part (d), shift the graph in (a) to the
left 5 units, stretch it vertically by a factor of
2, and then shift it upward 4 units.

35.

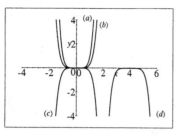

For part (b), shrink the graph in (a) vertically by a factor of $\frac{1}{3}$; for part (c), shrink the graph in (a) vertically by a factor of $\frac{1}{3}$, and reflect it about the x-axis; for part (d), shift the graph in (a) to the left 4 units, shrink vertically by a factor of $\frac{1}{3}$, and then reflect it about the x-axis.

37. $f(-x) = (-x)^{-2} = x^{-2} = f(x)$. Thus $f(x)$ is even.

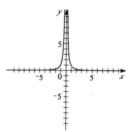

39. $f(-x) = (-x)^2 + (-x) = x^2 - x$. Thus $f(-x) \neq f(x)$. Also, $f(-x) \neq -f(x)$, so $f(x)$ is neither odd nor even.

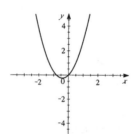

41. $f(-x) = (-x)^3 - (-x) = -x^3 + x$
     $= -(x^3 - x) = -f(x)$. Thus $f(x)$ is odd.

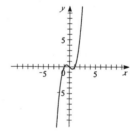

43. $f(-x) = 1 - \sqrt[3]{(-x)} = 1 + \sqrt[3]{x}$. Thus $f(-x) \neq f(x)$. Also $f(-x) \neq -f(x)$, so $f(x)$ is neither odd nor even.

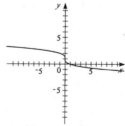

45. Since $f(x) = x^2 - 4 < 0$, for $-2 < x < 2$, the graph of $y = g(x)$ is found by sketching the graph of $y = f(x)$ for $x \leq -2$ and $x \geq 2$, then reflecting about the x-axis the part of the graph of $y = f(x)$ for $-2 < x < 2$.

47. (a) $f(x) = 4x - x^2$

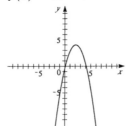

(b) $f(x) = |4x - x^2|$

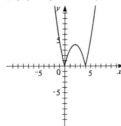

49. f even implies $f(-x) = f(x)$; g even implies $g(-x) = g(x)$; f odd implies $f(-x) = -f(x)$; and g odd implies $g(-x) = -g(x)$

If f and g are both even, then $(f + g)(-x) = f(-x) + g(-x) = f(x) + g(x) = (f + g)(x)$ and $f + g$ is even.

If f and g are both odd, then $(f + g)(-x) = f(-x) + g(-x) = -f(x) - g(x) = -(f + g)(x)$ and $f + g$ is odd.

If f odd and g even, then $(f + g)(-x) = f(-x) + g(-x) = -f(x) + g(x)$, which is neither odd nor even.

51. $f(x) = x^n$ is even when n is an even integer and $f(x) = x^n$ is odd when n is an odd integer.

These names were chosen because polynomials with only terms with odd powers are odd functions, and polynomials with only terms with even powers are even functions.

Exercises 4.6

1. $y = x^2 - 8x$

    Vertex: $y = x^2 - 8x = x^2 - 8x + 16 - 16 = (x-4)^2 - 16$. Vertex is at $(4, -16)$.

    x-intercepts: $y = 0 \Rightarrow 0 = x^2 - 8x = x(x-8)$. So $x = 0$ or $x = 8$. The x-intercepts are at $x = 0$ and $x = 8$.

    y-intercepts: $x = 0 \Rightarrow y = 0$. The y-intercept is at $y = 0$.

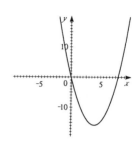

3. $y = 2x^2 - 6x$

    Vertex: $y = 2x^2 - 6x = 2(x^2 - 3x) = 2\left[x^2 - 3x + \left(\frac{3}{2}\right)^2\right] - \frac{9}{2}$

    $= 2\left(x - \frac{3}{2}\right)^2 - \frac{9}{2}$. Vertex is at $\left(\frac{3}{2}, -\frac{9}{2}\right)$.

    x-intercepts: $y = 0 \Rightarrow 0 = 2x^2 - 6x = 2x(x-3) \Rightarrow x = 0$ or $x = 3$. The x-intercepts are at $x = 0$ and $x = 3$.

    y-intercepts: $x = 0 \Rightarrow y = 0$. The y-intercept is at $y = 0$.

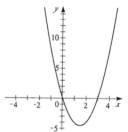

5. $y = x^2 + 4x + 1$

    Vertex: $y = x^2 + 4x + 1 = x^2 + 4x + 4 - 4 + 1 = (x+2)^2 - 3$. Vertex is at $(-2, -3)$.

    x-intercepts: $y = 0 \Rightarrow 0 = x^2 + 4x + 1$. Using the quadratic formula, $x = \frac{-4 \pm \sqrt{12}}{2} = \frac{-4 \pm 2\sqrt{3}}{2} = -2 \pm \sqrt{3}$. The x-intercepts are at $x = -2 - \sqrt{3}$ and $x = -2 + \sqrt{3}$.

    y-intercepts: $x = 0 \Rightarrow y = 1$. The y-intercept is at $y = 1$.

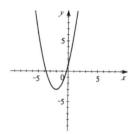

7. $y = x^2 + 6x + 8$

    Vertex: $y = x^2 + 6x + 8 = (x^2 + 6x) + 8 = (x^2 + 6x + 9) + 8 - 9 = (x+3)^2 - 1$. Vertex is at $(-3, -1)$.

    x-intercepts: $y = 0 \Rightarrow 0 = x^2 + 6x + 8 (x+2)(x+4) \Rightarrow x = -2$ or $x = -4$. The x-intercepts are at $x = -2$ and $x = -4$.

    y-intercepts: $x = 0 \Rightarrow y = 8$. The y-intercept is at $y = 8$.

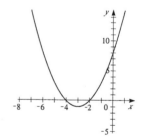

9. $y = 2x^2 + 4x + 3$

    Vertex: $y = 2x^2 + 4x + 3 = 2(x^2 + 2x) + 3 = 2(x^2 + 2x + 1) + 3 - 2 = 2(x+1)^2 + 1$. Vertex is at $(-1, 1)$.

    x-intercepts: $y = 0 \Rightarrow 0 = 2x^2 + 4x + 3 = 2(x+1)^2 + 1 \Leftrightarrow 2(x+1)^2 = -1$. Since this last equation has no real solution, there are no x-intercepts.

    y-intercepts: $x = 0 \Rightarrow y = 3$. The y-intercept is at $y = 3$.

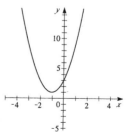

11. $y = 2x^2 - 20x + 57$

Vertex: $y = 2x^2 - 20x + 57 = 2(x^2 - 10x) + 57$
$= 2(x^2 - 10x + 25) + 57 - 50 = 2(x - 5)^2 + 7$. Vertex is at $(5, 7)$.

x-intercepts: $y = 0 \Rightarrow 0 = 2x^2 - 20x + 57 = 2(x-5)^2 + 7 \Leftrightarrow$
$2(x - 5)^2 = -7$. Since this last equation has no real solution, there are
no x-intercepts.

y-intercepts: $x = 0 \Rightarrow y = 57$. The y-intercept is at $y = 57$.

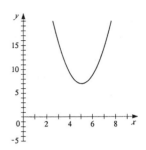

13. $y = -4x^2 - 16x + 3$

Vertex: $y = -4x^2 - 16x + 3 = -4(x^2 + 4x) + 3$
$= -4(x^2 + 4x + 4) + 3 + 16 = -4(x + 2)^2 + 19$. Vertex is at
$(-2, 19)$.

x-intercepts: $y = 0 \Rightarrow 0 = -4x^2 - 16x + 3 = -4(x + 2)^2 + 19$
$\Leftrightarrow 4(x + 2)^2 = 19 \Leftrightarrow (x + 2)^2 = \frac{19}{4} \Rightarrow$
$x + 2 = \pm\sqrt{\frac{19}{4}} = \pm\frac{\sqrt{19}}{2} \Leftrightarrow x = -2 \pm \frac{\sqrt{19}}{2}$. The x-intercepts are
at $x = -2 - \frac{\sqrt{19}}{2}$ and $x = -2 + \frac{\sqrt{19}}{2}$.

y-intercepts: $x = 0 \Rightarrow y = 3$. The y-intercept is at $y = 3$.

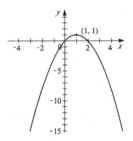

15. $f(x) = 2x - x^2 = -(x^2 - 2x) = -(x^2 - 2x + 1) + 1$
$= -(x - 1)^2 + 1$.

Therefore, the maximum value is $f(1) = 1$.

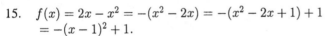

17. $f(x) = x^2 + 2x - 1 = (x^2 + 2x) - 1 = (x^2 + 2x + 1) - 1 - 1$
$= (x + 1)^2 - 2$.

Therefore, the minimum value is $f(-1) = -2$.

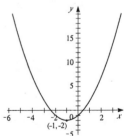

19. $f(x) = -x^2 - 3x + 3 = -(x^2 + 3x) + 3 = -\left(x^2 + 3x + \frac{9}{4}\right) + 3 + \frac{9}{4}$
$= -\left(x + \frac{3}{2}\right)^2 + \frac{21}{4}$.

Therefore, the maximum value is $f\left(-\frac{3}{2}\right) = \frac{21}{4}$.

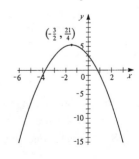

21. $g(x) = 3x^2 - 12x + 13 = 3(x^2 - 4x) + 13$
     $= 3(x^2 - 4x + 4) + 13 - 12 = 3(x - 2)^2 + 1.$

     Therefore, the minimum value is $g(2) = 1.$

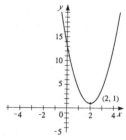

23. $h(x) = 1 - x - x^2 = -(x^2 + x) + 1 = -\left(x^2 + x + \frac{1}{4}\right) + 1 + \frac{1}{4}$
     $= -\left(x + \frac{1}{2}\right)^2 + \frac{5}{4}.$

     Therefore, the maximum value is $h\left(-\frac{1}{2}\right) = \frac{5}{4}.$

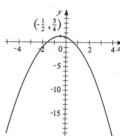

25. $f(x) = x^2 + x + 1 = (x^2 + x) + 1 = \left(x^2 + x + \frac{1}{4}\right) + 1 + \frac{1}{4} = \left(x + \frac{1}{2}\right)^2 + \frac{3}{4}.$
     Therefore, the minimum value is $f\left(-\frac{1}{2}\right) = \frac{3}{4}.$

27. $f(t) = 100 - 49t - 7t^2 = -7(t^2 + 7t) + 100 = -7\left(t^2 + 7t + \frac{49}{4}\right) + 100 + \frac{343}{4}$
     $= -7\left(t + \frac{7}{2}\right)^2 + \frac{743}{4}.$
     Therefore, the maximum value is $f\left(-\frac{7}{2}\right) = \frac{743}{4}.$

29. $f(s) = s^2 - 1.2s + 16 = (s^2 - 1.2s) + 16 = (s^2 - 1.2s + 0.36) + 16 - 0.36$
     $= (s - 0.6)^2 + 15.64.$

     Therefore, the minimum value is $f(0.6) = 15.64.$

31. $h(x) = \frac{1}{2}x^2 + 2x - 6 = \frac{1}{2}(x^2 + 4x) - 6 = \frac{1}{2}(x^2 + 4x + 4) - 6 - 2 = \frac{1}{2}(x + 2)^2 - 8.$

     Therefore, the minimum value is $h(-2) = -8.$

33. $f(x) = 3 - x - \frac{1}{2}x^2 = -\frac{1}{2}(x^2 + 2x) + 3 = -\frac{1}{2}(x^2 + 2x + 1) + 3 + \frac{1}{2} = -\frac{1}{2}(x + 1) + \frac{7}{2}.$
     Therefore, the maximum value is $f(-1) = \frac{7}{2}.$

35. Since the vertex is at $(1, -2)$, the function is of the form $f(x) = a(x - 1)^2 - 2$. Substituting the
     point $(4, 16)$, we get $16 = a(4 - 1)^2 - 2 \iff 16 = 9a - 2 \iff 9a = 18 \iff a = 2$. So
     the function is $f(x) = 2(x - 1)^2 - 2 = 2x^2 - 4x.$

37. $f(x) = -x^2 + 4x - 3 = -(x^2 - 4x) - 3 = -(x^2 - 4x + 4) - 3 + 4 = -(x - 2)^2 + 1$. So the
     domain of $f(x)$ is $(-\infty, \infty)$. Since $f(x)$ has a maximum value of 1, the range is $(-\infty, 1].$

39. $y = f(t) = 40t - 16t^2 = -16\left(t^2 - \frac{5}{2}\right) = -16\left[t^2 - \frac{5}{2}t + \left(\frac{5}{4}\right)^2\right] + 16\left(\frac{5}{4}\right)^2 = -16\left(t - \frac{5}{4}\right)^2 + 25.$
     Thus the maximum height attained by the ball is $f\left(\frac{5}{4}\right) = 25$ feet.

41. $R(x) = 80x - 0.4x^2 = -0.4(x^2 - 200x) = -0.4(x^2 - 200x + 10,000) + 4,000$
     $= -0.4(x - 100)^2 + 4,000.$
     So revenue is maximized at $\$4,000$ when 100 units are sold.

43. $E(n) = \frac{2}{3}n - \frac{1}{90}n^2 = -\frac{1}{90}(n^2 - 60n) = -\frac{1}{90}(n^2 - 60n + 900) + 10 = -\frac{1}{90}(n - 30)^2 + 10$. Since the maximum of the function occurs when $n = 30$, the viewer should watch the commercial 30 times for maximum effectiveness.

45. (a) $f(x) = x^2 + 1.79x - 3.21$ is shown in the viewing rectangle on the right. The minimum value is $f(x) \approx -4.01$.

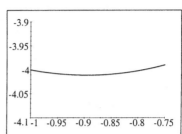

   (b) $f(x) = x^2 + 1.79x - 3.21$
   $$= \left[x^2 + 1.79x + \left(\frac{1.79}{2}\right)^2\right] - 3.21 - \left(\frac{1.79}{2}\right)^2$$
   $$= (x + 0.895)^2 - 4.011025$$

   Therefore, the exact minimum of $f(x)$ is -4.011025.

47. In the viewing rectangle on the right, we see that $f(x) = x^3 - x$ has a local minimum and a local maximum. Smaller x and y ranges (shown in the viewing rectangles below) show that $f(x)$ has a local maximum of ≈ 0.38 when $x \approx -0.58$ and a local minimum of ≈ -0.38 when $x \approx 0.58$.

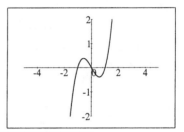

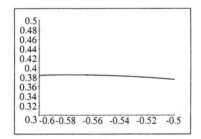

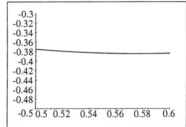

49. In the viewing rectangle on the right, we see that $g(x) = x^4 - 2x^3 - 11x^2$ has two local minimums and a local maximum. The local maximum is $g(x) = 0$ when $x = 0$. Smaller x and y ranges (shown in the viewing rectangles below) show that local minima are $g(x) \approx -13.61$ when $x \approx -1.71$ and $g(x) \approx -73.32$ when $x \approx 3.21$.

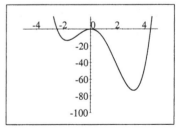

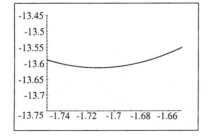

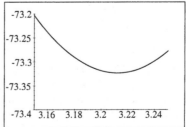

51. In the first viewing rectangle below, we see that $U(x) = x\sqrt{6-x}$ only has a local maximum. Smaller x and y ranges in the second viewing rectangle below show that $U(x)$ has a local maximum of ≈ 5.66 when $x \approx 4.00$.

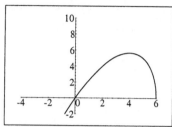 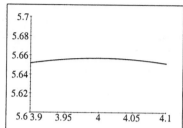

53. In the viewing rectangle on the right, we see that $V(x) = \dfrac{1 - x^2}{x^3}$ has a local minimum and a local maximum. Smaller x and y ranges (shown in the viewing rectangles below) show that $V(x)$ has a local maximum of ≈ 0.38 when $x \approx -1.73$ and a local minimum of ≈ -0.38 when $x \approx 1.73$.

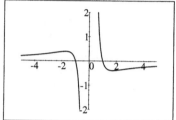

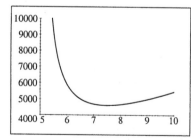 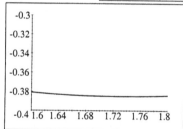

55. In the first viewing rectangle below, we see the general location of the minimum of $E(v) = 2.73v^3 \dfrac{10}{v - 5}$. In the second viewing rectangle, we isolate the minimum, and from this graph, we see that energy is minimized when $v \approx 7.5$ mi/h.

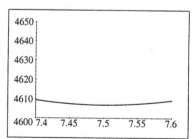

57. In the first viewing rectangle on the next page, we see the general location of the minimum of $V = 999.87 - 0.06426T + 0.0085043T^2 - 0.0000679T^3$ is around $T = 4$. In the second viewing rectangle, we isolate the minimum, and from this graph, we see that the minimum volume of 1 kg of water occurs at $T \approx 3.96°$ C.

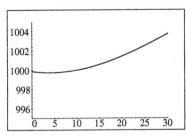   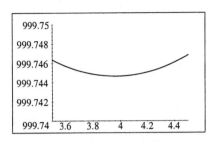

59. $f(x) = 3 + 4x^2 - x^4$. Substituting $t = x^2$, we have $f(\sqrt{t}) = 3 + 4t - t^2 = -(t^2 - 4t) + 3$
     $= -(t^2 - 4t + 4) + 3 + 4 = -(t - 2)^2 + 7$. Therefore, the function $f(\sqrt{t}) = f(x)$ has a
     maximum value of 7.

61. (a) If $x = a$ is a local maximum of $f(x)$ then $f(a) \geq f(x) \geq 0$ for all x around $x = a$. So
          $[g(a)]^2 \geq [g(x)]^2$ and thus $g(a) \geq g(x)$. Similarly, if $x = b$ is a local minimum of $f(x$, then
          $f(x) \geq f(b) \geq 0$, for all x around $x = b$. So $[g(x)]^2 \geq [g(b)]^2$ and thus $g(x) \geq g(b)$.

     (b) Using the distance formula, $g(x) = \sqrt{(x - 3)^2 + (x^2 - 0)^2} = \sqrt{x^4 + x^2 - 6x + 9}$.

     (c) Let $f(x) = x^4 + x^2 - 6x + 9$. In the viewing rectangle we see that $f(x)$ has a minimum at
          $x = 1$. Thus $g(x)$ also has a minimum at $x = 1$ and this minimum value is
          $g(1) = \sqrt{1^4 + 1^2 - 6(1) + 9} = \sqrt{5}$.

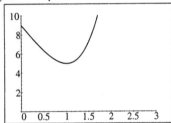

Exercises 4.7

1. $f(x) = x^2 - x$ has domain $(-\infty, \infty)$. $g(x) = x + 5$ has domain $(-\infty, \infty)$. The intersection of the domains of f and g is $(-\infty, \infty)$.

$(f + g)(x) = (x^2 - x) + (x + 5) = x^2 + 5$, and the domain is $(-\infty, \infty)$.

$(f - g)(x) = (x^2 - x) - (x + 5) = x^2 - 2x - 5$, and the domain is $(-\infty, \infty)$.

$(fg)(x) = (x^2 - x)(x + 5) = x^3 + 4x^2 - 5x$, and the domain is $(-\infty, \infty)$.

$\left(\dfrac{f}{g}\right)(x) = \dfrac{x^2 - x}{x + 5}$, and the domain is $\{\, x \mid x \neq -5 \,\}$.

3. $f(x) = \sqrt{1 + x}$, has domain $[-1, \infty)$. $g(x) = \sqrt{1 - x}$, has domain $(-\infty, 1]$. The intersection of the domains of f and g is $[-1, 1]$.

$(f + g)(x) = \sqrt{1 + x} + \sqrt{1 - x}$, and the domain is $[-1, 1]$.

$(f - g)(x) = \sqrt{1 + x} - \sqrt{1 - x}$, and the domain is $[-1, 1]$.

$(fg)(x) = \sqrt{1 + x} \cdot \sqrt{1 - x} = \sqrt{1 - x^2}$, and the domain is $[-1, 1]$.

$\left(\dfrac{f}{g}\right)(x) = \dfrac{\sqrt{1 + x}}{\sqrt{1 - x}} = \sqrt{\dfrac{1 + x}{1 - x}}$, and the domain is $[-1, 1)$.

5. $f(x) = \dfrac{2}{x}$, has domain $x \neq 0$. $g(x) = -\dfrac{2}{x + 4}$, has domain $x \neq -4$. The intersection of the domains of f and g is $x \neq 0, -4$, in interval notation, this is $(-\infty, -4) \cup (-4, 0) \cup (0, \infty)$.

$(f + g)(x) = \dfrac{2}{x} + \left(-\dfrac{2}{x + 4}\right) = \dfrac{2}{x} - \dfrac{2}{x + 4} = \dfrac{8}{x(x + 4)}$, and the domain is $(-\infty, -4) \cup (-4, 0) \cup (0, \infty)$.

$(f - g)(x) = \dfrac{2}{x} - \left(-\dfrac{2}{x + 4}\right) = \dfrac{2}{x} + \dfrac{2}{x + 4} = \dfrac{4x + 8}{x(x + 4)}$, and the domain is $(-\infty, -4) \cup (-4, 0) \cup (0, \infty)$.

$(fg)(x) = \dfrac{2}{x} \cdot \left(-\dfrac{2}{x + 4}\right) = -\dfrac{4}{x(x + 4)}$, and the domain is $(-\infty, -4) \cup (-4, 0) \cup (0, \infty)$.

$\left(\dfrac{f}{g}\right)(x) = \dfrac{\dfrac{2}{x}}{-\dfrac{2}{x + 4}} = -\dfrac{x + 4}{x}$, and the domain is $(-\infty, -4) \cup (-4, 0) \cup (0, \infty)$.

7. $f(x) = \sqrt{x} + \sqrt{1 - x}$. The domain of $\sqrt{x}$ is $[0, \infty]$, and the domain of $\sqrt{1 - x}$ is $(-\infty, 1]$. Thus the domain is $(-\infty, 1] \cap [0, \infty) = [0, 1]$.

9. $h(x) = (x + 1)^2(2x - 8)^{1/4}$. Since $1/4$ is an even root, $2x - 8 \geq 0 \iff x \geq 4$. So the domain is $[4, \infty)$.

11.

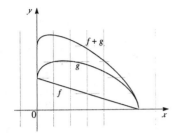

13.

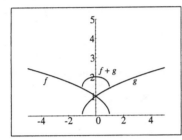

15.

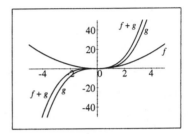

17. (a) $f(g(0)) = f(2 - (0)^2) = f(2) = 3(2) - 5 = 1$

(b) $g(f(0)) = g(3(0) - 5) = g(-5) = 2 - (-5)^2 = -23$

19. (a) $(f \circ g)(-2) = f(g(-2)) = f(2 - (-2)^2) = f(-2) = 3(-2) - 5 = -11$

(b) $(g \circ f)(-2) = g(f(-2)) = g(3(-2) - 5) = g(-11) = 2 - (-11)^2 = -119$

21. (a) $(f \circ g)(x) = f(g(x)) = f(2 - x^2) = 3(2 - x^2) - 5 = 6 - 3x^2 - 5 = 1 - 3x^2$

(b) $(g \circ f)(x) = g(f(x)) = g(3x - 5) = 2 - (3x - 5)^2 = 2 - (9x^2 - 30x + 25)$
$= -9x^2 + 30x - 23$

23. $f(g(2)) = f(5) = 4$

25. $(g \circ f)(4) = g(f(4)) = g(2) = 5$

27. $(g \circ g)(-2) = g(g(-2)) = g(1) = 4$

29. $f(x) = 2x + 3$, has domain $(-\infty, \infty)$; $g(x) = 4x - 1$, has domain $(-\infty, \infty)$.
$(f \circ g)(x) = f(4x - 1) = 2(4x - 1) + 3 = 8x + 1$, and the domain is $(-\infty, \infty)$.
$(g \circ f)(x) = g(2x + 3) = 4(2x + 3) - 1 = 8x + 11$, and the domain is $(-\infty, \infty)$.
$(f \circ f)(x) = f(2x + 3) = 2(2x + 3) + 3 = 4x + 9$, and the domain is $(-\infty, \infty)$.
$(g \circ g)(x) = g(4x - 1) = 4(4x - 1) - 1 = 16x - 5$, and the domain is $(-\infty, \infty)$.

31. $f(x) = x^2$, has domain $(-\infty, \infty)$; $g(x) = x + 1$, has domain $(-\infty, \infty)$.
$(f \circ g)(x) = f(x + 1) = (x + 1)^2 = x^2 + 2x + 1$, and the domain is $(-\infty, \infty)$.
$(g \circ f)(x) = g(x^2) = (x^2) + 1 = x^2 + 1$, and the domain is $(-\infty, \infty)$.
$(f \circ f)(x) = f(x^2) = (x^2)^2 = x^4$, and the domain is $(-\infty, \infty)$.
$(g \circ g)(x) = g(x + 1) = (x + 1) + 1 = x + 2$, and the domain is $(-\infty, \infty)$.

33. $f(x) = \dfrac{1}{x}$, has domain $\{x \mid x \neq 0\}$; $g(x) = 2x + 4$, has domain $(-\infty, \infty)$.

$(f \circ g)(x) = f(2x + 4) = \dfrac{1}{2x + 4}$. $(f \circ g)(x)$ is defined for $2x + 4 \neq 0 \quad \Leftrightarrow \quad x \neq -2$. So the domain is $\{ x \mid x \neq -2 \} = (-\infty, -2) \cup (-2, \infty)$.

$(g \circ f)(x) = g\left(\dfrac{1}{x}\right) = 2\left(\dfrac{1}{x}\right) + 4 = \dfrac{2}{x} + 4$, the domain is $\{x \mid x \neq 0\} = (-\infty, 0) \cup (0, \infty)$.

$(f \circ f)(x) = f\left(\dfrac{1}{x}\right) = \dfrac{1}{\left(\dfrac{1}{x}\right)} = x$. $(f \circ f)(x)$ is defined whenever both $f(x)$ and $f(f(x))$ are

defined; that is, whenever $\{x \mid x \neq 0\} = (-\infty, 0) \cup (0, \infty)$.

$(g \circ g)(x) = g(2x + 4) = 2(2x + 4) + 4 = 4x + 8 + 4 = 4x + 12$, and the domain is $(-\infty, \infty)$.

35. $f(x) = |x|$, has domain $(-\infty, \infty)$; $g(x) = 2x + 3$, has domain $(-\infty, \infty)$

$(f \circ g)(x) = f(2x + 4) = |2x + 3|$, and the domain is $(-\infty, \infty)$.

$(g \circ f)(x) = g(|x|) = 2|x| + 3$, and the domain is $(-\infty, \infty)$.

$(f \circ f)(x) = f(|x|) = ||x|| = |x|$, and the domain is $(-\infty, \infty)$.

$(g \circ g)(x) = g(2x + 3) = 2(2x + 3) + 3 = 4x + 6 + 3 = 4x + 9$. Domain is $(-\infty, \infty)$.

37. $f(x) = \dfrac{x}{x + 1}$, has domain $\{x \mid x \neq -1\}$; $g(x) = 2x - 1$, has domain $(-\infty, \infty)$

$(f \circ g)(x) = f(2x - 1) = \dfrac{2x - 1}{(2x - 1) + 1} = \dfrac{2x - 1}{2x}$, and the domain is

$\{x \mid x \neq 0\} = (-\infty, 0) \cup (0, \infty)$.

$(g \circ f)(x) = g\left(\dfrac{x}{x + 1}\right) = 2\left(\dfrac{x}{x + 1}\right) - 1 = \dfrac{2x}{x + 1} - 1$, and the domain is

$\{x \mid x \neq -1\} = (-\infty, -1) \cup (-1, \infty)$

$(f \circ f)(x) = f\left(\dfrac{x}{x + 1}\right) = \dfrac{\dfrac{x}{x + 1}}{\dfrac{x}{x + 1} + 1} \cdot \dfrac{x + 1}{x + 1} = \dfrac{x}{x + x + 1} = \dfrac{x}{2x + 1}$. $(f \circ f)(x)$ is defined

whenever both $f(x)$ and $f(f(x))$ are defined; that is, whenever $\{x \mid x \neq -1\}$ and $2x + 1 \neq 0 \quad \Rightarrow$ $\{x \mid x \neq -\frac{1}{2}\}$ which is $(-\infty, -1) \cup (-1, -\frac{1}{2}) \cup (-\frac{1}{2}, \infty)$.

$(g \circ g)(x) = g(2x - 1) = 2(2x - 1) - 1 = 4x - 2 - 1 = 4x - 3$, and the domain is $(-\infty, \infty)$.

39. $f(x) = \sqrt[3]{x}$, has domain $(-\infty, \infty)$; $g(x) = \sqrt[4]{x}$, has domain $[0, \infty)$.

$(f \circ g)(x) = f(\sqrt[4]{x}) = \sqrt[3]{\sqrt[4]{x}} = \sqrt[12]{x}$. $(f \circ g)(x)$ is defined whenever both $g(x)$ and $f(g(x))$ are defined. Since $f(x)$ has no restriction, the domain is $[0, \infty)$.

$(g \circ f)(x) = g(\sqrt[3]{x}) = \sqrt[4]{\sqrt[3]{x}} = \sqrt[12]{x}$. $(g \circ f)(x)$ is defined whenever both $f(x)$ and $g(f(x))$ are defined; that is, whenever $x \geq 0$. So the domain is $[0, \infty)$.

$(f \circ f)(x) = f(\sqrt[3]{x}) = \sqrt[3]{\sqrt[3]{x}} = \sqrt[9]{x}$. $(f \circ f)(x)$ is defined whenever both $f(x)$ and $f(f(x))$ are defined. Since $f(x)$ is defined everywhere, the domain is $(-\infty, \infty)$.

$(g \circ g)(x) = g(\sqrt[4]{x}) = \sqrt[4]{\sqrt[4]{x}} = \sqrt[16]{x}$. $(g \circ g)(x)$ is defined whenever both $g(x)$ and $g(g(x))$ are defined; that is, whenever $x \geq 0$. So the domain is $[0, \infty)$.

41. $(f \circ g \circ h)(x) = f(g(h(x))) = f(g(x-1)) = f(\sqrt{x-1}) = \sqrt{x-1} - 1$

43. $(f \circ g \circ h)(x) = f(g(h(x))) = f\left(g(\sqrt{x})\right) = f(\sqrt{x} - 5) = (\sqrt{x} - 5)^4 + 1$

45. $F(x) = (x-9)^5$. Let $f(x) = x^5$ and $g(x) = x - 9$, then $F(x) = (f \circ g)(x)$.

47. $G(x) = \dfrac{x^2}{x^2+4}$. Let $f(x) = \dfrac{x}{x+4}$ and $g(x) = x^2$, then $G(x) = (f \circ g)(x)$.

49. $H(x) = |1 - x^3|$. Let $f(x) = |x|$ and $g(x) = 1 - x^3$, then $H(x) = (f \circ g)(x)$.

For Exercises 51 and 53 there are several possible solutions, only one of which is shown.

51. $F(x) = \dfrac{1}{x^2+1}$. Let $f(x) = \dfrac{1}{x}$, $g(x) = x + 1$, and $h(x) = x^2$, then $F(x) = (f \circ g \circ h)(x)$.

53. $G(x) = (4 + \sqrt[3]{x})^9$. Let $f(x) = x^9$, $g(x) = 4 + x$, and $h(x) = \sqrt[3]{x}$, then $G(x) = (f \circ g \circ h)(x)$.

55. Let r be the radius of the circular ripple in centimeters. Since the ripple travels at a speed of 60 cm/s, the distance traveled in t seconds is the radius, so $r = 60t$. Therefore, the area of the circle can be written as $A(t) = \pi r^2 = \pi(60t)^2 = 3600\pi t^2$ cm^2.

57. Let t be the time since the plane flew over the radar station.

    (a) Let s be the distance in miles between the plane and the radar station, and let d be the horizontal distance that the plane has flown. Using the Pythagorean theorem,
$$s = f(d) = \sqrt{1 + d^2}.$$

    (b) Since *distance* = *rate* × *time* we have $d = g(t) = 350t$.

    (c) $s(t) = (f \circ g)(t) = f(350t) = \sqrt{1 + (350t)^2} = \sqrt{1 + 122{,}500t^2}$.

59. Yes. If $f(x) = m_1 x + b_1$ and $g(x) = m_2 x + b_2$, then
$(f \circ g)(x) = f(m_2 x + b_2) = m_1(m_2 x + b_2) + b_1 = m_1 m_2 x + m_1 b_2 + b_1$, which is a linear function, because it is of the form $y = mx + b$. The slope is $m_1 m_2$.

61. If $g(x)$ is even, then $h(-x) = f(g(-x)) = f(g(x)) = h(x)$. So yes, h is always an even function.

If $g(x)$ is odd, then h is not necessarily an odd function. For example, if we let $f(x) = x - 1$ and $g(x) = x^3$, g is an odd function, but $h(x) = (f \circ g)(x) = f(x^3) = x^3 - 1$ is not an odd function.

If $g(x)$ is odd and f is also odd, then $h(-x) = (f \circ g)(-x) = f(g(-x)) = f(-g(x)) = -f(g(x))$ $= -(f \circ g)(x) = -h(x)$. So in this case, h is also an odd function.

If $g(x)$ is odd and f is even, then $h(-x) = (f \circ g)(-x) = f(g(-x)) = f(-g(x)) = f(g(x))$ $= (f \circ g)(x) = h(x)$, so in this case, h is an even function.

Exercises 4.8

1. By the Horizontal Line Test, f is not one-to-one.

3. By the Horizontal Line Test, f is one-to-one.

5. By the Horizontal Line Test, f is not one-to-one.

7. $f(x) = 7x - 3$. If $x_1 \neq x_2$, then $7x_1 \neq 7x_2$ and $7x_1 - 3 \neq 7x_2 - 3$. So f is a one-to-one function.

9. $g(x) = \sqrt{x}$. If $x_1 \neq x_2$, then $\sqrt{x_1} \neq \sqrt{x_2}$ because two different numbers cannot have the same square root. Therefore, g is a one-to-one function.

11. $h(x) = x^3 + 1$. If $x_1 \neq x_2$, then $x_1^3 \neq x_2^3$ and $x_1^3 + 1 \neq x_2^3 + 1$. So f is a one-to-one function.

13. $f(x) = x^4 + 5$. Every nonzero number and its negative have the same fourth power. For example, $(-1)^4 = 1 = (1)^4$, so $f(-1) = f(1)$. Thus f is not a one-to-one function.

15. $f(x) = \dfrac{1}{x^2}$. Every nonzero number and its negative have the same square. For example, $\frac{1}{(-1)^2} = 1 = \frac{1}{(1)^2}$, so $f(-1) = f(1)$. Thus f is not a one-to-one function.

17. (a) $f(2) = 7$. Since f is one-to-one, $f^{-1}(7) = 2$.
    (b) $f^{-1}(3) = -1$. Since f is one-to-one, $f(-1) = 3$.

19. $f(x) = 5 - 2x$. Since f is one-to-one and $f(1) = 5 - 2(1) = 3$, then $f^{-1}(3) = 1$. (Find 1 by solving the equation $5 - 2x = 3$.)

21. $f(g(x)) = f(x - 3) = (x - 3) + 3 = x$, for all x.
    $g(f(x)) = g(x + 3) = (x + 3) - 3 = x$, for all x. Thus f and g are inverses of each other.

23. $f(g(x)) = f\left(\dfrac{x+5}{2}\right) = 2\left(\dfrac{x+5}{2}\right) - 5 = x + 5 - 5 = x$, for all x.

    $g(f(x)) = g(2x - 5) = \dfrac{(2x - 5) + 5}{2} = x$, for all x. Thus f and g are inverses of each other.

25. $f(g(x)) = f\left(\dfrac{1}{x}\right) = \dfrac{1}{1/x} = x$, for all x. Since $f(x) = g(x)$, we also have $g(f(x)) = x$. Thus f and g are inverses of each other.

27. $f(g(x)) = f\left(\sqrt{x+4}\right) = \left(\sqrt{x+4}\right)^2 - 4 = x + 4 - 4 = x$, for all $x \geq -4$.
    $g(f(x)) = g(x^2 - 4) = \sqrt{(x^2 - 4) + 4} = \sqrt{x^2} = x$, for all $x \geq 0$. Thus f and g are inverses of each other.

29. $f(g(x)) = f\left(\dfrac{1}{x} + 1\right) = \dfrac{1}{\left(\dfrac{1}{x} + 1\right) - 1} = x$, for all $x \neq 0$.

    $g(f(x)) = g\left(\dfrac{1}{x-1}\right) = \dfrac{1}{\left(\dfrac{1}{x-1}\right)} + 1 = (x - 1) + 1 = x$, for all $x \neq 1$. Thus f and g are inverses of each other.

31. $f(x) = 2x + 1.$ $y = 2x + 1$ $\Leftrightarrow$ $2x = y - 1$ $\Leftrightarrow$ $x = \frac{1}{2}(y - 1).$ So $f^{-1}(x) = \frac{1}{2}(x - 1).$

33. $f(x) = 4x + 7.$ $y = 4x + 7$ $\Leftrightarrow$ $4x = y - 7$ $\Leftrightarrow$ $x = \frac{1}{4}(y - 7).$ So $f^{-1}(x) = \frac{1}{4}(x - 7).$

35. $f(x) = \dfrac{x}{2}.$ $y = \dfrac{x}{2}$ $\Leftrightarrow$ $x = 2y.$ So $f^{-1}(x) = 2x.$

37. $f(x) = \dfrac{1}{x + 2}.$ $y = \dfrac{1}{x + 2}$ $\Leftrightarrow$ $x + 2 = \dfrac{1}{y}$ $\Leftrightarrow$ $x = \dfrac{1}{y} - 2.$ So $f^{-1}(x) = \dfrac{1}{x} - 2.$

39. $f(x) = \dfrac{1 + 3x}{5 - 2x}.$ $y = \dfrac{1 + 3x}{5 - 2x}$ $\Leftrightarrow$ $y(5 - 2x) = 1 + 3x$ $\Leftrightarrow$ $5y - 2xy = 1 + 3x$ $\Leftrightarrow$

$3x + 2xy = 5y - 1$ $\Leftrightarrow$ $x(3 + 2y) = 5y - 1$ $\Leftrightarrow$ $x = \dfrac{5y - 1}{2y + 3}.$ So $f^{-1}(x) = \dfrac{5x - 1}{2x + 3}.$

41. $f(x) = \sqrt{2 + 5x}, x \geq -\frac{2}{5}.$ $y = \sqrt{2 + 5x}, y \geq 0$ $\Leftrightarrow$ $y^2 = 2 + 5x$ $\Leftrightarrow$ $5x = y^2 - 2$ $\Leftrightarrow$
$x = \frac{1}{5}(y^2 - 2)$ and $y \geq 0.$ So $f^{-1}(x) = \frac{1}{5}(x^2 - 2), x \geq 0.$

43. $f(x) = 4 - x^2, x \geq 0.$ $y = 4 - x^2$ $\Leftrightarrow$ $x^2 = 4 - y$ $\Leftrightarrow$ $x = \sqrt{4 - y}.$ So
$f^{-1}(x) = \sqrt{4 - x}.$ Note: $x \geq 0$ $\Rightarrow$ $f(x) \leq 4.$

45. $f(x) = 4 + \sqrt[3]{x}.$ $y = 4 + \sqrt[3]{x}$ $\Leftrightarrow$ $\sqrt[3]{x} = y - 4$ $\Leftrightarrow$ $x = (y - 4)^3.$ So $f^{-1}(x) = (x - 4)^3.$

47. $f(x) = 1 + \sqrt{1 + x}.$ $y = 1 + \sqrt{1 + x}, y \geq 1$ $\Leftrightarrow$ $\sqrt{1 + x} = y - 1$ $\Leftrightarrow$ $1 + x = (y - 1)^2$
$\Leftrightarrow$ $x = (y - 1)^2 - 1 = y^2 - 2y.$ So $f^{-1}(x) = x^2 - 2x, x \geq 1.$

49. $f(x) = x^4, x \geq 0.$ $y = x^4, y \geq 0$ $\Leftrightarrow$ $x = \sqrt[4]{y}.$ So $f^{-1}(x) = \sqrt[4]{x}, x \geq 0.$

51. (a)   (b)

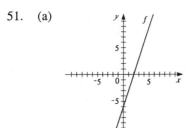   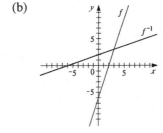

(c) $f(x) = 3x - 6.$ $y = 3x - 6$ $\Leftrightarrow$ $3x = y + 6$ $\Leftrightarrow$ $x = \frac{1}{3}(y + 6).$ So
$f^{-1}(x) = \frac{1}{3}(x + 6).$

53. (a)   (b)

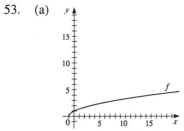   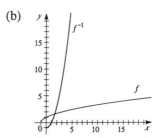

(c) $f(x) = \sqrt{x + 1}, x \geq -1.$ $y = \sqrt{x + 1}, y \geq 0$ $\Leftrightarrow$ $y^2 = x + 1$ $\Leftrightarrow$ $x = y^2 - 1$ and
$y \geq 0.$ So $f^{-1}(x) = x^2 - 1, x \geq 0.$

55. $f(x) = x^3 - x$. Using a graphing device and
     the Horizontal Line Test, we see that f is not a
     one-to-one function. For example,
     $f(0) = 0 = f(-1)$.

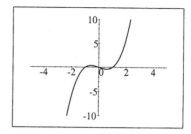

57. $f(x) = \dfrac{x + 12}{x - 6}$. Using a graphing device and
     the Horizontal Line Test, we see that f is a
     one-to-one function.

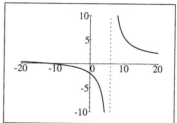

59. $f(x) = |x| - |x - 6|$. Using a graphing
     device and the Horizontal Line Test, we see
     that f is not a one-to-one function. For
     example $f(0) = -6 = f(-2)$.

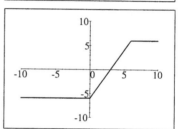

61. If we restrict the domain of $f(x)$ to $[0, \infty)$, then $y = 4 - x^2 \quad \Leftrightarrow \quad x^2 = 4 - y \quad \Rightarrow$
     $x = \sqrt{4 - y}$ (since $x \geq 0$, we take the positive square root). So $f^{-1}(x) = \sqrt{4 - x}$.
     If we restrict the domain of $f(x)$ to $(-\infty, 0]$, then $y = 4 - x^2 \quad \Leftrightarrow \quad x^2 = 4 - y \quad \Rightarrow$
     $x = -\sqrt{4 - y}$ (since $x \leq 0$, we take the negative square root). So $f^{-1}(x) = -\sqrt{4 - x}$.

63. If we restrict the domain of $h(x)$ to $[-2, \infty)$, then $y = (x + 2)^2 \quad \Rightarrow \quad x + 2 = \sqrt{y}$ (since
     $x \geq -2$, we take the positive square root) $\quad \Leftrightarrow \quad x = -2 + \sqrt{y}$. So $h^{-1}(x) = -2 + \sqrt{x}$.
     If we restrict the domain of $h(x)$ to $(-\infty, -2]$, then $y = (x + 2)^2 \quad \Rightarrow \quad x + 2 = -\sqrt{y}$ (since
     $x \leq -2$, we take the negative square root) $\quad \Leftrightarrow \quad x = -2 - \sqrt{y}$. So $h^{-1}(x) = -2 - \sqrt{x}$.

65.

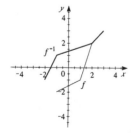

67. $f(x) = 7 + 2x$. $y = 7 + 2x \quad \Leftrightarrow \quad 2x = y - 7 \quad \Leftrightarrow \quad x = \dfrac{y - 7}{2}$. So $f^{-1}(x) = \dfrac{x - 7}{2}$. f^{-1} is
     the number of toppings on a pizza that costs $f(x)$ dollars.

69. (a) $f(x) = \dfrac{2x + 1}{5}$ is "multiply by 2, add 1, and then divide by 5". So the reverse is "multiply by

5, subtract 1 , and then divide by 2" or $f^{-1}(x) = \dfrac{5x - 1}{2}$. Check:

$$f \circ f^{-1}(x) = f\left(\frac{5x - 1}{2}\right)$$

$$= \frac{2\left(\dfrac{5x - 1}{2}\right) + 1}{5}$$

$$= \frac{5x - 1 + 1}{5} = \frac{5x}{5} = x$$

$$f^{-1} \circ f(x) = f\left(\frac{2x + 1}{5}\right)$$

$$= \frac{5\left(\dfrac{2x + 1}{5}\right) - 1}{2}$$

$$= \frac{2x + 1 - 1}{2} = \frac{2x}{2} = x$$

(b) $f(x) = 3 - \dfrac{1}{x} = \dfrac{-1}{x} + 3$ is "take the negative reciprocal and add 3". Since the reverse of
"take the negative reciprocal" is "take the negative reciprocal ", $f^{-1}(x)$ is "subtract 3 and take
the negative reciprocal ", that is, $f^{-1}(x) = \dfrac{-1}{x - 3}$. Check:

$$f \circ f^{-1}(x) = f\left(\frac{-1}{x - 3}\right)$$

$$= 3 - \frac{1}{\dfrac{-1}{x - 3}}$$

$$= 3 - \left(1 \cdot \frac{x - 3}{-1}\right)$$

$$= 3 + x - 3 = x$$

$$f^{-1} \circ f(x) = f\left(3 - \frac{1}{x}\right)$$

$$= \frac{-1}{\left(3 - \dfrac{1}{x}\right) - 3}$$

$$= \frac{-1}{-\dfrac{1}{x}} = -1 \cdot \frac{x}{-1} = x$$

(c) $f(x) = \sqrt{x^3 + 2}$ is "cube, add 2, and then take the square root 3". So the reverse is "square,
subtract 2, then take cube root " or $f^{-1}(x) = \sqrt[3]{x^2 - 2}$. Domain for $f(x)$ is $[-\sqrt[3]{2}, \infty)$;
domain for $f^{-1}(x)$ is $[0, \infty)$. Check:

$$f \circ f^{-1}(x) = f\left(\sqrt[3]{x^2 - 2}\right)$$

$$= \sqrt{\left(\sqrt[3]{x^2 - 2}\right)^3 + 2}$$

$$= \sqrt{x^2 - 2 + 2}$$

$$= \sqrt{x^2} = x \text{ (for the domains)}$$

$$f^{-1} \circ f(x) = f\left(\sqrt{x^3 + 2}\right)$$

$$= \sqrt[3]{\left(\sqrt{x^3 + 2}\right)^2 - 2}$$

$$= \sqrt[3]{x^3 + 2 - 2}$$

$$= \sqrt[3]{x^3} = x \text{ (for the domains)}$$

No; in a function like $f(x) = 3x - 2$, the variable occurs only once and it easy to see how to reverse

these operations step by step. But in $f(x) = \dfrac{3x - 2}{x + 7}$, you apply two different operations to the

variable x and then find the quotient of each of these values, so it is not possible to reverse the
operations step by step.

71. (a) We find $g^{-1}(x)$: $y = 2x + 1 \quad \Leftrightarrow \quad 2x = y - 1 \quad \Leftrightarrow \quad x = \frac{1}{2}(y - 1)$. So
$g^{-1}(x) = \frac{1}{2}(x - 1)$. Thus
$$f(x) = h \circ g^{-1}(x) = h\left(\tfrac{1}{2}(x - 1)\right) = 4\left[\tfrac{1}{2}(x - 1)\right]^2 + 4\left[\tfrac{1}{2}(x - 1)\right] + 7$$
$$= x^2 - 2x + 1 + 2x - 2 + 7 = x^2 + 6.$$

(b) $f \circ g = h$ $\Leftrightarrow$ $f^{-1} \circ f \circ g = f^{-1} \circ h$ $\Leftrightarrow$ $I \circ g = f^{-1} \circ h$ $\Leftrightarrow$ $g = f^{-1} \circ h$. Note
that we compose with f^{-1} on the left on each side of the equation. We find f^{-1}: $y = 3x + 5$
$\Leftrightarrow$ $3x = y - 5$ $\Leftrightarrow$ $x = \frac{1}{3}(y - 5)$. So $f^{-1}(x) = \frac{1}{3}(x - 5)$. Thus
$g(x) = f^{-1} \circ h(x) = f^{-1}(3x^2 + 3x + 2) = \frac{1}{3}[(3x^2 + 3x + 2) - 5]$
$= \frac{1}{3}[3x^2 + 3x - 3] = x^2 + x - 1$.

Review Exercises for Chapter 4

1. $f(x) = x^2 - x + 1$; $f(0) = (0)^2 - (0) + 1 = 1$; $f(2) = (2)^2 - (2) + 1 = 3$;
   $f(-2) = (-2)^2 - (-2) + 1 = 7$; $f(a) = (a)^2 - (a) + 1 = a^2 - a + 1$;
   $f(-a) = (-a)^2 - (-a) + 1 = a^2 + a + 1$;
   $f(x+1) = (x+1)^2 - (x+1) + 1 = x^2 + 2x + 1 - x - 1 + 1 = x^2 + x + 1$;
   $f(2x) = (2x)^2 - (2x) + 1 = 4x^2 - 2x + 1$;
   $2f(x) - 2 = 2(x^2 - x + 1) - 2 = 2x^2 - 2x + 2 - 2 = 2x^2 - 2x$.

3. (a) $f(-2) = -1$. $f(2) = 2$.

   (b) The domain of f is $[-4, 5]$.

   (c) The range of f is $[-4, 4]$.

   (d) f is increasing on $[-4, -2]$ and $[-1, 4]$; f is decreasing on $[-2, -1]$ and $[4, 5]$.

   (e) f is not a one-to-one, for example, $f(-2) = -1 = f(0)$. There are many more examples.

5. Domain: We must have $x + 3 \geq 0$ $\Leftrightarrow$ $x \geq -3$. In interval notation, the domain is $[-3, \infty)$.
   Range: For x in the domain of f, we have $x \geq -3$ $\Leftrightarrow$ $x + 3 \geq 0$ $\Leftrightarrow$ $\sqrt{x+3} \geq 0$ $\Leftrightarrow$
   $f(x) \geq 0$. So the range is $[0, \infty)$.

7. $f(x) = 7x + 15$. The domain is all real numbers, $(-\infty, \infty)$.

9. $f(x) = \sqrt{x+4}$. We require $x + 4 \geq 0$ $\Leftrightarrow$ $x \geq -4$. Thus the domain is $[-4, \infty)$.

11. $f(x) = \frac{1}{x} + \frac{1}{x+1} + \frac{1}{x+2}$. The denominators cannot equal 0, therefore the domain is
    $\{x | x \neq 0, -1, -2\}$.

13. $h(x) = \sqrt{4 - x} + \sqrt{x^2 - 1}$. We require the expression inside the radicals be nonnegative. So
    $4 - x \geq 0$ $\Leftrightarrow$ $4 \geq x$; also $x^2 - 1 \geq 0$ $\Leftrightarrow$ $(x-1)(x+1) \geq 0$. Using the methods from
    Chapter 2, we have:

Interval	$(-\infty, -1)$	$(-1, 1)$	$(1, \infty)$
Sign of $x - 1$	$-$	$-$	$+$
Sign of $x + 1$	$-$	$+$	$+$
Sign of $(x-1)(x+1)$	$+$	$-$	$+$

Thus the domain is $(-\infty, 4] \cap \{ (-\infty, -1] \cup [1, \infty)\} = (-\infty, -1] \cup [1, 4]$.

15. $f(x) = 1 - 2x$

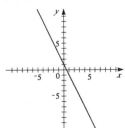

17. $f(t) = 1 - \frac{1}{2}t^2$

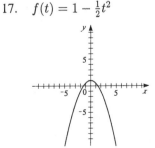

19. $f(x) = x^2 - 6x + 6$

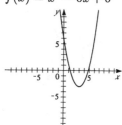

21. $y = 1 - \sqrt{x}$

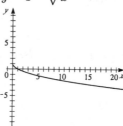

23. $y = \frac{1}{2}x^3$

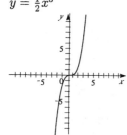

25. $h(x) = \sqrt[3]{x}$

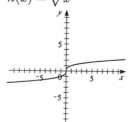

27. $g(x) = \frac{1}{x^2}$

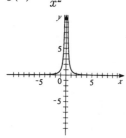

29. $f(x) = \begin{cases} 1 - x & \text{if } x < 0 \\ 1 & \text{if } x \geq 0 \end{cases}$

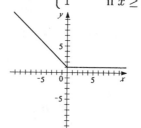

31. $f4(x) = \begin{cases} x + 6 & \text{if } x < -2 \\ x^2 & \text{if } x \geq -2 \end{cases}$

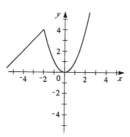

33. $f(x) = 6x^3 - 15x^2 + 4x - 1$
     (i) $[-2, 2]$ by $[-2, 2]$
     (ii) $[-8, 8]$ by $[-8, 8]$

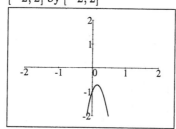

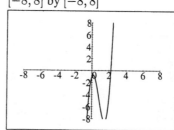

(iii) $[-4, 4]$ by $[-12, 12]$

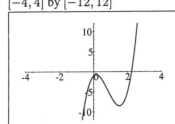

(iv) $[-100, 100]$ by $[-100, 100]$

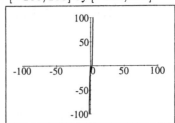

From the graphs, we see that the viewing rectangle in (iii) produces the most appropriate graph.

35. $f(x) = x^2 + 25x + 173$
$= (x^2 + 25x + \frac{625}{4}) + 173 - \frac{625}{4} = (x + \frac{25}{2})^2 + \frac{67}{4}$.
We use the viewing rectangle $[-30, 0]$ by $[0, 40]$.

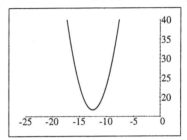

37. $y = \dfrac{x}{\sqrt{x^2 + 16}}$. Since $\sqrt{x^2 + 16} \geq \sqrt{x^2} = |x|$, it

follows that y should behave like $\dfrac{x}{|x|}$. Thus we use the

viewing rectangle $[-20, 20]$ by $[-2, 2]$.

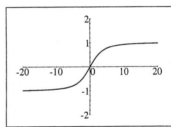

39. $f(x) = \sqrt{x^3 - 4x + 1}$. The domain consists of all x
where $x^3 - 4x + 1 \geq 0$. Using a graphing device, the
domain is approximately $[-2.1, 0.2] \cup [1.9, \infty)$.

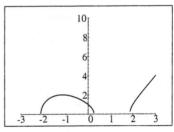

41. Since M varies directly as z we have $M = kz$. Substituting $M = 120$ when $z = 15$, we find
$120 = k(15) \quad \Leftrightarrow \quad k = 8$. Therefore, $M = 8z$.

43. (a) The intensity I varies inversely as the square of the distance d, so $I = \dfrac{k}{d^2}$.

(b) Substituting $I = 1000$ when $d = 8$, we get $1000 = \frac{k}{(8)^2} \quad \Leftrightarrow \quad k = 64,000$.

(c) From parts (a) and (b), we have $I = \dfrac{64000}{d^2}$. Substituting $d = 20$, we get $I = \dfrac{64000}{(20)^2} = 160$
candles.

45. Let v be the terminal velocity of the parachutist in mi/h and w be his weight in pounds. Since the
terminal velocity is directly proportional to the square root of the weight, we have $v = k\sqrt{w}$.

Substituting $v = 9$ when $w = 160$, we solve for k. This gives $9 = k\sqrt{160}$ $\Leftrightarrow$ $k = \frac{9}{\sqrt{160}} \approx 0.712$. Thus $v = 0.712\sqrt{w}$. When $w = 240$, the terminal velocity is $v = 0.712\sqrt{240} \approx 11$ mi/h.

47. Average rate of change $= \dfrac{f(2) - f(0)}{2 - 0} = \dfrac{[(2)^2 + 3(2)] - [0^2 + 3(0)]}{2} = \dfrac{4 + 6 - 0}{2} = 5.$

49. Average rate of change $= \dfrac{f(3 + h) - f(3)}{(3 + h) - 3} = \dfrac{\left[\frac{1}{3+h}\right] - \left[\frac{1}{3}\right]}{h} \cdot \dfrac{3(3 + h)}{3(3 + h)} = \dfrac{3 - (3 + h)}{3h(3 + h)}$
$= \dfrac{-h}{3h(3 + h)} = \dfrac{-1}{3(3 + h)}.$

51. $f(x) = x^3 - 4x^2$ is graphed in the viewing rectangle $[-5, 5]$ by $[-20, 10]$. $f(x)$ is increasing on $(-\infty, 0]$ and $[2.67, \infty)$. It is decreasing on $[0, 2.67]$

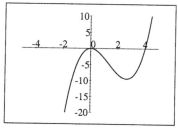

53. (a) $y = f(x) + 8$. Shift the graph of $f(x)$ upward 8 units.

   (b) $y = f(x + 8)$. Shift the graph of $f(x)$ to the left 8 units.

   (c) $y = 1 + 2f(x)$. Stretch the graph of $f(x)$ vertically by a factor of 2, then shift it upward 1 unit.

   (d) $y = f(x - 2) - 2$. Shift the graph of $f(x)$ to the right 2 units, then downward 2 units.

   (e) $y = f(-x)$. Reflect the graph of $f(x)$ about the y-axis.

   (f) $y = -f(-x)$. Reflect the graph of $f(x)$ first about the y-axis, then reflect about the x-axis.

   (g) $y = -f(x)$. Reflect the graph of $f(x)$ about the x-axis.

   (h) $y = f^{-1}(x)$. Reflect the graph of $f(x)$ about the line $y = x$.

55. (a) $f(x) = 2x^5 - 3x^2 + 2$.

       $f(-x) = 2(-x)^5 - 3(-x)^2 + 2 = -2x^5 - 3x^2 + 2$. Since $f(x) \neq f(-x)$, f is not even.
       $-f(x) = -2x^5 + 3x^2 - 2$. Since $-f(x) \neq f(-x)$, f is not odd.

   (b) $f(x) = x^3 - x^7$.

       $f(-x) = (-x)^3 - (-x)^7 = -(x^3 - x^7) = -f(x)$, hence f is odd.

   (c) $f(x) = \dfrac{1 - x^2}{1 + x^2}$. $f(-x) = \dfrac{1 - (-x)^2}{1 + (-x)^2} = \dfrac{1 - x^2}{1 + x^2} = f(x)$. Since $f(x) = f(-x)$, f is even.

   (d) $f(x) = \dfrac{1}{x + 2}$. $f(-x) = \dfrac{1}{(-x) + 2} = \dfrac{1}{2 - x}$. $-f(x) = -\dfrac{1}{x + 2}$. Since $f(x) \neq f(-x)$, f is not even, and since $f(-x) \neq -f(x)$, f is not odd.

57. $f(x) = x^2 + 4x + 1 = (x^2 + 4x + 4) + 1 - 4 = (x + 2)^2 - 3$

59. $g(x) = 2x^2 + 4x - 5 = 2(x^2 + 2x) - 5 = 2(x^2 + 2x + 1) - 5 - 2 = 2(x + 1)^2 - 7$. So the minimum value is $g(-1) = -7$.

61. $h(t) = -16t^2 + 48t + 32 = -16(t^2 - 3t) + 32 = -16(t^2 - 3t + \frac{9}{4}) + 32 + 36$
$$= -16(t^2 - 3t + \frac{9}{4}) + 68 = -16(t - \frac{3}{2})^2 + 68$$
The stone reaches a maximum height of 68 feet.

63. $f(x) = 3.3 + 1.6x - 2.5x^3$. In the first
viewing rectangle, $[-2, 2]$ by $[-4, 8]$, we see
that $f(x)$ has a local maximum and a local
minimum.

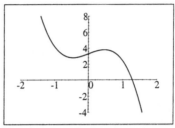

In the next viewing rectangle, $[0.4, 0.5]$ by $[3.78, 3.80]$, we isolate the local maximum value as
approximately 3.79 when $x \approx 0.46$. In the last viewing rectangle, $[-0.5, -0.4]$ by $[2.80, 2.82]$, we
isolate the local minimum value as 2.81 when $x \approx -0.46$.

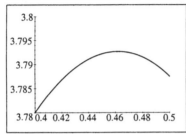
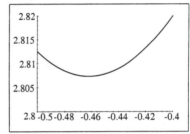

65. $f(x) = x^2 - 3x + 2$ and $g(x) = 4 - 3x$.
  (a) $(f + g)(x) = (x^2 - 3x + 2) + (4 - 3x) = x^2 - 6x + 6$
  (b) $(f - g)(x) = (x^2 - 3x + 2) - (4 - 3x) = x^2 - 2$
  (c) $(fg)(x) = (x^2 - 3x + 2)(4 - 3x) = 4x^2 - 12x + 8 - 3x^3 + 9x^2 - 6x$
       $= -3x^3 + 13x^2 - 18x + 8$
  (d) $\left(\frac{f}{g}\right)(x) = \dfrac{x^2 - 3x + 2}{4 - 3x}, \ x \neq \dfrac{4}{3}$
  (e) $(f \circ g)(x) = f(4 - 3x) = (4 - 3x)^2 - 3(4 - 3x) + 2 = 16 - 24x + 9x^2 - 12 + 9x + 2$
       $= 9x^2 - 15x + 6$
  (f) $(g \circ f)(x) = g(x^2 - 3x + 2) = 4 - 3(x^2 - 3x + 2) = -3x^2 + 9x - 2$

67. $f(x) = 3x - 1$ and $g(x) = 2x - x^2$.

  $(f \circ g)(x) = f(2x - x^2) = 3(2x - x^2) - 1 = -3x^2 + 6x - 1$, and the domain is $(-\infty, \infty)$.

  $(g \circ f)(x) = g(3x - 1) = 2(3x - 1) - (3x - 1)^2 = 6x - 2 - 9x^2 + 6x - 1 = -9x^2 + 12x - 3$,
  and the domain is $(-\infty, \infty)$

  $(f \circ f)(x) = f(3x - 1) = 3(3x - 1) - 1 = 9x - 4$, and the domain is $(-\infty, \infty)$.

  $(g \circ g)(x) = g(2x - x^2) = 2(2x - x^2) - (2x - x^2)^2 = 4x - 2x^2 - 4x^2 + 4x^3 - x^4$

  $= -x^4 + 4x^3 - 6x^2 + 4x$, and domain is $(-\infty, \infty)$.

69. $f(x) = \sqrt{1 - x}, \ g(x) = 1 - x^2$ and $h(x) = 1 + \sqrt{x}$.

$$(f \circ g \circ h)(x) = f(g(h(x))) = f\left(g\left(1 + \sqrt{x}\right)\right) = f\left(1 - \left(1 + \sqrt{x}\right)^2\right) = f(1 - (1 + 2\sqrt{x} + x)) =$$
$$f(-x - 2\sqrt{x}) = \sqrt{1 - (-x - 2\sqrt{x})} = \sqrt{1 + 2\sqrt{x} + x} = \sqrt{(1 + \sqrt{x})^2} = 1 + \sqrt{x}$$

71. $f(x) = 3 + x^3$. If $x_1 \neq x_2$, then $x_1^3 \neq x_2^3$ (unequal numbers have unequal cubes), and therefore $3 + x_1^3 \neq 3 + x_2^3$. Thus f is a one-to-one function.

73. $h(x) = \dfrac{1}{x^4}$. Since the fourth powers of a number and its negative are equal, h is not one-to-one. For example, $h(-1) = \dfrac{1}{(-1)^4} = 1$ and $h(1) = \dfrac{1}{(1)^4} = 1$, so $h(-1) = h(1)$.

75. $p(x) = 3.3 + 1.6x - 2.5x^3$. Using a graphing device and the Horizontal Line Test, we see that p is not a one-to-one function.

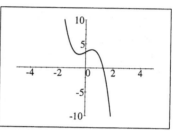

77. $f(x) = 3x - 2 \quad \Leftrightarrow \quad y = 3x - 2 \quad \Leftrightarrow \quad 3x = y + 2 \quad \Leftrightarrow \quad x = \frac{1}{3}(y + 2)$. So $f^{-1}(x) = \frac{1}{3}(x + 2)$.

79. $f(x) = (x + 1)^3 \quad \Leftrightarrow \quad y = (x + 1)^3 \quad \Leftrightarrow \quad x + 1 = \sqrt[3]{y} \quad \Leftrightarrow \quad x = -1 + \sqrt[3]{y}$. So $f^{-1}(x) = -1 + \sqrt[3]{x}$.

81. $f(x) = x^2 - 4, \ x \geq 0$.

(a)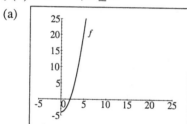

(b)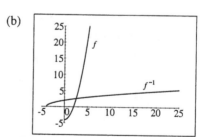

(c) $f(x) = x^2 - 4, \ x \geq 0 \quad \Leftrightarrow \quad y = x^2 - 4, \ y \geq -4 \quad \Leftrightarrow \quad x^2 = y + 4 \quad \Leftrightarrow \quad x = \sqrt{y + 4}$. So $f^{-1}(x) = \sqrt{x + 4}, \ x \geq -4$.

Chapter 4 Test

1. By the Vertical Line Test, figures (a) and (b) are graphs of functions. By the Horizontal Line Test, only figure (a) is the graph of a one-to-one function.

2. (a) $f(4) = \frac{\sqrt{4}}{4-1} = \frac{2}{3}$; $f(6) = \frac{\sqrt{6}}{6-1} = \frac{\sqrt{6}}{5}$; $f(a+1) = \frac{\sqrt{a+1}}{(a+1)-1} = \frac{\sqrt{a+1}}{a}$.

   (b) $f(x) = \frac{\sqrt{x}}{x-1}$. Our restrictions are that the input to the radical is nonnegative, and the denominator must not be equal to zero. Thus $x \geq 0$ and $x \neq 1$. In interval notation, the domain is $[0, 1) \cup (1, \infty)$.

3. (a) $M = k\dfrac{wh^2}{L}$.

   (b) Substituting $w = 4$, $h = 6$, $L = 12$, and $M = 4800$, we have $4800 = k\frac{(4)(6^2)}{12}$ $\Leftrightarrow$ $k = 400$. Thus $M = 400\dfrac{wh^2}{L}$.

   (c) Now if $L = 10$, $w = 3$, and $h = 10$, then $M = 400\frac{(3)(10^2)}{10} = 12,000$. So the beam can support $12,000$ pounds.

4. Average rate of change $= \dfrac{f(2+h) - f(2)}{(2+h) - 2} = \dfrac{[(2+h)^2 + 3(2+h)] - [2^2 + 3(2)]}{h}$

   $= \dfrac{4 + 4h + h^2 + 6 + 3h - 4 - 6}{h} = \dfrac{7h + h^2}{h} = \dfrac{h(7+h)}{h} = 7 + h.$

5. (a) $f(x) = x^3$

   (b) $g(x) = (x-1)^3 - 2$. To obtain the graph of g, shift the graph of f to the right 1 unit and downward 2 units.

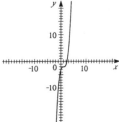

6. (a) $y = f(x-3) + 2$. Shift the graph of $f(x)$ to the right 3 units, then shift the graph upward 2 units.

   (b) $y = f(-x)$. Reflect the graph of $f(x)$ about the y-axis.

7. (a) $f(x) = 2x^2 - 8x + 13$

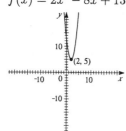

(b) $f(x) = 2x^2 - 8x + 13$
$$= 2(x^2 - 4x) + 13$$
$$= 2(x^2 - 4x + 4) + 13 - 8$$
$$= 2(x - 2)^2 + 5.$$

Thus the minimum value of f is
$f(2) = 5$.

8. (a) $f(-2) = 1 - (-2)^2 = 1 - 4 = -3$
    (since $-2 \le 0$).

    $f(1) = 2(1) + 1 = 2 + 1 = 3$ (since
    $1 > 0$).

(b)

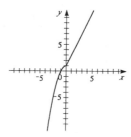

9. $f(x) = x^2 + 2x - 1;\ g(x) = 2x - 3.$

(a) $(f \circ g)(x) = f(g(x)) = f(2x - 3) = (2x - 3)^2 + 2(2x - 3) - 1$
     $= 4x^2 - 12x + 9 + 4x - 6 - 1 = 4x^2 - 8x + 2$

(b) $(g \circ f)(x) = g(f(x)) = g(x^2 + 2x - 1) = 2(x^2 + 2x - 1) - 3 = 2x^2 + 4x - 2 - 3$
     $= 2x^2 + 4x - 5$

(c) $f(g(2)) = f(1) = (1)^2 + 2(1) - 1 = 2.$ (We have used the fact that $g(2) = 2(2) - 3 = 1.$)

(d) $g(f(2)) = g(7) = 2(7) - 3 = 11.$ (We have used the fact that $f(2) = 2^2 + 2(2) - 1 = 7.$)

(e) $(g \circ g \circ g)(x) = g(g(g(x))) = g(g(2x - 3)) = g(4x - 9) = 2(4x - 9) - 3 = 8x - 18 - 3$
     $= 8x - 21.$ (We have used the fact that $g(2x - 3) = 2(2x - 3) - 3 = 4x - 6 - 3 = 4x - 9.$)

10. (a) $f(x) = \sqrt{3 - x},\ x \le 3 \quad \Leftrightarrow$
    $y = \sqrt{3 - x} \quad \Leftrightarrow \quad y^2 = 3 - x \quad \Leftrightarrow$
    $x = 3 - y^2.$ Thus $f^{-1}(x) = 3 - x^2,$
    $x \ge 0.$

(b) $f(x) = \sqrt{3 - x},\ x \le 3$ and
     $f^{-1}(x) = 3 - x^2,\ x \ge 0$

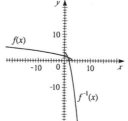

11. (a) The domain of f is $[0, 6]$, and
    the range of f is $[1, 7]$.

(b)

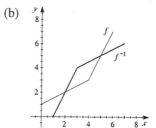

(c) Average rate of change $= \dfrac{f(6) - f(2)}{6 - 2} = \dfrac{7 - 2}{4} = \dfrac{5}{4}$.

12. (a) $f(x) = 3x^4 - 14x^2 + 5x - 3$. The graph is shown in the viewing rectangle $[-10, 10]$ by $[-30, 10]$.

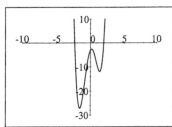

(b) No, by the Horizontal Line Test.

(c) The local minimum is approximately -27.18 when $x \approx -1.61$. Shown is the viewing rectangle $[-1.65, -1.55]$ by $[-27.5, -27]$.

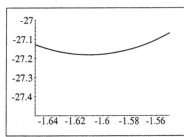

The local maximum is approximately -2.55 when $x \approx 0.18$. Shown is the viewing rectangle $[0.15, 0.25]$ by $[-2.6, -2.5]$.

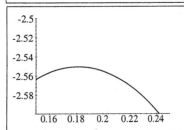

The local minimum is approximately -11.93 when $x \approx 1.43$. Shown is the viewing rectangle $[1.4, 1.5]$ by $[-12, -11.9]$.

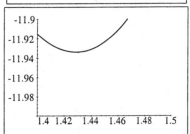

(d) Using the graph in part (a) and the local minimum, -27.18, found in part (c), we see that the range is $[-27.18, \infty)$

(e) Using the information from part (c) and the graph in part (a), $f(x)$ is increasing on the intervals $[-1.61, 0.18]$ and $[1.43, \infty)$ and decreasing on the intervals $(-\infty, -1.61]$ and $[0.18, 1.43]$.

Focus on Modeling

1. Let w be the width of the building lot. Then the length of the lot is $3w$. So the area of the building lot is $A(w) = 3w^2$, $w > 0$.

3. Let w be the width of the base of the rectangle. Then the height of the rectangle is $\frac{1}{2}w$. Thus the volume of the box is given by the function $V(w) = \frac{1}{2}w^3$, $w > 0$.

5. Let P be the perimeter of the rectangle and y be the length of the other side. Since $P = 2x + 2y$ and the perimeter is 20, we have $2x + 2y = 20 \quad \Leftrightarrow \quad x + y = 10 \quad \Leftrightarrow \quad y = 10 - x$. Since area is $A = xy$, substituting gives $A(x) = x(10 - x) = 10x - x^2$, and since A must be positive, the domain is $0 < x < 10$.

7. Let h be the height of an altitude of the equilateral triangle whose side has length x, see diagram. Thus the area is given by $A = \frac{1}{2}xh$. By the Pythagorean
Theorem, $h^2 + \left(\frac{1}{2}x\right)^2 = x^2 \quad \Leftrightarrow \quad h^2 + \dfrac{x^2}{4} = x^2 \quad \Leftrightarrow \quad h^2 = \dfrac{3x^2}{4} \quad \Leftrightarrow$
$h = \frac{\sqrt{3}x}{2}$. Substituting into the area of a triangle, we get
$A(x) = \frac{1}{2}xh = \frac{1}{2}x\left(\frac{\sqrt{3}x}{2}\right) = \frac{\sqrt{3}x^2}{4}$, $x > 0$.

9. We solve for r in the formula for the area of a circle. This gives $A = \pi r^2 \quad \Leftrightarrow \quad r^2 = \frac{A}{\pi} \quad \Rightarrow$
$r = \sqrt{\frac{A}{\pi}}$, so the model is $r(A) = \sqrt{\frac{A}{\pi}}$, $A > 0$.

11. Let h be the height of the box in feet. The volume of the box is $V = 60$. Then $x^2h = 60 \quad \Leftrightarrow$
$h = \dfrac{60}{x^2}$. The surface area, A, of the box is sum of the area of the 4 sides and the area of the base.
Thus $A = 4xh + x^2 = 4x\left(\dfrac{60}{x^2}\right) + x^2 = \dfrac{240}{x} + x^2$, so the model is $A(x) = \dfrac{240}{x} + x^2$, $x > 0$.

13. Let d_1 be the distance traveled south by the first ship and d_2 be the distance traveled east by the second ship. Since the first ship travels south for t hours at 15 mi/h, $d_1 = 15t$ and, similarly, $d_2 = 20t$. Since the ships are traveling at right angles to each other, we can apply the Pythagorean Theorem to get $D^2 = d_1^2 + d_2^2 = (15t)^2 + (20t)^2 = 225t^2 + 400t^2 = 625t^2$. Thus $D(t) = 25t$, $t \geq 0$.

15. Let x be the length of the base, l be the length of the equal sides, and h be the height in centimeters. Since the perimeter is 8, $2l + x = 8 \quad \Leftrightarrow$
$2l = 8 - x \quad \Leftrightarrow \quad l = \frac{1}{2}(8 - x)$. By the Pythagorean Theorem,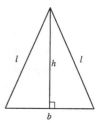
$h^2 + \left(\frac{1}{2}x\right)^2 = l^2 \quad \Leftrightarrow \quad h^2 = l^2 - \left(\frac{1}{2}x\right)^2 \quad \Leftrightarrow \quad h = \sqrt{l^2 - \frac{1}{4}x^2}$.
Therefore the area of the triangle is
$A = \frac{1}{2} \cdot x \cdot h = \frac{1}{2} \cdot x\sqrt{l^2 - \frac{1}{4}x^2} = \dfrac{x}{2}\sqrt{\frac{1}{4}(8 - x)^2 - \frac{1}{4}x^2}$

$= \dfrac{x}{4}\sqrt{64 - 16x + x^2 - x^2} = \dfrac{x}{4}\sqrt{64 - 16x} = \dfrac{x}{4} \cdot 4\sqrt{4 - x} = x\sqrt{4 - x}$,
so the model is $A(x) = x\sqrt{4 - x}$, $0 < x < 4$.

17. Let w be the length of the rectangle. By the Pythagorean Theorem, $\left(\frac{1}{2}w\right)^2 + h^2 = 10^2$ $\Leftrightarrow$ $\frac{w^2}{4} + h^2 = 10^2$ $\Leftrightarrow$ $w^2 = 4(100 - h^2)$ $\Leftrightarrow$ $w = 2\sqrt{100 - h^2}$ (since $w > 0$). Therefore, the area of the rectangle is $A = wh = 2h\sqrt{100 - h^2}$, so the model is $A(h) = 2h\sqrt{100 - h^2}$, $0 < h < 10$.

19. (a) We complete the table.

First number	Second number	Product
1	18	18
2	17	34
3	16	48
4	15	60
5	14	70
6	13	78
7	12	84
8	11	88
9	10	90
10	9	90
11	8	88

From the table we conclude that the numbers is still increasing, the numbers whose product is a maximum should both be 9.5.

(b) Let x be one number: then $19 - x$ is the other number, and so the product, p, is
$p(x) = x(19 - x) = 19x - x^2$.

(c) $p(x) = 19x - x^2 = -(x^2 - 19x) = -\left[x^2 - 19x + \left(\frac{19}{2}\right)^2\right] + \left(\frac{19}{2}\right)^2 = -(x - 9.5)^2 + 90.25$.
So the product is maximized when the numbers are both 9.5.

21. Let x and y be the two numbers. Since their sum is -24, we have $x + y = -24$ $\Leftrightarrow$ $y = -x - 24$. The product of the two numbers is $P = xy = x(-x - 24) = -x^2 - 24x$, which we wish to maximize. So $P = -x^2 - 24x = -(x^2 + 24x) = -(x^2 + 24x + 144) + 144$ $= -(x + 12)^2 + 144$. Thus the maximum product is 144, and it occurs when $x = -12$ and $y = -(-12) - 24 = -12$. Thus the two numbers are -12 and -12.

23. (a) Let x be the width of the field (in feet) and l be the length of the field (in feet). Since the farmer has 2400 ft of fencing we must have $2x + l = 2400$.

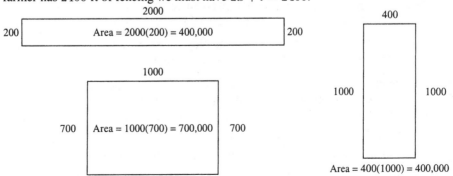

Width	Length	Area
200	2000	400,000
300	1800	540,000
400	1600	640,000
500	1400	700,000
600	1200	720,000
700	1000	700,000
800	800	640,000

It appears that the feild of largest area is about 600 ft. × 1200 ft.

(b) Let x be the width of the field (in feet) and l be the length of the field (in feet). Since the farmer has 2400 ft of fencing we must have $2x + l = 2400$ $\Leftrightarrow$ $l = 2400 - 2x$. The area of the fenced-in field is given by
$$A(x) = l \cdot x = (2400 - 2x)x = -2x^2 + 2400x = -2(x^2 - 1200x).$$

(c) The area is $A(x) = -2(x^2 - 1200x + 600^2) + 2(600^2) = -2(x - 600)^2 + 720000$. So the maximum area occurs when $x = 600$ feet and $l = 2400 - 2(600) = 1200$ feet.

25. Let x be the length of the fence along the road. If the area is 600, we have $600 = x \cdot$ length, and the width of the garden is $\dfrac{600}{x}$. Then the cost of the fence is given by the function

$C(x) = 5(x) + 3\left[x + 2 \cdot \dfrac{600}{x}\right] = 8x + \dfrac{3600}{x}$. The function $y = C(x)$ is shown in the first viewing rectangle below. In the second viewing rectangle, we isolate the minimum, and we see that the amount of cost of the material is minimized when x is 21.21 ft. So the garden should be 21.21 ft by $\dfrac{600}{21.21} = 28.28$ ft.

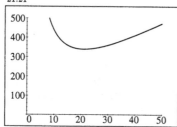
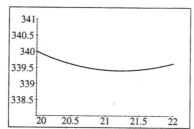

27. Let x represent the number of dollars that the ticket prices are lowered. So the new ticket price is $10 - x$, and the number of tickets sold is $n = 27000 + 3000x$. Thus the revenue is
$R(x) = (10 - x)(27000 + 3000x) = 270000 + 3000x - 3000x^2 = -3000(x^2 - x) + 270000$
$= -3000\left(x^2 - x + \frac{1}{4}\right) + 270000 + 750 = -3000\left(x - \frac{1}{2}\right)^2 + 270750$. The revenue is maximized when $x = \frac{1}{2}$, and so the price should be set at $10 - \frac{1}{2} = \$9.50$.

29. Let h be the height in feet of the straight portion of the window. The circumference of the semicircle is $C = \frac{1}{2}\pi x$. Since the perimeter of the window is 30 feet, we have $x + 2h + \frac{1}{2}\pi x = 30$. Solving for h, we get $2h = 30 - x - \frac{1}{2}\pi x$ $\Leftrightarrow$ $h = 15 - \frac{1}{2}x - \frac{1}{4}\pi x$. The area of the window is
$A(x) = xh + \frac{1}{2}\pi\left(\frac{1}{2}x\right)^2 = x\left(15 - \frac{1}{2}x - \frac{1}{4}\pi x\right) + \frac{1}{8}\pi x^2 = 15x - \frac{1}{2}x^2 - \frac{1}{4}\pi x^2 + \frac{1}{8}\pi x^2$
$= 15x - \frac{1}{2}x^2 - \frac{1}{8}\pi x^2 = 15x - \frac{1}{8}(\pi + 4)x^2 = -\frac{1}{8}(\pi + 4)\left[x^2 - \frac{120}{\pi+4}x\right]$
$= -\frac{1}{8}(\pi + 4)\left[x^2 - \frac{120}{\pi+4}x + \left(\frac{60}{\pi+4}\right)^2\right] + \frac{450}{\pi+4} = -\frac{1}{8}(\pi + 4)\left(x - \frac{60}{\pi+4}\right)^2 + \frac{450}{\pi+4}$. The area is maximized when $x = \frac{60}{\pi+4} \approx 8.40$, and hence $h \approx 15 - \frac{1}{2}(8.40) - \frac{1}{4}\pi(8.40) \approx 4.20$.

31. Let h be the height of the box in feet. The volume of the box is $V = 12$. Then $x^2h = 12$ ⇔ $h = \dfrac{12}{x^2}$. The surface area, A, of the box is sum of the area of the 4 sides and the area of the base. Thus the surface area of the box is given by the formula

$A(x) = 4xh + x^2 = 4x\left(\dfrac{12}{x^2}\right) + x^2 = \dfrac{48}{x} + x^2, x > 0$. The function $y = A(x)$ is shown in the first

viewing rectangle below. In the second viewing rectangle, we isolate the minimum, and we see that the amount of material is minimized when x (the length and width) is 2.88 ft. Then the height is

$h = \dfrac{12}{x^2} \approx 1.44$ ft.

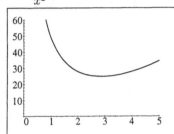
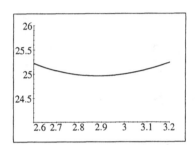

33. Let w be the width of the pen and l be the length in meters. We use the area to establish a relationship between w and l. Since the area is 100 m², we have $l \cdot w = 100$ ⇔ $l = \dfrac{100}{w}$. So the amount of fencing used is $F = 2l + 2w = 2\left(\dfrac{100}{w}\right) + 2w = \dfrac{200 + 2w^2}{w}$. Using a graphing device, we first graph F in the viewing rectangle, $[0, 40]$ by $[0, 100]$, and locate the approximate location of the minimum value. In the second viewing rectangle, $[8, 12]$ by $[39, 41]$, we see that the minimum value of F occurs when $w = 10$. Therefore the pen should be a square with side 10 m.

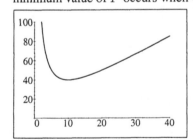
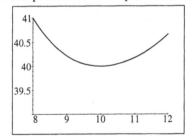

35. Let x be the distance from point B to C, in miles. Then the distance from A to C is $\sqrt{x^2 + 25}$, and the energy used in flying from A to C then C to D is $f(x) = 14\sqrt{x^2 + 25} + 10(12 - x)$. By using a graphing device, the energy expenditure is minimized when the distance from B to C is about 5.1 miles.

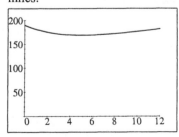
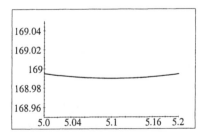

Chapter Five

Exercises 5.1

1. $P(x) = x^3 - 8$

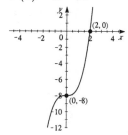

3. $P(x) = -x^4 + 16$

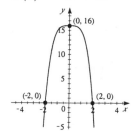

5. $P(x) = -(x-1)^4 + 1$

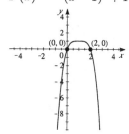

7. $P(x) = 4(x-2)^5 - 4$

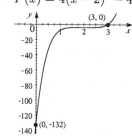

9. $P(x) = (x-3)(x+1)$

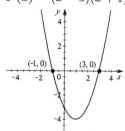

11. $P(x) = x(x-2)(x+1)$

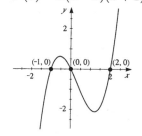

13. $P(x) = (x-3)(x+2)(3x-2)$

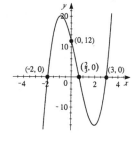

15. $P(x) = (x-1)^2(x-3)$

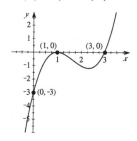

17. $P(x) = \frac{1}{12}(x+2)^2(x-3)^2$

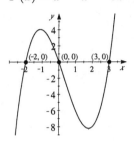

19. $P(x) = x^3(x+2)(x-3)^2$

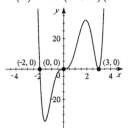

21. $P(x) = x^3 - x^2 - 6x$

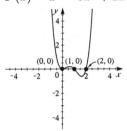

23. $P(x) = -x^3 + x^2 + 12x$

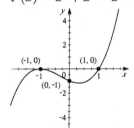

25. $P(x) = x^4 - 3x^3 + 2x^2$

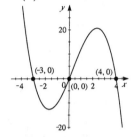

27. $P(x) = x^3 + x^2 - x - 1$

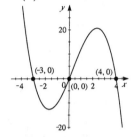

29. $P(x) = 2x^3 - x^2 - 18x + 9$

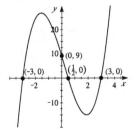

31. $P(x) = x^4 - 2x^3 - 8x + 16$

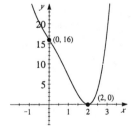

33. $P(x) = x^4 - 3x^2 - 4$

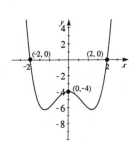

35. III   37. V   39. VI

41. $P(x) = 3x^3 - x^2 + 5x + 1$; $Q(x) = 3x^3$.
     Since P has odd degree and positive leading coefficient, it has the following end behavior:
     $y \to \infty$ as $x \to \infty$ and $y \to -\infty$ as $x \to -\infty$.

On the large viewing rectangle, the
graphs of P and Q look almost the same.

On the small viewing rectangle, the graphs of
P and Q have different intercepts.

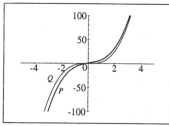

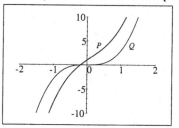

43. $P(x) = x^4 - 7x^2 + 5x + 5$; $Q(x) = x^4$.
     Since P has even degree and positive leading coefficient, it has the following end behavior:
     $y \to \infty$ as $x \to \infty$ and $y \to \infty$ as $x \to -\infty$.

On the large viewing rectangle, the
graphs of P and Q look almost the same.

On the small viewing rectangle, the graphs of
P and Q look very different and have
different intercepts..

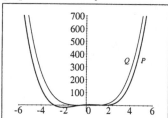

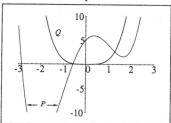

45. $P(x) = x^{11} - 9x^9$; $Q(x) = x^{11}$.
     Since P has odd degree and positive leading coefficient, it has the following end behavior:
     $y \to \infty$ as $x \to \infty$ and $y \to -\infty$ as $x \to -\infty$.

On the large viewing rectangle, the
graphs of P and Q look like they have the
same end behavior.

On the small viewing rectangle, the graphs of
P and Q look very different and seem to have
different end behavior.

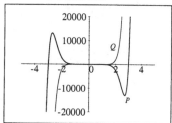

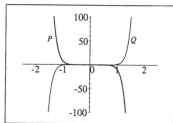

47. $y = -x^2 + 8x$, $[-4, 12]$ by $[-50, 30]$.
y-intercept: 0
Local maximum: $(4, 16)$

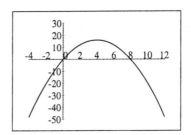

49. $y = x^3 - 12x + 9$, $[-5, 5]$ by $[-30, 30]$.
Local maximum: $(-2, 25)$
Local minimum: $(2, -7)$

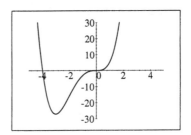

51. $y = x^4 + 4x^3$, $[-5, 5]$ by $[-30, 30]$.
Local minimum: $(-3, -27)$

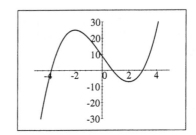

53. $y = 3x^5 - 5x^3 + 3$, $[-3, 3]$ by $[-5, 10]$.
Local maximum: $(-1, 5)$
Local minimum: $(1, 1)$

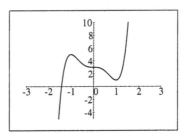

55. $y = -2x^2 + 3x + 5$.
One local maximum at $(0.75, 6.13)$.

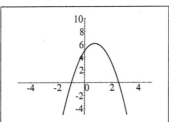

57. $y = x^3 - x^2 - x$.

One local maximum at $(-0.33, 0.19)$ and one
local minimum at $(1.00, -1.00)$.

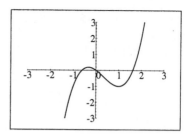

59. $y = x^4 - 5x^2 + 4$.

One local maximum at $(0, 4)$ and two local
minima at $(-1.58, -2.25)$ and $(1.58, -2.25)$.

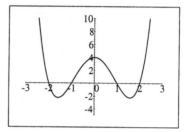

61. $y = (x - 2)^5 + 32$.

No extrema.

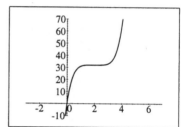

63. $y = x^8 - 3x^4 + x$.

One local maximum at $(0.44, 0.33)$ and two
local minima at $(1.09, -1.15)$ and
$(-1.12, -3.36)$.

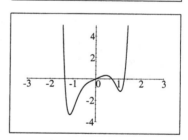

65. $P(x) = cx^3$; $c = 1, 2, 5$, and $\frac{1}{2}$. Increasing
the value of c stretches the graph vertically.

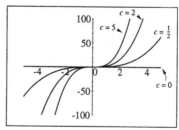

67. $P(x) = x^4 + c$; $c = -1, 0, 1$, and 2.
Increasing the value of c moves the graph up.

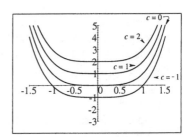

69. $P(x) = x^4 - cx$; $c = 0, 1, 8$, and 27.
Increasing the value of c causes a deeper dip
in the graph, in the fourth quadrant, and moves
the positive x-intercept to the right.

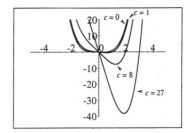

71. (a)

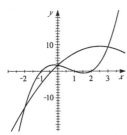

(b)   The two graphs appear to intersect at 3 points.

(c) $x^3 - 2x^2 - x + 2 = -x^2 + 5x + 2 \quad \Leftrightarrow \quad x^3 - x^2 - 6x = 0 \quad \Leftrightarrow \quad x(x^2 - x - 6) = 0$
$\Leftrightarrow \quad x(x - 3)(x + 2) = 0$. Then either $x = 0$, $x = 3$, or $x = -2$. If $x = 0$, then $y = 2$; if
$x = 3$ then $y = 8$; if $x = -2$, then $y = -12$. Hence the points where the two graphs intersect
are $(0, 2)$, $(3, 8)$, and $(-2, -12)$.

73. (a)   Since f and g are odd, $f(-x) = -f(x)$ and $g(-x) = -g(x)$. Therefore,
$(f + g)(-x) = f(-x) + g(-x) = -f(x) - g(x) = -[f(x) + g(x)] = -(f + g)(x)$, and so
the function $f + g$ is also odd.

(b)   Since f and g are even, $f(-x) = f(x)$ and $g(-x) = g(x)$. Therefore,
$(f + g)(-x) = f(-x) + g(-x) = f(x) + g(x) = (f + g)(x)$, and so the function $f + g$ is
also even.

(c)   Since f is odd and g is even, $f(-x) = -f(x)$ and $g(-x) = g(x)$. Therefore,
$(f + g)(-x) = f(-x) + g(-x) = -f(x) + g(x) \neq -[f(x) + g(x)]$ (since $g(x) \neq -g(x)$ if x
is a point for which $g(x) \neq 0$), and so the function $f + g$ is not odd. Also,
$-f(x) + g(x) \neq f(x) + g(x)$ (for a similar reason), and so the function $f + g$ is not even.
Thus, $f + g$ is neither even nor odd.

(d)   Let $P(x)$ be a polynomial containing only odd powers of x. Then each term of $P(x)$ can be
written as Cx^{2n+1}, for some constant C and integer n. Since $C(-x)^{2n+1} = -Cx^{2n+1}$, each
term of $P(x)$ is an odd function. Thus by part (a), $P(x)$ is an odd function.

(e) Let $P(x)$ be a polynomial containing only even powers of x. Then each term of $P(x)$ can be written as Cx^{2n}, for some constant C and integer n. Since $C(-x)^{2n} = Cx^{2n}$, each term of $P(x)$ is an even function. Thus by part (b), $P(x)$ is an even function.

(f) Since $P(x)$ contains both even and odd powers of x, we can write it in the form $P(x) = R(x) + Q(x)$, where $R(x)$ contains all the even-powered terms in $P(x)$ and $Q(x)$ contains all the odd. By part (d), $Q(x)$ is an odd function, and by part (e), $R(x)$ is an even function. Thus, since neither $Q(x)$ nor $R(x)$ are constantly 0 (by assumption), by part (c), $P(x) = R(x) + Q(x)$ is neither even nor odd.

(g) $P(x) = x^5 + 6x^3 - x^2 - 2x + 5 = (x^5 + 6x^3 - 2x) + (-x^2 + 5) = P_O(x) + P_E(x)$ where $P_O(x) = x^5 + 6x^3 - 2x$ and $P_E(x) = -x^2 + 5$. Since $P_O(x)$ contains only odd powers of x, it is an odd function, and since $P_E(x)$ contains only even powers of x, it is an even function.

75. (a) Let h = height of the box. Then the total length of all 12 edges is $8x + 4h = 144$ in. Thus, $8x + 4h = 144 \Leftrightarrow 2x + h = 36 \Leftrightarrow h = 36 - 2x$. The volume of the box is equal to (area of base) $\times$ (height) $= (x^2) \times (36 - 2x) = -2x^3 + 36x^2$. Therefore, the volume of the box is $V(x) = -2x^3 + 36x^2 = 2x^2(18 - x)$.

(b) Since the length of the base is x, we must have $x > 0$. Likewise, the height must be positive so $36 - 2x > 0 \Leftrightarrow x < 18$. Putting these together, we get that the domain of V is $0 < x < 18$.

(c) Using the domain from part (b), we graph V in the viewing rectangle $[0, 18]$ by $[0, 2000]$. The maximum volume is $V = 1728$ when $x = 12$.

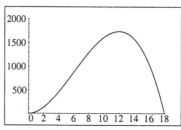

77. $P(t) = 120t - 0.4t^4 + 1000$

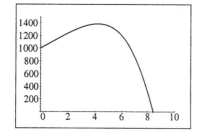

(a) A maximum population of approximately 1380 is attained after 4.22 months.

(b) The rabbit population disappears after approximately 8.42 months.

79. (a) From the graph, $P(x) = x^3 - 4x = x(x - 2)(x + 2)$ has three x-intercepts and two local extrema.

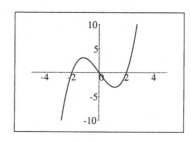

(b) From the graph,
$Q(x) = x^3 + 4x = x(x^2 + 4)$ has
one x-intercept and no local extrema.

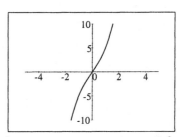

(c) For the x-intercepts of $P(x) = x^3 - ax$, we solve $x^3 - ax = 0$. Then we have $x(x^2 - a) = 0$
$\Leftrightarrow \quad x = 0$ or $x^2 = a$. If $x^2 = a$, then $x = \pm\sqrt{a}$. So P has 3 x-intercepts. Since
$P(x) = x(x^2 - a) = x(x + \sqrt{a})(x - \sqrt{a})$, by part (c) of problem 67, P has 2 local extrema.
For the x-intercepts of $Q(x) = x^3 + ax$, we solve $x^3 + ax = 0$. Then we have $x(x^2 + a) = 0$
$\Leftrightarrow \quad x = 0$ or $x^2 = -a$. The equation $x^2 = -a$ has no real solutions because $a > 0$. So Q
has 1 x-intercept. We now show that Q is always increasing and hence has no extrema. If
$x_1 < x_2$, then $ax_1 < ax_2$ (because $a > 0$) and $x_1^3 < x_2^3$. So we have $x_1^3 + ax_1 < x_2^3 + ax_2$, and
hence $Q(x_1) < Q(x_2)$. Thus Q is increasing, that is, its graph always rises, and so it has no
local extrema.

81. Since the polynomial shown has five zeros, it has at least five factors, and so the degree of the
polynomial is greater than or equal to 5.

83. No, it is not possible. Clearly a polynomial must have a local minimum between any two local
maxima.

Exercises 5.2

1.

$$\begin{array}{r} x\ +1 \\ x+3\overline{)x^2+4x-8} \\ \underline{x^2+3x} \\ x-8 \\ \underline{x+3} \\ -11 \end{array}$$

Thus the quotient is $x+1$ and the remainder is -11.

3.

$$\begin{array}{r} x^2\ +4x\ +22 \\ x-4\overline{)x^3+0x^2\ +6x+\ 5} \\ \underline{x^3-4x^2} \\ 4x^2\ +6x \\ \underline{4x^2-16x} \\ 22x+\ 5 \\ \underline{22x-88} \\ 93 \end{array}$$

Thus the quotient is $x^2-4x+22$ and the remainder is 93.

5.

$$\begin{array}{r} x\ +2 \\ x^2-2x+2\overline{)x^3+0x^2+6x+3} \\ \underline{x^3-2x^2+2x} \\ 2x^2+4x+3 \\ \underline{2x^2-4x+4} \\ 8x-1 \end{array}$$

Thus the quotient is $x+2$, and the remainder is $8x-1$.

7.

$$\begin{array}{r} 3x\ +1 \\ 2x^2+0x+5\overline{)6x^3+2x^2+22x+0} \\ \underline{6x^3\ \ \ \ \ +15x} \\ 2x^2\ +7x+0 \\ \underline{2x^2\ \ \ \ \ +5} \\ 7x-5 \end{array}$$

Thus the quotient is $3x+1$, and the remainder is $7x-5$.

9.

$$\begin{array}{r} x^4\ \ \ \ \ \ \ \ \ \ \ +1 \\ x^2+1\overline{)x^6+0x^5+x^4+0x^3+x^2+0x+1} \\ \underline{x^6\ \ \ \ \ +x^4} \\ 0\ \ \ \ +x^2\ \ \ \ +1 \\ \underline{x^2\ \ \ \ +1} \\ 0 \end{array}$$

Thus the quotient is x^4+1, and the remainder is 0.

11. The synthetic division table for this problem takes the following form.

$$\begin{array}{r|rrr} 3 & 1 & -5 & 4 \\ & & 3 & -6 \\ \hline & 1 & -2 & -2 \end{array}$$

Thus the quotient is $x-2$, and the remainder is -2.

13. The synthetic division table for this problem takes the following form.

$$
\begin{array}{r|rrr}
6 & 3 & 5 & 0 \\
  &   & 18 & 138 \\
\hline
  & 3 & 23 & 138
\end{array}
$$
Thus the quotient is $3x + 23$, and the remainder is 138.

15. Since $x + 2 = x - (-2)$, the synthetic division table for this problem takes the following form.

$$
\begin{array}{r|rrrr}
-2 & 1 & 2 & 2 & 1 \\
   &   & -2 & 0 & -4 \\
\hline
   & 1 & 0 & 2 & -3
\end{array}
$$
Thus the quotient is $x^2 + 2$, and the remainder is -3.

17. Since $x + 3 = x - (-3)$ and $x^3 - 8x + 2 = x^3 + 0x^2 - 8x + 2$, the synthetic division table for this problem takes the following form.

$$
\begin{array}{r|rrrr}
-3 & 1 & 0 & -8 & 2 \\
   &   & -3 & 9 & -3 \\
\hline
   & 1 & -3 & 1 & -1
\end{array}
$$
Thus the quotient is $x^2 - 3x + 1$, and the remainder is -1.

19. Since $x^5 + 3x^3 - 6 = x^5 + 0x^4 + 3x^3 + 0x^2 + 0x - 6$, the synthetic division table for this problem takes the following form.

$$
\begin{array}{r|rrrrrr}
1 & 1 & 0 & 3 & 0 & 0 & -6 \\
  &   & 1 & 1 & 4 & 4 & 4 \\
\hline
  & 1 & 1 & 4 & 4 & 4 & -2
\end{array}
$$
Thus the quotient is $x^4 + x^3 + 4x^2 + 4x + 4$, and the remainder is -2.

21. The synthetic division table for this problem takes the following form.

$$
\begin{array}{r|rrrr}
\frac{1}{2} & 2 & 3 & -2 & 1 \\
            &   & 1 & 2 & 0 \\
\hline
            & 2 & 4 & 0 & 1
\end{array}
$$
Thus the quotient is $2x^2 + 4x$, and the remainder is 1.

23. Since $x^3 - 27 = x^3 + 0x^2 + 0x - 27$, the synthetic division table for this problem takes the following form.

$$
\begin{array}{r|rrrr}
3 & 1 & 0 & 0 & -27 \\
  &   & 3 & 9 & 27 \\
\hline
  & 1 & 3 & 9 & 0
\end{array}
$$
Thus the quotient is $x^2 + 3x + 9$, and the remainder is 0.

25. $P(x) = 4x^2 + 12x + 5$, $c = -1$

$$
\begin{array}{r|rrr}
-1 & 4 & 12 & 5 \\
   &   & -4 & -8 \\
\hline
   & 4 & 8 & -3
\end{array}
$$
Therefore by the Remainder Theorem, $P(-1) = -3$.

27. $P(x) = x^3 + 3x^2 - 7x + 6$, $c = 2$

$$
\begin{array}{r|rrrr}
2 & 1 & 3 & -7 & 6 \\
  &   & 2 & 10 & 6 \\
\hline
  & 1 & 5 & 3 & 12
\end{array}
$$
Therefore by the Remainder Theorem, $P(2) = 12$.

29. $P(x) = x^3 + 2x^2 - 7$, $c = -2$

$$
\begin{array}{r|rrrr}
-2 & 1 & 2 & 0 & -7 \\
   &   & -2 & 0 & 0 \\
\hline
   & 1 & 0 & 0 & -7
\end{array}
$$
Therefore by the Remainder Theorem, $P(-2) = -7$.

31. $P(x) = 5x^4 + 30x^3 - 40x^2 + 36x + 14, c = -7$

$$
\begin{array}{r|rrrrr}
-7 & 5 & 30 & -40 & 36 & 14 \\
   &   & -35 & 35 & 35 & -497 \\
\hline
   & 5 & -5 & -5 & 71 & -483
\end{array}
$$
Therefore by the Remainder Theorem, $P(-7) = -483$.

33. $P(x) = x^7 - 3x^2 - 1 = x^7 + 0x^6 + 0x^5 + 0x^4 + 0x^3 - 3x^2 + 0x - 1, c = 3$

$$
\begin{array}{r|rrrrrrrr}
3 & 1 & 0 & 0 & 0 & 0 & -3 & 0 & -1 \\
  &   & 3 & 9 & 27 & 81 & 243 & 720 & 2160 \\
\hline
  & 1 & 3 & 9 & 27 & 81 & 240 & 720 & 2159
\end{array}
$$
Therefore by the Remainder Theorem, $P(3) = 2159$.

35. $P(x) = 3x^3 + 4x^2 - 2x + 1, c = \frac{2}{3}$

$$
\begin{array}{r|rrrr}
\frac{2}{3} & 3 & 4 & -2 & 1 \\
  &   & 2 & 4 & \frac{4}{3} \\
\hline
  & 3 & 6 & 2 & \frac{7}{3}
\end{array}
$$
Therefore by the Remainder Theorem, $P\left(\frac{2}{3}\right) = \frac{7}{3}$.

37. $P(x) = x^3 + 2x^2 - 3x - 8, c = 0.1$

$$
\begin{array}{r|rrrr}
0.1 & 1 & 2 & -3 & -8 \\
  &   & 0.1 & 0.21 & -0.279 \\
\hline
  & 1 & 2.1 & -2.79 & -8.279
\end{array}
$$
Therefore by the Remainder Theorem, $P(0.1) = -8.279$.

39. $P(x) = x^3 - 3x^2 + 3x - 1, c = 1$

$$
\begin{array}{r|rrrr}
1 & 1 & -3 & 3 & -1 \\
  &   & 1 & -2 & 1 \\
\hline
  & 1 & -2 & 1 & 0
\end{array}
$$
Since the remainder is 0, $x - 1$ is a factor.

41. $P(x) = 2x^3 + 7x^2 + 6x - 5, c = \frac{1}{2}$

$$
\begin{array}{r|rrrr}
\frac{1}{2} & 2 & 7 & 6 & -5 \\
  &   & 1 & 4 & 5 \\
\hline
  & 2 & 8 & 10 & 0
\end{array}
$$
Since the remainder is 0, $x - \frac{1}{2}$ is a factor.

43. $P(x) = x^3 - x^2 - 11x + 15, c = 3$

$$
\begin{array}{r|rrrr}
3 & 1 & -1 & -11 & 15 \\
  &   & 3 & 6 & -15 \\
\hline
  & 1 & 2 & -5 & 0
\end{array}
$$
Since the remainder is 0, we know that 3 is a zero $x^3 - x^2 - 11x + 15 = (x - 3)(x^2 + 2x - 5)$.
Now $x^2 + 2x - 5 = 0$ when $x = \dfrac{-2 \pm \sqrt{2^2 + 4(1)(5)}}{2} = -1 \pm \sqrt{6}$. Hence, the zeros are
$-1 - \sqrt{6}, -1 + \sqrt{6}$, and 3.

45. Since the zeros are $x = -1$, $x = 1$, and $x = 3$, the factors are $x + 1$, $x - 1$, and $x - 3$. Thus
$P(x) = (x + 1)(x - 1)(x - 3) = x^3 - 3x^2 - x + 3$.

47. Since the zeros are $x = -1$, $x = 1$, $x = 3$, and $x = 5$, the factors are $x + 1$, $x - 1$, $x - 3$, and
$x - 5$. Thus $P(x) = (x + 1)(x - 1)(x - 3)(x - 5) = x^4 - 8x^3 + 14x^2 + 8x - 15$.

49. Since the zeros of the polynomial are $1, -2$, and 3, it follows that $P(x) = C(x - 1)(x + 2)(x - 3)$
$= C(x^3 - 2x^2 - 5x + 6) = Cx^3 - 2Cx^2 - 5Cx + 6C$. Since the coefficient of x^2 is to be 3,
$-2C = 3$ so $C = -\frac{3}{2}$. Therefore, $P(x) = -\frac{3}{2}(x^3 - 2x^2 - 5x + 6) = -\frac{3}{2}x^3 + 3x^2 + \frac{15}{2}x - 9$ is
the polynomial.

51. (A) By the Remainder Theorem the remainder when $P(x) = 6x^{1000} - 17x^{562} + 12x + 26$ is
    divided by $x + 1$ is
    $$P(-1) = 6(-1)^{1000} - 17(-1)^{562} + 12(-1) + 26 = 6 - 17 - 12 + 26 = 3.$$

    (B) If $x - 1$ is a factor of $Q(x) = x^{567} - 3x^{400} + x^9 + 2$, then $Q(1)$ must equal 0.
    $$Q(1) = (1)^{567} - 3(1)^{400} + (1)^9 + 2 = 1 - 3 + 1 + 2 = 1 \neq 0, \text{ so } x - 1 \text{ is not a factor.}$$

Exercises 5.3

1. $P(x) = x^3 - 4x^2 + 3$ has possible rational zeros ± 1 and ± 3.

3. $R(x) = 2x^5 + 3x^3 + 4x^2 - 8$ has possible rational zeros $\pm 1, \pm 2, \pm 4, \pm 8, \pm \frac{1}{2}$.

5. $P(x) = x^3 + 3x^2 - 4$. The possible rational zeros are $\pm 1, \pm 2, \pm 4$. $P(x)$ has 1 variation in sign and hence 1 positive real zero. $P(-x) = -x^3 + 3x^2 - 4$ has 2 variations in sign and hence 0 or 2 negative real zeros.

$$\begin{array}{r|rrr} 1 & 1 & 3 & 0 & -4 \\ & & 1 & 4 & 4 \\ \hline & 1 & 4 & 4 & 0 \end{array} \quad \Rightarrow \quad x = 1 \text{ is a zero.}$$

$P(x) = x^3 + 3x^2 - 4 = (x-1)(x^2 + 4x + 4) = (x-1)(x+2)^2$. Therefore, the zeros are $x = -2, 1$.

7. $P(x) = x^3 - 3x - 2$. The possible rational zeros are $\pm 1, \pm 2$. $P(x)$ has 1 variation in sign and hence 1 positive real zero. $P(-x) = -x^3 + 3x - 2$ has 2 variations in sign and hence 0 or 2 negative real zeros.

$$\begin{array}{r|rrr} 1 & 1 & 0 & -3 & -2 \\ & & 1 & 1 & -2 \\ \hline & 1 & 1 & -2 & -4 \end{array} \quad \Rightarrow \quad x = 1 \text{ is not a zero.}$$

$$\begin{array}{r|rrr} 2 & 1 & 0 & -3 & -2 \\ & & 2 & 4 & 2 \\ \hline & 1 & 2 & 1 & 0 \end{array} \quad \Rightarrow \quad x = 2 \text{ is a zero.}$$

$P(x) = x^3 - 3x - 2 = (x-2)(x^2 + 2x + 1) = (x-2)(x+1)^2$. Therefore, the zeros are $x = 2, -1$.

9. $P(x) = x^3 - 6x^2 + 12x - 8$. The possible rational zeros are $\pm 1, \pm 2, \pm 4, \pm 8$. $P(x)$ has 3 variations in sign and hence 1 or 3 positive real zeros. $P(-x) = -x^3 - 6x^2 - 12x - 8$ has no variations in sign and hence 0 negative real zeros.

$$\begin{array}{r|rrr} 1 & 1 & -6 & 12 & -8 \\ & & 1 & -5 & 7 \\ \hline & 1 & -5 & 7 & -1 \end{array} \quad \Rightarrow \quad x = 1 \text{ is not a zero.}$$

$$\begin{array}{r|rrr} 2 & 1 & -6 & 12 & -8 \\ & & 2 & -8 & 8 \\ \hline & 1 & -4 & 4 & 0 \end{array} \quad \Rightarrow \quad x = 2 \text{ is a zero.}$$

$P(x) = x^3 - 6x^2 + 12x - 8 = (x-2)(x^2 - 4x + 4) = (x-2)^3$. Therefore, the zero is $x = 2$.

11. $P(x) = x^3 - 4x^2 + x + 6$. The possible rational zeros are $\pm 1, \pm 2, \pm 3, \pm 6$. $P(x)$ has 2 variations in sign and hence 0 or 2 positive real zeros. $P(-x) = -x^3 - 4x^2 - x + 6$ has 1 variation in sign and hence 1 negative real zeros.

$$\begin{array}{r|rrr} -1 & 1 & -4 & 1 & 6 \\ & & -1 & 5 & -6 \\ \hline & 1 & -5 & 6 & 0 \end{array} \quad \Rightarrow \quad x + 1 \text{ is a factor.}$$

So $P(x) = x^3 - 4x^2 + x + 6 = (x+1)(x^2 - 5x + 6) = (x+1)(x-3)(x-2)$. Therefore, the zeros are $x = -1, 2, 3$.

13. $P(x) = x^3 + 3x^2 + 6x + 4$. The possible rational zeros are $\pm 1, \pm 2, \pm 4$. $P(x)$ has no variation in sign and hence no positive real zeros. $P(-x) = -x^3 + 3x^2 - 6x + 4$ has 3 variations in sign and hence 1 or 3 negative real zeros.

$$
\begin{array}{r|rrrr}
-1 & 1 & 3 & 6 & 4 \\
   &   & -1 & -2 & -4 \\
\hline
   & 1 & 2 & 4 & 0
\end{array}
\quad \Rightarrow \quad x + 1 \text{ is a factor.}
$$

So $P(x) = x^3 + 3x^2 + 6x + 4 = (x + 1)(x^2 + 2x + 4)$. Now, $Q(x) = x^2 + 2x + 4$ has no real zeros, since the discriminant of this quadratic is $b^2 - 4ac = (2)^2 - 4(1)(4) = -12 < 0$. Thus, the only real zero is $x = -1$.

15. **Method #1:** $P(x) = x^4 - 5x^2 + 4$. The possible rational zeros are $\pm 1, \pm 2, \pm 4$. $P(x)$ has 1 variation in sign and hence 1 positive real zero. $P(-x) = x^4 - 5x^2 + 4$ has 2 variations in sign and hence 0 or 2 negative real zeros.

$$
\begin{array}{r|rrrrr}
1 & 1 & 0 & -5 & 0 & 4 \\
  &   & 1 & 1 & -4 & -4 \\
\hline
  & 1 & 1 & -4 & -4 & 0
\end{array}
\quad \Rightarrow \quad x = 1 \text{ is a zero.}
$$

Thus $P(x) = x^4 - 5x^2 + 4 = (x - 1)(x^3 + x^2 - 4x - 4)$. Continuing with the quotient we have:

$$
\begin{array}{r|rrrr}
-1 & 1 & 1 & -4 & -4 \\
   &   & -1 & 0 & 4 \\
\hline
   & 1 & 0 & -4 & 0
\end{array}
\quad \Rightarrow \quad x = -1 \text{ is a zero.}
$$

$P(x) = x^4 - 5x^2 + 4 = (x - 1)(x + 1)(x^2 - 4) = (x - 1)(x + 1)(x - 2)(x + 2)$. Therefore, the zeros are $x = \pm 1, \pm 2$.

Method #2: Substituting $u = x^2$, the polynomial becomes $P(u) = u^2 - 5u + 4$, which factors: $u^2 - 5u + 4 = (u - 1)(u - 4) = (x^2 - 1)(x^2 - 4)$, so either $x^2 = 1$ or $x^2 = 4$. If $x^2 = 1$, then $x = \pm 1$; if $x^2 = 4$, then $x = \pm 2$. Therefore, the zeros are $x = \pm 1, \pm 2$.

17. $P(x) = x^4 + 6x^3 + 7x^2 - 6x - 8$. The possible rational zeros are $\pm 1, \pm 2, \pm 4, \pm 8$. $P(x)$ has 1 variation in sign and hence 1 positive real zero. $P(-x) = x^4 - 6x^3 + 7x^2 + 6x - 8$ has 3 variations in sign and hence 1 or 3 negative real zeros.

$$
\begin{array}{r|rrrrr}
1 & 1 & 6 & 7 & -6 & -8 \\
  &   & 1 & 7 & 14 & 8 \\
\hline
  & 1 & 7 & 14 & 8 & 0
\end{array}
\quad \Rightarrow \quad x = 1 \text{ is a zero, and there are no other positive zeros.}
$$

Thus $P(x) = x^4 + 6x^3 + 7x^2 - 6x - 8 = (x - 1)(x^3 + 7x^2 + 14x + 8)$. Continuing by factoring the quotient, we have:

$$
\begin{array}{r|rrrr}
-1 & 1 & 7 & 14 & 8 \\
   &   & -1 & -6 & -8 \\
\hline
   & 1 & 6 & 8 & 0
\end{array}
\quad \Rightarrow \quad x = -1 \text{ is a zero.}
$$

So $P(x) = x^4 + 6x^3 + 7x^2 - 6x - 8 = (x - 1)(x + 1)(x^2 + 6x + 8)$
$= (x - 1)(x + 1)(x + 2)(x + 4)$. Therefore, the zeros are $x = -4, -2, \pm 1$.

19. $P(x) = 4x^4 - 25x^2 + 36$ has possible rational zeros $\pm 1, \pm 2, \pm 3, \pm 4, \pm 6, \pm 9, \pm 12, \pm 18, \pm 36,$ $\pm \frac{1}{2}, \pm \frac{1}{4}, \pm \frac{3}{2}, \pm \frac{3}{4}, \pm \frac{9}{2}, \pm \frac{9}{4}$. Since $P(x)$ has 2 variations in sign, there are 0 or 2 positive real zeros. Since $P(-x) = 4x^4 - 25x^2 + 36$ has 2 variations in sign, there are 0 or 2 negative real zeros.

$$\begin{array}{r|rrrrr} 1 & 4 & 0 & -25 & 0 & 36 \\ & & 4 & 4 & -21 & -21 \\ \hline & 4 & 4 & -21 & -21 & 15 \end{array}$$

$$\begin{array}{r|rrrrr} 2 & 4 & 0 & -25 & 0 & 36 \\ & & 8 & 16 & -18 & -36 \\ \hline & 4 & 8 & -9 & -18 & 0 \end{array} \Rightarrow \quad x = 2 \text{ is a zero.}$$

$P(x) = (x-2)(4x^3 + 8x^2 - 9x - 18)$

$$\begin{array}{r|rrrr} 2 & 4 & 8 & -9 & -18 \\ & & 8 & 32 & 46 \\ \hline & 4 & 16 & 23 & 28 \end{array} \Rightarrow \quad \text{all positive, } x = 2 \text{ is an upper bound.}$$

$$\begin{array}{r|rrrr} \frac{1}{2} & 4 & 8 & -9 & -18 \\ & & 2 & 5 & -2 \\ \hline & 4 & 10 & -4 & -20 \end{array} \qquad \begin{array}{r|rrrr} \frac{1}{4} & 4 & 8 & -9 & -18 \\ & & 1 & \frac{9}{4} & -\frac{27}{16} \\ \hline & 4 & 9 & -\frac{27}{4} & -\frac{315}{16} \end{array}$$

$$\begin{array}{r|rrrr} \frac{3}{2} & 4 & 8 & -9 & -18 \\ & & 6 & 21 & 18 \\ \hline & 4 & 14 & 12 & 0 \end{array} \Rightarrow \quad x = \frac{3}{2} \text{ is a zero.}$$

$P(x) = (x-2)(2x-3)(2x^2 + 7x + 6) = (x-2)(2x-3)(2x+3)(x+2)$. Therefore, the zeros are $x = \pm 2, \pm \frac{3}{2}$.

Note: Since $P(x)$ has only even terms, factoring by substitution also works. Let $x^2 = u$; then $P(u) = 4u^2 - 25u + 36 = (u-4)(4u-9) = (x^2 - 4)(4x^2 - 9)$, which yields the same results.

21. $P(x) = x^4 + 8x^3 + 24x^2 + 32x + 16$. The possible rational zeros are $\pm 1, \pm 2, \pm 4, \pm 8, \pm 16$. $P(x)$ has no variations in sign and hence no positive real zero. $P(-x) = x^4 - 8x^3 + 24x^2 - 32x + 16$ has 4 variations in sign and hence 0 or 2 or 4 negative real zeros.

$$\begin{array}{r|rrrrr} -1 & 1 & 8 & 24 & 32 & 16 \\ & & -1 & -7 & -17 & -15 \\ \hline & 1 & 7 & 17 & 15 & 1 \end{array} \Rightarrow \quad x = -1 \text{ is not a zero.}$$

$$\begin{array}{r|rrrrr} -2 & 1 & 8 & 24 & 32 & 16 \\ & & -2 & -12 & -24 & -16 \\ \hline & 1 & 6 & 12 & 8 & 0 \end{array} \Rightarrow \quad x = -2 \text{ is a zero.}$$

Thus $P(x) = x^4 + 8x^3 + 24x^2 + 32x + 16 = (x+2)(x^3 + 6x^2 + 12x + 8)$. Continuing by factoring the quotient, we have:

$$\begin{array}{r|rrrr} -2 & 1 & 6 & 12 & 8 \\ & & -2 & -8 & -8 \\ \hline & 1 & 4 & 4 & 0 \end{array} \Rightarrow \quad x = -2 \text{ is a zero.}$$

Thus $P(x) = (x+2)^2(x^2 + 4x + 4) = (x+2)^4$. Therefore, the zero is $x = -2$

23. Factoring by grouping can be applied to this exercise. $4x^3 + 4x^2 - x - 1 = 4x^2(x+1) - (x+1)$ $= (x+1)(4x^2 - 1) = (x+1)(2x+1)(2x-1)$. Therefore, the zeros are $x = -1, \pm \frac{1}{2}$.

25. $P(x) = 4x^3 - 7x + 3$. The possible rational zeros are $\pm 1, \pm 3, \pm \frac{1}{2}, \pm \frac{3}{2}, \pm \frac{1}{4}, \pm \frac{3}{4}$. Since $P(x)$ has 2 variations in sign, there are 0 or 2 positive zeros. Since $P(-x) = -4x^3 + 7x + 3$ has 1 variation in sign, there is 1 negative zero.

$$\frac{1}{2} \begin{array}{|rrrr} 4 & 0 & -7 & 3 \\ & 2 & 1 & -3 \\ \hline 4 & 2 & -6 & 0 \end{array} \quad \Rightarrow \quad x = \tfrac{1}{2} \text{ is a zero.}$$

$P(x) = (x - \tfrac{1}{2})(4x^2 + 2x - 6) = (2x - 1)(2x^2 + x - 3) = (2x - 1)(x - 1)(2x + 3) = 0.$ Thus, the zeros are $x = -\tfrac{3}{2}, \tfrac{1}{2}, 1.$

27. $P(x) = 2x^4 - 7x^3 + 3x^2 + 8x - 4$. The possible rational zeros are $\pm 1, \pm 2, \pm 4, \pm\tfrac{1}{2}$. $P(x)$ has 3 variations in sign and hence 1 or 3 positive real zeros. $P(-x) = 2x^4 + 7x^3 + 3x^2 - 8x - 4$ has 1 variation in sign and hence 1 negative real zero.

$$1 \begin{array}{|rrrrr} 2 & -7 & 3 & 8 & -4 \\ & 2 & -5 & -2 & 6 \\ \hline 2 & -5 & -2 & 6 & 2 \end{array} \quad \Rightarrow \quad x = 1 \text{ is not a zero.}$$

$$\tfrac{1}{2} \begin{array}{|rrrrr} 2 & -7 & 3 & 8 & -4 \\ & 1 & -3 & 0 & 4 \\ \hline 2 & -6 & 0 & 8 & 0 \end{array} \quad \Rightarrow \quad x = \tfrac{1}{2} \text{ is a zero.}$$

Thus $P(x) = 2x^4 - 7x^3 + 3x^2 + 8x - 4 = \left(x - \tfrac{1}{2}\right)(2x^3 - 6x^2 + 8)$. Continuing by factoring the quotient, we have:

$$2 \begin{array}{|rrrr} 2 & -6 & 0 & 8 \\ & 4 & -4 & -8 \\ \hline 2 & -2 & -4 & 0 \end{array} \quad \Rightarrow \quad x = 2 \text{ is a zero.}$$

So $P(x) = \left(x - \tfrac{1}{2}\right)(x - 2)(2x^2 - 2x - 4) = 2\left(x - \tfrac{1}{2}\right)(x - 2)(x^2 - x - 2)$
$= 2\left(x - \tfrac{1}{2}\right)(x - 2)^2(x + 1)$. Thus, the zeros are $x = \tfrac{1}{2}, 2, -1.$

29. $P(x) = x^5 + 3x^4 - 9x^3 - 31x^2 + 36$. The possible rational zeros are $\pm 1, \pm 2, \pm 3, \pm 4, \pm 6, \pm 8,$ $\pm 9, \pm 12, \pm 18$. $P(x)$ has 2 variations in sign and hence 0 or 2 positive real zeros. $P(-x) = -x^5 + 3x^4 + 9x^3 - 31x^2 + 36$ has 3 variations in sign and hence 1 or 3 negative real zeros.

$$1 \begin{array}{|rrrrrr} 1 & 3 & -9 & -31 & 0 & 36 \\ & 1 & 4 & -5 & -36 & -36 \\ \hline 1 & 4 & -5 & -36 & -36 & 0 \end{array} \quad \Rightarrow \quad x = 1 \text{ is a zero.}$$

So $P(x) = x^5 + 3x^4 - 9x^3 - 31x^2 + 36 = (x - 1)(x^4 + 4x^3 - 5x^2 - 36x - 36)$. Continuing by factoring the quotient, we have:

$$1 \begin{array}{|rrrrr} 1 & 4 & -5 & -36 & -36 \\ & 1 & 5 & 0 & -36 \\ \hline 1 & 1 & 0 & -36 & -72 \end{array} \qquad 2 \begin{array}{|rrrrr} 1 & 4 & -5 & -36 & -36 \\ & 2 & 12 & 14 & -44 \\ \hline 1 & 6 & 7 & -22 & -80 \end{array}$$

$$3 \begin{array}{|rrrrr} 1 & 4 & -5 & -36 & -36 \\ & 3 & 21 & 48 & 36 \\ \hline 1 & 7 & 16 & 12 & 0 \end{array} \quad \Rightarrow \quad x = 3 \text{ is a zero.}$$

So $P(x) = (x - 1)(x - 3)(x^3 + 7x^2 + 16x + 12)$. Since we have 2 positive zeros, there are no more positive zeros, so we continue by factoring the quotient with possible negative zeros.

$$-1 \begin{array}{|rrrr} 1 & 7 & 16 & 12 \\ & -1 & -6 & -10 \\ \hline 1 & 6 & 10 & 2 \end{array} \qquad -2 \begin{array}{|rrrr} 1 & 7 & 16 & 12 \\ & -2 & -10 & -12 \\ \hline 1 & 5 & 6 & 0 \end{array} \quad \Rightarrow \quad x = -2 \text{ is a zero.}$$

Then $P(x) = (x-1)(x-3)(x+2)(x^2+5x+6) = (x-1)(x-3)(x+2)^2(x+3)$. Thus, the zeros are $x = 1, 3, -2, -3$.

31. $P(x) = 3x^5 - 14x^4 - 14x^3 + 36x^2 + 43x + 10$ has possible rational zeros $\pm 1, \pm 2, \pm 5, \pm 10, \pm \frac{1}{3}$, $\pm \frac{2}{3}, \pm \frac{5}{3}, \pm \frac{10}{3}$. Since $P(x)$ has 2 variations in sign, there are 0 or 2 positive real zeros. Since $P(-x) = -3x^5 - 14x^4 + 14x^3 + 36x^2 - 43x + 10$ has 3 variations in sign, there are 1 or 3 negative real zeros.

$$\begin{array}{r|rrrrrr}
1 & 3 & -14 & -14 & 36 & 43 & 10 \\
  &   & 3 & -11 & -25 & 11 & 54 \\
\hline
  & 3 & -11 & -25 & 11 & 54 & 64
\end{array}$$

$$\begin{array}{r|rrrrrr}
2 & 3 & -14 & -14 & 36 & 43 & 10 \\
  &   & 6 & -16 & -60 & -48 & -10 \\
\hline
  & 3 & -8 & -30 & -24 & -5 & 0
\end{array} \Rightarrow \quad x = 2 \text{ is a zero.}$$

$P(x) = (x-2)(3x^4 - 8x^3 - 30x^2 - 24x - 5)$

$$\begin{array}{r|rrrrr}
2 & 3 & -8 & -30 & -24 & -5 \\
  &   & 6 & -4 & -68 & -184 \\
\hline
  & 3 & -2 & -34 & -92 & -189
\end{array}$$

$$\begin{array}{r|rrrrr}
5 & 3 & -8 & -30 & -24 & -5 \\
  &   & 15 & 35 & 25 & 5 \\
\hline
  & 3 & 7 & 5 & 1 & 0
\end{array} \Rightarrow \quad x = 5 \text{ is a zero.}$$

$P(x) = (x-2)(x-5)(3x^3 + 7x^2 + 5x + 1)$

Since $3x^3 + 7x^2 + 5x + 1$ has no change in signs, there are no more positive zeros.

$$\begin{array}{r|rrrr}
-1 & 3 & 7 & 5 & 1 \\
   &   & -3 & -4 & -1 \\
\hline
   & 3 & 4 & 1 & 0
\end{array} \Rightarrow \quad x = -1 \text{ is a zero.}$$

$P(x) = (x-2)(x-5)(x+1)(3x^2 + 4x + 1) = (x-2)(x-5)(x+1)(x+1)(3x+1)$.
Therefore, the zeros are $x = -1, -\frac{1}{3}, 2, 5$.

33. $P(x) = x^3 + 4x^2 + 3x - 2$. The possible rational zeros are $\pm 1, \pm 2$. $P(x)$ has 1 variation in sign and hence 1 positive real zero. $P(-x) = -x^3 + 4x^2 - 3x - 2$ has 2 variations in sign and hence 0 or 2 negative real zeros.

$$\begin{array}{r|rrrr}
1 & 1 & 4 & 3 & -2 \\
  &   & 1 & 5 & 8 \\
\hline
  & 1 & 5 & 8 & 6
\end{array} \Rightarrow \quad x = 1 \text{ is an upper bound.}$$

$$\begin{array}{r|rrrr}
-1 & 1 & 4 & 3 & -2 \\
   &   & -1 & -3 & 0 \\
\hline
   & 1 & 3 & 0 & -2
\end{array}
\qquad
\begin{array}{r|rrrr}
-2 & 1 & 4 & 3 & -2 \\
   &   & -2 & -4 & 2 \\
\hline
   & 1 & 2 & -1 & 0
\end{array} \Rightarrow \quad x = -2 \text{ is a zero.}$$

So $P(x) = (x+2)(x^2 + 2x - 1)$. Using the quadratic formula on the second factor, we have:
$$x = \frac{-2 \pm \sqrt{2^2 - 4(1)(-1)}}{2(1)} = \frac{-2 \pm \sqrt{8}}{2} = \frac{-2 \pm 2\sqrt{2}}{2} = -1 \pm \sqrt{2}. \text{ Therefore, the zeros are}$$
$x = -2, -1 + \sqrt{2}, -1 - \sqrt{2}$.

35. $P(x) = x^4 - 6x^3 + 4x^2 + 15x + 4$. The possible rational zeros are $\pm 1, \pm 2, \pm 4$. $P(x)$ has 2 variations in sign and hence 0 or 2 positive real zeros. $P(-x) = x^4 + 6x^3 + 4x^2 - 15x + 4$ has 2 variations in sign and hence 0 or 2 negative real zeros.

$$1 \overline{\smash{\big)}\,\begin{array}{rrrrr} 1 & -6 & 4 & 15 & 4 \\ & 1 & -5 & -1 & 14 \\ \hline 1 & -5 & -1 & 14 & 18 \end{array}} \qquad 2 \overline{\smash{\big)}\,\begin{array}{rrrrr} 1 & -6 & 4 & 15 & 4 \\ & 2 & -8 & -8 & 14 \\ \hline 1 & -4 & -4 & 7 & 18 \end{array}}$$

$$4 \overline{\smash{\big)}\,\begin{array}{rrrrr} 1 & -6 & 4 & 15 & 4 \\ & 4 & -8 & -16 & -4 \\ \hline 1 & -2 & -4 & -1 & 0 \end{array}} \Rightarrow \quad x = 4 \text{ is a zero.}$$

So $P(x) = (x - 4)(x^3 - 2x^2 - 4x - 1)$. Continuing by factoring the quotient, we have:

$$4 \overline{\smash{\big)}\,\begin{array}{rrrr} 1 & -2 & -4 & -1 \\ & 4 & 8 & 16 \\ \hline 1 & 2 & 4 & 15 \end{array}} \Rightarrow \quad x = 4 \text{ is an upper bound.}$$

$$-1 \overline{\smash{\big)}\,\begin{array}{rrrr} 1 & -2 & -4 & -1 \\ & -1 & 3 & 1 \\ \hline 1 & -3 & -1 & 0 \end{array}} \Rightarrow \quad x = -1 \text{ is a zero.}$$

So $P(x) = (x - 4)(x + 1)(x^2 - 3x - 1)$. Using the quadratic formula on the third factor, we have:
$$x = \frac{-(-3) \pm \sqrt{(-3)^2 - 4(1)(-1)}}{2(1)} = \frac{3 \pm \sqrt{13}}{2}. \text{ Therefore, the zeros are } x = 4, -1, \frac{3 \pm \sqrt{13}}{2}.$$

37. $P(x) = x^4 - 7x^3 + 14x^2 - 3x - 9$. The possible rational zeros are $\pm 1, \pm 3, \pm 9$. $P(x)$ has 3 variations in sign and hence 1 or 3 positive real zeros. $P(-x) = x^4 + 7x^3 + 14x^2 + 3x - 4$ has 1 variation in sign and hence 1 negative real zero.

$$1 \overline{\smash{\big)}\,\begin{array}{rrrrr} 1 & -7 & 14 & -3 & -9 \\ & 1 & -6 & 8 & 5 \\ \hline 1 & -6 & 8 & 5 & 4 \end{array}} \qquad 3 \overline{\smash{\big)}\,\begin{array}{rrrrr} 1 & -7 & 14 & -3 & -9 \\ & 3 & -12 & 6 & 9 \\ \hline 1 & -4 & 2 & 3 & 0 \end{array}} \Rightarrow \quad x = 3 \text{ is a zero.}$$

So $P(x) = (x - 3)(x^3 - 4x^2 + 2x + 3)$. Since the constant term of the second term is 3, ± 9 are no longer possible zeros. Continuing by factoring the quotient, we have:

$$3 \overline{\smash{\big)}\,\begin{array}{rrrr} 1 & -4 & 2 & 3 \\ & 3 & -3 & -3 \\ \hline 1 & -1 & -1 & 0 \end{array}} \Rightarrow \quad x = 3 \text{ is a zero again.}$$

So $P(x) = (x - 3)^2(x^2 - x - 1)$. Using the quadratic formula on the second factor, we have:
$$x = \frac{-(-1) \pm \sqrt{(-1)^2 - 4(1)(-1)}}{2(1)} = \frac{1 \pm \sqrt{5}}{2}. \text{ Therefore, the zeros are } x = 3, \frac{1 \pm \sqrt{5}}{2}.$$

39. $P(x) = 4x^3 - 6x^2 + 1$. The possible rational zeros are $\pm 1, \pm \frac{1}{2}, \pm \frac{1}{4}$. $P(x)$ has 2 variations in sign and hence 0 or 2 positive real zeros. $P(-x) = -4x^3 - 6x^2 + 1$ has 1 variation in sign and hence 1 negative real zero.

$$1 \overline{\smash{\big)}\,\begin{array}{rrrr} 4 & -6 & 0 & 1 \\ & 4 & -2 & -2 \\ \hline 4 & -2 & -2 & -1 \end{array}} \qquad \tfrac{1}{2} \overline{\smash{\big)}\,\begin{array}{rrrr} 4 & -6 & 0 & 1 \\ & 2 & -2 & -1 \\ \hline 4 & -4 & -2 & 0 \end{array}} \Rightarrow \quad x = \tfrac{1}{2} \text{ is a zero.}$$

So $P(x) = \left(x - \frac{1}{2}\right)(4x^2 - 4x - 2)$. Using the quadratic formula on the second factor, we have:
$$x = \frac{-(-4) \pm \sqrt{(-4)^2 - 4(4)(-2)}}{2(4)} = \frac{4 \pm \sqrt{48}}{8} = \frac{4 \pm 4\sqrt{3}}{8} = \frac{1 \pm \sqrt{3}}{2}. \text{ Therefore, the zeros}$$

are $x = \frac{1}{2}, \frac{1 \pm \sqrt{3}}{2}$.

41. $P(x) = 2x^4 + 15x^3 + 17x^2 + 3x - 1$. The possible rational zeros are $\pm 1, \pm\frac{1}{2}$. $P(x)$ has 1 variation in sign and hence 1 positive real zero. $P(-x) = 2x^4 - 15x^3 + 17x^2 - 3x - 1$ has 3 variations in sign and hence 1 or 3 negative real zeros.

$$
\frac{1}{2} \begin{array}{|rrrrr} 2 & 15 & 17 & 3 & -1 \\ & 1 & 8 & \frac{25}{2} & \frac{31}{4} \\ \hline 2 & 16 & 25 & \frac{31}{2} & \frac{27}{4} \end{array} \quad \Rightarrow \quad x = \tfrac{1}{2} \text{ is an upper bound.}
$$

$$
-\tfrac{1}{2} \begin{array}{|rrrrr} 2 & 15 & 17 & 3 & -1 \\ & & -1 & -7 & -5 & 1 \\ \hline 2 & 14 & 10 & -2 & 0 \end{array} \quad \Rightarrow \quad x = -\tfrac{1}{2} \text{ is a zero.}
$$

So $P(x) = \left(x + \tfrac{1}{2}\right)(2x^3 + 14x^2 + 10x - 2) = 2\left(x + \tfrac{1}{2}\right)(x^3 + 7x^2 + 5x - 1)$.

$$
-1 \begin{array}{|rrrr} 1 & 7 & 5 & -1 \\ & -1 & -6 & 1 \\ \hline 1 & 6 & -1 & 0 \end{array} \quad \Rightarrow \quad x = -1 \text{ is a zero.}
$$

So $P(x) = \left(x + \tfrac{1}{2}\right)(2x^3 + 14x^2 + 10x - 2) = 2\left(x + \tfrac{1}{2}\right)(x + 1)(x^2 + 6x - 1)$

Using the quadratic formula on the third factor, we have: $x = \dfrac{-(6) \pm \sqrt{(6)^2 - 4(1)(-1)}}{2(1)}$

$= \frac{-6 \pm \sqrt{40}}{2} = \frac{-6 \pm 2\sqrt{10}}{2} = -3 \pm \sqrt{10}$. Therefore, the zeros are $x = -1, -\tfrac{1}{2}, -3 \pm \sqrt{10}$.

43. (a) $P(x) = x^3 - 3x^2 - 4x + 12$ has possible rational zeros $\pm 1, \pm 2, \pm 3, \pm 4, \pm 6, \pm 12$.

$$
1 \begin{array}{|rrrr} 1 & -3 & -4 & 12 \\ & 1 & -2 & -6 \\ \hline 1 & -2 & -6 & 6 \end{array} \qquad\qquad 2 \begin{array}{|rrrr} 1 & -3 & -4 & 12 \\ & 2 & -2 & -12 \\ \hline 1 & -1 & -6 & 0 \end{array} \quad \Rightarrow \quad x = 2 \text{ is a zero.}
$$

So $P(x) = (x - 2)(x^2 - x - 6) = (x - 2)(x + 2)(x - 3)$. The real zeros of P are $-2, 2, 3$.

(b)

45. (a) $P(x) = -x^3 + 2x^2 + 15x - 36$ has possible rational zeros $\pm 1, \pm 2, \pm 3, \pm 4, \pm 6, \pm 9, \pm 12, \pm 18$.

$$
1 \begin{array}{|rrrr} -1 & 2 & 15 & -36 \\ & -1 & 1 & 16 \\ \hline -1 & 1 & 16 & -20 \end{array} \qquad\qquad 2 \begin{array}{|rrrr} -1 & 2 & 15 & -36 \\ & -2 & 0 & 30 \\ \hline -1 & 0 & 15 & -6 \end{array}
$$

$$
3 \begin{array}{|rrrr} -1 & 2 & 15 & -36 \\ & -3 & -3 & -36 \\ \hline -1 & -1 & 12 & 0 \end{array} \quad \Rightarrow \quad x = 3 \text{ is a zero.}
$$

So $P(x) = (x - 3)(-x^2 - x + 12) = (x - 3)(-x + 3)(x + 4) = -(x - 3)^2(x + 4)$. The real zeros of P are $-4, 3$.

(b)

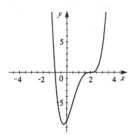

47. (a) $P(x) = x^4 - 5x^3 + 6x^2 + 4x - 8$ has possible rational zeros $\pm 1, \pm 2, \pm 4, \pm 8$.

$$
\begin{array}{r|rrrrr}
1 & 1 & -5 & 6 & 4 & -8 \\
  &   & 1 & -4 & 2 & 6 \\
\hline
  & 1 & -4 & 2 & 6 & -2
\end{array}
\qquad
\begin{array}{r|rrrrr}
2 & 1 & -5 & 6 & 4 & -8 \\
  &   & 2 & -6 & 0 & 8 \\
\hline
  & 1 & -3 & 0 & 4 & 0
\end{array}
\ \Rightarrow \quad x = 2 \text{ is a zero.}
$$

So $P(x) = (x-2)(x^3 - 3x^2 + 4)$, and the possible rational zeros are restricted to $-1, \pm 2$, ± 4.

$$
\begin{array}{r|rrrr}
2 & 1 & -3 & 0 & 4 \\
  &   & 2 & -2 & -4 \\
\hline
  & 1 & -1 & -2 & 0
\end{array}
\ \Rightarrow \quad x = 2 \text{ is a zero again.}
$$

$P(x) = (x-2)^2(x^2 - x - 2) = (x-2)^2(x-2)(x+1) = (x-2)^3(x+1)$. So the real zeros of P are -1 and 2.

(b)

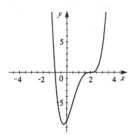

49. (a) $P(x) = x^5 - x^4 - 5x^3 + x^2 + 8x + 4$ has possible rational zeros $\pm 1, \pm 2, \pm 4$.

$$
\begin{array}{r|rrrrrr}
1 & 1 & -1 & -5 & 1 & 8 & 4 \\
  &   & 1 & 0 & -5 & -4 & 4 \\
\hline
  & 1 & 0 & -5 & -4 & 4 & 8
\end{array}
$$

$$
\begin{array}{r|rrrrrr}
2 & 1 & -1 & -5 & 1 & 8 & 4 \\
  &   & 2 & 2 & -6 & -10 & -4 \\
\hline
  & 1 & 1 & -3 & -5 & -2 & 0
\end{array}
\ \Rightarrow \quad x = 2 \text{ is a zero.}
$$

So $P(x) = (x-2)(x^4 + x^3 - 3x^2 - 5x - 2)$, and the possible rational zeros are restricted to $-1, \pm 2$.

$$
\begin{array}{r|rrrrr}
2 & 1 & 1 & -3 & -5 & -2 \\
  &   & 2 & 6 & 6 & 2 \\
\hline
  & 1 & 3 & 3 & 1 & 0
\end{array}
\ \Rightarrow \quad x = 2 \text{ is a zero again.}
$$

So $P(x) = (x-2)^2(x^3 + 3x^2 + 3x + 1)$, and the possible rational zeros are restricted to -1.

$$
\begin{array}{r|rrrr}
-1 & 1 & 3 & 3 & 1 \\
   &   & -1 & -2 & -1 \\
\hline
   & 1 & 2 & 1 & 0
\end{array}
\ \Rightarrow \quad x = -1 \text{ is a zero.}
$$

So $P(x) = (x-2)^2(x+1)(x^2+2x+1) = (x-2)^2(x+1)^3.$, and the real zeros of P are -1 and 2.

(b)

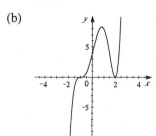

51. $P(x) = x^3 - x^2 - x - 3$. Since $P(x)$ has 1 variation in sign, P has 1 positive real zero. Since $P(-x) = -x^3 - x^2 + x - 3$ has 2 variations in sign, P has 2 or 0 negative real zeros. Thus, P has 1 or 3 real zeros.

53. $P(x) = 2x^6 + 5x^4 - x^3 - 5x - 1$. Since $P(x)$ has 1 variation in sign, P has 1 positive real zero. Since $P(-x) = 2x^6 + 5x^4 + x^3 + 5x - 1$ has 1 variation in sign, P has 1 negative real zero. Therefore, P has 2 real zeros.

55. $P(x) = x^5 + 4x^3 - x^2 + 6x$. Since $P(x)$ has 2 variations in sign, P has 2 or 0 positive real zeros. Since $P(-x) = -x^5 - 4x^3 - x^2 - 6x$ has 0 variations in sign, P has 0 negative real zeros. Therefore, P has a total of 1 or 3 real zeros (since $x = 0$ is a zero, but is neither positive nor negative).

57. $P(x) = 2x^3 + 5x^2 + x - 2; \ a = -3, b = 1$

$$
\begin{array}{r|rrrr}
-3 & 2 & 5 & 1 & -2 \\
   &   & -6 & 3 & -12 \\
\hline
   & 2 & -1 & 4 & -14
\end{array}
\qquad \text{alternating signs} \ \Rightarrow \ \text{lower bound.}
$$

$$
\begin{array}{r|rrrr}
1 & 2 & 5 & 1 & -2 \\
  &   & 2 & 7 & 8 \\
\hline
  & 2 & 7 & 8 & 6
\end{array}
\qquad \text{all non-negative} \ \Rightarrow \ \text{upper bound.}
$$

Therefore $a = -3, b = 1$ are lower and upper bounds, respectively.

59. $P(x) = 8x^3 + 10x^2 - 39x + 9; \ a = -3, b = 2$

$$
\begin{array}{r|rrrr}
-3 & 8 & 10 & -39 & 9 \\
   &   & -24 & 42 & -9 \\
\hline
   & 8 & -14 & 3 & 0
\end{array}
\qquad \text{alternating signs} \ \Rightarrow \ \text{lower bound.}
$$

$$
\begin{array}{r|rrrr}
2 & 8 & 10 & -39 & 9 \\
  &   & 16 & 52 & 26 \\
\hline
  & 8 & 26 & 13 & 35
\end{array}
\qquad \text{all non-negative} \ \Rightarrow \ \text{upper bound.}
$$

Therefore $a = -3, b = 2$ are lower and upper bounds, respectively. Note that $x = -3$ is also a zero.

There are many possible solutions to Exercises 61 and 63 since we are only asked to find 'an upper bound' and 'a lower bound'.

61. $P(x) = x^3 - 3x^2 + 4$ and use the Upper and Lower Bounds Theorem:

$$
\begin{array}{r|rrrr}
-1 & 1 & -3 & 0 & 4 \\
   &   & -1 & 4 & -4 \\
\hline
   & 1 & -4 & 4 & 0
\end{array}
\qquad \text{alternating signs} \ \Rightarrow \ \text{lower bound.}
$$

$$
\begin{array}{r|rrrr}
3 & 1 & -3 & 0 & 4 \\
  &   & 3 & 0 & 0 \\
\hline
  & 1 & 0 & 0 & 4
\end{array}
\qquad \text{all non-negative} \quad \Rightarrow \quad \text{upper bound.}
$$

Therefore -1 is a lower bound (and a zero) and 3 is an upper bound.

63. $P(x) = x^4 - 2x^3 + x^2 - 9x + 2$.

$$
\begin{array}{r|rrrrr}
1 & 1 & -2 & 1 & -9 & 2 \\
  &   & 1 & -1 & 0 & -9 \\
\hline
  & 1 & -1 & 0 & -9 & -7
\end{array}
\qquad
\begin{array}{r|rrrrr}
2 & 1 & -2 & 1 & -9 & 2 \\
  &   & 2 & 0 & 2 & -14 \\
\hline
  & 1 & 0 & 1 & -7 & -12
\end{array}
$$

$$
\begin{array}{r|rrrrr}
3 & 1 & -2 & 1 & -9 & 2 \\
  &   & 3 & 3 & 12 & 9 \\
\hline
  & 1 & 1 & 4 & 3 & 11
\end{array}
\qquad \text{all positive} \quad \Rightarrow \quad \text{upper bound.}
$$

$$
\begin{array}{r|rrrrr}
-1 & 1 & -2 & 1 & -9 & 2 \\
   &   & -1 & 3 & -4 & 13 \\
\hline
   & 1 & -3 & 4 & -13 & 15
\end{array}
\qquad \text{alternating signs} \quad \Rightarrow \quad \text{lower bound.}
$$

Therefore -1 is a lower bound and 3 is an upper bound.

65. $P(x) = 2x^4 + 3x^3 - 4x^2 - 3x + 2$.

$$
\begin{array}{r|rrrrr}
1 & 2 & 3 & -4 & -3 & 2 \\
  &   & 2 & 5 & 1 & -2 \\
\hline
  & 2 & 5 & 1 & -2 & 0
\end{array}
\quad \Rightarrow \quad x = 1 \text{ is a zero.}
$$

$P(x) = (x - 1)(2x^3 + 5x^2 + x - 2)$

$$
\begin{array}{r|rrrr}
-1 & 2 & 5 & 1 & -2 \\
   &   & -2 & -3 & 2 \\
\hline
   & 2 & 3 & -2 & 0
\end{array}
\quad \Rightarrow \quad x = -1 \text{ is a zero.}
$$

$P(x) = (x - 1)(x + 1)(2x^2 + 3x - 2) = (x - 1)(x + 1)(2x - 1)(x + 2)$. Therefore, the zeros are $x = -2, \frac{1}{2}, \pm 1$.

67. <u>Method #1</u>: $P(x) = 4x^4 - 21x^2 + 5$ has 2 variations in sign, so by Descartes' rule of signs there are either 2 or 0 positive zeros. If we replace x with $(-x)$, the function does not change, so there are either 2 or 0 negative zeros. Possible rational zeros are $\pm 1, \pm \frac{1}{2}, \pm \frac{1}{4}, \pm 5, \pm \frac{5}{2}, \pm \frac{5}{4}$. By inspection, ± 1 and ± 5 are not zeros, so we must look for non-integer solutions:

$$
\begin{array}{r|rrrrr}
\frac{1}{2} & 4 & 0 & -21 & 0 & 5 \\
            &   & 2 & 1 & -10 & -5 \\
\hline
            & 4 & 2 & -20 & -10 & 0
\end{array}
\quad \Rightarrow \quad x = \frac{1}{2} \text{ is a zero.}
$$

$P(x) = \left(x - \frac{1}{2}\right)(4x^3 + 2x^2 - 20x - 10)$, continuing with the quotient, we have:

$$
\begin{array}{r|rrrr}
-\frac{1}{2} & 4 & 2 & -20 & -10 \\
             &   & -2 & 0 & 10 \\
\hline
             & 4 & 0 & -20 & 0
\end{array}
\quad \Rightarrow \quad x = -\frac{1}{2} \text{ is a zero.}
$$

$P(x) = \left(x - \frac{1}{2}\right)\left(x + \frac{1}{2}\right)(4x^2 - 20) = 0$. If $4x^2 - 20 = 0$, then $x = \pm\sqrt{5}$. Thus the zeros are $x = \pm\frac{1}{2}, \pm\sqrt{5}$.

<u>Method #2</u>: Substituting $u = x^2$, the equation becomes $4u^2 - 21u + 5 = 0$, which factors: $4u^2 - 21u + 5 = (4u - 1)(u - 5) = (4x^2 - 1)(x^2 - 5)$. Then either we have $x^2 = 5$, so that $x = \pm\sqrt{5}$, or we have $x^2 = \frac{1}{4}$, so that $x = \pm\sqrt{\frac{1}{4}} = \pm\frac{1}{2}$. Thus the zeros are $x = \pm\frac{1}{2}, \pm\sqrt{5}$.

69. $P(x) = x^5 - 7x^4 + 9x^3 + 23x^2 - 50x + 24$. The possible rational zeros are $\pm 1, \pm 2, \pm 3, \pm 4, \pm 6$, $\pm 8, \pm 12, \pm 24$. $P(x)$ has 4 variations in sign and hence 0, 2, or 4 positive real zeros. $P(-x) = -x^5 - 7x^4 - 9x^3 + 23x^2 + 50x + 24$ has 1 variation in sign, and hence 1 negative real zero.

$$
\begin{array}{r|rrrrrr}
1 & 1 & -7 & 9 & 23 & -50 & 24 \\
  &   & 1 & -6 & 3 & 26 & -24 \\
\hline
  & 1 & -6 & 3 & 26 & -24 & 0
\end{array}
\quad \Rightarrow \quad x = 1 \text{ is a zero.}
$$

$P(x) = (x - 1)(x^4 - 6x^3 + 3x^2 + 26x - 24)$; continuing with the quotient, we try 1 again.

$$
\begin{array}{r|rrrrr}
1 & 1 & -6 & 3 & 26 & -24 \\
  &   & 1 & -5 & -2 & 24 \\
\hline
  & 1 & -5 & -2 & 24 & 0
\end{array}
\quad \Rightarrow \quad x = 1 \text{ is a zero again.}
$$

$P(x) = (x - 1)^2(x^3 - 5x^2 - 2x + 24)$; continuing with the quotient, we start by trying 1 again.

$$
\begin{array}{r|rrrr}
1 & 1 & -5 & -2 & 24 \\
  &   & 1 & -4 & -6 \\
\hline
  & 1 & -4 & -6 & 18
\end{array}
\qquad
\begin{array}{r|rrrr}
2 & 1 & -5 & -2 & 24 \\
  &   & 2 & -6 & -16 \\
\hline
  & 1 & -3 & -8 & 8
\end{array}
$$

$$
\begin{array}{r|rrrr}
3 & 1 & -5 & -2 & 24 \\
  &   & 3 & -6 & -24 \\
\hline
  & 1 & -2 & -8 & 0
\end{array}
\quad \Rightarrow \quad x = 3 \text{ is a zero.}
$$

$P(x) = (x - 1)^2(x - 3)(x^2 - 2x - 8) = (x - 1)^2(x - 3)(x - 4)(x + 2)$. Therefore, the zeros are $x = -2, 1, 3, 4$.

71. $P(x) = x^3 - x - 2$. The only possible rational zeros of $P(x)$ are ± 1 and ± 2.

$$
\begin{array}{r|rrrr}
1 & 1 & 0 & -1 & -2 \\
  &   & 1 & 1 & 0 \\
\hline
  & 1 & 1 & 0 & -2
\end{array}
\qquad
\begin{array}{r|rrrr}
2 & 1 & 0 & -1 & -2 \\
  &   & 2 & 4 & 6 \\
\hline
  & 1 & 2 & 3 & 4
\end{array}
\qquad
\begin{array}{r|rrrr}
-1 & 1 & 0 & -1 & -2 \\
   &   & -1 & 1 & 0 \\
\hline
   & 1 & -1 & 0 & -2
\end{array}
$$

Since the row that contains -1 alternates between non-negative and non-positive, -1 is a lower bound and there is no need to try -2. Therefore, $P(x)$ does not have any rational zeros.

73. $P(x) = 3x^3 - x^2 - 6x + 12$ has possible rational zeros $\pm 1, \pm 2, \pm 3, \pm 4, \pm 6, \pm 12, \pm \frac{1}{3}, \pm \frac{2}{3}, \pm \frac{4}{3}$.

$$
\begin{array}{r|rrrr}
 & 3 & -1 & -6 & 12 \\
\hline
1 & 3 & 2 & -4 & 8 \\
2 & 3 & 5 & 4 & 20 \\
\frac{1}{3} & 3 & 0 & -6 & 10 \\
\frac{2}{3} & 3 & 1 & -\frac{16}{3} & \frac{76}{9} \\
\frac{4}{3} & 3 & 3 & -2 & \frac{28}{3} \\
-1 & 3 & -4 & -2 & 14 \\
-2 & 3 & -7 & 8 & -4 \\
-\frac{1}{3} & 3 & -2 & -\frac{16}{3} & \frac{124}{9} \\
-\frac{2}{3} & 3 & -3 & -4 & \frac{44}{3} \\
-\frac{4}{3} & 3 & -5 & \frac{2}{3} & \frac{100}{9}
\end{array}
$$

all positive $\Rightarrow$ $x = 2$ is an upper bound

alternating signs $\Rightarrow$ $x = -2$ is a lower bound

Therefore, there are no rational zeros.

75. $P(x) = x^3 - 3x^2 - 4x + 12$; $[-4, 4]$ by $[-15, 15]$.
The possible rational zeros are $\pm 1, \pm 2, \pm 3, \pm 4, \pm 6, \pm 12$.
By observing the graph of P, the rational zeros are
$x = -2, 2, 3$.

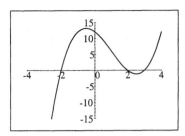

77. $P(x) = 2x^4 - 5x^3 - 14x^2 + 5x + 12$; $[-2, 5]$ by $[-40, 40]$.
The possible rational zeros are $\pm 1, \pm 2, \pm 3, \pm 4, \pm 6$,
$\pm 12, \pm \frac{1}{2}, \pm \frac{3}{2}$.
By observing the graph of P, the zeros are $x = -\frac{3}{2}, -1, 1, 4$.

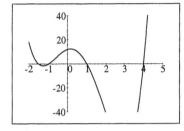

79. $x^4 - x - 4 = 0$. Possible rational solutions are $\pm 1, \pm 2, \pm 4$.

$$
\begin{array}{r|rrrrr}
1 & 1 & 0 & 0 & -1 & -4 \\
  &   & 1 & 1 & 1 & 0 \\
\hline
  & 1 & 1 & 1 & 0 & -4
\end{array}
\qquad
\begin{array}{r|rrrrr}
2 & 1 & 0 & 0 & -1 & -4 \\
  &   & 2 & 4 & 8 & 14 \\
\hline
  & 1 & 2 & 4 & 7 & 10
\end{array}
\Rightarrow \quad x = 2
$$

is an upper bound.

$$
\begin{array}{r|rrrrr}
-1 & 1 & 0 & 0 & -1 & -4 \\
   &   & -1 & 1 & -1 & 2 \\
\hline
   & 1 & -1 & 1 & -2 & -2
\end{array}
\qquad
\begin{array}{r|rrrrr}
-2 & 1 & 0 & 0 & -1 & -4 \\
   &   & -2 & 4 & -8 & 18 \\
\hline
   & 1 & -2 & 4 & -9 & 14
\end{array}
\Rightarrow \quad x = -2
$$

is a lower bound.

Therefore, we graph the function in the viewing rectangle
$[-2, 2]$ by $[-5, 20]$ and see there are two solutions.

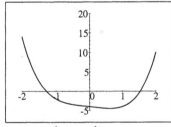

Viewing rectangle: $[-1.3, -1.25]$
by $[-0.1, 0.1]$. Solution $x \approx -1.28$.

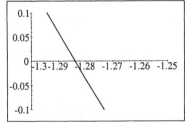

Viewing rectangle: $[1.5.1.6]$ by
$[-0.1, 0.1]$. Solution $x \approx 1.53$.

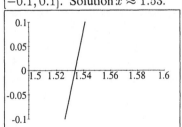

Thus the solutions are $x \approx -1.28, 1.53$.

81. $4.00x^4 + 4.00x^3 - 10.96x^2 - 5.88x + 9.09 = 0.$

$$
\begin{array}{r|rrrrr}
1 & 4 & 4 & -10.96 & -5.88 & 9.09 \\
  &   & 4 & 8 & -2.96 & -8.84 \\
\hline
  & 4 & 8 & -2.96 & -8.84 & 0.25
\end{array}
$$

$$
\begin{array}{r|rrrrr}
2 & 4 & 4 & -10.96 & -5.88 & 9.09 \\
  &   & 8 & 24 & 26.08 & 40.40 \\
\hline
  & 4 & 12 & 13.04 & 20.2 & 49.49
\end{array}
$$
$\Rightarrow \quad x = 2$ is an upper bound.

$$
\begin{array}{r|rrrrr}
-2 & 4 & 4 & -10.96 & -5.88 & 9.09 \\
   &   & -8 & 8 & 5.92 & -0.08 \\
\hline
   & 4 & -4 & -2.96 & 0.04 & 9.01
\end{array}
$$

$$
\begin{array}{r|rrrrr}
-3 & 4 & 4 & -10.96 & -5.88 & 9.09 \\
   &   & -12 & 24 & -39.12 & 135 \\
\hline
   & 4 & -8 & 13.04 & -45 & 144.09
\end{array}
$$
$\Rightarrow \quad x = -3$ is a lower bound.

Therefore, we graph the function in the viewing rectangle $[-3, 2]$ by $[-10, 40]$. There appear to be two solutions.

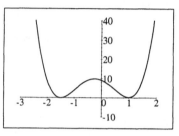

Viewing rectangle: $[-1.6, -1.4]$ by $[-0.1, 0.1]$. Solution $x \approx -1.50$.

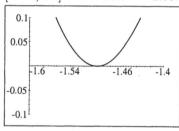

Viewing rectangle: $[0.8, 1.2]$ by $[0, 1]$. The graph comes close but does not go through the x-axis. Thus there is no solution here.

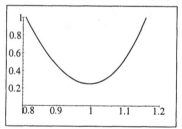

Therefore, the only solution is $x \approx -1.50$.

83. Let $r = $ the radius of the silo. The volume of the hemispherical roof is $\frac{1}{2}\left(\frac{4}{3}\pi r^3\right) = \frac{2}{3}\pi r^3$. The volume of the cylindrical section is $\pi(r^2)(30) = 30\pi r^2$. Because the total volume of the silo is 15000 ft^3, we get the following equation: $\frac{2}{3}\pi r^3 + 30\pi r^2 = 15000 \quad \Leftrightarrow$ $\frac{2}{3}\pi r^3 + 30\pi r^2 - 15000 = 0 \quad \Leftrightarrow \quad \pi r^3 + 45\pi r^2 - 22500 = 0.$ Using a graphing device, we first graph the polynomial in the viewing rectangle $[0, 15]$ by $[-10000, 10000]$. The solution, $r \approx 11.28$ ft., is shown in the viewing rectangle $[11.2, 11.4]$ by $[-1, 1]$.

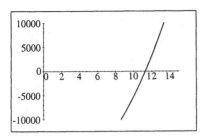 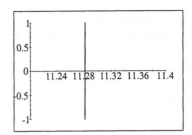

85. $h(t) = 11.60t - 12.41t^2 + 6.20t^3 - 1.58t^4 + 0.20t^5 - 0.01t^6$ is shown in the viewing rectangle $[0, 10]$ by $[0, 6]$.

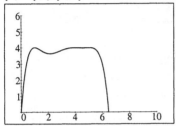

(a) It started to snow again.

(b) No, $h(t) \le 4$.

(c) The function $h(t)$ is shown in the viewing rectangle $[6, 6.5]$ by $[0, 0.5]$. The x-intercept of the function is a little less than 6.5, which means that the snow melted just before midnight on Saturday night.

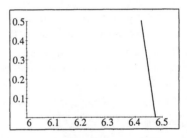

87. Let $r =$ the radius of the cone and cylinder. Let $h =$ the height of the cone. Since the height and diameter are equal, we get $h = 2r$. So the volume of the cylinder is $V_1 = \pi r^2 \cdot (cylinder\ height)$ $= 20\pi r^2$, and the volume of the cone is $V_2 = \frac{1}{3}\pi r^2 h = \frac{1}{3}\pi r^2(2r) = \frac{2}{3}\pi r^3$. Since the total volume is $\frac{500\pi}{3}$, it follows that $\frac{2}{3}\pi r^3 + 20\pi r^2 = \frac{500\pi}{3}$ $\Leftrightarrow$ $r^3 + 30r^2 - 250 = 0$. By Descartes' Rule of Signs, there is 1 positive zero.

r	$r^3 + 30r^2 - 250$
1	-219
2	-122
3	47
2.7	-11.617
2.8	7.152
2.765	1.440932999

Therefore, the radius should be 2.76 m (correct to two decimals).

89. Let b be the width of the base, and let l be the length of the box. Then the length plus girth is $l + 4b = 108$, and the volume is $V = lb^2 = 2200$. Solving the first equation for l and substituting this value into the second equation yields $l = 108 - 4b$ $\Rightarrow V = (108 - 4b)b^2 = 2200$ $\Leftrightarrow$ $4b^3 - 108b^2 + 2200 = 0$ $\Leftrightarrow$ $4(b^3 - 27b^2 + 550) = 0$. Now $P(b) = b^3 - 27b^2 + 550$ has two

variations in sign, so there are 0 or 2 positive real zeros. We also observe that since $l > 0$, $b < 27$, so $b = 27$ is an upper bound. Thus the possible positive rational real zeros are 1, 2, 3, 10, 11, 22, 25.

$$
\begin{array}{r|rrrr}
1 & 1 & -27 & 0 & 550 \\
  &   & 1 & -26 & -26 \\
\hline
  & 1 & -26 & -26 & 524
\end{array}
\qquad
\begin{array}{r|rrrr}
2 & 1 & -27 & 0 & 550 \\
  &   & 2 & -50 & -100 \\
\hline
  & 1 & -25 & -50 & 450
\end{array}
$$

$$
\begin{array}{r|rrrr}
5 & 1 & -27 & 0 & 550 \\
  &   & 5 & -110 & -550 \\
\hline
  & 1 & -22 & -110 & 0
\end{array}
\qquad \Rightarrow \quad b = 5 \text{ is a zero.}
$$

$P(b) = (b - 5)(b^2 - 22b - 110)$. The other zeros are $b = \dfrac{22 \pm \sqrt{484 - 4(1)(-110)}}{2}$

$= \dfrac{22 \pm \sqrt{924}}{2} = \dfrac{22 \pm 30.397}{2}$. The positive answer from this factor is $b \approx 26.20$. Thus we have two possible solutions, $b = 5$ or $b \approx 26.20$. If $b = 5$, then $l = 108 - 4(5) = 88$; if $b \approx 26.20$, then $l = 108 - 4(26.20) = 3.20$. Thus the length of the box is either 88 in. or 3.20 in.

91. Suppose $-b$ is a negative lower bound for the real zeros of $P(x)$. Then clearly b is an upper bound for $P_1(x) = P(-x)$. Thus, as in Exercise 90, we can write $P_1(x) = (x - b) \cdot Q(x) + r$, where $r > 0$ and the coefficients of Q are all non-negative, and
$P(x) = P_1(-x) = (-x - b) \cdot Q(-x) + r = (x + b) \cdot [-Q(-x)] + r$. Since the coefficients of $Q(x)$ are all non-negative, the coefficients of $-Q(-x)$ will be alternately non-positive and non-negative, which proves the second part of the Upper and Lower Bounds Theorem.

93. (a) An odd degree polynomial must have a real zero. The end behavior of such a polynomial requires that the graph of the polynomial heads off in opposite directions as $x \to \infty$ and $x \to -\infty$. Thus the graph must cross the x-axis.

   (b) There are many possibilities one of which is $P(x) = x^4 + 1$.

   (c) $P(x) = x(x - \sqrt{2})(x + \sqrt{2}) = x^3 - 2x$.

   (d) $P(x) = (x - \sqrt{2})(x + \sqrt{2})(x - \sqrt{3})(x + \sqrt{3}) = x^4 - 5x^2 + 6$.

   If a polynomial with integer coefficients has no real zeroes, then the polynomial must have even degree.

95. (a) Using the cubic formula:

$$
x = \sqrt[3]{\frac{-2}{2} + \sqrt{\frac{2^2}{4} + \frac{(-3)^3}{27}}} + \sqrt[3]{\frac{-2}{2} - \sqrt{\frac{2^2}{4} + \frac{(-3)^3}{27}}} = \sqrt[3]{-1} + \sqrt[3]{-1} = -1 - 1 = -2.
$$

$$
\begin{array}{r|rrrr}
-2 & 1 & 0 & -3 & 2 \\
   &   & -2 & 4 & -2 \\
\hline
   & 1 & -2 & 1 & 0
\end{array}
$$

   So $(x + 2)(x^2 - 2x + 1) = (x + 2)(x - 1)^2 = 0 \quad \Rightarrow \quad x = -2, 1$.

   Using the methods from this section, we have
$$
\begin{array}{r|rrrr}
1 & 1 & 0 & -3 & 2 \\
  &   & 1 & 1 & -2 \\
\hline
  & 1 & 1 & -2 & 0
\end{array}
$$

   So $x^3 - 3x + 2 = (x - 1)(x^2 + x - 2) = (x - 1)^2(x + 2) = 0 \quad \Leftrightarrow \quad x = -2, 1$.

   Since this factors easily, the factoring method was easier.

(b) Using the cubic formula:

$$x = \sqrt[3]{\frac{-(-54)}{2} + \sqrt{\frac{(-54)^2}{4} + \frac{(-27)^3}{27}}} + \sqrt[3]{\frac{-(-54)}{2} - \sqrt{\frac{(-54)^2}{4} + \frac{(-27)^3}{27}}}$$

$$= \sqrt[3]{\frac{54}{2} + \sqrt{27^2 - 27^2}} + \sqrt[3]{\frac{54}{2} - \sqrt{27^2 - 27^2}} = \sqrt[3]{27} + \sqrt[3]{27} = 3 + 3 = 6.$$

```
6 | 1    0   -27   -54
  |      6    36    54
  -----------------------
    1    6     9     0
```

$$x^3 - 27x - 54 = (x - 6)(x^2 + 6x + 9) = (x - 6)(x + 3)^2 = 0 \quad \Rightarrow \quad x = -3, 6.$$

Using methods from this section,

```
-1 | 1    0   -27   -54          -2 | 1    0   -27   -54
   |     -1     1    26             |     -2     4    46
   ------------------------         ------------------------
     1   -1   -26   -28               1   -2   -23    -8
```

```
-3 | 1    0   -27   -54
   |     -3     9    54
   ------------------------
     1   -3   -18     0
```

So $x^3 - 27x - 54 = = (x + 3)(x^2 - 3x - 18) = (x - 6)(x + 3)^2 = 0 \quad \Leftrightarrow \quad x = -3, 6.$

Since this factors easily, the factoring method was easier.

(c) Using the cubic formula:

$$x = \sqrt[3]{\frac{-4}{2} + \sqrt{\frac{4^2}{4} + \frac{3^3}{27}}} + \sqrt[3]{\frac{-4}{2} - \sqrt{\frac{4^2}{4} + \frac{3^3}{27}}}$$

$$= \sqrt[3]{-2 + \sqrt{4 + 1}} + \sqrt[3]{-2 - \sqrt{4 + 1}} = \sqrt[3]{-2 + \sqrt{5}} + \sqrt[3]{-2 - \sqrt{5}}.$$

From the graphing calculator, we see that $P(x) = x^3 + 3x + 4$ has one zero.

Using methods from this section, $P(x)$ has possible rational zeros $\pm 1, \pm 2, \pm 4$.

```
1 | 1    0    3    4
  |      1    1    4
  ---------------------
    1    1    4    4     ⇒     1 is an upper bound.
```

```
-1 | 1    0    3    4
   |     -1    1   -4
   ---------------------
     1   -1    4    0     ⇒ x = -1 is a zero.
```

$P(x) = x^3 + 3x + 4 = (x + 1)(x^2 - x + 4)$. Using the quadratic formula we have:

$$x = \frac{-(-1) \pm \sqrt{(-1)^2 - 4(1)(4)}}{2} = \frac{1 \pm \sqrt{-15}}{2} \text{ which is not a real number.}$$

Since it is not easy to see that $\sqrt[3]{-2 + \sqrt{5}} + \sqrt[3]{-2 - \sqrt{5}} = -1$, we see that the factoring method was *much* easier.

Exercises 5.4

1. $P(x) = x^2 + 9 = (x - 3i)(x + 3i)$. The zeros of P are $3i$ (multiplicity 1) and $-3i$ (multiplicity 1).

3. $Q(x) = x^2 + 2x + 2$. Using the quadratic formula $\quad x = \dfrac{-(2) \pm \sqrt{(2)^2 - 4(1)(2)}}{2(1)} = \dfrac{-2 \pm \sqrt{-4}}{2}$
    $= \dfrac{-2 \pm 2i}{2} = -1 \pm i$. The zeros of Q are $-1 - i$ (multiplicity 1) and $-1 + i$ (multiplicity 1).

5. $P(x) = x^3 + 4x = x(x^2 + 4) = x(x - 2i)(x + 2i)$. The zeros of P are 0, $2i$, and $-2i$ (all multiplicity 1).

7. $Q(x) = x^4 - 1 = (x^2 - 1)(x^2 + 1) = (x - 1)(x + 1)(x^2 + 1) = (x - 1)(x + 1)(x - i)(x + i)$.
    The zeros of Q are 1, -1, i, and $-i$ (all multiplicity 1).

9. $P(x) = 16x^4 - 81 = (4x^2 - 9)(4x^2 + 9) = (2x - 3)(2x + 3)(2x - 3i)(2x + 3i)$. The zeros of P
    are $\frac{3}{2}$, $-\frac{3}{2}$, $\frac{3}{2}i$, and $-\frac{3}{2}i$ (all multiplicity 1).

11. $P(x) = x^3 + x^2 + 9x + 9 = x^2(x + 1) + 9(x + 1) = (x + 1)(x^2 + 9) = (x + 1)(x - 3i)(x + 3i)$.
    The zeros of P are -1, $3i$, and $-3i$ (all multiplicity 1).

13. $Q(x) = x^4 + 2x^2 + 1 = (x^2 + 1)^2 = (x - i)^2(x + i)^2$. The zeros of Q are i and $-i$ (both multiplicity 2).

15. $P(x) = x^4 + 3x^2 - 4 = (x^2 - 1)(x^2 + 4) = (x - 1)(x + 1)(x - 2i)(x + 2i)$. The zeros of P are
    1, -1, $2i$, and $-2i$ (all multiplicity 1).

17. $P(x) = x^5 + 6x^3 + 9x = x(x^4 + 6x^2 + 9) = x(x^2 + 3)^2 = x(x - \sqrt{3}\,i)^2(x + \sqrt{3}\,i)^2$. The zeros
    of P are 0 (multiplicity 1), $\sqrt{3}\,i$ (multiplicity 2), and $-\sqrt{3}\,i$ (multiplicity 2).

19. Since $1 + i$ and $1 - i$ are conjugates, the factorization of the polynomial must be
    $P(x) = a(x - [1 + i])(x - [1 - i]) = a(x^2 - 2x + 2)$. If we let $a = 1$, we get
    $P(x) = x^2 - 2x + 2$.

21. Since $2i$ and $-2i$ are conjugates, the factorization of the polynomial must be
    $Q(x) = b(x - 3)(x - 2i)(x + 2i]) = b(x - 3)(x^2 + 4) = b(x^3 - 3x^2 + 4x - 12)$. If we let $b = 1$,
    we get $Q(x) = x^3 - 3x^2 + 4x - 12$.

23. Since i is a zero, by the Conjugate Roots Theorem, $-i$ is also a zero. So the factorization of the
    polynomial must be $P(x) = a(x - 2)(x - i)(x + i) = a(x^3 - 2x^2 + x - 2)$. If we let $a = 1$, we
    get $P(x) = x^3 - 2x^2 + x - 2$.

25. Since the zeros are $1 - 2i$ and 1 (with multiplicity two), by the Conjugate Roots Theorem, the other
    zero is $1 + 2i$. So a factorization is
    $R(x) = c(x - [1 - 2i])(x - [1 + 2i])(x - 1)^2 = c([x - 1] + 2i)([x - 1] - 2i)(x - 1)^2$
    $= c([x - 1]^2 - [2i]^2)(x^2 - 2x + 1) = c(x^2 - 2x + 1 + 4)(x^2 - 2x + 1)$
    $= c(x^2 - 2x + 5)(x^2 - 2x + 1) = c(x^4 - 2x^3 + x^2 - 2x^3 + 4x^2 - 2x + 5x^2 - 10x + 5)$
    $= c(x^4 - 4x^3 + 10x^2 - 12x + 5)$. If we let $c = 1$ we get $R(x) = x^4 - 4x^3 + 10x^2 - 12x + 5$.

27. Since the zeros are i and $1 + i$, by the Conjugate Roots Theorem, the other zeros are $-i$ and $1 - i$.
    So a factorization is

$T(x) = C(x - i)(x + i)(x - [1 + i])(x - [1 - i]) = C(x^2 - i^2)([x - 1] - i)([x - 1] + i)$
$= C(x^2 + 1)(x^2 - 2x + 1 - i^2) = C(x^2 + 1)(x^2 - 2x + 2) = C(x^4 - 2x^3 + 2x^2 + x^2 - 2x + 2)$
$= C(x^4 - 2x^3 + 3x^2 - 2x + 2) = Cx^4 - 2Cx^3 + 3Cx^2 - 2Cx + 2C.$

Since the constant coefficient is 12, it follows that $2C = 12 \iff C = 6$, and so
$T(x) = 6(x^4 - 2x^3 + 3x^2 - 2x + 2) = 6x^4 - 12x^3 + 18x^2 - 12x + 12.$

29. $P(x) = x^3 + 2x^2 + 4x + 8 = x^2(x + 2) + 4(x + 2) = (x + 2)(x^2 + 4)$
     $= (x + 2)(x - 2i)(x + 2i)$. Thus the zeros are -2 and $\pm 2i$.

31. $P(x) = x^3 - 2x^2 + 2x - 1$. By inspection, $P(1) = 1 - 2 + 2 - 1 = 0$, and hence $x = 1$ is a zero.

$$\begin{array}{r|rrrr} 1 & 1 & -2 & 2 & -1 \\ & & 1 & -1 & 1 \\ \hline & 1 & -1 & 1 & 0 \end{array}$$

Thus $P(x) = (x - 1)(x^2 - x + 1)$. So $x = 1$ or $x^2 - x + 1 = 0$.

Using the quadratic formula, we have $x = \dfrac{1 \pm \sqrt{1 - 4(1)(1)}}{2} = \dfrac{1 \pm i\sqrt{3}}{2}$. Hence, the zeros are 1 and $\dfrac{1 \pm i\sqrt{3}}{2}$.

33. $P(x) = x^3 - 3x^2 + 3x - 2.$

$$\begin{array}{r|rrrr} 2 & 1 & -3 & 3 & -2 \\ & & 2 & -2 & 2 \\ \hline & 1 & -1 & 1 & 0 \end{array}$$

Thus $P(x) = (x - 2)(x^2 - x + 1)$. So $x = 2$ or $x^2 - x + 1 = 0$

Using the quadratic formula we have $x = \dfrac{1 \pm \sqrt{1 - 4(1)(1)}}{2} = \dfrac{1 \pm i\sqrt{3}}{2}$. Hence, the zeros are 2, and $\dfrac{1 \pm i\sqrt{3}}{2}$

35. $P(x) = 2x^3 + 7x^2 + 12x + 9$ has possible rational zeros $\pm 1, \pm 3, \pm 9, \pm\frac{1}{2}, \pm\frac{3}{2}, \pm\frac{9}{2}$. Since all coefficients are positive, there are no positive real zeros.

$$\begin{array}{r|rrrr} -1 & 2 & 7 & 12 & 9 \\ & & -2 & -5 & -7 \\ \hline & 2 & 5 & 7 & 2 \end{array} \qquad \begin{array}{r|rrrr} -2 & 2 & 7 & 12 & 9 \\ & & -4 & -6 & -12 \\ \hline & 2 & 3 & 6 & -3 \end{array}$$

There is a zero between -1 and -2.

$$\begin{array}{r|rrrr} -\frac{3}{2} & 2 & 7 & 12 & 9 \\ & & -3 & -6 & -9 \\ \hline & 2 & 4 & 6 & 0 \end{array} \Rightarrow \quad x = -\frac{3}{2} \text{ is a zero.}$$

$P(x) = \left(x + \frac{3}{2}\right)(2x^2 + 4x + 6) = 2\left(x + \frac{3}{2}\right)(x^2 + 2x + 3)$. Now $x^2 + 2x + 3$ has zeros

$x = \dfrac{-2 \pm \sqrt{4 - 4(3)(1)}}{2} = \dfrac{-2 \pm 2\sqrt{-2}}{2} = -1 \pm i\sqrt{2}$. Hence, the zeros are $-\frac{3}{2}$ and $-1 \pm i\sqrt{2}$.

37. $P(x) = x^4 + x^3 + 7x^2 + 9x - 18$. Since $P(x)$ has one change in sign, we are guaranteed a positive zero, and since $P(-x) = x^4 - x^3 + 7x^2 - 9x - 18$, there are 1 or 3 negative zeros.

$$\begin{array}{r|rrrrr} 1 & 1 & 1 & 7 & 9 & -18 \\ & & 1 & 2 & 9 & 18 \\ \hline & 1 & 2 & 9 & 18 & 0 \end{array} \Rightarrow \quad \text{Therefore, } P(x) = (x - 1)(x^3 + 2x^2 + 9x + 18).$$

Continuing with the quotient, we try negative zeros.

$$
\begin{array}{r|rrrr}
-1 & 1 & 2 & 9 & 18 \\
   &   & -1 & -1 & -8 \\
\hline
   & 1 & 1 & 8 & 10
\end{array}
$$

$$
\begin{array}{r|rrrr}
-2 & 1 & 2 & 9 & 18 \\
   &   & -2 & 0 & -18 \\
\hline
   & 1 & 0 & 9 & 0
\end{array}
$$

$P(x) = (x - 1)(x + 2)(x^2 + 9) = (x - 1)(x + 2)(x - 3i)(x + 3i)$. Therefore, the zeros are 1, -2, and $\pm 3i$.

39. We see a pattern and utilize it to factor by grouping. This gives
$P(x) = x^5 - x^4 + 7x^3 - 7x^2 + 12x - 12 = x^4(x - 1) + 7x^2(x - 1) + 12(x - 1)$
$= (x - 1)(x^4 + 7x^2 + 12) = (x - 1)(x^2 + 3)(x^2 + 4)$
$= (x - 1)(x - i\sqrt{3})(x + i\sqrt{3})(x - 2i)(x + 2i)$. Therefore, the zeros are 1, $\pm i\sqrt{3}$, and $\pm 2i$.

41. $P(x) = x^4 - 6x^3 + 13x^2 - 24x + 36$ has possible rational zeros $\pm 1, \pm 2, \pm 3, \pm 4, \pm 6, \pm 9, \pm 12,$ ± 18. $P(x)$ has 4 variations in signs and $P(-x)$ has no variation in signs.

$$
\begin{array}{r|rrrrr}
1 & 1 & -6 & 13 & -24 & 36 \\
  &   & 1 & -5 & 8 & -16 \\
\hline
  & 1 & -5 & 8 & -16 & 20
\end{array}
\qquad
\begin{array}{r|rrrrr}
2 & 1 & -6 & 13 & -24 & 36 \\
  &   & 2 & -8 & 10 & -28 \\
\hline
  & 1 & -4 & 5 & -14 & 8
\end{array}
$$

$$
\begin{array}{r|rrrrr}
3 & 1 & -6 & 13 & -24 & 36 \\
  &   & 3 & -9 & 12 & -36 \\
\hline
  & 1 & -3 & 4 & -12 & 0
\end{array}
\quad \Rightarrow \quad x = 3 \text{ is a zero.}
$$

Continuing:

$$
\begin{array}{r|rrrr}
3 & 1 & -3 & 4 & -12 \\
  &   & 3 & 0 & 12 \\
\hline
  & 1 & 0 & 4 & 0
\end{array}
\quad \Rightarrow \quad x = 3 \text{ is a zero.}
$$

$P(x) = (x - 3)^2(x^2 + 4) = (x - 3)^2(x - 2i)(x + 2i)$. Therefore, the zeros are 3, and $\pm 2i$.

43. $P(x) = x^5 - 3x^4 + 12x^3 - 28x^2 + 27x - 9$ has possible rational zeros $\pm 1, \pm 3, \pm 9$. $P(x)$ has 4 variations in sign and $P(-x)$ has 1 variation in sign.

$$
\begin{array}{r|rrrrrr}
1 & 1 & -3 & 12 & -28 & 27 & -9 \\
  &   & 1 & -2 & 10 & -18 & 9 \\
\hline
  & 1 & -2 & 10 & -18 & 9 & 0
\end{array}
\quad \Rightarrow \quad x = 1 \text{ is a zero.}
$$

$$
\begin{array}{r|rrrrr}
1 & 1 & -2 & 10 & -18 & 9 \\
  &   & 1 & -1 & 9 & -9 \\
\hline
  & 1 & -1 & 9 & -9 & 0
\end{array}
\quad \Rightarrow \quad x = 1 \text{ is a zero.}
$$

$$
\begin{array}{r|rrrr}
1 & 1 & -1 & 9 & -9 \\
  &   & 1 & 0 & 9 \\
\hline
  & 1 & 0 & 9 & 0
\end{array}
\quad \Rightarrow \quad x = 1 \text{ is a zero.}
$$

$P(x) = (x - 1)^3(x^2 + 9) = (x - 1)^3(x - 3i)(x + 3i)$. Therefore, the zeros are 1 and $\pm 3i$.

45. (a) $P(x) = x^3 - 5x^2 + 4x - 20 = x^2(x - 5) + 4(x - 5) = (x - 5)(x^2 + 4)$.

(b) $P(x) = (x - 5)(x - 2i)(x + 2i)$.

47. (a) $P(x) = x^4 + 8x^2 - 9 = (x^2 - 1)(x^2 + 9) = (x - 1)(x + 1)(x^2 + 9)$.

(b) $P(x) = (x - 1)(x + 1)(x - 3i)(x + 3i)$.

49. (a) $P(x) = x^6 - 64 = (x^3 - 8)(x^3 + 8) = (x - 2)(x^2 + 2x + 4)(x + 2)(x^2 - 2x + 4)$.

     (b) $P(x) = (x - 2)(x + 2)(x + 1 - i\sqrt{3})(x + 1 + i\sqrt{3})(x - 1 - i\sqrt{3})(x - 1 + i\sqrt{3})$.

51. (a) $x^4 - 2x^3 - 11x^2 + 12x = x(x^3 - 2x^2 - 11x + 12) = 0$. We first find the bounds for our
          viewing rectangle.

     $$\begin{array}{r|rrrr}
      & 1 & -2 & -11 & 12 \\
     \hline
     5 & 1 & 3 & 4 & 32 \\
     -4 & 1 & -6 & 13 & -50
     \end{array}$$
     $\Rightarrow \quad x = 5$ is an upper bound.
     $\Rightarrow \quad x = -4$ is a lower bound.

     The viewing rectangle $[-4, 5]$ by $[-50, 10]$ shows that $P(x) = x^4 - 2x^3 - 11x^2 + 12x$ has 4
     real solutions. Since this matches the degree of $P(x)$, $P(x)$ has no imaginary solutions.

     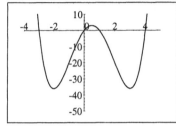

     (b) $x^4 - 2x^3 - 11x^2 + 12x - 5 = 0$. We use the same bounds for our viewing rectangle, $[-4, 5]$
          by $[-50, 10]$, and see that $R(x) = x^4 - 2x^3 - 11x^2 + 12x - 5$ has 2 real solutions. Since the
          degree of $R(x)$ is 4, $R(x)$ must have 2 imaginary solutions.

     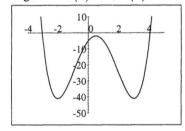

     (c) $x^4 - 2x^3 - 11x^2 + 12x + 40 = 0$. We use the same bounds for our viewing rectangle, $[-4, 5]$
          by $[-10, 50]$, and see that $T(x) = x^4 - 2x^3 - 11x^2 + 12x + 40$ has no real solutions. Since
          the degree of $T(x)$ is 4, $T(x)$ must have 4 imaginary solutions.

     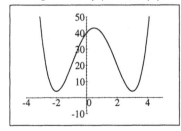

53. (a) $P(x) = x^2 - (1 + i)x + (2 + 2i)$. So
          $P(2i) = (2i)^2 - (1 + i)(2i) + 2 + 2i = -4 - 2i + 2 + 2 + 2i = 0$, and
          $P(1 - i) = (1 - i)^2 - (1 + i)(1 - i) + (2 + 2i) = 1 - 2i - 1 - 1 - 1 + 2 + 2i = 0$.
          Therefore, $2i$ and $1 - i$ are solutions of the equation $x^2 - (1 + i)x + (2 + 2i) = 0$. However,
          $P(-2i) = (-2i)^2 - (1 + i)(-2i) + 2 + 2i = -4 + 2i - 2 + 2 + 2i = -4 + 4i$, and

$P(1+i) = (1+i)^2 - (1+i)(1+i) + 2 + 2i = 2 + 2i$. Since, $P(-2i) \neq 0$ and $P(1+i) \neq 0$, $-2i$ and $1+i$ are not solutions.

(b)   This does not violate the Conjugate Roots Theorem because the coefficients of the polynomial $P(x)$ are not all real.

55.   Since P has real coefficients, the imaginary zeros come in pairs: $a \pm bi$ (by the Conjugate Roots Theorem), where $b \neq 0$. Thus there must be an even number of imaginary zeros. Since P is of odd degree, it has an odd number of zeros (counting multiplicity). It follows that P has at least one real zero.

Exercises 5.5

1. $r(x) = \dfrac{x-2}{x+3}$. When $x = 0$, we have $r(0) = -\frac{2}{3}$, so the y-intercept is $-\frac{2}{3}$. The numerator is 0 when $x = 2$, so the x-intercept is 2.

3. $t(x) = \dfrac{x^2 - x - 2}{x - 6}$. When $x = 0$, we have $t(0) = \frac{-2}{-6} = 3$, so the y-intercept is 3. The numerator is 0 when $x^2 - x - 2 = (x-2)(x+1) = 0$ or when $x = 2$ or $x = -1$, so the x-intercepts are 2 and -1.

5. $r(x) = \dfrac{x^2 - 9}{x^2}$. Since 0 is not in the domain of $r(x)$, there is no y-intercept. The numerator is 0 when $x^2 - 9 = (x-3)(x+3) = 0$ or when $x = \pm 3$, so the x-intercepts are ± 3.

7. From the graph, the x-intercept is 3, the y-intercept is 3, the vertical asymptote is $x = 2$, and the horizontal asymptote is $y = 2$.

9. $r(x) = \dfrac{5}{x+3}$. There is a vertical asymptote where $x + 3 = 0 \quad \Leftrightarrow \quad x = -3$. We have
$$r(x) = \frac{5}{x+3} = \frac{\dfrac{5}{x}}{1 + \dfrac{3}{x}} \to 0 \text{ as } x \to \pm\infty \text{, so the horizontal asymptote is } y = 0.$$

11. $t(x) = \dfrac{x^2}{x^2 - x - 6} = \dfrac{x^2}{(x-3)(x+2)} = \dfrac{1}{1 - \dfrac{1}{x} - \dfrac{6}{x^2}} \to 1 \text{ as } x \to \pm\infty.$ Hence, the horizontal asymptote is $y = 1$. The vertical asymptotes occur when $(x-3)(x+2) = 0 \quad \Leftrightarrow \quad x = 3$ or $x = -2$, and so the vertical asymptotes are $x = 3$ and $x = -2$.

13. $s(x) = \dfrac{6}{x^2 + 2}$. There is no vertical asymptote since $x^2 + 2$ is never 0. Since
$$s(x) = \frac{6}{x^2 + 2} = \frac{\dfrac{6}{x^2}}{1 + \dfrac{2}{x^2}} \to 0 \text{ as } x \to \pm\infty, \text{ the horizontal asymptote is } y = 0.$$

15. $r(x) = \dfrac{6x - 2}{x^2 + 5x - 6}$. A vertical asymptote occurs when $x^2 + 5x - 6 = (x+6)(x-1) = 0 \quad \Leftrightarrow \quad x = 1$ or $x = -6$. Because the degree of the denominator is greater than the degree of the numerator, the horizontal asymptote is $y = 0$.

17. $y = \dfrac{x^2 + 2}{x - 1}$. A vertical asymptote occurs when $x - 1 = 0 \quad \Leftrightarrow \quad x = 1$. There are no horizontal asymptotes because the degree of the numerator is greater than the degree of the denominator.

19. $r(x) = \dfrac{4}{x-2}$. When $x = 0$, $r(0) = \frac{4}{0-2} = -2$, and so the y-intercept is -2.

Since the numerator can never be zero, there is no x-intercept. Since
the degree of the numerator is less than the degree of the denominator,
the horizontal asymptote is $y = 0$. There is a vertical asymptote when
$x - 2 = 0 \quad \Leftrightarrow \quad x = 2$.

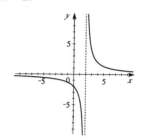

as $x \to$	2^+	2^-
sign of $y = \dfrac{4}{x-2}$	$\dfrac{(+)}{(+)}$	$\dfrac{(+)}{(-)}$
$y \to$	∞	$-\infty$

21. $r(x) = \dfrac{x-1}{x-2}$. When $x = 0$, $y = \frac{1}{2}$, so the y-intercept is $\frac{1}{2}$. When $y = 0$, $x - 1 = 0 \quad \Leftrightarrow \quad x = 1$,

so the x-intercept is 1. Since the degree of the numerator and
denominator are the same the horizontal asymptote is $y = \frac{1}{1} = 1$. A
vertical asymptote occurs when $x - 2 = 0 \quad \Leftrightarrow \quad x = 2$.

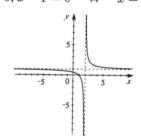

as $x \to$	2^+	2^-
sign of $y = \dfrac{x-1}{x-2}$	$\dfrac{(+)}{(+)}$	$\dfrac{(+)}{(-)}$
$y \to$	∞	$-\infty$

23. $y = \dfrac{4x-4}{x+2}$. When $x = 0$, $y = -2$, so the y-intercept is -2. When $y = 0$, $4x - 4 = 0 \quad \Leftrightarrow$

$x = 1$, so the x-intercept is 1. Since the degree of the numerator and
denominator are the same the horizontal asymptote is $y = \frac{4}{1} = 4$. A
vertical asymptote occurs when $x = -2$.

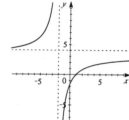

as $x \to$	-2^+	-2^-
sign of $y = \dfrac{4x-4}{x+2}$	$\dfrac{(-)}{(+)}$	$\dfrac{(-)}{(-)}$
$y \to$	$-\infty$	∞

25. $s(x) = \dfrac{4-3x}{x+7}$. When $x = 0$, $y = \frac{4}{7}$, so the y-intercept is $\frac{4}{7}$. The

x-intercepts occur when $y = 0 \quad \Leftrightarrow \quad 4 - 3x = 0 \quad \Leftrightarrow \quad x = \frac{4}{3}$.
A vertical asymptote occurs when $x = -7$. Since the degree of the
numerator and denominator are the same the horizontal asymptote is

$y = \frac{-3}{1} = -3$.

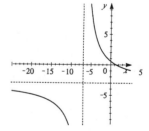

27. $r(x) = \dfrac{18}{(x-3)^2}$. When $x = 0$, $y = \frac{18}{9} = 2$, and so the y-intercept is

2. Since the numerator can never be zero, there is no x-intercept.
There is a vertical asymptote when $x - 3 = 0 \quad \Leftrightarrow \quad x = 3$, and
because the degree of the asymptote is $y = 0$.

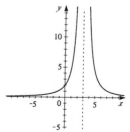

29. $s(x) = \dfrac{4x - 8}{(x - 4)(x + 1)}$. When $x = 0$, $y = \frac{-8}{(-4)(1)} = 2$, so the

y-intercept is 2. When $y = 0$, $4x + 8 = 0$ $\Leftrightarrow$ $x = -2$, so the
x-intercept is -2. The vertical asymptotes are $x = -1$ and $x = 4$, and
because the degree of the numerator is less than the degree of the
denominator, the horizontal asymptote is $y = 0$.

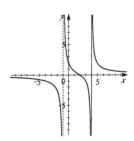

31. $s(x) = \dfrac{6}{x^2 - 5x - 6}$. When $x = 0$, $y = \frac{6}{-6} = -1$, so the y-intercept

is -1. Since the numerator is never zero, there is no x-intercept. The
vertical asymptotes occur when $x^2 - 5x - 6 = (x + 1)(x - 3)$ $\Leftrightarrow$
$x = -1$ and $x = 3$, and because the degree of the numerator is less
less than the degree of the denominator, the horizontal asymptote is
$y = 0$.

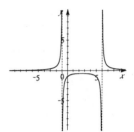

33. $t(x) = \dfrac{3x + 6}{x^2 + 2x - 8}$. When $x = 0$, $y = \frac{6}{-8} = -\frac{3}{4}$, so the y-intercept

y-intercept is $-\frac{3}{4}$. When $y = 0$, $3x + 6 = 0$ $\Leftrightarrow$ $x = -2$, so the
x-intercept is -2. The vertical asymptotes occur when
$x^2 + 2x - 8 = (x - 2)(x + 4) = 0$ $\Leftrightarrow$ $x = 2$ and $x = -4$.
Since the degree of the numerator is less than the degree of the
denominator, the horizontal asymptote is $y = 0$.

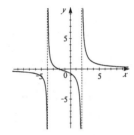

35. $r(x) = \dfrac{(x - 1)(x + 2)}{(x + 1)(x - 3)}$. When $x = 0$, $y = \frac{2}{3}$, so the y-intercept is $\frac{2}{3}$.

When $y = 0$, $(x - 1)(x + 2) = 0$ $\Rightarrow$ $x = -2, 1$, so, the
x-intercepts are -2 and 1. The vertical asymptotes are $x = -1$ and
$x = 3$, and because the degree of the numerator and denominator are
the same the horizontal asymptote is $y = \frac{1}{1} = 1$.

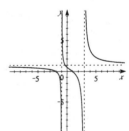

37. $r(x) = \dfrac{x^2 - 2x + 1}{x^2 + 2x + 1} = \dfrac{(x - 1)^2}{(x + 1)^2} = \left(\dfrac{x - 1}{x + 1}\right)^2$. When $x = 0$, $y = 1$,

so the y-intercept is 1. When $y = 0$, $x = 1$, so the x-intercept is 1. A
vertical asymptote occurs at $x + 1 = 0$ $\Leftrightarrow$ $x = -1$. Because the
degree of the numerator and denominator are the same the horizontal
asymptote is $y = \frac{1}{1} = 1$.

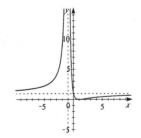

39. $r(x) = \dfrac{2x^2 + 10x - 12}{x^2 + x - 6} = \dfrac{2(x-1)(x+6)}{(x-2)(x+3)}$. When $x = 0$,

$y = \frac{2(-1)(6)}{(-2)(3)} = 2$, so the y-intercept is 2. When $y = 0$,

$2(x-1)(x+6) = 0 \Rightarrow x = -6, 1$, so the x-intercepts are -6

and 1. Vertical asymptotes occur when $(x-2)(x+3) = 0 \Leftrightarrow$

$x = -3$ or $x = 2$. Because the degree of the numerator and

denominator are the same the horizontal asymptote is $y = \frac{2}{1} = 2$.

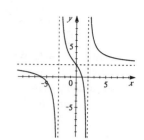

41. $y = \dfrac{x^2 - x - 6}{x^2 + 3x} = \dfrac{(x-3)(x+2)}{x(x+3)}$. The x-intercept occurs when $y = 0$

$\Leftrightarrow (x-3)(x+2) = 0 \Rightarrow x = -2, 3$, so the x-intercepts are

-2 and 3. There is no y-intercept because y is undefined when $x = 0$.

The vertical asymptotes are $x = 0$ and $x = -3$. Because the degree of

the numerator and denominator are the same, the horizontal asymptotes

is $y = \frac{1}{1} = 1$.

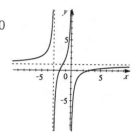

43. $r(x) = \dfrac{3x^2 + 6}{x^2 - 2x - 3} = \dfrac{3(x^2 + 2)}{(x-3)(x+1)}$. When $x = 0$, $y = -2$, so

the y-intercept is -2. Since the numerator can never equal zero, there

is -2. Since the numerator can never equal zero, there is no x-intercept.

Vertical asymptotes occur when $x = -1, 3$. Because the degree of

the numerator and denominator are the same, the horizontal

asymptote is.$y = \frac{3}{1} = 3$.

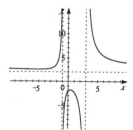

45. $s(x) = \dfrac{x^2 - 2x + 1}{x^3 - 3x^2} = \dfrac{(x-1)^2}{x^2(x-3)}$. Since $x = 0$ is not in the domain

of $s(x)$, there is no y-intercept. The x-intercept occurs when $y = 0$

$\Leftrightarrow x^2 - 2x + 1 = (x-1)^2 = 0 \Rightarrow x = 1$, so the x-intercept

is 1. Vertical asymptotes occur when $x = -1, 3$. Since the degree of

the numerator is less than the degree of the denominator, the horizontal

asymptote is $y = 0$.

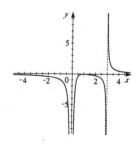

47. $r(x) = \dfrac{x^2}{x - 2}$. When $x = 0$, $y = 0$, so the graph passes through the

origin. There is a vertical asymptote when $x - 2 = 0 \Leftrightarrow x = 2$,

with $y \to \infty$ as $x \to 2^+$, and $y \to -\infty$ as $x \to 2^-$. Because the

degree of the numerator is greater than the degree of the denominator,

there is no horizontal asymptotes. By using long division, we see

that $y = x + 2 + \dfrac{4}{x - 2}$, so $y = x + 2$ is a slant asymptote.

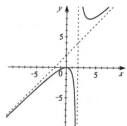

49. $r(x) = \dfrac{x^2 - 2x - 8}{x} = \dfrac{(x-4)(x+2)}{x}$. The vertical asymptote is
$x = 0$, thus, there is no y-intercept. If $y = 0$, then $(x-4)(x+2) = 0$
$\Rightarrow$ $x = -2, 4$, so the x-intercepts are -2 and 4. Because the degree
f the numerator is greater than the degree of the denominator, there
are no horizontal asymptotes. By using long division, we see that
$y = x - 2 - \dfrac{8}{x}$, so $y = x - 2$ is a slant asymptote.

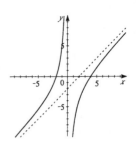

51. $r(x) = \dfrac{x^2 + 5x + 4}{x - 3} = \dfrac{(x+4)(x+1)}{x - 3}$. When $x = 0$, $y = -\frac{4}{3}$, so the
y-intercept is $-\frac{4}{3}$. When $y = 0$, $(x+4)(x+1) = 0$ $\Leftrightarrow$ $x = -4$,
-1, so the two x-intercepts are -4 and -1. A vertical asymptote
occurs when $x = 3$, with $y \to \infty$ as $x \to 3^{+}$, and $y \to -\infty$
as $x \to 3^{-}$. Using long division, we see that $y = x + 8 + \dfrac{28}{x - 3}$,
so $y = x + 8$ is a slant asymptote.

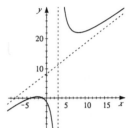

53. $r(x) = \dfrac{x^3 + x^2}{x^2 - 4} = \dfrac{x^2(x+1)}{(x-2)(x+2)}$. When $x = 0$, $y = 0$, so the
graph passes through the origin. Moreover, when $y = 0$, we have
$x^2(x+1) = 0$ $\Rightarrow$ $x = 0, -1$, so the x-intercepts are 0 and -1.
Vertical asymptotes occur when $x = \pm 2$, and because the degree of
the numerator is greater than the degree of the dominator there are
no horizontal asymptotes. Using long division, we see that

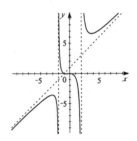

$y = x + 1 + \dfrac{4x + 4}{x^2 - 4}$, so $y = x + 1$ is a slant asymptote.

as $x \to$	-2^{-}	-2^{+}	2^{-}	2^{+}
sign of $y = \dfrac{x^2(x+1)}{(x-2)(x+2)}$	$\dfrac{(+)(-)}{(-)(-)}$	$\dfrac{(+)(-)}{(-)(+)}$	$\dfrac{(+)(+)}{(-)(+)}$	$\dfrac{(+)(+)}{(+)(+)}$
$y \to$	$-\infty$	∞	$-\infty$	∞

55. $f(x) = \dfrac{2x^2 + 6x + 6}{x + 3}$, $g(x) = 2x$. Vertical asymptote: $x = -3$.

$[-10, 5]$ by $[-20, 10]$ Graph of f.

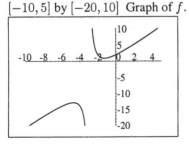

$[-20, 20]$ by $[-50, 50]$ Graph of f and g.

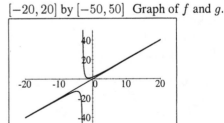

57. $f(x) = \dfrac{x^3 - 2x^2 + 16}{x - 2}$, $g(x) = x^2$. Vertical asymptote: $x = 2$.

$[-4, 8]$ by $[-30, 30]$ Graph of f.

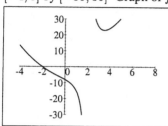

$[-20, 20]$ by $[-50, 50]$ Graph of f and g.

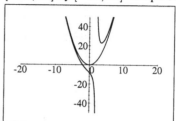

59. $f(x) = \dfrac{2x^2 - 5x}{2x + 3}$.

Vertical asymptote: $x = -1.5$

x-intercepts: $0, 2.5$

y-intercept: 0

Local maximum: $(-3.9, -10.4)$

Local minimum: $(0.9, -0.6)$

$$
\begin{array}{r}
x\ -\ 4 \\
2x + 3\ \overline{)\ 2x^2 - 5x} \\
\underline{2x^2 + 3x} \\
-8x \\
\underline{-8x - 12} \\
+12
\end{array}
$$

Using long division, we get $f(x) = x - 4 + \dfrac{12}{2x + 3}$. From the graph, we see that the end behavior of $f(x)$ is like the end behavior of $g(x) = x - 4$.

$[-10, 10]$ by $[-30, 30]$ Graph of f.

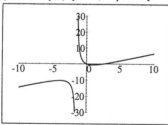

$[-20, 20]$ by $[-50, 50]$ Graph of f and g.

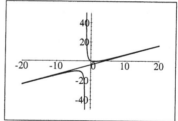

61. $f(x) = \dfrac{x^5}{x^3 - 1}$.

Vertical asymptote: $x = 1$

x-intercept: 0

y-intercept: 0

Local minimum: $(1.4, 3.1)$

$$
\begin{array}{r}
x^2 \\
x^3 - 1\ \overline{)\ x^5} \\
\underline{x^5 - x^2} \\
x^2
\end{array}
$$

Thus $y = x^2 + \dfrac{x^2}{x^3 - 1}$. From the graph we see that the end behavior of $f(x)$ is like the end behavior of $g(x) = x^2$.

[−5, 5] by [−10, 10] Graph of f.

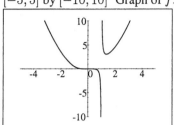

[−10, 10] by [−10, 10] Graph of f and g.

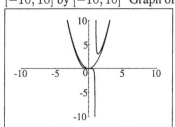

63. $f(x) = \dfrac{x^4 - 3x^3 + 6}{x - 3}$.

Vertical asymptote: $x = 3$
x-intercept: $1.6, 2.7$
y-intercept: -2
Local maximum: $(-0.4, -1.8)$, $(2.4, 3.8)$
Local minimum: $(0.6, -2.3)$, $(3.4, 54.3)$

$$x - 3\overline{)\begin{array}{l} x^3 \\ x^4 - 3x^3 + 6 \\ \underline{x^4 - 3x^3} \\ \phantom{x^4 - 3x^3 + {}} 6 \end{array}}$$

Thus $y = x^3 + \dfrac{6}{x - 3}$. From the graph we see that the end behavior of $f(x)$ is like the end behavior of $g(x) = x^3$.

[−10, 10] by [−20, 100] Graph of f.

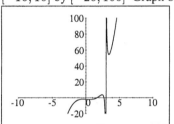

[−10, 10] by [−20, 100] Graph of f and g.

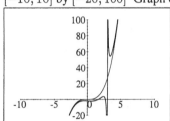

65. (a)

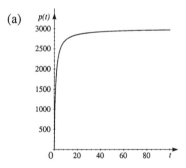

(b) $p(t) = \dfrac{3000t}{t + 1} = 3000 - \dfrac{3000}{t + 1}$.

So as $t \to \infty$, we have $p(t) \to 3000$.

67. [0, 20] by [0, 3]

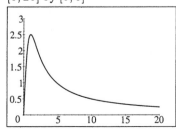

(a) The highest concentration of drug is 2.50 mg/L, and it is reached 1 hour after the drug is administered.

(b) The concentration of the drug in the bloodstream goes to 0.

(c) From the first viewing rectangle, we see that an approximate solution is near $t = 15$. Thus we graph $y = \dfrac{5t}{t^2 + 1}$ and $y = 0.3$ in the viewing rectangle $[14, 18]$ by $[0, 0.5]$. So it takes about 16.61 hours for the concentration to drop below 0.3 mg/L.

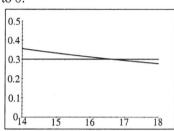

69. $P(v) = P_0 \left(\dfrac{s_0}{s_0 - v} \right) \quad \Rightarrow$

$P(v) = 440 \left(\dfrac{332}{332 - v} \right)$

$[0, 400]$ by $[0, 5000]$

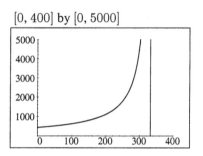

If the speed of the train approaches the speed of sound, the pitch of the whistle becomes very large. This would be experienced as a "sonic boom"— an effect seldom heard with trains.

71. Vertical asymptote $x = 3$: $p(x) = \dfrac{1}{x - 3}$

Vertical asymptote $x = 3$ and horizontal asymptote $y = 2$: $r(x) = \dfrac{2x}{x - 3}$.

Vertical asymptotes $x = 1$ and $x = -1$, horizontal asymptote 0, and x-intercept 4:

$q(x) = \dfrac{x - 4}{(x - 1)(x + 1)}$.

Of course, other answers are possible.

73. (a) $r(x) = \dfrac{3x^2 - 3x - 6}{x - 2} = \dfrac{3(x - 2)(x + 1)}{x - 2} = 3(x + 1)$,

for $x \neq 2$. Therefore, $r(x) = 3x + 3$, $x \neq 2$. Since $3(2) + 3 = 9$, the graph is the line $y = 3x + 3$ with the point $(2, 9)$ removed.

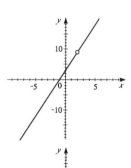

(b) $s(x) = \dfrac{x^2 + x - 20}{x + 5} = \dfrac{(x - 4)(x + 5)}{x + 5} = x - 4$, for

$x \neq -5$. Therefore, $s(x) = x - 4$, $x \neq -5$. Since $(-5) - 4 = -9$, the graph is the line $y = x - 4$ with the point $(-5, -9)$ removed.

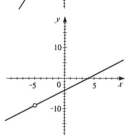

$t(x) = \dfrac{2x^2 - x - 1}{x - 1} = \dfrac{(2x + 1)(x - 1)}{x - 1} = 2x + 1$, for

$x \neq 1$. Therefore, $t(x) = 2x + 1$, $x \neq 1$. Since
$2(1) + 1 = 3$, the graph is the line $y = 2x + 1$ with
the point $(1, 3)$ removed.

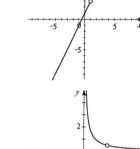

$u(x) = \dfrac{x - 2}{x^2 - 2x} = \dfrac{x - 2}{x(x - 2)} = \dfrac{1}{x}$, for $x \neq 2$.

Therefore, $u(x) = \dfrac{1}{x}$, $x \neq 2$. When $x = 2$, $\dfrac{1}{x} = \dfrac{1}{2}$,

the graph is the curve $y = \dfrac{1}{x}$ with the point $(2, \tfrac{1}{2})$

removed.

Review Exercises for Chapter 5

1. $P(x) = (x-2)^3 + 8$

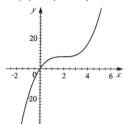

3. $P(x) = x^3 - 9x = x(x-3)(x+3)$

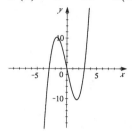

5. $P(x) = x^3 - 5x^2 - 4x + 20$
   $= (x-5)(x-2)(x+2)$

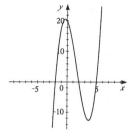

7. $P(x) = 2x^3 + x^2 - 18x - 9.$
   x-intercepts: $-3, -0.5, 3$
   y-intercept: -9
   local maximum: $(-1.9, 15.1)$
   local minimum: $(1.6, -27.1)$
   $y \to \infty$ as $x \to \infty$;
   $y \to -\infty$ as $x \to -\infty$

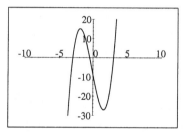

9. $P(x) = x^5 + x^2 - 5.$
   x-intercept: 1.3
   y-intercept: -5
   local maximum: $(-0.7, -4.7)$
   local minimum: $(0, -5)$
   $y \to \infty$ as $x \to \infty$;
   $y \to -\infty$ as $x \to -\infty$

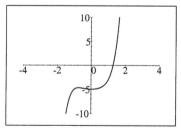

11.
$$
\begin{array}{r|rrrr}
3 & 1 & -1 & 1 & -11 \\
  &   & 3 & 6 & 21 \\
\hline
  & 1 & 2 & 7 & 10
\end{array}
$$

Therefore, $Q(x) = x^2 + 2x + 7$, and $R(x) = 10$.

13.
$$
\begin{array}{r}
x \quad - 3 \\
x^2 + 2x - 5 \overline{\smash{\big)}\, x^3 \;- x^2 - 11x \;+6} \\
\underline{x^3 + 2x^2 \;- 5x} \\
-3x^2 \;- 6x \;+6 \\
\underline{-3x^2 \;- 6x + 15} \\
-9
\end{array}
$$

Therefore, $Q(x) = x - 3$, and $R(x) = -9$.

15.
$$
\begin{array}{r|rrrrr}
-5 & 1 & 0 & -25 & 4 & 15 \\
& & -5 & 25 & 0 & -20 \\
\hline
& 1 & -5 & 0 & 4 & -5
\end{array}
$$

Therefore, $Q(x) = x^3 - 5x^2 + 4$, and $R(x) = -5$.

17.
$$
\begin{array}{r|rrrrr}
2 & 1 & 1 & -2 & -3 & -1 \\
& & 2 & 6 & 8 & 10 \\
\hline
& 1 & 3 & 4 & 5 & 9
\end{array}
$$

Therefore, $Q(x) = x^3 + 3x^2 + 4x + 5$, and $R(x) = 9$.

19. $P(x) = 2x^3 - 9x^2 - 7x + 13$; find $P(5)$.
$$
\begin{array}{r|rrrr}
5 & 2 & -9 & -7 & 13 \\
& & 10 & 5 & -10 \\
\hline
& 2 & 1 & -2 & 3
\end{array}
$$
Therefore, $P(5) = 3$.

21. $\frac{1}{2}$ is a zero of $P(x) = 2x^4 + x^3 - 5x^2 + 10x - 4$ if $P\left(\frac{1}{2}\right) = 0$.
$$
\begin{array}{r|rrrrr}
\frac{1}{2} & 2 & 1 & -5 & 10 & -4 \\
& & 1 & 1 & -2 & 4 \\
\hline
& 2 & 2 & -4 & 8 & 0
\end{array}
$$
Since $P\left(\frac{1}{2}\right) = 0$, $\frac{1}{2}$ is a zero of the polynomial.

23. $P(x) = x^{500} + 6x^{201} - x^2 - 2x + 4$. The remainder from dividing $P(x)$ by $x - 1$ is $P(1) = (1)^{500} + 6(1)^{201} - (1)^2 - 2(1) + 4 = 8$.

25. (a) $x^5 - 6x^3 - x^2 + 2x + 18 = 0$ has possible rational zeros $\pm 1, \pm 2, \pm 3, \pm 6, \pm 9, \pm 18$.

(b) Since $P(x)$ has 2 variations in sign, there are either 0 or 2 positive real zeros. Since $P(-x) = -x^5 + 6x^3 - x^2 - 2x + 18$ has 3 variations in sign, there are 1 or 3 negative real zeros.

27. Since the zeros are $-\frac{1}{2}$, 2, and 3, a factorization is $P(x) = C\left(x + \frac{1}{2}\right)(x - 2)(x - 3)$
$= \frac{C}{2}(2x + 1)(x^2 - 5x + 6) = \frac{C}{2}(2x^3 - 10x^2 + 12x + x^2 - 5x + 6) = \frac{C}{2}(2x^3 - 9x^2 + 7x + 6)$.
Since the constant coefficient is 12, $\frac{C}{2}(6) = 12 \quad \Leftrightarrow \quad C = 4$, and so the polynomial is $P(x) = 4x^3 - 18x^2 + 14x + 12$.

29. No, there is no polynomial of degree 4 with integer coefficients that has zeros i, $2i$, $3i$ and $4i$. Since the imaginary zeros of polynomial equations with real coefficients come in complex conjugate pairs, there would have to be 8 zeros, which is impossible for a polynomial of degree 4.

31. $P(x) = x^3 - 3x^2 - 13x + 15$ has possible rational zeros $\pm 1, \pm 3, \pm 5, \pm 15$.

$$\begin{array}{r|rrrr} 1 & 1 & -3 & -13 & 15 \\ & & 1 & -2 & -15 \\ \hline & 1 & -2 & -15 & 0 \end{array} \quad \Rightarrow \quad x = 1 \text{ is a zero.}$$

So $P(x) = x^3 - 3x^2 - 13x + 15 = (x-1)(x^2 - 2x - 15) = (x-1)(x-5)(x+3)$. Therefore, the zeros are -3, 1, and 5.

33. $P(x) = x^4 + 6x^3 + 17x^2 + 28x + 20$ has possible rational zeros $\pm 1, \pm 2, \pm 4, \pm 5, \pm 10, \pm 20$. Since all of the coefficients are positive, there are no positive real zeros.

$$\begin{array}{r|rrrrr} -1 & 1 & 6 & 17 & 28 & 20 \\ & & -1 & -5 & -12 & -16 \\ \hline & 1 & 5 & 12 & 16 & 4 \end{array}$$

$$\begin{array}{r|rrrrr} -2 & 1 & 6 & 17 & 28 & 20 \\ & & -2 & -8 & -18 & -20 \\ \hline & 1 & 4 & 9 & 10 & 0 \end{array} \quad \Rightarrow \quad x = -2 \text{ is a zero.}$$

$P(x) = x^4 + 6x^3 + 17x^2 + 28x + 20 = (x+2)(x^3 + 4x^2 + 9x + 10)$. Continuing with the quotient, we have

$$\begin{array}{r|rrrr} -2 & 1 & 4 & 9 & 10 \\ & & -2 & -4 & -10 \\ \hline & 1 & 2 & 5 & 0 \end{array} \quad \Rightarrow \quad x = -2 \text{ is a zero.}$$

Thus $P(x) = x^4 + 6x^3 + 17x^2 + 28x + 20 = (x+2)^2(x^2 + 2x + 5)$. Now $x^2 + 2x + 5 = 0$ when $x = \frac{-2 \pm \sqrt{4 - 4(5)(1)}}{2} = \frac{-2 \pm 4i}{2} = -1 \pm 2i$. Thus, the zeros are -2 (multiplicity 2) and $-1 \pm 2i$.

35. $P(x) = x^5 - 3x^4 - x^3 + 11x^2 - 12x + 4$ has possible rational zeros $\pm 1, \pm 2, \pm 4$.

$$\begin{array}{r|rrrrrr} 1 & 1 & -3 & -1 & 11 & -12 & 4 \\ & & 1 & -2 & -3 & 8 & -4 \\ \hline & 1 & -2 & -3 & 8 & -4 & 0 \end{array} \quad \Rightarrow \quad x = 1 \text{ is a zero.}$$

$P(x) = x^5 - 3x^4 - x^3 + 11x^2 - 12x + 4 = (x-1)(x^4 - 2x^3 - 3x^2 + 8x - 4)$. Continuing with the quotient, we have

$$\begin{array}{r|rrrrr} 1 & 1 & -2 & -3 & 8 & -4 \\ & & 1 & -1 & -4 & 4 \\ \hline & 1 & -1 & -4 & 4 & 0 \end{array} \quad \Rightarrow \quad x = 1 \text{ is a zero.}$$

So $x^5 - 3x^4 - x^3 + 11x^2 - 12x + 4 = (x-1)^2(x^3 - x^2 - 4x + 4) = (x-1)^3(x^2 - 4)$
$= (x-1)^3(x-2)(x+2)$. Therefore, the zeros are 1 (multiplicity 3), -2, and 2.

37. $P(x) = x^6 - 64 = (x^3 - 8)(x^3 + 8) = (x-2)(x^2 + 2x + 4)(x+2)(x^2 - 2x + 4)$. Now using the quadratic formula to find the zeros of $x^2 + 2x + 4$, we have
$x = \frac{-2 \pm \sqrt{4 - 4(4)(1)}}{2} = \frac{-2 \pm 2\sqrt{3}i}{2} = -1 \pm \sqrt{3}i$, and using the quadratic formula to find the zeros of
$x^2 - 2x + 4$, we have $x = \frac{2 \pm \sqrt{4 - 4(4)(1)}}{2} = \frac{2 \pm 2\sqrt{3}i}{2} = 1 \pm \sqrt{3}i$. Therefore, the zeros are 2, -2, $1 \pm \sqrt{3}i$, and $-1 \pm \sqrt{3}i$.

39. $P(x) = 6x^4 - 18x^3 + 6x^2 - 30x + 36 = 6(x^4 - 3x^3 + x^2 - 5x + 6)$ has possible rational zeros $\pm 1, \pm 2, \pm 3, \pm 6$.

$$\begin{array}{r|rrrrr} 1 & 6 & -18 & 6 & -30 & 36 \\ & & 6 & -12 & -6 & -36 \\ \hline & 6 & -12 & -6 & -36 & 0 \end{array} \quad \Rightarrow \quad x = 1 \text{ is a zero.}$$

So $P(x) = 6x^4 - 18x^3 + 6x^2 - 30x + 36 = (x - 1)(6x^3 - 12x^2 - 6x - 36)$
$= 6(x - 1)(x^3 - 2x^2 - x - 6)$. Continuing with the quotient we have

$$
\begin{array}{r|rrrr}
1 & 1 & -2 & -1 & -6 \\
  &   & 1 & -1 & -2 \\
\hline
  & 1 & -1 & -2 & -8 \\
\end{array}
$$

$$
\begin{array}{r|rrrr}
2 & 1 & -2 & -1 & -6 \\
  &   & 2 & 0 & -2 \\
\hline
  & 1 & 0 & -1 & -8 \\
\end{array}
$$

$$
\begin{array}{r|rrrr}
3 & 1 & -2 & -1 & -6 \\
  &   & 3 & 3 & 6 \\
\hline
  & 1 & 1 & 2 & 0 \\
\end{array} \Rightarrow \quad x = 3 \text{ is a zero.}
$$

So, $P(x) = 6x^4 - 18x^3 + 6x^2 - 30x + 36 = 6(x - 1)(x - 3)(x^2 + x + 2)$. Now, $x^2 + x + 2 = 0$
when $x = \frac{-1 \pm \sqrt{1 - 4(1)(2)}}{2} = \frac{-1 \pm \sqrt{7}i}{2}$, and so the zeros are 1, 3, and $\frac{-1 \pm \sqrt{7}i}{2}$.

41. $2x^2 = 5x + 3 \quad \Leftrightarrow \quad 2x^2 - 5x - 3 = 0$.
The solutions are $x = -0.5, 3$.

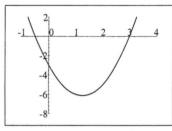

43. $x^4 - 3x^3 - 3x^2 - 9x - 2 = 0$ has solutions
$x \approx -0.24, 4.24$.

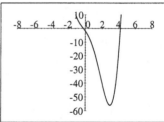

45. $r(x) = \dfrac{3x - 12}{x + 1}$. When $x = 0$, we have $r(0) = \frac{-12}{1} = -12$, so the
y-intercept is -12. Since $y = 0$, when $3x - 12 = 0 \quad \Leftrightarrow \quad x = 4$,
the x-intercept is 4. The vertical asymptote is $x = -1$. Because the
degree of the denominator and numerator are the same, the
horizontal asymptote is $y = \frac{3}{1} = 3$.

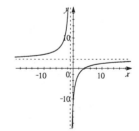

47. $r(x) = \dfrac{x - 2}{x^2 - 2x - 8} = \dfrac{x - 2}{(x + 2)(x - 4)}$. When $x = 0$, we have
$r(0) = \frac{-2}{-8} = \frac{1}{4}$, so the y-intercept is $\frac{1}{4}$. When $y = 0$, we have
$x - 2 = 0 \quad \Leftrightarrow \quad x = 2$, so the x-intercepts is 2. The vertical
asymptotes occur when $x = -2$ and $x = 4$. The horizontal
than the degree of the numerator.

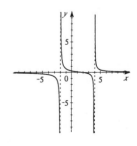

49. $r(x) = \dfrac{x^2 - 9}{2x^2 + 1} = \dfrac{(x+3)(x-3)}{2x^2 + 1}$. When $x = 0$, we have

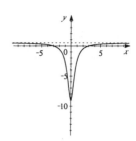

$r(0) = \frac{-9}{1}$, so the y-intercept is -9. When $y = 0$, we have
$x^2 - 9 = 0 \quad \Leftrightarrow \quad x = \pm 3$ so the x-intercepts are -3 and 3. Since
$2x^2 + 1 > 0$, the denominator is never zero so there are no vertical
asymptotes. The horizontal asymptote is at $y = \frac{1}{2}$ because the
degree of the denominator and numerator are the same.

51. $r(x) = \dfrac{x - 3}{2x + 6}$.

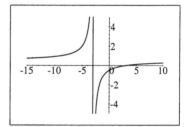

From the graph we see that
x-intercept: 3
y-intercept: -0.5
Vertical asymptote: $x = -3$
Horizontal asymptote: $y = 0.5$
No local extrema.

53. $r(x) = \dfrac{x^3 + 8}{x^2 - x - 2}$.

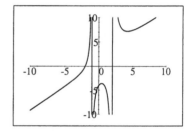

From the graph we see that
x-intercept: -2
y-intercept: -4
Vertical asymptote: $x = -1, x = 2$
Horizontal asymptote is $y = 0.5$.
Local maximum is $(0.425, -3.599)$.
Local minimum is $(4.216, 7.175)$.

By using long division, we see that $f(x) = x + 1 + \dfrac{10 - x}{x^2 - x - 2}$, so f has a slant asymptote
$y = x + 1$.

55. The graphs $y = x^4 + x^2 + 24x$ and $y = 6x^3 + 20$ intersect when $x^4 + x^2 + 24x = 6x^3 + 20 \quad \Leftrightarrow$
$x^4 - 6x^3 + x^2 + 24x - 20 = 0$. The possible rational zeros are $\pm 1, \pm 2, \pm 4, \pm 5, \pm 10, \pm 20$.

$$
\begin{array}{r|rrrrr}
1 & 1 & -6 & 1 & 24 & -20 \\
  &   & 1 & -5 & -4 & 20 \\
\hline
  & 1 & -5 & -4 & 20 & 0
\end{array} \quad \Rightarrow \quad x = 1 \text{ is a zero.}
$$

So $x^4 - 6x^3 + x^2 + 24x - 20 = (x - 1)(x^3 - 5x^2 - 4x + 20) = 0$. Continuing with the quotient:

$$
\begin{array}{r|rrrr}
1 & 1 & -5 & -4 & 20 \\
  &   & 1 & -4 & -8 \\
\hline
  & 1 & -4 & -8 & 12
\end{array}
\qquad
\begin{array}{r|rrrr}
2 & 1 & -5 & -4 & 20 \\
  &   & 2 & -6 & -20 \\
\hline
  & 1 & -3 & -10 & 0
\end{array} \quad \Rightarrow \quad x = 2 \text{ is a zero.}
$$

So $x^4 - 6x^3 + x^2 + 24x - 20 = (x - 1)(x - 2)(x^2 - 3x - 10)$
$= (x - 1)(x - 2)(x - 5)(x + 2) = 0$. Hence, the points of intersection are $(1, 26)$, $(2, 68)$,
$(5, 770)$, and $(-2, -28)$.

Chapter 5 Test

1. $f(x) = x^3 - x^2 - 9x + 9$
$\quad = (x-1)(x^2 - 9)$
$\quad = (x-1)(x-3)(x+3).$

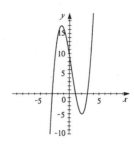

2. (a)
$$\begin{array}{r|rrrr} 2 & 1 & 0 & -4 & 2 & 5 \\ & & 2 & 4 & 0 & 4 \\ \hline & 1 & 2 & 0 & 2 & 9 \end{array}$$

   Therefore, quotient is $Q(x) = x^3 + 2x^2 + 2$, and the remainder is $R(x) = 9$.

   (b)
$$\begin{array}{r} x^3 + 2x^2 \qquad + \frac{1}{2} \\ 2x^2 - 1 \overline{\smash{)}\; 2x^5 + 4x^4 - x^3 - \; x^2 + 0x + 7} \\ \underline{2x^5 \qquad\quad -x^3} \\ 4x^4 \qquad\quad - \; x^2 \\ \underline{4x^4 \qquad\quad - 2x^2} \\ x^2 \qquad + 7 \\ \underline{x^2 \qquad - \frac{1}{2}} \\ 7\frac{1}{2} \text{ or } \frac{15}{2} \end{array}$$

   Therefore, quotient is $Q(x) = x^3 + 2x^2 + \frac{1}{2}$, and the remainder is $R(x) = \frac{15}{2}$.

3. (a) Possible rational zeros are: $\pm 1, \pm 3, \pm\frac{1}{2}, \pm\frac{3}{2}$.

   (b)
$$\begin{array}{r|rrrrr} 1 & 2 & -7 & 1 & 7 & -3 \\ & & 2 & -5 & -4 & 3 \\ \hline & 2 & -5 & -4 & 3 & 0 \end{array} \Rightarrow \quad x = 1 \text{ is a zero.}$$

   $P(x) = (x-1)(2x^3 - 5x^2 - 4x + 3)$

$$\begin{array}{r|rrrr} 1 & 2 & -5 & -4 & 3 \\ & & 2 & -3 & -7 \\ \hline & 2 & -3 & -7 & -4 \end{array} \qquad \begin{array}{r|rrrr} 3 & 2 & -5 & -4 & 3 \\ & & 6 & 3 & -3 \\ \hline & 2 & 1 & -1 & 0 \end{array} \Rightarrow \quad x = 3 \text{ is a zero.}$$

   $P(x) = (x-1)(x-3)(2x^2 + x - 1) = (x-1)(x-3)(x-\frac{1}{2})(x+1)$

   (c) The zeros of P are $x = \pm 1, 3, \frac{1}{2}$.

4. $P(x) = x^3 - x^2 - 4x - 6$. Possible rational zeros are: $\pm 1, \pm 2, \pm 3, \pm 6$.

$$\begin{array}{r|rrrr} 1 & 1 & -1 & -4 & -6 \\ & & 1 & 0 & -4 \\ \hline & 1 & 0 & -4 & -10 \end{array} \qquad \begin{array}{r|rrrr} 2 & 1 & -1 & -4 & -6 \\ & & 2 & 2 & -4 \\ \hline & 1 & 1 & -2 & -10 \end{array}$$

$$\begin{array}{r|rrrr} 3 & 1 & -1 & -4 & -6 \\ & & 3 & 6 & 6 \\ \hline & 1 & 2 & 2 & 0 \end{array} \quad \Rightarrow \quad x = 3 \text{ is a zero.}$$

So $P(x) = (x-3)(x^2 + 2x + 2)$. Using the quadratic formula on the second factor, we have

$$x = \frac{-2 \pm \sqrt{2^2 - 4(1)(2)}}{2(1)} = \frac{-2 \pm \sqrt{-4}}{2} = \frac{-2 \pm 2\sqrt{-1}}{2} = -1 \pm i. \text{ So zeros of } P(x) \text{ are } 3,$$

$-1 - i$, and $-1 + i$.

5. $P(x) = x^4 - 2x^3 + 5x^2 - 8x + 4$. The possible rational zeros of P are: $\pm 1, \pm 2$, and ± 4. Since there are four changes in sign, P has 4, 2, or 0 positive real zeros.

$$\begin{array}{r|rrrrr} 1 & 1 & -2 & 5 & -8 & 4 \\ & & 1 & -1 & 4 & -4 \\ \hline & 1 & -1 & 4 & -4 & 0 \end{array}$$

So $P(x) = (x-1)(x^3 - x^2 + 4x - 4)$. Factoring the second factor by grouping we have:
$P(x) = (x-1)[x^2(x-1) + 4(x-1)] = (x-1)(x^2+4)(x-1) = (x-1)^2(x-2i)(x+2i)$.

6. Since $3i$ is a zero of $P(x)$, $-3i$ is also a zero of $P(x)$. And since -1 is a zero of multiplicity 2,
$P(x) = (x+1)^2(x-3i)(x+3i) = (x^2 + 2x + 1)(x^2 + 9) = x^4 + 2x^3 + 10x^2 + 18x + 9$.

7. $P(x) = 2x^4 - 7x^3 + x^2 - 18x + 3$.

(a) Since $P(x)$ has 4 variations in sign, $P(x)$ can have 4, 2, or 0 positive real zeros. Since $P(-x) = 2x^4 + 7x^3 + x^2 + 18x + 3$ has no variations in sign, there are no negative real zeros.

(b)
$$\begin{array}{r|rrrrr} 4 & 2 & -7 & 1 & -18 & 3 \\ & & 8 & 4 & 20 & 8 \\ \hline & 2 & 1 & 5 & 2 & 11 \end{array}$$

Since the last row contains no negative entries, 4 is an upper bound for the real zeros of $P(x)$.

$$\begin{array}{r|rrrrr} -1 & 2 & -7 & 1 & -18 & 3 \\ & & -1 & 9 & -10 & 28 \\ \hline & 2 & -9 & 10 & -28 & 31 \end{array}$$

Since the last row alternates in sign, -1 is a lower bound for the real zeros of $P(x)$.

(c) Using the upper and lower limit from part b, we graph $P(x)$ in the viewing rectangle $[-1, 4]$ by $[-1, 1]$. The two real zeros are 0.17 and 3.93.

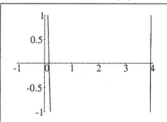

8. $r(x) = \dfrac{2x-1}{x^2 - x - 2}$, $s(x) = \dfrac{x^3 + 27}{x^2 + 4}$, $t(x) = \dfrac{x^3 - 9x}{x+2}$ and $u(x) = \dfrac{x^2 + x - 6}{x^2 - 25}$.

(a) $r(x)$ has the horizontal asymptote $y = 0$ because the degree of the denominator is greater than the degree of the numerator. $u(x)$ has the horizontal asymptote $y = \frac{1}{1} = 1$ because the degree of the numerator and the denominator are the same.

(b) The degree of the numerator of $s(x)$ is one more than the degree of the denominator, so $s(x)$ has a slant asymptote.

(c)   The denominator of s(x) is never 0, so s(x) has no vertical asymptotes.

(d) $u(x) = \dfrac{x^2 + x - 6}{x^2 - 25} = \dfrac{(x + 3)(x - 2)}{(x - 5)(x + 5)}$. When $x = 0$, we have $u(x) = \dfrac{-6}{-25} = \dfrac{6}{25}$, so the

y-intercept is $y = \frac{6}{25}$. When $y = 0$, we have $x = -3$ or $x = 2$, so the x-intercepts are -3 and
2. The vertical asymptotes are $x = -5$ and $x = 5$. The horizontal asymptote occurs at
$y = \frac{1}{1} = 1$ because the degree of the denominator and numerator are the same.

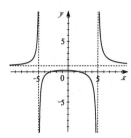

(e)
$$
\begin{array}{r}
x^2 \;\;\; - 2x \;\;\; - 5 \\
x + 2 \overline{\smash{\big)}\, x^3 + 0x^2 - 9x \;\; + 0} \\
\underline{x^3 + 2x^2 } \\
-2x^2 - 9x \\
\underline{-2x^2 - 4x } \\
-5x \;\; + 0 \\
\underline{-5x - 10} \\
-10
\end{array}
$$

Thus $P(x) = x^2 - 2x - 5$ and $t(x) = \dfrac{x^3 - 9x}{x + 2}$ have the same end behavior.

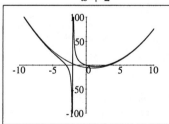

Focus on Problem Solving

1. The network is not traversable, since there are more than two vertices of odd order.

    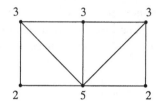

3. Since there are exactly two vertices of odd order, the network is traversable. Any path that traverses this network must start at one odd vertex and end at the other. For example, we may choose the path shown.

    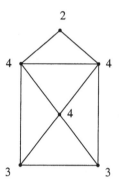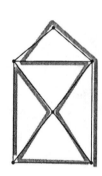

5. The diagram shows that a museum tour is possible if there are 5 rooms to a side. For an $n \times n$ museum, where n is an <u>odd</u> number, a similar right-left path works. If n is <u>even</u>, an argument similar to the one in the text (WBWB$\cdots$WB) shows that such a tour is impossible.

    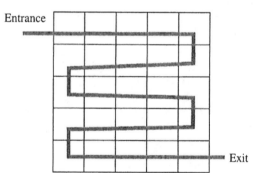

7. Consider the line joining B to A. Each time we cross the curve we move from outside to inside or from inside to outside. So we count the number of crossings in going from B to A along the line segment. B is clearly outside the curve, so if the number n of crossings is odd, A is inside; if n is even, A is outside. In the case illustrated, $n = 5$, so A is inside.

9. The equation will have four roots, so let the other root be k. Then $x^4 + ax^2 + bx + c = (x - 1)(x - 2)(x - 3)(x - k)$. The coefficient of x^3 on the right will be $-1 - 2 - 3 - k = -6 - k$. But this must equal the coefficient of x^3 on the left which is 0. Therefore $k = -6$. Equating constant coefficients gives $c = (-1)(-2)(-3)(-k)$ $= (-1)(-2)(-3)(6) = -36$.

11. Color the grid as a checkerboard, that is, with squares alternating black and white. Now each domino covers two squares, one black and one white. For the dominoes to cover the grid without overlapping, an equal number of black and white squares must be on the grid. However, the grid has 32 of one color and 30 of the other color. Therefore, it is impossible to cover the grid without overlapping the dominoes.

13. $f_0(x) = \dfrac{1}{1-x}$ and $f_{n+1} = f_0 \circ f_n$ for $n = 0, 1, 2, \ldots$

$$f_1(x) = f_0\left(\frac{1}{1-x}\right) = \frac{1}{1 - \frac{1}{1-x}} = \frac{1-x}{1(1-x)-1} = \frac{1-x}{-x} = \frac{x-1}{x},$$

$$f_2(x) = f_0\left(\frac{x-1}{x}\right) = \frac{1}{1 - \left(\frac{x-1}{x}\right)} = \frac{x}{x-(x-1)} = x,$$

$$f_3(x) = f_0(x) \quad \Rightarrow \quad f_4(x) = f_1(x) \quad \Rightarrow \quad f_5(x) = f_2(x) \quad \Rightarrow \quad f_6(x) = f_3(x) = f_0(x).$$

Thus $f_n(x) = \begin{cases} f_0(x) & \text{if } n = 3m \\ f_1(x) & \text{if } n = 3m+1 \\ f_2(x) & \text{if } n = 3m+2 \end{cases}$ for any natural number m.

Since $100 = 3 \cdot 33 + 1$, $f_{100}(x) = f_1(x)$ and $f_{100}(3) = f_1(3) = \frac{3-1}{3} = \frac{2}{3}$.

15. $f(x) = x - [\![x]\!]$

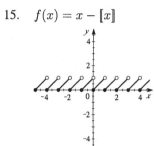

17. (a) Let n, $n+1$, $n+2$, and $n+3$ be four consecutive integers. Thus,
         $n(n+1)(n+2)(n+3) + 1 = n^4 + 6n^3 + 11n^2 + 6n + 1$. To show that this is a perfect
         square, we start by looking at some cases of n. For $n = 1$, we have
         $n(n+1)(n+2)(n+3) + 1 = 1 \cdot 2 \cdot 3 \cdot 4 + 1 = 25 = 5^2$, and for $n = 2$, we have
         $n(n+1)(n+2)(n+3) + 1 = 2 \cdot 3 \cdot 4 \cdot 5 + 1 = 121 = 11^2$. We next need to find a relation
         between n and the perfect square. After some experimentation we find that $5 = (1)(4) + 1$,
         and $11 = (2)(5) + 1$. Thus the perfect square is $[n(n+3) + 1]^2 = (n^2 + 3n + 1)^2$
         $= n^4 + 6n^3 + 11n^2 + 6n + 1$. Putting these together we can check that
         $n(n+1)(n+2)(n+3) + 1 = (n^2 + 3n + 1)^2$.

     (b) Let n, $n+1$, and $n+2$ be three consecutive integers. Then the product of three consecutive
         integers plus the middle is $n(n+1)(n+2) + (n+1) = n^3 + 3n^2 + 2n + n + 1$
         $= n^3 + 3n^2 + 3n + 1 = (n+1)^3$.

19. We follow the hint and put the original rectangle in the coordinate plane which has been tiled in a
     checkerboard pattern of black and white squares, each $\frac{1}{2}$ unit by $\frac{1}{2}$ unit. Since each of the smaller
     rectangles has at least one side of integer length, the area of each of the smaller rectangles is
     composed of equal amounts of black and white. So when we add all the smaller rectangles, the total
     area is composed of equal amounts of black and white. Thus the original rectangle is composed of
     equal amounts of black and white, so at least one side of the original rectangle must have integer
     length.

Chapter Six
Exercises 6.1

1. $f(x) = 2^x$

x	y
-4	$\frac{1}{16}$
-2	$\frac{1}{4}$
0	1
2	4
4	16

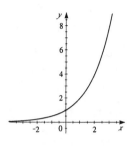

3. $f(x) = 6^x$

x	y
-2	$\frac{1}{36}$
-1	$\frac{1}{6}$
0	1
1	6
2	36

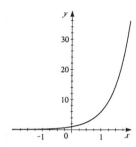

5. $f(x) = \left(\frac{1}{3}\right)^x$

x	y
-2	9
-1	3
0	1
1	$\frac{1}{3}$
2	$\frac{1}{9}$

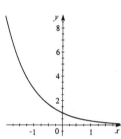

7. $g(x) = \left(\frac{1}{4}\right)^x$

x	y
-2	16
-1	4
0	1
1	$\frac{1}{4}$
2	$\frac{1}{16}$

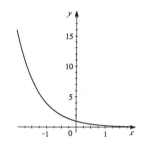

9. $y = 4^x$ and $y = 7^x$.

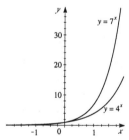

11. From the graph, $f(2) = a^2 = 9$, so $a = 3$. Thus $f(x) = 3^x$.

13. From the graph, $f(2) = a^2 = \frac{1}{16}$, so $a = \frac{1}{4}$. Thus $f(x) = \left(\frac{1}{4}\right)^x$.

15. III   17. I   19. II

21. $f(x) = -3^x$

The graph of f is obtained by reflecting the graph of $y = 3^x$ about the x-axis.

Domain: $(-\infty, \infty)$

Range: $(-\infty, 0)$

Asymptote: $y = 0$

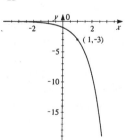

23. $g(x) = 2^x - 3$

The graph of g is obtained by shifting the graph of $y = 2^x$
downward 3 units.

Domain: $(-\infty, \infty)$

Range: $(-3, \infty)$

Asymptote: $y = -3$

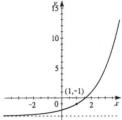

25. $h(x) = 4 + \left(\frac{1}{2}\right)^x$

The graph of h is obtained by shifting the graph of $y = \left(\frac{1}{2}\right)^x$
upward 4 units.

Domain: $(-\infty, \infty)$

Range: $(4, \infty)$

Asymptote: $y = 4$

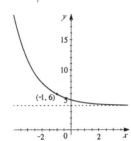

27. $f(x) = 10^{x+3}$

The graph of f is obtained by shifting the graph of $y = 10^x$ to
the left 3 units.

Domain: $(-\infty, \infty)$

Range: $(0, \infty)$

Asymptote: $y = 0$

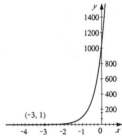

29. $f(x) = -3^{-x}$

The graph of f is obtained by reflecting the graph of $y = 3^x$
about the y-axis and then reflecting about the x-axis.

Domain: $(-\infty, \infty)$

Range: $(-\infty, 0)$

Asymptote: $y = 0$

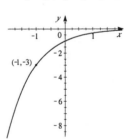

31. $f(x) = 5^{-2x}$

The graph of f is obtained by reflecting the graph of $y = 5^x$
about the y-axis and by shrinking it horizontally by a factor of 2.

Domain: $(-\infty, \infty)$

Range: $(0, \infty)$

Asymptote: $y = 0$

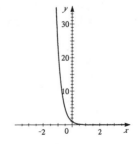

33. $f(x) = 5 - 2^{x-1}$

The graph of f is obtained by shifting the graph of $y = 2^x$ to the
right 1 unit, reflecting about the x-axis, then shifting upward 5
units.

Domain: $(-\infty, \infty)$

Range: $(-\infty, 5)$

Asymptote: $y = 5$

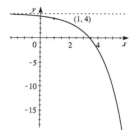

35. $f(x) = 2^{|x|}$

Note that $f(x) = \begin{cases} 2^x & x \geq 0 \\ 2^{-x} & x < 0. \end{cases}$

So, for $x \geq 0$ the graph of f is just the graph of $y = 2^x$; for
$x < 0$ the graph of f is obtained by reflecting the graph of
$y = 2^x$ about the y-axis.

Domain: $(-\infty, \infty)$

Range: $[1, \infty)$

Asymptote: None

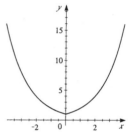

37. Using the points $(0, 3)$ and $(2, 12)$, we have $f(0) = Ca^0 = 3 \iff C = 3$. We also have
$f(2) = 3a^2 = 12 \iff a^2 = 4 \iff a = 2$ (recall that for an exponential function $f(x) = a^x$
we require $a > 0$). Thus $f(x) = 3 \cdot 2^x$.

39. (a)

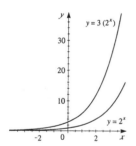

(b) Since $g(x) = 3(2^x) = 3f(x)$ and $f(x) > 0$, the
height of the graph of $g(x)$ is always three times
the height of the graph of $f(x) = 2^x$, so the
graph of g is steeper than the graph of f.

41. $f(x) = 10^x$, so $\dfrac{f(x+h) - f(x)}{h} = \dfrac{10^{x+h} - 10^x}{h} = \dfrac{10^x \cdot 10^h - 10^x}{h} = 10^x \dfrac{10^h - 1}{h}$.

43. (a) From the graphs in (i) - (iii) below, we see that the graph of f ultimately increases much more
quickly than the graph of g.

(i) $[0, 5]$ by $[0, 20]$

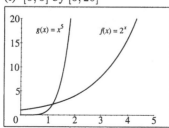

(ii) $[0, 25]$ by $[0, 10^7]$

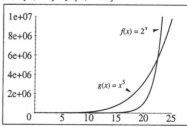

(iii) $[0, 50]$ by $[0, 10^8]$

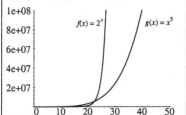

(b) From the graphs in parts (i) and (ii), we see that the approximate solutions are $x \approx 1.2$ and $x \approx 22.4$.

45.

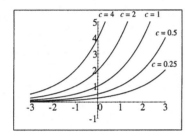

The larger the value of c, the more rapidly the graph of $f(x) = c\,2^x$ increases. Also, some students might notice that the graphs are just shifted horizontally 1 unit. This is because of our choice of c; each c in this exercise is of the form 2^k. So $f(x) = 2^k \cdot 2^x = 2^{x+k}$.

47. $y = 2^{1/x}$

Vertical Asymptote: $x = 0$

Horizontal Asymptote: $y = 1$

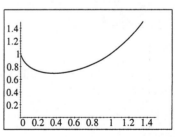

49. $g(x) = x^x$. Notice that $g(x)$ is only defined for $x \geq 0$. The graph of $g(x)$ is shown in the viewing rectangle $[0, 1.5]$ by $[0, 1.5]$. From the graphs, we see that there is a local minimum ≈ 0.69 when $x \approx 0.37$.

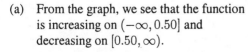

51. $y = 10^{x - x^2}$

(a) From the graph, we see that the function is increasing on $(-\infty, 0.50]$ and decreasing on $[0.50, \infty)$.

(b) From the graph, we see that the range is approximately $(0, 1.78]$.

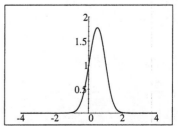

53. Calculating the pay for method (b), we have $pay = 2 + 2^2 + 2^3 + \cdots + 2^{30} > 2^{30}$ cents $= \$10,737,418.24$. Since this is much more than method (a), method (b) is more profitable.

Exercises 6.2

1.　$f(x) = 3e^x$

x	$f(x) = 3e^x$
-2	0.406006
-1.5	0.66939
-1	1.10364
-0.5	1.81959
0	3
0.5	4.94616
1	8.15485
1.5	13.4451
2	22.1672

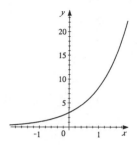

3.　$y = -e^x$

The graph of $y = -e^x$ is obtained from the graph of $y = e^x$ by reflecting it about the x-axis.

Domain: $(-\infty, \infty)$

Range: $(-\infty, 0)$

Asymptote: $y = 0$

5.　$y = e^{-x} - 1$

The graph of $y = e^{-x} - 1$ is obtained from the graph of $y = e^x$ by reflecting it about the y-axis then shifting downward 1 unit.

Domain: $(-\infty, \infty)$

Range: $(-1, \infty)$

Asymptote: $y = -1$

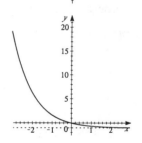

7.　$y = e^{x-2}$

The graph of $y = e^{x-2}$ is obtained from the graph of $y = e^x$ by shifting it to the right 2 units.

Domain: $(-\infty, \infty)$

Range: $(0, \infty)$

Asymptote: $y = 0$

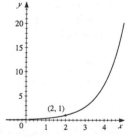

9.　$P = 10,000$, $r = 0.10$, and $n = 2$. So $A(t) = 10,000\left(1 + \frac{0.10}{2}\right)^{2t} = 10,000 \cdot 1.05^{2t}$.

(a)　$A(5) = 10,000 \cdot 1.05^{10} \approx 16,288.95$, and so the value of the investment is \$16,288.95.

(b)　$A(10) = 10,000 \cdot 1.05^{20} \approx 26,532.98$, and so the value of the investment is \$26,532.98.

(c)　$A(15) = 10,000 \cdot 1.05^{30} \approx 43,219.42$, and so the value of the investment is \$43,219.42.

11. $P = 3000$ and $r = 0.09$. Then we have $A(t) = 3000\left(1 + \dfrac{0.09}{n}\right)^{nt}$, and so

$$A(5) = 3000\left(1 + \dfrac{0.09}{n}\right)^{5n}.$$

   (a)   If $n = 1$, $A(5) = 3000\left(1 + \frac{0.09}{1}\right)^{5} = 3000 \cdot 1.09^5 \approx \$4{,}615.87$.

   (b)   If $n = 2$, $A(5) = 3000\left(1 + \frac{0.09}{2}\right)^{10} = 3000 \cdot 1.045^{10} \approx \$4{,}658.91$.

   (c)   If $n = 12$, $A(5) = 3000\left(1 + \frac{0.09}{12}\right)^{60} = 3000 \cdot 1.0075^{60} \approx \$4{,}697.04$.

   (d)   If $n = 52$, $A(5) = 3000\left(1 + \frac{0.09}{52}\right)^{260} \approx \$4{,}703.11$.

   (e)   If $n = 365$, $A(5) = 3000\left(1 + \frac{0.09}{365}\right)^{1825} \approx \$4{,}704.68$.

   (f)   If $n = 24 \cdot 365 = 8760$, $A(5) = 3000\left(1 + \frac{0.09}{8760}\right)^{43800} \approx \$4{,}704.93$.

   (g)   If interest is compounded continuously, $A(5) = 3000 \cdot e^{0.45} \approx \$4{,}704.94$.

13.   We find the effective rate, with $P = 1$, and $t = 1$. So $A = \left(1 + \frac{r}{n}\right)^{n}$

   (i) $n = 2, r = 0.085$; $A(2) = \left(1 + \frac{0.085}{2}\right)^{2} = (1.0425)^2 \approx 1.0868$.

   (ii) $n = 4, r = 0.0825$; $A(4) = \left(1 + \frac{0.0825}{4}\right)^{4} = (1.020625)^4 \approx 1.0851$.

   (iii)   continuous compounding, $r = 0.08$; $A(1) = e^{0.08} \approx 1.0833$.

   Since (i) is larger than the others, the best investment is the one at 8.5% compounded semiannually.

15.   We must solve for P in the equation $10000 = P\left(1 + \frac{0.09}{2}\right)^{2(3)} = P(1.045)^6 \quad \Leftrightarrow$
       $10000 = 1.3023P \quad \Leftrightarrow \quad P = 7678.96$. Thus the present value is \$7,678.96.

17. $n(t) = 500\,e^{0.45t}$

   (a)   The growth rate is 45%.

   (b) $n(0) = 500\,e^0 = 500$. Thus the initial population was 500 bacteria.

   (c) $n(5) = 500\,e^{0.45(5)} = 500\,e^{2.25} \approx 4744$. Thus the culture will contain 4744 bacteria at $t = 5$.

19.   (a) $r = 0.08$ and $n(0) = 18000$. Thus the population is given by the formula $n(t) = 18000\,e^{0.08t}$.

   (b) $t = 2008 - 2000 = 8$. Then we have $n(8) = 18000\,e^{0.08(8)} = 18000\,e^{0.64} \approx 34{,}137$. Thus
           there should be 34,137 foxes in the region by the year 2008.

   (c)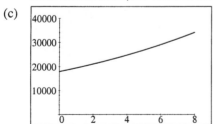

21. $n(t) = n_0\,e^{rt}$; $n_0 = 110$ million, $t = 2020 - 1995 = 25$

   (a) $r = 0.03$; $n(25) = 110{,}000{,}000\,e^{0.03(25)} = 110{,}000{,}000\,e^{0.75} \approx 232{,}870{,}000$. Thus at a 3%
           growth rate, the projected population will be approximately 233 million people by the year
           2020.

(b) $r = 0.02$; $n(25) = 110{,}000{,}000\, e^{0.02(25)} = 110{,}000{,}000\, e^{0.50} \approx 181{,}359{,}340$. Thus at a 2% growth rate, the projected population will be approximately 181 million people by the year 2020.

23. $n(t) = n_0\, e^{rt}$; $n_0 = 5$ billion, $t = 1995 - 1987 = 8$, $r = 0.02$. Thus

$n(8) = (5 \text{ billion})e^{0.02(8)} \approx 5.87$ billion.

25. $r = 2.20$ so $n(t) = n_0\, e^{2.2t}$

(a) $n(2) = n_0\, e^{2.2(2)} \approx 40000 \quad \Leftrightarrow \quad n_0\,(81.45) \approx 40000 \quad \Leftrightarrow \quad n_0 \approx 491$. Thus, about 500 bacteria were initially introduced into the food.

(b) $n(3) = 491\, e^{2.2(3)} = 491\, e^{6.6} \approx 360{,}932$. Thus there will be approximately 361,000 bacteria in 3 hours.

27. (a) $m(0) = 6\, e^{-0.087(0)} \approx 6$ grams

(b) $m(20) = 6\, e^{-0.087(20)} \approx 6(0.1755) = 1.053$. Thus approximately 1 gram of radioactive iodine remains after 20 days.

29. (a) $Q(5) = 15\left(1 - e^{-0.04(5)}\right) \approx 15(0.1813)$
$= 2.7345$. Thus approximately 2.7 lb of salt are in the barrel after 5 minutes.

(b) $Q(10) = 15\left(1 - e^{-0.04(10)}\right) \approx 15(0.3297)$
$= 4.946$. Thus approximately 4.9 lb of salt are in the barrel after 10 minutes.

(c)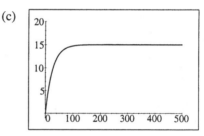

(d) 15 lb. Yes, since 50 gal $\times$ 0.3 lb/gal $= 15$ lb.

31. (a) Substituting $n_0 = 50$ and $t = 12$, we have $n(12) = \dfrac{300}{0.05 + \left(\frac{300}{50} - 0.05\right)e^{-0.55(12)}} \approx 5164$.

(b)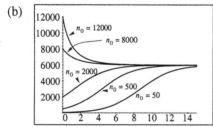

(c) The population approaches a size of 6000 rabbits.

33.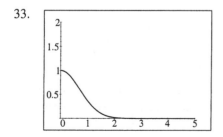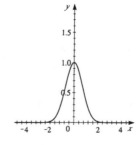

35.

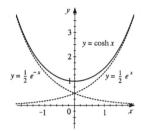

37. $\cosh(-x) = \dfrac{e^{-x} + e^{-(-x)}}{2} = \dfrac{e^{-x} + e^{x}}{2} = \dfrac{e^{x} + e^{-x}}{2} = \cosh(x).$

39. $[\cosh(x)]^2 - [\sinh(x)]^2 = \left(\dfrac{e^x + e^{-x}}{2}\right)^2 - \left(\dfrac{e^x - e^{-x}}{2}\right)^2$

$= \frac{1}{4}(e^{2x} + 2 + e^{-2x}) - \frac{1}{4}(e^{2x} - 2 + e^{-2x}) = \frac{2}{4} + \frac{2}{4} = 1.$

41.

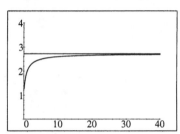

Note from the graph, that $y = \left(1 + \frac{1}{x}\right)^x$ approaches e as x get large.

43. (a)

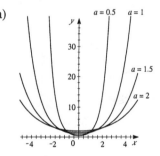

(b) As a increases the curve $y = \dfrac{a}{2}\left(e^{x/a} + e^{-x/a}\right)$ flattens out and the y intercept increases.

45. $f(x) = x\,e^{-x}$

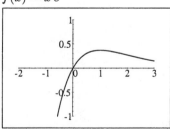

The graph of $f(x)$ is shown in the viewing rectangle $[-2, 3]$ by $[-1, 1]$. From the graph, we see that there is a local maximum when $x \approx 1.00$, and the maximum value is approximately 0.37.

Exercises 6.3

1.   (a) $2^5 = 32$ (b) $5^0 = 1$

3.   (a) $4^{1/2} = 2$ (b) $2^{-4} = \frac{1}{16}$

5.   (a) $e^x = 5$ (b) $e^5 = y$

7.   (a) $\log_2 8 = 3$ (b) $\log_{10} 0.001 = -3$

9.   (a) $\log_4 0.125 = -\frac{3}{2}$ (b) $\log_7 343 = 3$

11.   (a) $\ln 2 = x$ (b) $\ln y = 3$

13.   (a) $\log_5 5^4 = 4$ (b) $\log_4 64 = \log_4 4^3 = 3$

     (c) $\log_9 9 = 1$

15.   (a) $\log_8 64 = \log_8 8^2 = 2$ (b) $\log_7 49 = \log_7 7^2 = 2$

     (c) $\log_7 7^{10} = 10$

17.   (a) $\log_3 \left(\frac{1}{27}\right) = \log_3 3^{-3} = -3$ (b) $\log_{10} \sqrt{10} = \log_{10} 10^{1/2} = \frac{1}{2}$

     (c) $\log_5 0.2 = \log_5 \left(\frac{1}{5}\right) = \log_5 5^{-1} = -1$

19.   (a) $2^{\log_2 37} = 37$ (b) $3^{\log_3 8} = 8$

     (c) $e^{\ln \sqrt{5}} = \sqrt{5}$

21.   (a) $\log_8 0.25 = \log_8 8^{-2/3} = -\frac{2}{3}$ (b) $\ln e^4 = 4$

     (c) $\ln \left(\frac{1}{e}\right) = \ln e^{-1} = -1$

23.   (a) $\log_2 x = 5$ $\Leftrightarrow$ $x = 2^5 = 32$ (b) $x = \log_2 16 = \log_2 2^4 = 4$

25.   (a) $\log_{10} x = 2$ $\Leftrightarrow$ $x = 10^2 = 100$ (b) $\log_5 x = 2$ $\Leftrightarrow$ $x = 5^2 = 25$

27.   (a) $\log_x 16 = 4$ $\Leftrightarrow$ $x^4 = 16$ $\Leftrightarrow$ $x = 2$

     (b) $\log_x 8 = \frac{3}{2}$ $\Leftrightarrow$ $x^{3/2} = 8$ $\Leftrightarrow$ $x = 8^{2/3} = 4$

29.   (a) $\log 2 \approx 0.3010$ (b) $\log 35.2 \approx 1.5465$

     (c) $\log\left(\frac{2}{3}\right) \approx -0.1761$

31.   (a) $\ln 5 \approx 1.6094$ (b) $\ln 25.3 \approx 3.2308$

     (c) $\ln\left(1 + \sqrt{3}\right) \approx 1.0051$

33.   Since the point $(5, 1)$ is on the graph, we have $1 = \log_a 5$ $\Leftrightarrow$ $a^1 = 5$. Thus the function is $y = \log_5 x$.

35.   Since the point $\left(3, \frac{1}{2}\right)$ is on the graph, we have $\frac{1}{2} = \log_a 3$ $\Leftrightarrow$ $a^{1/2} = 3$ $\Leftrightarrow$ $a = 9$. Thus the function is $y = \log_9 x$.

37.   II   39.   III   41.   VI

43. The graph of $y = \log_4 x$ is obtained from the graph of $y = 4^x$ by reflecting it about the line $y = x$.

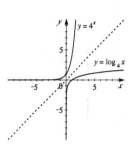

45. $f(x) = \log_2(x - 4)$

The graph of f is obtained from the graph of $y = \log_2 x$ by shifting it to the right 4 units.

Domain: $(4, \infty)$

Range: $(-\infty, \infty)$

Vertical asymptote: $x = 4$

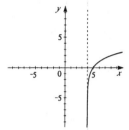

47. $g(x) = \log_5(-x)$

The graph of g is obtained from the graph of $y = \log_5 x$ by reflecting it about the y-axis.

Domain: $(-\infty, 0)$

Range: $(-\infty, \infty)$

Vertical asymptote: $x = 0$

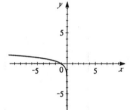

49. $y = 2 + \log_3 x$

The graph of $y = 2 + \log_3 x$ is obtained from the graph of $y = \log_3 x$ by shifting it upward 2 units.

Domain: $(0, \infty)$

Range: $(-\infty, \infty)$

Vertical asymptote: $x = 0$

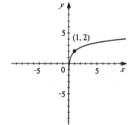

51. $y = 1 - \log_{10} x$

The graph of $y = 1 - \log_{10} x$ is obtained from the graph of $y = \log_{10} x$ by reflecting it about the x-axis, and then shifting it upward 1 unit.

Domain: $(0, \infty)$

Range: $(-\infty, \infty)$

Vertical asymptote: $x = 0$

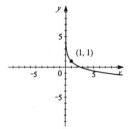

53. $y = |\ln x|$

Note that $y = \begin{cases} \ln x & x \geq 1 \\ -\ln x & 0 < x < 1. \end{cases}$

The graph of $y = |\ln x|$ is obtained from the graph of $y = \ln x$
by reflecting the part of the graph for $0 < x < 1$ about the
x-axis.

Domain: $(0, \infty)$

Range: $[0, \infty)$

Vertical asymptote: $x = 0$

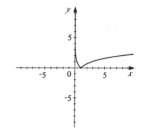

55. $f(x) = \log_{10}(2 + 5x)$. We require that $2 + 5x > 0 \quad \Leftrightarrow \quad 5x > -2 \quad \Leftrightarrow \quad x > -\frac{2}{5}$, so the
domain is $\left(-\frac{2}{5}, \infty\right)$.

57. $g(x) = \log_3(x^2 - 1)$. We require that $x^2 - 1 > 0 \quad \Leftrightarrow \quad x^2 > 1 \quad \Rightarrow \quad x < -1$ or $x > 1$, so the
domain is $(-\infty, -1) \cup (1, \infty)$.

59. $h(x) = \ln x + \ln(2 - x)$. We require that $x > 0$ and $2 - x > 0 \quad \Leftrightarrow \quad x > 0$ and $x < 2 \quad \Leftrightarrow$
$0 < x < 2$, so the domain is $(0, 2)$.

61. $y = \log_{10}(1 - x^2)$

Domain: $(-1, 1)$

Vertical asymptote: $x = -1$ and $x = 1$

Local maximum $y = 0$ at $x = 0$

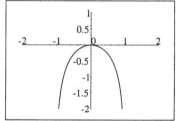

63. $y = x + \ln x$

Domain: $(0, \infty)$

Vertical asymptote: $x = 0$

No local extrema.

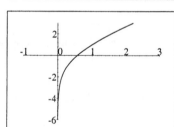

65. $y = \dfrac{\ln x}{x}$

Domain: $(0, \infty)$

Vertical asymptote: $x = 0$

Horizontal asymptote: $y = 0$

Local maximum $y \approx 0.37$ at $x \approx 2.72$

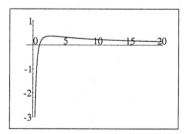

67. The graph of $g(x) = \sqrt{x}$ grows faster than the graph of $f(x) = \ln x$.

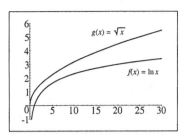

69. (a)

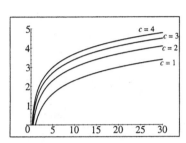

(b) Notice that $f(x) = \log(cx) = \log c + \log x$, so as c increases, the graph of $f(x) = \log(cx)$ is shifted upward $\log c$ units.

71. (a) $f(x) = \log_2(\log_{10}x)$. Since the domain of $\log_2 x$ is the positive real numbers, we have: $\log_{10}x > 0 \quad \Leftrightarrow \quad x > 10^0 = 1$. Thus the domain of $f(x)$ is $(1, \infty)$.

(b) $y = \log_2(\log_{10}x) \quad \Leftrightarrow \quad 2^y = \log_{10}x \quad \Leftrightarrow \quad 10^{2^y} = x$. Thus $f^{-1}(x) = 10^{2^x}$.

73. (a) $f(x) = \dfrac{2^x}{1+2^x}$. $y = \dfrac{2^x}{1+2^x} \quad \Leftrightarrow \quad y + y\,2^x = 2^x \quad \Leftrightarrow \quad y = 2^x - y\,2^x = 2^x(1-y) \quad \Leftrightarrow$

$2^x = \dfrac{y}{1-y} \quad \Leftrightarrow \quad x = \log_2\left(\dfrac{y}{1-y}\right)$. Thus $f^{-1}(x) = \log_2\left(\dfrac{x}{1-x}\right)$.

(b) $\dfrac{x}{1-x} > 0$. Solving this using the methods from Chapter 2, we start with the endpoints of the potential intervals, 0 and 1.

Interval	$(-\infty,0)$	$(0,1)$	$(1,\infty)$
Sign of x	$-$	$+$	$+$
Sign of $1-x$	$+$	$+$	$-$
Sign of $\dfrac{x}{1-x}$	$-$	$+$	$-$

Thus the domain of $f^{-1}(x)$ is $(0,1)$.

75. $\log(\log 10^{100}) = \log 100 = 2$

$\log(\log(\log 10^{\text{googol}})) = \log(\log(\text{googol})) = \log(\log 10^{100}) = \log(100) = 2$

77. The numbers between 1000 and 9999 (inclusive) each have 4 digits, while $\log 1000 = 3$ and $\log 10{,}000 = 4$. Since $[\![\,\log x\,]\!] = 3$ for all integers x where $1000 \le x < 10000$, the number of digits is $[\![\,\log x\,]\!] + 1$. Likewise, if x is an integer where $10^{n-1} \le x < 10^n$, then x would have n digits and $[\![\,\log x\,]\!] = n - 1$. Since $[\![\,\log x\,]\!] = n - 1 \quad \Leftrightarrow \quad n = [\![\,\log x\,]\!] + 1$, the number of digits in x is $[\![\,\log x\,]\!] + 1$.

Exercises 6.4

1. $\log_2[x(x-1)] = \log_2 x + \log_2(x-1)$

3. $\log 7^{23} = 23 \log 7$

5. $\log_2(AB^2) = \log_2 A + \log_2 B^2 = \log_2 A + 2\log_2 B$

7. $\log_3(x\sqrt{y}) = \log_3 x + \log_3 \sqrt{y} = \log_3 x + \frac{1}{2}\log_3 y$

9. $\log_5 \sqrt[3]{x^2+1} = \frac{1}{3}\log_5(x^2+1)$

11. $\ln\sqrt{ab} = \frac{1}{2}\ln ab = \frac{1}{2}(\ln a + \ln b)$

13. $\log\left(\dfrac{x^3 y^4}{z^6}\right) = \log\left(x^3 y^4\right) - \log z^6 = 3\log x + 4\log y - 6\log z$

15. $\log_2\left(\dfrac{x(x^2+1)}{\sqrt{x^2-1}}\right) = \log_2 x + \log_2(x^2+1) - \dfrac{1}{2}\log_2(x^2-1)$

17. $\ln\left(x\sqrt{\dfrac{y}{z}}\right) = \ln x + \dfrac{1}{2}\ln\left(\dfrac{y}{z}\right) = \ln x + \dfrac{1}{2}(\ln y - \ln z)$

19. $\log\sqrt[4]{x^2+y^2} = \frac{1}{4}\log(x^2+y^2)$

21. $\log\sqrt{\dfrac{x^2+4}{(x^2+1)(x^3-7)^2}} = \frac{1}{2}\log\dfrac{x^2+4}{(x^2+1)(x^3-7)^2} = \frac{1}{2}[\log(x^2+4) - \log(x^2+1)(x^3-7)^2]$

$= \frac{1}{2}[\log(x^2+4) - \log(x^2+1) - 2\log(x^3-7)]$

23. $\ln\dfrac{z^4\sqrt{x}}{\sqrt[3]{y^2+6y+17}} = \ln(z^4\sqrt{x}) - \ln\sqrt[3]{y^2+6y+17} = 4\ln z + \dfrac{1}{2}\ln x - \dfrac{1}{3}\ln(y^2+6y+17)$

25. $\log_5\sqrt{125} = \log_5 5^{3/2} = \frac{3}{2}$

27. $\log 2 + \log 5 = \log 10 = 1$

29. $\log_4 192 - \log_4 3 = \log_4 \frac{192}{3} = \log_4 64 = \log_4 4^3 = 3$

31. $\ln 6 - \ln 15 + \ln 20 = \ln\frac{6}{15} + \ln 20 = \ln(\frac{2}{5}\cdot 20) = \ln 8$

33. $10^{2\log 4} = (10^{\log 4})^2 = 4^2 = 16$

35. $\log_3 5 + 5\log_3 2 = \log_3 5 + \log_3 2^5 = \log_3(5\cdot 2^5) = \log_3 160$

37. $\log_2 A + \log_2 B - 2\log_2 C = \log_2(AB) - \log_2(C^2) = \log_2\left(\dfrac{AB}{C^2}\right)$

39. $4\log x - \frac{1}{3}\log(x^2+1) + 2\log(x-1) = \log x^4 - \log\sqrt[3]{x^2+1} + \log(x-1)^2$

$= \log\left(\dfrac{x^4}{\sqrt[3]{x^2+1}}\right) + \log(x-1)^2 = \log\left(\dfrac{x^4(x-1)^2}{\sqrt[3]{x^2+1}}\right)$

41. $\ln 5 + 2\ln x + 3\ln(x^2 + 5) = \ln(5x^2) + \ln(x^2 + 5)^3 = \ln[5x^2(x^2 + 5)^3]$

43. $\frac{1}{3}\log(2x + 1) + \frac{1}{2}[\log(x - 4) - \log(x^4 - x^2 - 1)] =$

$\log\sqrt[3]{2x + 1} + \frac{1}{2}\log\dfrac{x - 4}{x^4 - x^2 - 1} = \log\left(\sqrt[3]{2x + 1} \cdot \sqrt{\dfrac{x - 4}{x^4 - x^2 - 1}}\right)$

45. $\log_2 7 = \frac{\log 7}{\log 2} \approx 2.807355$

47. $\log_3 11 = \frac{\log 11}{\log 3} \approx 2.182658$

49. $\log_7 3.58 = \frac{\log 3.58}{\log 7} \approx 0.655407$

51. $\log_4 322 = \frac{\log 322}{\log 4} \approx 4.165458$

53. $\log_3 x = \dfrac{\log_e x}{\log_e 3} = \dfrac{\ln x}{\ln 3} = \dfrac{1}{\ln 3}\ln x.$

The graph of $y = \dfrac{1}{\ln 3}\ln x$ is shown in the viewing rectangle $[-1, 4]$ by $[-3, 2]$.

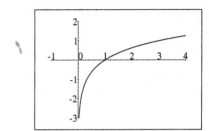

55. $\log e = \frac{\ln e}{\ln 10} = \frac{1}{\ln 10}$

57. $-\ln(x - \sqrt{x^2 - 1}) = \ln\left(\dfrac{1}{x - \sqrt{x^2 - 1}}\right) = \ln\left(\dfrac{1}{x - \sqrt{x^2 - 1}} \cdot \dfrac{x + \sqrt{x^2 - 1}}{x + \sqrt{x^2 - 1}}\right)$

$= \ln\left(\dfrac{x + \sqrt{x^2 - 1}}{x^2 - (x^2 - 1)}\right) = \ln\left(x + \sqrt{x^2 - 1}\right)$

59. The error is on the first line: $\log 0.1 < 0$, so $2\log 0.1 < \log 0.1$.

Exercises 6.5

1. $5^x = 16$ $\Leftrightarrow$ $\log 5^x = \log 16$ $\Leftrightarrow$ $x \log 5 = \log 16$ $\Leftrightarrow$ $x = \frac{\log 16}{\log 5} = 1.7227$

3. $2^{1-x} = 3$ $\Leftrightarrow$ $\log 2^{1-x} = \log 3$ $\Leftrightarrow$ $(1-x) \log 2 = \log 3$ $\Leftrightarrow$ $1 - x = \frac{\log 3}{\log 2}$ $\Leftrightarrow$
   $x = 1 - \frac{\log 3}{\log 2} \approx -0.5850$

5. $3 e^x = 10$ $\Leftrightarrow$ $e^x = \frac{10}{3}$ $\Leftrightarrow$ $x = \ln\left(\frac{10}{3}\right) \approx 1.2040$

7. $e^{1-4x} = 2$ $\Leftrightarrow$ $1 - 4x = \ln 2$ $\Leftrightarrow$ $-4x = -1 + \ln 2$ $\Leftrightarrow$ $x = \frac{1 - \ln 2}{4} = 0.0767$

9. $4 + 3^{5x} = 8$ $\Leftrightarrow$ $3^{5x} = 4$ $\Leftrightarrow$ $\log 3^{5x} = \log 4$ $\Leftrightarrow$ $5x \log 3 = \log 4$ $\Leftrightarrow$ $5x = \frac{\log 4}{\log 3}$
   $\Leftrightarrow$ $x = \frac{\log 4}{5 \log 3} \approx 0.2524$

11. $8^{0.4x} = 5$ $\Leftrightarrow$ $\log 8^{0.4x} = \log 5$ $\Leftrightarrow$ $0.4x \log 8 = \log 5$ $\Leftrightarrow$ $0.4x = \frac{\log 5}{\log 8}$ $\Leftrightarrow$
    $x = \frac{\log 5}{0.4 \log 8} \approx 1.9349$

13. $5^{-x/100} = 2$ $\Leftrightarrow$ $\log 5^{-x/100} = \log 2$ $\Leftrightarrow$ $-\frac{x}{100} \log 5 = \log 2$ $\Leftrightarrow$
    $x = -\frac{100 \log 2}{\log 5} \approx -43.0677$

15. $e^{2x+1} = 200$ $\Leftrightarrow$ $2x + 1 = \ln 200$ $\Leftrightarrow$ $2x = -1 + \ln 200$ $\Leftrightarrow$ $x = \frac{-1 + \ln 200}{2} \approx 2.1492$

17. $5^x = 4^{x+1}$ $\Leftrightarrow$ $\log 5^x = \log 4^{x+1}$ $\Leftrightarrow$ $x \log 5 = (x+1) \log 4 = x \log 4 + \log 4$ $\Leftrightarrow$
    $x \log 5 - x \log 4 = \log 4$ $\Leftrightarrow$ $x(\log 5 - \log 4) = \log 4$ $\Leftrightarrow$ $x = \frac{\log 4}{\log 5 - \log 4} \approx 6.2126$

19. $2^{3x+1} = 3^{x-2}$ $\Leftrightarrow$ $\log 2^{3x+1} = \log 3^{x-2}$ $\Leftrightarrow$ $(3x+1) \log 2 = (x-2) \log 3$ $\Leftrightarrow$
    $3x \log 2 + \log 2 = x \log 3 - 2 \log 3$ $\Leftrightarrow$ $3x \log 2 - x \log 3 = -\log 2 - 2 \log 3$ $\Leftrightarrow$
    $x(3 \log 2 - \log 3) = -(\log 2 + 2 \log 3)$ $\Leftrightarrow$ $s = -\frac{\log 2 + 2 \log 3}{3 \log 2 - \log 3} \approx -2.9469$

21. $\dfrac{50}{1 + e^{-x}} = 4$ $\Leftrightarrow$ $50 = 4 + 4 e^{-x}$ $\Leftrightarrow$ $46 = 4 e^{-x}$ $\Leftrightarrow$ $11.5 = e^{-x}$ $\Leftrightarrow$ $\ln 11.5 = -x$
    $\Leftrightarrow$ $x = -\ln 11.5 \approx -2.4423$

23. $100(1.04)^{2t} = 300$ $\Leftrightarrow$ $1.04^{2t} = 3$ $\Leftrightarrow$ $\log 1.04^{2t} = \log 3$ $\Leftrightarrow$ $2t \log 1.04 = \log 3$ $\Leftrightarrow$
    $t = \frac{\log 3}{2 \log 1.04} \approx 14.0055$

25. $x^2 2^x - 2^x = 0$ $\Leftrightarrow$ $2^x(x^2 - 1) = 0$ $\Rightarrow$ $2^x = 0$ (never) or $x^2 - 1 = 0$. If $x^2 - 1 = 0$, then
    $x^2 = 1$ $\Rightarrow$ $x = \pm 1$. So the only solutions are $x = \pm 1$.

27. $4x^3 e^{-3x} - 3x^4 e^{-3x} = 0$ $\Leftrightarrow$ $x^3 e^{-3x}(4 - 3x) = 0$ $\Rightarrow$ $x = 0$ or $e^{-3x} = 0$ (never) or
    $4 - 3x = 0$. If $4 - 3x = 0$, then $3x = 4$ $\Leftrightarrow$ $x = \frac{4}{3}$. So the solutions are $x = 0$ and $x = \frac{4}{3}$.

29. $e^{2x} - 3e^x + 2 = 0$ $\Leftrightarrow$ $(e^x - 1)(e^x - 2) = 0$ $\Rightarrow$ $e^x - 1 = 0$ or $e^x - 2 = 0$. If $e^x - 1 = 0$,
    then $e^x = 1$ $\Leftrightarrow$ $x = \ln 1 = 0$. If $e^x - 2 = 0$, then $e^x = 2$ $\Leftrightarrow$ $x = \ln 2 \approx 0.6931$. So the
    solutions are $x = 0$ and $x \approx 0.6931$.

31. $e^{4x} + 4e^{2x} - 21 = 0$ $\Leftrightarrow$ $(e^{2x} + 7)(e^{2x} - 3) = 0$ $\Rightarrow$ $e^{2x} = -7$ or $e^{2x} = 3$. Now $e^{2x} = -7$
    has no solution, since $e^{2x} > 0$ for all x. But we _can_ solve $e^{2x} = 3$ $\Leftrightarrow$ $2x = \ln 3$ $\Leftrightarrow$
    $x = \frac{1}{2} \ln 3 \approx 0.5493$. So the only solution is $x \approx 0.5493$.

33. $\ln x = 10$ $\Leftrightarrow$ $x = e^{10} \approx 22026$

35. $\log x = -2 \quad \Leftrightarrow \quad x = 10^{-2} = 0.01$

37. $\log(3x + 5) = 2 \quad \Leftrightarrow \quad 3x + 5 = 10^2 = 100 \quad \Leftrightarrow \quad 3x = 95 \quad \Leftrightarrow \quad x = \frac{95}{3} \approx 31.6667$

39. $2 - \ln(3 - x) = 0 \quad \Leftrightarrow \quad 2 = \ln(3 - x) \quad \Leftrightarrow \quad e^2 = 3 - x \quad \Leftrightarrow \quad x = 3 - e^2 \approx -4.3891$

41. $\log_2 3 + \log_2 x = \log_2 5 + \log_2(x - 2) \quad \Leftrightarrow \quad \log_2(3x) = \log_2(5x - 10) \quad \Leftrightarrow \quad 3x = 5x - 10$
$\Leftrightarrow \quad 2x = 10 \quad \Leftrightarrow \quad x = 5$

43. $\log x + \log(x - 1) = \log(4x) \quad \Leftrightarrow \quad \log[x(x - 1)] = \log(4x) \quad \Leftrightarrow \quad x^2 - x = 4x \quad \Leftrightarrow$
$x^2 - 5x = 0 \quad \Leftrightarrow \quad x(x - 5) = 0 \quad \Rightarrow \quad x = 0$ or $x = 5$. So the *possible* solutions are $x = 0$ and
$x = 5$. However, when $x = 0$, $\log x$ is undefined. Thus the only solution is $x = 5$.

45. $\log_5(x + 1) - \log_5(x - 1) = 2 \quad \Leftrightarrow \quad \log_5\left(\dfrac{x + 1}{x - 1}\right) = 2 \quad \Leftrightarrow \quad \dfrac{x + 1}{x - 1} = 5^2 \quad \Leftrightarrow$
$x + 1 = 25x - 25 \quad \Leftrightarrow \quad 24x = 26 \quad \Leftrightarrow \quad x = \frac{13}{12}$

47. $\log_9(x - 5) + \log_9(x + 3) = 1 \quad \Leftrightarrow \quad \log_9[(x - 5)(x + 3)] = 1 \quad \Leftrightarrow \quad (x - 5)(x + 3) = 9^1$
$\Leftrightarrow \quad x^2 - 2x - 24 = 0 \quad \Leftrightarrow \quad (x - 6)(x + 4) = 0 \quad \Rightarrow \quad x = 6$ or -4. However, $x = -4$ is
inadmissible, so $x = 6$ is the only solution.

49. $\log(x + 3) = \log x + \log 3 \quad \Leftrightarrow \quad \log(x + 3) = \log(3x) \quad \Leftrightarrow \quad x + 3 = 3x \quad \Leftrightarrow \quad 2x = 3 \quad \Leftrightarrow$
$x = \frac{3}{2}$

51. $2^{2/\log_5 x} = \frac{1}{16} \quad \Leftrightarrow \quad \log_2 2^{2/\log_5 x} = \log_2\left(\frac{1}{16}\right) \quad \Leftrightarrow \quad \dfrac{2}{\log_5 x} = -4 \quad \Leftrightarrow \quad \log_5 x = -\frac{1}{2} \quad \Leftrightarrow$
$x = 5^{-1/2} = \frac{1}{\sqrt{5}} \approx 0.4472$

53. $15e^{-0.087t} = 5 \quad \Leftrightarrow \quad e^{-0.087t} = \frac{1}{3} \quad \Leftrightarrow \quad -0.087t = \ln\left(\frac{1}{3}\right) = -\ln 3 \quad \Leftrightarrow \quad t = \frac{\ln 3}{0.087} \approx 12.6277$.
So only 5 grams remain after approximately 13 days.

55. (a) $I = \frac{60}{13}\left(1 - e^{-13t/5}\right) \quad \Leftrightarrow \quad \frac{13}{60}I = 1 - e^{-13t/5} \quad \Leftrightarrow \quad e^{-13t/5} = 1 - \frac{13}{60}I \quad \Leftrightarrow$
$-\frac{13}{5}t = \ln\left(1 - \frac{13}{60}I\right) \quad \Leftrightarrow \quad t = -\frac{5}{13}\ln\left(1 - \frac{13}{60}I\right)$.

   (b)   Substituting $I = 2$, we have $t = -\frac{5}{13}\ln\left[1 - \frac{13}{60}(2)\right] \approx 0.218$ seconds.

57. $\ln x = 3 - x \quad \Leftrightarrow \quad \ln x + x - 3 = 0$. Let
$f(x) = \ln x + x - 3$. We need to solve the equation
$f(x) = 0$. From the graph of f, we get $x \approx 2.21$.

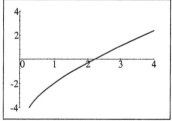

59. $x^3 - x = \log_{10}(x + 1) \quad \Leftrightarrow \quad x^3 - x - \log_{10}(x + 1) = 0$.
Let $f(x) = x^3 - x - \log_{10}(x + 1)$. We need to solve the
equation $f(x) = 0$. From the graph of f, we get $x = 0$
or $x \approx 1.14$.

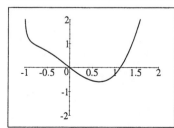

61. $e^x = -x$ $\Leftrightarrow$ $e^x + x = 0$. Let $f(x) = e^x + x$. We need to
     solve the equation $f(x) = 0$. From the graph of f, we get
     $x \approx -0.57$.

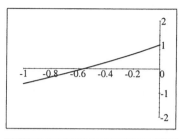

63. $4^{-x} = \sqrt{x}$ $\Leftrightarrow$ $4^{-x} - \sqrt{x} = 0$. Let $f(x) = 4^{-x} - \sqrt{x}$.
     We need to solve the equation $f(x) = 0$. From the graph of f,
     we get $x \approx 0.36$.

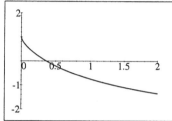

65. $\log(x - 2) + \log(9 - x) < 1$ $\Leftrightarrow$ $\log[(x - 2)(9 - x)] < 1$ $\Leftrightarrow$ $\log(-x^2 + 11x - 18) < 1$
     $\Rightarrow$ $-x^2 + 11x - 18 < 10^1$ $\Leftrightarrow$ $0 < x^2 - 11x + 28$ $\Leftrightarrow$ $0 < (x - 7)(x - 4)$. Also, since
     the domain of a logarithm is positive we must have $0 < -x^2 + 11x - 18$ $\Leftrightarrow$
     $0 < (x - 2)(9 - x)$. Using the methods from Chapter 3 with the endpoints 2, 4, 7, 9 for the
     intervals, we have:

Interval	$(-\infty, 2)$	$(2, 4)$	$(4, 7)$	$(7, 9)$	$(9, \infty)$
Sign of $x - 7$	$-$	$-$	$-$	$+$	$+$
Sign of $x - 4$	$-$	$-$	$+$	$+$	$+$
Sign of $x - 2$	$-$	$+$	$+$	$+$	$+$
Sign of $9 - x$	$+$	$+$	$+$	$+$	$-$
Sign of $(x - 7)(x - 4)$	$+$	$+$	$-$	$+$	$+$
Sign of $(x - 2)(9 - x)$	$-$	$+$	$+$	$+$	$-$

Thus the solution is $(2, 4) \cup (7, 9)$.

67. $2 < 10^x < 5$ $\Leftrightarrow$ $\log 2 < x < \log 5$ $\Leftrightarrow$ $0.3010 < x < 0.6990$. Hence the solution to the
     inequality is approximately the interval $(0.3010, 0.6990)$.

69. Since $x - 1 > 0$ $\Leftrightarrow$ $x > 1$, we can take the log of both sides of the equation
     $(x - 1)^{\log(x-1)} = 100(x - 1)$. Then we have $\log\left[(x - 1)^{\log(x-1)}\right] = \log[100(x - 1)]$ $\Leftrightarrow$
     $[\log(x - 1)][\log(x - 1)] = 2 + \log(x - 1)$ $\Leftrightarrow$ $[\log(x - 1)]^2 - \log(x - 1) - 2 = 0$ $\Leftrightarrow$
     $[\log(x - 1) + 1][\log(x - 1) - 2] = 0$ $\Rightarrow$

$$\log(x - 1) + 1 = 0 \qquad \text{or} \qquad \log(x - 1) - 2 = 0$$
$$\log(x - 1) = -1 \qquad\qquad\qquad \log(x - 1) = 2$$
$$x - 1 = \tfrac{1}{10} \qquad\qquad\qquad\qquad x - 1 = 10^2$$
$$x = \tfrac{11}{10} \qquad\qquad\qquad\qquad\quad x = 101$$

The solutions are $x = 1.1$ or $x = 101$.

71. $4^x - 2^{x+1} = 3$ $\Leftrightarrow$ $2^{2x} - 2 \cdot 2^x = 3$ $\Leftrightarrow$ $(2^x)^2 - 2 \cdot 2^x - 3 = 0$ $\Leftrightarrow$
     $(2^x - 3)(2^x + 1) = 0$ $\Leftrightarrow$ $2^x = 3$ or $2^x = -1$. Since $2^x > 0$ for all x, $2^x = -1$ has no solution,
     but $2^x = 3$ $\Leftrightarrow$ $x \log 2 = \log 3$ $\Leftrightarrow$ $x = \frac{\log 3}{\log 2} \approx 1.58$. So the only solution is $x \approx 1.58$.

73.   Notice that $\log\left(x^{1/\log x}\right) = \dfrac{1}{\log x} \log x = 1$, so $x^{1/\log x} = 10^1$ for all $x > 0$. So $x^{1/\log x} = 5$ has no

solution, and $x^{1/\log x} = k$ has a solution only when $k = 10$. This is verified by the graph of
$f(x) = x^{1/\log x}$.

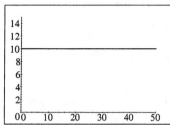

Exercises 6.6

1. (a) $A(3) = 10000\left(1 + \frac{0.085}{4}\right)^{4(3)} = 10000(1.02125^{12}) = 12870.19$. Thus the amount after 3 years is \$12,870.19.

   (b) $20000 = 10000\left(1 + \frac{0.085}{4}\right)^{4t} = 10000(1.02125^{4t}) \quad \Leftrightarrow \quad 2 = 1.02125^{4t} \quad \Leftrightarrow$
   $\log 2 = 4t \log 1.02125 \quad \Leftrightarrow \quad t = \frac{\log 2}{4 \log 1.02125} \approx 8.24$ years. Thus the investment will double in about 8.24 years.

3. $8000 = 5000\left(1 + \frac{0.095}{4}\right)^{4t} = 5000(1.02375^{4t}) \quad \Leftrightarrow \quad 1.6 = 1.02375^{4t} \quad \Leftrightarrow$
   $\log 1.6 = 4t \log 1.02375 \quad \Leftrightarrow \quad t = \frac{\log 1.6}{4 \log 1.02375} \approx 5.0059$ years. The investment will increase to \$8000 in approximately 5 years.

5. $2 = e^{0.085t} \quad \Leftrightarrow \quad \ln 2 = 0.085t \quad \Leftrightarrow \quad t = \frac{\ln 2}{0.085} \approx 8.15$ years. Thus the investment will double in about 8.15 years.

7. $r_{\text{eff}} = \left(1 + \frac{r}{n}\right)^n - 1$. Here $r = 0.08$ and $n = 12$, so $r_{\text{eff}} = \left(1 + \frac{0.08}{12}\right)^{12} - 1 = (1.0066667)^{12} - 1$
   $= 1.08299 - 1 \approx 0.083$. So the effective yield is about 8.3%.

9. (a) $n(0) = 500$.

   (b) The relative growth rate is $0.45 = 45\%$.

   (c) $n(3) = 500e^{0.45(3)} \approx 1929$.

   (d) $10000 = 500\, e^{0.45t} \quad \Leftrightarrow \quad 20 = e^{0.45t} \quad \Leftrightarrow \quad 0.45t = \ln 20 \quad \Leftrightarrow \quad t = \frac{\ln 20}{0.45} \approx 6.66$ hours, or 6 hours 40 minutes.

11. (a) $n(t) = 112000e^{0.04t}$.

    (b) $t = 2000 - 1994 = 6$ and $n(6) = 112000\, e^{0.04(6)} \approx 142380$. The projected population is about 142,000.

    (c) $200000 = 112000\, e^{0.04t} \quad \Leftrightarrow \quad \frac{25}{14} = e^{0.04t} \quad \Leftrightarrow \quad 0.04t = \ln\left(\frac{25}{14}\right) \quad \Leftrightarrow$
    $t = 25 \ln\left(\frac{25}{14}\right) \approx 14.5$. Since $1994 + 14.5 = 2008.5$, the population will reach 200,000 during the year 2008.

13. (a) The deer population in 1996 was 20,000.

    (b) Using the model $n(t) = 20000\, e^{rt}$ and the point $(4, 31000)$, we have $31000 = 20000\, e^{4r} \quad \Leftrightarrow$
    $1.55 = e^{4r} \quad \Leftrightarrow \quad 4r = \ln 1.55 \quad \Leftrightarrow \quad r = \frac{1}{4} \ln 1.55 \approx 0.1096$. Thus $n(t) = 20000\, e^{0.1096t}$

    (c) $n(8) = 20000\, e^{0.1096(8)} \approx 48218$, so the projected deer population in 2004 is about 48,000.

    (d) $100000 = 20000\, e^{0.1096t} \quad \Leftrightarrow \quad 5 = e^{0.1096t} \quad \Leftrightarrow \quad 0.1096t = \ln 5 \quad \Leftrightarrow \quad t = \frac{\ln 5}{0.1096}$
    ≈ 14.63. Since $1996 + 14.63 = 2010.63$, the deer population will reach 100,000 during the year 2010.

15. (a) Using the formula $n(t) = n_0\, e^{rt}$ with $n_0 = 8600$ and $n(1) = 10000$, we solve for r, giving
    $10000 = n(1) = 8600\, e^r \quad \Leftrightarrow \quad \frac{50}{43} = e^r \quad \Leftrightarrow \quad r = \ln\left(\frac{50}{43}\right) \approx 0.1508$. Thus
    $n(t) = 8600\, e^{0.1508\, t}$.

    (b) $n(2) = 8600\, e^{0.1508(2)} \approx 11627$. Thus the number of bacteria after two hours is about 11,600.

(c) $17200 = 8600e^{0.1508t}$ $\Leftrightarrow$ $2 = e^{0.1508t}$ $\Leftrightarrow$ $0.1508t = \ln 2$ $\Leftrightarrow$ $t = \frac{\ln 2}{0.1508} \approx 4.596$.
Thus the number of bacteria will double in about 4.6 hours.

17.   (a) $2n_0 = n_0 e^{0.02t}$ $\Leftrightarrow$ $2 = e^{0.02t}$ $\Leftrightarrow$ $0.02t = \ln 2$ $\Leftrightarrow$ $t = 50 \ln 2 \approx 34.65$. So we have
$t = 1995 + 34.65 = 2029.65$, and hence at the current growth rate the population will double
by the year 2029.

(b) $3n_0 = n_0 e^{0.02t}$ $\Leftrightarrow$ $3 = e^{0.02t}$ $\Leftrightarrow$ $0.02t = \ln 3$ $\Leftrightarrow$ $t = 50 \ln 3 \approx 54.93$. So we
have $t = 1995 + 54.93 = 2049.93$, and hence at the current growth rate the population will
triple by the year 2049.

19. $n(t) = n_0 e^{2t}$. When $n_0 = 1$, the critical level is $n(24) = e^{2(24)} = e^{48}$. We solve the equation
$e^{48} = n_0 e^{2t}$, where $n_0 = 10$. This gives $e^{48} = 10 e^{2t}$ $\Leftrightarrow$ $48 = \ln 10 + 2t$ $\Leftrightarrow$
$2t = 48 - \ln 10$ $\Leftrightarrow$ $t = \frac{1}{2}(48 - \ln 10) \approx 22.85$ hours.

21.   (a)   Using $m(t) = m_0 e^{-rt}$ with $m_0 = 10$ and $h = 30$, we have $r = \dfrac{\ln 2}{h} = \dfrac{\ln 2}{30} \approx 0.023$. Thus
$m(t) = 10e^{-0.023t}$.

(b) $m(80) = 10e^{-0.023(80)} \approx 1.6$ grams.

(c) $2 = 10e^{-0.023t}$ $\Leftrightarrow$ $\frac{1}{5} = e^{-0.023t}$ $\Leftrightarrow$ $\ln\left(\frac{1}{5}\right) = -0.023t$ $\Leftrightarrow$ $t = \frac{-\ln 5}{-0.023} \approx 70$ years.

23.   By the formula in the text, $m(t) = m_0 e^{-rt}$ where $r = \frac{\ln 2}{h}$, so $m(t) = 50e^{-\frac{\ln 2}{28} \cdot t}$. We need to solve
for t in the equation $32 = 50e^{-\frac{\ln 2}{28}t}$. This gives $e^{-\frac{\ln 2}{28}t} = \dfrac{32}{50}$ $\Leftrightarrow$ $-\dfrac{\ln 2}{28}t = \ln\left(\dfrac{32}{50}\right)$ $\Leftrightarrow$
$t = -\frac{28}{\ln 2} \cdot \ln\left(\frac{32}{50}\right) \approx 18.03$, so it takes about 18 years.

25.   By the formula for radioactive decay, we have $m(t) = m_0 e^{-rt}$, where $r = \frac{\ln 2}{h}$, in other words
$m(t) = m_0 e^{-\frac{\ln 2}{h} \cdot t}$. In this exercise we have to solve for h in the equation $200 = 250e^{-\frac{\ln 2}{h} \cdot 48}$
$\Leftrightarrow$ $0.8 = e^{-\frac{\ln 2}{h} \cdot 48}$ $\Leftrightarrow$ $\ln(0.8) = -\frac{\ln 2}{h} \cdot 48$ $\Leftrightarrow$ $h = -\frac{\ln 2}{\ln 0.8} \cdot 48 \approx 149.1$ hours. So the
half-life is approximately 149 hours.

27.   By the formula in the text, $m(t) = m_0 e^{-\frac{\ln 2}{h} \cdot t}$, so we have $0.65 = 1 \cdot e^{-\frac{\ln 2}{5730} \cdot t}$ $\Leftrightarrow$
$\ln(0.65) = -\dfrac{\ln 2}{5730}t$ $\Leftrightarrow$ $t = -\dfrac{5730 \ln 0.65}{\ln 2} \approx 3561$. Thus the artifact is about 3560 years old.

29.   (a) $T(0) = 65 + 145 e^{-0.05(0)} = 65 + 145 = 210°F$.

(b) $T(10) = 65 + 145 e^{-0.05(10)} \approx 152.9$. Thus the temperature after 10 minutes is about 153°F.

(c) $100 = 65 + 145 e^{-0.05t}$ $\Leftrightarrow$ $35 = 145 e^{-0.05t}$ $\Leftrightarrow$ $0.2414 = e^{-0.05t}$ $\Leftrightarrow$
$\ln 0.2414 = -0.05t$ $\Leftrightarrow$ $t = -\dfrac{\ln 0.2414}{0.05} \approx 28.4$. Thus the temperature will be 100°F in
about 28 minutes.

31.   Using Newton's Law of Cooling, $T(t) = T_s + D_0 e^{-kt}$ with $T_s = 75$ and $D_0 = 185 - 75 = 110$. So
$T(t) = 75 + 110 e^{-kt}$.

(a)   Since $T(30) = 150$, we have $T(30) = 75 + 110 e^{-30k} = 150$ $\Leftrightarrow$ $110 e^{-30k} = 75$ $\Leftrightarrow$
$e^{-30k} = \frac{15}{22}$ $\Leftrightarrow$ $-30k = \ln\left(\frac{15}{22}\right)$ $\Leftrightarrow$ $k = -\frac{1}{30} \ln\left(\frac{15}{22}\right)$. Thus we have
$T(45) = 75 + 110 e^{(45/30) \ln(15/22)} \approx 136.9$, and so the temperature of the turkey after 45
minutes is about 137° F.

(b) The temperature will be $100°\,$F when $75 + 110\, e^{(t/30)\ln(15/22)} = 100$ $\Leftrightarrow$

$e^{(t/30)\ln(15/22)} = \frac{25}{110} = \frac{5}{22}$ $\Leftrightarrow$ $\left(\frac{t}{30}\right)\ln\left(\frac{15}{22}\right) = \ln\left(\frac{5}{22}\right)$ $\Leftrightarrow$ $t = 30\dfrac{\ln\left(\frac{5}{22}\right)}{\ln\left(\frac{15}{22}\right)} \approx 116.1$. So the

temperature will be $100°\,$F after 116 minutes.

33. (a) $\text{pH} = -\log\left[\text{H}^+\right] = -\log(5.0 \times 10^{-3}) \approx 2.3$

(b) $\text{pH} = -\log\left[\text{H}^+\right] = -\log(3.2 \times 10^{-4}) \approx 3.5$

(c) $\text{pH} = -\log\left[\text{H}^+\right] = -\log(5.0 \times 10^{-9}) \approx 8.3$

35. (a) $\text{pH} = -\log\left[\text{H}^+\right] = 3.0$ $\Leftrightarrow$ $\left[\text{H}^+\right] = 10^{-3}\,\text{M}$

(b) $\text{pH} = -\log\left[\text{H}^+\right] = 6.5$ $\Leftrightarrow$ $\left[\text{H}^+\right] = 10^{-6.5} \approx 3.2 \times 10^{-7}\,\text{M}$

37. $4.0 \times 10^{-7} \le \left[\text{H}^+\right] \le 1.6 \times 10^{-5}$ $\Leftrightarrow$ $\log(4.0 \times 10^{-7}) \le \log\left[\text{H}^+\right] \le \log(1.6 \times 10^{-5})$ $\Leftrightarrow$ $-\log(4.0 \times 10^{-7}) \ge \text{pH} \ge -\log(1.6 \times 10^{-5})$ $\Leftrightarrow$ $6.4 \ge \text{pH} \ge 4.8$. Therefore the range of pH readings for cheese is approximately 4.8 to 6.4.

39. Let I_0 be the intensity of the smaller earthquake and I_1 the intensity of the larger earthquake. Then $I_1 = 20\, I_0$. Notice that $M_0 = \log\left(\frac{I_0}{S}\right) = \log I_0 - \log S$ and $M_1 = \log\left(\frac{I_1}{S}\right) = \log\left(\frac{20\,I_0}{S}\right)$ $= \log 20 + \log I_0 - \log S$. Then $M_1 - M_0 = \log 20 + \log I_0 - \log S - \log I_0 + \log S = \log 20 \approx 1.3$. Therefore the magnitude is 1.3 times larger.

41. Let the subscript A represent the Alaska earthquake and SF represent the San Francisco earthquake. Then $M_A = \log\left(\frac{I_A}{S}\right) = 8.6$ $\Leftrightarrow$ $I_A = S \cdot 10^{8.6}$; also, $M_{SF} = \log\left(\frac{I_{SF}}{S}\right) = 8.3$ $\Leftrightarrow$ $I_{SF} = S \cdot 10^{8.6}$. So $\dfrac{I_A}{I_{SF}} = \dfrac{S \cdot 10^{8.6}}{S \cdot 10^{8.3}} = 10^{0.3} \approx 1.995$, and hence the Alaskan earthquake was roughly twice as intense as the San Francisco earthquake.

43. Let the subscript MC represent the Mexico City earthquake, and T represent the Tangshan earthquake. We have $\dfrac{I_T}{I_{MC}} = 1.26$ $\Leftrightarrow$ $\log 1.26 = \log\dfrac{I_T}{I_{MC}} = \log\dfrac{I_T/S}{I_{MC}/S} = \log\dfrac{I_T}{S} - \log\dfrac{I_{MC}}{S}$ $= M_T - M_{MC}$. Therefore $M_T = M_{MC} + \log 1.26 \approx 8.1 + 0.1 = 8.2$. Thus the magnitude of the Tangshan earthquake was roughly 8.2.

45. $98 = 10\log\left(\frac{I}{10^{-12}}\right)$ $\Leftrightarrow$ $\log(I \cdot 10^{12}) = 9.8$ $\Leftrightarrow$ $\log I = 9.8 - \log 10^{12} = -2.2$ $\Leftrightarrow$ $I = 10^{-2.2} \approx 6.3 \times 10^{-3}$. So the intensity was 6.3×10^{-3} watts/m^2.

47. (a) $\beta_1 = 10\log\left(\dfrac{I_1}{I_0}\right)$ and $I_1 = \dfrac{k}{d_1^2}$ $\Leftrightarrow$ $\beta_1 = 10\log\left(\dfrac{k}{d_1^2\, I_0}\right) = 10\left[\log\left(\dfrac{k}{I_0}\right) - 2\log d_1\right]$

$= 10\log\left(\dfrac{k}{I_0}\right) - 20\log d_1$. Similarly, $\beta_2 = 10\log\left(\dfrac{k}{I_0}\right) - 20\log d_2$. Substituting the

expression for β_1, gives $\beta_2 = 10\log\left(\dfrac{k}{I_0}\right) - 20\log d_1 + 20\log d_1 - 20\log d_2$

$= \beta_1 + 20\log d_1 - 20\log d_2 = \beta_1 + 20\log\left(\dfrac{d_1}{d_2}\right)$.

(b) $\beta_1 = 120$, $d_1 = 2$, and $d_2 = 10$. Then $\beta_2 = \beta_1 + 20\log\left(\dfrac{d_1}{d_2}\right) = 120 + 20\log\left(\dfrac{2}{10}\right)$

$= 120 + 20\log 0.2 \approx 106$, and so the intensity level at 10 m is approximately 106 dB.

Review Exercises for Chapter 6

1. $f(x) = \dfrac{1}{2^x}$.
    Domain: $(-\infty, \infty)$
    Range: $(0, \infty)$
    Asymptote: $y = 0$.

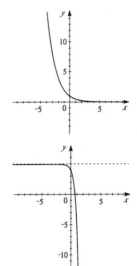

3. $y = 5 - 10^x$.
    Domain: $(-\infty, \infty)$
    Range: $(-\infty, 5)$
    Asymptote: $y = 5$.

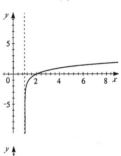

5. $f(x) = \log_3(x - 1)$.
    Domain: $(1, \infty)$
    Range: $(-\infty, \infty)$
    Asymptote: $x = 1$.

7. $y = 2 - \log_2 x$.
    Domain: $(0, \infty)$
    Range: $(-\infty, \infty)$
    Asymptote: $x = 0$.

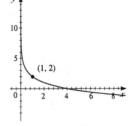

9. $F(x) = e^x - 1$.
    Domain: $(-\infty, \infty)$
    Range: $(-1, \infty)$
    Asymptote: $y = -1$.

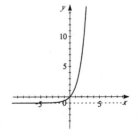

11. $y = 2 \ln x$.
   Domain: $(0, \infty)$
   Range: $(-\infty, \infty)$
   Asymptote: $x = 0$.

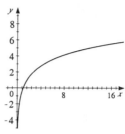

13. $f(x) = 10^{x^2} + \log(1 - 2x)$. Since $\log u$ is defined only for $u > 0$, we require $1 - 2x > 0 \quad \Leftrightarrow$
   $-2x > -1 \quad \Leftrightarrow \quad x < \frac{1}{2}$, and so the domain is $\left(-\infty, \frac{1}{2}\right)$.

15. $\log_2 1024 = 10 \quad \Leftrightarrow \quad 2^{10} = 1024$

17. $\log x = y \quad \Leftrightarrow \quad 10^y = x$

19. $2^6 = 64 \quad \Leftrightarrow \quad \log_2 64 = 6$

21. $10^x = 74 \quad \Leftrightarrow \quad \log_{10} 74 = x \quad \Leftrightarrow \quad \log 74 = x$

23. $\log_2 128 = \log_2(2^7) = 7$

25. $10^{\log 45} = 45$

27. $\ln(e^6) = 6$

29. $\log_3 \frac{1}{27} = \log_3 3^{-3} = -3$

31. $\log_5 \sqrt{5} = \log_5 5^{1/2} = \frac{1}{2}$

33. $\log 25 + \log 4 = \log(25 \cdot 4) = \log 10^2 = 2$

35. $\log_2(16^{23}) = \log_2(2^4)^{23} = \log_2 2^{92} = 92$

37. $\log_8 6 - \log_8 3 + \log_8 2 = \log_8\left(\frac{6}{3} \cdot 2\right) = \log_8 4 = \log_8 8^{2/3} = \frac{2}{3}$

39. $\log(AB^2C^3) = \log A + 2\log B + 3\log C$

41. $\ln\sqrt{\frac{x^2 - 1}{x^2 + 1}} = \frac{1}{2}\ln\left(\frac{x^2 - 1}{x^2 + 1}\right) = \frac{1}{2}[\ln(x^2 - 1) - \ln(x^2 + 1)]$

43. $\log_5\left(\frac{x^2(1 - 5x)^{3/2}}{\sqrt{x^3 - x}}\right) = \log_5 x^2(1 - 5x)^{3/2} - \log_5\sqrt{x(x^2 - 1)}$

   $= 2\log_5 x + \frac{3}{2}\log_5(1 - 5x) - \frac{1}{2}\log_5(x^3 - x)$

45. $\log 6 + 4\log 2 = \log 6 + \log 2^4 = \log(6 \cdot 2^4) = \log 96$

47. $\frac{3}{2}\log_2(x - y) - 2\log_2(x^2 + y^2) = \log_2(x - y)^{3/2} - \log_2(x^2 + y^2)^2 = \log_2\left(\frac{(x - y)^{3/2}}{(x^2 + y^2)^2}\right)$

49. $\log(x - 2) + \log(x + 2) - \frac{1}{2}\log(x^2 + 4) = \log[(x - 2)(x + 2)] - \log\sqrt{x^2 + 4} =$
   $\log\left(\frac{x^2 - 4}{\sqrt{x^2 + 4}}\right)$

51. $\log_2(1-x) = 4 \quad \Leftrightarrow \quad 1-x = 2^4 \quad \Leftrightarrow \quad x = 1 - 2^4 = -15$

53. $5^{5-3x} = 26 \quad \Leftrightarrow \quad \log_5 26 = 5 - 3x \quad \Leftrightarrow \quad 3x = 5 - \log_5 26 \quad \Leftrightarrow \quad x = \frac{1}{3}(5 - \log_5 26) \approx 0.99$

55. $e^{3x/4} = 10 \quad \Leftrightarrow \quad \ln e^{3x/4} = \ln 10 \quad \Leftrightarrow \quad \frac{3x}{4} = \ln 10 \quad \Leftrightarrow \quad x = \frac{4}{3} \ln 10 \approx 3.07$

57. $\log x + \log(x+1) = \log 12 \quad \Leftrightarrow \quad \log[x(x+1)] = \log 12 \quad \Leftrightarrow \quad x(x+1) = 12 \quad \Leftrightarrow$
$x^2 + x - 12 = 0 \quad \Leftrightarrow \quad (x+4)(x-3) = 0 \quad \Rightarrow \quad x = 3 \text{ or } -4$. Because $\log x$ and $\log(x+1)$
are undefined at $x = -4$, it follows that $x = 3$ is the only solution.

59. $x^2 e^{2x} + 2x\, e^{2x} = 8\, e^{2x} \quad \Leftrightarrow \quad e^{2x}(x^2 + 2x - 8) = 0 \quad \Leftrightarrow \quad x^2 + 2x - 8 = 0$ (since $e^{2x} > 0$ for
all x) $\quad \Leftrightarrow \quad (x+4)(x-2) = 0 \quad \Leftrightarrow \quad x = 2, -4$

61. $5^{-2x/3} = 0.63 \quad \Leftrightarrow \quad \dfrac{-2x}{3} \log 5 = \log 0.63 \quad \Leftrightarrow \quad x = -\dfrac{3\log 0.63}{2\log 5} \approx 0.430618$

63. $5^{2x+1} = 3^{4x-1} \quad \Leftrightarrow \quad (2x+1)\log 5 = (4x-1)\log 3 \quad \Leftrightarrow \quad 2x\log 5 + \log 5 = 4x\log 3 - \log 3$
$\Leftrightarrow \quad x(2\log 5 - 4\log 3) = -\log 3 - \log 5 \quad \Leftrightarrow \quad x = \dfrac{\log 3 + \log 5}{4\log 3 - 2\log 5} \approx 2.303600$

65. $y = e^{x/(x+2)}$.
Vertical Asymptote: $x = -2$
Horizontal Asymptote: $y = 2.72$
No maximum or minimum.

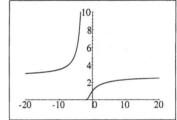

67. $y = \log(x^3 - x)$.
Vertical Asymptotes: $x = -1, x = 0, x = 1$
Horizontal Asymptote: none
Local maximum ≈ -0.41 when $x \approx -0.58$

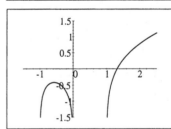

69. $3\log x = 6 - 2x$.
We graph $y = 3\log x$ and $y = 6 - 2x$ in the same
viewing rectangle. The solution occurs where the
two graphs intersect. From the graphs, we see that the
solution is $x \approx 2.42$.

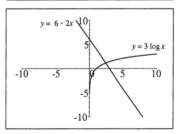

71. $\ln x > x - 2$. We graph the function $f(x) = \ln x - x + 2$,
     and we see that the graph lies above the x-axis for
     $0.16 < x < 3.15$. So the approximate solution of the
     given inequality is $0.16 < x < 3.15$.

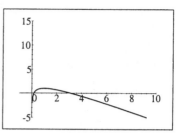

73. $f(x) = e^x - 3e^{-x} - 4x$.
     We graph the function $f(x)$, and we see that the
     function is increasing on $(-\infty, 0]$ and $[1.10, \infty)$
     and that it is decreasing on $[0, 1.10]$.

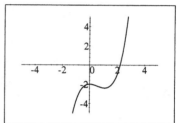

75. $\log_4 15 = \frac{\log 15}{\log 4} = 1.953445$

77. Notice that $\log_4 258 > \log_4 256 = \log_4 4^4 = 4$ and so $\log_4 258 > 4$. Also
     $\log_5 620 < \log_5 625 = \log_5 5^4 = 4$ and so $\log_5 620 < 4$. Then $\log_4 258 > 4 > \log_5 620$ and so
     $\log_4 258$ is larger.

79. $P = 12{,}000$, $r = 0.10$, and $t = 3$. Then $A = P(1 + \frac{r}{n})^{nt}$.

     (a) For $n = 2$, $A = 12{,}000(1 + \frac{0.10}{2})^{2(3)} = 12{,}000(1.05^6) \approx \$16{,}081.15$.

     (b) For $n = 12$, $A = 12{,}000(1 + \frac{0.10}{12})^{12(3)} \approx \$16{,}178.18$.

     (c) For $n = 365$, $A = 12{,}000(1 + \frac{0.10}{365})^{365(3)} \approx \$16{,}197.64$.

     (d) For $n = \infty$, $A = P e^{rt} = 12{,}000\, e^{0.10(3)} \approx \$16{,}198.31$.

81. (a) Using the model $n(t) = n_0\, e^{rt}$, with $n_0 = 30$ and $r = 0.15$, we have the formula
          $n(t) = 30\, e^{0.15t}$.

     (b) $n(4) = 30\, e^{0.15(4)} \approx 55$.

     (c) $500 = 30\, e^{0.15t} \quad \Leftrightarrow \quad \frac{50}{3} = e^{0.15t} \quad \Leftrightarrow \quad 0.15t = \ln\left(\frac{50}{3}\right) \quad \Leftrightarrow \quad t = \frac{1}{0.15}\ln\left(\frac{50}{3}\right) \approx 18.76$.
          So the stray cat population will reach 500 in about 19 years.

83. (a) From the formula for radioactive decay, we have $m(t) = 10e^{-rt}$, where $r = -\dfrac{\ln 2}{2.7 \times 10^5}$. So
          after 1000 years the amount remaining is $m(1000) = 10 \cdot e^{-\frac{\ln 2}{2.7\times10^5} \cdot 1000} = 10e^{-\frac{\ln 2}{2.7\times10^2}}$
          $= 10e^{-\frac{\ln 2}{270}} \approx 9.97$. Therefore the amount remaining is about 9.97 mg.

     (b) We solve for t in the equation $7 = 10e^{-\frac{\ln 2}{2.7\times10^5} \cdot t}$. We have $7 = 10\, e^{-\frac{\ln 2}{2.7\times10^5} t} \quad \Leftrightarrow$
          $0.7 = e^{-\frac{\ln 2}{2.7\times10^5} \cdot t} \quad \Leftrightarrow \quad \ln 0.7 = -\dfrac{\ln 2}{2.7 \times 10^5} \cdot t \quad \Leftrightarrow$
          $t = -\dfrac{\ln 0.7}{\ln 2} \cdot 2.7 \times 10^5 \approx 138{,}934.75$. Thus it takes about 139,000 years.

85. (a) From the formula for radioactive decay, $r = \dfrac{\ln 2}{1590} \approx 0.0004359$ and $n(t) = 150 \cdot e^{-0.0004359\,t}$.

   (b) $n(1000) = 150 \cdot e^{-0.0004359 \cdot 1000} \approx 97.00$, and so the amount remaining is about 97.00 mg.

   (c) Find t so that $50 = 150 \cdot e^{-0.0004359\,t}$. We have $50 = 150 \cdot e^{-0.0004359\,t}$ $\Leftrightarrow$ $\frac{1}{3} = e^{-0.0004359\,t}$
   $\Leftrightarrow$ $t = -\frac{1}{0.0004359}\ln\left(\frac{1}{3}\right) \approx 2520$. Thus only 50 mg remain after about 2520 years.

87. (a) Using $n_0 = 1500$ and $n(5) = 3200$ in the formula $n(t) = n_0\,e^{rt}$, we have
   $3200 = n(5) = 1500\,e^{5r}$ $\Leftrightarrow$ $e^{5r} = \frac{32}{15}$ $\Leftrightarrow$ $5r = \ln\left(\frac{32}{15}\right)$ $\Leftrightarrow$ $r = \frac{1}{5}$
   $\ln\left(\frac{32}{15}\right) \approx 0.1515$. Thus $n(t) = 1500 \cdot e^{0.1515t}$.

   (b) We have $t = 1999 - 1988 = 11$ so $n(11) = 1500e^{0.1515 \cdot 11} \approx 7940$. Thus in 1999 the bird population should be about 7940 birds.

89. $[\text{H}^+] = 1.3 \times 10^{-8}$ M. Then $\text{pH} = -\log[\text{H}^+] = -\log(1.3 \times 10^{-8}) \approx 7.9$, and so fresh egg whites are basic.

91. Let I_0 be the intensity of the smaller earthquake and I_1 be the intensity of the larger earthquake. Then $I_1 = 35\,I_0$. Since $M = \log\left(\frac{I}{S}\right)$, we have $M_0 = \log\left(\frac{I_0}{S}\right) = 6.5$ and $M_1 = \log\left(\frac{I_1}{S}\right)$
   $= \log\left(\frac{35\,I_0}{S}\right) = \log 35 + \log\left(\frac{I_0}{S}\right) = \log 35 + M_0 = \log 35 + 6.5 \approx 8.04$. So the magnitude on the Richter scale of the larger earthquake is approximately 8.0.

Chapter 6 Test

1. $y = 4^x$ and $y = \log_4 x$.

    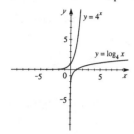

2. $f(x) = \log(x + 2)$.
    Domain: $(-2, \infty)$
    Range: $(-\infty, \infty)$
    Vertical asymptote: $x = -2$.

    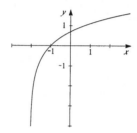

3. (a) $\log_3 \sqrt{27} = \log_3 (3^3)^{\frac{1}{2}} \log_3 3^{3/2} = \frac{3}{2}$

    (b) $\log_2 56 - \log_2 7 = \log_2 \left(\frac{56}{7}\right) = \log_2 8 = \log_2 2^3 = 3$

    (c) $\log_8 4 = \log_8 8^{2/3} = \frac{2}{3}$

    (d) $\log_6 4 + \log_6 9 = \log_6 (4 \cdot 9) = \log_6 6^2 = 2$

4. $\log \sqrt{\dfrac{x^2 - 1}{x^3 (y^2 + 1)^5}} = \frac{1}{2} \log \left(\dfrac{x^2 - 1}{x^3 (y^2 + 1)^5}\right) = \frac{1}{2} \left[\log(x^2 - 1) - (3 \log x + 5 \log(y^2 + 1))\right]$

    $= \frac{1}{2} \left[\log(x^2 - 1) - 3 \log x - 5 \log(y^2 + 1)\right]$

5. $\ln x - 2 \ln(x^2 + 1) + \frac{1}{2} \ln(3 - x^4) = \ln\left(x\sqrt{3 - x^4}\right) - \ln(x^2 + 1)^2 = \ln\left(\dfrac{x\sqrt{3 - x^4}}{(x^2 + 1)^2}\right)$

6. (a) $2^{x-1} = 10 \quad \Leftrightarrow \quad \log 2^{x-1} = \log 10 = 1 \quad \Leftrightarrow \quad (x - 1) \log 2 = 1 \quad \Leftrightarrow \quad x - 1 = \frac{1}{\log 2}$
        $\Leftrightarrow \quad x = 1 + \frac{1}{\log 2} \approx 4.32$

    (b) $5 \ln(3 - x) = 4 \quad \Leftrightarrow \quad \ln(3 - x) = \frac{4}{5} \quad \Leftrightarrow \quad e^{\ln(3-x)} = e^{4/5} \quad \Leftrightarrow \quad 3 - x = e^{4/5} \quad \Leftrightarrow$
        $x = 3 - e^{4/5} \approx 0.77$

    (c) $10^{x+3} = 6^{2x} \quad \Leftrightarrow \quad \log 10^{x+3} = \log 6^{2x} \quad \Leftrightarrow \quad x + 3 = 2x \log 6 \quad \Leftrightarrow \quad 2x \log 6 - x = 3$
        $\Leftrightarrow \quad x(2 \log 6 - 1) = 3 \quad \Leftrightarrow \quad x = \dfrac{3}{2 \log 6 - 1} \approx 5.39$

    (d) $\log_2 (x + 2) + \log_2 (x - 1) = 2 \quad \Leftrightarrow \quad \log_2 ((x + 2)(x - 1)) = 2 \quad \Leftrightarrow \quad x^2 + x - 2 = 2^2$
        $\Leftrightarrow \quad x^2 + x - 6 = 0 \quad \Leftrightarrow \quad (x + 3)(x - 2) = 0 \quad \Rightarrow \quad x = -3 \text{ or } x = 2.$ However, both
        logarithms are undefined at $x = -3$, so the only solution is $x = 2$.

7. (a) From the formula for population growth, we have $8000 = 1000e^{r \cdot 1}$ ⟺ $8 = e^r$ ⟺ $r = \ln 8 \approx 2.07944$. Thus $n(t) = 1000e^{2.07944\,t}$.

   (b) $n(1.5) = 1000e^{2.07944(1.5)} \approx 22{,}627$

   (c) $15000 = 1000e^{2.07944\,t}$ ⟺ $15 = e^{2.07944\,t}$ ⟺ $\ln 15 = 2.07944t$ ⟺ $t = \frac{\ln 15}{2.07944} \approx 1.3$. Thus the population will reach 15,000 after approximately 1.3 hours.

   (d)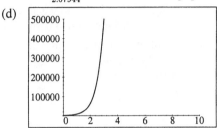

8. (a) $A(t) = 12{,}000\left(1 + \dfrac{0.056}{12}\right)^{12t}$, where t is in years.

   (b) $A(t) = 12{,}000\left(1 + \dfrac{0.056}{365}\right)^{365t}$. So $A(3) = 12{,}000\left(1 + \dfrac{0.056}{365}\right)^{365(3)} = \$14{,}195.06$.

   (c) $A(t) = 12{,}000\left(1 + \dfrac{0.056}{2}\right)^{2t} = 12{,}000(1.028)^{2t}$. So $20{,}000 = 12{,}000(1.028)^{2t}$ ⟺ $1.6667 = 1.028^{2t}$ ⟺ $\ln 1.6667 = \ln 1.028^{2t}$ ⟺ $\ln 1.6667 = 2t\ln 1.028$ ⟺ $t = \frac{\ln 1.6667}{2\ln 1.028} \approx 9.25$ years.

9. $f(x) = \dfrac{e^x}{x^3}$

   (a)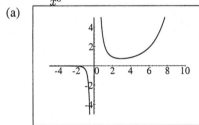

   (b) Vertical asymptote: $x = 0$; Horizontal asymptote: $y = 0$.

   (c) Local minimum ≈ 0.74 when $x \approx 3.00$

   (d) Range $\approx (-\infty, 0) \cup [0.74, \infty)$

   (e) $\dfrac{e^x}{x^3} = 2x + 1$.

   We see that the graphs intersect at $x \approx -0.85, 0.96,$ and 9.92.

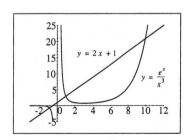

Focus on Modeling

1. (a)

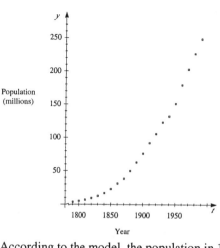

Year

(b) Using a graphing calculator, we obtain
the model $y = ab^t$, where
$a = 4.041807 \times 10^{-16}$ and
$b = 1.021003194$, and y is the
population (in millions) in the year t.

(c) Substituting $t = 2000$ into the model of
part (b), we get $y = ab^{2000} \approx 457.9$
million.

(d) According to the model, the population in 1965 should have been about $y = ab^{1965} \approx 221.2$
million.

(e) The values given by the model are clearly much too large. This means that an exponential
model is NOT appropriate for these data.

3. (a) Yes.

(b)

Year	Health Expenditures (billions of dollars)	
t	E	$\ln E$
1960	27.1	3.29953
1970	74.3	4.30811
1980	251.1	5.52585
1985	434.5	6.07420
1987	506.2	6.22693
1989	623.9	6.43599
1990	696.6	6.54621
1991	755.8	6.62778
1992	820.3	6.70967
1993	884.2	6.78468

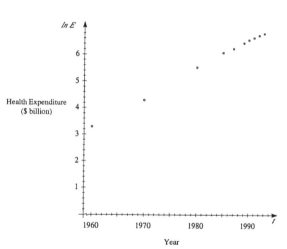

Year

Yes, the scatter plot appears to be linear.

(c) Let T be the number of years elapsed since 1960. Then $\ln E = 3.3016 + 0.10769T$.

(d) $E = e^{3.3016 + 0.10769T} = 27.156\, e^{0.10769\,T}$ (where $T = t - 1960$).

(e) When $t = 1996$, then $T = 36$, so the estimated 1996 health-care expenditures are
$27.156\, e^{0.10769 \cdot (36)} = 1310.9$ billion dollars.

5. (a) Let t be the number of years elapsed since 1970, and let y be the number of millions of tons of lead emissions in year t. Using a graphing calculator, we obtain the exponential model $y = ab^t$, where $a = 301.813054$ and $b = 0.819745$.

    (b) Using a graphing calculator, we obtain the fourth-degree polynomial model $y = at^4 + bt^3 + ct^2 + dt + e$, where $a = -0.002430$, $b = 0.135159$, $c = -2.014322$, $d = -4.055294$, and $e = 199.092227$.

    (c) The exponential model is shown as a solid curve and the polynomial model as a dotted curve in the graph to the right. From the graph, the polynomial model appears to fit the data better.

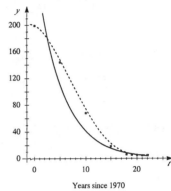

Lead Emissions (million metric tons)

Years since 1970

    (d) Exponential model: 1972 $(t = 2)$ $y = 202.8$
                               1982 $(t = 12)$ $y = 27.8$

         Polynomial model: 1972 $y = 184.0$
                               1982 $y = 43.5$

Chapter Seven
Exercises 7.1

1. $40° = 40° \cdot \frac{\pi}{180°}$ rad $= \frac{2\pi}{9}$ rad ≈ 0.698 rad

3. $72° = 72° \cdot \frac{\pi}{180°}$ rad $= \frac{2\pi}{5}$ rad ≈ 1.257 rad

5. $45° = 45° \cdot \frac{\pi}{180°}$ rad $= \frac{\pi}{4}$ rad ≈ 0.785 rad

7. $765° = 765° \cdot \frac{\pi}{180°}$ rad $= \frac{17\pi}{4}$ rad ≈ 13.352 rad

9. $36° = 36° \cdot \frac{\pi}{180°}$ rad $= \frac{\pi}{5}$ rad ≈ 0.628 rad

11. $-\frac{7\pi}{2} = -\frac{7\pi}{2} \cdot \frac{180°}{\pi} = -630°$

13. $2 = 2 \cdot \frac{180°}{\pi} = \frac{360°}{\pi} \approx 114.6°$

15. $\frac{2\pi}{9} = \frac{2\pi}{9} \cdot \frac{180°}{\pi} = 40°$

17. $\frac{\pi}{5} = \frac{\pi}{5} \cdot \frac{180°}{\pi} = 36°$

19. $300°$ is coterminal with: $300° + 360° = 660°$, $300° + 720° = 1020°$, $300° - 360° = -60°$, $300° - 720° = -420°$.

21. $\frac{3\pi}{4}$ is coterminal with: $\frac{3\pi}{4} + 2\pi = \frac{11\pi}{4}$, $\frac{3\pi}{4} + 4\pi = \frac{19\pi}{4}$, $\frac{3\pi}{4} - 2\pi = -\frac{5\pi}{4}$, $\frac{3\pi}{4} - 4\pi = -\frac{13\pi}{4}$.

23. $-\frac{\pi}{4}$ is coterminal with: $-\frac{\pi}{4} + 2\pi = \frac{7\pi}{4}$, $-\frac{\pi}{4} + 4\pi = \frac{15\pi}{4}$, $-\frac{\pi}{4} - 2\pi = -\frac{9\pi}{4}$, $-\frac{\pi}{4} - 4\pi = -\frac{17\pi}{4}$

25. Since $430° - 70° = 360°$, the angles are coterminal.

27. Since $\frac{17\pi}{6} - \frac{5\pi}{6} = \frac{12\pi}{6} = 2\pi$; the angles are coterminal.

29. Since $875° - 155° = 720° = 2 \times 360°$, the angles are coterminal.

31. Since $733° - 2 \cdot 360° = 13°$, the angles $733°$ and $13°$ are coterminal.

33. Since $2223° - 6 \cdot 360° = 63°$, the angles $2223°$ and $63°$ are coterminal.

35. Since $-800° + 3 \cdot 360° = 280°$, the angles $-800°$ and $280°$ are coterminal.

37. Since $\frac{12\pi}{5} - 2\pi = \frac{2\pi}{5}$, the angles $\frac{12\pi}{5}$ and $\frac{2\pi}{5}$ are coterminal.

39. Since $87\pi - 43 \cdot 2\pi = \pi$, the angles 87π and π are coterminal.

41. Since $\frac{17\pi}{4} - 2 \cdot 2\pi = \frac{\pi}{4}$, the angles $\frac{17\pi}{4}$ and $\frac{\pi}{4}$ are coterminal.

43. Using the formula $s = \theta r$, the length of the arc is $s = \left(220 \cdot \frac{\pi}{180}\right) \cdot 5 = \frac{55\pi}{9} \approx 19.2$.

45. Solving for r we have $r = \frac{s}{\theta}$, so the radius of the circle is $r = \frac{8}{2} = 4$.

47. Using the formula $s = \theta r$, the length of the arc is $s = 2 \cdot 2 = 4$ mi.

49. Solving for θ we have $\theta = \frac{s}{r}$, so the measure of the central angle is $\theta = \frac{100}{50} = 2$ rad. Converting to degrees we have $\theta = 2 \cdot \frac{180°}{\pi} \approx 114.6°$

51. Solving for r we have $r = \frac{s}{\theta}$, so the radius of the circle is $r = \frac{6}{\pi/6} = \frac{36}{\pi} \approx 11.46$ m.

53. We find the measure of the angle in degrees and then convert to radians. $\theta = 40.5° - 25.5° = 15°$ and $15 \cdot \frac{\pi}{180}$ rad $= \frac{\pi}{12}$ rad. Then using the formula $s = \theta r$, we have $s = \frac{\pi}{12} \cdot 3960 = 330\pi \approx 1036.725$ and so the distance between the two cities is roughly 1037 mi.

55. In one day, the earth travels $\frac{1}{365}$ of its orbit which is $\frac{2\pi}{365}$ rad. Then $s = \theta r = \frac{2\pi}{365} \cdot 93,000,000 \approx 1,600,911.3$ and so the distance traveled is approximately 1.6 million miles.

57. The central angle is 1 minute $= \frac{1}{60}° = \frac{1}{60} \cdot \frac{\pi}{180}$ rad $= \frac{\pi}{10,800}$ rad. Then $s = \theta r = \frac{\pi}{10,800} \cdot 3960 \approx 1.152$ and so a nautical mile is approximately 1.15 mi.

59. $A = \frac{1}{2}r^2\theta = \frac{1}{2} \cdot 10^2 \cdot 1 = 50$ m^2

61. $\theta = 2$ rad, $A = 16$ m^2. Since $A = \frac{1}{2}r^2\theta$, we have $r = \sqrt{2A/\theta} = \sqrt{2 \cdot 16/2} = \sqrt{16} = 4$ m.

63. Since the area of the circle is 72 cm^2, the radius of the circle is $r = \sqrt{A/\pi} = \sqrt{72/\pi}$. Then the area of the sector is $A = \frac{1}{2}r^2\theta = \frac{1}{2} \cdot \frac{72}{\pi} \cdot \frac{\pi}{6} = 6$ cm^2.

Exercises 7.2

1. $\sin\theta = \frac{4}{5}$, $\cos\theta = \frac{3}{5}$, $\tan\theta = \frac{4}{3}$, $\csc\theta = \frac{5}{4}$, $\sec\theta = \frac{5}{3}$, $\cot\theta = \frac{3}{4}$

3. The remaining side is obtained by the Pythagorean Theorem: $\sqrt{7^2 - 4^2} = \sqrt{33}$. Then $\sin\theta = \frac{4}{7}$, $\cos\theta = \frac{\sqrt{33}}{7}$, $\tan\theta = \frac{4}{\sqrt{33}} = \frac{4\sqrt{33}}{33}$, $\csc\theta = \frac{7}{4}$, $\sec\theta = \frac{7}{\sqrt{33}} = \frac{7\sqrt{33}}{33}$, $\cot\theta = \frac{\sqrt{33}}{4}$

5. $c = \sqrt{3^2 + 2^2} = \sqrt{13}$

   (a) $\sin\alpha = \cos\beta = \frac{2}{\sqrt{13}} = \frac{2\sqrt{13}}{13}$

   (b) $\tan\alpha = \cot\beta = \frac{2}{3}$

   (c) $\sec\alpha = \csc\beta = \frac{\sqrt{13}}{3}$

7. Since $\sin 30° = \dfrac{x}{25}$, we have $x = 25\sin 30° = 25 \cdot \frac{1}{2} = \frac{25}{2}$.

9. Since $\sin 60° = \dfrac{x}{13}$, we have $x = 13\sin 60° = 13 \cdot \frac{\sqrt{3}}{2} = \frac{13\sqrt{3}}{2}$.

11. Since $\tan 36° = \dfrac{12}{x}$, we have $x = \frac{12}{\tan 36°} \approx 16.51658$.

13. $\dfrac{x}{28} = \cos\theta \;\Leftrightarrow\; x = 28\cos\theta$, and $\dfrac{y}{28} = \sin\theta \;\Leftrightarrow\; y = 28\sin\theta$

15. $\sin\theta = \frac{3}{5}$. Then the third side is
    $x = \sqrt{5^2 - 3^2} = 4$. The other five ratios are
    $\cos\theta = \frac{4}{5}$, $\tan\theta = \frac{3}{4}$, $\csc\theta = \frac{5}{3}$, $\sec\theta = \frac{5}{4}$,
    and $\cot\theta = \frac{4}{3}$.

17. $\cot\theta = 1$. Then the third side is
    $r = \sqrt{1^2 + 1^2} = \sqrt{2}$. The other five ratios
    are $\sin\theta = \frac{1}{\sqrt{2}} = \frac{\sqrt{2}}{2}$, $\cos\theta = \frac{1}{\sqrt{2}} = \frac{\sqrt{2}}{2}$,
    $\tan\theta = 1$, $\csc\theta = \sqrt{2}$, and $\sec\theta = \sqrt{2}$.

19. $\sec\theta = 7$. The third side is
    $y = \sqrt{7^2 - 1^2} = \sqrt{48} = 4\sqrt{3}$.
    The other five ratios are $\sin\theta = \frac{4\sqrt{3}}{7}$, $\cos\theta = \frac{1}{7}$,
    $\tan\theta = 4\sqrt{3}$, $\csc\theta = \frac{7}{4\sqrt{3}} = \frac{7\sqrt{3}}{12}$, and $\cot\theta = \frac{1}{4\sqrt{3}} = \frac{\sqrt{3}}{12}$.

21. $\sin\frac{\pi}{6} + \cos\frac{\pi}{6} = \frac{1}{2} + \frac{\sqrt{3}}{2} = \frac{1+\sqrt{3}}{2}$

23. $\sin 30° \cos 60° + \sin 60° \cos 30° = \frac{1}{2} \cdot \frac{1}{2} + \frac{\sqrt{3}}{2} \cdot \frac{\sqrt{3}}{2} = \frac{1}{4} + \frac{3}{4} = 1$

25. $(\cos 30°)^2 - (\sin 30°)^2 = \left(\frac{\sqrt{3}}{2}\right)^2 - \left(\frac{1}{2}\right)^2 = \frac{3}{4} - \frac{1}{4} = \frac{1}{2}$

27. This is an isosceles right triangle, so the
     other leg $= 16\tan 45° = 16$,
     hypotenuse $= \frac{16}{\sin 45°} = 16\sqrt{2} \approx 22.63$, and
     other angle $= 90° - 45° = 45°$

29. The other leg $= 35\tan 52° = 44.80$,
     hypotenuse $= \frac{35}{\cos 52°} = 56.85$, and
     other angle $= 90° - 52° = 52°$.

31. $\sin\theta \approx \frac{1}{2.24} \approx 0.45.$ $\cos\theta \approx \frac{2}{2.24} \approx 0.89$,
     $\tan\theta = \frac{1}{2}$, $\csc\theta \approx 2.24$, $\sec\theta \approx \frac{2.24}{2} \approx 1.12$,
     $\cot\theta \approx 2.00$.

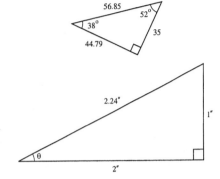

33. Let h be the height, in feet, of the Empire State Building. Then $\tan 11° = \dfrac{h}{5280}$ $\Leftrightarrow$
     $h = 5280 \cdot \tan 11° \approx 1026$ ft.

35. (a) Let h be the distance, in miles, that the beam has diverged. Then $\tan 0.5° = \dfrac{h}{240{,}000}$ $\Leftrightarrow$
          $h = 240{,}000 \cdot \tan 0.5° \approx 2100$ mi.

     (b) Since the deflection is about 2100 mi whereas the radius of the moon is about 1000 mi, the
          beam will not strike the moon.

37. Let h represent the height, in feet, that the ladder reaches on the building. Then $\sin 72° = \dfrac{h}{20}$ $\Leftrightarrow$
     $h = 20 \cdot \sin 72° \approx 19$ ft.

39. Let θ be the angle of elevation of the sun. Then $\tan\theta = \frac{96}{120} = 0.8$ $\Leftrightarrow$
     $\theta = \tan^{-1} 0.8 \approx 0.675 \approx 38.7°$.

41. Let h be the height, in feet, of the kite above the ground. Then $\sin 50° = \dfrac{h}{450}$ $\Leftrightarrow$
     $h = 450 \cdot \sin 50° \approx 345$ ft.

43. Let h_1 be the height of the window in feet and h_2 be the height from the window to the top of the
     tower. Then $\tan 25° = \dfrac{h_1}{325}$ $\Leftrightarrow$ $h_1 = 325 \cdot \tan 25° \approx 152$ ft. Also, $\tan 39° = \dfrac{h_2}{325}$ $\Leftrightarrow$
     $h_2 = 325 \cdot \tan 39° \approx 263$ ft. Therefore, the height of the window is approximately 152 ft and the
     height of the tower is approximately $152 + 263 = 415$ ft.

45. Let d_1 be the distance, in feet, between a point directly below the plane and one car, and d_2 be the
     distance, in feet, between the same point and the other car. Then $\tan 52° = \dfrac{d_1}{5150}$ $\Leftrightarrow$
     $d_1 = 5150 \cdot \tan 52° \approx 6591.7$ ft. Also, $\tan 38° = \dfrac{d_2}{5150}$ $\Leftrightarrow$ $d_2 = 5150 \cdot \tan 38° \approx 4023.6$ ft.
     So the distance between the two cars is now about 2570 ft.

47. Let x be the horizontal distance, in feet, between a point on the ground directly below the top of the mountain and the point on the plain closest to the mountain. Let h be the height, in feet, of the mountain. Then $\tan 35° = \dfrac{h}{x}$ and $\tan 32° = \dfrac{h}{x + 1000}$. So $h = x \tan 35° = (x + 1000)\tan 32°$

$\Leftrightarrow \quad x = \dfrac{1000 \cdot \tan 32°}{\tan 35° - \tan 32°} \approx 8294.2$. Thus $h \approx 8294.2 \cdot \tan 35° \approx 5808$ ft.

49. Let d be the distance, in miles, from the earth to the sun. Then $\tan 89.95° = \dfrac{d}{240,000} \quad \Leftrightarrow$

$d = 240,000 \cdot \tan 89.95° \approx 91.7$ million miles.

51. Let r represent the radius, in miles, of the earth. Then $\sin 60.276° = \dfrac{r}{r + 600} \quad \Leftrightarrow$

$(r + 600)\sin 60.276° = r \quad \Leftrightarrow \quad 600 \cdot \sin 60.276° = r(1 - \sin 60.276°) \quad \Leftrightarrow$

$r = \dfrac{600 \cdot \sin 60.276°}{1 - \sin 60.276°} \approx 3960.099$. So the radius of the earth is approximately 3960 mi.

53. $x = \dfrac{100}{\tan 60°} + \dfrac{100}{\tan 30°} \approx 230.9$

55. Let h be the length of the shared side. Then $\sin 60° = \dfrac{50}{h} \quad \Leftrightarrow \quad h = \dfrac{50}{\sin 60°} \approx 57.735 \quad \Leftrightarrow$

$\sin 65° = \dfrac{h}{x} \quad \Leftrightarrow \quad x = \dfrac{h}{\sin 65°} \approx 63.7$

57. $\sin \theta = \dfrac{1}{a} \quad \Leftrightarrow \quad a = \sin \theta, \tan \theta = \dfrac{b}{1} \quad \Leftrightarrow \quad b = \tan \theta, \cos \theta = \dfrac{1}{c} \quad \Leftrightarrow \quad c = \sec \theta, \cos \theta = \dfrac{d}{1}$

$\Leftrightarrow \quad d = \cos \theta$

Exercises 7.3

1. (a) $225°$: reference angle is $225° - 180° = 45°$
   (b) $-35°$: reference angle is $35°$
   (c) $181°$: reference angle is $181° - 180° = 1°$

3. (a) $335°$: reference angle is $360° - 335° = 25°$
   (b) $-95°$: reference angle is $180° - 95° = 85°$
   (c) $165°$: reference angle is $180° - 165° = 15°$

5. (a) $\frac{17\pi}{3}$: reference angle is $6\pi - \frac{17\pi}{3} = \frac{\pi}{3}$
   (b) $-\frac{\pi}{4}$: reference angle is $\frac{\pi}{4}$
   (c) 3: reference angle is $\pi - 3 \approx 0.14$

7. $\sin 150° = \sin 30° = \frac{1}{2}$

9. $\sin 135° = \sin 45° = \frac{1}{\sqrt{2}} = \frac{\sqrt{2}}{2}$

11. $\sin(-60°) = -\sin 60° = -\frac{\sqrt{3}}{2}$

13. $\csc(-630°) = \csc 90° = \dfrac{1}{\sin 90°} = 1$

15. $\cos 570° = -\cos 30° = -\frac{\sqrt{3}}{2}$

17. $\tan 750° = \tan 30° = \frac{1}{\sqrt{3}} = \frac{\sqrt{3}}{3}$

19. $\sin\left(\frac{2\pi}{3}\right) = \sin\left(\frac{\pi}{3}\right) = \frac{\sqrt{3}}{2}$

21. $\sin\left(\frac{3\pi}{2}\right) = -\sin\left(\frac{\pi}{2}\right) = -1$

23. $\cos\left(-\frac{7\pi}{3}\right) = \cos\left(\frac{\pi}{3}\right) = \frac{1}{2}$

25. $\sec\left(\frac{17\pi}{3}\right) = \sec\left(\frac{\pi}{3}\right) = \dfrac{1}{\cos\left(\frac{\pi}{3}\right)} = 2$

27. $\cot\left(-\frac{\pi}{4}\right) = -\cot\left(\frac{\pi}{4}\right) = \frac{-1}{\tan\left(\frac{\pi}{4}\right)} = -1$

29. $\tan\left(\frac{5\pi}{2}\right) = \tan\left(\frac{\pi}{2}\right)$ which is undefined

31. Since $\sin\theta < 0$ and $\cos\theta < 0$, we have θ in quadrant III.

33. $\sec\theta > 0 \;\Rightarrow\; \cos\theta > 0$. Also $\tan\theta < 0 \;\Rightarrow\; \frac{\sin\theta}{\cos\theta} < 0 \;\Leftrightarrow\; \sin\theta < 0$ (since $\cos\theta > 0$). Since $\sin\theta < 0$ and $\cos\theta > 0$, we get θ is in quadrant IV.

35. $\sec^2\theta = 1 + \tan^2\theta \;\Leftrightarrow\; \tan^2\theta = \dfrac{1}{\cos^2\theta} - 1 \;\Leftrightarrow\; \tan\theta = \sqrt{\dfrac{1}{\cos^2\theta} - 1} = \sqrt{\dfrac{1 - \cos^2\theta}{\cos^2\theta}}$

$= \dfrac{\sqrt{1 - \cos^2\theta}}{|\cos\theta|} = \dfrac{\sqrt{1 - \cos^2\theta}}{-\cos\theta}$ (since $\cos\theta < 0$ in quadrant III, $|\cos\theta| = -\cos\theta$). Thus

$\tan\theta = -\dfrac{\sqrt{1 - \cos^2\theta}}{\cos\theta}$.

37. $\cos^2\theta + \sin^2\theta = 1 \;\Leftrightarrow\; \cos\theta = \sqrt{1 - \sin^2\theta}$ ($\cos\theta > 0$ in quadrant IV)

39. $\sec^2\theta = 1 + \tan^2\theta \;\Leftrightarrow\; \sec\theta = -\sqrt{1 + \tan^2\theta}$ ($\sec\theta < 0$ in quadrant II)

41. $\sin\theta = \frac{3}{5}$. Then $x = -\sqrt{5^2 - 3^2} = -\sqrt{16} = -4$, since θ in quadrant II. Thus, $\cos\theta = -\frac{4}{5}$, $\tan\theta = -\frac{3}{4}$, $\csc\theta = \frac{5}{3}$, $\sec\theta = -\frac{5}{4}$, $\cot\theta = -\frac{4}{3}$.

43. $\tan\theta = -\frac{3}{4}$. Then $r = \sqrt{3^2 + 4^2} = 5$, and so $\sin\theta = -\frac{3}{5}$, $\cos\theta = \frac{4}{5}$, $\csc\theta = -\frac{5}{3}$, $\sec\theta = \frac{5}{4}$, $\cot\theta = -\frac{4}{3}$.

45. $\csc\theta = 2$. Then $\sin\theta = \frac{1}{2}$ and $x = \sqrt{2^2 - 1^2} = \sqrt{3}$. So $\sin\theta = \frac{1}{2}$, $\cos\theta = \frac{\sqrt{3}}{2}$, $\tan\theta = \frac{1}{\sqrt{3}} = \frac{\sqrt{3}}{3}$, $\sec\theta = \frac{2}{\sqrt{3}} = \frac{2\sqrt{3}}{3}$, $\cot\theta = \sqrt{3}$.

47. $\cos\theta = -\frac{2}{7}$. Then $y = \sqrt{7^2 - 2^2} = \sqrt{45} = 3\sqrt{5}$, and so $\sin\theta = \frac{3\sqrt{5}}{7}$, $\tan\theta = -\frac{3\sqrt{5}}{2}$, $\csc\theta = \frac{7}{3\sqrt{5}} = \frac{7\sqrt{5}}{15}$, $\sec\theta = -\frac{7}{2}$, $\cot\theta = -\frac{2}{3\sqrt{5}} = -\frac{2\sqrt{5}}{15}$.

49. (a) $\sin 2\theta = \sin\left(2 \cdot \frac{\pi}{3}\right) = \sin\frac{2\pi}{3} = \sin\frac{\pi}{3} = \frac{\sqrt{3}}{2}$, while $2\sin\theta = 2\sin\frac{\pi}{3} = 2 \cdot \frac{\sqrt{3}}{2} = \sqrt{3}$.

     (b) $\sin\frac{1}{2}\theta = \sin\left(\frac{1}{2} \cdot \frac{\pi}{3}\right) = \sin\frac{\pi}{6} = \frac{1}{2}$, while $\dfrac{\sin\theta}{2} = \frac{1}{2} \cdot \sin\frac{\pi}{3} = \frac{1}{2} \cdot \frac{\sqrt{3}}{2} = \frac{\sqrt{3}}{4}$.

     (c) $\sin^2\theta = \left(\sin\frac{\pi}{3}\right)^2 = \left(\frac{\sqrt{3}}{2}\right)^2 = \frac{3}{4}$, while $\sin(\theta^2) = \sin\left(\frac{\pi}{3}\right)^2 = \sin\frac{\pi^2}{9} \approx 0.88967$.

51. $a = 10$, $b = 22$, and $\theta = 10°$. Then $A = \frac{1}{2}ab\sin\theta = \frac{1}{2} \cdot 10 \cdot 22 \cdot \sin 10° = 110\sin 10° \approx 19.1$

53. $A = 16$, $a = 5$, and $b = 7$. So $\sin\theta = \dfrac{2A}{ab} = \frac{2 \cdot 16}{5 \cdot 7} = \frac{32}{35}$ $\Leftrightarrow$ $\theta = \sin^{-1}\left(\frac{32}{35}\right) \approx 66.1°$.

55. $\sin^2\theta + \cos^2\theta = 1$ $\Leftrightarrow$ $(\sin^2\theta + \cos^2\theta) \cdot \dfrac{1}{\cos^2\theta} = 1 \cdot \dfrac{1}{\cos^2\theta}$ $\Leftrightarrow$ $\tan^2\theta + 1 = \sec^2\theta$

57. Since $\sin 4° = 0.0697564737$ while $\sin 4 = -0.7568024953$. Your partner has found $\sin 4°$ instead of $\sin 4$ radians.

Exercises 7.4

1. $\angle C = 180° - 52° - 70° = 58°.$ $x = \dfrac{23.1 \cdot \sin 52°}{\sin 58°} \approx 21.5$

3. $\angle C = 180° - 37.5° - 28.1° = 114.4°.$ $x = \dfrac{90 \cdot \sin 114.4°}{\sin 37.5°} \approx 134.6$

5. $\sin \theta = \dfrac{24 \cdot \sin 120°}{30} \approx 0.693$ $\Leftrightarrow$ $\angle C \approx \sin^{-1} 0.693 \approx 44°.$

7. $\angle C = 180° - 46° - 20° = 114°.$ Then $a = \dfrac{65 \cdot \sin 46°}{\sin 114°} \approx 51$ and $b = \dfrac{65 \cdot \sin 20°}{\sin 114°} \approx 24.$

9. $\angle C = 180° - 50° - 68° = 62°.$ Then

$a = \dfrac{230 \cdot \sin 50°}{\sin 62°} \approx 200,$ and

$b = \dfrac{230 \cdot \sin 68°}{\sin 62°} \approx 242.$

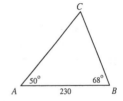

11. $\angle B = 180° - 30° - 65° = 85°.$ Then

$a = \dfrac{10 \cdot \sin 30°}{\sin 85°} \approx 5.0,$ and

$c = \dfrac{10 \cdot \sin 65°}{\sin 85°} \approx 9.$

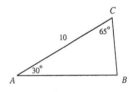

13. $\angle A = 180° - 51° - 29° = 100°.$ Then

$a = \dfrac{44 \cdot \sin 100°}{\sin 29°} \approx 89,$ and

$c = \dfrac{44 \cdot \sin 51°}{\sin 29°} \approx 71.$

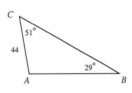

15. Since $\angle A > 90°$ there is only one triangle. $\sin B = \dfrac{15 \cdot \sin 110°}{28} \approx 0.503$ $\Leftrightarrow$

$\angle B \approx \sin^{-1} 0.503 \approx 30°.$ Then $\angle C \approx 180° - 110° - 30° = 40°,$ and so $c = \dfrac{28 \cdot \sin 40°}{\sin 110°} \approx 19.$

Thus $\angle B \approx 30°,$ $\angle C \approx 40°,$ and $c \approx 19.$

17. $\angle A = 125°$ is the largest angle but since side a is not the longest side, there can be no such triangle.

19. $\sin C = \dfrac{30 \cdot \sin 25°}{25} \approx 0.507$ $\Leftrightarrow$ $\angle C_1 \approx \sin^{-1} 0.507 \approx 30.47°$ or

$\angle C_2 \approx 180° - 39.47° \approx 149.53°.$

If $\underline{\angle C_1 = 30.47°},$ then $\angle A_1 \approx 180° - 25° - 30.47° = 124.53°$ and $a_1 = \dfrac{25 \cdot \sin 124.53°}{\sin 25°} \approx 48.73.$

If $\underline{\angle C_2 = 149.53°},$ then $\angle A_2 \approx 180° - 25° - 149.53° = 5.47°$ and $a_2 = \dfrac{25 \cdot \sin 5.47°}{\sin 25°} \approx 5.64.$

Thus the two triangles are: $\angle A_1 \approx 125°,$ $\angle C_1 \approx 30°,$ and $a_1 \approx 49;$ or $\angle A_2 \approx 5°,$ $\angle C_2 \approx 150°,$ and $a_2 \approx 5.6.$

21. $\sin B = \dfrac{100 \cdot \sin 50°}{50} \approx 1.532$. Since $|\sin \theta| \leq 1$ for all θ, there can be no such angle B, and thus no such triangle.

23. $\angle C = 180° - 82° - 52° = 46°$, so by the Law of Sines, $\dfrac{AC}{\sin 52°} = \dfrac{AB}{\sin 46°} \quad \Leftrightarrow$

$AC = \dfrac{AB \cdot \sin 52°}{\sin 46°}$, so substituting we have $AC = \dfrac{200 \cdot \sin 52°}{\sin 46°} \approx 219$ ft.

25. (a) Let a be the distance from satellite to the tracking station A in miles. Then the subtended angle at the satellite is $\angle C = 180° - 93° - 84.2° = 2.8°$, and so $a = \dfrac{50 \cdot \sin 84.2°}{\sin 2.8°} \approx 1018$ mi.

(b) Let d be the distance above the ground in miles. Then $d = 1018.3 \cdot \sin 87° \approx 1017$ mi.

27. Let x be the length of the cable as shown in the figure on the right. Since $\alpha = 12°$, other angles in $\triangle ABC$ are $\beta = 90° + 58° = 148°$, and $\gamma = 180° - (12° + 148°) = 20°$.

Thus, $\dfrac{x}{\sin 148°} = \dfrac{100}{\sin 20°} \quad \Leftrightarrow$

$x = 100 \cdot \dfrac{\sin 148°}{\sin 20°} \approx 155$ ft.

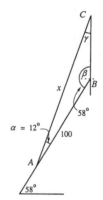

29. (a) In $\triangle ABC$, $\angle B = 180° - \beta$, so $\angle C = 180° - \alpha - (180° - \beta) = \beta - \alpha$. By the Law of Sines, $\dfrac{BC}{\sin \alpha} = \dfrac{AB}{\sin(\beta - \alpha)} \quad \Rightarrow \quad BC = AB \cdot \dfrac{\sin \alpha}{\sin(\beta - \alpha)} = d \cdot \dfrac{\sin \alpha}{\sin(\beta - \alpha)}$.

(b) From part (a) we know that $BC = d \cdot \dfrac{\sin \alpha}{\sin(\beta - \alpha)}$. But, $\sin \beta = \dfrac{h}{BC} \quad \Leftrightarrow \quad BC = \dfrac{h}{\sin \beta}$.

Therefore, $BC = d \cdot \dfrac{\sin \alpha}{\sin(\beta - \alpha)} = \dfrac{h}{\sin \beta} \quad \Rightarrow \quad h = d \cdot \dfrac{\sin \alpha \cdot \sin \beta}{\sin(\beta - \alpha)}$.

(c) $h = d \cdot \dfrac{\sin \alpha \cdot \sin \beta}{\sin(\beta - \alpha)} = 800 \cdot \dfrac{\sin 25° \sin 29°}{\sin 4°} \approx 800 \cdot \dfrac{0.20489}{0.06976} \approx 2350$ ft.

31. Label the diagram as shown to the right, and let the hill's angle of elevation be α. Then applying the Law of Sines to $\triangle ABC$, $\dfrac{\sin \beta}{120} = \dfrac{\sin 8°}{30} \quad \Leftrightarrow \quad \sin \beta = 4 \cdot \sin 8° \approx 0.55669 \quad \Rightarrow$
$\beta \approx \sin^{-1} 0.55669 \approx 33.8°$. But from $\triangle ABD$,
$\angle BAD + \angle B = (\alpha + 8°) + \beta = 90°$, so
$\alpha \approx 90° - 8° - 33.8° = 48.2°$

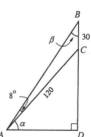

33. (a) $\sin B = \dfrac{20 \cdot \sin 40°}{15} \approx 0.857 \quad \Leftrightarrow \quad \angle B_1 \approx \sin^{-1} 0.857 \approx 58.99°$ or $\angle B_2 \approx 180° - 58.99° \approx 121.01°$.

If $\angle B_1 = 30.47°$, then $\angle C_1 \approx 180° - 15° - 58.99° = 106.01°$ and

$$c_1 = \frac{15 \cdot \sin 106.01°}{\sin 40°} \approx 22.43.$$

If $\angle B_2 = 121.01°$, then $\angle C_2 \approx 180° - 15° - 121.01° = 43.99°$ and

$$c_2 = \frac{15 \cdot \sin 43.99}{\sin 40°} \approx 16.21. \text{ Thus there are two triangles.}$$

(b) By the area formula given in Section 7.3, $\dfrac{\text{Area of } \triangle ABC}{\text{Area of } \triangle A'B'C'} = \dfrac{\frac{1}{2}ab \cdot \sin C}{\frac{1}{2}ab \cdot \sin C'} = \dfrac{\sin C}{\sin C'}$, since a

and b are the same in both triangles.

35.

$a \geq b$, 1 solution

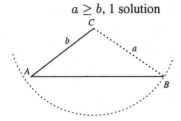

$b > a > b \sin A$, 2 solutions

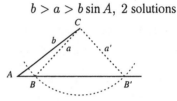

$a = b \sin A$, 1 solution

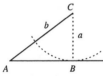

$a < b \sin A$, 0 solutions

$\angle A = 30°$, $b = 100$, $\sin A = \frac{1}{2}$. Using the criteria above,

If $a \geq b = 100$ then there is one triangle.

If $100 > a > 100 \sin 30° = 50$, then there are two possible triangles.

If $a = 50$, then there is one (right) triangle.

And if $a < 50$, then no triangles are possible.

Exercises 7.5

1. $x^2 = 10^2 + 20^2 - 2 \cdot 10 \cdot 20 \cdot \cos 35° = 100 + 400 - 400 \cos 35° \approx 172.339$ and so
$x \approx \sqrt{172.239} \approx 13.$

3. $37.83^2 = 68.01^2 + 42.15^2 - 2 \cdot 68.01 \cdot 42.15 \cdot \cos \theta.$ Then $\cos \theta = \frac{37.83^2 - 68.01^2 - 42.15^2}{-2 \cdot 68.01 \cdot 42.15} \approx 0.867$
$\Leftrightarrow \quad \theta \approx \cos^{-1} 0.867 \approx 29.89°.$

5. $x^2 = 24^2 + 30^2 - 2 \cdot 24 \cdot 30 \cdot \cos 30° = 576 + 900 - 1440 \cos 30° \approx 228.923$ and so
$x \approx \sqrt{228.923} \approx 15.$

7. $c^2 = 10^2 + 18^2 - 2 \cdot 10 \cdot 18 \cdot \cos 120° = 100 + 324 - 360 \cos 120° = 604$ and so
$c \approx \sqrt{604} \approx 24.576.$ Then $\sin A \approx \dfrac{18 \cdot \sin 120°}{24.576} \approx 0.634295 \quad \Leftrightarrow$
$\angle A \approx \sin^{-1} 0.634295 \approx 39.4°,$ and $\angle B \approx 180° - 120° - 39.4° = 20.6°.$

9. $c^2 = 3^2 + 4^2 - 2 \cdot 3 \cdot 4 \cdot \cos 53° = 9 + 16 - 24 \cos 53° \approx 10.556 \quad \Leftrightarrow \quad c \approx \sqrt{10.556} \approx 3.2.$
Then $\sin B = \dfrac{4 \cdot \sin 53°}{3.25} \approx 0.983 \quad \Leftrightarrow \quad \angle B \approx \sin^{-1} 0.983 \approx 79°$ and
$\angle A \approx 180° - 53° - 79° = 48°.$

11. $20^2 = 25^2 + 22^2 - 2 \cdot 25 \cdot 22 \cdot \cos A \quad \Leftrightarrow \quad \cos A = \dfrac{20^2 - 25^2 - 22^2}{-2 \cdot 25 \cdot 22} \approx 0.644 \quad \Leftrightarrow$
$\angle A \approx \cos^{-1} 0.644 \approx 50°.$ Then $\sin B \approx \dfrac{25 \cdot \sin 49.9°}{20} \approx 0.956 \quad \Leftrightarrow \quad \angle B \approx \sin^{-1} 0.956 \approx 73°,$
and so $\angle C \approx 180° - 50° - 73° = 57°.$

13. $\sin C = \dfrac{162 \cdot \sin 40°}{125} \approx 0.833 \quad \Leftrightarrow \quad \angle C_1 \approx \sin^{-1} 0.833 \approx 56.4°$ or
$\angle C_2 \approx 180° - 56.4° \approx 123.6°.$

If $\underline{\angle C_1 \approx 56.4°}$, then $\angle A_1 \approx 180° - 40° - 56.4° = 83.6°$ and $a_1 = \dfrac{125 \cdot \sin 83.6°}{\sin 40°} \approx 193$

If $\underline{\angle C_2 \approx 123.6°}$, then $\angle A_2 \approx 180° - 40° - 123.6° = 16.4°$ and $a_2 = \dfrac{125 \cdot \sin 16.4°}{\sin 40°} \approx 54.9.$

Thus the two triangles are: $\angle A \approx 83.6°, \angle C \approx 56.4°,$ and $a \approx 193;$ or $\angle A \approx 16.4°, \angle C \approx 123.6°,$
and $a \approx 55.$

15. $\sin B = \dfrac{65 \cdot \sin 55°}{50} \approx 1.065.$ Since $|\sin \theta| \le 1$ for all θ, there is no such $\angle B$, and hence no such
triangle.

17. $\angle B = 180° - 35° - 85° = 60°.$ Then $x = \dfrac{3 \cdot \sin 35°}{\sin 60°} \approx 2.$

19. $x = \dfrac{50 \cdot \sin 30°}{\sin 100°} \approx 25.4.$

21. $b^2 = 110^2 + 130^2 - 2(110)(130) \cdot \cos 38° = 12{,}100 + 16{,}900 - 28{,}600 \cos 38° \approx 6462.892$ and so
$b \approx 80.4.$ Therefore $\sin \theta \approx \dfrac{130 \cdot \sin 38°}{80.4} \approx 0.996 \quad \Leftrightarrow \quad \theta \approx 84.6°$

23. $x^2 = 38^2 + 48^2 - 2 \cdot 38 \cdot 48 \cdot \cos 30° = 1444 + 2304 - 3648 \cos 30° \approx 588.739$ and so $x \approx 24.3.$

25. Let c be the distance across the lake, in miles. Then $c^2 = 2.82^2 + 3.56^2 - 2(2.82)(3.56) \cdot \cos 40.3°$
≈ 5.313 $\Leftrightarrow$ $c \approx 2.30$ mi.

27. In half an hour, the faster car travels 25 miles while the slower car travels 15 miles. The distance between them is given by the Law of Cosines: $d^2 = 25^2 + 15^2 - 2(25)(15) \cdot \cos 65°$ $\Rightarrow$
$d = \sqrt{25^2 + 15^2 - 2(25)(15) \cdot \cos 65°} = 5\sqrt{25 + 9 - 30 \cdot \cos 65°} \approx 23.1$ mi.

29. The pilot travels a distance of $625 \times 1.5 = 937.5$ miles in her original direction and $625 \times 2 = 1250$ miles in the new direction. Since she makes a course correction of $10°$ to the right, the included angle is $180° - 10° = 170°$. From the figure below, we use the Law of Cosines to get the expression $d^2 = 937.5^2 + 1250^2 - 2(937.5)(1250) \cdot \cos 170° \approx 4,749,549.42$, so $d \approx 2179$ miles. Thus, the pilot's distance from her original position is approximately 2179 miles.

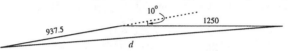

31. The largest angle is the one opposite the longest side; call this angle θ. Then by the Law of Cosines,
$44^2 = 36^2 + 22^2 - 2(36)(22) \cdot \cos \theta$ $\Leftrightarrow$ $\cos \theta = \dfrac{36^2 + 22^2 - 44^2}{2(36)(22)} = -0.09848$ $\Rightarrow$
$\theta \approx \cos^{-1}(-0.09848) \approx 96°$.

33. Let d be the distance between the kites; then $d^2 \approx 380^2 + 420^2 - 2(380)(420) \cdot \cos 30°$ $\Rightarrow$
$d \approx \sqrt{380^2 + 420^2 - 2(380)(420) \cdot \cos 30°} \approx 211$ ft.

35. <u>Solution #1:</u> From the figure on the right, we see that $\gamma = 106°$ and
$\sin 74° = \dfrac{3400}{b}$ $\Leftrightarrow$ $b = \dfrac{3400}{\sin 74°} \approx 3537$. Thus,
$x^2 = 800^2 + 3537^2 - 2(800)(3537) \cdot \cos 106°$ $\Rightarrow$
$x = \sqrt{800^2 + 3537^2 - 2(800)(3537) \cdot \cos 106°}$ $\Rightarrow$ $x \approx 3835$ ft.

<u>Solution #2:</u> Notice that $\tan 74° = \dfrac{3400}{a}$ $\Leftrightarrow$ $a = \dfrac{3400}{\tan 74°} \approx 974.9$.
By the Pythagorean theorem, $x^2 = (a + 800)^2 + 3400^2$. So,
$x = \sqrt{(974.9 + 800)^2 + 3400^2} \approx 3835$ ft.

37. Label the centers of the circles A, B, and C, as in the figure to the right. By
the Law of Cosines, $\cos A = \dfrac{AB^2 + AC^2 - BC^2}{2(AB)(AC)} = \dfrac{9^2 + 10^2 - 11^2}{2(9)(10)} = \dfrac{1}{3}$

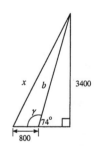

$\Rightarrow$ $\angle A \approx 70.53°$. Now, by the Law of Sines,
$\dfrac{\sin 70.53°}{11} = \dfrac{\sin B}{AC} = \dfrac{\sin C}{AB}$. So $\sin B = \dfrac{10}{11} \cdot \sin 70.53° \approx 0.85710$
$\Rightarrow$ $B \approx \sin^{-1} 0.85710 \approx 58.99°$ and $\sin C = \dfrac{9}{11} \cdot \sin 70.53° \approx 0.77139$ $\Rightarrow$
$C \approx \sin^{-1} 0.77139 \approx 50.48°$. The area of $\triangle ABC$ is
$\frac{1}{2}(AB)(AC)\sin A = \frac{1}{2}(9)(10)(\sin 70.53°) \approx 42.426$. The area of sector A is given by
$S_A = \pi R^2 \cdot \dfrac{\theta}{360°} = \pi(4)^2 \cdot \dfrac{70.53°}{360°} \approx 9.848$. Similarly, the areas of sectors B and C are
$S_B \approx 12.870$, $S_C \approx 15.859$. Thus, the area enclosed between the circles is
$A = \triangle ABC - S_A - S_B - S_C$ $\Rightarrow$ $A \approx 42.426 - 9.848 - 12.870 - 15.859 \approx 3.85$ cm^2.

39. First notice that $\angle DBC = 180° - 20° - 95° = 65°$ and $\angle DAC = 180° - 60° - 45° = 75°$. From $\triangle ACD$ we get $\dfrac{AC}{\sin 45°} = \dfrac{20}{\sin 75°}$ $\Leftrightarrow$ $AC = \dfrac{20\sin 45°}{\sin 75°} \approx 14.6°$. From $\triangle BCD$ we get $\dfrac{BC}{\sin 95°} = \dfrac{20}{\sin 65°}$ $\Leftrightarrow$ $BC = \dfrac{20\sin 95°}{\sin 65°} \approx 22.0$. By applying the Law of Cosines to $\triangle ABC$ we get $(AB)^2 = (AC)^2 + (BC)^2 - 2(AC)(BC)\cos 40°$ $\approx (14.6)^2 + (22.0)^2 - 2 \cdot 14.6 \cdot 22.0 \cdot \cos 40° \approx 205$, so $AB \approx \sqrt{205} \approx 14.3$m. Therefore, the distance between the points A and B is approximately 14.3 m.

41. By Heron's formula, $A = \sqrt{s(s-a)(s-b)(s-c)}$, where $s = \dfrac{a+b+c}{2} = \dfrac{112+148+190}{2}$ $= 225$. Thus, $A = \sqrt{225(225-112)(225-148)(225-190)} \approx 8277.7$ sq. ft. Since the land value is \$20 per square foot, the value of the lot is approximately $8277.7 \times \$20 = \$165{,}554$.

43. Using the Law of Cosines, $b^2 = a^2 + c^2 - 2ac\cos B$ $\Leftrightarrow$ $(10.5)^2 = (13.2)^2 + (18.0)^2 - 2(13.2)(18.0)\cos B$ $\Leftrightarrow$ $-387.99 = -475.2\cos B$ $\Leftrightarrow$ $\cos B \approx 0.816$ $\Leftrightarrow$ $B \approx 35.3°$. (The small difference in the two measurements is due to round off error.) There is only one possible answer with the Law of Cosines because cosine is uniquely defined between $0°$ and $180°$.

Review Exercises for Chapter 7

1. (a) $70° = 70 \cdot \frac{\pi}{180} = \frac{7\pi}{18} \approx 1.22$ rad   (b) $420° = 420 \cdot \frac{\pi}{180} = \frac{7\pi}{3} \approx 7.33$ rad

   (c) $-240° = -240 \cdot \frac{\pi}{180} = -\frac{4\pi}{3} \approx -4.19$ rad

   (d) $-40° = -40 \cdot \frac{\pi}{180} = -\frac{2\pi}{9} \approx -0.70$ rad

3. (a) $\frac{7\pi}{2}$ rad $= \frac{7\pi}{2} \cdot \frac{180}{\pi} = 630°$ (b) $-\frac{\pi}{3}$ rad $= -\frac{\pi}{3} \cdot \frac{180}{\pi} = -60°$

   (c) $\frac{7\pi}{4}$ rad $= \frac{7\pi}{4} \cdot \frac{180}{\pi} = 315°$ (d) 2.1 rad $= 2.1 \cdot \frac{180}{\pi} = \frac{378}{\pi} \approx 120.3°$

5. $r = 8$ m, $\theta = 1$ rad. Then $s = r\theta = 8 \cdot 1 = 8$ m.

7. $s = 100$ ft, $\theta = 70° = 70 \cdot \frac{\pi}{180} = \frac{7\pi}{18}$ rad. Then $r = \frac{s}{\theta} = 100 \cdot \frac{18}{7\pi} = \frac{1800}{7\pi} \approx 82$ ft.

9. $r = 3960$ miles, $s = 2450$ miles. Then $\theta = \dfrac{s}{r} = \frac{2450}{3960} \approx 0.619$ rad $= 0.619 \cdot \frac{\pi}{180}° \approx 35.448°$ and so the angle is approximately $35.4°$.

11. $A = \frac{1}{2}r^2\theta = \frac{1}{2}(200)^2\left(52° \times \frac{\pi}{180°}\right) \approx 18{,}151$ ft^2

13. $r = \sqrt{5^2 + 7^2} = \sqrt{74}$. Then $\sin\theta = \frac{5}{\sqrt{74}}$, $\cos\theta = \frac{7}{\sqrt{74}}$, $\tan\theta = \frac{5}{7}$, $\csc\theta = \frac{\sqrt{74}}{5}$, $\sec\theta = \frac{\sqrt{74}}{7}$, and $\cot\theta = \frac{7}{5}$.

15. $\dfrac{x}{5} = \cos 40°$ $\Leftrightarrow$ $x = 5\cos 40° \approx 3.83$, and $\dfrac{y}{5} = \sin 40°$ $\Leftrightarrow$ $y = 5\sin 40° \approx 3.21$

17. $\dfrac{1}{x} = \sin 20°$ $\Leftrightarrow$ $x = \dfrac{1}{\sin 20°} \approx 2.92$, and $\dfrac{x}{y} = \cos 20°$ $\Leftrightarrow$ $y = \dfrac{x}{\cos 20°} \approx \dfrac{2.924}{0.9397} \approx 3.11$

19. $w = 3\sin 20° \approx 1.026$,
    $v = 3\cos 20° \approx 2.819$.

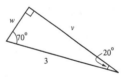

21. $\tan\theta = \dfrac{1}{a}$ $\Leftrightarrow$ $a = \dfrac{1}{\tan\theta} = \cot\theta$, $\sin\theta = \dfrac{1}{b}$ $\Leftrightarrow$ $b = \dfrac{1}{\sin\theta} = \csc\theta$

23. One side of the hexagon together with radial line segments through its endpoints forms a triangle with two sides of length 8 m and subtended angle $60°$. Let x be the length of one such side (in m). So by the Law of Cosines
$x^2 = 8^2 + 8^2 - 2 \cdot 8 \cdot 8 \cdot \cos 60° = 64$ $\Leftrightarrow$ $x = 8$. Thus the perimeter of the hexagon is $6x = 6 \cdot 8 = 48$ m

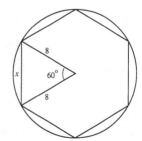

25. Let r represent the radius, in miles, of the moon. Then $\tan\frac{\theta}{2} = \dfrac{r}{r + |AB|}$, $\theta = 0.518°$ $\Leftrightarrow$
$r = (r + 236{,}900) \cdot \tan 0.259°$ $\Leftrightarrow$ $r(1 - \tan 0.259°) = 236{,}900 \cdot \tan 0.259°$ $\Leftrightarrow$
$r = \dfrac{236{,}900 \cdot \tan 0.259°}{1 - \tan 0.259°} \approx 1076$ and so the radius of the moon is roughly 1076 miles.

27. $\sin 315° = -\sin 45° = -\frac{1}{\sqrt{2}} = -\frac{\sqrt{2}}{2}$

29. $\tan(-135°) = \tan 45° = 1$

31. $\cot(-\frac{22\pi}{3}) = \cot(\frac{2\pi}{3}) = \cot\frac{\pi}{3} = -\frac{1}{\sqrt{3}} = -\frac{\sqrt{3}}{3}$

33. $\cos 585° = \cos 225° = -\cos 45° = -\frac{1}{\sqrt{2}} = -\frac{\sqrt{2}}{2}$

35. $\csc\frac{8\pi}{3} = \csc\frac{2\pi}{3} = \csc\frac{\pi}{3} = \frac{2}{\sqrt{3}} = \frac{2\sqrt{3}}{3}$

37. $\cot(-390°) = \cot(-30°) = -\cot 30° = -\sqrt{3}$

39. $r = \sqrt{(-5)^2 + 12^2} = \sqrt{169} = 13$. Then $\sin\theta = \frac{12}{13}$, $\cos\theta = -\frac{5}{13}$, $\tan\theta = -\frac{12}{5}$, $\csc\theta = \frac{13}{12}$, $\sec\theta = -\frac{13}{5}$, and $\cot\theta = -\frac{5}{12}$.

41. $y - \sqrt{3}x + 1 = 0 \iff y = \sqrt{3}x - 1$, so the slope of the line is $m = \sqrt{3}$. Then $\tan\theta = m = \sqrt{3} \iff \theta = 60°$.

43. $\sec^2\theta = 1 + \tan^2\theta \iff \tan^2\theta = \frac{1}{\cos^2\theta} - 1 \iff \tan\theta = \sqrt{\frac{1}{\cos^2\theta} - 1} = \sqrt{\frac{1 - \cos^2\theta}{\cos^2\theta}}$

$= \frac{\sqrt{1 - \cos^2\theta}}{|\cos\theta|} = \frac{\sqrt{1 - \cos^2\theta}}{-\cos\theta}$ (since $\cos\theta < 0$ in quadrant III, $|\cos\theta| = -\cos\theta$). Thus

$\tan\theta = -\frac{\sqrt{1 - \cos^2\theta}}{\cos\theta}$.

45. $\tan^2\theta = \frac{\sin^2\theta}{\cos^2\theta} = \frac{\sin^2\theta}{1 - \sin^2\theta}$

47. $\tan\theta = \frac{\sqrt{7}}{3}$, $\sec\theta = \frac{4}{3}$. Then $\cos\theta = \frac{3}{4}$ and $\sin\theta = \tan\theta \cdot \cos\theta = \frac{\sqrt{7}}{4}$, $\csc\theta = \frac{4}{\sqrt{7}} = \frac{4\sqrt{7}}{7}$, and $\cot\theta = \frac{3}{\sqrt{7}} = \frac{3\sqrt{7}}{7}$.

49. $\sin\theta = \frac{3}{5}$. Since $\cos\theta < 0$, θ is in quadrant II. Thus, $x = -\sqrt{5^2 - 3^2} = -\sqrt{16} = -4$. Therefore, $\cos\theta = -\frac{4}{5}$, $\tan\theta = -\frac{3}{4}$, $\csc\theta = \frac{5}{3}$, $\sec\theta = -\frac{5}{4}$, $\cot\theta = -\frac{4}{3}$.

51. $\tan\theta = -\frac{1}{2}$. $\sec^2\theta = 1 + \tan^2\theta = 1 + \frac{1}{4} = \frac{5}{4} \iff \cos^2\theta = \frac{4}{5} \implies \cos\theta = -\sqrt{\frac{4}{5}} = -\frac{2}{\sqrt{5}}$

since $\cos\theta < 0$ in quadrant II. But $\tan\theta = \frac{\sin\theta}{\cos\theta} = -\frac{1}{2} \iff \sin\theta = -\frac{1}{2}\cos\theta$

$= -\frac{1}{2}\left(-\frac{2}{\sqrt{5}}\right) = \frac{1}{\sqrt{5}}$. Therefore, $\sin\theta + \cos\theta = \frac{1}{\sqrt{5}} + \left(-\frac{2}{\sqrt{5}}\right) = -\frac{1}{\sqrt{5}} = -\frac{\sqrt{5}}{5}$.

53. By the Pythagorean identity, $\sin^2\theta + \cos^2\theta = 1$ for any angle θ.

55. $\angle B = 180° - 30° - 80° = 70°$, and so by the Law of Sines, $x = \dfrac{10 \cdot \sin 30°}{\sin 70°} \approx 5.32$

57. $x^2 = 100^2 + 210^2 - 2 \cdot 100 \cdot 210 \cdot \cos 40° \approx 21926.133 \iff x \approx 148.07$

59. $\sin B = \dfrac{20 \cdot \sin 60°}{70} \approx 0.247 \iff \angle B \approx \sin^{-1} 0.247 \approx 14.33°$. Then

$\angle C \approx 180° - 60° - 14.33° = 105.67°$, and so $x \approx \dfrac{70 \cdot \sin 105.67°}{\sin 60°} \approx 77.82$.

61. After 2 hours the ships have traveled distances $d_1 = 40$ mi and $d_2 = 56$ mi. The subtended angle is $180° - 32° - 42° = 106°$. Let d be the distance between the two ships in miles. Then by the Law of Cosines $d^2 = 40^2 + 56^2 - 2(40)(56) \cdot \cos 106° \approx 5970.855 \quad \Leftrightarrow \quad d \approx 77.3$ miles.

63. Let d be the distance, in miles, between the points A and B. Then by the Law of Cosines $d^2 = 3.2^2 + 5.6^2 - 2(3.2)(5.6) \cdot \cos 42° \approx 14.966 \quad \Leftrightarrow \quad d \approx 3.9$ mi.

65. Let h represent the height, in m, of the cliff, and d the horizontal distance to the cliff. The third horizontal angle is $180° - 69.4° - 51.6° = 59°$ and so $d = \dfrac{200 \cdot \sin 51.6°}{\sin 59°} \approx 182.857$. Then $\tan 33.1° = \dfrac{h}{d} \quad \Leftrightarrow \quad h = d \tan 33.1° \approx 182.857 \tan 33.1° \approx 119.2$ m.

67. By Heron's formula, $A = \sqrt{s(s-a)(s-b)(s-c)}$, where $s = \dfrac{a+b+c}{2} = \dfrac{5+6+8}{2} = 9.5$. Thus, $A = \sqrt{9.5(9.5-5)(9.5-6)(9.5-8)} \approx 14.98$.

Chapter 7 Test

1. $300° = 300 \cdot \frac{\pi}{180} = \frac{5\pi}{3}$ rad. $-18° = -18 \cdot \frac{\pi}{180} = -\frac{\pi}{10}$ rad.

2. $\frac{5\pi}{6}$ rad $= \frac{5\pi}{6} \cdot \frac{180}{\pi} = 150°$ 2.4 rad $= 2.4 \cdot \frac{180}{\pi} = \frac{432}{\pi} \approx 137.5°$

3. (a) $\sin 405° = \sin 45° = \frac{1}{\sqrt{2}} = \frac{\sqrt{2}}{2}$ (b) $\tan(-150°) = \tan 30° = \frac{1}{\sqrt{3}} = \frac{\sqrt{3}}{3}$

    (c) $\sec \frac{5\pi}{3} = \sec \frac{\pi}{3} = 2$ (d) $\csc \frac{5\pi}{2} = \csc \frac{\pi}{2} = 1$

4. $r = \sqrt{3^2 + 2^2} = \sqrt{13}$. Then $\tan\theta + \sin\theta = \frac{2}{3} + \frac{2}{\sqrt{13}} = \frac{2(\sqrt{13}+3)}{3\sqrt{13}} = \frac{26+6\sqrt{13}}{39}$

5. $\sin\theta = \dfrac{a}{24} \quad \Leftrightarrow \quad a = 24 \sin\theta$. Also, $\cos\theta = \dfrac{b}{24} \quad \Leftrightarrow \quad b = 24 \cos\theta$.

6. $\cos\theta = -\frac{1}{3}$ and θ is in quadrant III, so $r = 3$, $x = -1$, and $y = -\sqrt{3^2 - 1^2} = -2\sqrt{2}$. Then
    $\tan\theta \cot\theta + \csc\theta = \tan\theta \cdot \dfrac{1}{\tan\theta} + \csc\theta = 1 - \dfrac{3}{2\sqrt{2}} = \dfrac{2\sqrt{2}-3}{2\sqrt{2}} = \dfrac{4-3\sqrt{2}}{4}$.

7. $\sin\theta = \frac{5}{13}$, $\tan\theta = -\frac{5}{12}$. Then $\sec\theta = \dfrac{1}{\cos\theta} = \dfrac{1}{\cos\theta} \cdot \dfrac{\sin\theta}{\sin\theta} = \tan\theta \cdot \dfrac{1}{\sin\theta} = -\frac{5}{12} \cdot \frac{13}{5} = -\frac{13}{12}$.

8. $\sec^2\theta = 1 + \tan^2\theta \quad \Leftrightarrow \quad \tan\theta = \pm\sqrt{\sec^2\theta - 1}$. Thus, $\tan\theta = -\sqrt{\sec^2\theta - 1}$ since $\tan\theta < 0$ in quadrant II.

9. $\tan 73° = \dfrac{h}{6} \quad \Rightarrow \quad h = 6 \tan 73° \approx 19.6$ ft.

10. By the Law of Cosines $x^2 = 10^2 + 12^2 - 2(10)(12) \cdot \cos 48° \approx 8.409 \quad \Leftrightarrow \quad x \approx 9.1$

11. $\angle C = 180° - 52° - 69° = 59°$. Then by the Law of Sines $x = \dfrac{230 \cdot \sin 69°}{\sin 59°} \approx 250.5$

12. Let h be the height of the shorter altitude. Then $\tan 20° = \dfrac{h}{50} \quad \Leftrightarrow \quad h = 50 \tan 20°$ and
    $\tan 28° = \dfrac{x+h}{50} \quad \Leftrightarrow \quad x + h = 50 \tan 28° \quad \Leftrightarrow \quad x = 50 \tan 28° - h = 50 \tan 28° - 50 \tan 20°$
    ≈ 8.4.

13. Let $\angle A$ and $\angle X$ be the other angles in the triangle. Then $\sin A = \dfrac{15 \cdot \sin 108°}{28} \approx 0.509 \quad \Leftrightarrow$
    $\angle A \approx 30.63°$. Then $\angle X \approx 180° - 108° - 30.63° \approx 41.37°$, and so $x \approx \dfrac{28 \cdot \sin 41.37°}{\sin 108°} \approx 19.5$

14. $A(\text{sector}) = \frac{1}{2}r^2\theta = \frac{1}{2} \cdot 10^2 \cdot 72 \cdot \frac{\pi}{180} = 50 \cdot \frac{72\pi}{180}$
    $A(\text{triangle}) = \frac{1}{2}r \cdot r\sin\theta = \frac{1}{2} \cdot 10^2 \sin 72°$. Thus the area of the shaded region is
    $A(\text{shaded}) = A(\text{sector}) - A(\text{triangle}) = 50\left(\frac{72\pi}{180} - \sin 72°\right) \approx 15.3\text{m}^2$.

15. The shaded region is bounded by two pieces: one piece is part of the triangle, the other is part of the circle. The first part has length $l = \sqrt{10^2 + 10^2 - 2(10)(10) \cdot \cos 72°} = 10\sqrt{2 - 2 \cdot \cos 72°}$. The

second has length $s = 10 \cdot 72° \cdot \frac{\pi}{180°} = 4\pi$. Thus, the perimeter of the shaded region is $p = l + s = 10\sqrt{2 - 2\cos 72°} + 4\pi \approx 24.3$ m.

16. If θ is the angle opposite the longest side, then by the Laws of Cosines $\cos\theta = \frac{9^2 + 13^2 - 20^2}{2(9)(20)} = -0.6410$. Therefore, $\theta = \cos^{-1}(-0.6410) \approx 129.9°$.

17. From question 16, $\theta = 130°$, so the area of the triangle is $\mathcal{A} = \frac{1}{2}(9)(13) \cdot \sin 130° \approx 44.9$ units^2.

    Another way to find the area is to use Heron's formula: $\mathcal{A} = \sqrt{s(s - a)(s - b)(s - c)}$, where $s = \frac{a + b + c}{2} = \frac{9 + 13 + 20}{2} = 21$. Thus, $\mathcal{A} = \sqrt{21(21 - 20)(21 - 13)(21 - 9)}$ $= \sqrt{2016} \approx 44.9$ units^2.

18. Label the figure as shown to the right. Now $\angle\beta = 85° - 75° = 10°$, so by the Law of Sines, $\dfrac{x}{\sin 75°} = \dfrac{100}{\sin 10°} \quad\Leftrightarrow$

    $x = 100 \cdot \dfrac{\sin 75°}{\sin 10°}$. Since $\sin 85° = \dfrac{h}{x} \quad\Leftrightarrow$

    $h = x\sin 85° = 100 \cdot \dfrac{\sin 75°}{\sin 10°} \sin 85° \approx 554.$

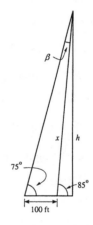

Focus on Problem Solving

1. Tetrahedron: faces are all triangles and three faces meet at each vertex. Thus $V = \dfrac{3F}{3} = F$ and $E = \dfrac{3F}{2}$. Substituting into Euler's formula we have $F - \dfrac{3F}{2} + F = 2$ $\Leftrightarrow$ $\frac{1}{2}F = 2$ $\Leftrightarrow$ $F = 4$. So $V = 4$ and $E = \frac{12}{2} = 6$.

    Octahedron: faces are all triangles and four faces meet at each vertex. Thus $V = \dfrac{3F}{4}$ and $E = \dfrac{3F}{2}$. Substituting into Euler's formula we have $F - \dfrac{3F}{2} + \dfrac{3F}{4} = 2$ $\Leftrightarrow$ $\frac{1}{4}F = 2$ $\Leftrightarrow$ $F = 8$. So $V = \frac{24}{4} = 6$ and $E = \frac{24}{2} = 12$.

    Cube: faces are all rectangles and three faces meet at each vertex. Thus $V = \dfrac{4F}{3}$ and $E = \dfrac{4F}{2} = 2F$. Substituting into Euler's formula we have $F - 2F + \dfrac{4F}{3} = 2$ $\Leftrightarrow$ $\frac{1}{3}F = 2$ $\Leftrightarrow$ $F = 6$. So $V = \frac{24}{3} = 8$ and $E = 12$.

    Dodecahedron: faces are all pentagon and three faces meet at each vertex. Thus $V = \dfrac{5F}{3}$ and $E = \dfrac{5F}{2}$. Substituting into Euler's formula we have $F - \dfrac{5F}{2} + \dfrac{5F}{3} = 2$ $\Leftrightarrow$ $\frac{1}{6}F = 2$ $\Leftrightarrow$ $F = 12$. So $V = \frac{60}{3} = 20$ and $E = \frac{60}{2} = 30$.

    Icosahedron is done in the text.

3. In order to tile the plane with regular polygons, a whole number of polygons must meet at each corner point (so the sum of the angles at each corner is 360°). Since the interior angle of a regular pentagon is 108°, it is impossible for a whole number of pentagons to meet at a corner (because 360° is not evenly divisible by 108°). The plane can be tiled with regular hexagons, as shown is the figure on the right. For $n \geq 7$, the interior angle of a regular n-gon is strictly between 120° and 180°, so it is impossible for a whole number of such n-gons to meet at a corner (because 360° is not evenly divisible by any such number). Thus it is only possible to tile the plane with equilateral triangles, squares, and regular hexagons.

5. The belt is tangent to the pulley at the point where the belt leaves the pulley. Thus the length of the belt between adjacent tangent points is the same as the length between the centers of the corresponding pulleys. The length of the belt around the pulleys can be pieced together to form a circle of radius 1. Thus the length of the belt is $L = P + 2\pi(1) = P + 2\pi$. (See the following figure.)

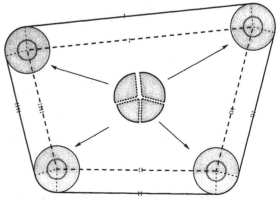

7. The area we seek is the shaded region in the figure on the left. It consists of two sections (like that shown in the figure on the right) with the overlap being the parallelogram $ABCD$. Now $AB = AC = AD = CB = CA = CD = 1$ since all are radii. So $\triangle ABC$ and $\triangle ADC$ are equilateral triangles whose area are each $\frac{1}{2}bh = \frac{1}{2} \cdot 1 \cdot \cos 60° = \frac{\sqrt{3}}{4}$. Thus the area of parallelogram $ABCD$ is $\frac{\sqrt{3}}{2}$ Since $\angle BAD = \angle BCD = 120°$, the area of each sector is $\frac{1}{3}\pi$. Thus the area we seek is $2 \cdot \frac{1}{3}\pi - \frac{\sqrt{3}}{2} = \frac{4\pi - 3\sqrt{3}}{6} \approx 1.2$.

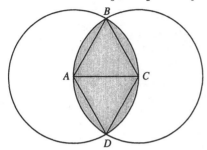 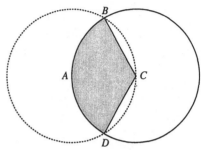

9. We use the identity $\sin(2\pi - x) = -\sin x$ to deduce $\sin\left(\frac{\pi}{100}\right) = -\sin\left(\frac{199\pi}{100}\right)$, $\sin\left(\frac{2\pi}{100}\right) = -\sin\left(\frac{198\pi}{100}\right)$, $\dots$, and in general, $\sin\left(\frac{k\pi}{100}\right) = -\sin\left(\frac{(200-k)\pi}{100}\right)$, for $k = 1, 2, \dots, 199$. Thus the sum is $\sin\frac{\pi}{100} + \dots + \sin\frac{199\pi}{100} = 0$.

11. Imagine that the walls of the room are folded down as in the diagram. Notice that the diagram shows two points labeled B. So one of the straight lines from A to B gives the shortest path. If $a \le b$ then the length of the shortest path is $\sqrt{(a+c)^2 + b^2}$ and if $a \ge b$, then the length of the shortest path is $\sqrt{a^2 + (b+c)^2}$.

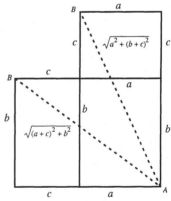

13. $x^2 + y^2 = 4z + 3$. We consider all possible cases for x and y even or odd.

Case 1: x, y are both odd. Since the square of an odd number is odd, and the sum of two odd numbers is even, the left hand side is even but the right hand side is odd. Thus the left hand side cannot equal the right hand side.

Case 2: x, y are both even. Since the square of an even number is even, and the sum of two even numbers is even, the left hand side is even but the right hand side is odd. Thus the left hand side cannot equal the right hand side.

Case 3: x odd and y even. Let $x = 2n + 1$ and let $y = 2m$ for some integers n and m. Then $(2n+1)^2 + (2m)^2 = 4z + 3 \iff 4n^2 + 4n + 1 + 4m^2 = 4z + 3 \iff 4(n^2 + n + m^2) + 1 = 4z + 3$. However, when the left hand side is divided by 4 the remainder is 1, but when the right hand side is divided by 4 the remainder is 3. Thus the left hand side cannot equal the right hand side.

Case 4: x even and y odd. This is handled in the same way as case 3.

Thus there are no integer solutions to $x^2 + y^2 = 4z + 3$.

15. (a) $13 = 2^2 + 3^2$; $41 = 4^2 + 5^2$

(b) $(a^2 + b^2)(c^2 + d^2) = a^2 c^2 + b^2 d^2 + a^2 d^2 + b^2 c^2$
$$= a^2 c^2 + 2abcd + b^2 d^2 + a^2 d^2 - 2abcd + b^2 c^2$$
$$= (ac + bd)^2 + (ad - bc)^2$$

(c) We first factor $533 = 13 \cdot 41$.

Let $a = 2$, $b = 3$, $c = 4$, $d = 5$, then by the formula in part (b)
$533 = 13 \cdot 41 = (2^2 + 3^2)(4^2 + 5^2) = (2 \cdot 4 + 3 \cdot 5)^2 + (2 \cdot 5 - 3 \cdot 4)^2 = 23^2 + (-2)^2$
$= 23^2 + 2^2$. So $533 = 23^2 + 2^2$.

Let $a = 2$, $b = 3$, $c = 5$, $d = 4$, then by the formula in part (b)
$533 = 13 \cdot 41 = (2^2 + 3^2)(5^2 + 4^2) = (2 \cdot 5 + 3 \cdot 4)^2 + (2 \cdot 4 - 3 \cdot 5)^2 = 22^2 + (-7)^2$
$= 22^2 + 7^2$. So $533 = 22^2 + 7^2$.

Chapter Eight
Exercises 8.1

1. Since $\left(\frac{3}{5}\right)^2 + \left(\frac{4}{5}\right)^2 = \frac{9}{25} + \frac{16}{25} = 1$, $P\left(\frac{3}{5}, \frac{4}{5}\right)$ lies on the unit circle.

3. Since $\left(-\frac{1}{3}\right)^2 + \left(\frac{2\sqrt{2}}{3}\right)^2 = \frac{1}{9} + \frac{8}{9} = 1$, $P\left(-\frac{1}{3}, \frac{2\sqrt{2}}{3}\right)$ lies on the unit circle.

5. $\left(\frac{3}{5}\right)^2 + y^2 = 1 \iff y^2 = 1 - \frac{9}{25} \iff y^2 = \frac{16}{25} \iff y = \pm\frac{4}{5}$. Since $P(x, y)$ is in quadrant I, y is positive, so the point is $P\left(\frac{3}{5}, \frac{4}{5}\right)$.

7. $\left(\frac{2}{3}\right)^2 + y^2 = 1 \iff y^2 = 1 - \frac{4}{9} \iff y^2 = \frac{5}{9} \iff y = \pm\frac{\sqrt{5}}{3}$. Since y is negative, the point is $P\left(\frac{2}{3}, -\frac{\sqrt{5}}{3}\right)$.

9. $\left(\frac{\sqrt{2}}{3}\right)^2 + y^2 = 1 \iff y^2 = 1 - \frac{2}{9} \iff y^2 = \frac{7}{9} \iff y = \pm\frac{\sqrt{7}}{3}$. Since $P(x, y)$ is in quadrant IV, y is negative; thus the point is $P\left(\frac{\sqrt{2}}{3}, -\frac{\sqrt{7}}{3}\right)$.

11. Quadrant I

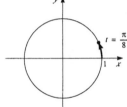

13. Quadrant II

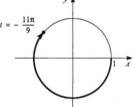

15. Quadrant III

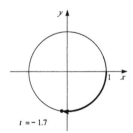

17. $P(x, y) = \left(\frac{\sqrt{2}}{2}, \frac{\sqrt{2}}{2}\right)$

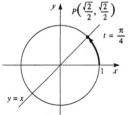

19. $P(x, y) = \left(\frac{1}{2}, -\frac{\sqrt{3}}{2}\right)$

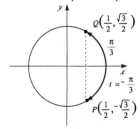

21. $P(x, y) = (-1, 0)$

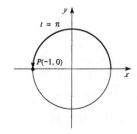

23. $P(x,y) = \left(-\frac{1}{2}, \frac{\sqrt{3}}{2}\right)$

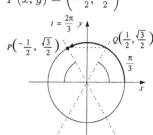

25. $P(x,y) = \left(-\frac{\sqrt{2}}{2}, -\frac{\sqrt{2}}{2}\right)$

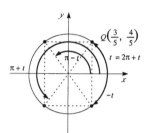

27. Let $Q(x,y) = \left(\frac{3}{5}, \frac{4}{5}\right)$ be the terminal point determined by t.

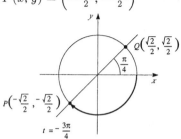

  (a) $\pi - t$ determines the point $P(-x,y) = \left(-\frac{3}{5}, \frac{4}{5}\right)$

  (b) $-t$ determines the point $P(x,-y) = \left(\frac{3}{5}, -\frac{4}{5}\right)$

  (c) $\pi + t$ determines the point $P(-x,-y) = \left(-\frac{3}{5}, -\frac{4}{5}\right)$

  (d) $2\pi + t$ determines the point $P(x,y) = \left(\frac{3}{5}, \frac{4}{5}\right)$

29. **(a)** $\bar{t} = \frac{5\pi}{4} - \pi = \frac{\pi}{4}$ **(b)** $\bar{t} = \frac{7\pi}{3} - 2\pi = \frac{\pi}{3}$

     (c) $\bar{t} = \frac{4\pi}{3} - \pi = \frac{\pi}{3}$ **(d)** $\bar{t} = \frac{\pi}{6}$

31. **(a)** Since $-\frac{11\pi}{5}$ is coterminal with $-\frac{11\pi}{5} + 2\pi = -\frac{\pi}{5}$, we have $\bar{t} = -\left(-\frac{\pi}{5}\right) = \frac{\pi}{5}$

     (b) $\bar{t} = \frac{13\pi}{6} - 2\pi = \frac{\pi}{6}$ **(c)** $\bar{t} = \frac{7\pi}{3} - 2\pi = \frac{\pi}{3}$ **(d)** $\bar{t} = \pi - \frac{5\pi}{6} = \frac{\pi}{6}$

33. **(a)** $\bar{t} = \pi - \frac{3\pi}{4} = \frac{\pi}{4}$ **(b)** $P\left(-\frac{\sqrt{2}}{2}, \frac{\sqrt{2}}{2}\right)$

35. **(a)** $\bar{t} = \pi - \frac{2\pi}{3} = \frac{\pi}{3}$ **(b)** $P\left(-\frac{1}{2}, -\frac{\sqrt{3}}{2}\right)$

37. **(a)** $\bar{t} = \frac{13\pi}{4} - 3\pi = \frac{\pi}{4}$ **(b)** $P\left(-\frac{\sqrt{2}}{2}, -\frac{\sqrt{2}}{2}\right)$

39. **(a)** $\bar{t} = \frac{7\pi}{6} - \pi = \frac{\pi}{6}$ **(b)** $P\left(-\frac{\sqrt{3}}{2}, -\frac{1}{2}\right)$

41. **(a)** $\bar{t} = 4\pi - \frac{11\pi}{3} = \frac{\pi}{3}$ **(b)** $P\left(\frac{1}{2}, \frac{\sqrt{3}}{2}\right)$

43. **(a)** $\bar{t} = \frac{16\pi}{3} - 5\pi = \frac{\pi}{3}$ **(b)** $P\left(-\frac{1}{2}, -\frac{\sqrt{3}}{2}\right)$

45. $t = 1 \quad \Rightarrow \quad (0.5, 0.8)$ **47.** $t = -1.1 \quad \Rightarrow \quad (0.5, -0.9)$

49. The distances PQ and PR are equal because they both subtend arcs of length $\frac{\pi}{3}$. Since $P(x,y)$ is a point on the unit circle, $x^2 + y^2 = 1$. Now, $d(P,Q) = \sqrt{(x-x)^2 + (y-(-y))^2} = 2y$ and $d(R,S) = \sqrt{(x-0)^2 + (y-1)^2} = \sqrt{x^2 + y^2 - 2y + 1} = \sqrt{2 - 2y}$ (using the fact that $x^2 + y^2 = 1$). Setting these equal gives $2y = \sqrt{2 - 2y} \quad \Rightarrow \quad 4y^2 = 2 - 2y \quad \Leftrightarrow \quad 4y^2 + 2y - 2 = 0 \quad \Leftrightarrow \quad 2(2y-1)(y+1) = 0$. So $y = -1$ or $y = \frac{1}{2}$. Since P is in quadrant I, $y = \frac{1}{2}$ is the only viable solution. Again using $x^2 + y^2 = 1$ we have $x^2 + \left(\frac{1}{2}\right)^2 = 1 \quad \Leftrightarrow \quad x^2 = \frac{3}{4} \quad \Rightarrow \quad x = \pm\frac{\sqrt{3}}{2}$. Again, since P is in quadrant I the coordinates must be $\left(\frac{\sqrt{3}}{2}, \frac{1}{2}\right)$.

Exercises 8.2

1. (a) $\sin 0 = 0$ (b) $\cos 0 = 1$

3. (a) $\sin(-\pi) = 0$ (b) $\cos(-\pi) = -1$

5. (a) $\sin\frac{\pi}{2} = 1$ (b) $\sin\frac{3\pi}{2} = -1$

7. (a) $\cos\frac{\pi}{2} = 0$ (b) $\cos\frac{5\pi}{2} = 0$

9. (a) $\cos\frac{7\pi}{3} = \frac{1}{2}$ (b) $\sec\frac{7\pi}{3} = 2$

11. (a) $\cos\frac{\pi}{3} = \frac{1}{2}$ (b) $\cos\left(-\frac{\pi}{3}\right) = \frac{1}{2}$

13. (a) $\tan\frac{\pi}{6} = \frac{1}{\sqrt{3}} = \frac{\sqrt{3}}{3}$ (b) $\tan\left(-\frac{\pi}{6}\right) = -\frac{1}{\sqrt{3}} = -\frac{\sqrt{3}}{3}$

15. (a) $\sec\frac{11\pi}{3} = \sec\frac{5\pi}{3} = 2$ (b) $\csc\frac{11\pi}{3} = \csc\frac{5\pi}{3} = -\frac{2}{\sqrt{3}} = -\frac{2\sqrt{3}}{3}$

17. (a) $\sin\frac{9\pi}{4} = \sin\frac{\pi}{4} = \frac{1}{\sqrt{2}} = \frac{\sqrt{2}}{2}$ (b) $\csc\frac{9\pi}{4} = \csc\frac{\pi}{4} = \sqrt{2}$

19. (a) $\tan\left(-\frac{\pi}{4}\right) = -1$ (b) $\cot\left(-\frac{\pi}{4}\right) = -1$

21. $t = 0 \quad\Rightarrow\quad \sin t = 0$, $\cos t = 1$, $\tan t = 0$, $\sec t = 1$, $\csc t$ and $\cot t$ are undefined.

23. $t = \pi \quad\Rightarrow\quad \sin t = 0$, $\cos t = -1$, $\tan t = 0$, $\sec t = -1$, $\csc t$ and $\cot t$ are undefined.

25. $\left(\frac{3}{5}\right)^2 + \left(\frac{4}{5}\right)^2 = \frac{9}{25} + \frac{16}{25} = 1$. So $\sin t = \frac{4}{5}$, $\cos t = \frac{3}{5}$, and $\tan t = \frac{\frac{4}{5}}{\frac{3}{5}} = \frac{4}{3}$.

27. $\left(\frac{6}{7}\right)^2 + \left(-\frac{\sqrt{13}}{7}\right)^2 = \frac{36}{49} + \frac{13}{49} = 1$. So $\sin t = -\frac{\sqrt{13}}{7}$, $\cos t = \frac{6}{7}$, and $\tan t = \frac{-\frac{\sqrt{13}}{7}}{\frac{6}{7}} = -\frac{\sqrt{13}}{6}$.

29. $\left(\frac{40}{41}\right)^2 + \left(\frac{9}{41}\right)^2 = \frac{1600}{1681} + \frac{81}{1681} = 1$. So $\sin t = \frac{9}{41}$, $\cos t = \frac{40}{41}$, and $\tan t = \frac{\frac{9}{41}}{\frac{40}{41}} = \frac{9}{40}$.

31. $\left(-\frac{5}{13}\right)^2 + \left(-\frac{12}{13}\right)^2 = \frac{25}{169} + \frac{144}{169} = 1$. So $\sin t = -\frac{12}{13}$, $\cos t = -\frac{5}{13}$, and $\tan t = \frac{-\frac{12}{13}}{-\frac{5}{13}} = \frac{12}{5}$.

33. (a) 0.8 (b) 0.84147

35. (a) 0.9 (b) 0.93204

37. (a) 1.0 (b) 1.02964

39. (a) -0.6 (b) -0.57482

41. $\tan t \cdot \cos t = \sin t$ is positive in quadrant II.

43. $\dfrac{\tan t \cdot \sin t}{\cot t} = \tan t \cdot \dfrac{1}{\cot t} \cdot \sin t = \tan t \cdot \tan t \cdot \sin t = \tan^2 t \cdot \sin t$. Since $\tan^2 t$ is always positive and $\sin t$ is negative in quadrant III, the expression is negative in quadrant III.

45. quadrant II 47. quadrant II

49. $\sin t = \sqrt{1 - \cos^2 t}$

51. $\tan t = \dfrac{\sin t}{\cos t} = \dfrac{\sin t}{\sqrt{1 - \sin^2 t}}$

53. $\sec t = -\sqrt{1 + \tan^2 t}$

55. $\tan t = \sqrt{\sec^2 t - 1}$

57. $\tan^2 t = \dfrac{\sin^2 t}{\cos^2 t} = \dfrac{\sin^2 t}{1 - \sin^2 t}$

59. $\sin t = \frac{3}{5}$ and t is in quadrant II, so the terminal point determined by t is $P\left(x, \frac{3}{5}\right)$. Since P is on the unit circle $x^2 + \left(\frac{3}{5}\right)^2 = 1$. Solving for x gives $x = \pm\sqrt{1 - \frac{9}{25}} = \pm\sqrt{\frac{16}{25}} = \pm\frac{4}{5}$. Since t is in quadrant III, $x = -\frac{4}{5}$. Thus the terminal point is $P\left(-\frac{4}{5}, \frac{3}{5}\right)$. Thus, $\cos t = -\frac{4}{5}$, $\tan t = -\frac{3}{4}$, $\csc t = \frac{5}{3}$, $\sec t = -\frac{5}{4}$, $\cot t = -\frac{4}{3}$.

61. $\tan t = -\frac{3}{4}$ and $\cos t > 0$, so t is in quadrant IV. Since $\sec^2 t = \tan^2 t + 1$ we have $\sec^2 t = \left(-\frac{3}{4}\right)^2 + 1 = \frac{9}{16} + 1 = \frac{25}{16}$. Thus $\sec t = \pm\sqrt{\frac{25}{16}} = \pm\frac{5}{4}$. Since $\cos t > 0$, we have $\cos t = \dfrac{1}{\sec t} = \dfrac{1}{\frac{5}{4}} = \frac{4}{5}$. Let $P\left(\frac{4}{5}, y\right)$. Since $\tan t \cdot \cos t = \sin t$ we have $\sin t = \left(-\frac{3}{4}\right)\left(\frac{4}{5}\right) = -\frac{3}{5}$. Thus, the terminal point determined by t is $P\left(\frac{4}{5}, -\frac{3}{5}\right)$, and so $\sin t = -\frac{3}{5}$, $\cos t = \frac{4}{5}$, $\csc t = -\frac{5}{3}$, $\sec t = \frac{5}{4}$, $\cot t = -\frac{4}{3}$.

63. $\sec t = 2$ and $\sin t < 0$, so t is in quadrant IV. Thus, $\cos t = \frac{1}{2}$ and the terminal point determined by t is $P\left(\frac{1}{2}, y\right)$. Since P is on the unit circle $\left(\frac{1}{2}\right)^2 + y^2 = 1$. Solving for y gives $y = \pm\sqrt{1 - \frac{1}{4}} = \pm\sqrt{\frac{3}{4}} = \pm\frac{\sqrt{3}}{2}$. Since t is in quadrant IV, $y = -\frac{\sqrt{3}}{2}$. Thus the terminal point is $P\left(\frac{1}{2}, -\frac{\sqrt{3}}{2}\right)$, and so $\sin t = -\frac{\sqrt{3}}{2}$, $\cos t = \frac{1}{2}$, $\tan t = -\sqrt{3}$, $\csc t = -\frac{2}{\sqrt{3}} = -\frac{2\sqrt{3}}{3}$, $\cot t = -\frac{1}{\sqrt{3}} = -\frac{\sqrt{3}}{3}$.

65. $\sin t = -\frac{1}{4}$, $\sec t < 0$, so t is in quadrant III. So the terminal point determined by t is $P\left(x, -\frac{1}{4}\right)$. Since P is on the unit circle $x^2 + \left(-\frac{1}{4}\right)^2 = 1$. Solving for x gives $x = \pm\sqrt{1 - \frac{1}{16}} = \pm\sqrt{\frac{15}{16}} = \pm\frac{\sqrt{15}}{4}$. Since t is in quadrant III, $x = -\frac{\sqrt{15}}{4}$. Thus, the terminal point determined by t is $P\left(-\frac{\sqrt{15}}{4}, -\frac{1}{4}\right)$, and so $\cos t = -\frac{\sqrt{15}}{4}$, $\tan t = \frac{1}{\sqrt{15}} = \frac{\sqrt{15}}{15}$, $\csc t = -4$, $\sec t = -\frac{4}{\sqrt{15}} = -\frac{4\sqrt{15}}{15}$, $\cot t = \sqrt{15}$.

67. $f(-x) = (-x)^2 \sin(-x) = -x^2 \sin x = -f(x)$ so f is odd.

69. $f(-x) = \sin(-x)\cos(-x) = -\sin x \cos x = -f(x)$ so f is odd.

71. $f(-x) = |-x|\cos(-x) = |x|\cos x = f(x)$ so f is even.

73. $f(-x) = (-x)^3 + \cos(-x) = -x^3 + \cos x$ which is neither $f(x)$ nor $-f(x)$, so f is neither even nor odd.

75. Notice that if $P(t) = (x, y)$, then $P(t + \pi) = (-x, -y)$. Thus,
   (a) $\sin(t + \pi) = -y$, and $\sin t = y$. Therefore, $\sin(t + \pi) = -\sin t$.
   (b) $\cos(t + \pi) = -x$, and $\cos t = x$. Therefore, $\cos(t + \pi) = -\cos t$.
   (c) $\tan(t + \pi) = \dfrac{\sin(t + \pi)}{\cos(t + \pi)} = \dfrac{-y}{-x} = \dfrac{y}{x} = \dfrac{\sin t}{\cos t} = \tan t$.

Exercises 8.3

1. $y = 2 + \sin x$

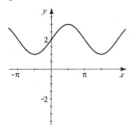

3. $y = 1 - \cos x$

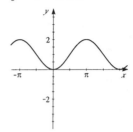

5. $y = 2 \cos x$

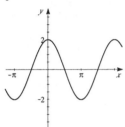

7. $y = 3 + 3 \cos x$

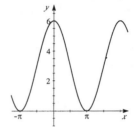

9. $y = |\sin x|$

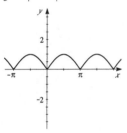

11. $y = 3 \sin 3x$

amplitude = 3, period = $\frac{2\pi}{3}$

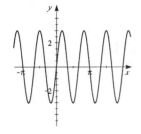

13. $y = 10 \sin \frac{1}{2} x$

amplitude = 10, period = 4π

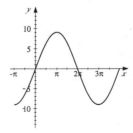

15. $y = -\cos \frac{1}{3}x$

amplitude $= 1$, period $= 6\pi$

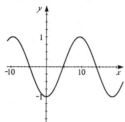

17. $y = 3\cos 3\pi x$

amplitude $= 3$, period $= \frac{2}{3}$

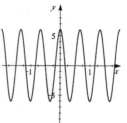

19. $y = \cos(x - \frac{\pi}{2})$

amplitude $= 1$, period $= 2\pi$,
phase shift $= \frac{\pi}{2}$

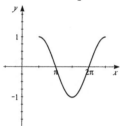

21. $y = -2\sin\left(x - \frac{\pi}{6}\right)$

amplitude $= 2$, period $= 2\pi$,
phase shift $= \frac{\pi}{6}$

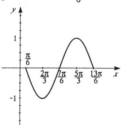

23. $y = 5\cos(3x - \frac{\pi}{4}) = 5\cos 3(x - \frac{\pi}{12})$
amplitude $= 5$, period $= \frac{2\pi}{3}$,
phase shift $= \frac{\pi}{12}$

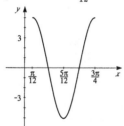

25. $y = 2\sin(\frac{2}{3}x - \frac{\pi}{6}) = 2\sin \frac{2}{3}(x - \frac{\pi}{4})$
amplitude $= 2$, period $= 3\pi$,
phase shift $= \frac{\pi}{4}$

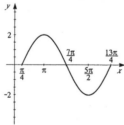

27. $y = 3\cos \pi(x + \frac{1}{2})$
amplitude $= 3$, period $= 2$,
phase shift $= -\frac{1}{2}$

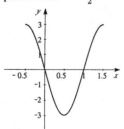

29. $y = -\frac{1}{2}\cos(2x - \frac{\pi}{3}) = -\frac{1}{2}\cos 2(x - \frac{\pi}{6})$
amplitude $= \frac{1}{2}$, period $= \pi$,
phase shift $= \frac{\pi}{6}$

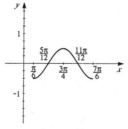

31. $y = \sin(3x + \pi) = \sin 3\left(x + \frac{\pi}{3}\right)$

amplitude $= 1$, period $= \frac{2\pi}{3}$,

phase shift $= -\frac{\pi}{3}$

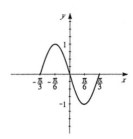

33. (a) amplitude $= a = 4$, period $= \frac{2\pi}{k} = 2\pi$, phase shift $= b = 0$

(b) $y = a \sin k(x - b) = 4 \sin x$

35. (a) amplitude $= a = 3$, period $= \frac{2\pi}{k} = 4\pi$, phase shift $= b = 0$

(b) $y = 3 \sin \frac{1}{2}x$

37. (a) amplitude $= a = \frac{1}{2}$, period $= \frac{2\pi}{k} = \pi$, phase shift $= b = -\frac{\pi}{3}$

(b) $y = -\frac{1}{2} \cos 2\left(x + \frac{\pi}{3}\right)$

39. $f(x) = \cos 100x$,

[-0.1, 0.1] by [-1.5, 1.5]

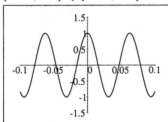

41. $f(x) = \sin \frac{x}{40}$,

[-250, 250] by [-1.5, 1.5]

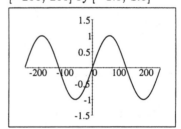

43. $y = \tan 25x$, [-0.2, 0.2] by [-3, 3]

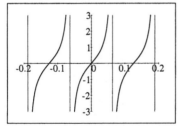

45. $y = e^{\sin 20x}$, [-0.5, 0.5] by [$0, 3$]

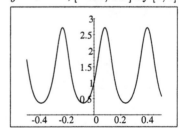

47. $f(x) = x$, $g(x) = \sin x$, [-6.28, 6.28] by [-7, 7]

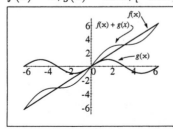

49. $y = x^2$, $y = -x^2$, $y = x^2 \sin x$

$y = x^2 \sin x$ is a sine curve that lies between
the graphs of $y = x^2$ and $y = -x^2$.

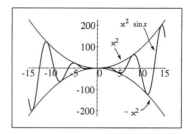

51. $y = e^x$, $y = -e^x$, $y = e^x \sin 5\pi x$

$y = e^x \sin 5\pi x$ is a sine curve that lies between
the graphs of $y = e^x$ and $y = -e^x$.

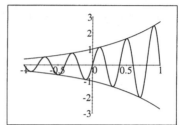

53. $y = \cos 3\pi x$, $y = -\cos 3\pi x$,
$y = \cos 3\pi x \cos 21\pi x$

$y = \cos 3\pi x \cos 21\pi x$ is a cosine curve that
lies between the graphs of $y = \cos 3\pi x$ and
$y = -\cos 3\pi x$.

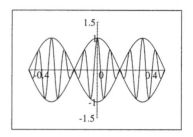

55. $y = |\sin x|$
(a) $[-6.28, 6.28]$ by $[-1.5, 1.5]$

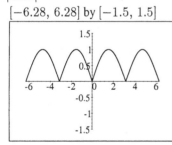

(b)   period $= \pi$

(c)   function is even

57. $y = e^{\sin x}$
(a) $[-6.28, 6.28]$ by $[-0.5, 3]$

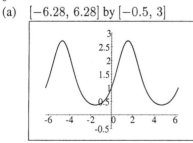

(b)   period $= 2\pi$

(c)   function is neither even nor odd

59. $y = \sin^2 x$

(a) $[-6.28, 6.28]$ by $[-0.5, 1.5]$

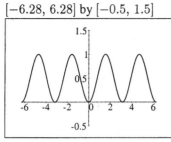

(b)   period = π

(c)   function is even

61. $y = \sin x + \sin 2x$. The period is 2π, so we graph the function over one period, $(-\pi, \pi)$. Maximum value 1.76 when $x \approx 0.94 + 2n\pi$, minimum value -1.76 when $x \approx -0.94 + 2n\pi$, for $n = 0, \pm 1, \pm 2, \ldots$.

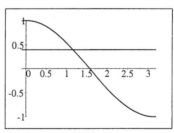

63. $y = 2\sin x + \sin^2 x$. he period is 2π, so we graph the function over one period, $(-\pi, \pi)$. Maximum value 3.00 when $x \approx 1.57 + 2n\pi$, minimum value -1.00 when $x \approx -1.57 + 2n\pi$, for $n = 0, \pm 1, \pm 2, \ldots$.

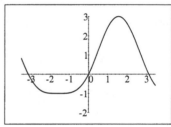

65. $\cos x = 0.4$, $x \in [0, \pi]$

$x \approx 1.16$

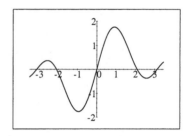

67. $\csc x = 3$, $x \in [0, \pi]$

$x \approx 0.34, 2.80$

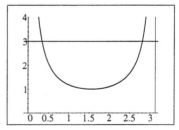

69. $f(x) = \dfrac{1 - \cos x}{x}$ (c)

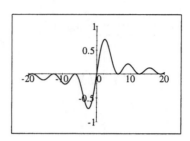

   (a) Since $f(-x) = \dfrac{1 - \cos(-x)}{-x}$

        $= \dfrac{1 - \cos x}{-x} = -f(x)$, the function is

        odd.

   (b) The x-intercepts occur when
        $1 - \cos x = 0 \quad \Leftrightarrow \quad \cos x = 1 \quad \Leftrightarrow$
        $x = 0, \pm 2\pi, \pm 4\pi, \pm 6\pi, \ldots$

   (d) As $x \to \pm\infty,\, f(x) \to 0$.

   (e) As $x \to 0,\, f(x) \to 0$.

71. From the graph we see that the amplitude is 5 and the period is 2π. In Exercise 75 in Section 8.2, we proved the reduction formulas $\sin(x + \pi) = -\sin x$. So starting with $y = -5\sin x$ we may also represent the curve by:

     $y = -5\sin(x - b)$ where $b = 2n\pi$, where n is an integer;

     $y = 5\sin(x - b)$ where $b = (2n + 1)\pi$, where n is an integer;

     We also proved the identity $\cos(x + \pi) = -\cos x$. So starting with $y = -5\cos\left(x - \frac{\pi}{2}\right)$ we may also represent the curve by:

     $y = -5\cos(x - b)$ where $b = \frac{\pi}{2} + 2n\pi$, where n is an integer;

     $y = 5\cos(x - b)$ where $b = \frac{\pi}{2} + (2n + 1)\pi$, where n is an integer.

Exercises 8.4

1. $y = 3\tan x$, period $= \pi$

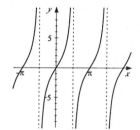

3. $y = \frac{1}{2}\tan x$, period $= \pi$

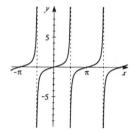

5. $y = 4\cot x$, period $= \pi$

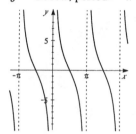

7. $y = 2\csc x$, period $= 2\pi$

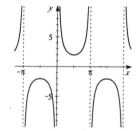

9. $y = 4\sec x$, period $= 2\pi$

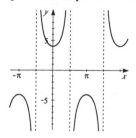

11. $y = \tan\left(x + \frac{\pi}{2}\right)$, period $= \pi$

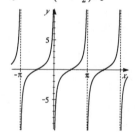

13. $y = \csc\left(x + \frac{\pi}{2}\right)$, period $= 2\pi$

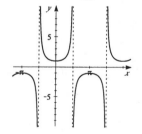

15. $y = \cot\left(x + \frac{\pi}{4}\right)$, period $= \pi$

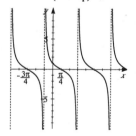

17. $y = \frac{1}{2}\sec\left(x - \frac{\pi}{6}\right)$, period $= 2\pi$

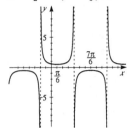

19. $y = \tan 2x$, period $= \frac{\pi}{2}$

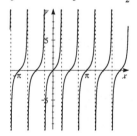

21. $y = \tan(\pi x)$, period $= \frac{\pi}{\pi} = 1$

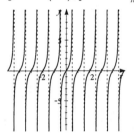

23. $y = \sec 2x$, period $= \frac{2\pi}{2} = \pi$

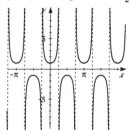

25. $y = \csc 2x$, period $= \frac{2\pi}{2} = \pi$

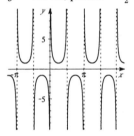

27. $y = 2\tan 3x$, period $= \frac{\pi}{3}$

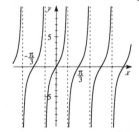

29. $y = 5\csc 3x$, period $= \frac{2\pi}{3}$

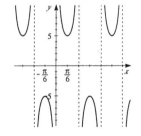

31. $y = \tan 2\left(x + \frac{\pi}{2}\right)$, period $= \frac{\pi}{2}$

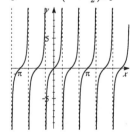

33. $y = \tan 2(x - \pi) = \tan 2x$, period $= \frac{\pi}{2}$

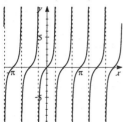

35. $y = \cot(2x - \frac{\pi}{2}) = \cot 2(x - \frac{\pi}{4})$, period $= \frac{\pi}{2}$

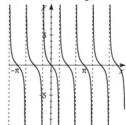

37. $y = 2\csc(\pi x - \frac{\pi}{3}) = 2\csc \pi(x - \frac{1}{3})$,
    period $= \frac{2\pi}{\pi} = 2$

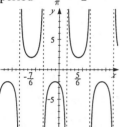

39. $y = 5\sec(3x - \frac{\pi}{2}) = 5\sec 3(x - \frac{\pi}{6})$,
    period $= \frac{2\pi}{3}$

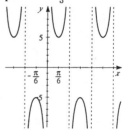

41. $y = \tan(\frac{2}{3}x - \frac{\pi}{6}) = \tan \frac{2}{3}(x - \frac{\pi}{4})$,
    period $= \dfrac{\pi}{\left(\frac{2}{3}\right)} = \dfrac{3\pi}{2}$

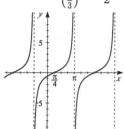

43. $y = 3\sec\pi(x + \frac{1}{2})$, period $= \frac{2\pi}{\pi} = 2$

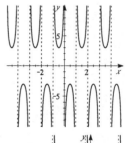

45. $y = -2\tan(2x - \frac{\pi}{3}) = -2\tan 2(x - \frac{\pi}{6})$,
    period $= \frac{\pi}{2}$

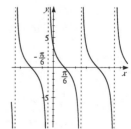

47. (a) If f is periodic with period p, then by the definition of a period, $f(x + p) = f(x)$ for all x in
        the domain of f. Therefore, $\dfrac{1}{f(x + p)} = \dfrac{1}{f(x)}$ for all $f(x) \neq 0$. Thus, $\dfrac{1}{f}$ is also periodic
        with period p.

    (b) Since $\sin x$ has period 2π, it follows from part (a) that $\csc x = \dfrac{1}{\sin x}$ also has period 2π.

        Similarly, since $\cos x$ has period 2π, we conclude $\sec x = \dfrac{1}{\cos x}$ also has period 2π.

Review Exercises for Chapter 8

1. (a) Since $\left(\frac{1}{2}\right)^2 + \left(\frac{\sqrt{3}}{2}\right)^2 = \frac{1}{4} + \frac{3}{4} = 1$, the point $P\left(\frac{1}{2}, \frac{\sqrt{3}}{2}\right)$ lies on the unit circle.

    (b) $\sin t = \frac{\sqrt{3}}{2}$, $\cos t = \frac{1}{2}$, $\tan t = \frac{\frac{\sqrt{3}}{2}}{\frac{1}{2}} = \sqrt{3}$

3. $t = \frac{\pi}{3}$

    (a) $\bar{t} = \frac{\pi}{3}$

    (b) $P\left(\frac{1}{2}, \frac{\sqrt{3}}{2}\right)$

    (c) $\sin t = \frac{\sqrt{3}}{2}$, $\cos t = \frac{1}{2}$, $\tan t = \sqrt{3}$, $\csc t = \frac{2\sqrt{3}}{3}$, $\sec t = 2$, and $\cot t = \frac{\sqrt{3}}{3}$.

5. $t = \frac{11\pi}{4}$

    (a) $\bar{t} = 3\pi - \frac{11\pi}{4} = \frac{\pi}{4}$

    (b) $P\left(-\frac{\sqrt{2}}{2}, \frac{\sqrt{2}}{2}\right)$

    (c) $\sin t = \frac{\sqrt{2}}{2}$, $\cos t = -\frac{\sqrt{2}}{2}$, $\tan t = -1$, $\csc t = \sqrt{2}$, $\sec t = -\sqrt{2}$, $\cot t = -1$.

7. (a) $\sin \frac{3\pi}{4} = \sin \frac{\pi}{4} = \frac{\sqrt{2}}{2}$ (b) $\cos \frac{3\pi}{4} = -\cos \frac{\pi}{4} = -\frac{\sqrt{2}}{2}$

9. (a) $\sin 1.1 \approx 0.89121$ (b) $\cos 1.1 \approx 0.45360$

11. (a) $\cos \frac{9\pi}{2} = \cos \frac{\pi}{2} = 0$ (b) $\sec \frac{9\pi}{2}$ is undefined

13. (a) $\tan \frac{5\pi}{2}$ is undefined   (b) $\cot \frac{5\pi}{2} = \cot \frac{\pi}{2} = 0$

15. (a) $\tan \frac{\pi}{8} \approx 0.41421$ (b) $\cot \frac{\pi}{8} \approx 2.41421$

17. $\dfrac{\tan t}{\cos t} = \dfrac{\frac{\sin t}{\cos t}}{\cos t} = \dfrac{\sin t}{\cos^2 t} = \dfrac{\sin t}{1 - \sin^2 t}$

19. $\tan t = \dfrac{\sin t}{\cos t} = \dfrac{\sin t}{\pm\sqrt{1 - \sin^2 t}} = \dfrac{\sin t}{\sqrt{1 - \sin^2 t}}$ (since t is in quadrant IV, $\cos t$ is positive)

21. $\sin t = \frac{5}{13}$, $\cos t = -\frac{12}{13}$. Then $\tan t = \dfrac{\frac{5}{13}}{-\frac{12}{13}} = -\frac{5}{12}$, $\csc t = \frac{13}{5}$, $\sec t = -\frac{13}{12}$, and $\cot t = -\frac{12}{5}$.

23. $\cot t = -\frac{1}{2}$, $\csc t = \frac{\sqrt{5}}{2}$. Since $\csc t = \dfrac{1}{\sin t}$, we know $\sin t = \frac{2}{\sqrt{5}} = \frac{2\sqrt{5}}{5}$. Now $\cot t = \dfrac{\cos t}{\sin t}$, so

    $\cos t = \sin t \cdot \cot t = \frac{2\sqrt{5}}{5} \cdot \left(-\frac{1}{2}\right) = -\frac{\sqrt{5}}{5}$, and $\tan t = \dfrac{1}{\left(-\frac{1}{2}\right)} = -2$ while

    $\sec t = \dfrac{1}{\cos t} = \dfrac{1}{\left(-\frac{\sqrt{5}}{5}\right)} = -\frac{5}{\sqrt{5}} = -\sqrt{5}$.

25. $\tan t = \frac{1}{4}$, t is in quadrant III $\Rightarrow$ $\sec t + \cot t = -\sqrt{\tan^2 t + 1} + \dfrac{1}{\tan t} = -\sqrt{\left(\frac{1}{4}\right)^2 + 1} + 4$

    $= -\sqrt{\frac{17}{16}} + 4 = 4 - \frac{\sqrt{17}}{4} = \frac{16 - \sqrt{17}}{4}$

27. $\cos t = \frac{3}{5}$, t is in quadrant I $\Rightarrow$ $\tan t + \sec t = \dfrac{\sin t}{\cos t} + \dfrac{1}{\cos t} = \dfrac{\sqrt{1 - \cos^2 t}}{\cos t} + \dfrac{1}{\cos t}$

$$= \dfrac{\sqrt{1 - \left(\frac{3}{5}\right)^2}}{\frac{3}{5}} + \dfrac{5}{3} = \dfrac{\sqrt{\frac{16}{25}}}{\frac{3}{5}} + \dfrac{5}{3} = \dfrac{4}{5} \cdot \dfrac{5}{3} + \dfrac{5}{3} = \dfrac{9}{3} = 3$$

29. $y = 10 \cos \frac{1}{2} x$

     (a)   amplitude $= 10$, period $= \dfrac{2\pi}{\left(\frac{1}{2}\right)} = 4\pi,$ (b)

         phase shift $= 0$

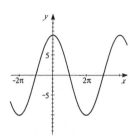

31. $y = -\sin \frac{1}{2} x$

     (a)   amplitude $= 1$, period $= \dfrac{2\pi}{\left(\frac{1}{2}\right)} = 4\pi,$ (b)

         phase shift $= 0$

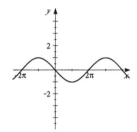

33. $y = 3 \sin(2x - 2) = 3 \sin 2(x - 1)$

     (a)   amplitude $= 3$, period $= \frac{2\pi}{2} = \pi,$ (b)

         phase shift $= 1$

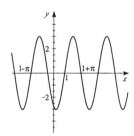

35. $y = -\cos\left(\frac{\pi}{2} x + \frac{\pi}{6}\right) = -\cos \frac{\pi}{2}\left(x + \frac{1}{3}\right)$

     (a)   amplitude $= 1$, period $= \dfrac{2\pi}{\left(\frac{\pi}{2}\right)} = 4,$ (b)

         phase shift $= -\frac{1}{3}$

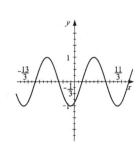

37.   From the graph we see that the amplitude is 5, the period is $\frac{\pi}{2}$, and there is no phase shift. Therefore, the function is $y = 5 \sin 4x$.

39. From the graph we see that the amplitude is $\frac{1}{2}$, the period is 1, and there is a phase shift of $-\frac{1}{3}$. Therefore, the function is $y = \frac{1}{2}\sin 2\pi(x + \frac{1}{3})$.

41. $y = 3\tan x$, period $= \pi$

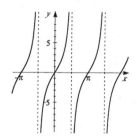

43. $y = 2\cot\left(x - \frac{\pi}{2}\right)$, period $= \pi$

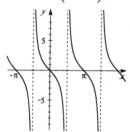

45. $y = 4\csc(2x - \pi) = 4\csc 2(x - \frac{\pi}{2})$,

   period $= \frac{2\pi}{2} = \pi$

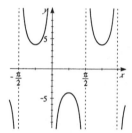

47. $y = \tan\left(\frac{1}{2}x - \frac{\pi}{8}\right) = \tan\frac{1}{2}(x - \frac{\pi}{4})$,

   period $= \dfrac{\pi}{\left(\frac{1}{2}\right)} = 2\pi$

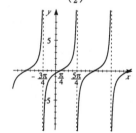

49. (a) $y = |\cos x|$

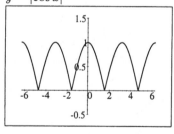

   (b) function has period π
   (c) function is even

51. (a) $y = \cos(2^{0.1x})$

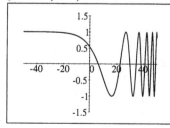

   (b) function is not periodic
   (c) function is neither even nor odd

53. (a) $y = e^{-|x|}\cos 3x$

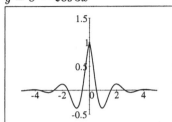

(b) function is not periodic

(c) function is even

55. $y = x$, $y = -x$, $y = x\sin x$

$y = x\sin x$ is a sine function whose graph lies between those of $y = x$ and $y = -x$.

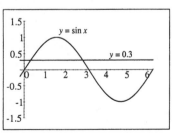

57. $y = x$, $y = \sin 4x$, $y = x + \sin 4x$

$y = x + \sin 4x$ is the sum of the two functions $y = x$ and $y = \sin 4x$.

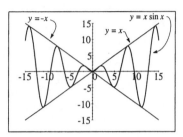

59. $y = \cos x + \sin 2x$

Since the period is 2π, we graph over the interval $-\pi$ to π. The maximum value is 1.76 when $x \approx 0.63 \pm 2n\pi$, the minimum value is -1.76 when $x \approx 2.51 \pm 2n\pi$, n an integer.

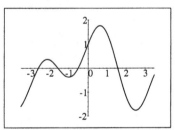

61. We want to find solutions to $\sin x = 0.3$ in the interval $[0, 2\pi]$, so we plot the functions $y = \sin x$ and $y = 0.3$ and look for their intersection. We see that $x \approx 0.305$ or $x \approx 2.837$.

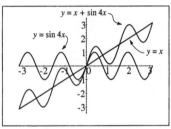

63. $f(x) = \dfrac{\sin^2 x}{x}$

(a) The function is odd.

(b) The graph intersects the x-axis at $x = 0$, $\pm\pi$, $\pm 2\pi$, $\pm 3\pi$, ...

(c)

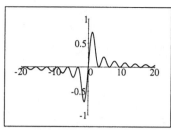

(d)   As $x \to \pm\infty$, $f(x) \to 0$.

(e)   As $x \to 0$, $f(x) \to 0$.

Chapter 8 Test

1. Since $P(x, y)$ lies on the unit circle, $x^2 + y^2 = 1 \Rightarrow y = \pm\sqrt{1 - \left(\frac{12}{13}\right)^2} = \pm\sqrt{\frac{25}{169}} = \pm\frac{5}{13}$.
   But $P(x, y)$ lies in the fourth quadrant. Therefore y is negative $\Rightarrow y = -\frac{5}{13}$.

2. Since P is on the unit circle, $x^2 + y^2 = 1 \Leftrightarrow x^2 = 1 - y^2$. Thus, $x^2 = 1 - \left(\frac{4}{5}\right)^2 = \frac{9}{25}$, and so
   $x = \pm\frac{3}{5}$. From the diagram, x is clearly negative $\Rightarrow x = -\frac{3}{5}$. Therefore, P is the point
   $\left(-\frac{3}{5}, \frac{4}{5}\right)$.

   (a) $\sin t = \frac{4}{5}$.

   (b) $\cos(-t) = \cos t = -\frac{3}{5}$.

   (c) $\tan(\pi - t) = \tan(-t) = -\tan t = \dfrac{-\frac{4}{5}}{-\frac{3}{5}} = \frac{4}{3}$.

   (d) $\sec(\pi + t) = \dfrac{1}{\cos(\pi + t)} = \dfrac{1}{-\cos t} = \dfrac{1}{-\left(-\frac{3}{5}\right)} = \frac{5}{3}$.

3. $\tan t = \dfrac{\sin t}{\cos t} = \dfrac{\sin t}{\pm\sqrt{1 - \sin^2 t}}$. But t is in quadrant II $\Rightarrow \cos t$ is negative, so we choose the
   negative square root. Thus, $\tan t = \dfrac{\sin t}{-\sqrt{1 - \sin^2 t}}$.

4. $\cos t = -\frac{8}{17}$, t in quadrant III $\Rightarrow \tan t \cdot \cot t + \csc t = 1 + \dfrac{1}{-\sqrt{1 - \cos^2 t}}$ (since t in quadrant
   III) $= 1 - \dfrac{1}{-\sqrt{1 - \frac{64}{289}}} = 1 - \dfrac{1}{\frac{15}{17}} = -\dfrac{2}{15}$.

5. $y = 2\cos 3x$

   (a) amplitude $= 2$, period $= \frac{2\pi}{3}$, no phase
   shift.

   (b)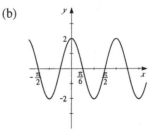

6. $y = \sin(\frac{1}{2}x - \frac{\pi}{6}) = \sin\frac{1}{2}\left(x - \frac{\pi}{3}\right)$

   (a) amplitude $= 1$, period $= \dfrac{2\pi}{\left(\frac{1}{2}\right)} = 4\pi$,

   phase shift $= \frac{\pi}{3}$.

   (b)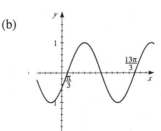

7. $y = -\csc 2x$, period $= \frac{2\pi}{2} = \pi$

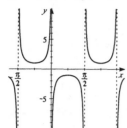

8. $y = \tan 2\left(x - \frac{\pi}{4}\right)$, period $= \frac{\pi}{2}$

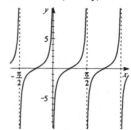

9. From the graph we see that the amplitude is 2 and the phase shift is $-\frac{\pi}{3}$. Also, the period is π so $\frac{2\pi}{k} = \pi \quad \Rightarrow \quad k = \frac{2\pi}{\pi} = 2$. Thus, the function is $y = 2\sin 2\left(x + \frac{\pi}{3}\right)$.

10. $y = \dfrac{\cos x}{1 + x^2}$

    (a)

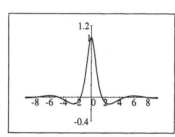

    (b) function is even
    (c) minimum value -0.11 when $x \approx \pm 2.54$, maximum value 1 when $x = 0$.

11. $y = e^{-\sin 3x}$

    (a)

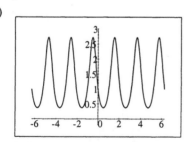

    (b) function is neither even nor odd
    (c) The function has period $\frac{2\pi}{3}$. The minimum value is 0.37 when $x = \frac{\pi}{6} + \frac{2n\pi}{3}$, and the maximum value is 2.72 when

    $$x = -\frac{\pi}{6} + \frac{2n\pi}{3}, \text{ for } n \text{ an integer.}$$

12. $y_1 = \cos(\sin x)$, $y_2 = \sin(\cos x)$

    (a)

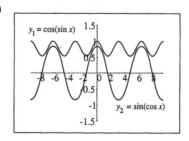

    (b) y_1 has period π, while y_2 has period 2π
    (c) $\sin(\cos x) < \cos(\sin x)$ for all x.

Focus on Modeling

1. The graph resembles a sine wave with an amplitude of 5, a period of $\frac{2}{5}$, and no phase shift.
   Therefore, $a = 5$, $\dfrac{2\pi}{\omega} = \frac{2}{5}$ $\Leftrightarrow$ $\omega = 5\pi$, and the formula is $d(t) = 5\sin 5\pi t$.

3. $a = 21$, $f = \frac{1}{12}$ cycle/hour $\Rightarrow$ $\dfrac{\omega}{2\pi} = \frac{1}{12}$
   $\Leftrightarrow$ $\omega = \frac{\pi}{6}$. So, $y = 21\sin\left(\frac{\pi}{6}t\right)$ (assuming
   the tide is at mean level and rising when
   $t = 0$).

   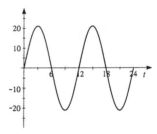

5. Since the mass travels from its highest point (compressed spring) to its lowest point in $\frac{1}{2}$ s, it
   completes half a period in $\frac{1}{2}$ s. So, $\frac{1}{2}$(one period) $= \frac{1}{2}$ s $\Rightarrow$ $\frac{1}{2} \cdot \dfrac{2\pi}{\omega} = \frac{1}{2}$ $\Leftrightarrow$ $\omega = 2\pi$. Also,
   $a = 5$. So $y = 5\cos 2\pi t$.

7. Since the ferris wheel has a radius of 10 m and the bottom of the wheel is 1 m above the ground, the
   minimum height is 1 m and the maximum height is 21 m. Then $a = 10$ and $\dfrac{2\pi}{\omega} = 20$ sec $\Leftrightarrow$
   $\omega = \frac{\pi}{10}$, and so $y = 11 + 10\sin\frac{\pi}{10}t$, where t is in seconds.

9. $a = 0.2$, $\dfrac{2\pi}{\omega} = 10$ $\Leftrightarrow$ $\omega = \frac{\pi}{5}$. Then $y = 3.8 + 0.2\sin\left(\frac{\pi}{5}t\right)$.

11. $E_0 = 155$, $\dfrac{\omega}{2\pi} = 60$ $\Leftrightarrow$ $\omega = 120\pi$. Thus $E(t) = E_0\cos\omega t = 155\cos 120\pi t$.

13. (a) The maximum voltage is the amplitude, that is, $V_{max} = a = 45$ V.

    (b) From the graph we see that 4 cycles are completed every 0.1 seconds, or equivalently, 40 cycles
        are completed every second, so $f = 40$.

    (c) The number of revolutions per second of the armature is the frequency, that is, $\dfrac{\omega}{2\pi} = f = 40$.

    (d) $a = 45$, $f = \dfrac{\omega}{2\pi} = 40$ $\Leftrightarrow$ $\omega = 80\pi$. Then $V(t) = 45\cos 80\pi t$.

15. $k = 1$, $c = 0.9$, and $\dfrac{\omega}{2\pi} = \frac{1}{2}$ $\Leftrightarrow$ $\omega = \pi$. Since $f(0) = 0$, $f(t) = e^{-0.9t}\sin\pi t$.

17. $\dfrac{ke^{-ct}}{ke^{-c(t+3)}} = 4$ $\Leftrightarrow$ $e^{-ct+c(t+3)} = 4$ $\Leftrightarrow$ $e^{3c} = 4$ $\Leftrightarrow$ $3c = \ln 4$ $\Leftrightarrow$ $c = \frac{1}{3}\ln 4 \approx 0.46$.

Chapter Nine
Exercises 9.1

1. $\cos x \tan x = \cos x \cdot \dfrac{\sin x}{\cos x} = \sin x$

3. $\sec^2 x - \tan^2 x = \dfrac{1}{\cos^2 x} - \dfrac{\sin^2 x}{\cos^2 x} = \dfrac{1 - \sin^2 x}{\cos^2 x} = \dfrac{\cos^2 x}{\cos^2 x} = 1$

5. $\cos u + \tan u \sin u = \cos u + \dfrac{\sin u}{\cos u} \sin u = \dfrac{\cos^2 u}{\cos u} + \dfrac{\sin^2 u}{\cos u} = \dfrac{1}{\cos u} = \sec u$

7. $\dfrac{\cos x \sec x}{\cot x} = \dfrac{\cos x \frac{1}{\cos x}}{\cot x} = \dfrac{1}{\cot x} = \tan x$

9. $\dfrac{1 + \sin y}{1 + \csc y} = \dfrac{1 + \sin y}{1 + \dfrac{1}{\sin y}} = \dfrac{1 + \sin y}{\dfrac{\sin y + 1}{\sin y}} = \dfrac{1 + \sin y}{1} \cdot \dfrac{\sin y}{\sin y + 1} = \sin y$

11. $\dfrac{\sec^2 x - 1}{\sec^2 x} = \dfrac{\tan^2 x}{\sec^2 x} = \dfrac{\sin^2 x}{\cos^2 x} \cdot \cos^2 x = \sin^2 x$

   Alternative approach: $\dfrac{\sec^2 x - 1}{\sec^2 x} = 1 - \dfrac{1}{\sec^2 x} = 1 - \cos^2 x = \sin^2 x$

13. $\dfrac{1 + \csc x}{\cos x + \cot x} = \dfrac{1 + \dfrac{1}{\sin x}}{\cos x + \dfrac{\cos x}{\sin x}} = \dfrac{1 + \dfrac{1}{\sin x}}{\cos x + \dfrac{\cos x}{\sin x}} \cdot \dfrac{\sin x}{\sin x} = \dfrac{\sin x + 1}{\cos x(\sin x + 1)} = \dfrac{1}{\cos x} = \sec x$

15. $\dfrac{1 + \sin u}{\cos u} + \dfrac{\cos u}{1 + \sin u} = \dfrac{(1 + \sin u)^2 + \cos^2 u}{\cos u(1 + \sin u)} = \dfrac{1 + 2\sin u + \sin^2 u + \cos^2 u}{\cos u(1 + \sin u)} = \dfrac{1 + 2\sin u + 1}{\cos u(1 + \sin u)}$

   $= \dfrac{2 + 2\sin u}{\cos u(1 + \sin u)} = \dfrac{2(1 + \sin u)}{\cos u(1 + \sin u)} = \dfrac{2}{\cos u} = 2\sec u$

17. $\dfrac{2 + \tan^2 x}{\sec^2 x} - 1 = \dfrac{1 + 1 + \tan^2 x}{\sec^2 x} - 1 = \dfrac{1}{\sec^2 x} + \dfrac{1 + \tan^2 x}{\sec^2 x} - 1 = \dfrac{1}{\sec^2 x} + \dfrac{\sec^2 x}{\sec^2 x} - 1$

   $= \dfrac{1}{\sec^2 x} + 1 - 1 = \dfrac{1}{\sec^2 x} = \cos^2 x$

19. $\tan \theta + \cos(-\theta) + \tan(-\theta) = \tan \theta + \cos \theta - \tan \theta = \cos \theta$

21. $\sin \theta \cot \theta = \sin \theta \dfrac{\cos \theta}{\sin \theta} = \cos \theta$

23. $\dfrac{\cos u \sec u}{\tan u} = \cos u \dfrac{1}{\cos u} \cot u = \cot u$

25. $\dfrac{\tan y}{\csc y} = \dfrac{\sin y}{\cos y} \sin y = \dfrac{\sin^2 y}{\cos y} = \dfrac{1 - \cos^2 y}{\cos y} = \sec y - \cos y$

27. $\sin B + \cos B \cot B = \sin B + \cos B \dfrac{\cos B}{\sin B} = \dfrac{\sin^2 B + \cos^2 B}{\sin B} = \dfrac{1}{\sin B} = \csc B$

29. $\cot(-\alpha)\cos(-\alpha) + \sin(-\alpha) = -\dfrac{\cos\alpha}{\sin\alpha}\cos\alpha - \sin\alpha = \dfrac{-\cos^2\alpha - \sin^2\alpha}{\sin\alpha} = \dfrac{-1}{\sin\alpha} = -\csc\alpha$

31. $(1 - \sin x)(1 + \sin x) = 1 - \sin^2 x = \cos^2 x$

33. $(1 - \cos\beta)(1 + \cos\beta) = 1 - \cos^2\beta = \sin^2\beta = \dfrac{1}{\csc^2\beta}$

35. $\dfrac{(\sin x + \cos x)^2}{\sin^2 x - \cos^2 x} = \dfrac{(\sin x + \cos x)^2}{(\sin x + \cos x)(\sin x - \cos x)} = \dfrac{\sin x + \cos x}{\sin x - \cos x}$

$= \dfrac{(\sin x + \cos x)(\sin x - \cos x)}{(\sin x - \cos x)(\sin x - \cos x)} = \dfrac{\sin^2 x - \cos^2 x}{(\sin x - \cos x)^2}$

37. $\dfrac{\sec t - \cos t}{\sec t} = \dfrac{\dfrac{1}{\cos t} - \cos t}{\dfrac{1}{\cos t}} = \dfrac{\dfrac{1}{\cos t} - \cos t}{\dfrac{1}{\cos t}} \cdot \dfrac{\cos t}{\cos t} = \dfrac{1 - \cos^2 t}{1} = \sin^2 t$

39. $\dfrac{1}{1 - \sin^2 y} = \dfrac{1}{\cos^2 y} = \sec^2 y = 1 + \tan^2 y$

41. $(\cot x - \csc x)(\cos x + 1) = \cot x \cos x + \cot x - \csc x \cos x - \csc x$

$= \dfrac{\cos^2 x}{\sin x} + \dfrac{\cos x}{\sin x} - \dfrac{\cos x}{\sin x} - \dfrac{1}{\sin x} = \dfrac{\cos^2 x - 1}{\sin x} = \dfrac{-\sin^2 x}{\sin x} = -\sin x$

43. $(1 - \cos^2 x)(1 + \cot^2 x) = \sin^2 x\left(1 + \dfrac{\cos^2 x}{\sin^2 x}\right) = \sin^2 x + \cos^2 x = 1$

45. $2\cos^2 x - 1 = 2(1 - \sin^2 x) - 1 = 2 - 2\sin^2 x - 1 = 1 - 2\sin^2 x$

47. $\dfrac{1 - \cos\alpha}{\sin\alpha} = \dfrac{1 - \cos\alpha}{\sin\alpha} \cdot \dfrac{1 + \cos\alpha}{1 + \cos\alpha} = \dfrac{1 - \cos^2\alpha}{\sin\alpha(1 + \cos\alpha)} = \dfrac{\sin^2\alpha}{\sin\alpha(1 + \cos\alpha)} = \dfrac{\sin\alpha}{1 + \cos\alpha}$

49. $\dfrac{\sin x - 1}{\sin x + 1} = \dfrac{\sin x - 1}{\sin x + 1} \cdot \dfrac{\sin x + 1}{\sin x + 1} = \dfrac{\sin^2 x - 1}{(\sin x + 1)^2} = \dfrac{-\cos^2 x}{(\sin x + 1)^2}$

51. $\dfrac{(\sin t + \cos t)^2}{\sin t \cos t} = \dfrac{\sin^2 t + 2\sin t\cos t + \cos^2 t}{\sin t \cos t} = \dfrac{\sin^2 t + \cos^2 t}{\sin t \cos t} + \dfrac{2\sin t \cos t}{\sin t \cos t} = \dfrac{1}{\sin t\cos t} + 2$

$= 2 + \sec t\cos t$

53. $\dfrac{1 + \tan^2 u}{1 - \tan^2 u} = \dfrac{1 + \dfrac{\sin^2 u}{\cos^2 u}}{1 - \dfrac{\sin^2 u}{\cos^2 u}} = \dfrac{1 + \dfrac{\sin^2 u}{\cos^2 u}}{1 - \dfrac{\sin^2 u}{\cos^2 u}} \cdot \dfrac{\cos^2 u}{\cos^2 u} = \dfrac{\cos^2 u + \sin^2 u}{\cos^2 u - \sin^2 u} = \dfrac{1}{\cos^2 u - \sin^2 u}$

55. $\dfrac{\sec x}{\sec x - \tan x} = \dfrac{\sec x}{\sec x - \tan x} \cdot \dfrac{\sec x + \tan x}{\sec x + \tan x} = \dfrac{\sec x(\sec x + \tan x)}{\sec^2 x - \tan^2 x} = \dfrac{\sec x(\sec x + \tan x)}{1}$

$= \sec x(\sec x + \tan x)$

57. $\sec v - \tan v = (\sec v - \tan v) \cdot \dfrac{\sec v + \tan v}{\sec v + \tan v} = \dfrac{\sec^2 v - \tan^2 v}{\sec v + \tan v} = \dfrac{1}{\sec v + \tan v}$

59. $\dfrac{\sin x + \cos x}{\sec x + \csc x} = \dfrac{\sin x + \cos x}{\dfrac{1}{\cos x} + \dfrac{1}{\sin x}} = \dfrac{\sin x + \cos x}{\dfrac{\sin x + \cos x}{\cos x \sin x}} = (\sin x + \cos x)\dfrac{\cos x \sin x}{\sin x + \cos x} = \cos x \sin x$

61. $\dfrac{\csc x - \cot x}{\sec x - 1} = \dfrac{\dfrac{1}{\sin x} - \dfrac{\cos x}{\sin x}}{\dfrac{1}{\cos x} - 1} = \dfrac{\dfrac{1}{\sin x} - \dfrac{\cos x}{\sin x}}{\dfrac{1}{\cos x} - 1} \cdot \dfrac{\sin x \cos x}{\sin x \cos x} = \dfrac{\cos x(1 - \cos x)}{\sin x(1 - \cos x)} = \dfrac{\cos x}{\sin x} = \cot x$

63. $\tan^2 u - \sin^2 u = \dfrac{\sin^2 u}{\cos^2 u} - \dfrac{\sin^2 u \cos^2 u}{\cos^2 u} = \dfrac{\sin^2 u}{\cos^2 u}(1 - \cos^2 u) = \tan^2 u \sin^2 u$

65. $\sec^4 x - \tan^4 x = (\sec^2 x - \tan^2 x)(\sec^2 x + \tan^2 x) = 1\,(\sec^2 x + \tan^2 x) = \sec^2 x + \tan^2 x$

67. $\dfrac{\sin\theta - \csc\theta}{\cos\theta - \cot\theta} = \dfrac{\sin\theta - \dfrac{1}{\sin\theta}}{\cos\theta - \dfrac{\cos\theta}{\sin\theta}} = \dfrac{\dfrac{\sin^2\theta - 1}{\sin\theta}}{\dfrac{\cos\theta\sin\theta - \cos\theta}{\sin\theta}} = \dfrac{\cos^2\theta}{\cos\theta(\sin\theta - 1)} = \dfrac{\cos\theta}{\sin\theta - 1}$

69. $\dfrac{\cos^2 t + \tan^2 t - 1}{\sin^2 t} = \dfrac{-\sin^2 t + \tan^2 t}{\sin^2 t} = -1 + \dfrac{\sin^2 t}{\cos^2 t}\cdot\dfrac{1}{\sin^2 t} = -1 + \sec^2 t = \tan^2 t$

71. $\dfrac{1}{\sec x + \tan x} + \dfrac{1}{\sec x - \tan x} = \dfrac{\sec x - \tan x + \sec x + \tan x}{(\sec x + \tan x)(\sec x - \tan x)} = \dfrac{2\sec x}{\sec^2 x - \tan^2 x}$

$= \dfrac{2\sec x}{1} = 2\sec x$

73. $(\tan x + \cot x)^2 = \tan^2 x + 2\tan x \cot x + \cot^2 x = \tan^2 x + 2 + \cot^2 x = (\tan^2 x + 1) + (\cot^2 x + 1)$
$= \sec^2 x + \csc^2 x$

75. $\dfrac{\sec u - 1}{\sec u + 1} = \dfrac{\frac{1}{\cos u} - 1}{\frac{1}{\cos u} + 1} \cdot \dfrac{\cos u}{\cos u} = \dfrac{1 - \cos u}{1 + \cos u}$

77. $\dfrac{\sin^3 x + \cos^3 x}{\sin x + \cos x} = \dfrac{(\sin x + \cos x)(\sin^2 x - \sin x \cos x + \cos^2 x)}{\sin x + \cos x} = \sin^2 - \sin x \cos x + \cos^2 x$

$= 1 - \sin x \cos x$

79. $\dfrac{1 + \sin x}{1 - \sin x} = \dfrac{1 + \sin x}{1 - \sin x} \cdot \dfrac{1 + \sin x}{1 + \sin x} = \dfrac{(1 + \sin x)^2}{1 - \sin^2 x} = \dfrac{(1 + \sin x)^2}{\cos^2 x} = \left(\dfrac{1 + \sin x}{\cos x}\right)^2$

$= (\tan x + \sec x)^2$

81. $(\tan x + \cot x)^4 = \left(\dfrac{\sin x}{\cos x} + \dfrac{\cos x}{\sin x}\right)^4 = \left(\dfrac{\sin^2 x + \cos^2 x}{\sin x \cos x}\right)^4 = \left(\dfrac{1}{\sin x \cos x}\right)^4 = \sec^4 x \csc^4 x$

83. $x = \sin\theta$; then $\dfrac{x}{\sqrt{1 - x^2}} = \dfrac{\sin\theta}{\sqrt{1 - \sin^2\theta}} = \dfrac{\sin\theta}{\sqrt{\cos^2\theta}} = \dfrac{\sin\theta}{\cos\theta} = \tan\theta$ (since $\cos\theta \geq 0$ for

$0 \leq \theta \leq \frac{\pi}{2}$).

85. $x = \sec\theta$; then $\sqrt{x^2 - 1} = \sqrt{\sec^2\theta - 1} = \sqrt{(\tan^2\theta + 1) - 1} = \sqrt{\tan^2\theta} = \tan\theta$ (since $\tan\theta \geq 0$
for $0 \leq \theta < \frac{\pi}{2}$)

87. $x = 3\sin\theta$; then $\sqrt{9 - x^2} = \sqrt{9 - (3\sin\theta)^2} = \sqrt{9 - 9\sin^2\theta} = \sqrt{9(1 - \sin^2\theta)} = 3\sqrt{\cos^2\theta}$
   $= 3\cos\theta$ (since $\cos\theta \geq 0$ for $0 \leq \theta < \frac{\pi}{2}$).

89. Choose $x = \frac{\pi}{2}$. Then $\sin 2x = \sin\pi = 0$ whereas $2\sin x = 2\sin\frac{\pi}{2} = 2$.

91. Choose $\theta = \frac{\pi}{4}$. Then $\sec^2\theta + \csc^2\theta = (\sqrt{2})^2 + (\sqrt{2})^2 = 4 \neq 1$.

93. $f(x) = \cos^2 x - \sin^2 x$, $g(x) = 1 - 2\sin^2 x$. From the graph this appears to be an identity.

   <u>Proof:</u> $f(x) = \cos^2 x - \sin^2 x = \cos^2 x + \sin^2 x - 2\sin^2 x$

   $= 1 - 2\sin^2 x = g(x)$. Since $f(x) = g(x)$ for all x, this is an identity.

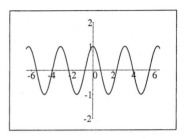

95. $f(x) = (\sin x + \cos x)^2$, $g(x) = 1$

   From the graph this does not appear to be an identity. In order to show this, we can set $x = \frac{\pi}{4}$. Then we have

   $f\left(\frac{\pi}{4}\right) = \left(\frac{1}{\sqrt{2}} + \frac{1}{\sqrt{2}}\right)^2 = \left(\frac{2}{\sqrt{2}}\right)^2 = \left(\sqrt{2}\right)^2 = 2 \neq 1 = g\left(\frac{\pi}{4}\right)$.
   Since $f\left(\frac{\pi}{4}\right) \neq g\left(\frac{\pi}{4}\right)$, this is not an identity.

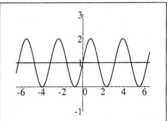

97. We label the triangle as shown in the figure to the right. Since $w = \frac{\pi}{2}$ and $u + v + w = \pi$ we must have $u + v = \frac{\pi}{2} \quad \Leftrightarrow \quad v = \frac{\pi}{2} - u$. Next we express all six trigonometric function for each angle.

$\cos u = \frac{a}{c} = \sin v \quad \Rightarrow \quad \cos u = \sin\left(\frac{\pi}{2} - u\right)$

$\sin u = \frac{b}{c} = \cos v \quad \Rightarrow \quad \sin u = \cos\left(\frac{\pi}{2} - u\right)$

$\tan u = \frac{b}{a} = \cot v \quad \Rightarrow \quad \tan u = \cot\left(\frac{\pi}{2} - u\right)$

$\cot u = \frac{a}{b} = \tan v \quad \Rightarrow \quad \cot u = \tan\left(\frac{\pi}{2} - u\right)$

$\sec u = \frac{c}{a} = \csc v \quad \Rightarrow \quad \sec u = \csc\left(\frac{\pi}{2} - u\right)$

$\csc u = \frac{c}{b} = \sec v \quad \Rightarrow \quad \csc u = \sec\left(\frac{\pi}{2} - u\right)$

Exercises 9.2

1. $\sin 15° = \sin(60° - 45°) = \sin 60° \cos 45° - \cos 60° \sin 45° = \frac{\sqrt{3}}{2} \cdot \frac{1}{\sqrt{2}} - \frac{1}{2} \cdot \frac{1}{\sqrt{2}} = \frac{1}{2\sqrt{2}}(\sqrt{3} - 1)$
    $= \frac{1}{4}(\sqrt{6} - \sqrt{2})$

3. $\tan 105° = \tan(45° + 60°) = \dfrac{\tan 45° + \tan 60°}{1 - \tan 45° \tan 60°} = \dfrac{1 + \sqrt{3}}{1 - 1 \cdot \sqrt{3}} = \dfrac{1 + \sqrt{3}}{1 - \sqrt{3}} = -\dfrac{1}{2}(1 + \sqrt{3})^2$
    $= -2 - \sqrt{3}$

5. $\sin(\frac{11\pi}{12}) = \sin\frac{\pi}{12} = \sin(\frac{\pi}{4} - \frac{\pi}{6}) = \sin\frac{\pi}{4}\cos\frac{\pi}{6} - \cos\frac{\pi}{4}\sin\frac{\pi}{6} = \frac{1}{\sqrt{2}} \cdot \frac{\sqrt{3}}{2} - \frac{1}{\sqrt{2}} \cdot \frac{1}{2} = \frac{1}{4}(\sqrt{6} - \sqrt{2})$

7. $\sin 18° \cos 27° + \cos 18° \sin 27° = \sin(18° + 27°) = \sin 45° = \frac{1}{\sqrt{2}} = \frac{\sqrt{2}}{2}.$

9. $\dfrac{\tan 73° - \tan 13°}{1 + \tan 73° \tan 13°} = \tan(73° - 13°) = \tan 60° = \sqrt{3}$

11. $\tan(\frac{\pi}{2} - u) = \dfrac{\sin(\frac{\pi}{2} - u)}{\cos(\frac{\pi}{2} - u)} = \dfrac{\sin\frac{\pi}{2}\cos u - \cos\frac{\pi}{2}\sin u}{\cos\frac{\pi}{2}\cos u + \sin\frac{\pi}{2}\sin u} = \dfrac{1 \cdot \cos u - 0 \cdot \sin u}{0 \cdot \cos u + 1 \cdot \sin u} = \dfrac{\cos u}{\sin u} = \cot u$

13. $\sec(\frac{\pi}{2} - u) = \dfrac{1}{\cos(\frac{\pi}{2} - u)} = \dfrac{1}{\cos\frac{\pi}{2}\cos u + \sin\frac{\pi}{2}\sin u} = \dfrac{1}{0 \cdot \cos u + 1 \cdot \sin u} = \dfrac{1}{\sin u} = \csc u$

15. $\sin(x - \frac{\pi}{2}) = \sin x \cos\frac{\pi}{2} - \cos x \sin\frac{\pi}{2} = 0 \cdot \sin x - 1 \cdot \cos x = -\cos x$

17. $\sin(x - \pi) = \sin x \cos \pi - \cos x \sin \pi = -1 \cdot \sin x - 0 \cdot \cos x = -\sin x$

19. $\tan(x - \pi) = \dfrac{\tan x - \tan \pi}{1 + \tan x \tan \pi} = \dfrac{\tan x - 0}{1 + \tan x \cdot 0} = \tan x$

21. $\cos(x + \frac{\pi}{6}) + \sin(x - \frac{\pi}{3}) = \cos x \cos\frac{\pi}{6} - \sin x \sin\frac{\pi}{6} + \sin x \cos\frac{\pi}{3} - \cos x \sin\frac{\pi}{3}$
    $= \frac{\sqrt{3}}{2}\cos x - \frac{1}{2}\sin x + \frac{1}{2}\sin x - \frac{\sqrt{3}}{2}\cos x = 0$

23. $\sin(x + y) - \sin(x - y) = \sin x \cos y + \cos x \sin y - (\sin x \cos y - \cos x \sin y) = 2\cos x \sin y$

25. $\cot(x - y) = \dfrac{1}{\tan(x - y)} = \dfrac{1 + \tan x \tan y}{\tan x - \tan y} = \dfrac{1 + \frac{1}{\cot x}\frac{1}{\cot y}}{\frac{1}{\cot x} - \frac{1}{\cot y}} \cdot \dfrac{\cot x \cot y}{\cot x \cot y} = \dfrac{\cot x \cot y + 1}{\cot y - \cot x}$

27. $\tan x - \tan y = \dfrac{\sin x}{\cos x} - \dfrac{\sin y}{\cos y} = \dfrac{\sin x \cos y - \cos x \sin y}{\cos x \cos y} = \dfrac{\sin(x - y)}{\cos x \cos y}$

29. $\dfrac{\sin(x + y) - \sin(x - y)}{\cos(x + y) + \cos(x - y)} = \dfrac{\sin x \cos y + \cos x \sin y - (\sin x \cos y - \cos x \sin y)}{\cos x \cos y - \sin x \sin y + \cos x \cos y + \sin x \sin y}$
    $= \dfrac{2\cos x \sin y}{2\cos x \cos y} = \tan y$

31. $\sin(x + y + z) = \sin[(x + y) + z] = \sin(x + y)\cos z + \cos(x + y)\sin z$
    $= \cos z\,[\sin x \cos y + \cos x \sin y] + \sin z\,[\cos x \cos y - \sin x \sin y]$
    $= \sin x \cos y \cos z + \cos x \sin y \cos z + \cos x \cos y \sin z - \sin x \sin y \sin z$

33. $k = \sqrt{A^2 + B^2} = \sqrt{\left(-\sqrt{3}\right)^2 + 1^2} = \sqrt{4} = 2.$ $\sin\phi = \frac{1}{2}$ and $\cos\phi = \frac{-\sqrt{3}}{2}$ $\Rightarrow$ $\phi = \frac{5\pi}{6}.$

Therefore, $-\sqrt{3}\sin x + \cos x = k\sin(x + \phi) = 2\sin(x + \frac{5\pi}{6}).$

35. $k = \sqrt{A^2 + B^2} = \sqrt{5^2 + (-5)^2} = \sqrt{50} = 5\sqrt{2}.$ $\sin\phi = \frac{-5}{5\sqrt{2}} = -\frac{1}{\sqrt{2}}$ and $\cos\phi = \frac{5}{5\sqrt{2}} = \frac{1}{\sqrt{2}}$

$\Rightarrow$ $\phi = \frac{7\pi}{4}.$ Therefore, $5(\sin 2x - \cos 2x) = k\sin(2x + \phi) = 5\sqrt{2}\sin(2x + \frac{7\pi}{4}).$

37. $f(x) = \sin x + \cos x$ $\Rightarrow$
$k = \sqrt{1^2 + 1^2} = \sqrt{2},$ and ϕ satisfies
$\sin\phi = \cos\phi = \frac{1}{\sqrt{2}}$ $\Rightarrow$ $\phi = \frac{\pi}{4}.$

Thus, we can write $f(x) = k\sin(x + \phi)$
$= \sqrt{2}\sin(x + \frac{\pi}{4}).$ We see that this is a sine
curve with amplitude $= \sqrt{2},$ period $= 2\pi,$ and
phase shift $= -\frac{\pi}{4}.$

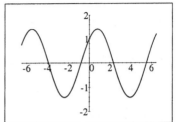

39. If $\beta - \alpha = \frac{\pi}{2},$ then $\beta = \alpha + \frac{\pi}{2}.$ Now if we let $y = x + \alpha,$ then $\sin(x + \alpha) + \cos(x + \alpha + \frac{\pi}{2})$

$= \sin y + \cos(y + \frac{\pi}{2}) = \sin y + (-\sin y) = 0.$ Therefore, $\sin(x + \alpha) + \cos(x + \beta) = 0.$

41. Let $\angle A$ and $\angle B$ be the two angles shown in the diagram below. Then $180° = \gamma + A + B,$
$90° = \alpha + A,$ and $90° = \beta + B.$ Subtracting the second and third equation from the first we get

$$
\begin{aligned}
180° &= \gamma + A + B \\
-90° &= -\alpha - A \\
-90° &= -\beta - B \\
\hline
0 &= \gamma - \alpha - \beta \qquad \Leftrightarrow \qquad \alpha + \beta = \gamma.
\end{aligned}
$$

$$\tan\gamma = \tan(\alpha + \beta) = \frac{\tan\alpha + \tan\beta}{1 - \tan\alpha\tan\beta} = \frac{\frac{4}{6} + \frac{3}{4}}{1 - \frac{4}{6} \cdot \frac{3}{4}} = \frac{\frac{8}{12} + \frac{9}{12}}{1 - \frac{1}{2}} = 2 \cdot \frac{17}{12} = \frac{17}{6}.$$

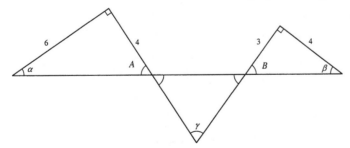

43. (a) $y = \sin^2\left(x + \frac{\pi}{4}\right) + \sin^2\left(x - \frac{\pi}{4}\right)$

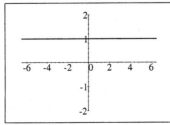

From the graph we see that the value of y
seems to always be equal to 1.

(b) $y = \sin^2\left(x + \frac{\pi}{4}\right) + \sin^2\left(x - \frac{\pi}{4}\right) = \left(\sin x \cos \frac{\pi}{4} + \cos x \sin \frac{\pi}{4}\right)^2 + \left(\sin x \cos \frac{\pi}{4} - \cos x \sin \frac{\pi}{4}\right)^2$

$= \left[\frac{1}{\sqrt{2}}(\sin x + \cos x)\right]^2 + \left[\frac{1}{\sqrt{2}}(\sin x - \cos x)\right]^2 = \frac{1}{2}[(\sin x + \cos x)^2 + (\sin x - \cos x)^2]$

$= \frac{1}{2}[(\sin^2 x + 2\sin x \cos x + \cos^2 x) + (\sin^2 x - 2\sin x \cos x + \cos^2 x)]$

$= \frac{1}{2}[(1 + 2\sin x \cos x) + (1 - 2\sin x \cos x)] = \frac{1}{2} \cdot 2 = 1$

45.   (a) $y = f_1(t) + f_2(t) = 5\sin t + 5\cos t = 5(\sin t + \cos t)$

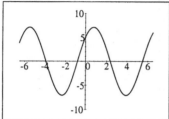

(b) $k = \sqrt{A^2 + B^2} = \sqrt{5^2 + 5^2} = 5\sqrt{2}$. Therefore, $\sin \phi = \frac{B}{\sqrt{A^2+B^2}} = \frac{5}{5\sqrt{2}} = \frac{1}{\sqrt{2}}$ and

$\cos \phi = \frac{A}{\sqrt{A^2+B^2}} = \frac{1}{\sqrt{2}}$. Thus, $\phi = \frac{\pi}{4}$.

47. $\tan(s + t) = \dfrac{\sin(s + t)}{\cos(s + t)} = \dfrac{\sin s \cos t + \cos s \sin t}{\cos s \cos t + \sin s \sin t} = \dfrac{\sin s \cos t + \cos s \sin t}{\cos s \cos t - \sin s \sin t} \cdot \dfrac{\frac{1}{\cos s \cos t}}{\frac{1}{\cos s \cos t}}$

$= \dfrac{\dfrac{\sin s}{\cos s} + \dfrac{\sin t}{\cos t}}{1 - \dfrac{\sin s}{\cos s} \cdot \dfrac{\sin t}{\cos t}} = \dfrac{\tan s + \tan t}{1 - \tan s \tan t}$

Exercises 9.3

1. $\sin x = \frac{5}{13}$, x in quadrant I $\Rightarrow$ $\cos x = \frac{12}{13}$ and $\tan x = \frac{5}{12}$.

    $\sin 2x = 2\sin x \cos x = 2\left(\frac{5}{13}\right)\left(\frac{12}{13}\right) = \frac{120}{169}$

    $\cos 2x = \cos^2 x - \sin^2 x = \left(\frac{12}{13}\right)^2 - \left(\frac{5}{13}\right)^2 = \frac{144-25}{169} = \frac{119}{169}$

    $\tan 2x = \dfrac{\sin 2x}{\cos 2x} = \dfrac{\frac{120}{169}}{\frac{119}{169}} = \frac{120}{169}\cdot\frac{169}{119} = \frac{120}{119}$

3. $\tan x = -\frac{4}{3}$. Then, $\sin x = \frac{4}{5}$ and $\cos x = -\frac{3}{5}$ (x is in quadrant II).

    $\sin 2x = 2\sin x \cos x = 2\cdot\frac{4}{5}\left(-\frac{3}{5}\right) = -\frac{24}{25}$

    $\cos 2x = \cos^2 x - \sin^2 x = \left(-\frac{3}{5}\right)^2 - \left(\frac{4}{5}\right)^2 = \frac{9-16}{25} = -\frac{7}{25}$

    $\tan 2x = \dfrac{\sin 2x}{\cos 2x} = \dfrac{-\frac{24}{25}}{-\frac{7}{25}} = \frac{24}{25}\cdot\frac{25}{7} = \frac{24}{7}$

5. $\sin x = -\frac{3}{5}$. Then, $\cos x = -\frac{4}{5}$ and $\tan x = \frac{3}{4}$ (x is in quadrant III).

    $\sin 2x = 2\sin x \cos x = 2(-\frac{3}{5})(-\frac{4}{5}) = \frac{24}{25}$

    $\cos 2x = \cos^2 x - \sin^2 x = (-\frac{4}{5})^2 - (-\frac{3}{5})^2 = \frac{16-9}{25} = \frac{7}{25}$

    $\tan 2x = \dfrac{\sin 2x}{\cos 2x} = \dfrac{\frac{24}{25}}{\frac{7}{25}} = \frac{24}{25}\cdot\frac{25}{7} = \frac{24}{7}$

7. $\sin^4 x = \left(\sin^2 x\right)^2 = \left(\dfrac{1-\cos 2x}{2}\right)^2 = \frac{1}{4} - \frac{1}{2}\cos 2x + \frac{1}{4}\cos^2 2x$

    $= \frac{1}{4} - \frac{1}{2}\cos 2x + \frac{1}{4}\cdot\dfrac{1+\cos 4x}{2} = \frac{1}{4} - \frac{1}{2}\cos 2x + \frac{1}{8} + \frac{1}{8}\cos 4x = \frac{3}{8} - \frac{1}{2}\cos 2x + \frac{1}{8}\cos 4x$

    $= \frac{1}{2}\left(\frac{3}{4} - \cos 2x + \frac{1}{4}\cos 4x\right)$

9. Since $\sin^4 x \cos^4 x = \left(\sin^2 x \cos^2 x\right)^2$ we use the result of Example 4, so

    $\sin^4 x \cos^4 x = \left(\frac{1}{8} - \frac{1}{8}\cos 4x\right)^2 = \frac{1}{64} - \frac{1}{32}\cos 4x + \frac{1}{64}\cos^2 4x$

    $= \frac{1}{64} - \frac{1}{32}\cos 4x + \frac{1}{64}\left(\dfrac{1+\cos 8x}{2}\right) = \frac{1}{64} - \frac{1}{32}\cos 4x + \frac{1}{128} + \frac{1}{128}\cos 8x$

    $= \frac{3}{128} - \frac{1}{32}\cos 4x + \frac{1}{128}\cos 8x = \frac{1}{32}\left(\frac{3}{4} - \cos 4x + \frac{1}{4}\cos 8x\right)$

11. Again use the result of Example 4, $\cos^2 x \sin^4 x = \left(\sin^2 x \cos^2 x\right)\sin^2 x$

    $= \left(\frac{1}{8} - \frac{1}{8}\cos 4x\right)\cdot\left(\frac{1}{2} - \dfrac{\cos 2x}{2}\right) = \frac{1}{16} - \frac{1}{16}\cos 2x - \frac{1}{16}\cos 4x + \frac{1}{16}\cos 2x \cos 4x$

13. $\sin 15° = \sqrt{\frac{1}{2}(1-\cos 30°)} = \sqrt{\frac{1}{2}\left(1-\frac{\sqrt{3}}{2}\right)} = \sqrt{\frac{1}{4}(2-\sqrt{3})} = \frac{1}{2}\sqrt{2-\sqrt{3}}$

15. $\cos 22.5° = \sqrt{\frac{1}{2}(1+\cos 45°)} = \sqrt{\frac{1}{2}\left(1+\frac{\sqrt{2}}{2}\right)} = \frac{1}{2}\sqrt{2+\sqrt{2}}$

17. $\sin \frac{\pi}{12} = \sqrt{\frac{1}{2}(1 - \cos \frac{\pi}{6})} = \sqrt{\frac{1}{2}(1 - \frac{\sqrt{3}}{2})} = \frac{1}{2}\sqrt{2 - \sqrt{3}}$

19. (a) $2 \sin 18° \cos 18° = \sin 36°$ (b) $2 \sin 3\theta \cos 3\theta = \sin 6\theta$

21. (a) $\cos^2 34° - \sin^2 34° = \cos 68°$ (b) $\cos^2 5\theta - \sin^2 5\theta = \cos 10\theta$

23. (a) $\dfrac{\sin 8°}{1 + \cos 8°} = \tan \frac{8°}{2} = \tan 4°$ (b) $\dfrac{1 - \cos 4\theta}{\sin 4\theta} = \tan \frac{4\theta}{2} = \tan 2\theta$

25. $\sin x = \frac{3}{5}$. Since x is in quadrant I, $\cos x = \frac{4}{5}$ and $\frac{x}{2}$ is also in quadrant I. Thus

$$\sin \frac{x}{2} = \sqrt{\frac{1}{2}(1 - \cos x)} = \sqrt{\frac{1}{2}(1 - \frac{4}{5})} = \frac{1}{\sqrt{10}} = \frac{\sqrt{10}}{10}$$

$$\cos \frac{x}{2} = \sqrt{\frac{1}{2}(1 + \cos x)} = \sqrt{\frac{1}{2}(1 + \frac{4}{5})} = \frac{3}{\sqrt{10}} = \frac{3\sqrt{10}}{10}$$

$$\tan \frac{x}{2} = \frac{\sin \frac{x}{2}}{\cos \frac{x}{2}} = \frac{1}{\sqrt{10}} \cdot \frac{\sqrt{10}}{3} = \frac{1}{3}$$

27. $\csc x = 3$. Then, $\sin x = \frac{1}{3}$ and since x is in quadrant II, $\cos x = -\frac{2\sqrt{2}}{3}$. Since $90° \le x \le 180°$, we have $45° \le \frac{x}{2} \le 90°$ and so $\frac{x}{2}$ is in quadrant I. Thus

$$\sin \frac{x}{2} = \sqrt{\frac{1}{2}(1 - \cos x)} = \sqrt{\frac{1}{2}(1 + \frac{2\sqrt{2}}{3})} = \sqrt{\frac{1}{6}(3 + 2\sqrt{2})}$$

$$\cos \frac{x}{2} = \sqrt{\frac{1}{2}(1 + \cos x)} = \sqrt{\frac{1}{2}(1 - \frac{2\sqrt{2}}{3})} = \sqrt{\frac{1}{6}(3 - 2\sqrt{2})}$$

$$\tan \frac{x}{2} = \frac{\sin \frac{x}{2}}{\cos \frac{x}{2}} = \sqrt{\frac{3 + 2\sqrt{2}}{3 - 2\sqrt{2}}} = 3 + 2\sqrt{2}$$

29. $\sec x = \frac{3}{2}$. Then $\cos x = \frac{2}{3}$ and since x is in quadrant IV, $\sin x = -\frac{\sqrt{5}}{3}$. Since $270° \le x \le 360°$, we have $135° \le \frac{x}{2} \le 180°$ and so $\frac{x}{2}$ is in quadrant II. Thus

$$\sin \frac{x}{2} = \sqrt{\frac{1}{2}(1 - \cos x)} = \sqrt{\frac{1}{2}(1 - \frac{2}{3})} = \frac{1}{\sqrt{6}} = \frac{\sqrt{6}}{6}$$

$$\cos \frac{x}{2} = -\sqrt{\frac{1}{2}(1 + \cos x)} = -\sqrt{\frac{1}{2}(1 + \frac{2}{3})} = \frac{-\sqrt{5}}{\sqrt{6}} = -\frac{\sqrt{30}}{6}$$

$$\tan \frac{x}{2} = \frac{\sin \frac{x}{2}}{\cos \frac{x}{2}} = \frac{1}{\sqrt{6}} \cdot \frac{\sqrt{6}}{-\sqrt{5}} = -\frac{1}{\sqrt{5}} = -\frac{\sqrt{5}}{5}$$

31. $\sin 2x \cos 3x = \frac{1}{2}[\sin(2x + 3x) + \sin(2x - 3x)] = \frac{1}{2}(\sin 5x - \sin x)$

33. $3\cos 4x \cos 7x = 3 \cdot \frac{1}{2}[\cos(4x + 7x) + \cos(4x - 7x)] = \frac{3}{2}(\cos 11x + \cos 3x)$

35. $\sin 5x + \sin 3x = 2 \sin\left(\frac{5x + 3x}{2}\right) \cos\left(\frac{5x - 3x}{2}\right) = 2 \sin 4x \cos x$

37. $\cos 4x - \cos 6x = -2 \sin\left(\frac{4x + 6x}{2}\right) \sin\left(\frac{4x - 6x}{2}\right) = -2 \sin 5x \sin(-x) = 2 \sin 5x \sin x$

39. $\sin 2x - \sin 7x = 2 \cos\left(\frac{2x + 7x}{2}\right) \sin\left(\frac{2x - 7x}{2}\right) = 2 \cos \frac{9x}{2} \sin \frac{-5x}{2} = -2 \cos \frac{9x}{2} \sin \frac{5x}{2}$

41. $2 \sin 52.5° \sin 97.5° = 2 \cdot \frac{1}{2}[\cos(52.5° - 97.5°) - \cos(52.5° + 97.5°)] = \cos(-45°) - \cos 150°$

$= \cos 45° - \cos 150° = \frac{\sqrt{2}}{2} + \frac{\sqrt{3}}{2} = \frac{1}{2}(\sqrt{2} + \sqrt{3})$

43. $\cos 37.5° \sin 7.5° = \frac{1}{2}(\sin 45° - \sin 30°) = \frac{1}{2}\left(\frac{\sqrt{2}}{2} - \frac{1}{2}\right) = \frac{1}{4}(\sqrt{2} - 1)$

45. $\cos 255° - \cos 195° = -2\sin\left(\frac{255°+195°}{2}\right)\sin\left(\frac{255°-195°}{2}\right) = -2\sin 225° \sin 30° = -2\left(-\frac{\sqrt{2}}{2}\right)\frac{1}{2} = \frac{\sqrt{2}}{2}$

47. $\cos^2 5x - \sin^2 5x = \cos(2 \cdot 5x) = \cos 10x$

49. $(\sin x + \cos x)^2 = \sin^2 x + 2\sin x \cos x + \cos^2 x = 1 + 2\sin x \cos x = 1 + \sin 2x$

51. $\dfrac{\sin 4x}{\sin x} = \dfrac{(2\sin 2x \cos 2x)}{\sin x} = \dfrac{2(2\sin x \cos x)(\cos 2x)}{\sin x} = 4\cos x \cos 2x$

53. $\dfrac{2(\tan x - \cot x)}{\tan^2 x - \cot^2 x} = \dfrac{2(\tan x - \cot x)}{(\tan x + \cot x)(\tan x - \cot x)} = \dfrac{2}{\tan x + \cot x} = \dfrac{2}{\dfrac{\sin x}{\cos x} + \dfrac{\cos x}{\sin x}}$

$= \dfrac{2}{\dfrac{\sin x}{\cos x} + \dfrac{\cos x}{\sin x}} \cdot \dfrac{\sin x \cos x}{\sin x \cos x} = \dfrac{2\sin x \cos x}{\sin^2 x + \cos^2 x} = 2\sin x \cos x = \sin 2x$

55. $\tan 3x = \tan(2x + x) = \dfrac{\tan 2x + \tan x}{1 - \tan 2x \tan x} = \dfrac{\dfrac{2\tan x}{1 - \tan^2 x} + \tan x}{1 - \dfrac{2\tan x}{1 - \tan^2 x}\tan x} = \dfrac{2\tan x + \tan x(1 - \tan^2 x)}{1 - \tan^2 x - 2\tan x \tan x}$

$= \dfrac{3\tan x - \tan^3 x}{1 - 3\tan^2 x}$

57. $\cos^4 x - \sin^4 x = (\cos^2 x + \sin^2 x)(\cos^2 x - \sin^2 x) = \cos^2 x - \sin^2 x = \cos 2x$

59. $\dfrac{\sin x + \sin 5x}{\cos x + \cos 5x} = \dfrac{2\sin 3x \cos 2x}{2\cos 3x \cos 2x} = \dfrac{\sin 3x}{\cos 3x} = \tan 3x$

61. $\dfrac{\sin 10x}{\sin 9x + \sin x} = \dfrac{2\sin 5x \cos 5x}{2\sin 5x \cos 4x} = \dfrac{\cos 5x}{\cos 4x}$

63. $\dfrac{\sin x + \sin y}{\cos x + \cos y} = \dfrac{2\sin\left(\frac{x+y}{2}\right)\cos\left(\frac{x-y}{2}\right)}{2\cos\left(\frac{x+y}{2}\right)\cos\left(\frac{x-y}{2}\right)} = \dfrac{\sin\left(\frac{x+y}{2}\right)}{\cos\left(\frac{x+y}{2}\right)} = \tan\left(\frac{x+y}{2}\right)$

65. $\sin 45° + \sin 15° = 2\sin\left(\frac{45°+15°}{2}\right)\cos\left(\frac{45°-15°}{2}\right) = 2\sin 30° \cos 15° = 2 \cdot \frac{1}{2} \cdot \cos 15° = \cos 15°$
$= \sin(90° - 15°) = \sin 75°$ (applying the cofunction identity)

67. $\dfrac{\sin x + \sin 2x + \sin 3x + \sin 4x + \sin 5x}{\cos x + \cos 2x + \cos 3x + \cos 4x + \cos 5x} = \dfrac{(\sin x + \sin 5x) + (\sin 2x + \sin 4x) + \sin 3x}{(\cos x + \cos 5x) + (\cos 2x + \cos 4x) + \cos 3x}$

$= \dfrac{2\sin 3x \cos 2x + 2\sin 3x \cos x + \sin 3x}{2\cos 3x \cos 2x + 2\cos 3x \cos x + \cos 3x} = \dfrac{\sin 3x(2\cos 2x + 2\cos x + 1)}{\cos 3x(2\cos 2x + 2\cos x + 1)} = \tan 3x$

69. (a) $f(x) = \dfrac{\sin 3x}{\sin x} - \dfrac{\cos 3x}{\cos x}$

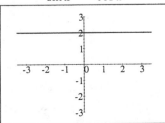

The function appears to have a constant value of 2 wherever it is defined.

(b) $f(x) = \dfrac{\sin 3x}{\sin x} - \dfrac{\cos 3x}{\cos x} = \dfrac{\sin 3x \cos x - \cos 3x \sin x}{\sin x \cos x} = \dfrac{\sin (3x - x)}{\sin x \cos x} = \dfrac{\sin 2x}{\sin x \cos x}$

$= \dfrac{2 \sin x \cos x}{\sin x \cos x} = 2$ for all x for which the function is defined.

71. (a) $y = \sin 6x + \sin 7x$

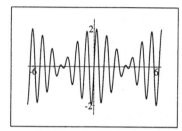

(b) By a sum to product identity
$y = \sin 6x + \sin 7x$

$= 2 \sin\left(\dfrac{6x + 7x}{2}\right) \cos\left(\dfrac{6x - 7x}{2}\right)$

$= 2 \cdot \sin\left(\tfrac{13}{2}x\right) \cdot \cos\left(-\tfrac{1}{2}x\right)$

$= 2 \sin\left(\tfrac{13}{2}x\right) \cos\left(\tfrac{1}{2}x\right).$

(c) $y = \sin 6x + \sin 7x$, with $y = 2 \cos\left(\tfrac{1}{2}x\right)$ and $y = -2 \cos\left(\tfrac{1}{2}x\right)$

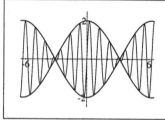

The graph of $y = f(x)$ lies between the other two graphs.

73. We first label the figure as shown on the right. Because the sheet of paper is folded over $\angle EAC = \angle CAB = \theta$. Thus $\angle BCA = \angle ACE = 90° - \theta$. It follows that $\angle ECD = 180° - \angle BCA - \angle ACE$ $= 180° - (90° - \theta) - (90° - \theta) = 2\theta$. Also, from the figure we see that $BC = L \sin \theta$ and $CE = L \sin \theta$ so $DC = EC \cos 2\theta = L \sin \theta \cos 2\theta$. Thus $6 = DB = DC + CB = L \sin \theta \cos 2\theta + L \sin \theta$ $= L \sin \theta (1 + \cos 2\theta) = L \sin \theta \cdot 2 \cos^2 \theta$. So $L = \dfrac{6}{2 \sin \theta \cos^2 \theta} = \dfrac{3}{\sin \theta \cos^2 \theta}.$

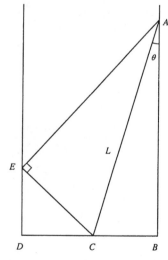

75. (a) $\cos 4x = \cos(2x + 2x) = 2 \cos^2 2x - 1 = 2 \left(2 \cos^2 x - 1\right)^2 - 1 = 8 \cos^4 x - 8 \cos^2 x + 1.$
         Thus the desired polynomial is $P(t) = 8t^4 - 8t^2 + 1.$

     (b) $\cos 5x = \cos(4x + x) = \cos 4x \cos x - \sin 4x \sin x$
          $= \cos x \left(8 \cos^4 x - 8 \cos^2 x + 1\right) - 2 \sin 2x \cos 2x \sin x$
          $= 8 \cos^5 x - 8 \cos^3 x + \cos x - 4 \sin x \cos x \left(2 \cos^2 x - 1\right) \sin x \qquad \text{(from part (a))}$
          $= 8 \cos^5 x - 8 \cos^3 x + \cos x - 4 \cos x \left(2 \cos^2 x - 1\right) \sin^2 x$

$$= 8\cos^5 x - 8\cos^3 x + \cos x - 4\cos x \,(2\cos^2 x - 1)(1 - \cos^2 x)$$
$$= 8\cos^5 x - 8\cos^3 x + \cos x + 8\cos^5 x - 12\cos^3 x + 4\cos x = 16\cos^5 x - 20\cos^3 x + 5\cos x.$$
Thus the desired polynomial is $P(t) = 16t^5 - 20t^3 + 5t$.

77. We find the area of $\triangle\,ABC$ in two different ways. First, let AB be the base and CD be the height. Since $\angle BOC = 2\theta$ we see that $CD = \sin 2\theta$. So

$$Area = \tfrac{1}{2}(base)(height) = \tfrac{1}{2}\cdot 2\cdot\sin 2\theta = \sin 2\theta.$$

On the other hand, in $\triangle\,ABC$ we see that $\angle C$ is a right angle. So, $BC = 2\sin\theta$ and $AC = 2\cos\theta$, and

$$Area = \tfrac{1}{2}(base)(height) = \tfrac{1}{2}\cdot(2\sin\theta)(2\cos\theta) = 2\sin\theta\cos\theta.$$

Comparing the two expressions for the area of $\triangle ABC$ we get $\sin 2\theta = 2\sin\theta\cos\theta$.

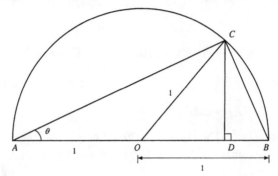

Note: The two facts from geometry used in this solution are given in the problem itself.

Exercises 9.4

1. (a) $\sin^{-1}\frac{1}{2} = \frac{\pi}{6}$ (b) $\cos^{-1}\frac{1}{2} = \frac{\pi}{3}$ (c) $\cos^{-1}2$ is not defined.

3. (a) $\sin^{-1}\frac{\sqrt{2}}{2} = \frac{\pi}{4}$ (b) $\cos^{-1}\frac{\sqrt{2}}{2} = \frac{\pi}{4}$ (c) $\sin^{-1}(-\frac{\sqrt{2}}{2}) = -\frac{\pi}{4}$

5. (a) $\sin^{-1}1 = \frac{\pi}{2}$ (b) $\cos^{-1}1 = 0$ (c) $\cos^{-1}(-1) = \pi$

7. (a) $\tan^{-1}\frac{\sqrt{3}}{3} = \frac{\pi}{6}$ (b) $\tan^{-1}(-\frac{\sqrt{3}}{3}) = -\frac{\pi}{6}$ (c) $\sin^{-1}(-2)$ is not defined.

9. (a) $\sin^{-1}(0.7688) \approx 0.87696$ (b) $\cos^{-1}(-0.5014) \approx 2.09601$

11. $\sin(\sin^{-1}\frac{1}{3}) = \frac{1}{3}$ 13. $\tan(\tan^{-1}10) = 10$

15. $\cos^{-1}(\cos\frac{\pi}{3}) = \frac{\pi}{3}$ 17. $\sin^{-1}\left[\sin\left(-\frac{\pi}{6}\right)\right] = -\frac{\pi}{6}$

19. $\tan^{-1}(\tan\frac{2\pi}{3}) = \tan^{-1}\left[\tan(\frac{2\pi}{3} - \pi)\right] = \tan^{-1}(\tan\frac{-\pi}{3}) = -\frac{\pi}{3}$ (since $\frac{2\pi}{3} > \frac{\pi}{2}$)

21. $\tan(\sin^{-1}\frac{1}{2}) = \tan\frac{\pi}{6} = \frac{\sqrt{3}}{3}$ 23. $\cos(\sin^{-1}\frac{\sqrt{3}}{2}) = \cos\frac{\pi}{3} = \frac{1}{2}$

25. $\tan^{-1}(2\sin\frac{\pi}{3}) = \tan^{-1}(2 \cdot \frac{\sqrt{3}}{2}) = \tan^{-1}\sqrt{3} = \frac{\pi}{3}$

27. Let $u = \cos^{-1}\frac{3}{5}$, so $\cos u = \frac{3}{5}$. Then from the triangle $\sin(\cos^{-1}\frac{3}{5}) = \sin u = \frac{4}{5}$.

29. Let $u = \tan^{-1}\frac{12}{5}$, so $\tan u = \frac{12}{5}$. Then from the triangle $\sin(\tan^{-1}\frac{12}{5}) = \sin u = \frac{12}{13}$.

31. Let $\theta = \sin^{-1}\frac{12}{13}$, so $\sin\theta = \frac{12}{13}$. Then from the triangle $\sec(\sin^{-1}\frac{12}{13}) = \sec\theta = \frac{13}{5}$.

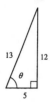

33. Let $u = \tan^{-1}2$, so $\tan u = 2$. Then from the triangle $\cos(\tan^{-1}2) = \cos u = \frac{1}{\sqrt{5}} = \frac{\sqrt{5}}{5}$.

35. Let $u = \cos^{-1}\frac{3}{5}$, so $\cos u = \frac{3}{5}$. From the triangle $\sin u = \frac{4}{5}$ so

$$\sin(2\cos^{-1}\frac{3}{5}) = \sin(2u) = 2\sin u \cos u = 2 \cdot \frac{4}{5} \cdot \frac{3}{5} = \frac{24}{25}$$

37. Let $u = \cos^{-1}\frac{1}{2}$ and $v = \sin^{-1}\frac{1}{2}$, so $\cos u = \frac{1}{2}$ and $\sin v = \frac{1}{2}$. From the triangle
$\sin(\sin^{-1}\frac{1}{2} + \cos^{-1}\frac{1}{2}) = \sin(v + u) = \sin v \cos u + \cos v \sin u$
$= \frac{1}{2} \cdot \frac{1}{2} + \frac{\sqrt{3}}{2} \cdot \frac{\sqrt{3}}{2} = 1.$
Another solution: $\sin(\sin^{-1}\frac{1}{2} + \cos^{-1}\frac{1}{2}) = \sin(\frac{\pi}{6} + \frac{\pi}{3}) = \sin\frac{\pi}{2} = 1.$

39. Let $u = \sin^{-1}x$, so $\sin u = x$. From the triangle $\cos(\sin^{-1}x) = \cos u = \sqrt{1 - x^2}$.

41. Let $u = \sin^{-1}x$, so $\sin u = x$. From the triangle $\tan(\sin^{-1}x) = \tan u = \dfrac{x}{\sqrt{1 - x^2}}$.

43. Let $u = \tan^{-1}x$, so $\tan u = x$. From the triangle
$\cos(2\tan^{-1}x) = \cos 2u - 1 = 2\cos^2 u - 1 = \dfrac{2}{1 + x^2} - 1 = \dfrac{1 - x^2}{1 + x^2}.$

45. Let $u = \cos^{-1}x$ and $v = \sin^{-1}x$, so $\cos u = x$ and $\sin v = x$. From the triangle
$\cos(\cos^{-1}x + \sin^{-1}x) = \cos(\cos^{-1}x)\cos(\sin^{-1}x) - \sin(\cos^{-1}x)\sin(\sin^{-1}x)$
$= \cos u \cos v - \sin u \sin v = x\sqrt{1 - x^2} - x\sqrt{1 - x^2} = 0.$

47. $\tan\theta = \dfrac{50}{s} \quad\Rightarrow\quad \theta = \tan^{-1}\left(\dfrac{50}{s}\right)$

49. (a) $y = \sin^{-1}x + \cos^{-1}x$. Note the domain of both $\sin^{-1}x$ and $\cos^{-1}x$ is $[-1, 1]$.

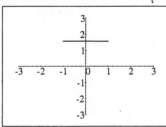

Conjecture: $y = \frac{\pi}{2}$ for $-1 \le x \le 1$.

(b) To prove this conjecture, let $\sin^{-1}x = u$
$\Leftrightarrow \quad \sin u = x$, $-\frac{\pi}{2} \le u \le \frac{\pi}{2}$. Using the cofunction identity $\sin u = \cos\left(\frac{\pi}{2} - u\right) = x$,
$\cos^{-1}x = \frac{\pi}{2} - u$. Therefore,
$\sin^{-1}x + \cos^{-1}x = u + \left(\frac{\pi}{2} - u\right) = \frac{\pi}{2}.$

51. (a) $\tan^{-1}x + \tan^{-1}2x = \frac{\pi}{4}$

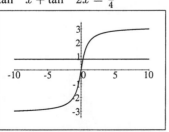

From the graph, $x \approx 0.28$

(b) To solve the equation exactly, we set $\tan^{-1}x + \tan^{-1}2x = \frac{\pi}{4}$ $\Leftrightarrow$

$\tan^{-1}x = \frac{\pi}{4} - \tan^{-1}2x$ $\Rightarrow$ $\tan(\tan^{-1}x) = \tan\left(\frac{\pi}{4} - \tan^{-1}2x\right)$ $\Rightarrow$

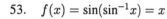

$x = \dfrac{\tan\frac{\pi}{4} - \tan\left(\tan^{-1}2x\right)}{1 + \tan\frac{\pi}{4} \cdot \tan\left(\tan^{-1}2x\right)} = \dfrac{1 - 2x}{1 + (1)2x}$ $\Rightarrow$ $x(1 + 2x) = 1 - 2x$ $\Leftrightarrow$

$x + 2x^2 = 1 - 2x$. Thus, $2x^2 + 3x - 1 = 0$. We use the quadratic formula to solve for x:

$x = \dfrac{-3 \pm \sqrt{3^2 - 4(2)(-1)}}{2(2)} = \dfrac{-3 \pm \sqrt{17}}{4}$. Substituting into the original equation, we see that

$x = \dfrac{-3 - \sqrt{17}}{4}$ is not a solution, and so $x = \dfrac{-3 + \sqrt{17}}{4} \approx 0.28$ is the only root.

53. $f(x) = \sin(\sin^{-1}x) = x$

Domain: $[-1, 1]$

Range: $[-1, 1]$

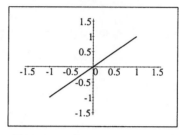

$g(x) = \sin^{-1}(\sin x)$

Domain: $[-\infty, \infty]$

Range: $[-\frac{\pi}{2}, \frac{\pi}{2}]$

Note $g(x) = x$ only for $x \in [-\frac{\pi}{2}, \frac{\pi}{2}]$

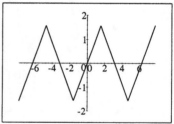

Exercises 9.5

1. $2\cos x - 1 = 0$ $\Leftrightarrow$ $2\cos x = 1$ $\Leftrightarrow$ $\cos x = \frac{1}{2}$. In the interval $[0, 2\pi)$ the solution are $x = \frac{\pi}{3}, \frac{5\pi}{3}$. Therefore, the solutions are $x = \frac{\pi}{3} + 2k\pi, \frac{5\pi}{3} + 2k\pi, k = 0, \pm1, \pm2, \ldots$.

3. $2\sin x - \sqrt{3} = 0$ $\Leftrightarrow$ $2\sin x - \sqrt{3}$ $\Leftrightarrow$ $\sin x = \frac{\sqrt{3}}{2}$. In the interval $[0, 2\pi)$ the solution are $x = \frac{\pi}{3}, \frac{2\pi}{3}$. Therefore, the solutions are $x = \frac{\pi}{3} + 2k\pi, \frac{2\pi}{3} + 2k\pi, k = 0, \pm1, \pm2, \ldots$.

5. $4\cos^2 x - 1 = 0$ $\Leftrightarrow$ $4\cos^2 x = 1$ $\Leftrightarrow$ $4\cos^2 x = 1$ $\Leftrightarrow$ $\cos^2 x = \frac{1}{4}$ $\Leftrightarrow$ $\cos x = \pm\frac{1}{2}$. In the interval $[0, 2\pi)$ the solutions are $x = \frac{\pi}{3}, \frac{2\pi}{3}, \frac{4\pi}{3}, \frac{5\pi}{3}$. So the solutions are $x = \frac{\pi}{3} + 2k\pi, \frac{2\pi}{3} + 2k\pi, \frac{4\pi}{3} + 2k\pi, \frac{5\pi}{3} + 2k\pi$ $\Leftrightarrow$ $x = \frac{\pi}{3} + k\pi, \frac{2\pi}{3} + k\pi, k = 0, \pm1, \pm2, \ldots$.

7. $\sec^2 x - 2 = 0$ $\Leftrightarrow$ $\sec^2 x = 2$ $\Leftrightarrow$ $\sec x = \pm\sqrt{2}$. In the interval $[0, 2\pi)$ the solutions are $x = \frac{\pi}{4}, \frac{3\pi}{4}, \frac{5\pi}{4}, \frac{7\pi}{4}$. So the solutions can be expressed as the odd multiples of $\frac{\pi}{4}$ which are $x = \frac{(2k+1)\pi}{4}, k = 0, \pm1, \pm2, \ldots$.

9. $\cos x (2\sin x + 1) = 0$ $\Leftrightarrow$ $\cos x = 0$ or $2\sin x + 1 = 0$ $\Leftrightarrow$ $\sin x = -\frac{1}{2}$. On $[0, 2\pi)$: $\cos x = 0$ $\Leftrightarrow$ $x = \frac{\pi}{2}, \frac{3\pi}{2}$ and $\sin x = -\frac{1}{2}$ $\Leftrightarrow$ $x = \frac{7\pi}{6}, \frac{11\pi}{6}$. Thus the solutions are $x = \frac{\pi}{2} + k\pi, x = \frac{7\pi}{6} + 2k\pi, x = \frac{11\pi}{6} + 2k\pi, k = 0, \pm1, \pm2, \ldots$.

11. $(\tan x + \sqrt{3})(\cos x + 2) = 0$ $\Leftrightarrow$ $\tan x + \sqrt{3} = 0$ or $\cos x + 2 = 0$. Since $|\cos x| \leq 1$ for all x, there is no solution for $\cos x + 2 = 0$. Hence, $\tan x + \sqrt{3} = 0$ $\Leftrightarrow$ $\tan x = -\sqrt{3}$ $\Leftrightarrow$ $x = -\frac{\pi}{3}$ on $(-\frac{\pi}{2}, \frac{\pi}{2})$. Thus the solutions are $x = -\frac{\pi}{3} + k\pi, k = 0, \pm1, \pm2, \ldots$.

13. $\cos x \sin x - 2\cos x = 0$ $\Leftrightarrow$ $\cos x (\sin x - 2) = 0$ $\Leftrightarrow$ $\cos x = 0$ or $\sin x - 2 = 0$. Since $|\sin x| \leq 1$ for all x, there is no solution for $\sin x - 2 = 0$. Hence, $\cos x = 0$ $\Leftrightarrow$ $x = \frac{\pi}{2} + 2k\pi, \frac{3\pi}{2} + 2k\pi$ $\Leftrightarrow$ $x = \frac{\pi}{2} + k\pi, k = 0, \pm1, \pm2, \ldots$.

15. $4\cos^2 x - 4\cos x + 1 = 0$ $\Leftrightarrow$ $(2\cos x - 1)^2 = 0$ $\Leftrightarrow$ $2\cos x - 1 = 0$ $\Leftrightarrow$ $\cos x = \frac{1}{2}$ $\Leftrightarrow$ $x = \frac{\pi}{3} + 2k\pi, \frac{5\pi}{3} + 2k\pi, k = 0, \pm1, \pm2, \ldots$.

17. $\sin^2 x = 2\sin x + 3$ $\Leftrightarrow$ $\sin^2 x - 2\sin x - 3 = 0$ $\Leftrightarrow$ $(\sin x - 3)(\sin x + 1) = 0$ $\Leftrightarrow$ $\sin x - 3 = 0$ or $\sin x + 1 = 0$. Since $|\sin x| \leq 1$ for all x, there is no solution for $\sin x - 3 = 0$. Hence $\sin x + 1 = 0$ $\Leftrightarrow$ $\sin x = -1$ $\Leftrightarrow$ $x = \frac{3\pi}{2} + 2k\pi, k = 0, \pm1, \pm2, \ldots$.

19. $\sin^2 x = 4 - 2\cos^2 x$ $\Leftrightarrow$ $\sin^2 x + \cos^2 x + \cos^2 x = 4$ $\Leftrightarrow$ $1 + \cos^2 x = 4$ $\Leftrightarrow$ $\cos^2 x = 3$. Since $|\cos x| \leq 1$ for all x, it follows that $\cos^2 x \leq 1$ and so there are no solutions for $\cos^2 x = 3$.

21. $2\sin 3x - 1 = 0$ $\Leftrightarrow$ $2\sin 3x = 1$ $\Leftrightarrow$ $\sin 3x = \frac{1}{2}$. In the interval $[0, 6\pi)$ the solution are $3x = \frac{\pi}{6}, \frac{5\pi}{6}, \frac{13\pi}{6}, \frac{17\pi}{6}, \frac{25\pi}{6}, \frac{29\pi}{6}$ $\Leftrightarrow$ $x = \frac{\pi}{18}, \frac{5\pi}{18}, \frac{13\pi}{18}, \frac{17\pi}{18}, \frac{25\pi}{18}, \frac{29\pi}{18}$. So $x = \frac{\pi}{18} + \frac{2k\pi}{3}, \frac{5\pi}{18} + \frac{2k\pi}{3}$, for any integer k.

23. $\cos \frac{x}{2} - 1 = 0$ $\Leftrightarrow$ $\cos \frac{x}{2} = 1$ $\Leftrightarrow$ $\frac{x}{2} = 2k\pi$ $\Leftrightarrow$ $x = 4k\pi$, for any integer k.

25. $\tan^5 x - 9\tan x = 0$ $\Leftrightarrow$ $\tan x (\tan^4 x - 9) = 0$ $\Leftrightarrow$ $\tan x = 0$ or $\tan^4 x = 9$ $\Leftrightarrow$ $\tan x = 0$ or $\tan x = \pm\sqrt{3}$ $\Leftrightarrow$ $x = 0, \pi, \frac{\pi}{3}, \frac{2\pi}{3}, \frac{4\pi}{3}, \frac{5\pi}{3}$ in $[0, 2\pi)$. Thus $x = \frac{k\pi}{3}$, for any integer k.

27. $4\sin x \cos x + 2\sin x - 2\cos x - 1 = 0$ $\Leftrightarrow$ $(2\sin x - 1)(2\cos x + 1) = 0$ $\Leftrightarrow$

$2\sin x - 1 = 0$ or $2\cos x + 1 = 0$ $\Leftrightarrow$ $\sin x = \frac{1}{2}$ or $\cos x = -\frac{1}{2}$ $\Leftrightarrow$ $x = \frac{\pi}{6}, \frac{5\pi}{6}, \frac{2\pi}{3}, \frac{4\pi}{3}$ in $[0, 2\pi)$. Thus $x = \frac{\pi}{6} + 2k\pi, \frac{5\pi}{6} + 2k\pi, \frac{2\pi}{3} + 2k\pi, \frac{4\pi}{3} + 2k\pi$, for any integer k.

29. $\cos^2 2x - \sin^2 2x = 0$ $\Leftrightarrow$ $(\cos 2x - \sin 2x)(\cos 2x + \sin 2x) = 0$ $\Leftrightarrow$ $\cos 2x = \pm\sin 2x$
    $\Leftrightarrow$ $\tan 2x = \pm 1$ $\Leftrightarrow$ $2x = \frac{\pi}{4}, \frac{3\pi}{4}, \frac{5\pi}{4}, \frac{7\pi}{4}, \frac{9\pi}{4}, \frac{11\pi}{4}, \frac{13\pi}{4}, \frac{15\pi}{4}$ in $[0, 4\pi)$ $\Leftrightarrow$ $x = \frac{\pi}{8}, \frac{3\pi}{8}, \frac{5\pi}{8},$
    $\frac{7\pi}{8}, \frac{9\pi}{8}, \frac{11\pi}{8}, \frac{13\pi}{8}, \frac{15\pi}{8}$ in $[0, 2\pi)$. So the solution can be expressed as the odd multiples of $\frac{\pi}{8}$ which are
    $x = \frac{\pi}{8} + \frac{k\pi}{4}$, for any integer k.

31. $2\cos 3x = 1$ $\Leftrightarrow$ $\cos 3x = \frac{1}{2}$ $\Rightarrow$ $3x = \frac{\pi}{3}, \frac{5\pi}{3}, \frac{7\pi}{3}, \frac{11\pi}{3}, \frac{13\pi}{3}, \frac{17\pi}{3}$ on $[0, 6\pi)$ $\Leftrightarrow$ $x = \frac{\pi}{9},$
    $\frac{5\pi}{9}, \frac{7\pi}{9}, \frac{11\pi}{9}, \frac{13\pi}{9}, \frac{17\pi}{9}$ on $[0, 2\pi)$.

33. $2\sin x \tan x - \tan x = 1 - 2\sin x$ $\Leftrightarrow$ $2\sin x \tan x - \tan x + 2\sin x - 1 = 0$ $\Leftrightarrow$
    $(2\sin x - 1)(\tan x + 1) = 0$ $\Leftrightarrow$ $2\sin x - 1 = 0$ or $\tan x + 1 = 0$ $\Leftrightarrow$ $\sin x = \frac{1}{2}$ or
    $\tan x = -1$ $\Leftrightarrow$ $x = \frac{\pi}{6}, \frac{5\pi}{6}$ or $x = \frac{3\pi}{4}, \frac{7\pi}{4}$. Thus, the solutions in $[0, 2\pi)$ are $\frac{\pi}{6}, \frac{3\pi}{4}, \frac{5\pi}{6}, \frac{7\pi}{4}$.

35. $\tan x - 3\cot x = 0$ $\Leftrightarrow$ $\dfrac{\sin x}{\cos x} - \dfrac{3\cos x}{\sin x} = 0$ $\Leftrightarrow$ $\dfrac{\sin^2 x - 3\cos^2 x}{\cos x \sin x} = 0$ $\Leftrightarrow$
    $\dfrac{\sin^2 x + \cos^2 x - 4\cos^2 x}{\cos x \sin x} = 0$ $\Leftrightarrow$ $\dfrac{1 - 4\cos^2 x}{\cos x \sin x} = 0$ $\Leftrightarrow$ $1 - 4\cos^2 x = 0$ $\Leftrightarrow$
    $4\cos^2 x = 1$ $\Leftrightarrow$ $\cos x = \pm\frac{1}{2}$ $\Leftrightarrow$ $x = \frac{\pi}{3}, \frac{2\pi}{3}, \frac{4\pi}{3}, \frac{5\pi}{3}$ in $[0, 2\pi)$.

37. $\tan 3x + 1 = \sec 3x$ $\Rightarrow$ $(\tan 3x + 1)^2 = \sec^2 3x$ $\Leftrightarrow$ $\tan^2 3x + 2\tan 3x + 1 = \sec^2 3x$ $\Leftrightarrow$
    $\sec^2 3x + 2\tan 3x = \sec^2 3x$ $\Leftrightarrow$ $2\tan 3x = 0$ $\Leftrightarrow$ $3x = 0, \pi, 2\pi, 3\pi, 4\pi, 5\pi$ in $[0, 6\pi)$
    $\Leftrightarrow$ $x = 0, \frac{\pi}{3}, \frac{2\pi}{3}, \pi, \frac{4\pi}{3}, \frac{5\pi}{3}$ in $[0, 2\pi)$. Since squaring both sides is an operation that can
    introduce extraneous solutions, we must check each of the possible solution in the original equation.
    We see that $x = \frac{\pi}{3}, \pi, \frac{5\pi}{3}$ are not solutions. Hence the only solutions are $x = 0, \frac{2\pi}{3}, \frac{4\pi}{3}$.

39. (a) $\cos x = 0.4$ $\Rightarrow$ $x = \cos^{-1} 0.4 \approx 1.15928$. The other solution is
    $2\pi - \cos^{-1} 0.4 \approx 5.12391$.

    (b) Since the period of cosine is 2π the solutions are $x \approx 1.15928 + 2k\pi$ and $x \approx 5.12391 + 2k\pi$,
    for any integer k.

41. (a) $\sec x = 5$ $\Leftrightarrow$ $\cos x = \frac{1}{5}$ $\Rightarrow$ $x = \cos^{-1} \frac{1}{5} \approx 1.36944$. The other solution is
    $2\pi - \cos^{-1} \frac{1}{5} \approx 4.91375$.

    (b) Since the period of secant is 2π the solutions are $x \approx 1.36943 \pm 2k\pi$ and $\approx 4.91375 \pm 2k\pi$,
    for any integer k.

43. (a) $5\sin^2 x - 1 = 0$ $\Rightarrow$ $\sin^2 x = \frac{1}{5}$ $\Rightarrow$ $\sin x = \pm\frac{1}{\sqrt{5}}$. Now $\sin x = \frac{1}{\sqrt{5}}$ $\Rightarrow$
    $x = \sin^{-1}\left(\frac{1}{\sqrt{5}}\right) \approx 0.46365$, and $x \approx \pi - 0.46365 \approx 2.67795$. Also, $\sin x = -\frac{1}{\sqrt{5}}$ $\Rightarrow$
    $x = \sin^{-1}\left(-\frac{1}{\sqrt{5}}\right) \approx -0.46365$, so in the interval $[0, 2\pi)$ $x \approx \pi - (-0.46365) \approx 3.60524$ and
    $x \approx 2\pi - 0.46365 \approx 5.81954$

    (b) Since $\sin(\pi + x) = -\sin x$, we can express the general solutions as $x \approx 0.46365 + k\pi$ and
    $x \approx 2.67795 + k\pi$, for any integer k.

45. (a) $3\sin^2 x - 7\sin x + 2 = 0$ $\Rightarrow$ $(3\sin x - 1)(\sin x - 2) = 0$ $\Rightarrow$ $3\sin x - 1 = 0$ or $\sin$
    $x - 2 = 0$. Since $|\sin x| \leq 1$, $\sin x - 2 = 0$ has no solution. Thus $3\sin x - 1 = 0$ $\Rightarrow$
    $\sin x = \frac{1}{3}$ $\Rightarrow$ $x \approx 0.33984$ and $x \approx \pi - 0.33984 \approx 2.80176$.

(b) Since the period for sine is 2π, the general solution is of the form $x \approx 0.33984 + 2k\pi$ and $x \approx 2.80176 + 2k\pi$, for any integer k.

47. $f(x) = 3\cos x + 1$; $g(x) = \cos x - 1$.
$f(x) = g(x)$ when $3\cos x + 1 = \cos x - 1$
$\Leftrightarrow \quad 2\cos x = -2 \quad \Leftrightarrow \quad \cos x = -1 \quad \Leftrightarrow$
$x = \pi + 2k\pi = (2k+1)\pi$. The points of
intersection are $([2k+1]\pi, -2)$, for any
integer k.

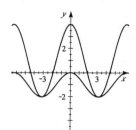

49. $f(x) = \tan x$; $g(x) = \sqrt{3}$. $f(x) = g(x)$ when
$\tan x = \sqrt{3} \quad \Leftrightarrow \quad x = \frac{\pi}{3} + k\pi$. The
intersection points are $(\frac{\pi}{3} + k\pi, \sqrt{3})$, for any
integer k.

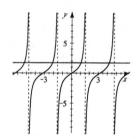

51. $\cos x \cos 3x - \sin x \sin 3x = 0 \quad \Leftrightarrow \quad \cos(x + 3x) = 0 \quad \Leftrightarrow \quad \cos 4x = 0 \quad \Leftrightarrow \quad 4x = \frac{\pi}{2}, \frac{3\pi}{2},$
$\frac{5\pi}{2}, \frac{7\pi}{2}, \frac{9\pi}{2}, \frac{11\pi}{2}, \frac{13\pi}{2}, \frac{15\pi}{2}$ in $[0, 8\pi) \quad \Leftrightarrow \quad x = \frac{\pi}{8}, \frac{3\pi}{8}, \frac{5\pi}{8}, \frac{7\pi}{8}, \frac{9\pi}{8}, \frac{11\pi}{8}, \frac{13\pi}{8}, \frac{15\pi}{8}$ in $[0, 2\pi)$.

53. $\sin 2x \cos x + \cos 2x \sin x = \frac{\sqrt{3}}{2} \quad \Leftrightarrow \quad \sin(2x + x) = \frac{\sqrt{3}}{2} \quad \Leftrightarrow \quad \sin 3x = \frac{\sqrt{3}}{2} \quad \Leftrightarrow \quad 3x = \frac{\pi}{3},$
$\frac{2\pi}{3}, \frac{7\pi}{3}, \frac{8\pi}{3}, \frac{13\pi}{3}, \frac{14\pi}{3}$ in $[0, 6\pi) \quad \Leftrightarrow \quad x = \frac{\pi}{9}, \frac{2\pi}{9}, \frac{7\pi}{9}, \frac{8\pi}{9}, \frac{13\pi}{9}, \frac{14\pi}{9}$ in $[0, 2\pi)$.

55. $\sin 2x + \cos x = 0 \quad \Leftrightarrow \quad 2\sin x \cos x + \cos x = 0 \quad \Leftrightarrow \quad \cos x (2\sin x + 1) = 0 \quad \Leftrightarrow$
$\cos x = 0$ or $\sin x = -\frac{1}{2} \quad \Leftrightarrow \quad x = \frac{\pi}{2}, \frac{7\pi}{6}, \frac{3\pi}{2}, \frac{11\pi}{6}$.

57. $\cos 2x + \cos x = 2 \quad \Leftrightarrow \quad 2\cos^2 x - 1 + \cos x - 2 = 0 \quad \Leftrightarrow \quad 2\cos^2 x + \cos x - 3 = 0$
$\Leftrightarrow \quad (2\cos x + 3)(\cos x - 1) = 0 \quad \Leftrightarrow \quad 2\cos x + 3 = 0$ or $\cos x - 1 = 0 \quad \Leftrightarrow$
$\cos x = -\frac{3}{2}$ (which is impossible) or $\cos x = 1 \quad \Leftrightarrow \quad x = 0$ in $[0, 2\pi)$.

59. $\sin x + \sin 3x = 0 \quad \Leftrightarrow \quad 2\sin 2x \cos(-x) = 0 \quad \Leftrightarrow \quad 2\sin 2x \cos x = 0 \quad \Leftrightarrow \quad \sin 2x = 0$ or
$\cos x = 0 \quad \Leftrightarrow \quad 2x = k\pi$ or $2x = \frac{k\pi}{2} \quad \Leftrightarrow \quad x = \frac{k\pi}{2}$, for any integer k.

61. $\cos 4x + \cos 2x = \cos x \quad \Leftrightarrow \quad 2\cos 3x \cos x = \cos x \quad \Leftrightarrow \quad \cos x (2\cos 3x - 1) = 0 \quad \Leftrightarrow$
$\cos x = 0$ or $\cos 3x = \frac{1}{2} \quad \Leftrightarrow \quad x = \frac{\pi(2k+1)}{2}$ or $3x = \frac{\pi}{3} + 2k\pi, \frac{5\pi}{3} + 2k\pi, \frac{7\pi}{3} + 2k\pi, \frac{11\pi}{3} + 2k\pi,$
$\frac{13\pi}{3} + 2k\pi, \frac{17\pi}{3} + 2k\pi \quad \Leftrightarrow \quad x = \frac{\pi(2k+1)}{2}, \frac{\pi}{9} + \frac{2k\pi}{3}, \frac{5\pi}{9} + \frac{2k\pi}{3}$.

63. $\sin 2x = x$

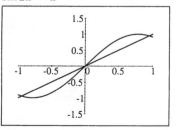

3 solutions: $x = 0$, $x \approx \pm 0.95$

65. $2^{\sin x} = x$

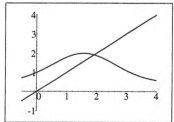

1 solution: $x \approx 1.92$

67. $\dfrac{\cos x}{1 + x^2} = x^2$

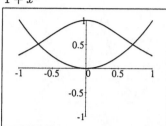

2 solutions: $x \approx \pm 0.71$

69. "A" is a true statement. Every identity is an equation. The difference between an identity and an equation is that an identity is true for all values in the domain, whereas an equation may only be true for certain values in the domain and false for others.

Exercises 9.6

1. $1 + i$. Then $\tan\theta = \frac{1}{1} = 1$ with θ in quadrant I $\Rightarrow$ $\theta = \frac{\pi}{4}$, and $r = \sqrt{1^2 + 1^2} = \sqrt{2}$. Hence,
$1 + i = \sqrt{2}\left(\cos\frac{\pi}{4} + i\sin\frac{\pi}{4}\right)$.

3. $\sqrt{2} - \sqrt{2}\,i$. Then $\tan\theta = \frac{\sqrt{2}}{\sqrt{2}} = -1$ with θ in quadrant IV $\Rightarrow$ $\theta = \frac{7\pi}{4}$, and $r = \sqrt{2+2} = 2$.
Hence, $\sqrt{2} - \sqrt{2}\,i = 2\left(\cos\frac{7\pi}{4} + i\sin\frac{7\pi}{4}\right)$.

5. $2\sqrt{3} - 2i$. Then $\tan\theta = \frac{-2}{2\sqrt{3}} = -\frac{1}{\sqrt{3}}$ with θ in quadrant IV $\Rightarrow$ $\theta = \frac{11\pi}{6}$, and
$r = \sqrt{12+4} = 4$. Hence, $2\sqrt{3} - 2i = 4\left(\cos\frac{11\pi}{6} + i\sin\frac{11\pi}{6}\right)$.

7. $-\sqrt{2}\,i$. Then $\theta = \frac{3\pi}{2}$, and $r = \sqrt{0+2} = \sqrt{2}$. Hence, $-\sqrt{2}\,i = \sqrt{2}\left(\cos\frac{3\pi}{2} + i\sin\frac{3\pi}{2}\right)$.

9. $5 + 5i$. Then $\tan\theta = \frac{5}{5} = 1$ with θ in quadrant I $\Rightarrow$ $\theta = \frac{\pi}{4}$, and $r = \sqrt{25+25} = 5\sqrt{2}$.
Hence, $5 + 5i = 5\sqrt{2}\left(\cos\frac{\pi}{4} + i\sin\frac{\pi}{4}\right)$.

11. $4\sqrt{3} - 4i$. Then $\tan\theta = \frac{-4}{4\sqrt{3}} = -\frac{1}{\sqrt{3}}$ with θ in quadrant IV $\Rightarrow$ $\theta = \frac{11\pi}{6}$, and
$r = \sqrt{48+16} = 8$. Hence, $4\sqrt{3} - 4i = 8\left(\cos\frac{11\pi}{6} + i\sin\frac{11\pi}{6}\right)$.

13. -20. Then $\theta = \pi$, and $r = 20$. Hence, $-20 = 20\left(\cos\pi + i\sin\pi\right)$.

15. $3 + 4i$. Then $\tan\theta = \frac{4}{3}$ with θ in quadrant I $\Rightarrow$ $\theta = \tan^{-1}\frac{4}{3}$, and $r = \sqrt{9+16} = 5$.   Hence,
$3 + 4i = 5\left[\cos(\tan^{-1}\frac{4}{3}) + i\sin(\tan^{-1}\frac{4}{3})\right]$.

17. $3i\,(1 + i) = -3 + 3i$. Then $\tan\theta = \frac{3}{-3} = -1$ with θ in quadrant II $\Rightarrow$ $\theta = \frac{3\pi}{4}$, and
$r = \sqrt{9+9} = 3\sqrt{2}$. Hence, $3i\,(1 + i) = 3\sqrt{2}\left(\cos\frac{3\pi}{4} + i\sin\frac{3\pi}{4}\right)$.

19. $4\left(\sqrt{3} + i\right) = 4\sqrt{3} + 4i$. Then $\tan\theta = \frac{4}{4\sqrt{3}} = \frac{1}{\sqrt{3}}$ with θ in quadrant I $\Rightarrow$ $\theta = \frac{\pi}{6}$, and
$r = \sqrt{48+16} = 8$. Hence, $4\left(\sqrt{3} + i\right) = 8\left(\cos\frac{\pi}{6} + i\sin\frac{\pi}{6}\right)$.

21. $2 + i$. Then $\tan\theta = \frac{1}{2}$ with θ in quadrant I $\Rightarrow$ $\theta = \tan^{-1}\frac{1}{2}$, and $r = \sqrt{4+1} = \sqrt{5}$. Hence,
$2 + i = \sqrt{5}\left[\cos(\tan^{-1}\frac{1}{2}) + i\sin(\tan^{-1}\frac{1}{2})\right]$.

23. $\sqrt{2} + \sqrt{2}\,i$. Then $\tan\theta = \frac{\sqrt{2}}{\sqrt{2}} = 1$ with θ in quadrant I $\Rightarrow$ $\theta = \frac{\pi}{4}$, and $r = \sqrt{2+2} = 2$.
Hence, $2 + \sqrt{2}\,i = 2\left(\cos\frac{\pi}{4} + i\sin\frac{\pi}{4}\right)$.

25. $z_1 = \cos\frac{\pi}{4} + i\sin\frac{\pi}{4}$, $z_2 = \cos\frac{3\pi}{4} + i\sin\frac{3\pi}{4}$

     $z_1 z_2 = \cos\left(\frac{\pi}{4} + \frac{3\pi}{4}\right) + i\sin\left(\frac{\pi}{4} + \frac{3\pi}{4}\right) = \cos\pi + i\sin\pi$

     $z_1/z_2 = \cos\left(\frac{\pi}{4} - \frac{3\pi}{4}\right) + i\sin\left(\frac{\pi}{4} - \frac{3\pi}{4}\right) = \cos\left(-\frac{\pi}{2}\right) + i\sin\left(-\frac{\pi}{2}\right) = \cos\frac{\pi}{2} - i\sin\frac{\pi}{2}$

27. $z_1 = 7\left(\cos\frac{9\pi}{7} + i\sin\frac{9\pi}{7}\right)$, $z_2 = 2\left(\cos\frac{3\pi}{7} + i\sin\frac{3\pi}{7}\right)$

     $z_1 z_2 = 7\cdot 2\left[\cos\left(\frac{9\pi}{7} + \frac{3\pi}{7}\right) + i\sin\left(\frac{9\pi}{7} + \frac{3\pi}{7}\right)\right] = 14\left(\cos\frac{12\pi}{7} + i\sin\frac{12\pi}{7}\right)$

     $z_1/z_2 = \frac{7}{2}\left[\cos\left(\frac{9\pi}{7} - \frac{3\pi}{7}\right) + i\sin\left(\frac{9\pi}{7} - \frac{3\pi}{7}\right)\right] = \frac{7}{2}\left(\cos\frac{6\pi}{7} + i\sin\frac{6\pi}{7}\right)$

29. $z_1 = 4\,(\cos 200° + i\sin 200°)$, $z_2 = 25\,(\cos 150° + i\sin 150°)$

     $z_1 z_2 = 4 \cdot 25[\cos(200° + 150°) + i\sin(200° + 150°)] = 100(\cos 350° + i\sin 350°)$

     $z_1/z_2 = \frac{4}{25}\,[\cos(200° - 150°) + i\sin(200° - 150°)] = \frac{4}{25}\,(\cos 50° + i\sin 50°)$

31. $z_1 = \sqrt{3} + i$. Then $\tan\theta_1 = \frac{1}{\sqrt{3}}$ with θ_1 in quadrant I $\quad\Rightarrow\quad \theta_1 = \frac{\pi}{6}$, and $r_1 = \sqrt{3+1} = 2$.

     $z_2 = 1 + \sqrt{3}\,i$. Then $\tan\theta_2 = \sqrt{3}$ with θ_2 in quadrant I $\quad\Rightarrow\quad \theta_2 = \frac{\pi}{3}$, and $r_1 = \sqrt{1+3} = 2$.

     Hence, $z_1 = 2\,(\cos\frac{\pi}{6} + i\sin\frac{\pi}{6})$ and $z_2 = 2\,(\cos\frac{\pi}{3} + i\sin\frac{\pi}{3})$.

     $z_1 z_2 = 2 \cdot 2\,[\cos(\frac{\pi}{6} + \frac{\pi}{3}) + i\sin(\frac{\pi}{6} + \frac{\pi}{3})] = 4\,(\cos\frac{\pi}{2} + i\sin\frac{\pi}{2})$

     $z_1/z_2 = \frac{2}{2}\,[\cos(\frac{\pi}{6} - \frac{\pi}{3}) + i\sin(\frac{\pi}{6} - \frac{\pi}{3})] = \cos(-\frac{\pi}{6}) + i\sin(-\frac{\pi}{6}) = \cos\frac{\pi}{6} - i\sin\frac{\pi}{6}$

     $1/z_1 = \frac{1}{2}[\cos(-\frac{\pi}{6}) + i\sin(-\frac{\pi}{6})] = \frac{1}{2}\,(\cos\frac{\pi}{6} - i\sin\frac{\pi}{6})$

33. $z_1 = 2\sqrt{3} - 2i$. Then $\tan\theta_1 = \frac{-2}{2\sqrt{3}} = -\frac{1}{\sqrt{3}}$ with θ_1 in quadrant IV $\quad\Rightarrow\quad \theta_1 = \frac{11\pi}{6}$, and

     $r_1 = \sqrt{12 + 4} = 4$.

     $z_2 = -1 + i$. Then $\tan\theta_2 = -1$ with θ_2 in quadrant II $\quad\Rightarrow\quad \theta_2 = \frac{3\pi}{4}$, and $r_2 = \sqrt{1+1} = \sqrt{2}$.

     Hence, $z_1 = 4\,(\cos\frac{11\pi}{6} + i\sin\frac{11\pi}{6})$ and $z_2 = \sqrt{2}\,(\cos\frac{3\pi}{4} + i\sin\frac{3\pi}{4})$.

     $z_1 z_2 = 4 \cdot \sqrt{2}\,[\cos(\frac{11\pi}{6} + \frac{3\pi}{4}) + i\sin(\frac{11\pi}{6} + \frac{3\pi}{4})] = 4\sqrt{2}\,(\cos\frac{7\pi}{12} + i\sin\frac{7\pi}{12})$

     $z_1/z_2 = \frac{4}{\sqrt{2}}\,[\cos(\frac{11\pi}{6} - \frac{3\pi}{4}) + i\sin(\frac{11\pi}{6} - \frac{3\pi}{4})] = 2\sqrt{2}\,(\cos\frac{13\pi}{12} + i\sin\frac{13\pi}{12})$

     $1/z_1 = \frac{1}{4}\,(\cos(-\frac{11\pi}{6}) + i\sin(-\frac{11\pi}{6})) = \frac{1}{4}\,(\cos\frac{11\pi}{6} - i\sin\frac{11\pi}{6})$

35. $z_1 = 5 + 5i$. Then $\tan\theta_1 = \frac{5}{5} = 1$ with θ_1 in quadrant I $\quad\Rightarrow\quad \theta_1 = \frac{\pi}{4}$, and

     $r_1 = \sqrt{25 + 25} = 5\sqrt{2}$.

     $z_2 = 4$. Then $\theta_2 = 0$, and $r_2 = 4$.

     Hence, $z_1 = 5\sqrt{2}\,(\cos\frac{\pi}{4} + i\sin\frac{\pi}{4})$ and $z_2 = 4\,(\cos 0 + i\sin 0)$.

     $z_1 z_2 = 5\sqrt{2} \cdot 4\,[\cos(\frac{\pi}{4} + 0) + i\sin(\frac{\pi}{4} + 0)] = 20\sqrt{2}\,(\cos\frac{\pi}{4} + i\sin\frac{\pi}{4})$

     $z_1/z_2 = \frac{5\sqrt{2}}{4}\,(\cos\frac{\pi}{4} + i\sin\frac{\pi}{4})$

     $1/z_1 = \frac{1}{5\sqrt{2}}\,(\cos(-\frac{\pi}{4}) + i\sin(-\frac{\pi}{4})) = \frac{\sqrt{2}}{10}\,(\cos\frac{\pi}{4} - i\sin\frac{\pi}{4})$

37. $z_1 = -20$. Then $\theta_1 = \pi$, and $r_1 = 20$.

     $z_2 = \sqrt{3} + i$. Then $\tan\theta_2 = \frac{1}{\sqrt{3}}$ with θ_2 in quadrant I $\quad\Rightarrow\quad \theta_2 = \frac{\pi}{6}$, and $r_2 = \sqrt{3+1} = 2$.

     Hence, $z_1 = 20\,(\cos\pi + i\sin\pi)$ and $z_2 = 2\,(\cos\frac{\pi}{6} + i\sin\frac{\pi}{6})$.

     $z_1 z_2 = 20 \cdot 2\,[\cos(\pi + \frac{\pi}{6}) + i\sin(\pi + \frac{\pi}{6})] = 40\,(\cos\frac{7\pi}{6} + i\sin\frac{7\pi}{6})$

     $z_1/z_2 = \frac{20}{2}\,[\cos(\pi - \frac{\pi}{6}) + i\sin(\pi - \frac{\pi}{6})] = 10\,(\cos\frac{5\pi}{6} + i\sin\frac{5\pi}{6})$

     $1/z_1 = \frac{1}{20}\,[\cos(-\pi) + i\sin(-\pi)] = \frac{1}{20}\,(\cos\pi - i\sin\pi)$

39.   From Exercise 1: $1 + i = \sqrt{2}\,(\cos\frac{\pi}{4} + i\sin\frac{\pi}{4})$. Thus,

     $(1 + i)^{20} = (\sqrt{2})^{20}\big(\cos 20(\frac{\pi}{4}) + i\sin 20(\frac{\pi}{4})\big) = (2^{1/2})^{20}\,(\cos 5\pi + i\sin 5\pi)$

     $= 2^{10}(-1 + 0i) = -1024$

41. $r = \sqrt{12 + 4} = 4$ and $\tan\theta = \frac{2}{2\sqrt{3}} = \frac{1}{\sqrt{3}}$ $\Rightarrow$ $\theta = \frac{\pi}{6}$. Thus $2\sqrt{3} + 2i = 4\left(\cos\frac{\pi}{6} + i\sin\frac{\pi}{6}\right)$.

So, $(2\sqrt{3} + 2i)^5 = 4^5\left(\cos\frac{5\pi}{6} + i\sin\frac{5\pi}{6}\right) = 1024\left(-\frac{\sqrt{3}}{2} + \frac{1}{2}i\right) = 512\left(-\sqrt{3} + i\right)$

43. $r = \sqrt{\frac{1}{2} + \frac{1}{2}} = 1$ and $\tan\theta = 1$ $\Rightarrow$ $\theta = \frac{\pi}{4}$. Thus $\frac{\sqrt{2}}{2} + \frac{\sqrt{2}}{2}i = \cos\frac{\pi}{4} + i\sin\frac{\pi}{4}$. Therefore,

$\left(\frac{\sqrt{2}}{2} + \frac{\sqrt{2}}{2}i\right)^{12} = \cos 12\left(\frac{\pi}{4}\right) + i\sin 12\left(\frac{\pi}{4}\right) = \cos 3\pi + i\sin 3\pi = -1$

45. $r = \sqrt{4 + 4} = 4\sqrt{2}$ and $\tan\theta = -1$ with θ in quadrant IV $\Rightarrow$ $\theta = \frac{7\pi}{4}$. Thus

$2 - 2i = 2\sqrt{2}\left(\cos\frac{7\pi}{4} + i\sin\frac{7\pi}{4}\right)$. So, $(2 - 2i)^8 = (2\sqrt{2})^8\left(\cos 14\pi + i\sin 14\pi\right)$

$= 4096\left(1 - 0i\right) = 4096$

47. $r = \sqrt{1 + 1} = \sqrt{2}$ and $\tan\theta = 1$ with θ in quadrant III $\Rightarrow$ $\theta = \frac{5\pi}{4}$. Thus

$-1 - i = \sqrt{2}\left(\cos\frac{5\pi}{4} + i\sin\frac{5\pi}{4}\right)$. Thus, $(-1 - i)^7 = (\sqrt{2})^7\left(\cos\frac{35\pi}{4} + i\sin\frac{35\pi}{4}\right)$

$= 8\sqrt{2}\left(\cos\frac{3\pi}{4} + i\sin\frac{3\pi}{4}\right) = 8\sqrt{2}\left(\frac{1}{\sqrt{2}} - i\frac{1}{\sqrt{2}}\right) = 8\left(-1 + i\right)$

49. $r = \sqrt{12 + 4} = 4$ and $\tan\theta = \frac{2}{2\sqrt{3}} = \frac{1}{\sqrt{3}}$ $\Rightarrow$ $\theta = \frac{\pi}{6}$. Thus $2\sqrt{3} + 2i = 4\left(\cos\frac{\pi}{6} + i\sin\frac{\pi}{6}\right)$.

So, $(2\sqrt{3} + 2i)^{-5} = \left(\frac{1}{4}\right)^5\left(\cos\frac{-5\pi}{6} + i\sin\frac{-5\pi}{6}\right) = \frac{1}{1024}\left(-\frac{\sqrt{3}}{2} - \frac{1}{2}i\right) = \frac{1}{2048}\left(-\sqrt{3} - i\right)$

51. $r = \sqrt{48 + 16} = 8$ and $\tan\theta = \frac{4}{4\sqrt{3}} = \frac{1}{\sqrt{3}}$ $\Rightarrow$ $\theta = \frac{\pi}{6}$. Thus

$4\sqrt{3} + 4i = 8\left(\cos\frac{\pi}{6} + i\sin\frac{\pi}{6}\right)$. So,

$(4\sqrt{3} + 4i)^{1/2} = \sqrt{8}\left[\cos\left(\frac{\pi/6 + 2k\pi}{2}\right) + i\sin\left(\frac{\pi/6 + 2k\pi}{2}\right)\right]$, for $k = 0, 1$.

Thus the two roots are $w_0 = 2\sqrt{2}\left(\cos\frac{\pi}{12} + i\sin\frac{\pi}{12}\right)$ and

$w_1 = 2\sqrt{2}\left(\cos\frac{13\pi}{12} + i\sin\frac{13\pi}{12}\right)$

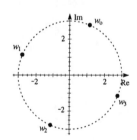

53. $-81i = 81\left(\cos\frac{3\pi}{2} + i\sin\frac{3\pi}{2}\right)$. Thus, $(-81i)^{1/4} = 81^{1/4}$

$\left[\cos\left(\frac{3\pi/2 + 2k\pi}{4}\right) + i\sin\left(\frac{3\pi/2 + 2k\pi}{4}\right)\right]$ for $k = 0, 1, 2, 3$. The four roots

are $w_0 = 3\left(\cos\frac{3\pi}{8} + i\sin\frac{3\pi}{8}\right)$, $w_1 = 3\left(\cos\frac{7\pi}{8} + i\sin\frac{7\pi}{8}\right)$,

$w_2 = 3\left(\cos\frac{11\pi}{8} + i\sin\frac{11\pi}{8}\right)$, and $w_3 = 3\left(\cos\frac{15\pi}{8} + i\sin\frac{15\pi}{8}\right)$

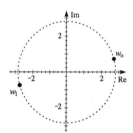

55. $1 = \cos 0 + i\sin 0$. Thus, $1^{1/8} = \cos\frac{2k\pi}{8} + i\sin\frac{2k\pi}{8}$, for $k = 0, 1, 2,$

$3, 4, 5, 6, 7$. So, the eight roots are $w_0 = \cos 0 + i\sin 0 = 1$,

$w_1 = \cos\frac{\pi}{4} + i\sin\frac{\pi}{4} = \frac{\sqrt{2}}{2} + i\frac{\sqrt{2}}{2}$, $w_2 = \cos\frac{\pi}{2} + i\sin\frac{\pi}{2} = i$,

$w_3 = \cos\frac{3\pi}{4} + i\sin\frac{3\pi}{4} = -\frac{\sqrt{2}}{2} + i\frac{\sqrt{2}}{2}$, $w_4 = \cos\pi + i\sin\pi = -1$,

$w_5 = \cos\frac{5\pi}{4} + i\sin\frac{5\pi}{4} = -\frac{\sqrt{2}}{2} - i\frac{\sqrt{2}}{2}$, $w_6 = \cos\frac{3\pi}{2} + i\sin\frac{3\pi}{2} = -i$,

and $w_7 = \cos\frac{7\pi}{4} + i\sin\frac{7\pi}{4} = \frac{\sqrt{2}}{2} - i\frac{\sqrt{2}}{2}$.

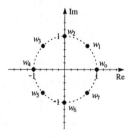

57. $i = \cos\frac{\pi}{2} + i\sin\frac{\pi}{2}$. So, $i^{1/3} = \cos\left(\frac{\pi/2+2k\pi}{3}\right) + i\sin\left(\frac{\pi/2+2k\pi}{3}\right)$ for

$k = 0, 1, 2$. Thus the three roots are $w_0 = \cos\frac{\pi}{6} + i\sin\frac{\pi}{6}$

$= \frac{\sqrt{3}}{2} + \frac{1}{2}i$, $w_1 = \cos\frac{5\pi}{6} + i\sin\frac{5\pi}{6} = -\frac{\sqrt{3}}{2} + \frac{1}{2}i$,

and $w_2 = \cos\frac{3\pi}{2} + i\sin\frac{3\pi}{2} = -i$.

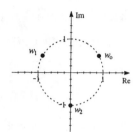

59. $-1 = \cos\pi + i\sin\pi$. Then $(-1)^{1/4} = \cos\left(\frac{\pi+2k\pi}{4}\right) + i\sin\left(\frac{\pi+2k\pi}{4}\right)$ for

$k = 0, 1, 2, 3$. So, the four roots are

$w_0 = \cos\frac{\pi}{4} + i\sin\frac{\pi}{4} = \frac{\sqrt{2}}{2} + i\frac{\sqrt{2}}{2}$, $w_1 = \cos\frac{3\pi}{4} + i\sin\frac{3\pi}{4}$

$= -\frac{\sqrt{2}}{2} + i\frac{\sqrt{2}}{2}$, $w_2 = \cos\frac{5\pi}{4} + i\sin\frac{5\pi}{4} = -\frac{\sqrt{2}}{2} - i\frac{\sqrt{2}}{2}$, and

$w_3 = \cos\frac{7\pi}{4} + i\sin\frac{7\pi}{4} = \frac{\sqrt{2}}{2} - i\frac{\sqrt{2}}{2}$.

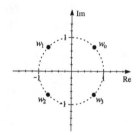

61. $z^4 + 1 = 0 \Leftrightarrow z = (-1)^{1/4} = \frac{\sqrt{2}}{2}(\pm 1 \pm i)$ (from Exercise 59).

63. $z^3 - 4\sqrt{3} - 4i = 0 \Leftrightarrow z = (4\sqrt{3} + 4i)^{1/3}$. Since $4\sqrt{3} + 4i = 8\left(\cos\frac{\pi}{6} + i\sin\frac{\pi}{6}\right)$,

$\left(4\sqrt{3} + 4i\right)^{1/3} = 8^{1/3}\left[\cos\left(\frac{\pi/6+2k\pi}{3}\right) + i\sin\left(\frac{\pi/6+2k\pi}{3}\right)\right]$, for $k = 0, 1, 2$. Thus there are three roots

$z = 2\left(\cos\frac{\pi}{18} + i\sin\frac{\pi}{18}\right)$, $z = 2\left(\cos\frac{13\pi}{18} + i\sin\frac{13\pi}{8}\right)$, $z = 2\left(\cos\frac{25\pi}{18} + i\sin\frac{25\pi}{18}\right)$

65. $z^3 + 1 = -i \Rightarrow z = (-1-i)^{1/3}$. Since $-1 - i = \sqrt{2}\left(\cos\frac{5\pi}{4} + i\sin\frac{5\pi}{4}\right)$,

$z = (-1-i)^{1/3} = 2^{1/6}\left[\cos\left(\frac{5\pi/4+2k\pi}{3}\right) + i\sin\left(\frac{5\pi/4+2k\pi}{3}\right)\right]$ for $k = 0, 1, 2$. Thus the three solutions to

this equation are $z = 2^{1/6}\left[\cos\frac{5\pi}{12} + i\sin\frac{5\pi}{12}\right]$, $2^{1/6}\left[\cos\frac{13\pi}{12} + i\sin\frac{13\pi}{12}\right]$, and $2^{1/6}\left[\cos\frac{21\pi}{12} + i\sin\frac{21\pi}{12}\right]$.

67. (a) $w = \cos\frac{2\pi}{n} + i\sin\frac{2\pi}{n}$ for a positive integer n. Then, $w^k = \cos\frac{2k\pi}{n} + i\sin\frac{2k\pi}{n}$. Now

$w^0 = \cos 0 + i\sin 0 = 1$ and for $k \neq 0$, $(w^k)^n = \cos 2k\pi + i\sin 2k\pi = 1$. So the nth roots of

1 are $\cos\frac{2k\pi}{n} + i\sin\frac{2k\pi}{n} = w^k$ for $k = 0, 1, 2, \ldots, n-1$. In other words, the nth roots of 1 are

$w^0, w^1, w^2, w^3, \ldots, w^{n-1}$ or $1, w, w^2, w^3, \ldots, w^{n-1}$.

(b) For $k = 0, 1, \ldots, n-1$, we have $(sw^k)^n = s^n(w^k)^n = z \cdot 1 = z$, so sw^k are nth roots of z for

$k = 0, 1, \ldots, n-1$.

69. The cube roots of 1 are $w^0 = 1$, $w^1 = \cos\frac{2\pi}{3} + i\sin\frac{2\pi}{3}$, $w^2 = \cos\frac{4\pi}{3} + i\sin\frac{4\pi}{3}$ so

$w^0 \cdot w^1 \cdot w^2 = (1)\left(\cos\frac{2\pi}{3} + i\sin\frac{2\pi}{3}\right)\left(\cos\frac{4\pi}{3} + i\sin\frac{4\pi}{3}\right) = \cos 2\pi + i\sin 2\pi = 1$.

The fourth roots of 1 are $w^0 = 1$, $w^1 = i$, $w^2 = -1$, $w^3 = -i$. So the product of the four fourth

roots of 1 is: $w^0 \cdot w^1 \cdot w^2 \cdot w^3 = (1) \cdot (i) \cdot (-1) \cdot (-i) = i^2 = -1$.

The fifth roots of 1 are $w^0 = 1$, $w^1 = \cos\frac{2\pi}{5} + i\sin\frac{2\pi}{5}$, $w^2 = \cos\frac{4\pi}{5} + i\sin\frac{4\pi}{5}$,

$w^3 = \cos\frac{6\pi}{5} + i\sin\frac{6\pi}{5}$, $w^4 = \cos\frac{8\pi}{5} + i\sin\frac{8\pi}{5}$. So the product of the five fifth roots of 1 is:

$1\left(\cos\frac{2\pi}{5} + i\sin\frac{2\pi}{5}\right)\left(\cos\frac{4\pi}{5} + i\sin\frac{4\pi}{5}\right)\left(\cos\frac{6\pi}{5} + i\sin\frac{6\pi}{5}\right)\left(\cos\frac{8\pi}{5} + i\sin\frac{8\pi}{5}\right) = \cos 4\pi + i\sin 4\pi = 1$.

The sixth roots of 1 are $w^0 = 1$, $w^1 = \cos\frac{\pi}{3} + i\sin\frac{\pi}{3}$, $w^2 = \cos\frac{2\pi}{3} + i\sin\frac{2\pi}{3}$, $w^3 = -1$,

$w^4 = \cos\frac{4\pi}{3} + i\sin\frac{4\pi}{3}$, $w^5 = \cos\frac{5\pi}{3} + i\sin\frac{5\pi}{3}$. So the product of the six sixth roots of 1 is:

$1(\cos\frac{\pi}{3} + i\sin\frac{\pi}{3})(\cos\frac{2\pi}{3} + i\sin\frac{2\pi}{3})(-1)(\cos\frac{4\pi}{3} + i\sin\frac{4\pi}{3})(\cos\frac{5\pi}{3} + i\sin\frac{5\pi}{3})$
$= \cos 5\pi + i\sin 5\pi = -1.$

The eight roots of 1 are $w^0 = 1$, $w^1 = \cos\frac{\pi}{4} + i\sin\frac{\pi}{4}$, $w^2 = i$, $w^3 = \cos\frac{3\pi}{4} + i\sin\frac{3\pi}{4}$, $w^4 = -1$, $w^5 = \cos\frac{5\pi}{4} + i\sin\frac{5\pi}{4}$, $w^6 = -i$, $w^7 = \cos\frac{7\pi}{4} + i\sin\frac{7\pi}{4}$. So the product of the eight eighth roots of 1 is: $1(\cos\frac{\pi}{4} + i\sin\frac{\pi}{4})i(\cos\frac{3\pi}{4} + i\sin\frac{3\pi}{4})(-1)(\cos\frac{5\pi}{4} + i\sin\frac{5\pi}{4})(-i)(\cos\frac{7\pi}{4} + i\sin\frac{7\pi}{4})$
$= i^2 \cdot (\cos 2\pi + i\sin 2\pi) = -1.$

The product of the nth roots of 1 is -1 if n is even and 1 if n is odd.

Proof. This proof requires one fact that will be proved in Chapter 12: The sum of the first m integers is $\frac{m(m+1)}{2}$.

Let $w = \cos\frac{2\pi}{n} + i\sin\frac{2\pi}{n}$. Then, $w^k = \cos\frac{2k\pi}{n} + i\sin\frac{2k\pi}{n}$ for $k = 0, 1, 2, \ldots, n-1$. The argument of the product of the n roots of unity can be found by adding the arguments of each w^k. So the argument of the product is $\theta = 0 + \frac{2(1)\pi}{n} + \frac{2(2)\pi}{n} + \frac{2(3)\pi}{n} + \cdots + \frac{2(n-2)\pi}{n} + \frac{2(n-1)\pi}{n}$
$= \frac{2\pi}{n}[0 + 1 + 2 + 3 + \cdots + (n-2) + (n-1)]$. Since this is the sum of the first $n-1$ integers, this sum is $\frac{2\pi}{n} \cdot \frac{(n-1)(n)}{2} = (n-1)\pi$. Thus the product of the n roots of unity is $\cos[(n-1)\pi] + i\sin[(n-1)\pi] = -1$ if n is even and 1 if n is odd.

Exercises 9.7

1. $2\mathbf{u} = 2\langle -2, 3 \rangle = \langle -4, 6 \rangle$

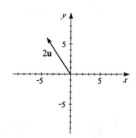

3. $\mathbf{u} + \mathbf{v} = \langle -2, 3 \rangle + \langle 3, 4 \rangle$
    $= \langle -2 + 3, 3 + 4 \rangle = \langle 1, 7 \rangle$

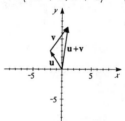

5. $\mathbf{v} - 2\mathbf{u} = \langle 3, 4 \rangle - 2\langle -2, 3 \rangle$
    $= \langle 3 - 2(-2), 4 - 2(3) \rangle = \langle 7, -2 \rangle$

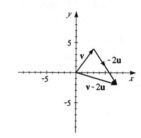

In Exercises 7 – 15, $\mathbf{v}$ represents the vector with initial point P and terminal point Q.

7. $P(2, 1), Q(5, 4).\ \mathbf{v} = \langle 5 - 2, 4 - 1 \rangle = \langle 3, 3 \rangle$

9. $P(1, 2), Q(4, 1).\ \mathbf{v} = \langle 4 - 1, 1 - 2 \rangle = \langle 3, -1 \rangle$

11. $P(3, 2), Q(8, 9).\ \mathbf{v} = \langle 8 - 3, 9 - 2 \rangle = \langle 5, 7 \rangle$

13. $P(5, 3), Q(1, 0).\ \mathbf{v} = \langle 1 - 5, 0 - 3 \rangle = \langle -4, -3 \rangle$

15. $P(-1, -1), Q(-1, 1).\ \mathbf{v} = \langle -1 - (-1), 1 - (-1) \rangle = \langle 0, 2 \rangle$

17. $\mathbf{u} = \langle 2, 7 \rangle,\ \mathbf{v} = \langle 3, 1 \rangle.\ 2\mathbf{u} = 2 \cdot \langle 2, 7 \rangle = \langle 4, 14 \rangle;\ -3\mathbf{v} = -3 \cdot \langle 3, 1 \rangle = \langle -9, -3 \rangle;$
     $\mathbf{u} + \mathbf{v} = \langle 2, 7 \rangle + \langle 3, 1 \rangle = \langle 5, 8 \rangle;\ 3\mathbf{u} - 4\mathbf{v} = \langle 6, 21 \rangle - \langle 12, 4 \rangle = \langle -6, 17 \rangle$

19. $\mathbf{u} = \langle 0, -1 \rangle,\ \mathbf{v} = \langle -2, 0 \rangle.\ 2\mathbf{u} = 2 \cdot \langle 0, -1 \rangle = \langle 0, -2 \rangle;\ -3\mathbf{v} = -3 \cdot \langle -2, 0 \rangle = \langle 6, 0 \rangle;$
     $\mathbf{u} + \mathbf{v} = \langle 0, -1 \rangle + \langle -2, 0 \rangle = \langle -2, -1 \rangle;\ 3\mathbf{u} - 4\mathbf{v} = \langle 0, -3 \rangle - \langle -8, 0 \rangle = \langle 8, -3 \rangle$

21. $\mathbf{u} = 2\mathbf{i},\ \mathbf{v} = 3\mathbf{i} - 2\mathbf{j}.\ 2\mathbf{u} = 2 \cdot 2\mathbf{i} = 4\mathbf{i};\ -3\mathbf{v} = -3(3\mathbf{i} - 2\mathbf{j}) = -9\mathbf{i} + 6\mathbf{j};$
     $\mathbf{u} + \mathbf{v} = 2\mathbf{i} + 3\mathbf{i} - 2\mathbf{j} = 5\mathbf{i} - 2\mathbf{j};\ 3\mathbf{u} - 4\mathbf{v} = 3 \cdot 2\mathbf{i} - 4(3\mathbf{i} - 2\mathbf{j}) = -6\mathbf{i} + 8\mathbf{j}$

23. $\mathbf{u} = 2\mathbf{i} + \mathbf{j},\ \mathbf{v} = 3\mathbf{i} - 2\mathbf{j}.$ Then $|\mathbf{u}| = \sqrt{2^2 + 1^2} = \sqrt{5};\ |\mathbf{v}| = \sqrt{3^2 + 2^2} = \sqrt{13};\ 2\mathbf{u} = 4\mathbf{i} + 2\mathbf{j};$
     $|2\mathbf{u}| = \sqrt{4^2 + 2^2} = 2\sqrt{5};\ \frac{1}{2}\mathbf{v} = \frac{3}{2}\mathbf{i} - \mathbf{j};\ |\frac{1}{2}\mathbf{v}| = \sqrt{\left(\frac{3}{2}\right)^2 + 1^2} = \frac{1}{2}\sqrt{13};\ \mathbf{u} + \mathbf{v} = 5\mathbf{i} - \mathbf{j};$
     $|\mathbf{u} + \mathbf{v}| = \sqrt{5^2 + 1^2} = \sqrt{26};\ \mathbf{u} - \mathbf{v} = 2\mathbf{i} + \mathbf{j} - 3\mathbf{i} + 2\mathbf{j} = -\mathbf{i} + 3\mathbf{j};$
     $|\mathbf{u} - \mathbf{v}| = \sqrt{1^2 + 3^2} = \sqrt{10};\ |\mathbf{u}| - |\mathbf{v}| = \sqrt{5} - \sqrt{13}$

25. $\mathbf{u} = \langle 10, -1 \rangle,\ \mathbf{v} = \langle -2, -2 \rangle.$ Then $|\mathbf{u}| = \sqrt{10^2 + 1^2} = \sqrt{101};\ |\mathbf{v}| = \sqrt{(-2)^2 + (-2)^2} = 2\sqrt{2};$
     $2\mathbf{u} = \langle 20, -2 \rangle;\ |2\mathbf{u}| = \sqrt{20^2 + 2^2} = \sqrt{404} = 2\sqrt{101};\ \frac{1}{2}\mathbf{v} = \langle -1, -1 \rangle;$

$\left|\frac{1}{2}\mathbf{v}\right| = \sqrt{(-1)^2 + (-1)^2} = \sqrt{2};\ \mathbf{u} + \mathbf{v} = \langle 8, -3\rangle;\ |\mathbf{u} + \mathbf{v}| = \sqrt{8^2 + 3^2} = \sqrt{73};\ \mathbf{u} - \mathbf{v} = \langle 12, 1\rangle;$
$|\mathbf{u} - \mathbf{v}| = \sqrt{12^2 + 1^2} = \sqrt{145};\ |\mathbf{u}| - |\mathbf{v}| = \sqrt{101} - 2\sqrt{2}$

In Exercises 27 – 33, x represents the horizontal component and y the vertical component.

27. $|\mathbf{v}| = 40$, direction $\theta = 30°$. $x = 40\cos 30° = 20\sqrt{3}$, $y = 40\sin 30° = 20$. Thus,
    $\mathbf{v} = x\mathbf{i} + y\mathbf{j} = 20\sqrt{3}\,\mathbf{i} + 20\mathbf{j}$

29. $|\mathbf{v}| = 1$, direction $\theta = 225°$. $x = \cos 225° = -\frac{1}{\sqrt{2}}$, $y = \sin 225° = -\frac{1}{\sqrt{2}}$. Thus,
    $\mathbf{v} = x\mathbf{i} + y\mathbf{j} = -\frac{1}{\sqrt{2}}\mathbf{i} - \frac{1}{\sqrt{2}}\mathbf{j} = -\frac{\sqrt{2}}{2}\mathbf{i} - \frac{\sqrt{2}}{2}\mathbf{j}$

31. $|\mathbf{v}| = 4$, direction $\theta = 10°$. $x = 4\cos 10° \approx 3.94$, $y = 4\sin 10° \approx 0.69$. Thus,
    $\mathbf{v} = x\mathbf{i} + y\mathbf{j} = (4\cos 10°)\mathbf{i} + (4\sin 10°)\mathbf{j} \approx 3.94\mathbf{i} + 0.69\mathbf{j}$

33. $|\mathbf{v}| = 30$, direction $\theta = 30°$. $x = 30\cos 30° = 30 \cdot \frac{\sqrt{3}}{2} \approx 25.98$, $y = 30\sin 30° = 15$. So, the
    horizontal component of force is $15\sqrt{3}$ lb and the vertical component is -15 lb.

35. $\mathbf{v} = \langle 3, 4\rangle$. The magnitude is $|\mathbf{v}| = \sqrt{3^2 + 4^2} = 5$. The direction is $\tan\theta = \frac{4}{3}$ $\Leftrightarrow$
    $\theta = \tan^{-1}\left(\frac{4}{3}\right) \approx 53.13°$

37. $\mathbf{v} = \langle -12, 5\rangle$. The magnitude is $|\mathbf{v}| = \sqrt{(-12)^2 + 5^2} = \sqrt{169} = 13$. The direction is θ where
    $\tan\theta = -\frac{5}{12}$ with θ in quadrant II $\Leftrightarrow$ $\theta = \tan^{-1}\left(-\frac{5}{12}\right) \approx 157.38°$

39. $\mathbf{v} = \mathbf{i} + \sqrt{3}\,\mathbf{j}$. The magnitude is $|\mathbf{v}| = \sqrt{1^2 + (\sqrt{3})^2} = 2$. The direction is θ where $\tan\theta = \sqrt{3}$
    with θ in quadrant I $\Leftrightarrow$ $\theta = \tan^{-1}\sqrt{3} = 60°$

41. (a) The true velocity of the jet is $\mathbf{v} = 425\,\mathbf{i} + 40\,\mathbf{j}$

    (b) The true speed of the jet is $|\mathbf{v}| = \sqrt{425^2 + 40^2} \approx 427$ mi/h, and the true direction is
        $\theta = \tan^{-1}\left(\frac{40}{425}\right) \approx 5.4°$ $\Rightarrow$ θ is N 84.6° E.

43. If the direction of the plane is N 30° W, the airplane speed is $\mathbf{u} = \langle u_x, u_y\rangle$ where $u_x = -765\cos 60°$
    $= -382.5$, and $u_y = 765\sin 60° \approx 662.51$. If the direction of the wind is N 30° E, the wind speed
    is $\mathbf{w} = \langle w_x, w_y\rangle$ where $w_x = 55\cos 60° = 27.5$, and $w_y = 55\sin 60° \approx 47.63$. Thus, the actual
    flight path is $\mathbf{v} = \mathbf{u} + \mathbf{w} = \langle -382.5 + 27.5, 662.51 + 47.63\rangle = \langle -355, 710.14\rangle$, and so the true
    speed is $|\mathbf{v}| = \sqrt{355^2 + 710.14^2} \approx 794$ mi/h, and the true direction is $\theta = \tan^{-1}\left(\frac{710.14}{-355}\right) \approx 116.6°$
    so θ is N 26.6° W.

45. (a) Let $\mathbf{r} = \langle r_x, r_y\rangle$ represent the velocity of the river and $\mathbf{b} = \langle b_x, b_y\rangle$ the velocity of the boat.
        Then, $r_x = 10$, $r_y = 0$, $b_x = 20\cos 60° = 10$, and $b_y = 20\sin 60° \approx 17.32$. The resultant
        velocity is then $\mathbf{w} \approx \langle 10 + 10, 0 + 17.32\rangle = \langle 20, 17.32\rangle = 20\,\mathbf{i} + 17.32\,\mathbf{j}$.

    (b) The true speed of the boat is $|\mathbf{w}| = \sqrt{20^2 + 17.32^2} \approx 26.5$ mi/h, and the true direction is
        $\theta = \tan^{-1}\left(\frac{17.32}{20}\right) \approx 40.9° \approx$ N 49.1° E.

47. Let $\mathbf{b} = \langle b_x, b_y\rangle$ represent the velocity of the boat and $\mathbf{w} = \langle w_x, w_y\rangle$ the velocity of the water. Then
    $\mathbf{b} = \langle 24\cos 18°, 24\sin 18°\rangle$, and $\mathbf{w} = \langle 0, w\rangle$ where w is the speed of the water. So the true velocity
    of the boat is $\mathbf{b} + \mathbf{w} = \langle 24\cos 18°, 24\sin 18° - w\rangle$. For the direction to be due east, we must have
    $24\sin 18° - w = 0$ $\Leftrightarrow$ $w = 7.42$ mi/h. Therefore, the true speed of the water is 7.4 mi/h. Since
    $\mathbf{b} + \mathbf{w} = \langle 24\cos 18°, 0\rangle$, the true speed of the boat is $|\mathbf{b} + \mathbf{w}| = 24\cos 18° \approx 22.8$ mi/h.

49. $\mathbf{F}_1 = \langle 2, 5 \rangle$, $\mathbf{F}_2 = \langle 3, -8 \rangle$.

   (a)   Then $\mathbf{F}_1 + \mathbf{F}_2 = \langle 2 + 3, 5 - 8 \rangle = \langle 5, -3 \rangle$.

   (b)   The additional force required is $\mathbf{F}_3 = \langle 0, 0 \rangle - \langle 5, -3 \rangle = \langle -5, 3 \rangle$.

51. $\mathbf{F}_1 = 4\mathbf{i} - \mathbf{j}$, $\mathbf{F}_2 = 3\mathbf{i} - 7\mathbf{j}$, $\mathbf{F}_3 = -8\mathbf{i} + 3\mathbf{j}$, and $\mathbf{F}_4 = \mathbf{i} + \mathbf{j}$.

   (a)   Then $\mathbf{F}_1 + \mathbf{F}_2 + \mathbf{F}_3 + \mathbf{F}_4 = (4 + 3 - 8 + 1)\mathbf{i} + (-1 - 7 + 3 + 1)\mathbf{j} = 0\mathbf{i} - 4\mathbf{j}$.

   (b)   The additional force required is $\mathbf{F}_5 = 0\mathbf{i} + 0\mathbf{j} - (0\mathbf{i} - 4\mathbf{j}) = 4\mathbf{j}$.

53. $\mathbf{F}_1 = \langle 10 \cos 60°, 10 \sin 60° \rangle = \langle 5, 5\sqrt{3} \rangle$, $\mathbf{F}_2 = \langle -8 \cos 30°, 8 \sin 30° \rangle = \langle -4\sqrt{3}, 4 \rangle$, and
   $\mathbf{F}_3 = \langle -6 \cos 20°, -6 \sin 20° \rangle \approx \langle -5.638, -2.052 \rangle$.

   (a) $\mathbf{F}_1 + \mathbf{F}_2 + \mathbf{F}_3 = \langle 5 - 4\sqrt{3} - 5.638, 5\sqrt{3} + 4 - 2.052 \rangle \approx \langle -7.57, 10.61 \rangle$.

   (b)   The additional force required is $\mathbf{F}_4 = \langle 0, 0 \rangle - \langle -7.57, 10.61 \rangle = \langle 7.57, -10.61 \rangle$.

55.   From the figure we see that $\mathbf{T}_1 = -|\mathbf{T}_1| \cos 50° \mathbf{i} + |\mathbf{T}_1| \sin 50° \mathbf{j}$ and
   $\mathbf{T}_2 = |\mathbf{T}_2| \cos 30° \mathbf{i} + |\mathbf{T}_2| \sin 30° \mathbf{j}$. Since $\mathbf{T}_1 + \mathbf{T}_2 = 100\mathbf{j}$ we get
   $-|\mathbf{T}_1| \cos 50° + |\mathbf{T}_2| \cos 30° = 0$ and $|\mathbf{T}_1| \sin 50° + |\mathbf{T}_2| \sin 30° = 100$. From the first equation
   $|\mathbf{T}_2| = |\mathbf{T}_1| \dfrac{\cos 50°}{\cos 30°}$ and substituting into the second equation gives

   $|\mathbf{T}_1| \sin 50° + |\mathbf{T}_1| \dfrac{\cos 50° \sin 30°}{\cos 30°} = 100 \quad \Leftrightarrow \quad |\mathbf{T}_1| (\sin 50° \cos 30° + \cos 50° \sin 30°)$

   $= 100 \cos 30° \quad \Leftrightarrow \quad |\mathbf{T}_1| \sin(50° + 30°) = 100 \cos 30° \quad \Leftrightarrow \quad |\mathbf{T}_1| = 100 \dfrac{\cos 30°}{\sin 80°} \approx 87.9416$.

   Similarly, solving for $|\mathbf{T}_1|$ in the first equation gives $|\mathbf{T}_1| = |\mathbf{T}_2| \dfrac{\cos 30°}{\cos 50°}$ and substituting gives

   $|\mathbf{T}_2| \dfrac{\cos 30° \sin 50°}{\cos 50°} + |\mathbf{T}_2| \sin 30° = 100 \quad \Leftrightarrow$

   $|\mathbf{T}_2| (\cos 30° \sin 50° + \cos 50° \sin 30°) = 100 \cos 50° \quad \Leftrightarrow \quad |\mathbf{T}_2| = \dfrac{100 \cos 50°}{\sin 80°} \approx 65.2704$. Thus

   $\mathbf{T}_1 \approx (-87.9416 \cos 50°)\mathbf{i} + (87.9416 \sin 50°)\mathbf{j} \approx -56.5\mathbf{i} + 67.4\mathbf{j}$ and

   $\mathbf{T}_2 \approx (65.2704 \cos 30°)\mathbf{i} + (65.2704 \sin 30°)\mathbf{j} \approx 56.5\mathbf{i} + 32.6\mathbf{j}$.

57.   When we add two (or more vectors) the resultant vector can be found by first placing the initial point
   of the second vector at the terminal point of the first vector. The resultant vector can then found by
   using the new terminal point of the second vector and the initial point of the first vector. When the n
   vectors are placed head to tail in the plane so that they form a polygon, the initial point and the
   terminal point are the same. Thus the sum of these n vectors is the zero vector.

Exercises 9.8

1. (a) $\mathbf{u} \cdot \mathbf{v} = \langle 2, 0 \rangle \cdot \langle 1, 1 \rangle = 2 + 0 = 2$

   (b) $\cos\theta = \dfrac{\mathbf{u} \cdot \mathbf{v}}{|\mathbf{u}|\,|\mathbf{v}|} = \dfrac{2}{2\cdot\sqrt{2}} = \dfrac{1}{\sqrt{2}} \;\Rightarrow\; \theta = 45°$

3. (a) $\mathbf{u} \cdot \mathbf{v} = \langle 2, 7 \rangle \cdot \langle 3, 1 \rangle = 6 + 7 = 13$

   (b) $\cos\theta = \dfrac{\mathbf{u} \cdot \mathbf{v}}{|\mathbf{u}|\,|\mathbf{v}|} = \dfrac{13}{\sqrt{53}\cdot\sqrt{10}} \;\Rightarrow\; \theta \approx 56°$

5. (a) $\mathbf{u} \cdot \mathbf{v} = \langle 3, -2 \rangle \cdot \langle 1, 2 \rangle = 3 + (-4) = -1$

   (b) $\cos\theta = \dfrac{\mathbf{u} \cdot \mathbf{v}}{|\mathbf{u}|\,|\mathbf{v}|} = \dfrac{-1}{\sqrt{13}\cdot\sqrt{5}} \;\Rightarrow\; \theta \approx 97°$

7. (a) $\mathbf{u} \cdot \mathbf{v} = \langle 0, -5 \rangle \cdot \langle -1, -\sqrt{3} \rangle = 0 + 5\sqrt{3} = 5\sqrt{3}$

   (b) $\cos\theta = \dfrac{\mathbf{u} \cdot \mathbf{v}}{|\mathbf{u}|\,|\mathbf{v}|} = \dfrac{5\sqrt{3}}{5\cdot 2} = \dfrac{\sqrt{3}}{2} \;\Rightarrow\; \theta = 30°$

9. $\mathbf{u} \cdot \mathbf{v} = -12 + 12 = 0 \;\Rightarrow\;$ vectors are orthogonal

11. $\mathbf{u} \cdot \mathbf{v} = -8 + 12 = 4 \neq 0 \;\Rightarrow\;$ vectors are not orthogonal

13. $\mathbf{u} \cdot \mathbf{v} = -24 + 24 = 0 \;\Rightarrow\;$ vectors are orthogonal

15. $\mathbf{u} \cdot \mathbf{v} + \mathbf{u} \cdot \mathbf{w} = \langle 2, 1 \rangle \cdot \langle 1, -3 \rangle + \langle 2, 1 \rangle \cdot \langle 3, 4 \rangle = 2 - 3 + 6 + 4 = 9$

17. $(\mathbf{u} + \mathbf{v}) \cdot (\mathbf{u} - \mathbf{v}) = [\langle 2, 1 \rangle + \langle 1, -3 \rangle] \cdot [\langle 2, 1 \rangle - \langle 1, -3 \rangle] = \langle 3, -2 \rangle \cdot \langle 1, 4 \rangle = 3 - 8 = -5$

19. $x = \dfrac{\mathbf{u} \cdot \mathbf{v}}{|\mathbf{v}|} = \dfrac{12 - 24}{5} = -\dfrac{12}{5}$

21. $x = \dfrac{\mathbf{u} \cdot \mathbf{v}}{|\mathbf{v}|} = \dfrac{0 - 24}{1} = -24$

23. $\mathbf{W} = \mathbf{F} \cdot \mathbf{d} = \langle 4, -5 \rangle \cdot \langle 3, 8 \rangle = -28$

25. $\mathbf{W} = \mathbf{F} \cdot \mathbf{d} = \langle 10, 3 \rangle \cdot \langle 4, -5 \rangle = 25$

27. $\mathbf{W} = \mathbf{F} \cdot \mathbf{d} = \langle 4, -7 \rangle \cdot \langle 4, 0 \rangle = 16$ ft-lb

29. The distance vector is $\mathbf{D} = \langle 200, 0 \rangle$ and the force vector is $\mathbf{F} = \langle 50\cos 30°, 50\sin 30° \rangle$. Hence, the work done is $\mathbf{W} = \mathbf{F} \cdot \mathbf{D} = \langle 200, 0 \rangle \cdot \langle 50\cos 30°, 50\sin 30° \rangle = 200 \cdot 50\cos 30° \approx 8660$ ft-lb.

31. Since the weight of the car is 2755 lb, the force exerted perpendicular to the earth is 2755 lb. Resolving this into a force $\mathbf{u}$ perpendicular to the driveway gives $|\mathbf{u}| = 2766\cos 65° \approx 1164$ lb. Thus, a force of about 1164 lb is required.

33. Since the force required parallel to the driveway is 80 lb and the weight of the package is 200 lb, it follows that $80 = 200\sin\theta$, where θ is the angle of inclination of the plane. Then $\theta = \sin^{-1}\left(\frac{80}{200}\right) \approx 23.58°$, and so the angle of inclination is approximately $23.6°$.

35. Let $\mathbf{u} = \langle u_1, u_2 \rangle$ and $\mathbf{v} = \langle v_1, v_2 \rangle$. Then
$$\mathbf{u} \cdot \mathbf{v} = \langle u_1, u_2 \rangle \cdot \langle v_1, v_2 \rangle = u_1 v_1 + u_2 v_2 = v_1 u_1 + v_2 u_2 = \langle v_1, v_2 \rangle \cdot \langle u_1, u_2 \rangle = \mathbf{v} \cdot \mathbf{u}$$

37. Let $\mathbf{u} = \langle u_1, u_2 \rangle$, $\mathbf{v} = \langle v_1, v_2 \rangle$, and $\mathbf{w} = \langle w_1, w_2 \rangle$. Then
$$(\mathbf{u} + \mathbf{v}) \cdot \mathbf{w} = (\langle u_1, u_2 \rangle + \langle v_1, v_2 \rangle) \cdot \langle w_1, w_2 \rangle = \langle u_1 + v_1, u_2 + v_2 \rangle \cdot \langle w_1, w_2 \rangle$$
$$= u_1 w_1 + v_1 w_1 + u_2 w_2 + v_2 w_2 = u_1 w_1 + u_2 w_2 + v_1 w_1 + v_2 w_2$$
$$= \langle u_1, u_2 \rangle \cdot \langle w_1, w_2 \rangle + \langle v_1, v_2 \rangle \cdot \langle w_1, w_2 \rangle = \mathbf{u} \cdot \mathbf{w} + \mathbf{v} \cdot \mathbf{w}$$

Review Exercises for Chapter 9

1. $\cos^2 x \csc x - \csc x = (1 - \sin^2 x) \csc x - \csc x = \csc x - \sin^2 x \csc x - \csc x$

$= -\sin^2 x \cdot \dfrac{1}{\sin x} = -\sin x$

3. $\dfrac{\cos^2 x - \tan^2 x}{\sin^2 x} = \dfrac{\cos^2 x}{\sin^2 x} - \dfrac{\tan^2 x}{\sin^2 x} = \cot^2 x - \dfrac{1}{\cos^2 x} = \cot^2 x - \sec^2 x$

5. $\dfrac{\cos^2 x}{1 - \sin x} = \dfrac{\cos x}{\dfrac{1}{\cos x}(1 - \sin x)} = \dfrac{\cos x}{\dfrac{1}{\cos x} - \dfrac{\sin x}{\cos x}} = \dfrac{\cos x}{\sec x - \tan x}$

7. $\sin^2 x \cot^2 x + \cos^2 x \tan^2 x = \sin^2 x \cdot \dfrac{\cos^2 x}{\sin^2 x} + \cos^2 x \cdot \dfrac{\sin^2 x}{\cos^2 x} = \cos^2 x + \sin^2 x = 1$

9. $\dfrac{\sin 2x}{1 + \cos 2x} = \dfrac{2 \sin x \cos x}{1 + 2\cos^2 x - 1} = \dfrac{2 \sin x \cos x}{2\cos^2 x} = \dfrac{2 \sin x}{2 \cos x} = \tan x$

11. $\tan \dfrac{x}{2} = \dfrac{1 - \cos x}{\sin x} = \dfrac{1}{\sin x} - \dfrac{\cos x}{\sin x} = \csc x - \cot x$

13. $\sin(x + y) \sin(x - y) = \frac{1}{2}\left[\cos((x+y) - (x-y)) - \cos((x+y) + (x-y))\right]$

$= \frac{1}{2}(\cos 2y - \cos 2x) = \frac{1}{2}[1 - 2\sin^2 y - (1 - 2\sin^2 x)] = \frac{1}{2}(2\sin^2 x - 2\sin^2 y) = \sin^2 x - \sin^2 y$

15. $1 + \tan x \tan \dfrac{x}{2} = 1 + \dfrac{\sin x}{\cos x} \cdot \dfrac{1 - \cos x}{\sin x} = 1 + \dfrac{1 - \cos x}{\cos x} = 1 + \dfrac{1}{\cos x} - 1 = \dfrac{1}{\cos x} = \sec x$

17. $(\cos \frac{x}{2} - \sin \frac{x}{2})^2 = \cos^2 \frac{x}{2} - 2\sin \frac{x}{2}\cos \frac{x}{2} + \sin^2 \frac{x}{2} = \sin^2 \frac{x}{2} + \cos^2 \frac{x}{2} - 2\sin \frac{x}{2}\cos \frac{x}{2}$

$= 1 - \sin(2 \cdot \frac{x}{2}) = 1 - \sin x$

19. $\dfrac{\sin 2x}{\sin x} - \dfrac{\cos 2x}{\cos x} = \dfrac{2 \sin x \cos x}{\sin x} - \dfrac{2\cos^2 x - 1}{\cos x} = 2\cos x - 2\cos x + \dfrac{1}{\cos x} = \sec x$

21. $\tan(x + \frac{\pi}{4}) = \dfrac{\tan x + \tan \frac{\pi}{4}}{1 - \tan x \tan \frac{\pi}{4}} = \dfrac{1 + \tan x}{1 - \tan x}$

23. (a) $f(x) = 1 - \left(\cos \frac{x}{2} - \sin \frac{x}{2}\right)^2$,

$g(x) = \sin x$

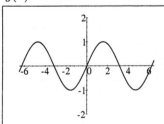

(b) The graphs suggest that $f(x) = g(x)$ is an identity. To prove this, expand $f(x)$ and simplify, using the double-angle formula for sine:

$f(x) = 1 - \left(\cos \frac{x}{2} - \sin \frac{x}{2}\right)^2$

$= 1 - \left[\cos^2 \frac{x}{2} - 2\cos \frac{x}{2}\sin \frac{x}{2} + \sin^2 \frac{x}{2}\right]$

$= 1 + 2\cos \frac{x}{2}\sin \frac{x}{2} - \left[\cos^2 \frac{x}{2} + \sin^2 \frac{x}{2}\right]$

$= 1 + \sin x - (1) = \sin x = g(x)$

25. (a) $f(x) = \tan x \tan \frac{x}{2}$, $g(x) = \dfrac{1}{\cos x}$

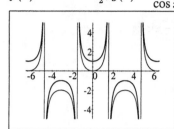

(b) The graphs suggest that $f(x) \neq g(x)$ in general. For example, choose $x = \frac{\pi}{3}$ and evaluate:

$$f(\tfrac{\pi}{3}) = \tan \tfrac{\pi}{3} \tan \tfrac{\pi}{6} = \sqrt{3} \cdot \tfrac{1}{\sqrt{3}} = 1, \text{ whereas}$$

$$g(\tfrac{\pi}{3}) = \frac{1}{\tfrac{1}{2}} = 2, \text{ so } f(x) \neq g(x).$$

27. (a) $f(x) = 2 \sin^2 3x + \cos 6x$

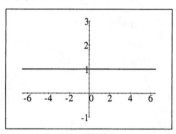

(b) The graph suggests that $f(x) = 1$ for all x. To prove this, we use the double angle formula to note that $\cos 6x = \cos[2(3x)] = 1 - 2\sin^2 3x$,

so $f(x) = 2\sin^2 3x + (1 - 2\sin^2 3x) = 1$.

29. $\cos x \sin x - \sin x = 0 \quad \Leftrightarrow \quad \sin x\,(\cos x - 1) = 0 \quad \Leftrightarrow \quad \sin x = 0$ or $\cos x = 1 \quad \Leftrightarrow$ $x = 0, \pi$ or $x = 0$. Therefore, the solutions are $x = 0$ and π.

31. $2\sin^2 x - 5\sin x + 2 = 0 \quad \Leftrightarrow \quad (2\sin x - 1)(\sin x - 2) = 0 \quad \Leftrightarrow \quad \sin x = \frac{1}{2}$ or $\sin x = 2$ (which is inadmissible) $\quad \Leftrightarrow \quad x = \frac{\pi}{6}, \frac{5\pi}{6}$. Thus, the solutions in $[0, 2\pi)$ are $x = \frac{\pi}{6}$ and $\frac{5\pi}{6}$.

33. $2\cos^2 x - 7\cos x + 3 = 0 \quad \Leftrightarrow \quad (2\cos x - 1)(\cos x - 3) = 0 \quad \Leftrightarrow \quad \cos x = \frac{1}{2}$ or $\cos x = 3$ (which is inadmissible) $\quad \Leftrightarrow \quad x = \frac{\pi}{3}, \frac{5\pi}{3}$. Therefore, the solutions in $[0, 2\pi)$ are $x = \frac{\pi}{3}, \frac{5\pi}{3}$.

35. Note: $x = \pi$ is not a solution because then the denominator is zero. $\dfrac{1 - \cos x}{1 + \cos x} = 3 \quad \Leftrightarrow$ $1 - \cos x = 3 + 3\cos x \quad \Leftrightarrow \quad -4\cos x = 2 \quad \Leftrightarrow \quad \cos x = -\frac{1}{2} \quad \Leftrightarrow \quad x = \frac{2\pi}{3}, \frac{4\pi}{3}$ in $[0, 2\pi)$.

37. (factor by grouping) $\tan^3 x + \tan^2 x - 3\tan x - 3 = 0 \quad \Leftrightarrow \quad (\tan x + 1)(\tan^2 x - 3) = 0 \quad \Leftrightarrow$ $\tan x = -1$ or $\tan x = \pm\sqrt{3} \quad \Leftrightarrow \quad x = \frac{3\pi}{4}, \frac{7\pi}{4}$ or $x = \frac{\pi}{3}, \frac{2\pi}{3}, \frac{4\pi}{3}, \frac{5\pi}{3}$. Therefore, the solutions in $[0, 2\pi)$ are $x = \frac{\pi}{3}, \frac{2\pi}{3}, \frac{3\pi}{4}, \frac{4\pi}{3}, \frac{5\pi}{3}, \frac{7\pi}{4}$.

39. $\tan \frac{1}{2}x + 2\sin 2x = \csc x \quad \Leftrightarrow \quad \dfrac{1 - \cos x}{\sin x} + 4\sin x \cos x = \dfrac{1}{\sin x} \quad \Leftrightarrow$ $1 - \cos x + 4\sin^2 x \cos x = 1 \quad \Leftrightarrow \quad 4\sin^2 x \cos x - \cos x = 0 \quad \Leftrightarrow \quad \cos x\,(4\sin^2 x - 1) = 0$ $\quad \Leftrightarrow \quad \cos x = 0$ or $\sin x = \pm\frac{1}{2} \quad \Leftrightarrow \quad x = \frac{\pi}{2}, \frac{3\pi}{2}$ or $x = \frac{\pi}{6}, \frac{5\pi}{6}, \frac{7\pi}{6}, \frac{11\pi}{6}$. Thus, the solutions in $[0, 2\pi)$ are $x = \frac{\pi}{6}, \frac{\pi}{2}, \frac{5\pi}{6}, \frac{7\pi}{6}, \frac{3\pi}{2}, \frac{11\pi}{6}$.

41. $\tan x + \sec x = \sqrt{3} \quad \Leftrightarrow \quad \dfrac{\sin x}{\cos x} + \dfrac{1}{\cos x} = \sqrt{3} \quad \Leftrightarrow \quad \sin x + 1 = \sqrt{3}\cos x \quad \Leftrightarrow$ $\sqrt{3}\cos x - \sin x = 1 \quad \Leftrightarrow \quad \frac{\sqrt{3}}{2}\cos x - \frac{1}{2}\sin x = \frac{1}{2} \quad \Leftrightarrow \quad \cos \frac{\pi}{6}\cos x - \sin \frac{\pi}{6}\sin x = \frac{1}{2} \quad \Leftrightarrow$ $\cos(x + \frac{\pi}{6}) = \frac{1}{2} \quad \Leftrightarrow \quad x + \frac{\pi}{6} = \frac{\pi}{3}, \frac{5\pi}{3} \quad \Leftrightarrow \quad x = \frac{\pi}{6}, \frac{3\pi}{2}$. However $x = \frac{3\pi}{2}$ is inadmissible because $\sec \frac{3\pi}{2}$ is undefined. Thus, the only solution is in $[0, 2\pi)$ is $x = \frac{\pi}{6}$.

43.   We graph $f(x) = \cos x$ and $g(x) = x^2 - 1$ in the viewing rectangle $[0,\ 6.5]$ by $[-2, 2]$. The two functions intersect at only one point, $x \approx 1.18$.

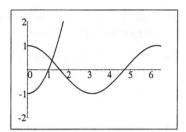

45.   Since $15°$ is in quadrant I: $\cos 15° = \sqrt{\dfrac{1 + \cos 30°}{2}} = \sqrt{\dfrac{2 + \sqrt{3}}{4}} = \dfrac{1}{2}\sqrt{2 + \sqrt{3}}$.

47. $\tan \dfrac{\pi}{8} = \dfrac{1 - \cos \frac{\pi}{4}}{\sin \frac{\pi}{4}} = \dfrac{1 - \frac{1}{\sqrt{2}}}{\frac{1}{\sqrt{2}}} = \left(1 - \dfrac{1}{\sqrt{2}}\right)\sqrt{2} = \sqrt{2} - 1$.

49. $\sin 5° \cos 40° + \cos 5° \sin 40° = \sin(5° + 40°) = \sin 45° = \dfrac{1}{\sqrt{2}} = \dfrac{\sqrt{2}}{2}$.

51. $\cos^2 \dfrac{\pi}{8} - \sin^2 \dfrac{\pi}{8} = \cos\left(2\left(\dfrac{\pi}{8}\right)\right) = \cos \dfrac{\pi}{4} = \dfrac{1}{\sqrt{2}} = \dfrac{\sqrt{2}}{2}$.

53. $\cos 37.5° \cos 7.5° = \dfrac{1}{2}\left[\cos 45° + \cos 30°\right] = \dfrac{1}{2}\left[\dfrac{\sqrt{2}}{2} + \dfrac{\sqrt{3}}{2}\right] = \dfrac{1}{4}\left(\sqrt{2} + \sqrt{3}\right)$. Note: We have used a product-to-sum identity.

In Exercises 55 – 59, x and y are in quadrant I, so we know that $\sec x = \dfrac{3}{2} \ \Rightarrow \ \cos x = \dfrac{2}{3}$, so $\sin x = \dfrac{\sqrt{5}}{3}$ and $\tan x = \dfrac{\sqrt{5}}{2}$. Also, $\csc y = 3 \ \Rightarrow \ \sin y = \dfrac{1}{3}$, and so $\cos y = \dfrac{2\sqrt{2}}{3}$, and $\tan y = \dfrac{1}{2\sqrt{2}} = \dfrac{\sqrt{2}}{4}$.

55. $\sin(x + y) = \sin x \cos y + \cos x \sin y = \dfrac{\sqrt{5}}{3} \cdot \dfrac{2\sqrt{2}}{3} + \dfrac{2}{3} \cdot \dfrac{1}{3} = \dfrac{2}{9}(1 + \sqrt{10})$.

57. $\tan(x + y) = \dfrac{\tan x + \tan y}{1 - \tan x \tan y} = \dfrac{\frac{\sqrt{5}}{2} + \frac{\sqrt{2}}{4}}{1 - \left(\frac{\sqrt{5}}{2}\right)\left(\frac{\sqrt{2}}{4}\right)} = \dfrac{\frac{\sqrt{5}}{2} + \frac{\sqrt{2}}{4}}{1 - \left(\frac{\sqrt{5}}{2}\right)\left(\frac{\sqrt{2}}{4}\right)} \cdot \dfrac{8}{8}$

       $= \dfrac{2\left(2\sqrt{5} + \sqrt{2}\right)}{8 - \sqrt{10}} \cdot \dfrac{8 + \sqrt{10}}{8 + \sqrt{10}} = \dfrac{2}{3}\left(\sqrt{2} + \sqrt{5}\right)$.

59. $\cos \dfrac{y}{2} = \sqrt{\dfrac{1 + \cos y}{2}} = \sqrt{\dfrac{1 + \left(\frac{2\sqrt{2}}{3}\right)}{2}} = \sqrt{\dfrac{3 + 2\sqrt{2}}{6}}$ (since cosine is positive in quadrant I).

61. $\sin^{-1}\left(\dfrac{\sqrt{3}}{2}\right) = \dfrac{\pi}{3}$.

63. $\cos(\tan^{-1}\sqrt{3}) = \cos \dfrac{\pi}{3} = \dfrac{1}{2}$.

65.   Let $u = \sin^{-1}\dfrac{2}{5}$ and so $\sin u = \dfrac{2}{5}$. Then from the triangle $\tan(\sin^{-1}\dfrac{2}{5}) = \tan u = \dfrac{2}{\sqrt{21}}$.

67. $\cos(2\sin^{-1}\dfrac{1}{3}) = 1 - 2\sin^2(\sin^{-1}\dfrac{1}{3}) = 1 - 2\left(\dfrac{1}{3}\right)^2 = \dfrac{7}{9}$.

69. Let $\theta = \tan^{-1}x$ $\Leftrightarrow$ $\tan\theta = x$. Then from the triangle on the right we have

$$\sin(\tan^{-1}x) = \sin\theta = \frac{x}{\sqrt{1+x^2}}$$

71. $\cos\theta = \frac{x}{3}$ $\Rightarrow$ $\theta = \cos^{-1}\left(\frac{x}{3}\right)$

73. The arc PQ subtends a central angle of 2θ. Thus, $s = r(2\theta) = 2r\theta$ $\Rightarrow$ $\theta = \frac{s}{2r}$. If we examine the triangles in the figure, we see that $\cos\theta = \frac{r}{r+h}$ $\Rightarrow$ $\cos\left(\frac{s}{2r}\right) = \frac{r}{r+h}$. Apply the inverse cosine function to each side gives $\cos^{-1}\left(\cos\left(\frac{s}{2r}\right)\right) = \cos^{-1}\left(\frac{r}{r+h}\right)$ $\Rightarrow$

$\frac{s}{2r} = \cos^{-1}\left(\frac{r}{r+h}\right)$. Therefore, $s = 2r\cos^{-1}\left(\frac{r}{r+h}\right) = 7920\cos^{-1}\left(\frac{3960}{3960+h}\right)$.

75. $4 + 4i$. Then $r = \sqrt{16+16} = 4\sqrt{2}$, and $\theta = \tan^{-1}\frac{4}{4} = \frac{\pi}{4}$ (in quadrant I). Thus,
    $4 + 4i = 4\sqrt{2}\left(\cos\frac{\pi}{4} + i\sin\frac{\pi}{4}\right)$.

77. $5 + 3i$. Then $r = \sqrt{25+9} = \sqrt{34}$, and $\theta = \tan^{-1}\frac{3}{5}$. Thus,
    $5 + 3i = \sqrt{34}\left[\cos(\tan^{-1}\frac{3}{5}) + i\sin(\tan^{-1}\frac{3}{5})\right]$.

79. $-1 + i$. Then $r = \sqrt{1+1} = \sqrt{2}$, and $\tan\theta = \frac{1}{-1}$ with θ in quadrant II $\Leftrightarrow$ $\theta = \frac{3\pi}{4}$. Thus,
    $-1 + i = \sqrt{2}\left(\cos\frac{3\pi}{4} + i\sin\frac{3\pi}{4}\right)$.

81. $1 - \sqrt{3}i$ has $r = \sqrt{1+3} = 2$ and $\tan\theta = \frac{-\sqrt{3}}{1} = -\sqrt{3}$ with θ in quadrant III $\Leftrightarrow$ $\theta = \frac{5\pi}{3}$.
    Therefore, $1 - \sqrt{3}i = 2\left(\cos\frac{5\pi}{3} + i\sin\frac{5\pi}{3}\right)$, and so $(1 - \sqrt{3}i)^4 = 2^4\left(\cos\frac{20\pi}{3} + i\sin\frac{20\pi}{3}\right)$
    $= 16\left(\cos\frac{2\pi}{3} + i\sin\frac{2\pi}{3}\right) = 16\left(-\frac{1}{2} + i\frac{\sqrt{3}}{2}\right) = 8(-1 + i\sqrt{3})$.

83. $\sqrt{3} + i$ has $r = \sqrt{3+1} = 2$ and $\tan\theta = \frac{1}{\sqrt{3}}$ with θ in quadrant I $\Leftrightarrow$ $\theta = \frac{\pi}{6}$. Therefore,
    $\sqrt{3} + i = 2\left(\cos\frac{\pi}{6} + i\sin\frac{\pi}{6}\right)$, and so $(\sqrt{3} + i)^{-4} = 2^{-4}\left(\cos\frac{-4\pi}{6} + i\sin\frac{-4\pi}{6}\right)$
    $= \frac{1}{16}\left(\cos\frac{2\pi}{3} - i\sin\frac{2\pi}{3}\right) = \frac{1}{16}\left(-\frac{1}{2} - i\frac{\sqrt{3}}{2}\right) = \frac{1}{32}(-1 - i\sqrt{3}) = -\frac{1}{32}(1 + i\sqrt{3})$.

85. $-16i$ has $r = 16$ and $\theta = \frac{3\pi}{2}$. Thus, $-16i = 16\left(\cos\frac{3\pi}{2} + i\sin\frac{3\pi}{2}\right)$ and so
    $(-16i)^{1/2} = 16^{1/2}\left[\cos\left(\frac{3\pi+4k\pi}{2}\right) + i\sin\left(\frac{3\pi+4k\pi}{2}\right)\right]$ for $k = 0, 1$. Thus the roots are
    $w_0 = 4\left(\cos\frac{3\pi}{4} + i\sin\frac{3\pi}{4}\right) = 4\left(-\frac{1}{\sqrt{2}} + i\frac{1}{\sqrt{2}}\right) = 2\sqrt{2}(-1 + i)$ and
    $w_1 = 4\left(\cos\frac{7\pi}{4} + i\sin\frac{7\pi}{4}\right) = 4\left(\frac{1}{\sqrt{2}} - i\frac{1}{\sqrt{2}}\right) = 2\sqrt{2}(1 - i)$.

87. $1 = \cos 0 + i\sin 0$. Then, $1^{1/6} = 1\left(\cos\frac{2k\pi}{6} + i\sin\frac{2k\pi}{6}\right)$ for $k = 0, 1, 2, 3, 4, 5$. Thus the six roots
    are $w_0 = 1(\cos 0 + i\sin 0) = 1$, $w_1 = 1\left(\cos\frac{\pi}{3} + i\sin\frac{\pi}{3}\right) = \frac{1}{2} + i\frac{\sqrt{3}}{2}$, $w_2 = 1\left(\cos\frac{2\pi}{3} + i\sin\frac{2\pi}{3}\right)$
    $= -\frac{1}{2} + i\frac{\sqrt{3}}{2}$, $w_3 = 1(\cos\pi + i\sin\pi) = -1$, $w_4 = 1\left(\cos\frac{4\pi}{3} + i\sin\frac{4\pi}{3}\right) = -\frac{1}{2} - i\frac{\sqrt{3}}{2}$, and
    $w_5 = 1\left(\cos\frac{5\pi}{3} + i\sin\frac{5\pi}{3}\right) = \frac{1}{2} - i\frac{\sqrt{3}}{2}$.

89. $\mathbf{u} = \langle -2, 3 \rangle$ and $\mathbf{v} = \langle 8, 1 \rangle$. Then $|\mathbf{u}| = \sqrt{(-2)^2 + 3^2} = \sqrt{13}$; $\mathbf{u} + \mathbf{v} = \langle -2 + 8, 3 + 1 \rangle = \langle 6, 4 \rangle$;
$\mathbf{u} - \mathbf{v} = \langle -2 - 8, 3 - 1 \rangle = \langle -10, 2 \rangle$; $2\mathbf{u} = \langle -4, 6 \rangle$;
and $3\mathbf{u} - 2\mathbf{v} = \langle -6, 9 \rangle - \langle 16, 2 \rangle = \langle -6 - 16, 9 - 2 \rangle = \langle -22, 7 \rangle$.

91. $P(0, 3)$ and $Q(3, -1)$. $\mathbf{u} = \langle 3 - 0, -1 - 3 \rangle = \langle 3, -4 \rangle = 3\mathbf{i} - 4\mathbf{j}$

93. Let $Q(x, y)$ be the terminal point. Then $\langle x - 5, y - 6 \rangle = \langle 5, -8 \rangle \quad \Leftrightarrow \quad x - 5 = 5$ and
$y - 6 = -8 \quad \Leftrightarrow \quad x = 10$ and $y = -2$. Therefore, the terminal point is $Q(10, -2)$.

95. (a) The resultant force $\mathbf{r}$ is the sum of the forces; to find this we first resolve the two vectors:
$\mathbf{v_1} = 2.0 \times 10^4$ lb N $50°$E $= \langle 2.0 \times 10^4 \cdot \cos 40°, 2.0 \times 10^4 \cdot \sin 40° \rangle \approx \langle 15321, 12856 \rangle$, and
$\mathbf{v_2} = 3.4 \times 10^4$ lb S $75°$E $= \langle 3.4 \times 10^4 \cdot \cos(-15°), 3.4 \times 10^4 \cdot \sin(-15°) \rangle$
$\approx \langle 32841, -8800 \rangle$. Thus $\mathbf{r} = \mathbf{v_1} + \mathbf{v_2} \approx \langle 15321, 12856 \rangle + \langle 32841, -8800 \rangle$
$= \langle 48163, 4056 \rangle \approx \langle 4.82, 0.41 \rangle \times 10^4 = (4.82\mathbf{i} + 0.41\mathbf{j}) \times 10^4$.

(b) The magnitude is $|\mathbf{r}| = \sqrt{48163^2 + 4056^2} \approx 4.8 \times 10^4$ lb. We have $\theta \approx \tan^{-1}\left(\frac{4056}{48163}\right) \approx 4.8°$.
Thus the direction of $\mathbf{r}$ is approximately N $85.2°$E.

97. $\mathbf{u} = \langle 4, -3 \rangle$ and $\mathbf{v} = \langle 9, -8 \rangle$. Then $|\mathbf{u}| = \sqrt{(4)^2 + (-3)^2} = \sqrt{25} = 5$;
$\mathbf{u} \cdot \mathbf{u} = \langle 4, -3 \rangle \cdot \langle 4, -3 \rangle = (4)(4) + (-3)(-3) = 16 + 9 = 25$;
$\mathbf{u} \cdot \mathbf{v} = \langle 4, -3 \rangle \cdot \langle 9, -8 \rangle = (4)(9) + (-3)(-8) = 36 + 24 = 60$.

99. $\mathbf{u} = -2\mathbf{i} + 2\mathbf{j}$ and $\mathbf{v} = \mathbf{i} + \mathbf{j}$. Then $|\mathbf{u}| = \sqrt{(-2)^2 + (2)^2} = \sqrt{8}$;
$\mathbf{u} \cdot \mathbf{u} = -2\mathbf{i} + 2\mathbf{j} \cdot -2\mathbf{i} + 2\mathbf{j} = (-2)(-2) + (2)(2) = 4 + 4 = 8$;
$\mathbf{u} \cdot \mathbf{v} = -2\mathbf{i} + 2\mathbf{j} \cdot \mathbf{i} + \mathbf{j} = (-2)(1) + (2)(1) = -2 + 2 = 0$.

101. $\mathbf{u} = \langle -4, 2 \rangle$ and $\mathbf{v} = \langle 3, 6 \rangle$. Since $\mathbf{u} \cdot \mathbf{v} = -12 + 12 = 0$, the vectors are orthogonal.

103. $\mathbf{u} = 2\mathbf{i} + \mathbf{j}$ and $\mathbf{v} = \mathbf{i} + 3\mathbf{j}$. Then $\cos\theta = \dfrac{\mathbf{u} \cdot \mathbf{v}}{|\mathbf{u}|\,|\mathbf{v}|} = \dfrac{2+3}{\sqrt{5}\sqrt{10}} = \dfrac{5}{5\sqrt{2}}$ Thus, $\theta = \cos^{-1}\dfrac{5}{5\sqrt{2}}$
$= \cos^{-1}\dfrac{1}{\sqrt{2}} = 45°$.

105. $\mathbf{u} = \langle 3, 1 \rangle$ and $\mathbf{v} = \langle 6, -1 \rangle$. Then the component of $\mathbf{u}$ along $\mathbf{v}$ is
$\dfrac{\mathbf{u} \cdot \mathbf{v}}{|\mathbf{v}|} = \dfrac{\langle 3,1 \rangle \cdot \langle 6,-1 \rangle}{\sqrt{\langle 6,-1 \rangle \cdot \langle 6,-1 \rangle}} = \dfrac{18-1}{\sqrt{36+1}} = \dfrac{17}{\sqrt{37}}$.

107. The displacement is $\mathbf{D} = \langle 7 - 1, -1 - 1 \rangle = \langle 6, -2 \rangle$. Thus, $W = \mathbf{F} \cdot \mathbf{D} = \langle 2, 9 \rangle \cdot \langle 6, -2 \rangle = -6$.

Chapter 9 Test

1. (a) $\dfrac{\tan x}{1 - \cos x} = \dfrac{\tan x}{1 - \cos x} \cdot \dfrac{1 + \cos x}{1 + \cos x} = \dfrac{\tan x\,(1 + \cos x)}{1 - \cos^2 x} = \dfrac{\dfrac{\sin x}{\cos x}\,(1 + \cos x)}{\sin^2 x}$

   $ = \dfrac{1}{\sin x} \cdot \dfrac{1 + \cos x}{\cos x} = \csc x\,(1 + \sec x)$

   (b) $\dfrac{2\tan x}{1 + \tan^2 x} = \dfrac{2\tan x}{\sec^2 x} = \dfrac{2\sin x}{\cos x} \cdot \cos^2 x = 2\sin x \cos x = \sin 2x$

2. (a) $2\cos^2 x + 5\cos x + 2 = 0 \;\Leftrightarrow\; (2\cos x + 1)(\cos x + 2) = 0 \;\Leftrightarrow\; \cos x = -\tfrac{1}{2}$ or
   $\cos x = -2$ (which is impossible). So, in the interval $[0,\,2\pi)$ the solutions are $x = \tfrac{2\pi}{3},\,\tfrac{4\pi}{3}$.

   (b) $\sin 2x - \cos x = 0 \;\Leftrightarrow\; 2\sin x \cos x - \cos x = 0 \;\Leftrightarrow\; \cos x\,(2\sin x - 1) = 0 \;\Leftrightarrow$
   $\cos x = 0$ or $\sin x = \tfrac{1}{2} \;\Leftrightarrow\; x = \tfrac{\pi}{2},\,\tfrac{3\pi}{2}$ or $x = \tfrac{\pi}{6},\,\tfrac{5\pi}{6}$. Therefore, the solutions in $[0,\,2\pi)$ are
   $x = \tfrac{\pi}{6},\,\tfrac{\pi}{2},\,\tfrac{5\pi}{6},\,\tfrac{3\pi}{2}$.

3. $5\cos 2x = 2 \;\Leftrightarrow\; \cos 2x = \tfrac{2}{5} \;\Leftrightarrow\; 2x = \cos^{-1} 0.4 \approx 1.159279$. The solution in $[0,\,4\pi)$ are
   $2x \approx 1.159279,\; 2\pi - 1.159279,\; 2\pi + 1.159279,\; 4\pi - 1.159279 \;\Leftrightarrow\; 2x \approx 1.159279,$
   $5.123906,\; 7.442465,\; 11.407091 \;\Leftrightarrow\; x \approx 0.57964,\, 2.56195,\, 3.72123,\, 5.70355$ in $[0,\,2\pi)$.

4. $\dfrac{x}{\sqrt{4 - x^2}} = \dfrac{2\sin\theta}{\sqrt{4 - (2\sin\theta)^2}} = \dfrac{2\sin\theta}{\sqrt{4 - 4\sin^2\theta}} = \dfrac{2\sin\theta}{2\sqrt{1 - \sin^2\theta}} = \dfrac{\sin\theta}{\sqrt{\cos^2\theta}} = \dfrac{\sin\theta}{|\cos\theta|} = \dfrac{\sin\theta}{\cos\theta}$
   $= \tan\theta$ (because $\cos\theta > 0$ for $-\tfrac{\pi}{2} < \theta < \tfrac{\pi}{2}$).

5. (a) $\sin 8° \cos 22° + \cos 8° \sin 22° = \sin(8° + 22°) = \sin 30° = \tfrac{1}{2}$.

   (b) $\sin 75° = \sin(45° + 30°) = \sin 45° \cos 30° + \cos 45° \sin 30° = \dfrac{\sqrt{2}}{2} \cdot \dfrac{\sqrt{3}}{2} + \dfrac{\sqrt{2}}{2} \cdot \dfrac{1}{2}$
   $ = \tfrac{1}{4}\big(\sqrt{6} + \sqrt{2}\big)$

   (c) $\sin \dfrac{\pi}{12} = \sin\left(\dfrac{\pi/6}{2}\right) = \sqrt{\dfrac{1 - \cos\frac{\pi}{6}}{2}} = \sqrt{\dfrac{1 - \frac{\sqrt{3}}{2}}{2}} = \sqrt{\dfrac{2 - \sqrt{3}}{4}} = \dfrac{1}{2}\sqrt{2 - \sqrt{3}}$

6. From the triangles we have $\cos(\alpha + \beta) = \cos\alpha \cos\beta - \sin\alpha \sin\beta = \dfrac{2}{\sqrt{5}} \cdot \dfrac{\sqrt{5}}{3} - \dfrac{1}{\sqrt{5}} \cdot \dfrac{2}{3}$
   $= \dfrac{2\sqrt{5} - 2}{3\sqrt{5}} = \dfrac{10 - 2\sqrt{5}}{15}$.

7. (a) $\sin 3x \cos 5x = \tfrac{1}{2}\left[\sin(3x + 5x) + \sin(3x - 5x)\right] = \tfrac{1}{2}\left(\sin 8x - \sin 2x\right)$

   (b) $\sin 2x - \sin 5x = 2\cos\left(\dfrac{2x + 5x}{2}\right)\sin\left(\dfrac{2x - 5x}{2}\right) = -2\cos\dfrac{7x}{2}\sin\dfrac{3x}{2}$

8. $\sin\theta = -\tfrac{4}{5}$. Since θ is in quadrant III, $\cos\theta = -\tfrac{3}{5}$. Then, $\tan\dfrac{\theta}{2} = \dfrac{1 - \cos u}{\sin u} = \dfrac{1 - (-3/5)}{-4/5}$
   $= -\dfrac{1 + (3/5)}{4/5} = -\dfrac{5 + 3}{4} = -2$.

9. $y = \sin x$ has domain $(-\infty, \infty)$ and $y = \sin^{-1}x$ has domain $[-1, 1]$.

$y = \sin x$ $\qquad\qquad\qquad\qquad$ $y = \sin^{-1}x$

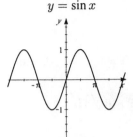

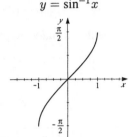

10. (a) $\tan\theta = \frac{x}{4} \quad\Rightarrow\quad \theta = \tan^{-1}\left(\frac{x}{4}\right)$

    (b) $\cos\theta = \frac{3}{x} \quad\Rightarrow\quad \theta = \cos^{-1}\left(\frac{3}{x}\right)$

11. Let $u = \tan^{-1}\frac{9}{40}$ so $\tan u = \frac{9}{40}$. From the triangle
$r = \sqrt{9^2 + 40^2} = 41$. So, $\cos(\tan^{-1}\frac{9}{40}) = \cos$
$u = \frac{40}{41}$.

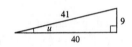

12. (a) $1 + \sqrt{3}\,i$ has $r = \sqrt{1 + 3} = 2$ and $\theta = \tan^{-1}(\sqrt{3}) = \frac{\pi}{3}$. So, in trigonometric form,
$1 + \sqrt{3}\,i = 2\left(\cos\frac{\pi}{3} + i\sin\frac{\pi}{3}\right)$.

    (b) $z = 1 + \sqrt{3}\,i = 2\left(\cos\frac{\pi}{3} + i\sin\frac{\pi}{3}\right) \quad\Rightarrow\quad z^9 = 2^9\left(\cos\frac{9\pi}{3} + i\sin\frac{9\pi}{3}\right)$
$= 512\left(\cos 3\pi + i\sin 3\pi\right) = 512\left(-1 + i\left(0\right)\right) = -512$.

13. $z_1 = 4\left(\cos\frac{7\pi}{12} + i\sin\frac{7\pi}{12}\right)$ and $z_2 = 2\left(\cos\frac{5\pi}{12} + i\sin\frac{5\pi}{12}\right)$.

    Then $z_1 z_2 = 4 \cdot 2\left[\cos\left(\frac{7\pi+5\pi}{12}\right) + i\sin\left(\frac{7\pi+5\pi}{12}\right)\right] = 8\left(\cos\pi + i\sin\pi\right) = -8$, and

    $z_1/z_2 = \frac{4}{2}\left[\cos\left(\frac{7\pi-5\pi}{12}\right) + i\sin\left(\frac{7\pi-5\pi}{12}\right)\right] = 2\left(\cos\frac{\pi}{6} + i\sin\frac{\pi}{6}\right) = 2\left(\frac{\sqrt{3}}{2} + \frac{1}{2}i\right) = \sqrt{3} + i$

14. $27i$ has $r = 27$ and $\theta = \frac{\pi}{2}$, so $27i = 27(\cos\frac{\pi}{2} + i\sin\frac{\pi}{2})$.

    Thus, $(27i)^{1/3} = \sqrt[3]{27}\left[\cos\left(\frac{(\pi/2)+2k\pi}{3}\right) + i\sin\left(\frac{(\pi/2)+2k\pi}{3}\right)\right]$

    $= 3\left[\cos\left(\frac{\pi+4k\pi}{6}\right) + i\sin\left(\frac{\pi+4k\pi}{6}\right)\right]$ for $k = 0, 1, 2$. Thus the three

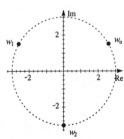

    roots are $w_0 = 3\left(\cos\frac{\pi}{6} + i\sin\frac{\pi}{6}\right) = 3\left(\frac{\sqrt{3}}{2} + \frac{1}{2}i\right) = \frac{3}{2}\left(\sqrt{3} + i\right)$,

    $w_1 = 3\left(\cos\frac{5\pi}{6} + i\sin\frac{5\pi}{6}\right) = 3\left(-\frac{\sqrt{3}}{2} + \frac{1}{2}i\right) = \frac{3}{2}\left(-\sqrt{3} + i\right)$, and

    $w_2 = 3\left(\cos\frac{9\pi}{6} + i\sin\frac{9\pi}{6}\right) = -3i$.

15. (a) Initial point $P(3, -1)$ and terminal point $Q(-3, 9)$. Therefore, $\mathbf{u} = \langle -3 - 3, 9 - (-1)\rangle$
$= \langle -6, 10\rangle = -6\mathbf{i} + 10\mathbf{j}$.

    (b) $\mathbf{u} = \langle -6, 10\rangle \quad\Rightarrow\quad |\mathbf{u}| = \sqrt{(-6)^2 + 10^2} = \sqrt{4 \cdot (9 + 25)} = 2\sqrt{34}$

16. $\mathbf{u} = \langle 1, 3\rangle$ and $\mathbf{v} = \langle -6, 2\rangle$

    (a) $\mathbf{u} - 3\mathbf{v} = \langle 1, 3\rangle - \langle -18, 6\rangle = \langle 19, -3\rangle$

(b) $|\mathbf{u} + \mathbf{v}| = |\langle 1 - 6, 3 + 2\rangle| = |\langle -5, 5\rangle| = \sqrt{(-5)^2 + 5^2} = 5\sqrt{2}$

(c) $\mathbf{u} \cdot \mathbf{v} = (1)(-6) + (3)(2) = -6 + 6 = 0$

(d) Since $\mathbf{u} \cdot \mathbf{v} = 0$, it is true that $\mathbf{u}$ and $\mathbf{v}$ are perpendicular.

17. Let $\mathbf{r}$ represent the velocity of the river and $\mathbf{b}$ the velocity of the boat. Then, $\mathbf{r} = \langle 8, 0\rangle$ and $\mathbf{b} = \langle 12\sin 30°, 12\cos 30°\rangle = \langle 6, 6\sqrt{3}\rangle$.

(a) The resultant (or true) velocity is $\mathbf{v} = \mathbf{r} + \mathbf{b} = \langle 14, 6\sqrt{3}\rangle = 14\,\mathbf{i} + 6\sqrt{3}\,\mathbf{j}$.

(b) $|\mathbf{v}| = \sqrt{14^2 + \left(6\sqrt{3}\right)^2} = \sqrt{196 + 108} = \sqrt{304} \approx 17.44$ and

$\theta = \tan^{-1}\left(\frac{6\sqrt{3}}{14}\right) = \tan^{-1}\left(\frac{3\sqrt{3}}{7}\right) \approx 36.6°$. Therefore, the true speed of the boat is approximately 17.4 mi/h and the true direction is approximately N 53.4° E.

18. $\cos\theta = \frac{\langle 3,2\rangle \cdot \langle 5,-1\rangle}{|\langle 3,2\rangle|\,|\langle 5,-1\rangle|} = \frac{15-2}{\sqrt{13}\,\sqrt{26}} = \frac{13}{13\sqrt{2}} = \frac{1}{\sqrt{2}} \quad \Rightarrow \quad \theta = \frac{\pi}{4}\ \text{(rad)} = 45°.$

19. The displacement vector is $\mathbf{D} = \langle 7, -13\rangle - \langle 2, 2\rangle = \langle 5, -15\rangle$. Therefore, the work done is $W = \mathbf{F} \cdot \mathbf{D} = \langle 3, -5\rangle \cdot \langle 5, -15\rangle = 15 + 75 = 90.$

Focus on Problem Solving

1. The solutions to the equation are the points of intersection of the graphs of the functions $y = \sin x$ and $y = \dfrac{x}{1{,}000{,}000}$. We first find the number of positive solutions. The graph of $y = \dfrac{x}{1{,}000{,}000}$ is a line with slope $m = \frac{1}{1{,}000{,}000}$. The graphs intersect at two points on each of the intervals $[0, 2\pi]$, $[2\pi, 4\pi]$, $[4\pi, 6\pi]$, ... This pattern continues as long as $\dfrac{x}{1{,}000{,}000} \le 1 \quad\Leftrightarrow\quad x \le 1{,}000{,}000$.

   So we need to know how many of the intervals described lie between 0 and 1,000,000. Since $\frac{1{,}000{,}000}{2\pi} \approx 159{,}154.94$ and since the points of intersection always occur in the first half of the positive intervals described above, it follows that the graphs intersect $159{,}155 \cdot 2 = 318{,}310$ times. Thus the given equation has 318,310 solutions for $x \ge 0$. Because of symmetry, it is clear that there are also 318,310 solutions for $x \le 0$. Since $x = 0$ is a solution counted in both cases, the total number of solutions of the given equation is $318{,}310 + 318{,}310 - 1 = 636{,}619$.

3. By similar triangles, $\dfrac{|AQ|}{|QC|} = \dfrac{|AP|}{x}$. So $\dfrac{50}{10} = \dfrac{30}{x} \quad\Leftrightarrow\quad$ $50x = 300 \quad\Leftrightarrow\quad x = 6$. Using the Pythagorean Theorem $y^2 + 6^2 = 10^2 \quad\Leftrightarrow\quad y = 8$ and so $|DE| = 16$ cm.

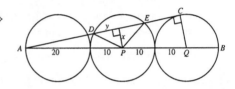

5. From Exercise 44 (b) in section 7.2, we know that $\tan\psi = \dfrac{m_2 - m_1}{1 + m_1 m_2}$, where ψ is the acute angle formed by two lines with slopes m_1 and m_2. In an equilateral triangle, all the angles are 60°, so $\tan 60° = \sqrt{3} = \dfrac{m_2 - m_1}{1 + m_1 m_2}$. Suppose that the triangle's vertices are located at (x_1, y_1), (x_2, y_2), and (x_3, y_3). Then $m_1 = \dfrac{y_1 - y_3}{x_1 - x_3}$ and $m_2 = \dfrac{y_2 - y_3}{x_2 - x_3}$, that is, m_1 and m_2 are rational numbers (since coordinates at a lattice point are integers). Thus, $\dfrac{m_2 - m_1}{1 + m_1 m_2}$ is also rational; but this contradicts $\tan 60° = \sqrt{3}$. Therefore, it is impossible for all three vertices to be at lattice points.

7. $\sin 2\theta = a \quad\Leftrightarrow\quad 2\sin\theta\cos\theta = a \quad\Leftrightarrow\quad \sin\theta\cos\theta = \frac{a}{2}$. Now, $(\sin\theta + \cos\theta)^2 = \sin^2\theta + 2\cos\theta\sin\theta + \cos^2\theta = 1 + 2\cos\theta\sin\theta = 1 + 2\cdot\frac{a}{2} = 1 + a$. Therefore $\sin\theta + \cos\theta = \sqrt{1 + a}$.

9. (a) Label the diagram as shown. Applying the Law of Sines to $\triangle ABC$, we get $\dfrac{\sin B}{b} = \dfrac{\sin 60°}{c}$ or $\sin B = \dfrac{b\sin 60°}{c}$. Similarly, applying the Law of Sines to $\triangle BCD$ gives $\dfrac{\sin B}{r} = \dfrac{\sin 120°}{c + d}$ or or $\sin B = \dfrac{r\sin 120°}{c + d}$. Since $\sin 120° = \sin 60°$,

   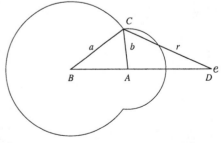

   we have: $\dfrac{b}{c} = \dfrac{r}{c + d} \quad\Leftrightarrow\quad \dfrac{b}{r} = \dfrac{c}{c + d}$ (*). Similarly, from $\triangle ADC$ and the Law of

Sines we have $\dfrac{\sin D}{b} = \dfrac{\sin 60°}{d}$ or $\sin D = \dfrac{b \sin 60°}{d}$, and from $\triangle BDC$ we have

$\dfrac{\sin D}{a} = \dfrac{\sin 120°}{c+d}$ or $\sin D = \dfrac{a \sin 120°}{c+d}$. Thus, $\dfrac{b \sin 60°}{d} = \dfrac{a \sin 120°}{c+d}$ or $\dfrac{b}{a} = \dfrac{d}{c+d}$.

Combining this with (*), we get $\dfrac{b}{r} + \dfrac{b}{a} = \dfrac{c}{c+d} + \dfrac{d}{c+d} = \dfrac{c+d}{c+d} = 1$. Solving for r,

we get $\dfrac{b}{r} = 1 - \dfrac{b}{a} = \dfrac{a-b}{a} \quad \Leftrightarrow \quad \dfrac{r}{b} = \dfrac{a}{a-b} \quad \Leftrightarrow \quad r = \dfrac{ab}{a-b}$.

(b) $r = \dfrac{4 \cdot 3}{4 - 3} = 12$ cm

(c)   If $a = b$, then r is infinite, and so the face is a plane.

11.   There are two possibilities: if $\sin x > 0$, then $|\sin x| = \sin x$, $0 < x < \pi$, and the equation becomes
$\sin x = \sin x + 2\cos x \quad \Leftrightarrow \quad 2\cos x = 0 \quad \Leftrightarrow \quad \cos x = 0 \quad \Leftrightarrow \quad x = \frac{\pi}{2}, \frac{3\pi}{2}$. The second root is
inadmissible since $\sin\frac{3\pi}{2} = -1 < 0$. Thus, if $\sin x > 0$ then $x = \frac{\pi}{2}$. On the other hand, if $\sin x < 0$
then $|\sin x| = -\sin x$, $\pi < x < 2\pi$, and the equation becomes $-\sin x = \sin x + 2\cos x \quad \Leftrightarrow$
$2\sin x = -2\cos x \quad \Leftrightarrow \quad \tan x = -1 \ (\cos x \neq 0) \quad \Leftrightarrow \quad x = \frac{3\pi}{4}, \frac{7\pi}{4}$. Since $\sin\frac{3\pi}{4} = \frac{1}{\sqrt{2}} > 0$, the
first root is inadmissible, and so $x = \frac{7\pi}{4}$.

13.   Using the Change of Base formula we have
$(\log_2 3)(\log_3 4)(\log_4 5)\cdots(\log_{31} 32) = \frac{\log 3}{\log 2} \cdot \frac{\log 4}{\log 3} \cdot \frac{\log 5}{\log 4} \cdot \ \ldots \ \cdot \frac{\log 31}{\log 30} \cdot \frac{\log 32}{\log 31} = \frac{\log 32}{\log 2} = \log_2 32$
$= \log_2 2^5 = 5$.

Chapter Ten
Exercises 10.1

1. $\begin{cases} x + y = 8 \\ x - 3y = 0 \end{cases}$ From the second equation, we have $x = 3y$, and substituting this into the first

   equation gives $3y + y = 8$ $\Leftrightarrow$ $4y = 8$ $\Leftrightarrow$ $y = 2$. Since $x = 3y$, we have $x = 6$ when
   $y = 2$. Thus the solution is $(6, 2)$.

3. $\begin{cases} y = x^2 \\ y = x + 6 \end{cases}$ Substituting $y = x^2$ into the second equation gives $x^2 = x + 6$ $\Leftrightarrow$

   $0 = x^2 - x - 6 = (x - 3)(x + 2)$ $\Rightarrow$ $x = 3$ or $x = -2$. So since $y = x^2$, the solutions are
   $(-2, 4)$ and $(3, 9)$.

5. $\begin{cases} x^2 + y^2 = 8 \\ x + y = 0 \end{cases}$ Solving the second equation for y gives $y = -x$, and substituting this into the first

   equation gives $x^2 + (-x)^2 = 8$

   $\Leftrightarrow$ $2x^2 = 8$ $\Leftrightarrow$ $x = \pm 2$. Thus the solutions are $(2, -2)$ and $(-2, 2)$.

7. $\begin{cases} 5x + 2y = 2 \\ 7x + 3y = 6 \end{cases}$ Multiplying the first equation by 3 and the second by -2 gives the system

   $\begin{cases} 15x + 6y = 6 \\ -14x - 6y = -12 \end{cases}$. Adding, we get $x = -6$, and substituting into the first equation in the

   original system gives $5(-6) + 2y = 2$ $\Leftrightarrow$ $2y = 2 + 30 = 32$ $\Leftrightarrow$ $y = 16$. The solution is
   $(-6, 16)$.

9. $\begin{cases} x^2 - 2y = 1 \\ x^2 + 5y = 29 \end{cases}$ Subtracting the first equation from the second equation gives $7y = 28$ $\Rightarrow$

   $y = 4$. Substituting $y = 4$ into the first equation of the original system gives $x^2 - 2(4) = 1$ $\Leftrightarrow$
   $x^2 = 9$ $\Leftrightarrow$ $x = \pm 3$. The solutions are $(3, 4)$ and $(-3, 4)$.

11. $\begin{cases} 3x^2 - y^2 = 11 \\ x^2 + 4y^2 = 8 \end{cases}$ Multiplying the first equation by 4 gives the system $\begin{cases} 12x^2 - 4y^2 = 44 \\ x^2 + 4y^2 = 8 \end{cases}$.

    Adding the equations gives $13x^2 = 52$ $\Leftrightarrow$ $x = \pm 2$. Thus, the solutions are $(2, 1)$, $(2, -1)$,
    $(-2, 1)$, and $(-2, -1)$.

13. $\begin{cases} y + x^2 = 4x \\ y + 4x = 16 \end{cases}$ Subtracting the second equation from the first equation gives $x^2 - 4x = 4x - 16$

    $\Leftrightarrow$ $x^2 - 8x + 16 = 0$ $\Leftrightarrow$ $(x - 4)^2 = 0$ $\Leftrightarrow$ $x = 4$. Substituting this value for x into
    either of the original equations gives $y = 0$. Therefore, the solution is $(4, 0)$.

15. $\begin{cases} x - 2y = 2 \\ y^2 - x^2 = 2x + 4 \end{cases}$ Now $x - 2y = 2$ $\Leftrightarrow$ $x = 2y + 2$. Substituting for x gives

    $y^2 - x^2 = 2x + 4$ $\Leftrightarrow$ $y^2 - (2y + 2)^2 = 2(2y + 2) + 4$ $\Leftrightarrow$
    $y^2 - 4y^2 - 8y - 4 = 4y + 4 + 4$ $\Leftrightarrow$ $y^2 + 4y + 4 = 0$ $\Leftrightarrow$ $(y + 2)^2 = 0$ $\Leftrightarrow$ $y = -2$.
    $x = 2(-2) + 2 = -2$. Thus the solution is $(-2, -2)$.

17. $\begin{cases} x - y = 4 \\ xy = 12 \end{cases}$ Now $x - y = 4 \quad \Leftrightarrow \quad x = 4 + y$. Substituting for x gives $xy = 12 \quad \Leftrightarrow$

$(4 + y)y = 12 \quad \Leftrightarrow \quad y^2 + 4y - 12 = 0 \quad \Leftrightarrow \quad (y + 6)(y - 2) = 0 \quad \Leftrightarrow \quad y = -6, y = 2$. Thus, the solutions are $(-2, -6)$ and $(6, 2)$.

19. $\begin{cases} x^2 y = 16 \\ x^2 + 4y + 16 = 0 \end{cases}$ Now $x^2 y = 16 \quad \Leftrightarrow \quad x^2 = \dfrac{16}{y}$. Substituting for x^2 gives

$\dfrac{16}{y} + 4y + 16 = 0 \quad \Rightarrow \quad 4y^2 + 16y + 16 = 0 \quad \Leftrightarrow \quad y^2 + 4y + 4 = 0 \quad \Leftrightarrow \quad (y + 2)^2 = 0$

$\Leftrightarrow \quad y = -2$. Therefore, $x^2 = \frac{16}{-2} = -8$, which has no real solution, and so the system has no solution.

21. $\begin{cases} x^2 + y^2 = 9 \\ x^2 - y^2 = 1 \end{cases}$ Adding the equations gives $2x^2 = 10 \quad \Leftrightarrow \quad x^2 = 5 \quad \Leftrightarrow \quad x = \pm\sqrt{5}$. Now

$x = \pm\sqrt{5} \quad \Rightarrow \quad y^2 = 9 - 5 = 4 \quad \Leftrightarrow \quad y = \pm 2$, and so the solutions are $(\sqrt{5}, 2)$, $(\sqrt{5}, -2)$, $(-\sqrt{5}, 2)$, and $(-\sqrt{5}, -2)$.

23. $\begin{cases} 2x^2 - 8y^3 = 19 \\ 4x^2 + 16y^3 = 34 \end{cases}$ Multiplying the first equation by 2 gives the system $\begin{cases} 4x^2 - 16y^3 = 38 \\ 4x^2 + 16y^3 = 34 \end{cases}$
Adding the two equations gives $8x^2 = 72 \quad \Leftrightarrow \quad x = \pm 3$, and then $2(9) - 8y^3 = 19 \quad \Leftrightarrow$
$y^3 = -\frac{1}{8} \quad \Leftrightarrow \quad y = -\frac{1}{2}$. Therefore, the solutions are $(3, -\frac{1}{2})$ and $(-3, -\frac{1}{2})$.

25. $\begin{cases} \dfrac{2}{x} - \dfrac{3}{y} = 1 \\ -\dfrac{4}{x} + \dfrac{7}{y} = 1 \end{cases}$ If we let $u = \dfrac{1}{x}$ and $v = \dfrac{1}{y}$, the system is equivalent to $\begin{cases} 2u - 3v = 1 \\ -4u + 7v = 1 \end{cases}$.

Multiplying the first equation by 4 gives the system $\begin{cases} 4u - 6v = 2 \\ -4u + 7v = 1 \end{cases}$. Adding the equations gives

$v = 3$, and $2u - 9 = 1 \quad \Leftrightarrow \quad u = 5$. Thus, the solution is $(\frac{1}{5}, \frac{1}{3})$.

27. $\begin{cases} y = 2x + 6 \\ y = -x + 5 \end{cases}$

The solution is $(-0.33, 5.33)$.

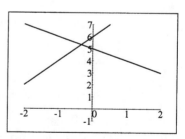

29. $\begin{cases} y = x^3 \\ y = 1 - x^4 \end{cases}$

The solutions are $(0.82, 0.55)$ and $(-1.38, -2.63)$.

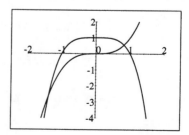

31. $\begin{cases} x^2 + y^2 = 25 \\ x + 3y = 2 \end{cases}$

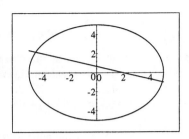

$\Leftrightarrow \begin{cases} y = \pm\sqrt{25 - x^2} \\ y = -\frac{1}{3}x + \frac{2}{3} \end{cases}$

The solutions are $(-4.51, 2.17)$ and $(4.91, -0.97)$.

33. $\begin{cases} \frac{x^2}{9} + \frac{y^2}{18} = 1 \\ y = -x^2 + 6x - 2 \end{cases}$

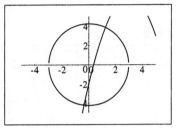

$\Leftrightarrow \begin{cases} y = \pm\sqrt{18 - 2x^2} \\ y = -x^2 + 6x - 2 \end{cases}$

The solutions are $(1.23, 3.87)$ and $(-0.35, -4.21)$.

35. $\begin{cases} x^4 + 16y^4 = 32 \\ x^2 + 2x + y = 0 \end{cases}$

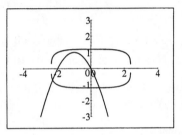

$\Leftrightarrow \begin{cases} y = \pm\dfrac{\sqrt[4]{32 - x^4}}{2} \\ y = -x^2 - 2x \end{cases}$

The solutions are $(-2.30, -0.70)$ and $(0.48, -1.19)$.

37. Let w and l be the lengths of the sides, in cm. Then we have the system $\begin{cases} lw = 180 \\ 2l + 2w = 54 \end{cases}$.

We solve the second equation for w giving, $w = 27 - l$, and substitute into the first equation to get $l(27 - l) = 180 \quad\Leftrightarrow\quad l^2 - 27l + 180 = 0 \quad\Leftrightarrow\quad (l - 15)(l - 12) = 0 \quad\Rightarrow\quad l = 15$ or $l = 12$. If $l = 15$, then $w = 27 - 15 = 12$, and if $l = 12$, then $w = 27 - 12 = 15$. Therefore, the dimensions of the rectangle are 12 cm by 15 cm.

39. Let l and w be the length and width, respectively, of the rectangle. Then, the system of equations is $\begin{cases} 2l + 2w = 70 \\ \sqrt{l^2 + w^2} = 25 \end{cases}$. Solving the first equation for l, we have $l = 35 - w$, and substituting into the second yields $\sqrt{l^2 + w^2} = 25 \quad\Leftrightarrow\quad l^2 + w^2 = 625 \quad\Leftrightarrow\quad (35 - w)^2 + w^2 = 625 \quad\Leftrightarrow\quad 1225 - 70w + w^2 + w^2 = 625 \quad\Leftrightarrow\quad 2w^2 - 70w + 600 = 0 \quad\Leftrightarrow\quad (w - 15)(w - 20) = 0 \quad\Rightarrow\quad w = 15$ or $w = 20$. So the dimensions of the rectangle are 15 by 20.

41. At the points where the rocket path and the hillside meet, we have $\begin{cases} y = \frac{1}{2}x \\ y = -x^2 + 401x \end{cases}$. Substituting for y in the second equation gives $\frac{1}{2}x = -x^2 + 401x \quad\Leftrightarrow\quad x^2 - \frac{801}{2}x = 0 \quad\Leftrightarrow\quad x\left(x - \frac{801}{2}\right) = 0 \quad\Rightarrow\quad x = 0, x = \frac{801}{2}$. When $x = 0$, the rocket has not left the pad. When $x = \frac{801}{2}$, then $y = \frac{1}{2}\left(\frac{801}{2}\right) = \frac{801}{4}$. So the rocket lands at the point $\left(\frac{801}{2}, \frac{801}{4}\right)$. The distance from the base of the hill is $\sqrt{\left(\frac{801}{2}\right)^2 + \left(\frac{801}{4}\right)^2} \approx 447.77$ meters.

43. $\begin{cases} x^2 + y^2 = 25 \\ x^2 - 3x + y^2 + y = 30 \end{cases}$ Subtracting the two equations, we get $3x - y = 25 - 30$ ⇔
$y = 3x + 5$. Since this is the equation of a line, and since any point satisfying both the original equations must satisfy this equation, this must be the line that contains the points of intersection. Thus the equation of the line passing through the intersection points is $y = 3x + 5$. (Notice, we did not have to find the points of intersection to determine the equation of the line that contains them.)

45. $\begin{cases} x^2 + xy = 1 \\ xy + y^2 = 3 \end{cases}$ Adding the equations gives

$x^2 + xy + xy + y^2 = 4$ ⇔ $x^2 + 2xy + y^2 = 4$ ⇔ $(x + y)^2 = 4$ ⇒ $x + y = \pm 2$. If $x + y = 2$, then from the first equation we get $x(x + y) = 1$ ⇒ $x \cdot 2 = 1$ ⇒ $x = \frac{1}{2}$, and so $y = 2 - \frac{1}{2} = \frac{3}{2}$. If $x + y = -2$, then from the first equation we get $x(x + y) = 1$ ⇒ $x \cdot (-2) = 1$ ⇒ $x = -\frac{1}{2}$, and so $y = -2 - \left(-\frac{1}{2}\right) = -\frac{3}{2}$. Thus the solutions are $\left(\frac{1}{2}, \frac{3}{2}\right)$ and $\left(-\frac{1}{2}, -\frac{3}{2}\right)$.

47. $\begin{cases} \log x + \log y = \frac{3}{2} \\ 2\log x - \log y = 0 \end{cases}$ Adding the two equations gives $3\log x = \frac{3}{2}$ ⇔ $\log x = \frac{1}{2}$ ⇔

$x = \sqrt{10}$. So $2\log 10^{1/2} - \log y = 0$ ⇔ $\log 10 - \log y = 0$ ⇔ $\log y = 1$ ⇔ $y = 10$. Thus the solution is $\left(\sqrt{10}, 10\right)$.

49. The point P is at an intersection of the circle of radius 26 centered at $A\,(22, 32)$ and the circle of radius 20 centered at $B\,(28, 20)$. We have

the system $\begin{cases} (x - 22)^2 + (y - 32)^2 = 26^2 \\ (x - 28)^2 + (y - 20)^2 = 20^2 \end{cases}$ ⇔

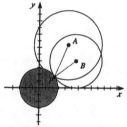

$\begin{cases} x^2 - 44x + 484 + y^2 - 64y + 1024 = 676 \\ x^2 - 56x + 784 + y^2 - 40y + 400 = 400 \end{cases}$ ⇔

$\begin{cases} x^2 - 44x + y^2 - 64y = -832 \\ x^2 - 56x + y^2 - 40y = -784 \end{cases}$. Subtracting the two equations, we get

$12x - 24y = -48$ ⇔ $x - 2y = -4$, which is the equation of a line. Solving for x, we have $x = 2y - 4$. Substituting into the first equation gives $(2y - 4)^2 - 44(2y - 4) + y^2 - 64y = -832$ ⇔ $4y^2 - 16y + 16 - 88y + 176 + y^2 - 64y = -832$ ⇔ $5y^2 - 168y + 192 = -832$ ⇔ $5y^2 - 168y + 1024 = 0$. Using the quadratic equation, we have
$y = \dfrac{168 \pm \sqrt{168^2 - 4(5)(1024)}}{2(5)} = \dfrac{168 \pm \sqrt{7744}}{10} = \dfrac{168 \pm 88}{10}$ ⇔ $y = 8$ or $y = 25.60$.
Since the y coordinate of the point P must be less than that of point A, we have $y = 8$. Then $x = 2(8) - 4 = 12$. So the coordinates of P are $(12, 8)$.
To solve graphically, we must solve each equation for y. This gives $(x - 22)^2 + (y - 32)^2 = 26^2$ ⇔ $(y - 32)^2 = 26^2 - (x - 22)^2$ ⇒ $y - 32 = \pm\sqrt{676 - (x - 22)^2}$ ⇔ $y = 32 \pm \sqrt{676 - (x - 22)^2}$. We use the function $y = 32 - \sqrt{676 - (x - 22)^2}$ because the intersection we at interested in is below the point A. Likewise, solving the second equation for y, we would get the function $y = 20 - \sqrt{400 - (x - 28)^2}$.
In a 3 dimensional situation, you would need the minimum of 3 satellites, since a point on the earth can be uniquely specified as the intersection of 3 spheres centered at the satellites.

Exercises 10.2

1. $\begin{cases} x + y = 3 \\ 2x - y = 0 \end{cases}$

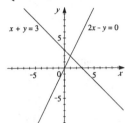

The solution is $x = 1, y = 2$.

3. $\begin{cases} 2x + 3y = 12 \\ x - y = 1 \end{cases}$

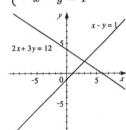

The solution is $x = 3, y = 2$.

5. $\begin{cases} 2x + 5y = 15 \\ 4x + 10y = 20 \end{cases}$

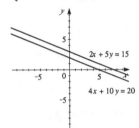

No solution. The lines are parallel, so there is no intersection.

7. $-x + y = 2 \quad \Leftrightarrow \quad y = x + 2$. Substituting for y into $4x - 3y = -3$ gives $4x - 3(x + 2) = -3$ $\Leftrightarrow \quad 4x - 3x - 6 = -3 \quad \Leftrightarrow \quad x = 3$, and so $y = (3) + 2 = 5$. Hence, the solution is $(x, y) = (3, 5)$.

9. $x + 2y = 7 \quad \Leftrightarrow \quad x = 7 - 2y$. Substituting for x into $5x - y = 2$ gives $5(7 - 2y) - y = 2 \quad \Leftrightarrow$ $35 - 10y - y = 2 \quad \Leftrightarrow \quad -11y = -33 \quad \Leftrightarrow \quad y = 3$, and so $x = 7 - 2(3) = 1$. Hence, the solution is $(1, 3)$.

11. $\frac{1}{2}x + \frac{1}{3}y = 2 \quad \Leftrightarrow \quad x + \frac{2}{3}y = 4 \quad \Leftrightarrow \quad x = 4 - \frac{2}{3}y$. Substituting for x into $\frac{1}{5}x - \frac{2}{3}y = 8$ gives $\frac{1}{5}(4 - \frac{2}{3}y) - \frac{2}{3}y = 8 \quad \Leftrightarrow \quad \frac{4}{5} - \frac{2}{15}y - \frac{10}{15}y = 8 \quad \Leftrightarrow \quad 12 - 2y - 10y = 120 \quad \Leftrightarrow \quad y = -9$, and so $x = 4 - \frac{2}{3}(-9) = 10$. Hence, the solution is $(10, -9)$.

13. Adding gives $3x + 2y = 8$
$$\underline{\quad x - 2y = 0 \quad}$$
$$4x \qquad = 8 \quad \Leftrightarrow \quad x = 2.$$

So $x - 2y = (2) - 2y = 0 \quad \Leftrightarrow \quad 2y = 2 \quad \Leftrightarrow \quad y = 1$. Thus, the solution is $(2, 1)$.

15. $\begin{cases} x + 4y = 8 \\ 3x + 12y = 2 \end{cases}$ Adding -3 times the first equation to the second equation gives
$$-3x - 12y = -24$$
$$\underline{\quad 3x + 12y = 2 \quad}$$
$$0 = -22, \text{ which is never true. Thus the system has no solution.}$$

17. $\begin{cases} 2x - 6y = 10 \\ -3x + 9y = -15 \end{cases}$ Adding 3 times the first equation to 2 times the second equation gives

$6x - 18y = 30$
$-6x + 18y = -30$

$\overline{ 0 = 0}$. Writing the equation in slope-intercept form, we have $2x - 6y = 10 \quad \Leftrightarrow$
$-6y = -2x + 10 \quad \Leftrightarrow \quad y = \frac{1}{3}x - \frac{5}{3}$, so a solution is any pair of the form $\left(x, \frac{1}{3}x - \frac{5}{3}\right)$, where x is any real number.

19. $\begin{cases} 6x + 4y = 12 \\ 9x + 6y = 18 \end{cases}$ Adding 3 times the first equation to -2 times the second equation gives

$18x + 12y = 36$
$-18x - 12y = -36$

$\overline{ 0 = 0}$. Writing the equation in slope-intercept form, we have $6x + 4y = 12 \quad \Leftrightarrow$
$4y = -6x + 12 \quad \Leftrightarrow \quad y = -\frac{3}{2}x + 3$, so a solution is any pair of the form $\left(x, -\frac{3}{2}x + 3\right)$, where x is any real number.

21. $\begin{cases} 8s - 3t = -3 \\ 5s - 2t = -1 \end{cases}$. Adding 2 times the first equation to 3 times the second equation gives

$16s - 6t = -6$
$15s - 6t = -3$

$\overline{ s = -3}$. So $8(-3) - 3t = -3 \quad \Leftrightarrow \quad -24 - 3t = -3 \quad \Leftrightarrow \quad t = -7$. Thus, the solution is $(s, t) = (-3, -7)$.

23. $\begin{cases} \frac{1}{2}x + \frac{3}{5}y = 3 \\ \frac{5}{3}x + 2y = 10 \end{cases}$. Adding 10 times the first equation to -3 times the second equation gives

$5x + 6y = 30$
$-5x - 6y = -30$

$\overline{ 0 = 0}$. Writing the equation in slope-intercept form, we have $\frac{1}{2}x + \frac{3}{5}y = 3 \quad \Leftrightarrow$
$\frac{3}{5}y = -\frac{1}{2}x + 3 \quad \Leftrightarrow \quad y = -\frac{5}{6}x + 5$, so a solution is any pair of the form $\left(x, -\frac{5}{6}x + 5\right)$, where x is any real number.

25. $\begin{cases} \dfrac{2x-5}{3} + \dfrac{y-1}{6} = \dfrac{1}{2} \\ \dfrac{x}{5} + \dfrac{3y-6}{12} = 1 \end{cases}$. We multiply the first equation by 6 and the second equation by 60 gives

the system $\begin{cases} 4x - 10 + y - 1 = 3 \\ 12x + 15y - 30 = 60 \end{cases}$, which simplifies to $\begin{cases} 4x + y = 14 \\ 12x + 15y = 90 \end{cases}$. Then adding 3 times the first equation to -1 times the second equation gives $\quad 12x + 3y = 42$

$\qquad\qquad\qquad\qquad\qquad\qquad\qquad\qquad\qquad -12x - 15y = -90$

$\qquad\qquad\qquad\qquad\qquad\qquad\qquad\qquad\qquad \overline{ -12y = -48} \quad \Leftrightarrow \quad y = 4.$

So $4x + (4) = 14 \quad \Leftrightarrow \quad 4x = 10 \quad \Leftrightarrow \quad x = \frac{5}{2}$. Thus, the solution is $\left(\frac{5}{2}, 4\right)$.

27. $\quad x - 2y = 1$
$\quad 2x + 2y = 1$

$\overline{ 3x \qquad = 2} \Leftrightarrow \quad x = \frac{2}{3}$.

So $\left(\frac{2}{3}\right) - 2y = 1 \quad \Leftrightarrow \quad 2y = -\frac{1}{3} \quad \Leftrightarrow \quad y = -\frac{1}{6}$. Thus the solution is $\left(\frac{2}{3}, -\frac{1}{6}\right)$.

29. $x + y = 0 \quad \times -1 \qquad -x - y = 0$
     $x + ay = 1 \quad \times 1 \quad \Rightarrow \quad x + ay = 1$
     $\qquad\qquad\qquad\qquad\qquad \overline{ay - y = 1} \quad \Leftrightarrow \quad y(a - 1) = 1 \quad \Leftrightarrow \quad y = \frac{1}{a-1}, a \neq 1.$

So $x + \left(\frac{1}{a-1}\right) = 0 \quad \Leftrightarrow \quad x = \frac{1}{1-a} = -\frac{1}{a-1}$. Thus the solution is $\left(-\frac{1}{a-1}, \frac{1}{a-1}\right)$.

31. $ax + by = 1 \quad \times -b \qquad -abx - b^2 y = -b$
     $bx + ay = 1 \quad \times a \quad \Rightarrow \quad abx + a^2 y = a$
     $\qquad\qquad\qquad\qquad\qquad \overline{(a^2 - b^2)y = a - b} \quad \Leftrightarrow \quad y = \frac{a-b}{a^2-b^2} = \frac{1}{a+b}, a^2 - b^2 \neq 0.$

So $ax + \frac{b}{a+b} = 1 \quad \Leftrightarrow \quad ax = \frac{a}{a+b} \quad \Leftrightarrow \quad x = \frac{1}{a+b}$. Thus the solution is $\left(\frac{1}{a+b}, \frac{1}{a+b}\right)$.

33. Let the two numbers be x and y. This gives $x + y = 34$
     $\qquad\qquad\qquad\qquad\qquad\qquad\qquad\qquad\qquad\quad x - y = 10$
     $\qquad\qquad\qquad\qquad\qquad\qquad\qquad\qquad \overline{2x \quad\;\; = 44} \quad \Leftrightarrow \quad x = 22.$

So $22 + y = 34 \quad \Leftrightarrow \quad y = 12$. Therefore, the two numbers are 22 and 12.

35. Let d be the number of dimes and q be the number of quarters. This gives
     $d + q = 14 \qquad\qquad \times -1 \qquad\quad -d - q = -14$
     $0.10d + 0.25q = 2.75 \quad \times 10 \quad \Rightarrow \quad d + 2.5q = 27.5$
     $\qquad\qquad\qquad\qquad\qquad\qquad\qquad\qquad \overline{1.5q = 13.5} \quad \Leftrightarrow \quad q = 9$

So $d + (9) = 14 \quad \Leftrightarrow \quad d = 5$. Thus the number of dimes is 5 and the number of quarters is 9.

37. Let x be the speed of the plane in still air and y be the speed of the wind. This gives
     $2x - 2y = 180 \qquad\quad \times -6 \qquad\quad 12x - 12y = 1080$
     $1.2x + 1.2y = 180 \quad\; \times 10 \quad \Rightarrow \quad 12x + 12y = 1800$
     $\qquad\qquad\qquad\qquad\qquad\qquad\qquad\qquad \overline{24x \qquad\quad = 2880} \quad \Leftrightarrow \quad x = 120.$

So $2(120) - 2y = 180 \quad \Leftrightarrow \quad -2y = -60 \quad \Leftrightarrow \quad y = 30$. Therefore, the speed of the plane is 120 mi/h and the wind speed is 30 mi/h.

39. Let x be the cycling speed and y be the running speed. (Remember to divide by 60 to convert minutes to decimal hours.) We have
     $0.5x + 0.5y = 12.5 \times -2 \qquad\qquad -x - y = -25$
     $0.75x + 0.2y = 16 \qquad \times 5 \quad \Rightarrow \quad 3.75x + y = 80$
     $\qquad\qquad\qquad\qquad\qquad\qquad\qquad\quad \overline{2.75x \qquad = 55} \quad \Leftrightarrow \quad x = 20.$

So $20 + y = 25 \quad \Leftrightarrow \quad y = 5$. Thus, the cycling speed is 20 mi/h and the running speed is 5 mi/h.

41. Let a be the grams of food A and b be the grams of food B. This gives
     $0.12a + 0.20b = 32 \qquad \times -250 \qquad -30a - 50b = -8000$
     $\qquad 100a + 50b = 22000 \qquad \times 1 \quad \Rightarrow \quad 100a + 50b = 22000$
     $\qquad\qquad\qquad\qquad\qquad\qquad\qquad\qquad\qquad \overline{70a \qquad\quad = 14000} \quad \Leftrightarrow \quad a = 200.$

So $0.12(200) + 0.20b = 32 \quad \Leftrightarrow \quad 0.20b = 8 \quad \Leftrightarrow \quad b = 40$. Thus, she should use 200 grams of food A and 40 grams of food B.

43. Let x and y be the sulfuric acid concentrations in the first and second containers.
     $300x + 600y = 900(0.15) \qquad \times -1 \qquad -300x - 600y = -135$
     $100x + 500y = 600(0.125) \qquad \times 3 \quad \Rightarrow \quad 300x + 1500y = 225$
     $\qquad\qquad\qquad\qquad\qquad\qquad\qquad\qquad\qquad\quad \overline{900y = 90} \qquad\quad \Leftrightarrow \quad y = 0.10.$

So $100x + 500(0.10) = 75$ $\Leftrightarrow$ $x = 0.25$. The concentrations of sulfuric acid are 25% in the first container and 10% in the second.

45. Let x be the amount he invests at 6% and let y be the amount he invests at 10%.
The ratio of the amounts invested gives $x = 2y$. Then the interest earned is $0.06x + 0.10y = 3520$
$\Leftrightarrow$ $6x + 10y = 352{,}000$. Substituting gives $6(2y) + 10y = 352{,}000$ $\Leftrightarrow$ $22y = 352{,}000$
$\Leftrightarrow$ $y = 16{,}000$. Then $x = 2(16{,}000) = 32{,}000$. He invests $32,000 at 6% and $16,000 at 10%.

47. Let x be the tens digit and y be the ones digit of the number.

$$\begin{array}{lll} x + y = 7 & \times 9 & 9x + 9y = 63 \\ 10y + x = 27 + 10x + y & \times 1 & 9x - 9y = -27 \\ \hline & & 18x \quad\quad = 36 \end{array}$$
$\Leftrightarrow$ $x = 2$.

So $2 + y = 7$ $\Leftrightarrow$ $y = 5$. Thus, the number is 25.

49. $\begin{cases} 0.21x + 3.17y = 9.51 & l_1 \\ 2.35x - 1.17y = 5.89 & l_2 \end{cases}$

The solution is approximately $(3.87, 2.74)$.

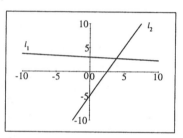

51. $\begin{cases} 2371x - 6552y = 13{,}591 & l_1 \\ 9815x + 992y = 618{,}555 & l_2 \end{cases}$

The solution is approximately $(61.00, 20.00)$.

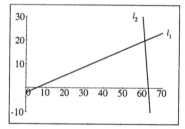

53. $\displaystyle\sum_{k=1}^{n} x_k = 1 + 2 + 3 + 5 + 7 = 18$,

$\displaystyle\sum_{k=1}^{n} y_k = 3 + 5 + 6 + 6 + 9 = 29$,

$\displaystyle\sum_{k=1}^{n} x_k y_k = 1(3) + 2(5) + 3(6) + 5(6) + 7(9) = 124$,

$\displaystyle\sum_{k=1}^{n} x_k^2 = 1^2 + 2^2 + 3^2 + 5^2 + 7^2 = 88$, $n = 5$. Thus we get the system

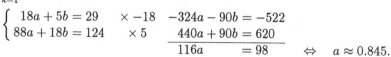

$$\begin{array}{lll} \begin{cases} 18a + 5b = 29 \\ 88a + 18b = 124 \end{cases} & \begin{array}{l} \times -18 \\ \times 5 \end{array} & \begin{array}{l} -324a - 90b = -522 \\ 440a + 90b = 620 \\ \hline 116a \quad\quad = 98 \end{array} \end{array}$$
$\Leftrightarrow$ $a \approx 0.845$.

Then $b = \frac{1}{5}[-18(0.845) + 29] \approx 2.758$. The regression line is $y = 0.845x + 2.758$.

Exercises 10.3

1. The equation $6x - 3y + 1000z - w = \sqrt{13}$ is linear.

3. The equation $e^2 x_1 + \pi x_2 - \sqrt{5} = x_3 - \frac{1}{2} x_4$ is linear.

5. (a) Yes, this matrix is in echelon form.
   (b) Yes, this matrix is in reduced echelon form.
   (c) $x = -3$
   $y = 5$

7. (a) Yes, this matrix is in echelon form.
   (b) No, this matrix is not in reduced echelon form, since the leading "one" in the second row does not have a zero above it.
   (c) $x + 2y + 8z = 0$
   $y + 3z = 2$
   $0 = 0$

9. (a) No, this matrix is not in echelon form, since the row of zeros is not at the bottom.
   (b) No, this matrix is not in reduced echelon form.
   (c) $x \qquad = 0$
   $0 = 0$
   $y + 5z = 1$

11. $\begin{bmatrix} 1 & -2 & 1 & 1 \\ 0 & 1 & 2 & 5 \\ 1 & 1 & 3 & 8 \end{bmatrix}$ $\xrightarrow[R_3 - R_1 \to R_3]{}$ $\begin{bmatrix} 1 & -2 & 1 & 1 \\ 0 & 1 & 2 & 5 \\ 0 & 3 & 2 & 7 \end{bmatrix}$ $\xrightarrow[R_3 - 3R_2 \to R_3]{}$ $\begin{bmatrix} 1 & -2 & 1 & 1 \\ 0 & 1 & 2 & 5 \\ 0 & 0 & -4 & -8 \end{bmatrix}$

Thus, $-4z = -8 \iff z = 2; y + 2(2) = 5 \iff y = 1$; and $x - 2(1) + (2) = 1 \iff x = 1$. Therefore, the solution is $(1, 1, 2)$.

13. $\begin{bmatrix} 1 & 1 & 1 & 2 \\ 2 & -3 & 2 & 4 \\ 4 & 1 & -3 & 1 \end{bmatrix}$ $\xrightarrow[\substack{R_2 - 2R_1 \to R_2 \\ R_3 - 4R_1 \to R_3}]{}$ $\begin{bmatrix} 1 & 1 & 1 & 2 \\ 0 & -5 & 0 & 0 \\ 0 & -3 & -7 & -7 \end{bmatrix}$ $\xrightarrow[R_3 - \frac{3}{5} R_2 \to R_3]{}$ $\begin{bmatrix} 1 & 1 & 1 & 2 \\ 0 & -5 & 0 & 0 \\ 0 & 0 & -7 & -7 \end{bmatrix}$

Thus, $-7z = -7 \iff z = 1; -5y = 0 \iff y = 0$; and $x + 0 + 1 = 2 \iff x = 1$. Therefore, the solution is $(x, y, z) = (1, 0, 1)$.

15. $\begin{bmatrix} 1 & 2 & -1 & -2 \\ 1 & 0 & 1 & 0 \\ 2 & -1 & -1 & -3 \end{bmatrix}$ $\xrightarrow[\substack{R_2 - R_1 \to R_2 \\ R_3 - 2R_1 \to R_3}]{}$ $\begin{bmatrix} 1 & 2 & -1 & -2 \\ 0 & -2 & 2 & 2 \\ 0 & -5 & 1 & 1 \end{bmatrix}$ $\xrightarrow[]{-\frac{1}{2} R_2}$

$\begin{bmatrix} 1 & 2 & -1 & -2 \\ 0 & 1 & 1 & 1 \\ 0 & -5 & 1 & 1 \end{bmatrix}$ $\xrightarrow[R_3 + 5R_2 \to R_3]{}$ $\begin{bmatrix} 1 & 2 & -1 & -2 \\ 0 & 1 & 1 & 1 \\ 0 & 0 & 6 & 6 \end{bmatrix}$

Thus, $6z = 6 \iff z = 1; y + (1) = 1 \iff y = 0$; and $x + 2(0) - (1) = -2 \iff x = -1$. Therefore, the solution is $(x, y, z) = (-1, 0, 1)$.

17. $\begin{bmatrix} 1 & 2 & -1 & 9 \\ 2 & 0 & -1 & -2 \\ 3 & 5 & 2 & 22 \end{bmatrix}$ $\xrightarrow[\substack{R_2 - 2R_1 \to R_2 \\ R_3 - 3R_1 \to R_3}]{}$ $\begin{bmatrix} 1 & 2 & -1 & 9 \\ 0 & -4 & 1 & -20 \\ 0 & -1 & 5 & -5 \end{bmatrix}$ $\xrightarrow[4R_3 - R_2 \to R_3]{}$

$$\begin{bmatrix} 1 & 2 & -1 & 9 \\ 0 & -4 & 1 & -20 \\ 0 & 0 & 19 & 0 \end{bmatrix}$$

Thus, $19x_3 = 0 \quad \Leftrightarrow \quad x_3 = 0$; $-4x_2 = -20 \quad \Leftrightarrow \quad x_2 = 5$; and $x_1 + 2(5) = 9 \quad \Leftrightarrow$
$x_1 = -1$. Therefore, the solution is $(x_1, x_2, x_3) = (-1, 5, 0)$.

19.
$$\begin{bmatrix} 2 & -3 & -1 & 13 \\ -1 & 2 & -5 & 6 \\ 5 & -1 & -1 & 49 \end{bmatrix} \xrightarrow[2R_3-5R_1 \to R_3]{2R_2+R_1 \to R_2} \begin{bmatrix} 2 & -3 & -1 & 13 \\ 0 & 1 & -11 & 25 \\ 0 & 13 & 3 & 33 \end{bmatrix} \xrightarrow{R_3-13R_2 \to R_3}$$
$$\begin{bmatrix} 2 & -3 & -1 & 13 \\ 0 & 1 & -11 & 25 \\ 0 & 0 & 146 & -292 \end{bmatrix}$$

Thus, $146z = -292 \quad \Leftrightarrow \quad z = -2$; $y - 11(-2) = 25 \quad \Leftrightarrow \quad y = 3$; and $2x - 3 \cdot 3 + 2 = 13$
$\Leftrightarrow \quad x = 10$. Therefore, the solution is $(10, 3, -2)$.

21.
$$\begin{bmatrix} 1 & 1 & 1 & 2 \\ 0 & 1 & -3 & 1 \\ 2 & 1 & 5 & 0 \end{bmatrix} \xrightarrow{R_3-2R_1 \to R_3} \begin{bmatrix} 1 & 1 & 1 & 2 \\ 0 & 1 & -3 & 1 \\ 0 & -1 & 3 & -4 \end{bmatrix} \xrightarrow{R_3+R_2 \to R_3} \begin{bmatrix} 1 & 1 & 1 & 3 \\ 0 & 1 & -3 & 1 \\ 0 & 0 & 0 & -3 \end{bmatrix}$$
The third row of the matrix states $0 = -3$, which is impossible. Hence, the system is inconsistent,
and there is no solution.

23.
$$\begin{bmatrix} 2 & -3 & -9 & -5 \\ 1 & 0 & 3 & 2 \\ -3 & 1 & -4 & -3 \end{bmatrix} \xrightarrow{R_1 \leftrightarrow R_2} \begin{bmatrix} 1 & 0 & 3 & 2 \\ 2 & -3 & -9 & -5 \\ -3 & 1 & -4 & -3 \end{bmatrix} \xrightarrow[R_3+3R_1 \to R_3]{R_2-2R_1 \to R_2}$$
$$\begin{bmatrix} 1 & 0 & 3 & 2 \\ 0 & -3 & -15 & -9 \\ 0 & 1 & 5 & 3 \end{bmatrix} \xrightarrow{-\frac{1}{3}R_2} \begin{bmatrix} 1 & 0 & 3 & 2 \\ 0 & 1 & 5 & 3 \\ 0 & 1 & 5 & 3 \end{bmatrix} \xrightarrow{R_3-R_2 \to R_3}$$
$$\begin{bmatrix} 1 & 0 & 3 & 2 \\ 0 & 1 & 5 & 3 \\ 0 & 0 & 0 & 0 \end{bmatrix}$$

Therefore, this system has infinitely many solutions, given by $x + 3z = 2 \quad \Leftrightarrow \quad x = 2 - 3z$, and
$y + 5z = 3 \quad \Leftrightarrow \quad y = 3 - 5z$. Hence, the solutions are $(2 - 3z, 3 - 5z, z)$, where z is any real
number.

25.
$$\begin{bmatrix} 1 & -1 & 3 & 3 \\ 4 & -8 & 32 & 24 \\ 2 & -3 & 11 & 4 \end{bmatrix} \xrightarrow[R_3-2R_1 \to R_3]{R_2-4R_1 \to R_2} \begin{bmatrix} 1 & -1 & 3 & 3 \\ 0 & -4 & 20 & 12 \\ 0 & -1 & 5 & -2 \end{bmatrix} \xrightarrow[R_3+R_2 \to R_3]{-\frac{1}{4}R_2}$$
$$\begin{bmatrix} 1 & -1 & 3 & 3 \\ 0 & 1 & -5 & -3 \\ 0 & 0 & 0 & -5 \end{bmatrix}$$
The third row of the matrix states $0 = -5$, which is impossible. Hence, the system is inconsistent,
and there is no solution.

27.
$$\begin{bmatrix} 1 & 4 & -2 & -3 \\ 2 & -1 & 5 & 12 \\ 8 & 5 & 11 & 30 \end{bmatrix} \xrightarrow[R_3-8R_1 \to R_3]{R_2-2R_1 \to R_2} \begin{bmatrix} 1 & 4 & -2 & -3 \\ 0 & -9 & 9 & 18 \\ 0 & -27 & 27 & 54 \end{bmatrix} \xrightarrow{R_3-3R_2 \to R_3}$$

$$\begin{bmatrix} 1 & 4 & -2 & -3 \\ 0 & -9 & 9 & 18 \\ 0 & 0 & 0 & 0 \end{bmatrix}$$

Therefore, this system has infinitely many solutions, given by $-9y + 9z = 18 \quad \Leftrightarrow \quad y = -2 + z$, and $x + 4(-2 + z) - 2z = -3 \quad \Leftrightarrow \quad x = 5 - 2z$. Hence, the solutions are $(5 - 2z, -2 + z, z)$, where z is any real number.

29. $\begin{bmatrix} 2 & 1 & -2 & 12 \\ -1 & -\frac{1}{2} & 1 & -6 \\ 3 & \frac{3}{2} & -3 & 18 \end{bmatrix} \xrightarrow[-R_1]{R_1 \leftrightarrow R_2} \begin{bmatrix} 1 & \frac{1}{2} & -1 & 6 \\ 2 & 1 & -2 & 12 \\ 3 & \frac{3}{2} & -3 & 18 \end{bmatrix} \xrightarrow[R_3 - 3R_1 \to R_3]{R_2 - 2R_1 \to R_2} \begin{bmatrix} 1 & \frac{1}{2} & -1 & 6 \\ 0 & 0 & 0 & 0 \\ 0 & 0 & 0 & 0 \end{bmatrix}$

Therefore, this system has infinitely many solutions, given by $x + \frac{1}{2} y - z = 6 \quad \Leftrightarrow \quad x = 6 - \frac{1}{2} y + z$. Hence, the solutions are $(6 - \frac{1}{2}y + z, y, z)$, where y and z are any real numbers.

31. $\begin{bmatrix} 4 & -3 & 1 & -8 \\ -2 & 1 & -3 & -4 \\ 1 & -1 & 2 & 3 \end{bmatrix} \xrightarrow{R_1 \leftrightarrow R_3} \begin{bmatrix} 1 & -1 & 2 & 3 \\ -2 & 1 & -3 & -4 \\ 4 & -3 & 1 & -8 \end{bmatrix} \xrightarrow[R_3 - 4R_1 \to R_3]{R_2 + 2R_1 \to R_2}$

$\begin{bmatrix} 1 & -1 & 2 & 3 \\ 0 & -1 & 1 & 2 \\ 0 & 1 & -7 & -20 \end{bmatrix} \xrightarrow{-R_2} \begin{bmatrix} 1 & -1 & 2 & 3 \\ 0 & 1 & -1 & -2 \\ 0 & 1 & -7 & -20 \end{bmatrix} \xrightarrow{R_3 - R_2 \to R_3} \begin{bmatrix} 1 & -1 & 2 & 3 \\ 0 & 1 & -1 & -2 \\ 0 & 0 & -6 & -18 \end{bmatrix}$

Therefore, $-6z = -18 \quad \Leftrightarrow \quad z = 3$; $y - (3) = -2 \quad \Leftrightarrow \quad y = 1$; and $x - (1) + 2(3) = 3 \quad \Leftrightarrow \quad x = -2$. Hence, the solution is $(-2, 1, 3)$.

33. $\begin{bmatrix} 1 & 2 & -3 & -5 \\ -2 & -4 & -6 & 10 \\ 3 & 7 & -2 & -13 \end{bmatrix} \xrightarrow[R_3 - 3R_1 \to R_3]{R_2 + 2R_1 \to R_2} \begin{bmatrix} 1 & 2 & -3 & -5 \\ 0 & 0 & -12 & 0 \\ 0 & 1 & 7 & 2 \end{bmatrix} \xrightarrow{R_2 \leftrightarrow R_3}$

$\begin{bmatrix} 1 & 2 & -3 & -5 \\ 0 & 1 & 7 & 2 \\ 0 & 0 & -12 & 0 \end{bmatrix}$

Therefore, $-12z = 0 \quad \Leftrightarrow \quad z = 0$; $y + 7(0) = 2 \quad \Leftrightarrow \quad y = 2$; and $x + 2(2) - 3(0) = -5 \quad \Leftrightarrow \quad x = -9$. Hence, the solution is $(-9, 2, 0)$.

35. $\begin{bmatrix} -1 & 2 & 1 & -3 & 3 \\ 3 & -4 & 1 & 1 & 9 \\ -1 & -1 & 1 & 1 & 0 \\ 2 & 1 & 4 & -2 & 3 \end{bmatrix} \xrightarrow{-R_1} \begin{bmatrix} 1 & -2 & -1 & 3 & -3 \\ 3 & -4 & 1 & 1 & 9 \\ -1 & -1 & 1 & 1 & 0 \\ 2 & 1 & 4 & -2 & 3 \end{bmatrix}$

$\xrightarrow[R_4 - 2R_1 \to R_4]{\substack{R_2 - 3R_1 \to R_2,\ R_3 + R_1 \to R_3}} \begin{bmatrix} 1 & -2 & -1 & 3 & -3 \\ 0 & 2 & 4 & -8 & 18 \\ 0 & -3 & 0 & 4 & -3 \\ 0 & 5 & 6 & -8 & 9 \end{bmatrix} \xrightarrow{\frac{1}{2} R_2}$

$\begin{bmatrix} 1 & -2 & -1 & 3 & -3 \\ 0 & 1 & 2 & -4 & 9 \\ 0 & -3 & 0 & 4 & -3 \\ 0 & 5 & 6 & -8 & 9 \end{bmatrix} \xrightarrow[R_4 - 5R_2 \to R_4]{R_3 + 3R_2 \to R_3} \begin{bmatrix} 1 & -2 & -1 & 3 & -3 \\ 0 & 1 & 2 & -4 & 9 \\ 0 & 0 & 6 & -8 & 24 \\ 0 & 0 & -4 & 12 & -36 \end{bmatrix}$

$\xrightarrow{3R_4 + 2R_3 \to R_4} \begin{bmatrix} 1 & -2 & -1 & 3 & -3 \\ 0 & 1 & 2 & -4 & 9 \\ 0 & 0 & 6 & -8 & 24 \\ 0 & 0 & 0 & 20 & -60 \end{bmatrix}$

Therefore, $20w = -60$ $\Leftrightarrow$ $w = -3$; $6z + 24 = 24$ $\Leftrightarrow$ $z = 0$; $y + 12 = 9$ $\Leftrightarrow$ $y = -3$; and $x + 6 - 9 = -3$ $\Leftrightarrow$ $x = 0$. Hence, the solution is $(0, -3, 0, -3)$.

37.
$$\begin{bmatrix} 1 & 1 & 2 & -1 & -2 \\ 0 & 3 & 1 & 2 & 2 \\ 1 & 1 & 0 & 3 & 2 \\ -3 & 0 & 1 & 2 & 5 \end{bmatrix} \xrightarrow[R_4+3R_1 \to R_4]{R_3-R_1 \to R_3} \begin{bmatrix} 1 & 1 & 2 & -1 & -2 \\ 0 & 3 & 1 & 2 & 2 \\ 0 & 0 & -2 & 4 & 4 \\ 0 & 3 & 7 & -1 & -1 \end{bmatrix} \xrightarrow{R_4-R_2 \to R_4}$$

$$\begin{bmatrix} 1 & 1 & 2 & -1 & -2 \\ 0 & 3 & 1 & 2 & 2 \\ 0 & 0 & -2 & 4 & 4 \\ 0 & 0 & 6 & -3 & -3 \end{bmatrix} \xrightarrow{R_4+3R_3 \to R_4} \begin{bmatrix} 1 & 1 & 2 & -1 & -2 \\ 0 & 3 & 1 & 2 & 2 \\ 0 & 0 & -2 & 4 & 4 \\ 0 & 0 & 0 & 9 & 9 \end{bmatrix}$$

Therefore, $9w = 9$ $\Leftrightarrow$ $w = 1$; $-2z + 4(1) = 4$ $\Leftrightarrow$ $z = 0$; $3y + (0) + 2(1) = 2$ $\Leftrightarrow$ $y = 0$; and $x + (0) + 2(0) - (1) = -2$ $\Leftrightarrow$ $x = -1$. Hence, the solution is $(-1, 0, 0, 1)$.

39.
$$\begin{bmatrix} 1 & 0 & 1 & 1 & 4 \\ 0 & 1 & -1 & 0 & -4 \\ 1 & -2 & 3 & 1 & 12 \\ 2 & 0 & -2 & 5 & -1 \end{bmatrix} \xrightarrow[R_4-2R_1 \to R_4]{R_3-R_1 \to R_3} \begin{bmatrix} 1 & 0 & 1 & 1 & 4 \\ 0 & 1 & -1 & 0 & -4 \\ 0 & -2 & 2 & 0 & 8 \\ 0 & 0 & -4 & 3 & -9 \end{bmatrix} \xrightarrow{R_3+2R_2 \to R_3}$$

$$\begin{bmatrix} 1 & 0 & 1 & 1 & 4 \\ 0 & 1 & -1 & 0 & -4 \\ 0 & 0 & 0 & 0 & 0 \\ 0 & 0 & -4 & 3 & -9 \end{bmatrix} \xrightarrow{R_3 \leftrightarrow -R_4} \begin{bmatrix} 1 & 0 & 1 & 1 & 4 \\ 0 & 1 & -1 & 0 & -4 \\ 0 & 0 & 4 & -3 & 9 \\ 0 & 0 & 0 & 0 & 0 \end{bmatrix}$$

Therefore, $4z - 3w = 9$ $\Leftrightarrow$ $4z = 9 + 3w$ $\Leftrightarrow$ $z = \frac{9}{4} + \frac{3}{4}w$; $y - \left(\frac{9}{4} + \frac{3}{4}w\right) = -4$ $\Leftrightarrow$ $y = \frac{-7}{4} + \frac{3}{4}$; and $x + \left(\frac{9}{4} + \frac{3}{4}w\right) + w = 4$ $\Leftrightarrow$ $x = \frac{7}{4} - \frac{7}{4}w$. Hence, the solutions are $\left(\frac{7}{4} + \frac{3}{4}w, \frac{-7}{4} + \frac{3}{4}w, \frac{9}{4} + \frac{3}{4}w, w\right)$, where w is any number.

41.
$$\begin{bmatrix} 1 & -1 & 0 & 1 & 0 \\ 3 & 0 & -1 & 2 & 0 \\ 1 & -4 & 1 & 2 & 0 \end{bmatrix} \xrightarrow[R_3-R_1 \to R_3]{R_2-3R_1 \to R_2} \begin{bmatrix} 1 & -1 & 0 & 1 & 0 \\ 0 & 3 & -1 & -1 & 0 \\ 0 & -3 & 1 & 1 & 0 \end{bmatrix} \xrightarrow{R_3+R_2 \to R_3}$$

$$\begin{bmatrix} 1 & -1 & 0 & 1 & 0 \\ 0 & 3 & -1 & -1 & 0 \\ 0 & 0 & 0 & 0 & 0 \end{bmatrix}$$

Therefore, the system has infinitely many solutions, given by $3y - z - w = 0$ $\Leftrightarrow$ $y = \frac{1}{3}(z + w)$; $x - \frac{1}{3}(z + w) + w = 0$ $\Leftrightarrow$ $x = \frac{1}{3}(z - 2w)$. So the solutions are $\left(\frac{1}{3}[z - 2w], \frac{1}{3}[z + w], z, w\right)$, where z and w are any real numbers.

43. Let x, y, z represent the number of VitaMax, Vitron, and VitaPlus pills taken daily. The matrix representation for the system of equations is:

$$\begin{bmatrix} 5 & 10 & 15 & 50 \\ 15 & 20 & 0 & 50 \\ 10 & 10 & 10 & 50 \end{bmatrix} \xrightarrow[\frac{1}{5}R_3]{\frac{1}{5}R_1, \frac{1}{5}R_2} \begin{bmatrix} 1 & 2 & 3 & 10 \\ 3 & 4 & 0 & 10 \\ 2 & 2 & 2 & 10 \end{bmatrix} \xrightarrow[R_3-2R_1 \to R_3]{R_2-3R_1 \to R_2} \begin{bmatrix} 1 & 2 & 3 & 10 \\ 0 & -2 & -9 & -20 \\ 0 & -2 & -4 & -10 \end{bmatrix}$$

$$\xrightarrow{R_3-R_2 \to R_3} \begin{bmatrix} 1 & 2 & 3 & 10 \\ 0 & -2 & -9 & -20 \\ 0 & 0 & 5 & 10 \end{bmatrix}$$

Thus, $5z = 10$ $\Leftrightarrow$ $z = 2$; $-2y - 18 = -20$ $\Leftrightarrow$ $y = 1$; and $x + 2 + 6 = 10$ $\Leftrightarrow$ $x = 2$. Hence, he should take 2 VitaMax, 1 Vitron, and 2 VitaPlus pills daily.

45. Let x, y, and z represent the distance, in miles, of the run, swim, and cycle parts of the race respectively. Then, since $time = \dfrac{distance}{speed}$, we get the following equations from the three contestants' race times:

$$\begin{cases} \left(\dfrac{x}{10}\right) + \left(\dfrac{y}{4}\right) + \left(\dfrac{z}{20}\right) = 2.5 \\ \left(\dfrac{x}{7.5}\right) + \left(\dfrac{y}{6}\right) + \left(\dfrac{z}{15}\right) = 3 \\ \left(\dfrac{x}{15}\right) + \left(\dfrac{y}{3}\right) + \left(\dfrac{z}{40}\right) = 1.75 \end{cases} \quad \Leftrightarrow \quad \begin{cases} 2x + 5y + z = 50 \\ 4x + 5y + 2z = 90 \\ 8x + 40y + 3z = 210 \end{cases},$$

which has the following matrix representation:

$$\begin{bmatrix} 2 & 5 & 1 & 50 \\ 4 & 5 & 2 & 90 \\ 8 & 40 & 3 & 210 \end{bmatrix} \xrightarrow[R_3 - 4R_1 \to R_3]{R_2 - 2R_1 \to R_2} \begin{bmatrix} 2 & 5 & 1 & 50 \\ 0 & -5 & 0 & -10 \\ 0 & 20 & -1 & 10 \end{bmatrix} \xrightarrow{R_3 + 4R_2 \to R_3}$$

$$\begin{bmatrix} 2 & 5 & 1 & 50 \\ 0 & -5 & 0 & -10 \\ 0 & 0 & -1 & -30 \end{bmatrix}.$$

Thus, $-z = -30 \;\Leftrightarrow\; z = 30$; $-5y = -10 \;\Leftrightarrow\; y = 2$; and $2x + 10 + 30 = 50 \;\Leftrightarrow\; x = 5$. So the race has a 5 mile run, 2 mile swim, and 30 mile cycle.

47. Let t be the number of tables produced, c the number of chairs, and a the number of armoires. Then, the system of equations is

$$\begin{cases} \frac{1}{2}t + c + a = 300 \\ \frac{1}{2}t + \frac{3}{2}c + a = 400 \\ t + \frac{3}{2}c + 2a = 590 \end{cases} \quad \Leftrightarrow \quad \begin{cases} t + 2c + 2a = 600 \\ t + 3c + 2a = 800 \\ 2t + 3c + 4a = 1180 \end{cases} \quad \text{and the matrix representation is:}$$

$$\begin{bmatrix} 1 & 2 & 2 & 600 \\ 1 & 3 & 2 & 800 \\ 2 & 3 & 4 & 1180 \end{bmatrix} \xrightarrow[R_3 - 2R_1 \to R_3]{R_2 - R_1 \to R_2} \begin{bmatrix} 1 & 2 & 2 & 600 \\ 0 & 1 & 0 & 200 \\ 0 & -1 & 0 & -20 \end{bmatrix} \xrightarrow{R_3 + R_2 \to R_3} \begin{bmatrix} 1 & 2 & 2 & 600 \\ 0 & 1 & 0 & 200 \\ 0 & 0 & 0 & 180 \end{bmatrix}.$$

The third row states $0 = 180$, which is impossible, and so the system is inconsistent. Therefore, it is impossible to use all of the available labor-hours.

49. (a) We begin by substituting $\left(\dfrac{x_0 + x_1}{2}, \dfrac{y_0 + y_1}{2}, \dfrac{z_0 + z_1}{2}\right)$ into the left side of the first equation which gives: $a_1\left(\dfrac{x_0 + x_1}{2}\right) + b_1\left(\dfrac{y_0 + y_1}{2}\right) + c_1\left(\dfrac{z_0 + z_1}{2}\right)$

$= \frac{1}{2}[(a_1 x_0 + b_1 y_0 + c_1 z_0) + (a_1 x_1 + b_1 y_1 + c_1 z_1)] = \frac{1}{2}[d_1 + d_1] = d_1$. Thus the given ordered triple satisfies the first equation. We can show that it satisfies the second and the third in exactly the same way. Thus it is a solution of the system.

(b) We have shown in part (a) that if the system has two different solutions, we can find a third one by averaging the two solutions. But then we can find a fourth and a fifth solution by averaging the new one with each of the previous two. Then we can find four more by repeating this process with these new solutions, and so on. Clearly this process can continue indefinitely, so there will be infinitely many solutions.

Exercises 10.4

In Exercises 1 – 21, the matrices A, B, C, D, E, F, and G are defined as follows:

$$A = \begin{bmatrix} 2 & -5 \\ 0 & 7 \end{bmatrix} \qquad B = \begin{bmatrix} 3 & \frac{1}{2} & 5 \\ 1 & -1 & 3 \end{bmatrix} \qquad C = \begin{bmatrix} 2 & -\frac{5}{2} & 0 \\ 0 & 2 & -3 \end{bmatrix} \qquad D = \begin{bmatrix} 7 & 3 \end{bmatrix}$$

$$E = \begin{bmatrix} 1 \\ 2 \\ 0 \end{bmatrix} \qquad F = \begin{bmatrix} 1 & 0 & 0 \\ 0 & 1 & 0 \\ 0 & 0 & 1 \end{bmatrix} \qquad G = \begin{bmatrix} 5 & -3 & 10 \\ 6 & 1 & 0 \\ -5 & 2 & 2 \end{bmatrix}$$

1. $B + C = \begin{bmatrix} 3 & \frac{1}{2} & 5 \\ 1 & -1 & 3 \end{bmatrix} + \begin{bmatrix} 2 & -\frac{5}{2} & 0 \\ 0 & 2 & -3 \end{bmatrix} = \begin{bmatrix} 5 & -2 & 5 \\ 1 & 1 & 0 \end{bmatrix}$

3. $C - B = \begin{bmatrix} 2 & -\frac{5}{2} & 0 \\ 0 & 2 & -3 \end{bmatrix} - \begin{bmatrix} 3 & \frac{1}{2} & 5 \\ 1 & -1 & 3 \end{bmatrix} = \begin{bmatrix} -1 & -3 & -5 \\ -1 & 3 & -6 \end{bmatrix}$

5. $3B + 2C = 3\begin{bmatrix} 3 & \frac{1}{2} & 5 \\ 1 & -1 & 3 \end{bmatrix} + 2\begin{bmatrix} 2 & -\frac{5}{2} & 0 \\ 0 & 2 & -3 \end{bmatrix} = \begin{bmatrix} 13 & -\frac{7}{2} & 15 \\ 3 & 1 & 3 \end{bmatrix}$

7. $2C - 6B = 2\begin{bmatrix} 2 & -\frac{5}{2} & 0 \\ 0 & 2 & -3 \end{bmatrix} - 6\begin{bmatrix} 3 & \frac{1}{2} & 5 \\ 1 & -1 & 3 \end{bmatrix} = \begin{bmatrix} -14 & -8 & -30 \\ -6 & 10 & -24 \end{bmatrix}$

9. AD is undefined because A (2×2) and D (1×2) have incompatible dimensions.

11. $BF = \begin{bmatrix} 3 & \frac{1}{2} & 5 \\ 1 & -1 & 3 \end{bmatrix}\begin{bmatrix} 1 & 0 & 0 \\ 0 & 1 & 0 \\ 0 & 0 & 1 \end{bmatrix} = \begin{bmatrix} 3 & \frac{1}{2} & 5 \\ 1 & -1 & 3 \end{bmatrix}$

13. $(DA)B = \begin{bmatrix} 7 & 3 \end{bmatrix}\begin{bmatrix} 2 & -5 \\ 0 & 7 \end{bmatrix}\begin{bmatrix} 3 & \frac{1}{2} & 5 \\ 1 & -1 & 3 \end{bmatrix} = \begin{bmatrix} 14 & -14 \end{bmatrix}\begin{bmatrix} 3 & \frac{1}{2} & 5 \\ 1 & -1 & 3 \end{bmatrix} = \begin{bmatrix} 28 & 21 & 28 \end{bmatrix}$

15. $GE = \begin{bmatrix} 5 & -3 & 10 \\ 6 & 1 & 0 \\ -5 & 2 & 2 \end{bmatrix}\begin{bmatrix} 1 \\ 2 \\ 0 \end{bmatrix} = \begin{bmatrix} -1 \\ 8 \\ -1 \end{bmatrix}$

17. $A^3 = \begin{bmatrix} 2 & -5 \\ 0 & 7 \end{bmatrix}\begin{bmatrix} 2 & -5 \\ 0 & 7 \end{bmatrix}\begin{bmatrix} 2 & -5 \\ 0 & 7 \end{bmatrix} = \begin{bmatrix} 4 & -45 \\ 0 & 49 \end{bmatrix}\begin{bmatrix} 2 & -5 \\ 0 & 7 \end{bmatrix} = \begin{bmatrix} 8 & -335 \\ 0 & 343 \end{bmatrix}$

19. B^2 is undefined because the dimensions (2×3 and 2×3) are incompatible.

21. $BF + FE$ is undefined because the dimensions, $(2 \times 3) \cdot (3 \times 3) = (2 \times 3)$ and $(3 \times 3) \cdot (3 \times 1) = (3 \times 1)$, are incompatible.

23. $\begin{cases} 2x - 5y = 7 \\ 3x + 2y = 4 \end{cases}$ written as a matrix equation is $\begin{bmatrix} 2 & -5 \\ 3 & 2 \end{bmatrix}\begin{bmatrix} x \\ y \end{bmatrix} = \begin{bmatrix} 7 \\ 4 \end{bmatrix}.$

25. $\begin{cases} 3x_1 + 2x_2 - x_3 + x_4 = 0 \\ x_1 - x_3 = 5 \\ 3x_2 + x_3 - x_4 = 4 \end{cases}$ written as a matrix equation is $\begin{bmatrix} 3 & 2 & -1 & 1 \\ 1 & 0 & -1 & 0 \\ 0 & 3 & 1 & -1 \end{bmatrix}\begin{bmatrix} x_1 \\ x_2 \\ x_3 \\ x_4 \end{bmatrix} = \begin{bmatrix} 0 \\ 5 \\ 4 \end{bmatrix}.$

27. $\begin{bmatrix} 6 & x \\ 1 & 0 \end{bmatrix} \begin{bmatrix} y & 2 \\ -1 & 2 \end{bmatrix} = \begin{bmatrix} 6y - x & 12 + 2x \\ y & 2 \end{bmatrix} = \begin{bmatrix} 4 & 16 \\ 1 & 2 \end{bmatrix}$. Thus we must solve the system

$$\begin{cases} 6y - x = 4 \\ 12 + 2x = 16 \\ y = 1 \\ 2 = 2 \end{cases}$$

So $y = 1$ and $12 + 2x = 16 \Leftrightarrow x = 2$. Since these values for x and y also satisfy the first two equations, the solution is $x = 2$, $y = 1$.

29. $2X - A = B \Leftrightarrow X = \frac{1}{2}(A + B) = \frac{1}{2}\left(\begin{bmatrix} 4 & 6 \\ 1 & 3 \end{bmatrix} + \begin{bmatrix} 2 & 5 \\ 3 & 7 \end{bmatrix}\right) = \frac{1}{2}\begin{bmatrix} 6 & 11 \\ 4 & 10 \end{bmatrix} = \begin{bmatrix} 3 & \frac{11}{2} \\ 2 & 5 \end{bmatrix}$.

31. $3X + B = C \Leftrightarrow 3X = C - B \Leftrightarrow X = \frac{1}{3}(C - B)$, which is undefined because the dimensions, (3×2) and (2×2), are incompatible.

33. (a) $BA = [\$0.90 \quad \$0.80 \quad \$1.10] \begin{bmatrix} 4000 & 1000 & 3500 \\ 400 & 300 & 200 \\ 700 & 500 & 9000 \end{bmatrix} = [\$4690 \quad \$1690 \quad \$13,210]$

   (b) The entries in the product matrix represent the total food sales in Santa Monica, Long Beach, and Anaheim respectively.

35. $A = \begin{bmatrix} 1 & 0 & 6 & -1 \\ 2 & \frac{1}{2} & 4 & 0 \end{bmatrix}$, $B = [1 \quad 7 \quad -9 \quad 2]$, and $C = \begin{bmatrix} 1 \\ 0 \\ -1 \\ -2 \end{bmatrix}$.

   ABC is undefined because the dimensions of A (2×4) and B (1×4) are not compatible.

   $ACB = \begin{bmatrix} -3 \\ -2 \end{bmatrix} [1 \quad 7 \quad -9 \quad 2] = \begin{bmatrix} -3 & -21 & 27 & -6 \\ -2 & -14 & 18 & -4 \end{bmatrix}$

   BAC is undefined because the dimensions of B (1×4) and A (2×4) are not compatible.

   BCA is undefined because the dimensions of C (4×1) and A (2×4) are not compatible.

   CAB is undefined because the dimensions of C (4×1) and A (2×4) are not compatible.

   CBA is undefined because the dimensions of B (1×4) and A (2×4) are not compatible.

37. No, from Exercise 36, $(A + B)^2 = (A + B)(A + B) = A^2 + AB + BA + B^2 \neq A^2 + 2AB + B^2$ unless $AB = BA$ which is, in general, not true, as we saw in Example 3.

39. $A = \begin{bmatrix} 1 & 1 \\ 1 & 1 \end{bmatrix}$;

   $A^2 = \begin{bmatrix} 1 & 1 \\ 1 & 1 \end{bmatrix} \begin{bmatrix} 1 & 1 \\ 1 & 1 \end{bmatrix} = \begin{bmatrix} 2 & 2 \\ 2 & 2 \end{bmatrix}$;

   $A^3 = A \cdot A^2 = \begin{bmatrix} 1 & 1 \\ 1 & 1 \end{bmatrix} \begin{bmatrix} 2 & 2 \\ 2 & 2 \end{bmatrix} = \begin{bmatrix} 4 & 4 \\ 4 & 4 \end{bmatrix}$;

   $A^4 = A \cdot A^3 = \begin{bmatrix} 1 & 1 \\ 1 & 1 \end{bmatrix} \begin{bmatrix} 4 & 4 \\ 4 & 4 \end{bmatrix} = \begin{bmatrix} 8 & 8 \\ 8 & 8 \end{bmatrix}$.

   From this pattern, we see that

   $A^n = \begin{bmatrix} 2^{n-1} & 2^{n-1} \\ 2^{n-1} & 2^{n-1} \end{bmatrix}$.

Exercises 10.5

1. $A = \begin{bmatrix} 7 & 4 \\ 3 & 2 \end{bmatrix} \quad \Leftrightarrow \quad A^{-1} = \frac{1}{14-12} \begin{bmatrix} 2 & -4 \\ -3 & 7 \end{bmatrix} = \begin{bmatrix} 1 & -2 \\ -\frac{3}{2} & \frac{7}{2} \end{bmatrix}.$

Then, $AA^{-1} = \begin{bmatrix} 7 & 4 \\ 3 & 2 \end{bmatrix} \begin{bmatrix} 1 & -2 \\ -\frac{3}{2} & \frac{7}{2} \end{bmatrix} = \begin{bmatrix} 1 & 0 \\ 0 & 1 \end{bmatrix},$

and $A^{-1}A = \begin{bmatrix} 1 & -2 \\ -\frac{3}{2} & \frac{7}{2} \end{bmatrix} \begin{bmatrix} 7 & 4 \\ 3 & 2 \end{bmatrix} = \begin{bmatrix} 1 & 0 \\ 0 & 1 \end{bmatrix}.$

3. $\begin{bmatrix} 5 & 3 \\ 3 & 2 \end{bmatrix}^{-1} = \frac{1}{10-9} \begin{bmatrix} 2 & -3 \\ -3 & 5 \end{bmatrix} = \begin{bmatrix} 2 & -3 \\ -3 & 5 \end{bmatrix}$

5. $\begin{bmatrix} 2 & 5 \\ -5 & -13 \end{bmatrix}^{-1} = \frac{1}{-26+25} \begin{bmatrix} -13 & -5 \\ 5 & 2 \end{bmatrix} = \begin{bmatrix} 13 & 5 \\ -5 & -2 \end{bmatrix}$

7. $\begin{bmatrix} 6 & -3 \\ -8 & 4 \end{bmatrix}^{-1} = \frac{1}{24-24} \begin{bmatrix} 4 & 3 \\ 8 & 6 \end{bmatrix}$, which is not defined, and so there is no inverse.

9. $\begin{bmatrix} 0.4 & -1.2 \\ 0.3 & 0.6 \end{bmatrix}^{-1} = \frac{1}{0.24+0.36} \begin{bmatrix} 0.6 & 1.2 \\ -0.3 & 0.4 \end{bmatrix} = \begin{bmatrix} 1 & 2 \\ -\frac{1}{2} & \frac{2}{3} \end{bmatrix}$

11. $\begin{bmatrix} 2 & 4 & 1 & 1 & 0 & 0 \\ -1 & 1 & -1 & 0 & 1 & 0 \\ 1 & 4 & 0 & 0 & 0 & 1 \end{bmatrix} \xrightarrow[2R_3-R_1 \to R_3]{2R_2+R_1 \to R_2} \begin{bmatrix} 2 & 4 & 1 & 1 & 0 & 0 \\ 0 & 6 & -1 & 1 & 2 & 0 \\ 0 & 4 & -1 & -1 & 0 & 2 \end{bmatrix}$

$\xrightarrow[3R_1-2R_2 \to R_1]{3R_3-2R_2 \to R_3} \begin{bmatrix} 6 & 0 & 5 & 1 & -4 & 0 \\ 0 & 6 & -1 & 1 & 2 & 0 \\ 0 & 0 & -1 & -5 & -4 & 6 \end{bmatrix} \xrightarrow[R_2-R_3 \to R_2]{R_1+5R_3 \to R_1}$

$\begin{bmatrix} 6 & 0 & 0 & -24 & -24 & 30 \\ 0 & 6 & 0 & 6 & 6 & -6 \\ 0 & 0 & -1 & -5 & -4 & 6 \end{bmatrix} \xrightarrow[\frac{1}{6}R_2, -R_3]{\frac{1}{6}R_1} \begin{bmatrix} 1 & 0 & 0 & -4 & -4 & 5 \\ 0 & 1 & 0 & 1 & 1 & -1 \\ 0 & 0 & 1 & 5 & 4 & -6 \end{bmatrix}$

Therefore, the inverse matrix is $\begin{bmatrix} -4 & -4 & 5 \\ 1 & 1 & -1 \\ 5 & 4 & -6 \end{bmatrix}$.

13. $\begin{bmatrix} 1 & 2 & 3 & 1 & 0 & 0 \\ 4 & 5 & -1 & 0 & 1 & 0 \\ 1 & -1 & -10 & 0 & 0 & 1 \end{bmatrix} \xrightarrow[R_3-R_1 \to R_3]{R_2-4R_1 \to R_2} \begin{bmatrix} 1 & 2 & 3 & 1 & 0 & 0 \\ 0 & -3 & -13 & -4 & 1 & 0 \\ 0 & -3 & -13 & -1 & 0 & 1 \end{bmatrix}$

$\xrightarrow{R_3-R_2 \to R_3} \begin{bmatrix} 1 & 2 & 3 & 1 & 0 & 0 \\ 0 & -3 & -13 & -4 & 1 & 0 \\ 0 & 0 & 0 & 3 & -1 & 1 \end{bmatrix}$

Since the left half of the last row consists entirely of zeros, there is no inverse matrix.

15.
$$\begin{bmatrix} 0 & -2 & 2 & 1 & 0 & 0 \\ 3 & 1 & 3 & 0 & 1 & 0 \\ 1 & -2 & 3 & 0 & 0 & 1 \end{bmatrix} \xrightarrow{R_1 \leftrightarrow R_3} \begin{bmatrix} 1 & -2 & 3 & 0 & 0 & 1 \\ 3 & 1 & 3 & 0 & 1 & 0 \\ 0 & -2 & 2 & 1 & 0 & 0 \end{bmatrix} \xrightarrow{R_2-3R_1 \to R_2}$$

$$\begin{bmatrix} 1 & -2 & 3 & 0 & 0 & 1 \\ 0 & 7 & -6 & 0 & 1 & -3 \\ 0 & -2 & 2 & 1 & 0 & 0 \end{bmatrix} \xrightarrow[R_2+3R_3 \to R_2]{R_1-R_3 \to R_1} \begin{bmatrix} 1 & 0 & 1 & -1 & 0 & 1 \\ 0 & 1 & 0 & 3 & 1 & -3 \\ 0 & -2 & 2 & 1 & 0 & 0 \end{bmatrix}$$

$$\xrightarrow{R_3+2R_2 \to R_3} \begin{bmatrix} 1 & 0 & 1 & -1 & 0 & 1 \\ 0 & 1 & 0 & 3 & 1 & -3 \\ 0 & 0 & 2 & 7 & 2 & -6 \end{bmatrix} \xrightarrow{\frac{1}{2}R_3} \begin{bmatrix} 1 & 0 & 1 & -1 & 0 & 1 \\ 0 & 1 & 0 & 3 & 1 & -3 \\ 0 & 0 & 1 & \frac{7}{2} & 1 & -3 \end{bmatrix}$$

$$\xrightarrow{R_1-R_3 \to R_1} \begin{bmatrix} 1 & 0 & 0 & -\frac{9}{2} & -1 & 4 \\ 0 & 1 & 0 & 3 & 1 & -3 \\ 0 & 0 & 1 & \frac{7}{2} & 1 & -3 \end{bmatrix}$$

Therefore, the inverse matrix is $\begin{bmatrix} -\frac{9}{2} & -1 & 4 \\ 3 & 1 & -3 \\ \frac{7}{2} & 1 & -3 \end{bmatrix}$.

17.
$$\begin{bmatrix} 1 & 2 & 0 & 3 & 1 & 0 & 0 & 0 \\ 0 & 1 & 1 & 1 & 0 & 1 & 0 & 0 \\ 0 & 1 & 0 & 1 & 0 & 0 & 1 & 0 \\ 1 & 2 & 0 & 2 & 0 & 0 & 0 & 1 \end{bmatrix} \xrightarrow[R_4-R_1 \to R_4]{R_3-R_2 \to R_3} \begin{bmatrix} 1 & 2 & 0 & 3 & 1 & 0 & 0 & 0 \\ 0 & 1 & 1 & 1 & 0 & 1 & 0 & 0 \\ 0 & 0 & -1 & 0 & 0 & -1 & 1 & 0 \\ 0 & 0 & 0 & -1 & -1 & 0 & 0 & 1 \end{bmatrix}$$

$$\xrightarrow[-R_4]{-R_3} \begin{bmatrix} 1 & 2 & 0 & 3 & 1 & 0 & 0 & 0 \\ 0 & 1 & 1 & 1 & 0 & 1 & 0 & 0 \\ 0 & 0 & 1 & 0 & 0 & 1 & -1 & 0 \\ 0 & 0 & 0 & 1 & 1 & 0 & 0 & -1 \end{bmatrix} \xrightarrow[R_2-R_3 \to R_2]{R_1-2R_2 \to R_1}$$

$$\begin{bmatrix} 1 & 0 & -2 & 1 & 1 & -2 & 0 & 0 \\ 0 & 1 & 0 & 1 & 0 & 0 & 1 & 0 \\ 0 & 0 & 1 & 0 & 0 & 1 & -1 & 0 \\ 0 & 0 & 0 & 1 & 1 & 0 & 0 & -1 \end{bmatrix} \xrightarrow[R_2-R_4 \to R_2]{R_1+2R_3 \to R_1}$$

$$\begin{bmatrix} 1 & 0 & 0 & 1 & 1 & 0 & -2 & 0 \\ 0 & 1 & 0 & 0 & -1 & 0 & 1 & 1 \\ 0 & 0 & 1 & 0 & 0 & 1 & -1 & 0 \\ 0 & 0 & 0 & 1 & 1 & 0 & 0 & -1 \end{bmatrix} \xrightarrow{R_1 \to R_1-R_4} \begin{bmatrix} 1 & 0 & 0 & 0 & 0 & 0 & -2 & 1 \\ 0 & 1 & 0 & 0 & -1 & 0 & 1 & 1 \\ 0 & 0 & 1 & 0 & 0 & 1 & -1 & 0 \\ 0 & 0 & 0 & 1 & 1 & 0 & 0 & -1 \end{bmatrix}$$

Therefore, the inverse matrix is $\begin{bmatrix} 0 & 0 & -2 & 1 \\ -1 & 0 & 1 & 1 \\ 0 & 1 & -1 & 0 \\ 1 & 0 & 0 & -1 \end{bmatrix}$.

19. $\begin{cases} 5x + 3y = 4 \\ 3x + 2y = 0 \end{cases}$ is equivalent to the matrix equation $\begin{bmatrix} 5 & 3 \\ 3 & 2 \end{bmatrix} \begin{bmatrix} x \\ y \end{bmatrix} = \begin{bmatrix} 4 \\ 0 \end{bmatrix}$.

Using the inverse from Exercise 3, $\begin{bmatrix} x \\ y \end{bmatrix} = \begin{bmatrix} 2 & -3 \\ -3 & 5 \end{bmatrix} \begin{bmatrix} 4 \\ 0 \end{bmatrix} = \begin{bmatrix} 8 \\ -12 \end{bmatrix}$.

Therefore, $x = 8$ and $y = -12$.

21. $\begin{cases} 2x + 5y = 2 \\ -5x - 13y = 20 \end{cases}$ is equivalent to the matrix equation $\begin{bmatrix} 2 & 5 \\ -5 & -13 \end{bmatrix} \begin{bmatrix} x \\ y \end{bmatrix} = \begin{bmatrix} 2 \\ 20 \end{bmatrix}$.

Using the inverse from Exercise 5, $\begin{bmatrix} x \\ y \end{bmatrix} = \begin{bmatrix} 13 & 5 \\ -5 & -2 \end{bmatrix} \begin{bmatrix} 2 \\ 20 \end{bmatrix} = \begin{bmatrix} 126 \\ -50 \end{bmatrix}$.

Therefore, $x = 126$ and $y = -50$.

23. $\begin{cases} 2x + 4y + z = 7 \\ -x + y - z = 0 \\ x + 4y = -2 \end{cases}$ is equivalent to the matrix equation $\begin{bmatrix} 2 & 4 & 1 \\ -1 & 1 & -1 \\ 1 & 4 & 0 \end{bmatrix} \begin{bmatrix} x \\ y \\ z \end{bmatrix} = \begin{bmatrix} 7 \\ 0 \\ -2 \end{bmatrix}$.

Using the inverse from Exercise 11, $\begin{bmatrix} x \\ y \\ z \end{bmatrix} = \begin{bmatrix} -4 & -4 & 5 \\ 1 & 1 & -1 \\ 5 & 4 & -6 \end{bmatrix} \begin{bmatrix} 7 \\ 0 \\ -2 \end{bmatrix} = \begin{bmatrix} -38 \\ 9 \\ 47 \end{bmatrix}$.

Therefore, $x = -38$, $y = 9$, and $z = 47$.

25. $\begin{cases} -2y + 2z = 12 \\ 3x + y + 3z = -2 \\ x - 2y + 3z = 8 \end{cases}$ is equivalent to the matrix equation $\begin{bmatrix} 0 & -2 & 2 \\ 3 & 1 & 3 \\ 1 & -2 & 3 \end{bmatrix} \begin{bmatrix} x \\ y \\ z \end{bmatrix} = \begin{bmatrix} 12 \\ -2 \\ 8 \end{bmatrix}$.

Using the inverse from Exercise 15, $\begin{bmatrix} x \\ y \\ z \end{bmatrix} = \begin{bmatrix} -\frac{9}{2} & -1 & 4 \\ 3 & 1 & -3 \\ \frac{7}{2} & 1 & -3 \end{bmatrix} \begin{bmatrix} 12 \\ -2 \\ 8 \end{bmatrix} = \begin{bmatrix} -20 \\ 10 \\ 16 \end{bmatrix}$.

Therefore, $x = -20$, $y = 10$, and $z = 16$.

27. We have $\begin{bmatrix} 3 & -2 \\ -4 & 3 \end{bmatrix}^{-1} = \frac{1}{9-8} \begin{bmatrix} 3 & 2 \\ 4 & 3 \end{bmatrix} = \begin{bmatrix} 3 & 2 \\ 4 & 3 \end{bmatrix}$.

Hence, $\begin{bmatrix} x & y & z \\ u & v & w \end{bmatrix} = \begin{bmatrix} 3 & 2 \\ 4 & 3 \end{bmatrix} \begin{bmatrix} 1 & 0 & -1 \\ 2 & 1 & 3 \end{bmatrix} = \begin{bmatrix} 7 & 2 & 3 \\ 10 & 3 & 5 \end{bmatrix}$.

29. (a) $\begin{bmatrix} 3 & 1 & 3 & 1 & 0 & 0 \\ 4 & 2 & 4 & 0 & 1 & 0 \\ 3 & 2 & 4 & 0 & 0 & 1 \end{bmatrix} \xrightarrow[\substack{R_3 - R_1 \to R_3 \\ R_1 \leftrightarrow R_2}]{} \begin{bmatrix} 4 & 2 & 4 & 0 & 1 & 0 \\ 3 & 1 & 3 & 1 & 0 & 0 \\ 0 & 1 & 1 & -1 & 0 & 1 \end{bmatrix} \xrightarrow[\substack{R_1 - R_2 \to R_1}]{}$

$\begin{bmatrix} 1 & 1 & 1 & -1 & 1 & 0 \\ 3 & 1 & 3 & 1 & 0 & 0 \\ 0 & 1 & 1 & -1 & 0 & 1 \end{bmatrix} \xrightarrow[\substack{R_2 - 3R_1 \to R_2}]{} \begin{bmatrix} 1 & 1 & 1 & -1 & 1 & 0 \\ 0 & -2 & 0 & 4 & -3 & 0 \\ 0 & 1 & 1 & -1 & 0 & 1 \end{bmatrix} \xrightarrow[\substack{R_1 - R_2 \to R_1, -\frac{1}{2}R_2 \\ R_3 - R_2 \to R_3}]{}$

$\begin{bmatrix} 1 & 0 & 1 & 1 & -\frac{1}{2} & 0 \\ 0 & 1 & 0 & -2 & \frac{3}{2} & 0 \\ 0 & 0 & 1 & 1 & -\frac{3}{2} & 1 \end{bmatrix} \xrightarrow[\substack{R_1 - R_3 \to R_1}]{} \begin{bmatrix} 1 & 0 & 0 & 0 & 1 & -1 \\ 0 & 1 & 0 & -2 & \frac{3}{2} & 0 \\ 0 & 0 & 1 & 1 & -\frac{3}{2} & 1 \end{bmatrix}$

Therefore, the inverse of the matrix is $\begin{bmatrix} 0 & 1 & -1 \\ -2 & \frac{3}{2} & 0 \\ 1 & -\frac{3}{2} & 1 \end{bmatrix}$.

(b) $\begin{bmatrix} A \\ B \\ C \end{bmatrix} = \begin{bmatrix} 0 & 1 & -1 \\ -2 & \frac{3}{2} & 0 \\ 1 & -\frac{3}{2} & 1 \end{bmatrix} \begin{bmatrix} 10 \\ 14 \\ 13 \end{bmatrix} = \begin{bmatrix} 1 \\ 1 \\ 2 \end{bmatrix}$

Therefore, he should feed the rats 1 oz of A, 1 oz of B, and 2 oz of C.

(c) $\begin{bmatrix} A \\ B \\ C \end{bmatrix} = \begin{bmatrix} 0 & 1 & -1 \\ -2 & \frac{3}{2} & 0 \\ 1 & -\frac{3}{2} & 1 \end{bmatrix} \begin{bmatrix} 9 \\ 12 \\ 10 \end{bmatrix} = \begin{bmatrix} 2 \\ 0 \\ 1 \end{bmatrix}$

Therefore, he should feed the rats 2 oz of A, 0 oz of B, and 1 oz of C.

(d) $\begin{bmatrix} A \\ B \\ C \end{bmatrix} = \begin{bmatrix} 0 & 1 & -1 \\ -2 & \frac{3}{2} & 0 \\ 1 & -\frac{3}{2} & 1 \end{bmatrix} \begin{bmatrix} 2 \\ 4 \\ 11 \end{bmatrix} = \begin{bmatrix} -7 \\ 2 \\ 7 \end{bmatrix}$

Since $A < 0$, there is no combination of foods giving the required supply.

31. (a) $\begin{cases} x + y + 2z = 675 \\ 2x + y + z = 600 \\ x + 2y + z = 625 \end{cases}$

(b) $\begin{bmatrix} 1 & 1 & 2 \\ 2 & 1 & 1 \\ 1 & 2 & 1 \end{bmatrix} \begin{bmatrix} x \\ y \\ z \end{bmatrix} = \begin{bmatrix} 675 \\ 600 \\ 625 \end{bmatrix}$

(c) $\begin{bmatrix} 1 & 1 & 2 & 1 & 0 & 0 \\ 2 & 1 & 1 & 0 & 1 & 0 \\ 1 & 2 & 1 & 0 & 0 & 1 \end{bmatrix} \xrightarrow[R_3 - R_1 \to R_3]{R_2 - 2R_1 \to R_2} \begin{bmatrix} 1 & 1 & 2 & 1 & 0 & 0 \\ 0 & -1 & -3 & -2 & 1 & 0 \\ 0 & 1 & -1 & -1 & 0 & 1 \end{bmatrix}$

$\xrightarrow[R_3 + R_2 \to R_3]{R_1 + R_2 \to R_1} \begin{bmatrix} 1 & 0 & -1 & -1 & 1 & 0 \\ 0 & -1 & -3 & -2 & 1 & 0 \\ 0 & 0 & -4 & -3 & 1 & 1 \end{bmatrix} \xrightarrow[-\frac{1}{4}R_3]{-R_2}$

$\begin{bmatrix} 1 & 0 & -1 & -1 & 1 & 0 \\ 0 & 1 & 3 & 2 & -1 & 0 \\ 0 & 0 & 1 & \frac{3}{4} & -\frac{1}{4} & -\frac{1}{4} \end{bmatrix} \xrightarrow[R_2 - 3R_3 \to R_2]{R_1 + R_3 \to R_1} \begin{bmatrix} 1 & 0 & 0 & -\frac{1}{4} & \frac{3}{4} & -\frac{1}{4} \\ 0 & 1 & 0 & -\frac{1}{4} & -\frac{1}{4} & \frac{3}{4} \\ 0 & 0 & 1 & \frac{3}{4} & -\frac{1}{4} & -\frac{1}{4} \end{bmatrix}$

Therefore, the inverse of the matrix is $\begin{bmatrix} -\frac{1}{4} & \frac{3}{4} & -\frac{1}{4} \\ -\frac{1}{4} & -\frac{1}{4} & \frac{3}{4} \\ \frac{3}{4} & -\frac{1}{4} & -\frac{1}{4} \end{bmatrix}$ and

$\begin{bmatrix} x \\ y \\ z \end{bmatrix} = \begin{bmatrix} -\frac{1}{4} & \frac{3}{4} & -\frac{1}{4} \\ -\frac{1}{4} & -\frac{1}{4} & \frac{3}{4} \\ \frac{3}{4} & -\frac{1}{4} & -\frac{1}{4} \end{bmatrix} \begin{bmatrix} 675 \\ 600 \\ 625 \end{bmatrix} = \begin{bmatrix} 125 \\ 150 \\ 200 \end{bmatrix}$.

Thus he earns \$125 on a standard set, \$150 on a deluxe set, and \$200 on a leather-bound set.

33. $\begin{bmatrix} e^x & -e^{2x} \\ e^{2x} & e^{3x} \end{bmatrix}^{-1} = \frac{1}{e^{4x} + e^{4x}} \begin{bmatrix} e^{3x} & e^{2x} \\ -e^{2x} & e^x \end{bmatrix} = \frac{1}{2} \begin{bmatrix} e^{-x} & e^{-2x} \\ -e^{-2x} & e^{-3x} \end{bmatrix}$

35. $\begin{bmatrix} \sin x & \cos x \\ -\cos x & \sin x \end{bmatrix}^{-1} = \frac{1}{\sin^2 x + \cos^2 x} \begin{bmatrix} \sin x & -\cos x \\ \cos x & \sin x \end{bmatrix} = \begin{bmatrix} \sin x & -\cos x \\ \cos x & \sin x \end{bmatrix}$

37. No, consider the following counterexample: $A = \begin{bmatrix} 0 & 1 \\ 0 & 0 \end{bmatrix}$ and $B = \begin{bmatrix} 0 & 2 \\ 0 & 0 \end{bmatrix}$. Then, $AB = O$, but neither $A = O$ nor $B = O$.

There are infinitely many matrices for which $A^2 = O$. One example is $A = \begin{bmatrix} 0 & 1 \\ 0 & 0 \end{bmatrix}$. Then, $A^2 = O$, but $A \neq O$.

Exercises 10.6

1. The matrix $[\,3\,]$ has determinant $|D| = 3$.

3. The matrix $\begin{bmatrix} 4 & 5 \\ 0 & -1 \end{bmatrix}$ has determinant $|D| = (4)(-1) - (5)(0) = -4$.

5. The matrix $[\,2 \quad 5\,]$ does not have a determinant because the matrix is not square.

7. The matrix $\begin{bmatrix} \frac{1}{2} & \frac{1}{8} \\ 1 & \frac{1}{2} \end{bmatrix}$ has determinant $|D| = \frac{1}{2} \cdot \frac{1}{2} - 1 \cdot \frac{1}{8} = \frac{1}{4} - \frac{1}{8} = \frac{1}{8}$.

In Exercises 9 – 13, the matrix is $A = \begin{bmatrix} 1 & 0 & \frac{1}{2} \\ -3 & 5 & 2 \\ 0 & 0 & 4 \end{bmatrix}$.

9. $M_{11} = 5 \cdot 4 - 0 \cdot 2 = 20$, $A_{11} = (-1)^2 \, M_{11} = 20$

11. $M_{12} = -3 \cdot 4 - 0 \cdot 2 = -12$, $A_{12} = (-1)^3 \, M_{12} = 12$

13. $M_{23} = 1 \cdot 0 - 0 \cdot 0 = 0$, $A_{23} = (-1)^5 \, M_{23} = 0$

15. $M = \begin{bmatrix} 1 & 3 & 7 \\ 2 & 0 & -1 \\ 0 & 2 & 6 \end{bmatrix}$. Therefore, expanding by the third row, $|M| = -2 \begin{vmatrix} 1 & 7 \\ 2 & -1 \end{vmatrix} + 6 \begin{vmatrix} 1 & 3 \\ 2 & 0 \end{vmatrix}$

$= -2(-1 - 14) + 6(0 - 6) = 30 - 36 = -6$. Since $|M| \neq 0$, the matrix has an inverse.

17. $M = \begin{bmatrix} 30 & 0 & 20 \\ 0 & -10 & -20 \\ 40 & 0 & 10 \end{bmatrix}$. Therefore, expanding by the first row,

$|M| = 30 \begin{vmatrix} -10 & -20 \\ 0 & 10 \end{vmatrix} + 20 \begin{vmatrix} 0 & -10 \\ 40 & 0 \end{vmatrix} = 30(-100 + 0) + 20(0 + 400)$

$= -3000 + 8000 = 5000$, and so M^{-1} exists.

19. $M = \begin{bmatrix} 1 & 3 & 3 & 0 \\ 0 & 2 & 0 & 1 \\ -1 & 0 & 0 & 2 \\ 1 & 6 & 4 & 1 \end{bmatrix}$. Therefore, expanding by the third row,

$|M| = -1 \begin{vmatrix} 3 & 3 & 0 \\ 2 & 0 & 1 \\ 6 & 4 & 1 \end{vmatrix} - 2 \begin{vmatrix} 1 & 3 & 3 \\ 0 & 2 & 0 \\ 1 & 6 & 4 \end{vmatrix} = 1 \begin{vmatrix} 3 & 3 \\ 6 & 4 \end{vmatrix} - 1 \begin{vmatrix} 3 & 3 \\ 2 & 0 \end{vmatrix} - 4 \begin{vmatrix} 1 & 3 \\ 1 & 4 \end{vmatrix} = -6 + 6 - 4 = -4$,

and so M^{-1} exists.

21. $|M| = \begin{vmatrix} 0 & 0 & 4 & 6 \\ 2 & 1 & 1 & 3 \\ 2 & 1 & 2 & 3 \\ 3 & 0 & 1 & 7 \end{vmatrix} = \begin{vmatrix} 0 & 0 & 4 & 6 \\ 2 & 1 & 1 & 3 \\ 0 & 0 & 1 & 0 \\ 3 & 0 & 1 & 7 \end{vmatrix}$, by replacing R_3 with $R_3 - R_2$. Then, expanding by the

third row, $|M| = 1 \begin{vmatrix} 0 & 0 & 6 \\ 2 & 1 & 3 \\ 3 & 0 & 7 \end{vmatrix} = 6 \begin{vmatrix} 2 & 1 \\ 3 & 0 \end{vmatrix} = 6(2 \cdot 0 - 3 \cdot 1) = -18.$

23. $M = \begin{bmatrix} 1 & 2 & 3 & 4 & 5 \\ 0 & 2 & 4 & 6 & 8 \\ 0 & 0 & 3 & 6 & 9 \\ 0 & 0 & 0 & 4 & 8 \\ 0 & 0 & 0 & 0 & 5 \end{bmatrix}$. Then, expanding by the fifth row, $|M| = 5 \begin{vmatrix} 1 & 2 & 3 & 4 \\ 0 & 2 & 4 & 6 \\ 0 & 0 & 3 & 6 \\ 0 & 0 & 0 & 4 \end{vmatrix}$

$= 5 \cdot 4 \begin{vmatrix} 1 & 2 & 3 \\ 0 & 2 & 4 \\ 0 & 0 & 3 \end{vmatrix} = 20 \cdot 3 \begin{vmatrix} 1 & 2 \\ 0 & 2 \end{vmatrix} = 60 \cdot 2 = 120.$

25. $B = \begin{bmatrix} 4 & 1 & 0 \\ -2 & -1 & 1 \\ 4 & 0 & 3 \end{bmatrix}$

(a) $|B| = 2 \begin{vmatrix} 1 & 0 \\ 0 & 3 \end{vmatrix} - 1 \begin{vmatrix} 4 & 0 \\ 4 & 3 \end{vmatrix} - 1 \begin{vmatrix} 4 & 1 \\ 4 & 0 \end{vmatrix} = 6 - 12 + 4 = -2$

(b) $|B| = -1 \begin{vmatrix} 4 & 1 \\ 4 & 0 \end{vmatrix} + 3 \begin{vmatrix} 4 & 1 \\ -2 & -1 \end{vmatrix} = 4 - 6 = -2$

(c) Yes, as expected, the results agree.

27. $\begin{cases} 2x - y = -9 \\ x + 2y = 8 \end{cases}$ Then, $|D| = \begin{vmatrix} 2 & -1 \\ 1 & 2 \end{vmatrix} = 5, |D_x| = \begin{vmatrix} -9 & -1 \\ 8 & 2 \end{vmatrix} = -10,$ and

$|D_y| = \begin{vmatrix} 2 & -9 \\ 1 & 8 \end{vmatrix} = 25.$ Hence, $x = \dfrac{|D_x|}{|D|} = \dfrac{-10}{5} = -2, y = \dfrac{|D_y|}{|D|} = \dfrac{25}{5} = 5,$ and so the solution

is $(-2, 5)$.

29. $\begin{cases} x - 6y = 3 \\ 3x + 2y = 1 \end{cases}$ Then, $|D| = \begin{vmatrix} 1 & -6 \\ 3 & 2 \end{vmatrix} = 20, |D_x| = \begin{vmatrix} 3 & -6 \\ 1 & 2 \end{vmatrix} = 12,$ and

$|D_y| = \begin{vmatrix} 1 & 3 \\ 3 & 1 \end{vmatrix} = -8.$ Hence, $x = \dfrac{|D_x|}{|D|} = \dfrac{12}{20} = 0.6, y = \dfrac{|D_y|}{|D|} = \dfrac{-8}{20} = -0.4,$ and so the solution

is $(0.6, -0.4)$.

31. $\begin{cases} 0.4x + 1.2y = 0.4 \\ 1.2x + 1.6y = 3.2 \end{cases}$ Then, $|D| = \begin{vmatrix} 0.4 & 1.2 \\ 1.2 & 1.6 \end{vmatrix} = -0.8, |D_x| = \begin{vmatrix} 0.4 & 1.2 \\ 3.2 & 1.6 \end{vmatrix} = -3.2,$ and

$|D_y| = \begin{vmatrix} 0.4 & 0.4 \\ 1.2 & 3.2 \end{vmatrix} = 0.8.$ Hence, $x = \dfrac{|D_x|}{|D|} = \dfrac{-3.2}{-0.8} = 4, y = \dfrac{|D_y|}{|D|} = \dfrac{0.8}{-0.8} = -1,$ and so the

solution is $(4, -1)$.

33. $\begin{cases} x - y + 2z = 0 \\ 3x + z = 11 \\ -x + 2y = 0 \end{cases}$ Then, expanding by the second row,

$|D| = \begin{vmatrix} 1 & -1 & 2 \\ 3 & 0 & 1 \\ -1 & 2 & 0 \end{vmatrix} = -3 \begin{vmatrix} -1 & 2 \\ 2 & 0 \end{vmatrix} - 1 \begin{vmatrix} 1 & -1 \\ -1 & 2 \end{vmatrix} = 12 - 1 = 11,$

$$|D_x| = \begin{vmatrix} 0 & -1 & 2 \\ 11 & 0 & 1 \\ 0 & 2 & 0 \end{vmatrix} = -11 \begin{vmatrix} -1 & 2 \\ 2 & 0 \end{vmatrix} = 44, \quad |D_y| = \begin{vmatrix} 1 & 0 & 2 \\ 3 & 11 & 1 \\ -1 & 0 & 0 \end{vmatrix} = 11 \begin{vmatrix} 1 & 2 \\ -1 & 0 \end{vmatrix} = 22,$$

and $|D_z| = \begin{vmatrix} 1 & -1 & 0 \\ 3 & 0 & 11 \\ -1 & 2 & 0 \end{vmatrix} = -11 \begin{vmatrix} 1 & -1 \\ -1 & 2 \end{vmatrix} = -11.$ Therefore, $x = \frac{44}{11} = 4$, $y = \frac{22}{11} = 2$,

$z = \frac{-11}{11} = -1$, and so the solution is $(4, 2, -1)$.

35. $\begin{cases} 2x_1 + 3x_2 - 5x_3 = 1 \\ x_1 + x_2 - x_3 = 2 \\ 2x_2 + x_3 = 8 \end{cases}$ Then, expanding by the third row,

$$|D| = \begin{vmatrix} 2 & 3 & -5 \\ 1 & 1 & -1 \\ 0 & 2 & 1 \end{vmatrix} = -2 \begin{vmatrix} 2 & -5 \\ 1 & -1 \end{vmatrix} + \begin{vmatrix} 2 & 3 \\ 1 & 1 \end{vmatrix} = -6 - 1 = -7,$$

$$|D_{x_1}| = \begin{vmatrix} 1 & 3 & -5 \\ 2 & 1 & -1 \\ 8 & 2 & 1 \end{vmatrix} = \begin{vmatrix} 1 & -1 \\ 2 & 1 \end{vmatrix} - 3 \begin{vmatrix} 2 & -1 \\ 8 & 1 \end{vmatrix} - 5 \begin{vmatrix} 2 & 1 \\ 8 & 2 \end{vmatrix} = 3 - 30 + 20 = -7,$$

$$|D_{x_2}| = \begin{vmatrix} 2 & 1 & -5 \\ 1 & 2 & -1 \\ 0 & 8 & 1 \end{vmatrix} = -8 \begin{vmatrix} 2 & -5 \\ 1 & -1 \end{vmatrix} + \begin{vmatrix} 2 & 1 \\ 1 & 2 \end{vmatrix} = -24 + 3 = -21, \text{ and}$$

$$|D_{x_3}| = \begin{vmatrix} 2 & 3 & 1 \\ 1 & 1 & 2 \\ 0 & 2 & 8 \end{vmatrix} = -2 \begin{vmatrix} 2 & 1 \\ 1 & 2 \end{vmatrix} + 8 \begin{vmatrix} 2 & 3 \\ 1 & 1 \end{vmatrix} = -6 - 8 = -14. \text{ Thus, } x_1 = \frac{-7}{-7} = 1,$$

$x_2 = \frac{-21}{-7} = 3$, $x_3 = \frac{-14}{-7} = 2$, and so the solution is $(1, 3, 2)$.

37. $\begin{cases} \frac{1}{3}x - \frac{1}{5}y + \frac{1}{2}z = \frac{7}{10} \\ -\frac{2}{3}x + \frac{2}{5}y + \frac{3}{2}z = \frac{11}{10} \\ x - \frac{4}{5}y + z = \frac{9}{5} \end{cases} \Leftrightarrow \begin{cases} 10x - 6y + 15z = 21 \\ -20x + 12y + 45z = 33 \\ 5x - 4y + 5z = 9 \end{cases}$

Then, $|D| = \begin{vmatrix} 10 & -6 & 15 \\ -20 & 12 & 45 \\ 5 & -4 & 5 \end{vmatrix} = 10 \begin{vmatrix} 12 & 45 \\ -4 & 5 \end{vmatrix} + 6 \begin{vmatrix} -20 & 45 \\ 5 & 5 \end{vmatrix} + 15 \begin{vmatrix} -20 & 12 \\ 5 & -4 \end{vmatrix}$

$= 2400 - 1950 + 300 = 750,$

$|D_x| = \begin{vmatrix} 21 & -6 & 15 \\ 33 & 12 & 45 \\ 9 & -4 & 5 \end{vmatrix} = 21 \begin{vmatrix} 12 & 45 \\ -4 & 5 \end{vmatrix} + 6 \begin{vmatrix} 33 & 45 \\ 9 & 5 \end{vmatrix} + 15 \begin{vmatrix} 33 & 12 \\ 9 & -4 \end{vmatrix}$

$= 5040 - 1440 - 3600 = 0,$

$|D_y| = \begin{vmatrix} 10 & 21 & 15 \\ -20 & 33 & 45 \\ 5 & 9 & 5 \end{vmatrix} = 10 \begin{vmatrix} 33 & 45 \\ 9 & 5 \end{vmatrix} - 21 \begin{vmatrix} -20 & 45 \\ 5 & 5 \end{vmatrix} + 15 \begin{vmatrix} -20 & 33 \\ 5 & 9 \end{vmatrix}$

$= -2400 + 6825 - 5175 = -750, \text{ and}$

$$|D_z| = \begin{vmatrix} 10 & -6 & 21 \\ -20 & 12 & 33 \\ 5 & -4 & 9 \end{vmatrix} = 10 \begin{vmatrix} 12 & 33 \\ -4 & 9 \end{vmatrix} + 6 \begin{vmatrix} -20 & 33 \\ 5 & 9 \end{vmatrix} + 21 \begin{vmatrix} -20 & 12 \\ 5 & -4 \end{vmatrix}$$

$$= 2400 - 2070 + 420 = 750.$$

Therefore, $x = 0$, $y = -1$, $z = 1$, and so the solution is $(0, -1, 1)$.

39. $\begin{cases} 3y + 5z = 4 \\ 2x - z = 10 \\ 4x + 7y = 0 \end{cases}$ Then, $|D| = \begin{vmatrix} 0 & 3 & 5 \\ 2 & 0 & -1 \\ 4 & 7 & 0 \end{vmatrix} = -3 \begin{vmatrix} 2 & -1 \\ 4 & 0 \end{vmatrix} + 5 \begin{vmatrix} 2 & 0 \\ 4 & 7 \end{vmatrix} = -12 + 70 = 58,$

$$|D_x| = \begin{vmatrix} 4 & 3 & 5 \\ 10 & 0 & -1 \\ 0 & 7 & 0 \end{vmatrix} = -7 \begin{vmatrix} 4 & 5 \\ 10 & -1 \end{vmatrix} = 378,$$

$$|D_y| = \begin{vmatrix} 0 & 4 & 5 \\ 2 & 10 & -1 \\ 4 & 0 & 0 \end{vmatrix} = 4 \begin{vmatrix} 4 & 5 \\ 10 & -1 \end{vmatrix} = -216, \text{ and}$$

$$|D_z| = \begin{vmatrix} 0 & 3 & 4 \\ 2 & 0 & 10 \\ 4 & 7 & 0 \end{vmatrix} = 4 \begin{vmatrix} 3 & 4 \\ 0 & 10 \end{vmatrix} - 7 \begin{vmatrix} 0 & 4 \\ 2 & 10 \end{vmatrix} = 120 + 56 = 176.$$

Thus, $x = \frac{189}{29}$, $y = -\frac{108}{29}$, and $z = \frac{88}{29}$, and so the solution is $\left(\frac{189}{29}, -\frac{108}{29}, \frac{88}{29} \right)$.

41. $\begin{cases} x + y + z + w = 0 \\ 2z \quad\quad + w = 0 \\ y - z \quad = 0 \\ x \quad + 2z \quad = 1 \end{cases}$ Then $|D| = \begin{vmatrix} 1 & 1 & 1 & 1 \\ 2 & 0 & 0 & 1 \\ 0 & 1 & -1 & 0 \\ 1 & 0 & 2 & 0 \end{vmatrix} = -1 \begin{vmatrix} 2 & 0 & 1 \\ 0 & -1 & 0 \\ 1 & 2 & 0 \end{vmatrix} - 1 \begin{vmatrix} 1 & 1 & 1 \\ 2 & 0 & 1 \\ 1 & 2 & 0 \end{vmatrix}$

$$= -\left(2 \begin{vmatrix} -1 & 0 \\ 2 & 0 \end{vmatrix} + 1 \begin{vmatrix} 0 & 1 \\ -1 & 0 \end{vmatrix} \right) - \left(-1 \begin{vmatrix} 2 & 1 \\ 1 & 0 \end{vmatrix} - 2 \begin{vmatrix} 1 & 1 \\ 2 & 1 \end{vmatrix} \right)$$

$$= -2(0) - 1(1) + 1(-1) + 2(-1) = -4,$$

$$|D_x| = \begin{vmatrix} 0 & 1 & 1 & 1 \\ 0 & 0 & 0 & 1 \\ 0 & 1 & -1 & 0 \\ 1 & 0 & 2 & 0 \end{vmatrix} = -1 \begin{vmatrix} 1 & 1 & 1 \\ 0 & 0 & 1 \\ 1 & -1 & 0 \end{vmatrix} = -1(-1) \begin{vmatrix} 1 & 1 \\ 1 & -1 \end{vmatrix} = -2,$$

$$|D_y| = \begin{vmatrix} 1 & 0 & 1 & 1 \\ 2 & 0 & 0 & 1 \\ 0 & 0 & -1 & 0 \\ 1 & 1 & 2 & 0 \end{vmatrix} = 1 \begin{vmatrix} 1 & 1 & 1 \\ 2 & 0 & 1 \\ 0 & -1 & 0 \end{vmatrix} = 1 \begin{vmatrix} 0 & 1 \\ -1 & 0 \end{vmatrix} - 2 \begin{vmatrix} 1 & 1 \\ -1 & 0 \end{vmatrix} = 1 - 2(1) = -1,$$

$$|D_z| = \begin{vmatrix} 1 & 1 & 0 & 1 \\ 2 & 0 & 0 & 1 \\ 0 & 1 & 0 & 0 \\ 1 & 0 & 1 & 0 \end{vmatrix} = -1 \begin{vmatrix} 1 & 1 & 1 \\ 2 & 0 & 1 \\ 0 & 1 & 0 \end{vmatrix} = -1 \begin{vmatrix} 0 & 1 \\ 1 & 0 \end{vmatrix} + 2 \begin{vmatrix} 1 & 1 \\ 1 & 0 \end{vmatrix} = -1(-1) + 2(-1) = -1, \text{ and}$$

$$|D_w| = \begin{vmatrix} 1 & 1 & 1 & 0 \\ 2 & 0 & 0 & 0 \\ 0 & 1 & -1 & 0 \\ 1 & 0 & 2 & 1 \end{vmatrix} = 1 \begin{vmatrix} 1 & 1 & 1 \\ 2 & 0 & 0 \\ 0 & 1 & -1 \end{vmatrix} = -2 \begin{vmatrix} 1 & 1 \\ 1 & -1 \end{vmatrix} = -2(-2) = 4. \text{ Hence, we have}$$

$x = \dfrac{|D_x|}{|D|} = \dfrac{-2}{-4} = \dfrac{1}{2}$; $y = \dfrac{|D_y|}{|D|} = \dfrac{-1}{-4} = \dfrac{1}{4}$; $z = \dfrac{|D_z|}{|D|} = \dfrac{-1}{-4} = \dfrac{1}{4}$; $w = \dfrac{|D_w|}{|D|} = \dfrac{4}{-4} = -1$, and the

solution is $\left(\dfrac{1}{2}, \dfrac{1}{4}, \dfrac{1}{4}, -1\right)$.

43.   (a)   Let $|M| = \begin{vmatrix} x_1 & y_1 & 1 \\ x_2 & y_2 & 1 \\ x & y & 1 \end{vmatrix}$. Then, expanding by the third column,

$$|M| = \begin{vmatrix} x_2 & y_2 \\ x & y \end{vmatrix} - \begin{vmatrix} x_1 & y_1 \\ x & y \end{vmatrix} + \begin{vmatrix} x_1 & y_1 \\ x_2 & y_2 \end{vmatrix} = x_2 y - x y_2 - x_1 y + x y_1 + x_1 y_2 - x_2 y_1$$

$= (x_2 - x_1)y + (-y_2 + y_1)x + x_1 y_2 - x_2 y_1$. So $|M| = 0 \iff$

$(x_2 - x_1)y + (-y_2 + y_1)x + x_1 y_2 - x_2 y_1 = 0 \iff$

$(x_2 - x_1)y = (y_2 - y_1)x - x_1 y_2 + x_2 y_1 \iff$

$(x_2 - x_1)y = (y_2 - y_1)x - x_1 y_2 + x_1 y_1 - x_1 y_1 + x_2 y_1 \iff$

$y = \dfrac{y_2 - y_1}{x_2 - x_1} x - \dfrac{x_1(y_2 - y_1)}{x_2 - x_1} + \dfrac{y_1(x_2 - x_1)}{x_2 - x_1} \iff y = \dfrac{y_2 - y_1}{x_2 - x_1}(x - x_1) + y_1 \iff$

$y - y_1 = \dfrac{y_2 - y_1}{x_2 - x_1}(x - x_1)$, which is the "two-point" form of the equation for the line passing

through the points (x_1, y_1) and (x_2, y_2).

  (b)   Using the result of part (a), the line has the equation

$\begin{vmatrix} 20 & 50 & 1 \\ -10 & 25 & 1 \\ x & y & 1 \end{vmatrix} = 0 \Rightarrow 20(25 - y) - 50(-10 - x) + (-10y - 25x) = 0 \iff$

$500 - 20y + 500 + 50x - 10y - 25x = 0 \iff 25x - 30y + 1000 = 0 \iff$

$5x - 6y + 200 = 0$.

45. $\begin{vmatrix} x & 12 & 13 \\ 0 & x - 1 & 23 \\ 0 & 0 & x - 2 \end{vmatrix} = 0 \iff (x - 2)\begin{vmatrix} x & 12 \\ 0 & x - 1 \end{vmatrix} = 0 \iff (x - 2) \cdot x(x - 1) = 0 \iff$

$x = 0, 1$ or 2.

47. $\begin{vmatrix} 1 & 0 & x \\ x^2 & 1 & 0 \\ x & 0 & 1 \end{vmatrix} = 0 \iff 1\begin{vmatrix} 1 & 0 \\ 0 & 1 \end{vmatrix} + x\begin{vmatrix} x^2 & 1 \\ x & 0 \end{vmatrix} = 0 \iff 1 - x^2 = 0 \iff x^2 = 1 \iff$

$x = \pm 1$.

49.   (a)   If A is a matrix with a row or column consisting entirely of zeros, then if we expand the
      determinant by this row or column, we get $|A| = 0 \cdot |A_{1j}| - 0 \cdot |A_{2j}| + \cdots - 0 \cdot |A_{nj}| = 0$.

  (b)   Use the principle that if matrix B is a square matrix obtained from A by adding a multiple of
      one row to another, or a multiple of one column to another, then $|A| = |B|$. If we let B be the
      matrix obtained by subtracting the two rows (or columns) that are the same, then matrix B will
      have a row or column that consists entirely of zeros. So $|B| = 0 \Rightarrow |A| = 0$.

  (c)   Again use the principle that if matrix B is a square matrix obtained from A by adding a
      multiple of one row to another, or a multiple of one column to another, then $|A| = |B|$. If we
      let B be the matrix obtained by subtracting the proper multiple of the row (or column) from the
      other similar row (or column), then matrix B will have a row or column that consists entirely of
      zeros. So $|B| = 0 \Rightarrow |A| = 0$.

Exercises 10.7

1. $x \leq 2$

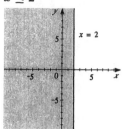

3. $y > x$

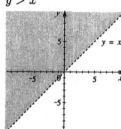

5. $y \geq 2x + 2$

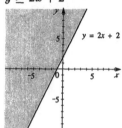

7. $2x - y \leq 8$

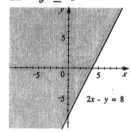

9. $4x + 5y < 25$

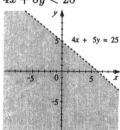

11. $y > x^2 + 1$

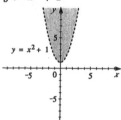

13. $\begin{cases} x + y \leq 4 \\ y \geq x \end{cases}$ The vertices occur where $\begin{cases} x + y = 4 \\ y = x \end{cases}$.
Substituting, we have $2x = 4 \iff x = 2$. Since $y = x$, the vertex is $(2, 2)$, and the solution set is not bounded.

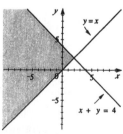

15. $\begin{cases} y < \frac{1}{4}x + 2 \\ y \geq 2x - 5 \end{cases}$ The vertex occurs where $\begin{cases} y = \frac{1}{4}x + 2 \\ y = 2x - 5 \end{cases}$.
Substituting for y gives $\frac{1}{4}x + 2 = 2x - 5 \iff \frac{7}{4}x = 7 \iff x = 4$, so $y = 3$. Hence, the vertex is $(4, 3)$, and the solution is not bounded.

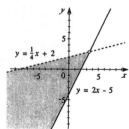

17. $\begin{cases} x \geq 0 \\ y \geq 0 \\ 3x + 5y \leq 15 \\ 3x + 2y \leq 9 \end{cases}$ From the graph, the points $(3, 0)$, $(0, 3)$ and

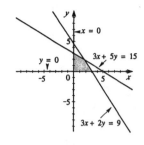

$(0, 0)$ are vertices, and the fourth vertex occurs where the lines $3x + 5y = 15$ and $3x + 2y = 9$ intersect. Subtracting these two equations gives $3y = 6 \quad \Leftrightarrow \quad y = 2$, and so $x = \frac{5}{3}$. Thus, the fourth vertex is $\left(\frac{5}{3}, 2\right)$, and the solution set is bounded.

19. $\begin{cases} y < 9 - x^2 \\ y \geq x + 3 \end{cases}$ The vertices occur where $\begin{cases} y = 9 - x^2 \\ y = x + 3 \end{cases}$.

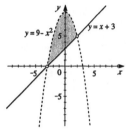

Substituting for y gives $9 - x^2 = x + 3 \quad \Leftrightarrow \quad x^2 + x - 6 = 0$ $\Leftrightarrow \quad (x - 2)(x + 3) = 0 \quad \Rightarrow \quad x = -3, x = 2$. Therefore, he vertices are $(-3, 0)$ and $(2, 5)$, and the solution set is bounded.

21. $\begin{cases} x^2 + y^2 \leq 4 \\ x - y > 0 \end{cases}$ The vertices occur where $\begin{cases} x^2 + y^2 = 4 \\ x - y = 0 \end{cases}$.

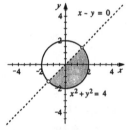

Since $x - y = 0 \quad \Leftrightarrow \quad x = y$, substituting for x gives $y^2 + y^2 = 4 \quad \Leftrightarrow \quad y^2 = 2 \quad \Rightarrow \quad y = \pm\sqrt{2}$, and $x = \pm\sqrt{2}$. Therefore, the vertices are $(-\sqrt{2}, -\sqrt{2})$ and $(\sqrt{2}, \sqrt{2})$, and the solution set is bounded.

23. $\begin{cases} x^2 - y \leq 0 \\ 2x^2 + y \leq 12 \end{cases}$ The vertices occur where $\begin{cases} x^2 - y = 0 \\ 2x^2 + y = 12 \end{cases}$

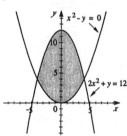

$\Leftrightarrow \begin{cases} 2x^2 - 2y = 0 \\ 2x^2 + y = 12 \end{cases}$. Subtracting the equations gives $3y = 12$ $\Leftrightarrow \quad y = 4$, and $x = \pm 2$. Thus, the vertices are $(2, 4)$ and $(-2, 4)$, and the solution set is bounded.

25. $\begin{cases} x + 2y \leq 14 \\ 3x - y \geq 0 \\ x - y \geq 2 \end{cases}$ We now find the vertices of the region by solving pairs of the corresponding equations.

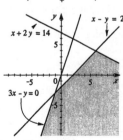

$\begin{cases} x + 2y = 14 \\ x - y = 2 \end{cases}$ Subtracting the second equation from the first gives $3y = 12 \quad \Leftrightarrow \quad y = 4$, and $x = 6$.

$\begin{cases} 3x - y = 0 \\ x - y = 2 \end{cases}$ Subtracting the second equation from the first gives $2x = -2 \quad \Leftrightarrow \quad x = -1$, and $y = -3$.

Therefore, the vertices are $(6, 4)$ and $(-1, -3)$, and the solution set is not bounded.

27. $\begin{cases} x \geq 0 \\ y \geq 0 \\ x \leq 5 \\ x + y \leq 7 \end{cases}$ The points of intersection are $(0,7)$, $(0,0)$, $(7,0)$,

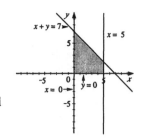

$(5,2)$, and $(5,0)$. However, the point $(7,0)$ is not in the solution set. Therefore, the vertices are $(0,7)$, $(0,0)$, $(5,0)$, and $(5,2)$, and the solution set is bounded.

29. $\begin{cases} y > x + 1 \\ x + 2y \leq 12 \\ x + 1 > 0 \end{cases}$ We now find the vertices of the region by solving pairs of the corresponding equations.

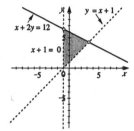

Using $x = -1$ and substituting for x in the line $y = x + 1$ gives the point $(-1, 0)$. Substituting for x in the line $x + 2y = 12$ gives the point $\left(-1, \frac{13}{2}\right)$.

$\begin{cases} y = x + 1 \\ x + 2y = 12 \end{cases}$ From the first equation, $x = y - 1$, and

substituting for x into the second equation gives $y - 1 + 2y = 12 \quad \Leftrightarrow \quad 3y = 13 \quad \Leftrightarrow \quad y = \frac{13}{3}$, and $x = \frac{10}{3}$. So the vertices are $(-1, 0)$, $\left(-1, \frac{13}{2}\right)$, and $\left(\frac{10}{3}, \frac{13}{3}\right)$, and none of these vertices are in the solution set. The solution set is bounded.

31. $\begin{cases} x^2 + y^2 \leq 8 \\ x \geq 2 \\ y \geq 0 \end{cases}$ The intersection points are $(2, \pm 2)$, $(2, 0)$, and

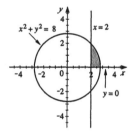

$(2\sqrt{2}, 0)$. However, since $(2, -2)$ is not part of the solution set, the vertices are $(2, 2)$, $(2, 0)$, and $(2\sqrt{2}, 0)$. The solution set is bounded.

33. $\begin{cases} x^2 + y^2 < 9 \\ x + y > 0 \\ x \leq 0 \end{cases}$ Substituting $x = 0$ into the equations $x^2 + y^2 = 9$

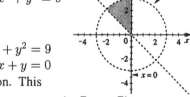

and $x + y = 0$ gives the vertices $(0, \pm 3)$ and $(0, 0)$.

To find the points of intersection for the equations $\begin{cases} x^2 + y^2 = 9 \\ x + y = 0 \end{cases}$

we solve for $x = -y$ and substitute into the first equation. This

gives $(-y)^2 + y^2 = 9 \quad \Rightarrow \quad y = \pm\frac{3\sqrt{2}}{2}$. The points $(0, -3)$ and $\left(\frac{3\sqrt{2}}{2}, -\frac{3\sqrt{2}}{2}\right)$ lie away from the

solution set, so the vertices are $(0, 0)$, $(0, 3)$, and $\left(-\frac{3\sqrt{2}}{2}, \frac{3\sqrt{2}}{2}\right)$. (Note that the vertices are not

solutions in this case.) The solution set is bounded.

35.   Let x be the number of fiction books published in a year, and y be the number of nonfiction books. Then, the following system of inequalities holds:

$$\begin{cases} x \geq 0 \\ y \geq 0 \\ x + y \leq 100 \\ y \geq 20 \\ x \geq y \end{cases}$$

From the graph, we see that the vertices are $(50, 50)$, $(80, 20)$ and $(20, 20)$.

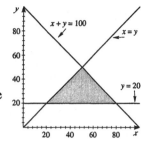

37.
$$\begin{cases} x + 2y > 4 \\ -x + y < 1 \\ x + 3y \leq 9 \\ x < 3 \end{cases}$$

We solve the system using the first method, that is, shading the solution to each inequality. However, since this text is published in black and white, we shade the region with lines perpendicular to the boundary. As you can see, as the number of inequalities in the system

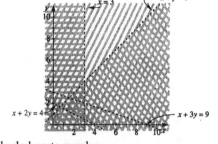

increases, it gets harder to locate the region where <u>all</u> the shaded parts overlap. In the second method, if a region is shaded, then it does <u>not</u> satisfy at least one equality. As a result, the region that is left unshaded satisfies each inequality, and is the solution to the system of inequalities. This makes it easier to locate the solution set.

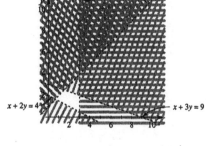

To finish, we find the vertices of the solution set. The line $x = 3$ intersects the line $x + 2y = 4$ at $\left(3, \frac{1}{2}\right)$ and the line $x + 3y = 9$ at $(3, 2)$. To find where the lines $-x + y = 1$ and $x + 2y = 4$ intersect, we add the two equations, which gives $3y = 5 \iff y = \frac{5}{3}$, and $x = \frac{2}{3}$. To find where the lines $-x + y = 1$ and $x + 3y = 9$ intersect, we add the two equations, which gives $4y = 10 \iff y = \frac{10}{4} = \frac{5}{2}$, and $x = \frac{3}{2}$. The vertices are $\left(3, \frac{1}{2}\right)$, $(3, 2)$, $\left(\frac{2}{3}, \frac{5}{3}\right)$, and $\left(\frac{3}{2}, \frac{5}{2}\right)$, and the solution set is bounded.

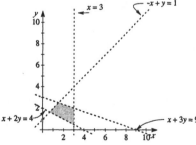

Exercises 10.8

1. $\dfrac{1}{(x-1)(x+2)} = \dfrac{A}{x-1} + \dfrac{B}{x+2}$

3. $\dfrac{x^2 - 3x + 5}{(x-2)^2(x+4)} = \dfrac{A}{x-2} + \dfrac{B}{(x-2)^2} + \dfrac{C}{x+4}$

5. $\dfrac{x^2}{(x-3)(x^2+4)} = \dfrac{A}{x-3} + \dfrac{Bx+C}{x^2+4}$

7. $\dfrac{x^3 - 4x^2 + 2}{(x^2+1)(x^2+2)} = \dfrac{Ax+B}{x^2+1} + \dfrac{Cx+D}{x^2+2}$

9. $\dfrac{x^3 + x + 1}{x(2x-5)^3(x^2+2x+5)^2} = \dfrac{A}{x} + \dfrac{B}{2x-5} + \dfrac{C}{(2x-5)^2} + \dfrac{D}{(2x-5)^3} +$

$$\dfrac{Ex+F}{x^2+2x+5} + \dfrac{Gx+H}{(x^2+2x+5)^2}$$

11. $\dfrac{5}{(x-1)(x+4)} = \dfrac{A}{x-1} + \dfrac{B}{x+4}$. Multiplying by $(x-1)(x+4)$, we get

$5 = A(x+4) + B(x-1) \quad \Leftrightarrow \quad 5 = Ax + 4A + Bx - B$. Thus

$\begin{cases} A+B = 0 \\ 4A - B = 5 \end{cases}$. Now $A + B = 0 \quad \Leftrightarrow \quad B = -A$, so substituting, we get $4A - (-A) = 5 \quad \Leftrightarrow$

$5A = 5 \quad \Leftrightarrow \quad A = 1$ and $B = -1$. The required partial fraction decomposition is

$\dfrac{5}{(x-1)(x+4)} = \dfrac{1}{x-1} - \dfrac{1}{x+4}$.

13. $\dfrac{12}{x^2-9} = \dfrac{12}{(x-3)(x+3)} = \dfrac{A}{x-3} + \dfrac{B}{x+3}$. Multiplying by $(x-3)(x+3)$, we get

$12 = A(x+3) + B(x-3) \quad \Leftrightarrow \quad 12 = Ax + 3A + Bx - 3B$. Thus

$\begin{cases} A+B = 0 \\ 3A - 3B = 12 \end{cases} \quad \Leftrightarrow \quad \begin{cases} A+B = 0 \\ A - B = 4 \end{cases}$. Adding we get $2A = 4 \quad \Leftrightarrow \quad A = 2$. So

$2 + B = 2 \quad \Leftrightarrow \quad$ and $B = -2$. The required partial fraction decomposition is

$\dfrac{12}{x^2-9} = \dfrac{2}{x-3} - \dfrac{2}{x+3}$.

15. $\dfrac{4}{x^2-4} = \dfrac{4}{(x-2)(x+2)} = \dfrac{A}{x-2} + \dfrac{B}{x+2}$. Multiplying by $x^2 - 4$, we get

$4 = A(x+2) + B(x-2) = (A+B)x + (2A - 2B)$, and so

$\begin{cases} A+B = 0 \\ 2A - 2B = 4 \end{cases} \quad \Leftrightarrow \quad \begin{cases} A+B = 0 \\ A - B = 2 \end{cases}$. Adding we get $2A = 2 \quad \Leftrightarrow \quad A = 1$, and $B = -1$.

Therefore, $\dfrac{4}{x^2-4} = \dfrac{1}{x-2} - \dfrac{1}{x+2}$.

17. $\dfrac{x+14}{x^2 - 2x - 8} = \dfrac{x+14}{(x-4)(x+2)} = \dfrac{A}{x-4} + \dfrac{B}{x+2}$. Hence,

$x + 14 = A(x+2) + B(x-4) = (A+B)x + (2A - 4B)$, and so

$$\begin{cases} A + B = 1 \\ 2A - 4B = 14 \end{cases} \Leftrightarrow \begin{cases} 2A + 2B = 2 \\ A - 2B = 7 \end{cases}.$$ Adding, we get $3A = 9 \Leftrightarrow A = 3$. So

$(3) + B = 1 \Leftrightarrow B = -2$. Therefore, $\dfrac{x + 14}{x^2 - 2x - 8} = \dfrac{3}{x - 4} - \dfrac{2}{x + 2}$.

19. $\dfrac{x}{8x^2 - 10x + 3} = \dfrac{x}{(4x - 3)(2x - 1)} = \dfrac{A}{4x - 3} + \dfrac{B}{2x - 1}$. Hence,

$x = A(2x - 1) + B(4x - 3) = (2A + 4B)x + (-A - 3B)$, and so

$$\begin{cases} 2A + 4B = 1 \\ -A - 3B = 0 \end{cases} \Leftrightarrow \begin{cases} 2A + 4B = 1 \\ -2A - 6B = 0 \end{cases}.$$ Adding, we get $-2B = 1 \Leftrightarrow B = -\frac{1}{2}$, and

$A = \frac{3}{2}$. Therefore, $\dfrac{x}{8x^2 - 10x + 3} = \dfrac{\frac{3}{2}}{4x - 3} - \dfrac{\frac{1}{2}}{2x - 1}$.

21. $\dfrac{9x^2 - 9x + 6}{2x^3 - x^2 - 8x + 4} = \dfrac{9x^2 - 9x + 6}{(x - 2)(x + 2)(2x - 1)} = \dfrac{A}{x - 2} + \dfrac{B}{x + 2} + \dfrac{C}{2x - 1}$. Thus,

$9x^2 - 9x + 6 = A(x + 2)(2x - 1) + B(x - 2)(2x - 1) + C(x - 2)(x + 2)$
$= A(2x^2 + 3x - 2) + B(2x^2 - 5x + 2) + C(x^2 - 4)$
$= (2A + 2B + C)x^2 + (3A - 5B)x + (-2A + 2B - 4C)$.

This leads to a system with the following matrix representation:

$$\begin{bmatrix} 2 & 2 & 1 & 9 \\ 3 & -5 & 0 & -9 \\ -2 & 2 & -4 & 6 \end{bmatrix} \begin{array}{c} R_2 + R_3 \to R_2 \\ R_1 + R_3 \to R_1 \end{array} \begin{bmatrix} 0 & 4 & -3 & 15 \\ 1 & -3 & -4 & -3 \\ -2 & 2 & -4 & 6 \end{bmatrix} \xrightarrow{R_1 \leftrightarrow R_2}$$

$$\begin{bmatrix} 1 & -3 & -4 & -3 \\ 0 & 4 & -3 & 15 \\ -2 & 2 & -4 & 6 \end{bmatrix} \xrightarrow{R_3 + 2R_1 \to R_3} \begin{bmatrix} 1 & -3 & -4 & -3 \\ 0 & 4 & -3 & 15 \\ 0 & -4 & -12 & 0 \end{bmatrix} \xrightarrow{R_3 + R_2 \to R_3}$$

$$\begin{bmatrix} 1 & -3 & -4 & -3 \\ 0 & 4 & -3 & 15 \\ 0 & 0 & -15 & 15 \end{bmatrix}.$$

Hence, $-15C = 15 \Leftrightarrow C = -1$; $4B + 3 = 15 \Leftrightarrow B = 3$; and $A - 9 + 4 = -3 \Leftrightarrow A = 2$.

Therefore, $\dfrac{9x^2 - 9x + 6}{2x^3 - x^2 - 8x + 4} = \dfrac{2}{x - 2} + \dfrac{3}{x + 2} - \dfrac{1}{2x - 1}$.

23. $\dfrac{x^2 + 1}{x^3 + x^2} = \dfrac{x^2 + 1}{x^2(x + 1)} = \dfrac{A}{x} + \dfrac{B}{x^2} + \dfrac{C}{x + 1}$. Hence,

$x^2 + 1 = Ax(x + 1) + B(x + 1) + Cx^2 = (A + C)x^2 + (A + B)x + B$, and so $B = 1$;

$A + 1 = 0 \Leftrightarrow A = -1$; and $-1 + C = 1 \Leftrightarrow C = 2$. Therefore, $\dfrac{x^2 + 1}{x^3 + x^2} = \dfrac{-1}{x} + \dfrac{1}{x^2} + \dfrac{2}{x + 1}$.

25. $\dfrac{2x}{4x^2 + 12x + 9} = \dfrac{2x}{(2x + 3)^2} = \dfrac{A}{2x + 3} + \dfrac{B}{(2x + 3)^2}$. Hence,

$2x = A(2x + 3) + B = 2Ax + (3A + B)$. So $2A = 2 \Leftrightarrow A = 1$; and $3(1) + B = 0 \Leftrightarrow B = -3$.

Therefore, $\dfrac{2x}{4x^2 + 12x + 9} = \dfrac{1}{2x + 3} - \dfrac{3}{(2x + 3)^2}$.

27. $\dfrac{4x^2 - x - 2}{x^4 + 2x^3} = \dfrac{4x^2 - x - 2}{x^3(x+2)} = \dfrac{A}{x} + \dfrac{B}{x^2} + \dfrac{C}{x^3} + \dfrac{D}{x+2}$. Hence,

$4x^2 - x - 2 = Ax^2(x+2) + Bx(x+2) + C(x+2) + Dx^3$
$\qquad\qquad = (A+D)x^3 + (2A+B)x^2 + (2B+C)x + 2C.$

So $2C = -2 \Leftrightarrow C = -1;\ 2B - 1 = -1 \Leftrightarrow B = 0;\ 2A + 0 = 4 \Leftrightarrow A = 2;$ and $2 + D = 0 \Leftrightarrow$
$D = -2.$ Therefore, $\dfrac{4x^2 - x - 2}{x^4 + 2x^3} = \dfrac{2}{x} - \dfrac{1}{x^3} - \dfrac{2}{x+2}.$

29. $\dfrac{-10x^2 + 27x - 14}{(x-1)^3(x+2)} = \dfrac{A}{x+2} + \dfrac{B}{x-1} + \dfrac{C}{(x-1)^2} + \dfrac{D}{(x-1)^3}.$ Thus,

$-10x^2 + 27x - 14 = A(x-1)^3 + B(x+2)(x-1)^2 + C(x+2)(x-1) + D(x+2)$
$= A(x^3 - 3x^2 + 3x - 1) + B(x+2)(x^2 - 2x + 1) + C(x^2 + x - 2) + D(x+2)$
$= A(x^3 - 3x^2 + 3x - 1) + B(x^3 - 3x + 2) + C(x^2 + x - 2) + D(x+2)$
$= (A+B)x^3 + (-3A+C)x^2 + (3A - 3B + C + D)x + (-A + 2B - 2C + 2D),$ which leads
to the following system of equations:

$$
\begin{bmatrix}
1 & 1 & 0 & 0 & 0 \\
-3 & 0 & 1 & 0 & -10 \\
3 & -3 & 1 & 1 & 27 \\
-1 & 2 & -2 & 2 & -14
\end{bmatrix}
\xrightarrow[\;R_4 + R_1 \to R_4\;]{R_2 + 3R_1 \to R_2,\; R_3 - 3R_1 \to R_3}
\begin{bmatrix}
1 & 1 & 0 & 0 & 0 \\
0 & 3 & 1 & 0 & -10 \\
0 & -6 & 1 & 1 & 27 \\
0 & 3 & -2 & 2 & -14
\end{bmatrix}
$$

$$
\xrightarrow[\;\overrightarrow{R_4 - R_2 \to R_4}\;]{R_3 + 2R_2 \to R_3}
\begin{bmatrix}
1 & 1 & 0 & 0 & 0 \\
0 & 3 & 1 & 0 & -10 \\
0 & 0 & 3 & 1 & 7 \\
0 & 0 & -3 & 2 & -4
\end{bmatrix}
\xrightarrow{R_4 + R_3 \to R_4}
\begin{bmatrix}
1 & 1 & 0 & 0 & 0 \\
0 & 3 & 1 & 0 & -10 \\
0 & 0 & 3 & 1 & 7 \\
0 & 0 & 0 & 3 & 3
\end{bmatrix}.
$$

Hence, $3D = 3 \Leftrightarrow D = 1;\ 3C + 1 = 7 \Leftrightarrow C = 2;\ 3B + 2 = -10 \Leftrightarrow B = -4;$ and $A - 4 = 0$
$\Leftrightarrow A = 4.$ Therefore, $\dfrac{-10x^2 + 27x - 14}{(x-1)^3(x+2)} = \dfrac{4}{x+2} - \dfrac{4}{x-1} + \dfrac{2}{(x-1)^2} + \dfrac{1}{(x-1)^3}.$

31. $\dfrac{3x^3 + 22x^2 + 53x + 41}{(x+2)^2(x+3)^2} = \dfrac{A}{x+2} + \dfrac{B}{(x+2)^2} + \dfrac{C}{x+3} + \dfrac{D}{(x+3)^2}.$ Thus,

$3x^3 + 22x^2 + 53x + 41 = A(x+2)(x+3)^2 + B(x+3)^2 + C(x+2)^2(x+3) + D(x+2)^2$
$= A(x^3 + 8x^2 + 21x + 18) + B(x^2 + 6x + 9) + C(x^3 + 7x^2 + 16x + 12) + D(x^2 + 4x + 4)$
$= (A+C)x^3 + (8A + B + 7C + D)x^2 + (21A + 6B + 16C + 4D)x +$
$(18A + 9B + 12C + 4D),$ so we must solve the system:

$$
\begin{bmatrix}
1 & 0 & 1 & 0 & 3 \\
8 & 1 & 7 & 1 & 22 \\
21 & 6 & 16 & 4 & 53 \\
18 & 9 & 12 & 4 & 41
\end{bmatrix}
\xrightarrow[\;R_4 - 18R_1 \to R_4\;]{R_2 - 8R_1 \to R_2,\; R_3 - 21R_1 \to R_3}
\begin{bmatrix}
1 & 0 & 1 & 0 & 3 \\
0 & 1 & -1 & 1 & -2 \\
0 & 6 & -5 & 4 & -10 \\
0 & 9 & -6 & 4 & -13
\end{bmatrix}
$$

$$
\xrightarrow[\;\overrightarrow{R_4 - 9R_2 \to R_4}\;]{R_3 - 6R_2 \to R_3}
\begin{bmatrix}
1 & 0 & 1 & 0 & 3 \\
0 & 1 & -1 & 1 & -2 \\
0 & 0 & 1 & -2 & 2 \\
0 & 0 & 3 & -5 & 5
\end{bmatrix}
\xrightarrow{R_4 - 3R_3 \to R_4}
\begin{bmatrix}
1 & 0 & 1 & 0 & 3 \\
0 & 1 & -1 & 1 & -2 \\
0 & 0 & 1 & -2 & 2 \\
0 & 0 & 0 & 1 & -1
\end{bmatrix}.
$$

Hence, $D = -1$; $\;\; C + 2 = 2 \Leftrightarrow C = 0$; $\;\; B - 0 - 1 = -2 \Leftrightarrow B = -1$; $\;$ and $A + 0 = 3 \Leftrightarrow$

$A = 3$. Therefore, $\dfrac{3x^3 + 22x^2 + 53x + 41}{(x+2)^2(x+3)^2} = \dfrac{3}{x+2} - \dfrac{1}{(x+2)^2} - \dfrac{1}{(x+3)^2}$.

33. $\dfrac{x-3}{x^3 + 3x} = \dfrac{x-3}{x(x^2+3)} = \dfrac{A}{x} + \dfrac{Bx+C}{x^2+3}$. Hence, $x - 3 = A(x^2 + 3) + Bx^2 + Cx$

$= (A+B)x^2 + Cx + 3A$. So $3A = -3 \Leftrightarrow A = -1$; $\;\; C = 1$; $\;$ and $-1 + B = 0 \Leftrightarrow B = 1$.

Therefore, $\dfrac{x-3}{x^3+3x} = -\dfrac{1}{x} + \dfrac{x+1}{x^2+3}$.

35. $\dfrac{2x^3 + 7x + 5}{(x^2 + x + 2)(x^2 + 1)} = \dfrac{Ax+B}{x^2+x+2} + \dfrac{Cx+D}{x^2+1}$. Thus,

$2x^3 + 7x + 5 = (Ax+B)(x^2+1) + (Cx+D)(x^2+x+2)$

$= Ax^3 + Ax + Bx^2 + B + Cx^3 + Cx^2 + 2Cx + Dx^2 + Dx + 2D$

$= (A+C)x^3 + (B+C+D)x^2 + (A+2C+D)x + (B+2D)$. We must solve the system:

$$\begin{bmatrix} 1 & 0 & 1 & 0 & 2 \\ 0 & 1 & 1 & 1 & 0 \\ 1 & 0 & 2 & 1 & 7 \\ 0 & 1 & 0 & 2 & 5 \end{bmatrix} \begin{matrix} \\ \\ {\scriptstyle R_3 - R_1 \rightarrow R_3} \\ {\scriptstyle R_4 - R_2 \rightarrow R_4} \end{matrix} \begin{bmatrix} 1 & 0 & 1 & 0 & 2 \\ 0 & 1 & 1 & 1 & 0 \\ 0 & 0 & 1 & 1 & 5 \\ 0 & 0 & -1 & 1 & 5 \end{bmatrix} \begin{matrix} \\ {\scriptstyle R_4 + R_3 \rightarrow R_4} \\ \longrightarrow \\ \\ \end{matrix} \begin{bmatrix} 1 & 0 & 1 & 0 & 2 \\ 0 & 1 & 1 & 1 & 0 \\ 0 & 0 & 1 & 1 & 5 \\ 0 & 0 & 0 & 2 & 10 \end{bmatrix}$$

Hence, $2D = 10 \Leftrightarrow D = 5$; $\;\; C + 5 = 5 \Leftrightarrow C = 0$; $\;\; B + 0 + 5 = 0 \Leftrightarrow B = -5$; $\;$ and $A + 0 = 2$

$\Leftrightarrow A = 2$. Therefore, $\dfrac{2x^3 + 7x + 5}{(x^2 + x + 2)(x^2 + 1)} = \dfrac{2x - 5}{x^2 + x + 2} + \dfrac{5}{x^2 + 1}$.

37. $\dfrac{x^4 + x^3 + x^2 - x + 1}{x(x^2+1)^2} = \dfrac{A}{x} + \dfrac{Bx+C}{x^2+1} + \dfrac{Dx+E}{(x^2+1)^2}$. Hence,

$x^4 + x^3 + x^2 - x + 1 = A(x^2+1)^2 + (Bx+C)x(x^2+1) + x(Dx+E)$

$= A(x^4 + 2x^2 + 1) + (Bx^2 + Cx)(x^2 + 1) + Dx^2 + Ex$

$= A(x^4 + 2x^2 + 1) + Bx^4 + Bx^2 + Cx^3 + Cx + Dx^2 + Ex$

$= (A+B)x^4 + Cx^3 + (2A+B+D)x^2 + (C+E)x + A$

Hence, $A = 1$; $\;\; 1 + B = 1 \Leftrightarrow B = 0$; $\;\; C = 1$; $\;\; 2 + 0 + D = 1 \Leftrightarrow D = -1$; $\;$ and $1 + E = -1$

$\Leftrightarrow E = -2$. Therefore, $\dfrac{x^4 + x^3 + x^2 - x + 1}{x(x^2+1)^2} = \dfrac{1}{x} + \dfrac{1}{x^2+1} - \dfrac{x+2}{(x^2+1)^2}$.

39. We must first get a proper rational function. Using long division, we have:

$$\begin{array}{r} x^2 \\ x^3 - 2x^2 + x - 2 \overline{\smash{\big)}\, x^5 - 2x^4 + x^3 + 0x^2 + x + 5} \\ \underline{x^5 - 2x^4 + x^3 - 2x^2} \\ 2x^2 + x + 5. \end{array}$$

Therefore, $\dfrac{x^5 - 2x^4 + x^3 + x + 5}{x^3 - 2x^2 + x - 2} = x^2 + \dfrac{2x^2 + x + 5}{x^3 - 2x^2 + x - 2} = x^2 + \dfrac{2x^2 + x + 5}{(x-2)(x^2+1)}$

$= x^2 + \dfrac{A}{x-2} + \dfrac{Bx+C}{x^2+1}$. Hence, $2x^2 + x + 5 = A(x^2+1) + (Bx+C)(x-2)$

$= Ax^2 + A + Bx^2 + Cx - 2Bx - 2C = (A+B)x^2 + (C-2B)x + (A-2C)$. Equating

coefficients, we get the system:

$$\begin{bmatrix} 1 & 1 & 0 & 2 \\ 0 & -2 & 1 & 1 \\ 1 & 0 & -2 & 5 \end{bmatrix} \xrightarrow{R_3 - R_1 \to R_3} \begin{bmatrix} 1 & 1 & 0 & 2 \\ 0 & -2 & 1 & 1 \\ 0 & -1 & -2 & 3 \end{bmatrix} \xrightarrow[\substack{R_2 - 2R_3 \to R_2 \\ -R_3 \leftrightarrow R_2}]{} \begin{bmatrix} 1 & 1 & 0 & 2 \\ 0 & 1 & 2 & -3 \\ 0 & 0 & 5 & -5 \end{bmatrix}$$

Therefore, $5C = -5 \quad \Leftrightarrow \quad C = -1, B - 2 = -3 \quad \Leftrightarrow \quad B = -1,$ and $A - 1 = 2 \quad \Leftrightarrow$

$A = 3.$ Therefore, $\dfrac{x^5 - 2x^4 + x^3 + x + 5}{x^3 - 2x^2 + x - 2} = x^2 + \dfrac{3}{x - 2} - \dfrac{x + 1}{x^2 + 1}.$

41. $\dfrac{ax + b}{x^2 - 1} = \dfrac{A}{x - 1} + \dfrac{B}{x + 1}.$ Hence, $ax + b = A(x + 1) + B(x - 1) = (A + B)x + (A - B).$ So

$\begin{cases} A + B = a \\ A - B = b \end{cases}.$ Adding, we get $2A = a + b \quad \Leftrightarrow \quad A = \dfrac{a + b}{2}.$ Substituting, we get

$B = a - A = \dfrac{2a}{2} - \dfrac{a + b}{2} = \dfrac{a - b}{2}.$ Therefore, $A = \dfrac{a + b}{2}$ and $B = \dfrac{a - b}{2}.$

43. (a) The expression $\dfrac{x}{x^2 + 1} + \dfrac{1}{x + 1}$ is already a partial fraction decomposition. The denominator in the first term is a quadratic which can't be factored and the degree of the numerator is less than 2. The denominator of the second term is linear and the numerator is a constant.

(b) The term $\dfrac{x}{(x + 1)^2}$ can be decomposed further, since the numerator and denominator both have linear factors. $\dfrac{x}{(x + 1)^2} = \dfrac{A}{x + 1} + \dfrac{B}{(x + 1)^2}.$ Hence, $x = A(x + 1) + B = Ax + (A + B).$ So $A = 1, B = -1,$ and $\dfrac{x}{(x + 1)^2} = \dfrac{1}{x + 1} + \dfrac{-1}{(x + 1)^2}.$

(c) The expression $\dfrac{1}{x + 1} + \dfrac{2}{(x + 1)^2}$ is already a partial fraction decomposition, since each numerator is constant.

(d) The expression $\dfrac{x + 2}{(x^2 + 1)^2}$ is already a partial fraction decomposition, since the denominator is the square of a quadratic which can't be factored and the degree of the numerator is less than 2.

Review Exercises for Chapter 10

1. $\begin{cases} 3x - y = 5 \\ 2x + y = 5 \end{cases}$ Adding, we get $5x = 10 \quad \Leftrightarrow \quad x = 2$.

So $2(2) + y = 5 \quad \Leftrightarrow \quad y = 1$. Thus the solution is $(2, 1)$.

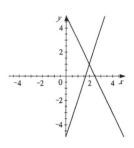

3. $\begin{cases} 2x - 7y = 28 \\ y = \frac{2}{7}x - 4 \end{cases} \quad \Leftrightarrow \quad \begin{cases} 2x - 7y = 28 \\ 2x - 7y = 28 \end{cases}$ Since these

equations represent the same line, any point on this line will satisfy the system. Thus the solution are $\left(x, \frac{2}{7}x - 4\right)$, where x is any real number.

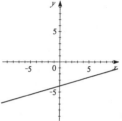

5. $\begin{cases} 2x - y = 1 \\ x + 3y = 10 \\ 3x + 4y = 15 \end{cases}$ Solving the first equation for y, we get

$y = -2x + 1$. Substituting into the second equation gives

$x + 3(-2x + 1) = 10 \quad \Leftrightarrow \quad -5x = 7 \quad \Leftrightarrow \quad x = -\frac{7}{5}$. So

$y = -\left(-\frac{7}{5}\right) + 1 = \frac{12}{5}$. Checking the point $\left(-\frac{7}{5}, \frac{12}{5}\right)$ in the third

equation we have $3\left(-\frac{7}{5}\right) + 4\left(\frac{12}{5}\right) \stackrel{?}{=} 15$ but $-\frac{21}{5} + \frac{48}{5} \neq 15$.

Thus, there is no solution, and the lines do not intersect at one point.

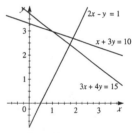

7. $\begin{cases} y = x^2 + 2x \\ y = 6 + x \end{cases}$ Substituting for y gives $6 + x = x^2 + 2x \quad \Leftrightarrow \quad x^2 + x - 6 = 0$. Factoring we

have $(x - 2)(x + 3) = 0$. Thus $x = 2$ or -3. If $x = 2$, then $y = 8$, and if $x = -3$, then $y = 3$.

Thus the solutions are $(-3, 3)$ and $(2, 8)$.

9. $\begin{cases} 3x + \dfrac{4}{y} = 6 \\ x - \dfrac{8}{y} = 4 \end{cases}$ Adding twice the first equation to the second equation gives $7x = 16 \quad \Leftrightarrow \quad x = \frac{16}{7}$. So $\frac{16}{7} - \frac{8}{y} = 4 \quad \Leftrightarrow \quad 16y - 56 = 28y \quad \Leftrightarrow \quad -12y = 56 \quad \Leftrightarrow \quad y = -\frac{14}{3}$. Thus the solution is $\left(\frac{16}{7}, -\frac{14}{3}\right)$.

11. $\begin{bmatrix} 1 & 1 & 2 & 6 \\ 2 & 0 & 5 & 12 \\ 1 & 2 & 3 & 9 \end{bmatrix} \xrightarrow[R_3 - R_1 \to R_3]{R_2 - 2R_1 \to R_2} \begin{bmatrix} 1 & 1 & 2 & 6 \\ 0 & -2 & 1 & 0 \\ 0 & 1 & 1 & 3 \end{bmatrix} \xrightarrow{R_3 \leftrightarrow R_2}$

$\begin{bmatrix} 1 & 1 & 2 & 6 \\ 0 & 1 & 1 & 3 \\ 0 & -2 & 1 & 0 \end{bmatrix} \xrightarrow{R_3 + 2R_2 \to R_3} \begin{bmatrix} 1 & 1 & 2 & 6 \\ 0 & 1 & 1 & 3 \\ 0 & 0 & 3 & 6 \end{bmatrix}$

Therefore, $3z = 6 \quad \Leftrightarrow \quad z = 2$; $y + 2 = 3 \quad \Leftrightarrow \quad y = 1$; and $x + 1 + 4 = 6 \quad \Leftrightarrow \quad x = 1$.

Hence, the solution is $(1, 1, 2)$.

13. $\begin{bmatrix} 1 & -2 & 3 & 1 \\ 2 & -1 & 1 & 3 \\ 2 & -7 & 11 & 2 \end{bmatrix}$ $\xrightarrow[R_3-2R_1 \to R_3]{R_2-2R_1 \to R_2}$ $\begin{bmatrix} 1 & -2 & 3 & 1 \\ 0 & 3 & -5 & 1 \\ 0 & -3 & 5 & 0 \end{bmatrix}$ $\xrightarrow{R_3+R_2 \to R_3}$

$\begin{bmatrix} 1 & -2 & 3 & 1 \\ 0 & 3 & -5 & 1 \\ 0 & 0 & 0 & 1 \end{bmatrix}$

Therefore, the system is inconsistent and has no solution.

15. $\begin{bmatrix} 1 & -3 & 1 & 4 \\ 4 & -1 & 15 & 5 \end{bmatrix}$ $\xrightarrow{R_2-4R_1 \to R_2}$ $\begin{bmatrix} 1 & -3 & 1 & 4 \\ 0 & 11 & 11 & -11 \end{bmatrix}$ $\xrightarrow{\frac{1}{11}R_2}$ $\begin{bmatrix} 1 & -3 & 1 & 4 \\ 0 & 1 & 1 & -1 \end{bmatrix}$

Thus, the system has infinitely many solutions given by $y + z = -1$ $\Leftrightarrow$ $y = -1 - z$, and $x + 3(1 + z) + z = 4$ $\Leftrightarrow$ $x = 1 - 4z$. Therefore, the solutions are $(1 - 4z, -1 - z, z)$, where z is any real number.

17. $\begin{bmatrix} -1 & 4 & 1 & 8 \\ 2 & -6 & 1 & -9 \\ 1 & -6 & -4 & -15 \end{bmatrix}$ $\xrightarrow{R_1 \leftrightarrow R_3}$ $\begin{bmatrix} 1 & -6 & -4 & -15 \\ 2 & -6 & 1 & -9 \\ -1 & 4 & 1 & 8 \end{bmatrix}$ $\xrightarrow[R_3+R_1 \to R_3]{R_2-2R_1 \to R_2}$

$\begin{bmatrix} 1 & -6 & -4 & -15 \\ 0 & 6 & 9 & 21 \\ 0 & -2 & -3 & -7 \end{bmatrix}$ $\xrightarrow[\frac{1}{6}R_2]{3R_3+R_2 \to R_3}$ $\begin{bmatrix} 1 & -6 & -4 & -15 \\ 0 & 1 & \frac{3}{2} & \frac{7}{2} \\ 0 & 0 & 0 & 0 \end{bmatrix}$

Thus, the system has infinitely many solutions given by $y + \frac{3}{2}z = \frac{7}{2}$ $\Leftrightarrow$ $y = \frac{7}{2} - \frac{3}{2}z$, and $x - 6(\frac{7}{2} - \frac{3}{2}z) - 4z = -15$ $\Leftrightarrow$ $x = 6 - 5z$. Therefore, the solutions are $(6 - 5z, \frac{7}{2} - \frac{3}{2}z, z)$, where z is any real number.

19. Let x be the amount in the 6% account, and y be the amount in the 7% account. Thus, the system is:

$$\begin{cases} y = 2x \\ 0.06x + 0.07y = 600 \end{cases}$$

Substituting gives $0.06x + 0.07(2x) = 600$ $\Leftrightarrow$ $0.2x = 600$ $\Leftrightarrow$ $x = 3000$. So $y = 2(3000) = 6000$. Hence, the man has $3,000 invested at 6% and $6,000 invested at 7%.

In Exercises 21 – 31:

$A = [2 \quad 0 \quad -1]$

$B = \begin{bmatrix} 1 & 2 & 4 \\ -2 & 1 & 0 \end{bmatrix}$

$C = \begin{bmatrix} \frac{1}{2} & 3 \\ 2 & \frac{3}{2} \\ -2 & 1 \end{bmatrix}$

$D = \begin{bmatrix} 1 & 4 \\ 0 & -1 \\ 2 & 0 \end{bmatrix}$

$E = \begin{bmatrix} 2 & -1 \\ -\frac{1}{2} & 1 \end{bmatrix}$

$F = \begin{bmatrix} 4 & 0 & 2 \\ -1 & 1 & 0 \\ 7 & 5 & 0 \end{bmatrix}$

$G = \begin{bmatrix} 5 \end{bmatrix}$

21. $A + B$ cannot be performed because the matrix dimensions (1×3 and 2×3) are not compatible.

23. $2C + 3D = 2\begin{bmatrix} \frac{1}{2} & 3 \\ 2 & \frac{3}{2} \\ -2 & 1 \end{bmatrix} + 3\begin{bmatrix} 1 & 4 \\ 0 & -1 \\ 2 & 0 \end{bmatrix} = \begin{bmatrix} 1 & 6 \\ 4 & 3 \\ -4 & 2 \end{bmatrix} + \begin{bmatrix} 3 & 12 \\ 0 & -3 \\ 6 & 0 \end{bmatrix} = \begin{bmatrix} 4 & 18 \\ 4 & 0 \\ 2 & 2 \end{bmatrix}$

25. $GA = \begin{bmatrix} 5 \end{bmatrix} [2 \quad 0 \quad -1] = [10 \quad 0 \quad -5]$

27. $BC = \begin{bmatrix} 1 & 2 & 4 \\ -2 & 1 & 0 \end{bmatrix} \begin{bmatrix} \frac{1}{2} & 3 \\ 2 & \frac{3}{2} \\ -2 & 1 \end{bmatrix} = \begin{bmatrix} -\frac{7}{2} & 10 \\ 1 & -\frac{9}{2} \end{bmatrix}$

29. $BF = \begin{bmatrix} 1 & 2 & 4 \\ -2 & 1 & 0 \end{bmatrix} \begin{bmatrix} 4 & 0 & 2 \\ -1 & 1 & 0 \\ 7 & 5 & 0 \end{bmatrix} = \begin{bmatrix} 30 & 22 & 2 \\ -9 & 1 & -4 \end{bmatrix}$

31. $(C + D)E = \left(\begin{bmatrix} \frac{1}{2} & 3 \\ 2 & \frac{3}{2} \\ -2 & 1 \end{bmatrix} + \begin{bmatrix} 1 & 4 \\ 0 & -1 \\ 2 & 0 \end{bmatrix} \right) \begin{bmatrix} 2 & -1 \\ -\frac{1}{2} & 1 \end{bmatrix}$

$= \begin{bmatrix} \frac{3}{2} & 7 \\ 2 & \frac{1}{2} \\ 0 & 1 \end{bmatrix} \begin{bmatrix} 2 & -1 \\ -\frac{1}{2} & 1 \end{bmatrix} = \begin{bmatrix} -\frac{1}{2} & \frac{11}{2} \\ \frac{15}{4} & -\frac{3}{2} \\ -\frac{1}{2} & 1 \end{bmatrix}$

33. $D = \begin{bmatrix} 1 & 4 \\ 2 & 9 \end{bmatrix}$. Then, $|D| = 1(9) - 2(4) = 1$, and so $D^{-1} = \begin{bmatrix} 9 & -4 \\ -2 & 1 \end{bmatrix}$.

35. $D = \begin{bmatrix} 4 & -12 \\ -2 & 6 \end{bmatrix}$. Then, $|D| = 4(6) - 2(12) = 0$, and so D has no inverse.

37. $D = \begin{bmatrix} 3 & 0 & 1 \\ 2 & -3 & 0 \\ 4 & -2 & 1 \end{bmatrix}$. Then, $|D| = 1 \begin{vmatrix} 2 & -3 \\ 4 & -2 \end{vmatrix} + 1 \begin{vmatrix} 3 & 0 \\ 2 & -3 \end{vmatrix} = -4 + 12 - 9 = -1$. So D^{-1} exists .

$\begin{bmatrix} 3 & 0 & 1 & 1 & 0 & 0 \\ 2 & -3 & 0 & 0 & 1 & 0 \\ 4 & -2 & 1 & 0 & 0 & 1 \end{bmatrix} \xrightarrow{R_1 - R_2 \to R_1} \begin{bmatrix} 1 & 3 & 1 & 1 & -1 & 0 \\ 2 & -3 & 0 & 0 & 1 & 0 \\ 4 & -2 & 1 & 0 & 0 & 1 \end{bmatrix} \begin{array}{l} R_2 - 2R_1 \to R_2 \\ \xrightarrow{} \\ R_3 - 4R_1 \to R_3 \end{array}$

$\begin{bmatrix} 1 & 3 & 1 & 1 & -1 & 0 \\ 0 & -9 & -2 & -2 & 3 & 0 \\ 0 & -14 & -3 & -4 & 4 & 1 \end{bmatrix} \begin{array}{l} -3R_2 \\ \xrightarrow{-2R_3} \end{array} \begin{bmatrix} 1 & 3 & 1 & 1 & -1 & 0 \\ 0 & 27 & 6 & 6 & -9 & 0 \\ 0 & 28 & 6 & 8 & -8 & -2 \end{bmatrix} \xrightarrow{R_3 - R_2 \to R_3}$

$\begin{bmatrix} 1 & 3 & 1 & 1 & -1 & 0 \\ 0 & 27 & 6 & 6 & -9 & 0 \\ 0 & 1 & 0 & 2 & 1 & -2 \end{bmatrix} \begin{array}{l} R_3 \leftrightarrow R_2 \\ \xrightarrow{\frac{1}{3}R_3} \end{array} \begin{bmatrix} 1 & 3 & 1 & 1 & -1 & 0 \\ 0 & 1 & 0 & 2 & 1 & -2 \\ 0 & 9 & 2 & 2 & -3 & 0 \end{bmatrix} \begin{array}{l} R_3 - 9R_2 \to R_3 \\ \xrightarrow{} \\ R_1 - 3R_2 \to R_1 \end{array}$

$\begin{bmatrix} 1 & 0 & 1 & -5 & -4 & 6 \\ 0 & 1 & 0 & 2 & 1 & -2 \\ 0 & 0 & 2 & -16 & -12 & 18 \end{bmatrix} \begin{array}{l} \frac{1}{2}R_3 \\ \xrightarrow{R_1 - R_3 \to R_1} \end{array} \begin{bmatrix} 1 & 0 & 0 & 3 & 2 & -3 \\ 0 & 1 & 0 & 2 & 1 & -2 \\ 0 & 0 & 1 & -8 & -6 & 9 \end{bmatrix}.$

Thus, $D^{-1} = \begin{bmatrix} 3 & 2 & -3 \\ 2 & 1 & -2 \\ -8 & -6 & 9 \end{bmatrix}$.

39. $\begin{bmatrix} 12 & -5 \\ 5 & -2 \end{bmatrix} \begin{bmatrix} x \\ y \end{bmatrix} = \begin{bmatrix} 10 \\ 17 \end{bmatrix}$. If we let $A = \begin{bmatrix} 12 & -5 \\ 5 & -2 \end{bmatrix}$, then $A^{-1} = \frac{1}{-24+25} \begin{bmatrix} -2 & 5 \\ -5 & 12 \end{bmatrix}$

$= \begin{bmatrix} -2 & 5 \\ -5 & 12 \end{bmatrix}$, and so $\begin{bmatrix} x \\ y \end{bmatrix} = \begin{bmatrix} -2 & 5 \\ -5 & 12 \end{bmatrix} \begin{bmatrix} 10 \\ 17 \end{bmatrix} = \begin{bmatrix} 65 \\ 154 \end{bmatrix}$. Therefore, the solution is $(65, 154)$.

41. $|D| = \begin{vmatrix} 2 & 7 \\ 6 & 16 \end{vmatrix} = 32 - 42 = -10; \quad |D_x| = \begin{vmatrix} 13 & 7 \\ 30 & 16 \end{vmatrix} = 208 - 210 = -2; \quad$ and

$|D_y| = \begin{vmatrix} 2 & 13 \\ 6 & 30 \end{vmatrix} = 60 - 78 = -18.$ Therefore, $x = \frac{-2}{-10} = \frac{1}{5}$ and $y = \frac{-18}{-10} = \frac{9}{5}$, and so the solution

is $\left(\frac{1}{5}, \frac{9}{5} \right)$.

43. $|D| = \begin{vmatrix} 2 & -1 & 5 \\ -1 & 7 & 0 \\ 5 & 4 & 3 \end{vmatrix} = 5 \begin{vmatrix} -1 & 7 \\ 5 & 4 \end{vmatrix} + 3 \begin{vmatrix} 2 & -1 \\ -1 & 7 \end{vmatrix} = -195 + 39 = -156;$

$|D_x| = \begin{vmatrix} 0 & -1 & 5 \\ 9 & 7 & 0 \\ -9 & 4 & 3 \end{vmatrix} = 5 \begin{vmatrix} 9 & 7 \\ -9 & 4 \end{vmatrix} + 3 \begin{vmatrix} 0 & -1 \\ 9 & 7 \end{vmatrix} = 495 + 27 = 522;$

$|D_y| = \begin{vmatrix} 2 & 0 & 5 \\ -1 & 9 & 0 \\ 5 & -9 & 3 \end{vmatrix} = 5 \begin{vmatrix} -1 & 9 \\ 5 & -9 \end{vmatrix} + 3 \begin{vmatrix} 2 & 0 \\ -1 & 9 \end{vmatrix} = -180 + 54 = -126;$ and

$|D_z| = \begin{vmatrix} 2 & -1 & 0 \\ -1 & 7 & 9 \\ 5 & 4 & -9 \end{vmatrix} = -9 \begin{vmatrix} 2 & -1 \\ 5 & 4 \end{vmatrix} - 9 \begin{vmatrix} 2 & -1 \\ -1 & 7 \end{vmatrix} = -117 - 117 = -234.$

Therefore, $x = \frac{522}{-156} = -\frac{87}{26}$, $y = \frac{-126}{-156} = \frac{21}{26}$, and $z = \frac{-234}{-156} = \frac{3}{2}$, and so the solution is $\left(-\frac{87}{26}, \frac{21}{26}, \frac{3}{2} \right)$.

45. $\begin{cases} x^2 + y^2 < 9 \\ x + y < 0 \end{cases}$ The vertices occur where $y = -x$. By

substitution, $x^2 + x^2 = 9 \Leftrightarrow x = \pm\frac{3}{\sqrt{2}}$, and so $y = \mp\frac{3}{\sqrt{2}}$.

Therefore, the vertices are $\left(\frac{3}{\sqrt{2}}, -\frac{3}{\sqrt{2}} \right)$ and $\left(-\frac{3}{\sqrt{2}}, \frac{3}{\sqrt{2}} \right)$ and

the solution set is bounded.

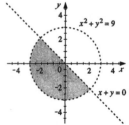

47. $\begin{cases} x \geq 0, \ y \geq 0 \\ x + 2y \leq 12 \\ y \leq x + 4 \end{cases}$ The intersection points are $(-4, 0)$, $(0, 4)$,

$\left(\frac{4}{3}, \frac{16}{3} \right)$, $(0, 6)$, $(0, 0)$, and $(12, 0)$. Since the points $(-4, 0)$ and
$(0, 6)$ are not in the solution set, the vertices are $(0, 4)$, $\left(\frac{4}{3}, \frac{16}{3} \right)$,
$(12, 0)$, and $(0, 0)$. The solution set is bounded.

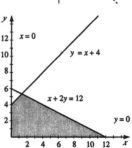

49. $\begin{bmatrix} -1 & 1 & 1 & a \\ 1 & -1 & 1 & b \\ 1 & 1 & -1 & c \end{bmatrix} \xrightarrow{-R_1} \begin{bmatrix} 1 & -1 & -1 & -a \\ 1 & -1 & 1 & b \\ 1 & 1 & -1 & c \end{bmatrix} \begin{matrix} R_2 - R_1 \to R_2 \\ R_3 - R_1 \to R_3 \end{matrix}$

$\begin{bmatrix} 1 & -1 & -1 & -a \\ 0 & 0 & 2 & a+b \\ 0 & 2 & 0 & a+c \end{bmatrix} \xrightarrow{R_2 \leftrightarrow R_3} \begin{bmatrix} 1 & -1 & -1 & -a \\ 0 & 2 & 0 & a+c \\ 0 & 0 & 2 & a+b \end{bmatrix}$

Thus, $z = \dfrac{a+b}{2}$; $y = \dfrac{a+c}{2}$; and $x - \dfrac{a+c}{2} - \dfrac{a+b}{2} = -a \Leftrightarrow x = \dfrac{b+c}{2}$. The solution is

$\left(\dfrac{b+c}{2}, \dfrac{a+c}{2}, \dfrac{a+b}{2} \right)$.

51. $\begin{bmatrix} 1 & 1 & 12 \\ k & -1 & 0 \\ -1 & 1 & 2k \end{bmatrix} \xrightarrow{R_3 + R_1 \to R_3} \begin{bmatrix} 1 & 1 & 12 \\ k & -1 & 0 \\ 0 & 2 & 12 + 2k \end{bmatrix}$

Thus, $2y = 12 + 2k \quad \Leftrightarrow \quad y = 6 + k$ and $x = \frac{6+k}{k}$. So $\frac{6+k}{k} + 6 + k = 12 \Rightarrow$
$6 + k + 6k + k^2 = 12k \quad \Leftrightarrow \quad k^2 - 5k + 6 = 0 \quad \Leftrightarrow \quad (k-3)(k-2) = 0 \Rightarrow k = 2$ or $k = 3$.

53. $\begin{cases} 0.32x + 0.43y = 0 \\ 7x - 12y = 341 \end{cases} \quad \Leftrightarrow \quad \begin{cases} y = -\dfrac{32}{43}x \\ y = \dfrac{7x - 341}{12} \end{cases}$

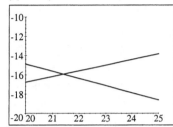

The solution is $(21.41, -15.93)$.

55. $\begin{cases} x - y^2 = 10 \\ x = \frac{1}{22}y + 12 \end{cases} \quad \Leftrightarrow \quad \begin{cases} y = \pm\sqrt{x - 10} \\ y = 22(x - 12) \end{cases}$

The solutions are $(11.94, -1.39)$ and $(12.07, re1.44)$.

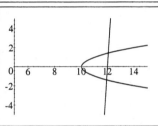

57. $\dfrac{3x+1}{x^2 - 2x - 15} = \dfrac{3x+1}{(x-5)(x+3)} = \dfrac{A}{x-5} + \dfrac{B}{x+3}$. Thus, $3x + 1 = A(x+3) + B(x-5)$
$= x(A+B) + (3A - 5B)$, and so

$\begin{cases} A + B = 3 \\ 3A - 5B = 1 \end{cases} \quad \Leftrightarrow \quad \begin{cases} -3A - 3B = -9 \\ 3A - 5B = 1 \end{cases}$. Adding, we have $-8B = -8 \quad \Leftrightarrow \quad B = 1$, and

$A = 2$. Hence, $\dfrac{3x+1}{x^2 - 2x - 15} = \dfrac{2}{x-5} + \dfrac{1}{x+3}$.

59. $\dfrac{2x-4}{x(x-1)^2} = \dfrac{A}{x} + \dfrac{B}{x-1} + \dfrac{C}{(x-1)^2}$. Then, $2x - 4 = A(x-1)^2 + Bx(x-1) + Cx$
$= Ax^2 - 2Ax + A + Bx^2 - Bx + Cx = x^2(A+B) + x(-2A - B + C) + A$. So $A = -4$;
$-4 + B = 0 \Leftrightarrow B = 4$; and $8 - 4 + C = 2 \Leftrightarrow C = -2$. Therefore,

$\dfrac{2x-4}{x(x-1)^2} = -\dfrac{4}{x} + \dfrac{4}{x-1} - \dfrac{2}{(x-1)^2}$.

Chapter 10 Test

1. Let w be the speed of the wind and a be the speed of the airplane in still air, in km per hour. Then the speed of the of the plane flying against the wind is $a - w$, and the speed of the plane flying with the wind is $a + w$. Using *distance = time × rate,* we get the system
$$\begin{cases} 600 = 2.5(a - w) \\ 300 = \frac{50}{60}(a + w) \end{cases} \quad \Leftrightarrow \quad \begin{cases} 240 = a - w \\ 360 = a + w \end{cases}.$$
Adding the two equations, we get $600 = 2a \quad \Leftrightarrow \quad a = 300.$ So $360 = 300 + w \quad \Leftrightarrow \quad w = 60.$ Thus the speed of the airplane in still air is 300 km/h and the speed of the wind is 60 km/h.

2. $\begin{cases} 3x - y = 10 \\ 2x + 5y = 1 \end{cases}$ Multiplying the first equation by 5 and then adding gives $17x = 51 \quad \Leftrightarrow$
$x = 3.$ So $3(3) - y = 10 \quad \Leftrightarrow \quad y = -1.$ Thus, the solution is $(3, -1)$, the system is linear, and it is neither inconsistent nor dependent.

3. $\begin{cases} x - y + 9z = -8 \\ \quad\ x - 4z = 7 \\ 3x - y + z = 5 \end{cases}$ This system is linear and has the following matrix representation.
$$\begin{bmatrix} 1 & -1 & 9 & -8 \\ 1 & 0 & -4 & 7 \\ 3 & -1 & 1 & 5 \end{bmatrix} \xrightarrow[R_3 \to R_3 - 3R_1]{R_2 \to R_2 - R_1} \begin{bmatrix} 1 & -1 & 9 & -8 \\ 0 & 1 & -13 & 15 \\ 0 & 2 & -26 & 29 \end{bmatrix} \xrightarrow{R_3 \to R_3 - 2R_2}$$
$$\begin{bmatrix} 1 & -1 & 9 & -8 \\ 0 & 1 & -13 & 15 \\ 0 & 0 & 0 & -1 \end{bmatrix}.$$ The third row states $0 = -1$, which is false.
Hence, the system is inconsistent.

4. $\begin{cases} 2x - y + z = 0 \\ 3x + 2y - 3z = 1 \\ x - 4y + 5z = -1 \end{cases}$ This system is linear and has the following matrix representation.
$$\begin{bmatrix} 2 & -1 & 1 & 0 \\ 3 & 2 & -3 & 1 \\ 1 & -4 & 5 & -1 \end{bmatrix} \xrightarrow{R_1 \leftrightarrow R_3} \begin{bmatrix} 1 & -4 & 5 & -1 \\ 3 & 2 & -3 & 1 \\ 2 & -1 & 1 & 0 \end{bmatrix} \xrightarrow[R_3 - 2R_1 \to R_3]{R_2 - 3R_1 \to R_2}$$
$$\begin{bmatrix} 1 & -4 & 5 & -1 \\ 0 & 14 & -18 & 4 \\ 0 & 7 & -9 & 2 \end{bmatrix} \xrightarrow{2R_3 - R_2 \to R_3} \begin{bmatrix} 1 & -4 & 5 & -1 \\ 0 & 14 & -18 & 4 \\ 0 & 0 & 0 & 0 \end{bmatrix}.$$
Then, $14\,y - 18z = 4 \quad \Leftrightarrow \quad y = \frac{2}{7} + \frac{9}{7}z;$ $x - 4\left(\frac{2}{7} + \frac{9}{7}z\right) + 5z = -1 \quad \Leftrightarrow \quad x = \frac{1}{7} + \frac{1}{7}z;$ and so the solutions are $\left(\frac{1}{7} + \frac{1}{7}z, \frac{2}{7} + \frac{9}{7}z, z\right)$ where z is any real number. The system is dependent.

5. $\begin{cases} 2x^2 + y^2 = 6 \\ 3x^2 - 4\,y = 11 \end{cases}$ Solving the second equation for x^2, we get $x^2 = \dfrac{11 + 4\,y}{3}.$ Substituting into the first equation gives $2\left(\dfrac{11 + 4\,y}{3}\right) + y^2 = 6 \quad \Leftrightarrow \quad 22 + 8\,y + 3\,y^2 = 18 \quad \Leftrightarrow$
$3y^2 + 8y + 4 = 0 \quad \Leftrightarrow \quad (3y + 2)(y + 2) = 0 \quad \Rightarrow \quad y = -\frac{2}{3},\ y = -2.$ Then $y = -\frac{2}{3} \quad \Rightarrow$
$x^2 = \frac{11 - \frac{8}{3}}{3} = \frac{25}{9} \quad \Rightarrow \quad x = \pm\frac{5}{3},$ and $y = -2 \quad \Rightarrow \quad x^2 = \frac{11 - 8}{3} = 1 \quad \Rightarrow \quad x = \pm 1.$
Therefore, the solutions are $\left(\frac{5}{3}, -\frac{2}{3}\right),\ \left(-\frac{5}{3}, -\frac{2}{3}\right),\ (1, -2)$ and $(-1, -2)$, and the system is nonlinear.

In problems 6 – 13, the matrices A, B, and C are defined as follows:

$$A = \begin{bmatrix} 2 & 3 \\ 2 & 4 \end{bmatrix} \qquad B = \begin{bmatrix} 2 & 4 \\ -1 & 1 \\ 3 & 0 \end{bmatrix} \qquad C = \begin{bmatrix} 1 & 0 & 4 \\ -1 & 1 & 2 \\ 0 & 1 & 3 \end{bmatrix}$$

6. $A + B$ is undefined because A is 2×2 and B is 3×2, so they have incompatible dimensions.

7. AB is undefined because A is 2×2 and B is 3×2, so they have incompatible dimensions.

8. $BA - 3B = \begin{bmatrix} 2 & 4 \\ -1 & 1 \\ 3 & 0 \end{bmatrix} \begin{bmatrix} 2 & 3 \\ 2 & 4 \end{bmatrix} - 3 \begin{bmatrix} 2 & 4 \\ -1 & 1 \\ 3 & 0 \end{bmatrix} = \begin{bmatrix} 12 & 22 \\ 0 & 1 \\ 6 & 9 \end{bmatrix} - \begin{bmatrix} 6 & 12 \\ -3 & 3 \\ 9 & 0 \end{bmatrix} = \begin{bmatrix} 6 & 10 \\ 3 & -2 \\ -3 & 9 \end{bmatrix}$

9. $CBA = \begin{bmatrix} 1 & 0 & 4 \\ -1 & 1 & 2 \\ 0 & 1 & 3 \end{bmatrix} \begin{bmatrix} 2 & 4 \\ -1 & 1 \\ 3 & 0 \end{bmatrix} \begin{bmatrix} 2 & 3 \\ 2 & 4 \end{bmatrix} = \begin{bmatrix} 14 & 4 \\ 3 & -3 \\ 8 & 1 \end{bmatrix} \begin{bmatrix} 2 & 3 \\ 2 & 4 \end{bmatrix} = \begin{bmatrix} 36 & 58 \\ 0 & -3 \\ 18 & 28 \end{bmatrix}$

10. $A = \begin{bmatrix} 2 & 3 \\ 2 & 4 \end{bmatrix} \Leftrightarrow A^{-1} = \dfrac{1}{8-6} \begin{bmatrix} 4 & -3 \\ -2 & 2 \end{bmatrix} = \begin{bmatrix} 2 & -\frac{3}{2} \\ -1 & 1 \end{bmatrix}$

11. B^{-1} does not exist since B is not a square matrix.

12. $|B|$ is not defined since B is not a square matrix.

13. $|C| = \begin{vmatrix} 1 & 0 & 4 \\ -1 & 1 & 2 \\ 0 & 1 & 3 \end{vmatrix} = 1 \begin{vmatrix} 1 & 2 \\ 1 & 3 \end{vmatrix} + 4 \begin{vmatrix} -1 & 1 \\ 0 & 1 \end{vmatrix} = 1 - 4 = -3$

14. The system $\begin{cases} 4x - 3y = 10 \\ 3x - 2y = 30 \end{cases}$ is equivalent to the matrix equation $\begin{bmatrix} 4 & -3 \\ 3 & -2 \end{bmatrix} \begin{bmatrix} x \\ y \end{bmatrix} = \begin{bmatrix} 10 \\ 30 \end{bmatrix}$.

We have $|D| = \begin{vmatrix} 4 & -3 \\ 3 & -2 \end{vmatrix} = 4(-2) - 3(-3) = 1$. So $D^{-1} = \begin{bmatrix} -2 & 3 \\ -3 & 4 \end{bmatrix}$, and

$\begin{bmatrix} x \\ y \end{bmatrix} = \begin{bmatrix} -2 & 3 \\ -3 & 4 \end{bmatrix} \begin{bmatrix} 10 \\ 30 \end{bmatrix} = \begin{bmatrix} 70 \\ 90 \end{bmatrix}$. Therefore, $x = 70$ and $y = 90$.

15. $\begin{cases} 2x - z = 14 \\ 3x - y + 5z = 0 \\ 4x + 2y + 3z = -2 \end{cases}$

Then, $|D| = \begin{vmatrix} 2 & 0 & -1 \\ 3 & -1 & 5 \\ 4 & 2 & 3 \end{vmatrix} = 2 \begin{vmatrix} -1 & 5 \\ 2 & 3 \end{vmatrix} - 1 \begin{vmatrix} 3 & -1 \\ 4 & 2 \end{vmatrix} = -26 - 10 = -36$;

$|D_x| = \begin{vmatrix} 14 & 0 & -1 \\ 0 & -1 & 5 \\ -2 & 2 & 3 \end{vmatrix} = 14 \begin{vmatrix} -1 & 5 \\ 2 & 3 \end{vmatrix} - 1 \begin{vmatrix} 0 & -1 \\ 4 & 2 \end{vmatrix} = -182 + 2 = -180$;

$|D_y| = \begin{vmatrix} 2 & 14 & -1 \\ 3 & 0 & 5 \\ 4 & -2 & 3 \end{vmatrix} = -3 \begin{vmatrix} 14 & -1 \\ -2 & 3 \end{vmatrix} - 5 \begin{vmatrix} 2 & 14 \\ 4 & -2 \end{vmatrix} = -120 + 300 = 180$; and

$$|D_z| = \begin{vmatrix} 2 & 0 & 14 \\ 3 & -1 & 0 \\ 4 & 2 & -2 \end{vmatrix} = 2\begin{vmatrix} -1 & 0 \\ 2 & -2 \end{vmatrix} + 14\begin{vmatrix} 3 & -1 \\ 2 & 2 \end{vmatrix} = 4 + 140 = 144.$$

Therefore, $x = \frac{-180}{-36} = 5$, $y = \frac{180}{-36} = -5$, $z = \frac{144}{-36} = -4$, and so the solution is $(5, -5, -4)$.

16. $|A| = \begin{vmatrix} 1 & 4 & 1 \\ 0 & 2 & 0 \\ 1 & 0 & 1 \end{vmatrix} = 2\begin{vmatrix} 1 & 1 \\ 1 & 1 \end{vmatrix} = 0$; $|B| = \begin{vmatrix} 1 & 4 & 0 \\ 0 & 2 & 0 \\ -3 & 0 & 1 \end{vmatrix} = 2\begin{vmatrix} 1 & 0 \\ -3 & 1 \end{vmatrix} = 2.$

Since $|A| = 0$, A does not have an inverse, and since $|B| \neq 0$, B does have an inverse.

$$\begin{bmatrix} 1 & 4 & 0 & 1 & 0 & 0 \\ 0 & 2 & 0 & 0 & 1 & 0 \\ -3 & 0 & 1 & 0 & 0 & 1 \end{bmatrix} \xrightarrow{R_3+3R_1 \rightarrow R_3} \begin{bmatrix} 1 & 4 & 0 & 1 & 0 & 0 \\ 0 & 2 & 0 & 0 & 1 & 0 \\ 0 & 12 & 1 & 3 & 0 & 1 \end{bmatrix} \begin{array}{l} \xrightarrow{R_1-2R_2 \rightarrow R_1} \\ \xrightarrow{R_3-6R_2 \rightarrow R_3} \end{array}$$

$$\begin{bmatrix} 1 & 0 & 0 & 1 & -2 & 0 \\ 0 & 2 & 0 & 0 & 1 & 0 \\ 0 & 0 & 1 & 3 & -6 & 1 \end{bmatrix} \xrightarrow{\frac{1}{2}R_2} \begin{bmatrix} 1 & 0 & 0 & 1 & -2 & 0 \\ 0 & 1 & 0 & 0 & \frac{1}{2} & 0 \\ 0 & 0 & 1 & 3 & -6 & 1 \end{bmatrix}.$$

Therefore, $B^{-1} = \begin{bmatrix} 1 & -2 & 0 \\ 0 & \frac{1}{2} & 0 \\ 3 & -6 & 1 \end{bmatrix}.$

17. $\begin{cases} x^2 - y + 5 \leq 0 \\ y \leq 5 + 2x \end{cases}$

Substituting $y = 5 + 2x$ into the first equation gives
$x^2 - (5 + 2x) + 5 = 0 \quad \Leftrightarrow \quad x^2 - 2x = 0 \quad \Leftrightarrow$
$x(x - 2) = 0 \quad \Leftrightarrow \quad x = 0$ or $x = 2$. If $x = 0$, then
$y = 5 + 2(0) = 5$, and if $x = 2$, then $y = 5 + 2(2) = 9$.
Thus the vertices are $(0, 5)$ and $(2, 9)$.

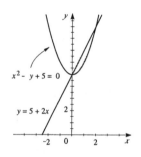

18. $\dfrac{4x - 1}{(x - 1)^2(x + 2)} = \dfrac{A}{x - 1} + \dfrac{B}{(x - 1)^2} + \dfrac{C}{x + 2}$. Thus,

$4x - 1 = A(x - 1)(x + 2) + B(x + 2) + C(x - 1)^2$
$= A(x^2 + x - 2) + B(x + 2) + C(x^2 - 2x + 1)$
$= (A + C)x^2 + (A + B - 2C)x + (-2A + 2B + C)$, which leads to the following system of

equations:

$\begin{cases} A + C = 0 \\ A + B - 2C = 4 \\ -2A + 2B + C = -1 \end{cases}$ So $\begin{bmatrix} 1 & 0 & 1 & 0 \\ 1 & 1 & -2 & 4 \\ -2 & 2 & 1 & -1 \end{bmatrix} \xrightarrow{R_1 \leftrightarrow R_2} \begin{bmatrix} 1 & 1 & -2 & 4 \\ 1 & 0 & 1 & 0 \\ -2 & 2 & 1 & -1 \end{bmatrix}$

$\begin{array}{l} \xrightarrow{R_2-R_1 \rightarrow R_2} \\ \xrightarrow{R_3+2R_1 \rightarrow R_3} \end{array} \begin{bmatrix} 1 & 1 & -2 & 4 \\ 0 & -1 & 3 & -4 \\ 0 & 4 & -3 & 7 \end{bmatrix} \xrightarrow{R_3+4R_2 \rightarrow R_3} \begin{bmatrix} 1 & 1 & -2 & 4 \\ 0 & -1 & 3 & -4 \\ 0 & 0 & 9 & -9 \end{bmatrix}.$

Therefore, $9C = -9 \Leftrightarrow C = -1$; $-B - 3 = -4 \Leftrightarrow B = 1$; and $A + 1 + 2 = 4 \Leftrightarrow A = 1$.

Therefore, $\dfrac{4x - 1}{(x - 1)^2(x + 2)} = \dfrac{1}{x - 1} + \dfrac{1}{(x - 1)^2} - \dfrac{1}{x + 2}.$

19. $\begin{cases} 2x^2 + y^2 = 16 \\ y = x^4 - 4x^3 + 6x^2 - 4x \end{cases}$ We graph the system $\begin{cases} y = \pm\sqrt{16 - 2x^2} \\ y = x^4 - 4x^3 + 6x^2 - 4x \end{cases}$.

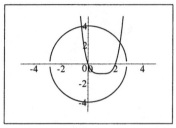

The solutions are approximately $(-0.49, 3.93)$ and $(2.34, 2.24)$.

Focus on Modeling

1.

Vertex	$M = 200 - x - y$	
$(0,2)$	$200 - (0) - (2) = 198$	← maximum value
$(0,5)$	$200 - (0) - (5) = 195$	← minimum value
$(4,0)$	$200 - (4) - (0) = 196$	

Thus the maximum value is 198, and the minimum value is 195.

3. $\begin{cases} x \geq 0, y \geq 0 \\ 2x + y \leq 10 \\ 2x + 4y \leq 28 \end{cases}$

The objective function is $P = 140 - x + 3y$.

From the graph, the vertices are $(0,0)$, $(5,0)$, $(2,6)$, and $(0,7)$.

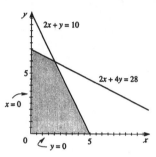

Vertex	$P = 140 - x + 3y$	
$(0,0)$	$140 - (0) + 3(0) = 140$	
$(5,0)$	$140 - (5) + 3(0) = 135$	← minimum value
$(2,6)$	$140 - (2) + 3(6) = 156$	
$(0,7)$	$140 - (0) + 3(7) = 161$	← maximum value

Thus the maximum value is 161, and the minimum value is 135.

5. Let t be the number of tables made daily and c be the number of chairs made daily. Then the data given can be summarized by the following table:

	Tables, t	Chairs, c	available time
carpentry	2 h	3 h	108 h
finishing	1 h	$\frac{1}{2}$ h	20 h
Profit	\$35	\$20	

Thus we wish to maximize the total profit, $P = 35t + 20c$, subject to the constraints:

$\begin{cases} 2t + 3c \leq 108 \\ t + \frac{1}{2}c \leq 20 \\ t \geq 0, c \geq 0. \end{cases}$

Thus we wish to maximize the total profit, $P = 35t + 20c$, subject to the constraints:

From the graph, the vertices occur at $(0,0)$, $(20,0)$, $(0,36)$, and $(3,34)$.

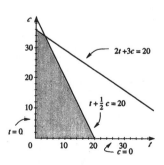

Vertex	$P = 35t + 20c$	
$(0,0)$	$35(0) + 20(0) = 0$	
$(20,0)$	$35(20) + 20(0) = 700$	
$(0,36)$	$35(0) + 20(36) = 720$	
$(3,34)$	$35(3) + 20(34) = 785$	← maximum value

Hence, 3 tables and 34 chairs should be produced daily for a maximum profit of \$785.

7. Let x be the number of crates of oranges and y be the number of crates of grapefruit. Then the data given can be summarized by the following table:

	oranges	grapefruit	available
volume	4 ft^3	6 ft^3	300 ft^3
weight	80 lb	100 lb	5600 lb
Profit	$2.50	$4.00	

In addition, $x \geq y$. Thus we wish to maximize the total profit, $P = 2.5x + 4y$, subject to the constraints:

$$\begin{cases} x \geq 0, y \geq 0 \\ x \geq y \\ 4x + 6y \leq 300 \\ 80x + 100y \leq 5600. \end{cases}$$

From the graph, the vertices occur at $(0,0)$, $(30, 30)$, $(45, 20)$, and $(70, 0)$.

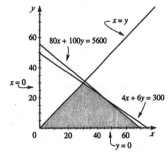

Vertex	$P = 2.5x + 4y$	
$(0,0)$	$2.5(0) + 4(0) = 0$	
$(30, 30)$	$2.5(30) + 4(30) = 195$	← maximum value
$(45, 20)$	$2.5(45) + 4(20) = 192.5$	
$(70, 0)$	$2.5(70) + 4(0) = 175$	

Thus, she should carry 30 crates of oranges and 30 crates of grapefruit for a maximum profit of $195.

9. Let x be the number of stereo sets shipped from Long Beach to Santa Monica and y be the number of stereo sets shipped from Long Beach to El Toro. Thus, $15 - x$ sets must be shipped to Santa Monica from Pasadena and $19 - y$ sets to El Toro from Pasadena. Thus $x \geq 0$, $y \geq 0$, $15 - x \geq 0$, $19 - y \geq 0$, $x + y \leq 24$, and $(15 - x) + (19 - y) \leq 18$. Simplifying, we get the inequalities (constraints):

$$\begin{cases} x \geq 0, y \geq 0 \\ x \leq 15, y \leq 19 \\ x + y \leq 24 \\ x + y \geq 16 \end{cases}$$

The objective function is the cost,
$C = 5x + 6y + 4(15 - x) + 5.5(19 - y) = x + 0.5y + 164.5$,
which we wish to minimize. From the graph, the vertices occur at $(0, 16)$, $(0, 19)$, $(5, 19)$, $(15, 9)$, and $(15, 1)$.

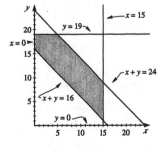

Vertex	$C = x + 0.5y + 164.5$	
$(0, 16)$	$(0) + 0.5(16) + 164.5 = 172.5$	← minimum value
$(0, 19)$	$(0) + 0.5(19) + 164.5 = 174$	
$(5, 19)$	$(5) + 0.5(19) + 164.5 = 179$	
$(15, 9)$	$(15) + 0.5(9) + 164.5 = 184$	
$(15, 1)$	$(15) + 0.5(1) + 164.5 = 180$	

The minimum cost is $172.50 and occurs when $x = 0$, $y = 16$. Hence, 0 sets should be shipped from Long Beach to Santa Monica, 16 from Long Beach to El Toro, 15 from Pasadena to Santa Monica, and 3 from Pasadena to El Toro.

11. Let x be the number of bags of standard mixtures and y be the number of bags of deluxe mixtures.
     Then the data can be summarized by the following table:

	standard	deluxe	available
cashews	100 g	150 g	15 kg
peanuts	200 g	50 g	20 kg
selling price	$1.95	$2.20	

Thus the total revenue, which we want to maximize, is given by $R = 1.95x + 2.25y$. We have the
following constraints:

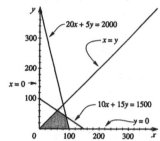

$$\begin{cases} x \geq 0, y \geq 0 \\ x \geq y \\ 0.1x + 0.15y \leq 15 \\ 0.2x + 0.05y \leq 20 \end{cases} \Leftrightarrow \begin{cases} x \geq 0, y \geq 0 \\ x \geq y \\ 10x + 15y \leq 1500 \\ 20x + 5y \leq 2000. \end{cases}$$

From the graph, the vertices occur at $(0,0)$, $(60,60)$, $(90,40)$, and
$(100,0)$.

Vertex	$R = 1.95x + 2.25y$
$(0,0)$	$1.96(0) + 2.25(0) = 0$
$(60,60)$	$1.95(60) + 2.25(60) = 252$
$(90,40)$	$1.95(90) + 2.25(40) = 265.5$ ← maximum revenue
$(100,0)$	$1.95(100) + 2.25(0) = 195$

Hence, he should pack 90 bags of standard and 40 bags of deluxe mixture for a maximum revenue of
$265.50.

13. Let x be the amount in municipal bonds and y be the amount in bank certificates, both in dollars.
     Then, $12000 - x - y$ is the amount in high-risk bonds. So our constraints can be stated as:

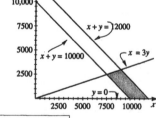

$$\begin{cases} x \geq 0, y \geq 0 \\ 12{,}000 - x - y \geq 0 \\ x \geq 3y \\ 12000 - x - y \leq 2000 \end{cases} \Leftrightarrow \begin{cases} x \geq 0, y \geq 0 \\ x + y \leq 12{,}000 \\ x \geq 3y \\ x + y \geq 10{,}000. \end{cases}$$

From the graph, the vertices occur at $(7500, 2500)$, $(10000, 0)$,
$(12000, 0)$, and $(9000, 3000)$. The objective function is
$P = 0.07x + 0.08y + 0.12(12000 - x - y)$
$= 1440 - 0.05x - 0.04y$, which we wish to maximize.

Vertex	$P = 1440 - 0.05x - 0.04y$
$(7500, 2500)$	$1440 - 0.05(7500) - 0.04(2500) = 965$ ← maximum value
$(10000, 0)$	$1440 - 0.05(10000) - 0.04(0) = 940$
$(12000, 0)$	$1440 - 0.05(12000) - 0.04(0) = 840$
$(9000, 3000)$	$1440 - 0.05(9000) - 0.04(3000) = 870$

Hence, she should invest $7500 in municipal bonds, $2500 in bank certificates, and the remaining
$2000 in high-risk bonds for a maximum yield of $965.

15. Let g be the number of games published and e be the number of educational programs published.
     Then the number of utility programs published is $36 - g - e$. Hence we wish to maximize profit,
     $P = 5000g + 8000e + 6000(36 - g - e) = 216{,}000 - 1000g + 2000e$, subject to the constraints:

$$\begin{cases} g \geq 4,\, e \geq 0 \\ 36 - g - e \geq 0 \\ 36 - g - e \leq 2e \end{cases} \quad \Leftrightarrow \quad \begin{cases} g \geq 4,\, e \geq 0 \\ g + e \leq 36 \\ g + 3e \geq 36. \end{cases}$$

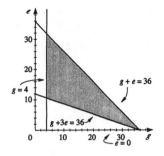

From the graph, the vertices are at $\left(4, \frac{32}{3}\right)$, $(4, 32)$, and $(36, 0)$.
The objective function is $P = 216{,}000 - 1000g + 2000e$.

Vertex	$P = 216{,}000 - 1000g + 2000e$	
$\left(4, \frac{32}{3}\right)$	$216{,}000 - 1000(4) + 2000\left(\frac{32}{3}\right) = 233{,}333.33$	
$(4, 32)$	$216{,}000 - 1000(4) + 2000(32) = 276{,}000$	← maximum value
$(36, 0)$	$216{,}000 - 1000(36) + 2000(0) = 180{,}000$	

So, they should publish 4 games, 32 educational programs, and 0 utility programs for a maximum profit of \$276,000 annually.

Chapter Eleven
Exercises 11.1

1. $y^2 = 4x$. Then $4p = 4 \quad \Leftrightarrow \quad p = 1$.
   Focus: $(1,0)$
   Directrix: $x = -1$
   Focal diameter: 4

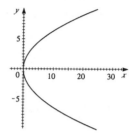

3. $x^2 = 9y$. Then $4p = 9 \quad \Leftrightarrow \quad p = \frac{9}{4}$.
   Focus: $\left(0, \frac{9}{4}\right)$
   Directrix: $y = -\frac{9}{4}$
   Focal diameter: 9

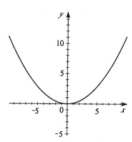

5. $y = 5x^2 \quad \Leftrightarrow \quad x^2 = \frac{1}{5}y$. Then $4p = \frac{1}{5} \quad \Leftrightarrow \quad p = \frac{1}{20}$.
   Focus: $\left(0, \frac{1}{20}\right)$
   Directrix: $y = -\frac{1}{20}$
   Focal diameter: $\frac{1}{5}$

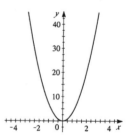

7. $x = -8y^2 \quad \Leftrightarrow \quad y^2 = -\frac{1}{8}x$. Then $4p = -\frac{1}{8} \quad \Leftrightarrow \quad p = -\frac{1}{32}$.
   Focus: $\left(-\frac{1}{32}, 0\right)$
   Directrix: $x = \frac{1}{32}$
   Focal diameter: $\frac{1}{8}$

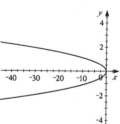

9. $x^2 + 6y = 0 \quad \Leftrightarrow \quad x^2 = -6y$. Then $4p = -6 \quad \Leftrightarrow \quad p = -\frac{3}{2}$.
   Focus: $\left(0, -\frac{3}{2}\right)$
   Directrix: $y = \frac{3}{2}$
   Focal diameter: 6

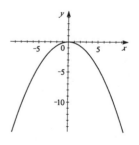

11. $5x + 3y^2 = 0 \quad \Leftrightarrow \quad y^2 = -\frac{5}{3}x.$ Then $4p = -\frac{5}{3} \quad \Leftrightarrow \quad p = -\frac{5}{12}.$

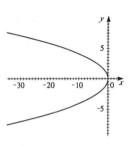

Focus: $\left(-\frac{5}{12}, 0\right)$

Directrix: $x = \frac{5}{12}$

Focal diameter: $\frac{5}{3}$

13. Since the focus is $(0, 2)$, $p = 2 \quad \Leftrightarrow \quad 4p = 8$. Hence, the equation of the parabola is $x^2 = 8y$.

15. Since the focus is $(-8, 0)$, $p = -8 \quad \Leftrightarrow \quad 4p = -32$. Hence, the equation of the parabola is $y^2 = -32x$.

17. Since the directrix is $x = 2$, $p = -2 \quad \Leftrightarrow \quad 4p = -8$. Hence, the equation of the parabola is $y^2 = -8x$.

19. Since the directrix is $y = -10$, $p = 10 \quad \Leftrightarrow \quad 4p = 40$. Hence, the equation of the parabola is $x^2 = 40y$.

21. The focus is on the positive x-axis, so the parabola opens horizontally with $2p = 2 \quad \Leftrightarrow \quad 4p = 4$. So the equation of the parabola is $y^2 = 4x$.

23. Since the parabola opens upward with focus 5 units from the vertex, the focus is $(5, 0)$. So $p = 5 \quad \Leftrightarrow \quad 4p = 20$. Thus the equation of the parabola is $x^2 = 20y$.

25. $p = -3 \quad \Leftrightarrow \quad 4p = -12$. Since the parabola opens downward, its equation is $x^2 = -12y$.

27. The focal diameter is $4p = \frac{3}{2} + \frac{3}{2} = 3$. Since the parabola opens to the left, its equation is $y^2 = -3x$.

29. The equation of the parabola has the form $y^2 = 4px$. Since the parabola passes through the point $(4, -2)$, $(-2)^2 = 4p(4) \quad \Leftrightarrow \quad 4p = 1$, and so the equation is $y^2 = x$.

31. The area of the shaded region is $width \times height = 4p \times p = 8$, and so $p^2 = 2 \quad \Leftrightarrow \quad p = -\sqrt{2}$ (because the parabola opens downward). Therefore, the equation is $x^2 = 4py = -4\sqrt{2}\,y \quad \Leftrightarrow \quad x^2 = -4\sqrt{2}\,y$.

33. (a) Since the focal diameter is 12 cm, $4p = 12$. Hence, the parabola has equation $y^2 = 12x$.

     (b) At a point 20 cm horizontally from the vertex, the parabola passes through the point $(20, y)$, and hence from part (a), $y^2 = 12(20) \quad \Leftrightarrow \quad y^2 = 240 \quad \Leftrightarrow \quad y = \pm 4\sqrt{15}$. Thus, $|CD| = 8\sqrt{15} \approx 31$ cm.

35. With the vertex at the origin, the top of one tower will be at the point $(300, 150)$. Inserting this point into the equation $x^2 = 4py$ gives $(300)^2 = 4p(150) \quad \Leftrightarrow \quad 90000 = 600p \quad \Leftrightarrow \quad p = 150$. So the equation of the parabolic part of the cables is $x^2 = 4(150)y \quad \Leftrightarrow \quad x^2 = 600y$.

37. (a) Since a parabola with directrix $y = -p$ has equation $x^2 = 4py$, we have the following.

     The directrix is $y = \frac{1}{2} \quad \Rightarrow \quad p = -\frac{1}{2}$, so the equation is $x^2 = 4\left(-\frac{1}{2}\right)y \quad \Leftrightarrow \quad x^2 = -2y$.

     The directrix is $y = 1 \quad \Rightarrow \quad p = -1$, so the equation is $x^2 = 4(-1)y \quad \Leftrightarrow \quad x^2 = -4y$.

     The directrix is $y = 4 \quad \Rightarrow \quad p = -4$, so the equation is $x^2 = 4(-4)y \quad \Leftrightarrow \quad x^2 = -16y$.

     The directrix is $y = 8 \quad \Rightarrow \quad p = -8$, so the equation is $x^2 = 4(-8)y \quad \Leftrightarrow \quad x^2 = -32y$.

(b)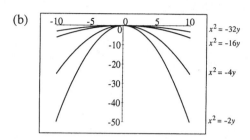

As the directrix moves further from the vertex, the parabola get flatter.

39.   Yes. If a cone intersects a plane that is parallel to a line on the cone, the resulting curve is a parabola (as shown on page 735 of the text).

Exercises 11.2

1. $\dfrac{x^2}{25} + \dfrac{y^2}{9} = 1$. This ellipse has $a = 5$, $b = 3$, and so
$c^2 = a^2 - b^2 = 16 \quad \Leftrightarrow \quad c = 4$.
Vertices: $(\pm 5, 0)$; foci: $(\pm 4, 0)$; eccentricity: $e = \frac{c}{a} = \frac{4}{5} = 0.8$;
length of the major axis: $2a = 10$; length of the minor axis: $2b = 6$

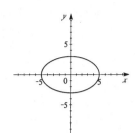

3. $9x^2 + 4y^2 = 36 \quad \Leftrightarrow \quad \dfrac{x^2}{4} + \dfrac{y^2}{9} = 1$. This ellipse has $a = 3$,
$b = 2$, and so $c^2 = 9 - 4 = 5 \quad \Leftrightarrow \quad c = \sqrt{5}$.
Vertices: $(0, \pm 3)$; foci: $(0, \pm\sqrt{5})$; eccentricity: $e = \frac{c}{a} = \frac{\sqrt{5}}{3}$;
length of the major axis: $2a = 6$; length of the minor axis: $2b = 4$

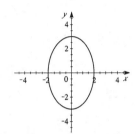

5. $x^2 + 4y^2 = 16 \quad \Leftrightarrow \quad \dfrac{x^2}{16} + \dfrac{y^2}{4} = 1$. This ellipse has $a = 4$,
$b = 2$, and so $c^2 = 16 - 4 = 12 \quad \Leftrightarrow \quad c = 2\sqrt{3}$.
Vertices: $(\pm 4, 0)$; foci: $(\pm 2\sqrt{3}, 0)$; eccentricity:
$e = \frac{c}{a} = \frac{2\sqrt{3}}{4} = \frac{\sqrt{3}}{2}$; length of the major axis: $2a = 8$; length of
the minor axis: $2b = 4$.

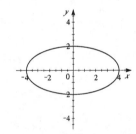

7. $2x^2 + y^2 = 3 \quad \Leftrightarrow \quad \dfrac{x^2}{\frac{3}{2}} + \dfrac{y^2}{3} = 1$. This ellipse has $a = \sqrt{3}$,
$b = \sqrt{\frac{3}{2}}$, and so $c^2 = 3 - \frac{3}{2} = \frac{3}{2} \quad \Leftrightarrow \quad c = \sqrt{\frac{3}{2}}$.
Vertices: $\left(0, \pm\sqrt{3}\right)$; foci: $\left(0, \pm\sqrt{\frac{3}{2}}\right)$; eccentricity:
$e = \frac{c}{a} = \dfrac{\sqrt{\frac{3}{2}}}{\sqrt{3}} = \dfrac{1}{\sqrt{2}} = \dfrac{\sqrt{2}}{2}$; length of the major axis: $2a = 2\sqrt{3}$
length of the minor axis: $2b = 2\sqrt{\frac{3}{2}} = \sqrt{6}$.

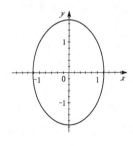

9. $x^2 + 4y^2 = 1 \quad \Leftrightarrow \quad \dfrac{x^2}{1} + \dfrac{y^2}{\frac{1}{4}} = 1$. This ellipse has $a = 1$, $b = \frac{1}{2}$,
and so $c^2 = 1 - \frac{1}{4} = \frac{3}{4} \quad \Leftrightarrow \quad c = \frac{\sqrt{3}}{2}$.
Vertices: $(\pm 1, 0)$; foci: $\left(\pm\frac{\sqrt{3}}{2}, 0\right)$; eccentricity:
$e = \frac{c}{a} = \frac{\sqrt{3}/2}{1} = \frac{\sqrt{3}}{2}$; length of the major axis: $2a = 2$; length
of the minor axis: $2b = 1$.

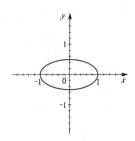

11. $\frac{1}{2}x^2 + \frac{1}{8}y^2 = \frac{1}{4}$ $\Leftrightarrow$ $2x^2 + \frac{1}{2}y^2 = 1$ $\Leftrightarrow$ $\dfrac{x^2}{\frac{1}{2}} + \dfrac{y^2}{2} = 1.$

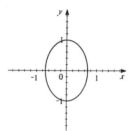

This ellipse has $a = \sqrt{2}$, $b = \frac{1}{\sqrt{2}}$, and so $c^2 = 2 - \frac{1}{2} = \frac{3}{2}$

$\Leftrightarrow$ $c = \sqrt{\frac{3}{2}}.$

Vertices: $(0, \pm\sqrt{2})$; foci: $\left(0, \pm\sqrt{\frac{3}{2}}\right)$; eccentricity:

$= \frac{c}{a} = \dfrac{\sqrt{\frac{3}{2}}}{\sqrt{2}} = \frac{1}{2}\sqrt{3}$; length of the major axis: $2a = 2\sqrt{2}$;

length of the minor axis: $2b = \sqrt{2}.$

13. $y^2 = 1 - 2x^2$ $\Leftrightarrow$ $2x^2 + y^2 = 1$ $\Leftrightarrow$ $\dfrac{x^2}{\frac{1}{2}} + \dfrac{y^2}{1} = 1.$ This

ellipse has $a = 1$, $b = \frac{1}{\sqrt{2}}$, and so $c^2 = 1 - \frac{1}{2} = \frac{1}{2}$ $\Leftrightarrow$ $c = \frac{1}{\sqrt{2}}.$

Vertices: $(0, \pm 1)$; foci: $\left(0, \pm\frac{1}{\sqrt{2}}\right)$; eccentricity:

$e = \frac{c}{a} = \dfrac{\frac{1}{\sqrt{2}}}{1} = \frac{1}{\sqrt{2}}$; length of the major axis: $2a = 2$; length of

the minor axis: $2b = \sqrt{2}.$

15. This ellipse has a horizontal major axis with $a = 5$ and $b = 4$, so the equation is $\dfrac{x^2}{(5)^2} + \dfrac{y^2}{(4)^2} = 1$

  $\Leftrightarrow$ $\dfrac{x^2}{25} + \dfrac{y^2}{16} = 1.$

17. This ellipse has a vertical major axis with $c = 2$ and $b = 2$. So $a^2 = c^2 + b^2 = 2^2 + 2^2 = 8$ $\Leftrightarrow$

$a = 2\sqrt{2}$. So the equation is $\dfrac{x^2}{(2)^2} + \dfrac{y^2}{\left(2\sqrt{2}\right)^2} = 1$ $\Leftrightarrow$ $\dfrac{x^2}{4} + \dfrac{y^2}{8} = 1.$

19. The foci are $(\pm 4, 0)$, and the vertices are $(\pm 5, 0)$. Thus, $c = 4$ and $a = 5$, and so $b^2 = 25 - 16 = 9$.

Therefore, the equation of the ellipse is $\dfrac{x^2}{25} + \dfrac{y^2}{9} = 1.$

21. The length of the major axis is $2a = 4$ $\Leftrightarrow$ $a = 2$, the length of the minor axis is $2b = 2$ $\Leftrightarrow$

$b = 1$, and the foci are on the y-axis. Therefore, the equation of the ellipse is $x^2 + \dfrac{y^2}{4} = 1.$

23. The foci are $(0, \pm 2)$, and the length of the minor axis is $2b = 6$ $\Leftrightarrow$ $b = 3$. Thus,

$a^2 = 4 + 9 = 13$. Since the foci are on the y-axis, the equation is $\dfrac{x^2}{9} + \dfrac{y^2}{13} = 1.$

25. The endpoints of the major axis are $(\pm 10, 0)$ $\Leftrightarrow$ $a = 10$, and the distance between the foci is

$2c = 6$ $\Leftrightarrow$ $c = 3$. Therefore, $b^2 = 100 - 9 = 91$, and so the equation of the ellipse is

$\dfrac{x^2}{100} + \dfrac{y^2}{91} = 1.$

27. The length of the major axis is 10, so $2a = 10$ $\Leftrightarrow$ $a = 5$, and the foci are on the x-axis, so the

form of the equation is $\dfrac{x^2}{25} + \dfrac{y^2}{b^2} = 1$. Since the ellipse passes through $\left(\sqrt{5}, 2\right)$, we know that

$\dfrac{\left(\sqrt{5}\right)^2}{25} + \dfrac{(2)^2}{b^2} = 1$ $\Leftrightarrow$ $\dfrac{5}{25} + \dfrac{4}{b^2} = 1$ $\Leftrightarrow$ $\dfrac{4}{b^2} = \dfrac{4}{5}$ $\Leftrightarrow$ $b^2 = 5$, and so the equation is

$\dfrac{x^2}{25} + \dfrac{y^2}{5} = 1$.

29. Since the foci are $(\pm 1.5, 0)$, we have $c = \dfrac{3}{2}$. Since the eccentricity is $0.8 = \dfrac{c}{a}$, we have $a = \dfrac{\frac{3}{2}}{\frac{4}{5}} = \dfrac{15}{8}$,

and so $b^2 = \dfrac{225}{64} - \dfrac{9}{4} = \dfrac{225 - 16 \cdot 9}{64} = \dfrac{81}{64}$. Therefore, the equation of the ellipse is $\dfrac{x^2}{\left(^{15}/_8\right)^2} + \dfrac{y^2}{^{81}/_{64}} = 1$

$\Leftrightarrow$ $\dfrac{64x^2}{225} + \dfrac{64y^2}{81} = 1$.

31. $\begin{cases} 4x^2 + y^2 = 4 \\ 4x^2 + 9y^2 = 36 \end{cases}$ Subtracting the first equation from the second

equation gives $8y^2 = 32$ $\Leftrightarrow$ $y^2 = 4$ $\Leftrightarrow$ $y = \pm 2$.
Substituting $y = \pm 2$ in the first equation gives $4x^2 + (\pm 2)^2 = 4$ $\Leftrightarrow$
$x = 0$, and so the points of intersection are $(0, \pm 2)$.

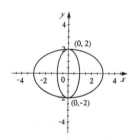

33. Using the perihelion, $a - c = 147{,}000{,}000$, while using the aphelion, $a + c = 153{,}000{,}000$.
Adding, we have $2a = 300{,}000{,}000$ $\Leftrightarrow$ $a = 150{,}000{,}000$. So $b^2 = a^2 - c^2$
$= (150 \times 10^6)^2 - (3 \times 10^6)^2 = 22{,}491 \times 10^{12} = 2.2491 \times 10^{16}$. Thus the equation of the orbit is
$\dfrac{x^2}{2.2500 \times 10^{16}} + \dfrac{y^2}{2.2491 \times 10^{16}} = 1$.

35. Using the perilune, $a - c = 1075 + 68 = 1143$, and using the apolune,
$a + c = 1075 + 195 = 1270$. Adding, we get $2a = 2413$ $\Leftrightarrow$ $a = 1206.5$. So
$c = 1270 - 1206.5$ $\Leftrightarrow$ $c = 63.5$. Therefore, $b^2 = (1206.5)^2 - (63.5)^2 = 1{,}451{,}610$. Since
$a^2 \approx 1{,}455{,}642$, the equation of Apollo 11's orbit is $\dfrac{x^2}{1{,}455{,}642} + \dfrac{y^2}{1{,}451{,}610} = 1$.

37. From the diagram, $a = 40$ and $b = 20$, and so the equation of the ellipse whose top half is the

window is $\dfrac{x^2}{1600} + \dfrac{y^2}{400} = 1$. Since the ellipse passes through the point $(25, h)$, by substituting, we

have $\dfrac{25^2}{1600} + \dfrac{h^2}{400} = 1$ $\Leftrightarrow$ $625 + 4y^2 = 1600$ $\Leftrightarrow$ $y = \dfrac{\sqrt{975}}{2} = \dfrac{5\sqrt{39}}{2} \approx 15.61$. Therefore,

the window is approximately 15.6 inches high at the specified point.

39. The foci are $(\pm c, 0)$, where $c^2 = a^2 - b^2$. The endpoints of one latus rectum are the points $(c, \pm k)$,

and the length is $2k$. Substituting this point into the equation, we get $\dfrac{c^2}{a^2} + \dfrac{k^2}{b^2} = 1$ $\Leftrightarrow$

$\dfrac{k^2}{b^2} = 1 - \dfrac{c^2}{a^2} = \dfrac{a^2 - c^2}{a^2}$ $\Leftrightarrow$ $k^2 = \dfrac{b^2(a^2 - c^2)}{a^2}$. Since $b^2 = a^2 - c^2$, the last equation becomes

$k^2 = \dfrac{b^4}{a^2}$ $\Rightarrow$ $k = \dfrac{b^2}{a}$. Thus, the length of the latus rectum is $2k = 2\left(\dfrac{b^2}{a}\right) = \dfrac{2b^2}{a}$.

41. (a) $\dfrac{x^2}{25} + \dfrac{y^2}{20} = 1 \iff \dfrac{y^2}{20} = 1 - \dfrac{x^2}{25} \iff y^2 = 20 - \dfrac{4x^2}{5} \implies y = \pm\sqrt{20 - \dfrac{4x^2}{5}}.$

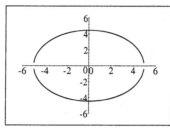

  (b) $6x^2 + y^2 = 36 \iff y^2 = 36 - 6x^2 \implies y = \pm\sqrt{36 - 6x^2}.$

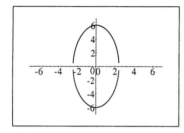

43. Have each friend hold one end of the string on the blackboard (the foci). Then use a piece of chalk and the string to draw the ellipse.

45. The shape drawn on the paper is *almost*, but not quite, an ellipse. For example, when the bottle has radius 1 unit and the compass legs are set 1 unit apart, then it can be shown that the equation of the resulting curve is $1 + y^2 = 2\cos x$. The graph of this curve differs very slightly from the ellipse with the same major and minor axis. This example shows that in mathematics, things are not always what they appear to be.

Exercises 11.3

1. The hyperbola $\dfrac{x^2}{4} - \dfrac{y^2}{16} = 1$ has $a = 2$, $b = 4$, and $c^2 = 16 + 4 \Rightarrow$
   $c = 2\sqrt{5}$.

   Vertices: $(\pm 2, 0)$; foci: $(\pm 2\sqrt{5}, 0)$; asymptotes: $y = \pm \frac{4}{2}x \iff$
   $y = \pm 2x$.

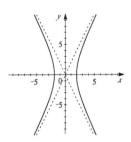

3. The hyperbola $\dfrac{y^2}{1} - \dfrac{x^2}{25} = 1$ has $a = 1$, $b = 5$, and $c^2 = 1 + 25 = 26$
   $\Rightarrow \quad c = \sqrt{26}$.

   Vertices: $(0, \pm 1)$; foci: $(0, \pm\sqrt{26})$; asymptotes: $y = \pm \frac{1}{5}x$.

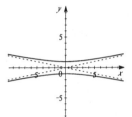

5. The hyperbola $x^2 - y^2 = 1$ has $a = 1$, $b = 1$, and $c^2 = 1 + 1 = 2$
   $\Rightarrow \quad c = \sqrt{2}$.

   Vertices: $(\pm 1, 0)$; foci: $(\pm\sqrt{2}, 0)$; asymptotes: $y = \pm x$.

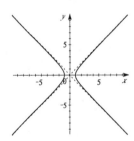

7. The hyperbola $25y^2 - 9x^2 = 225 \iff \dfrac{y^2}{9} - \dfrac{x^2}{25} = 1$ has $a = 3$,
   $b = 5$, and $c^2 = 25 + 9 = 34 \Rightarrow c = \sqrt{34}$.

   Vertices: $(0, \pm 3)$; foci: $(0, \pm\sqrt{34})$; asymptotes: $y = \pm \frac{3}{5}x$.

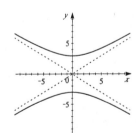

9. The hyperbola $x^2 - 4y^2 - 8 = 0 \iff \dfrac{x^2}{8} - \dfrac{y^2}{2} = 1$ has
   $a = \sqrt{8}$, $b = \sqrt{2}$, and $c^2 = 8 + 2 = 10 \Rightarrow c = \sqrt{10}$.

   Vertices: $(\pm 2\sqrt{2}, 0)$; foci: $(\pm\sqrt{10}, 0)$; asymptotes: $y = \pm \dfrac{\sqrt{2}}{\sqrt{8}} x$
   $\iff y = \pm \frac{1}{2}x$.

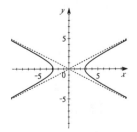

11. The hyperbola $4y^2 - x^2 = 1$ $\Leftrightarrow$ $\dfrac{y^2}{\frac{1}{4}} - x^2 = 1$ has $a = \dfrac{1}{2}$, $b = 1$,

and $c^2 = \frac{1}{4} + 1 = \frac{5}{4}$ $\Rightarrow$ $c = \frac{1}{2}\sqrt{5}$.

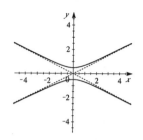

Vertices: $(0, \pm\frac{1}{2})$; foci: $(0, \pm\frac{1}{2}\sqrt{5})$; asymptotes: $y = \pm\dfrac{\frac{1}{2}}{1}x$ $\Leftrightarrow$
$y = \pm\frac{1}{2}x$.

13. From the graph, the foci are $(\pm 4, 0)$, and the vertices are $(\pm 2, 0)$, so $c = 4$ and $a = 2$. Thus,
$b^2 = 16 - 4 = 12$, and since the vertices are on the x-axis, the equation of the hyperbola is
$\dfrac{x^2}{4} - \dfrac{y^2}{12} = 1$.

15. From the graph, the vertices are $(0, \pm 4)$, the foci are on the y-axis, and the hyperbola passes through
the point $(3, -5)$. So the equation is of the form $\dfrac{y^2}{16} - \dfrac{x^2}{b^2} = 1$. Substituting the point $(3, -5)$, we
have $\dfrac{(-5)^2}{16} - \dfrac{(3)^2}{b^2} = 1$ $\Leftrightarrow$ $\dfrac{25}{16} - 1 = \dfrac{9}{b^2}$ $\Leftrightarrow$ $\dfrac{9}{16} = \dfrac{9}{b^2}$ $\Leftrightarrow$ $b^2 = 16$. Thus, the
equation of the hyperbola is $\dfrac{y^2}{16} - \dfrac{x^2}{16} = 1$.

17. The foci are $(\pm 5, 0)$, and the vertices are $(\pm 3, 0)$, so $c = 5$ and $a = 3$. Then $b^2 = 25 - 9 = 16$, and
since the vertices are on the x-axis, the equation of the hyperbola is $\dfrac{x^2}{9} - \dfrac{y^2}{16} = 1$.

19. The foci are $(0, \pm 2)$, and the vertices are $(0, \pm 1)$, so $c = 2$ and $a = 1$. Then $b^2 = 4 - 1 = 3$, and
since the vertices are on the y-axis, the equation is $y^2 - \dfrac{x^2}{3} = 1$.

21. The vertices are $(\pm 1, 0)$, and the asymptotes are $y = \pm 5x$, so $a = 1$. The asymptotes are $y = \pm\dfrac{b}{a}x$,
so $\frac{b}{1} = 5$ $\Leftrightarrow$ $b = 5$. Therefore, the equation of the hyperbola is $x^2 - \frac{y^2}{25} = 1$.

23. The foci are $(0, \pm 8)$, and the asymptotes are $y = \pm\frac{1}{2}x$, so $c = 8$. The asymptotes are $y = \pm\dfrac{a}{b}x$, so
$\dfrac{a}{b} = \frac{1}{2}$ and $b = 2a$. Since $a^2 + b^2 = c^2 = 64$, we have $a^2 + 4a^2 = 64$ $\Leftrightarrow$ $a^2 = \frac{64}{5}$ and
$b^2 = 4a^2 = \frac{256}{5}$. Thus, the equation of the hyperbola is $\dfrac{y^2}{64/5} - \dfrac{x^2}{256/5} = 1$ $\Leftrightarrow$
$\dfrac{5y^2}{64} - \dfrac{5x^2}{256} = 1$.

25. The asymptotes of the hyperbola are $y = \pm x$ so $b = a$. Since the hyperbola passes through the point
$(5, 3)$, its foci are on the x-axis, and the equation is of the form, $\dfrac{x^2}{a^2} - \dfrac{y^2}{a^2} = 1$, so it follows that
$\dfrac{25}{a^2} - \dfrac{9}{a^2} = 1$ $\Leftrightarrow$ $a^2 = 16 = b^2$. Therefore, the equation of the hyperbola is $\dfrac{x^2}{16} - \dfrac{y^2}{16} = 1$.

27. The foci are $(\pm 5, 0)$, and the length of the transverse axis is 6, so $c = 5$ and $2a = 6$ $\Leftrightarrow$ $a = 3$.
Thus, $b^2 = 25 - 9 = 16$, and the equation is $\dfrac{x^2}{9} - \dfrac{y^2}{16} = 1$.

29. (a) The hyperbola $x^2 - y^2 = 5$ $\Leftrightarrow$ $\dfrac{x^2}{5} - \dfrac{y^2}{5} = 1$ has $a = \sqrt{5}$ and $b = \sqrt{5}$. Thus, the asymptotes are $y = \pm x$, and their slopes are $m_1 = 1$ and $m_2 = -1$. Since $m_1 \times m_2 = -1$, the asymptotes are perpendicular.

(b) Since the asymptotes are perpendicular to each other, they must be of slope ± 1, so $a = b$.

Therefore, $c^2 = 2a^2$ $\Leftrightarrow$ $a^2 = \dfrac{c^2}{2}$, and since the vertices are on the x-axis, the equation is

$$\dfrac{x^2}{\frac{1}{2}c^2} - \dfrac{y^2}{\frac{1}{2}c^2} = 1 \quad \Leftrightarrow \quad x^2 - y^2 = \dfrac{c^2}{2}.$$

31. $\sqrt{(x+c)^2 + y^2} - \sqrt{(x-c)^2 + y^2} = \pm 2a$. Let us consider the positive case only. Then $\sqrt{(x+c)^2 + y^2} = 2a + \sqrt{(x-c)^2 + y^2}$, and squaring both sides gives $x^2 + 2cx + c^2 + y^2 = 4a^2 + 4a\sqrt{(x-c)^2 + y^2} + x^2 - 2cx + c^2 + y^2$ $\Leftrightarrow$ $4a\sqrt{(x-c)^2 + y^2} = 4cx - 4a^2$. Dividing by 4 and squaring both sides gives $a^2(x^2 - 2cx + c^2 + y^2) = c^2x^2 - 2a^2cx + a^4$ $\Leftrightarrow$ $a^2x^2 - 2a^2cx + a^2c^2 + a^2y^2 = c^2x^2 - 2a^2cx + a^4$ $\Leftrightarrow$ $a^2x^2 + a^2c^2 + a^2y^2 = c^2x^2 + a^4$. Rearranging the order, we have $c^2x^2 - a^2x^2 - a^2y^2 = a^2c^2 - a^4$ $\Leftrightarrow$ $(c^2 - a^2)x^2 - a^2y^2 = a^2(c^2 - a^2)$. The negative case gives the same result.

33. $d(AB) = 500 = 2c$ $\Leftrightarrow$ $c = 250$.

(a) Since $\Delta t = 2640$ and $v = 980$, then $\Delta d = d(PA) - d(PB) = v\,\Delta t$ $= (980 \text{ ft/}\mu s) \cdot (2640\ \mu s) = 2{,}587{,}200 \text{ ft} = 490 \text{ mi}$.

(b) $c = 250$, $2a = 490$ $\Leftrightarrow$ $a = 245$, and the foci are on the y-axis. Then, $b^2 = 250^2 - 245^2 = 2475$. Hence, the equation is $\dfrac{y^2}{60{,}025} - \dfrac{x^2}{2475} = 1$.

(c) Since P is due east of A, $c = 250$ is the y-coordinate of P. Therefore, P is at $(x, 250)$, and so $\dfrac{250^2}{245^2} - \dfrac{x^2}{2475} = 1$ $\Leftrightarrow$ $x^2 = 2475\left(\dfrac{250^2}{245^2} - 1\right) \approx 102.05$. Then, $x \approx 10.1$, and so P is approximately 10.1 miles from A.

35. (a) From the equation, we have $a^2 = k$ and $b^2 = 16 - k$. Thus, $c^2 = a^2 + b^2 = k + 16 - k = 16$ $\Rightarrow$ $c = \pm 4$. Thus the foci of the family of hyperbolas are $(0, \pm 4)$.

(b) $\dfrac{y^2}{k} - \dfrac{x^2}{16 - k} = 1$ $\Leftrightarrow$ $y^2 = k\left(1 + \dfrac{x^2}{16 - k}\right)$ $\Rightarrow$

$y = \pm\sqrt{k + \dfrac{kx^2}{16 - k}}$. For the top branch, we graph

$y = \sqrt{k + \dfrac{kx^2}{16 - k}}$, $k = 1, 4, 8, 12$.

As k increases, the hyperbola gets steeper and narrower, and the vertices move further apart.

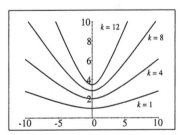

37. The wall is parallel to the axis of the cone of light coming from the top of the shade, so the intersection of the wall and the cone of light is a hyperbola. In the case of the flashlight, hold it parallel to the ground to form a hyperbola.

Exercises 11.4

1. The ellipse $\dfrac{(x-2)^2}{9} + \dfrac{(y-1)^2}{4} = 1$ is obtained from the ellipse

    $\dfrac{x^2}{9} + \dfrac{y^2}{4} = 1$ by shifting it 2 units to the right and 1 unit upward. So
    $a = 3$, $b = 2$, and $c = \sqrt{9-4} = \sqrt{5}$.

    Center: $(2,1)$; foci: $(2 \pm \sqrt{5}, 1)$; vertices: $(2 \pm 3, 1)$, so the vertices
    are $(-1,1)$ and $(5,1)$; length of the major axis: $2a = 6$; length of
    the minor axis: $2b = 4$.

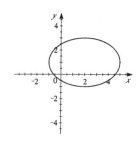

3. The ellipse $\dfrac{x^2}{9} + \dfrac{(y+5)^2}{25} = 1$ is obtained from the ellipse

    $\dfrac{x^2}{9} + \dfrac{y^2}{25} = 1$ by shifting it 5 units downward. So $a = 5$, $b = 3$, and
    $c = \sqrt{25-9} = 4$.

    Center: $(0,-5)$; foci: $(0, -5 \pm 4) = (0, -9)$ and $(0, -1)$; vertices:
    $(0, -5 \pm 5)$, so the vertices are $(0, -10)$ and $(0,0)$; length of the
    major axis: $2a = 10$; length of the minor axis: $2b = 6$.

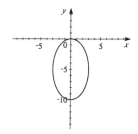

5. The parabola $(x-3)^2 = 8(y+1)$ is obtained from the parabola
    $x^2 = 8y$ by shifting it 3 units to the right and 1 unit down. So $4p = 8$
    $\Leftrightarrow \quad p = 2$.

    Vertex: $(3,-1)$; focus: $(3, -1+2) = (3,1)$; directrix:
    $y = -1 - 2 = -3$.

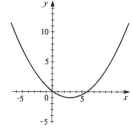

7. The parabola $-4\left(x + \frac{1}{2}\right)^2 = y \quad \Leftrightarrow \quad \left(x + \frac{1}{2}\right)^2 = -\frac{1}{4}y$ is

    obtained from the parabola $x^2 = -\frac{1}{4}y$ by shifting it $\frac{1}{2}$ unit to the

    left. So $4p = -\frac{1}{4} \quad \Leftrightarrow \quad p = -\frac{1}{16}$.

    Vertex: $\left(-\frac{1}{2}, 0\right)$; focus: $\left(-\frac{1}{2}, 0 - \frac{1}{16}\right) = \left(-\frac{1}{2}, -\frac{1}{16}\right)$; directrix:

    $y = 0 + \frac{1}{16} = \frac{1}{16}$.

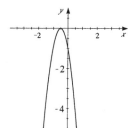

9. The hyperbola $\dfrac{(x+1)^2}{9} - \dfrac{(y-3)^2}{16} = 1$ is obtained from the

    hyperbola $\dfrac{x^2}{9} - \dfrac{y^2}{16} = 1$ by shifting it 1 unit to the left and 3 units

    up. So $a = 3$, $b = 4$, and $c = \sqrt{9+16} = 5$.

    Center: $(-1,3)$; foci: $(-1 \pm 5, 3)$, so the foci are $(-6,3)$ and $(4,3)$;

    vertices: $(-1 \pm 3, 3)$, so the vertices are $(-4,3)$ and $(2,3)$;

    asymptotes: $(y-3) = \pm\frac{4}{3}(x+1) \quad \Leftrightarrow \quad y = \pm\frac{4}{3}(x+1) + 3 \quad \Leftrightarrow \quad 3y = 4x + 13$ and
    $3y = -4x + 5$.

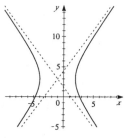

11. The hyperbola $y^2 - \dfrac{(x+1)^2}{4} = 1$ is obtained from hyperbola

$y^2 - \frac{x^2}{4} = 1$ by shifting it 1 to the unit left. So $a = 1$, $b = 2$, and

$c = \sqrt{1+4} = \sqrt{5}$.

Center: $(-1, 0)$; foci: $(-1, \pm\sqrt{5})$, so the foci are $(-1, -\sqrt{5})$ and

$(-1, \sqrt{5})$; vertices: $(-1, \pm 1)$, so the vertices are $(-1, -1)$ and

$(-1, 1)$; asymptotes: $y = \pm\frac{1}{2}(x+1)$ $\Leftrightarrow$ $y = \frac{1}{2}(x+1)$ and $y = -\frac{1}{2}(x+1)$.

13. This is a parabola that opens down with its vertex at $(0, 4)$, so its equation is of the form
$x^2 = a(y-4)$. Since $(1, 0)$ is a point on this parabola, we have $(1)^2 = a(0-4)$ $\Leftrightarrow$ $1 = -4a$
$\Leftrightarrow$ $a = -\frac{1}{4}$. Thus, the equation is $x^2 = -\frac{1}{4}(y-4)$.

15. This is an ellipse with the major axis parallel to the x-axis, with one vertex at $(0, 0)$, the other vertex
at $(10, 0)$, and one focus at $(8, 0)$. The center is at $\left(\frac{0+10}{2}, 0\right) = (5, 0)$, $a = 5$, and $c = 3$ (the distance
from one focus to the center). So $b^2 = a^2 - c^2 = 25 - 9 = 16$. Thus, the equation is
$\dfrac{(x-5)^2}{25} + \dfrac{y^2}{16} = 1$.

17. This is a hyperbola with center $(0, 1)$ and vertices $(0, 0)$ and $(0, 2)$. Since a is the distance form the
center to a vertex, we have $a = 1$. The slope of the given asymptote is 1, so $\frac{a}{b} = 1$ $\Leftrightarrow$ $b = 1$.
Thus, the equation of the hyperbola is $(y-1)^2 - x^2 = 1$.

19. $9x^2 - 36x + 4y^2 = 0$ $\Leftrightarrow$ $9(x^2 - 4x + 4) - 36 + 4y^2 = 0$ $\Leftrightarrow$

$9(x-2)^2 + 4y^2 = 36$ $\Leftrightarrow$ $\dfrac{(x-2)^2}{4} + \dfrac{y^2}{9} = 1$. This is an

ellipse that has $a = 3$, $b = 2$, and $c = \sqrt{9-4} = \sqrt{5}$.

Center: $(2, 0)$; foci: $(2, \pm\sqrt{5})$; vertices: $(2, \pm 3)$; length of major
axis: $2a = 6$; length of minor axis: $2b = 4$.

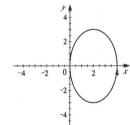

21. $x^2 - 4y^2 - 2x + 16y = 20$ $\Leftrightarrow$

$(x^2 - 2x + 1) - 4(y^2 - 4y + 4) = 20 + 1 - 16$ $\Leftrightarrow$

$(x-1)^2 - 4(y-2)^2 = 5$ $\Leftrightarrow$ $\dfrac{(x-1)^2}{5} - \dfrac{(y-2)^2}{\frac{5}{4}} = 1$. This

is a hyperbola that has $a = \sqrt{5}$, $b = \frac{1}{2}\sqrt{5}$, and $c = \sqrt{5 + \frac{5}{4}} = \frac{5}{2}$.

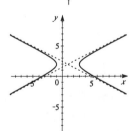

Center: $(1, 2)$; foci: $\left(1 \pm \frac{5}{2}, 2\right)$, so the foci are $\left(-\frac{3}{2}, 2\right)$ and $\left(\frac{7}{2}, 2\right)$; vertices: $(1 \pm \sqrt{5}, 2)$; asymptotes
$y - 2 = \pm\frac{1}{2}(x-1)$ $\Leftrightarrow$ $y = \pm\frac{1}{2}(x-1) + 2$ $\Leftrightarrow$ $y = \frac{1}{2}x + \frac{3}{2}$ and $y = -\frac{1}{2}x + \frac{5}{2}$.

23. $4x^2 + 25y^2 - 24x + 250y + 561 = 0$ $\Leftrightarrow$

$4(x^2 - 6x + 9) + 25(y^2 + 10y + 25) = -561 + 36 + 625$ $\Leftrightarrow$ $4(x-3)^2 + 25(y+5)^2 = 100$

$\Leftrightarrow$ $\dfrac{(x-3)^2}{25} + \dfrac{(y+5)^2}{4} = 1$. This is an ellipse that has $a = 5$, $b = 2$, and

$c = \sqrt{25 - 4} = \sqrt{21}$.

Center: $(3, -5)$; foci: $(3 \pm \sqrt{21}, -5)$; vertices: $(3 \pm 5, -5)$, so the
vertices are $(-2, -5)$ and $(8, -5)$; length of the major axis: $2a = 10$;
length of the minor axis: $2b = 4$.

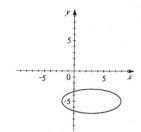

25. $16x^2 - 9y^2 - 96x + 288 = 0 \quad \Leftrightarrow \quad 16(x^2 - 6x) - 9y^2 + 288 = 0$
 $\Leftrightarrow \quad 16(x^2 - 6x + 9) - 9y^2 = 144 - 288 \quad \Leftrightarrow$
 $16(x - 3)^2 - 9y^2 = -144 \quad \Leftrightarrow \quad \dfrac{y^2}{16} - \dfrac{(x-3)^2}{9} = 1.$ This is a
 hyperbola that has $a = 4$, $b = 3$, and $c = \sqrt{16 + 9} = 5$.
 Center: $(3, 0)$; foci: $(3, \pm 5)$; vertices: $(3, \pm 4)$; asymptotes:
 $y = \pm \frac{4}{3}(x - 3) \quad \Leftrightarrow \quad y = \frac{4}{3}x - 4$ and $y = 4 - \frac{4}{3}x.$

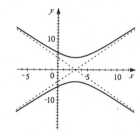

27. $x^2 + 16 = 4(y^2 + 2x) \quad \Leftrightarrow \quad x^2 - 8x - 4y^2 + 16 = 0 \quad \Leftrightarrow$
 $(x^2 - 8x + 16) - 4y^2 = -16 + 16 \quad \Leftrightarrow \quad 4y^2 = (x - 4)^2 \quad \Leftrightarrow$
 $y = \pm \frac{1}{2}(x - 4)$. Thus, the conic is degenerate, and its graph is a pair
 of lines, $y = \frac{1}{2}(x - 4)$ and $y = -\frac{1}{2}(x - 4).$

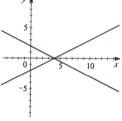

29. $3x^2 + 4y^2 - 6x - 24y + 39 = 0 \quad \Leftrightarrow$
 $3(x^2 - 2x) + 4(y^2 - 6y) = -39 \quad \Leftrightarrow$
 $3(x^2 - 2x + 1) + 4(y^2 - 6y + 9) = -39 + 3 + 36 \quad \Leftrightarrow$
 $3(x - 1)^2 + 4(y - 3)^2 = 0 \quad \Leftrightarrow \quad x = 1$ and $y = 3$. This is a
 degenerate conic whose graph is the point $(1, 3).$

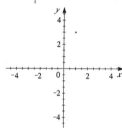

31. $4x^2 + y^2 + 4(x - 2y) + F = 0 \quad \Leftrightarrow \quad 4(x^2 + x) + (y^2 - 8y) = -F \quad \Leftrightarrow$
 $4(x^2 + x + \frac{1}{4}) + (y^2 - 8y + 16) = 16 + 1 - F \quad \Leftrightarrow \quad 4(x + \frac{1}{2})^2 + (y - 1)^2 = 17 - F.$
 (a) For an ellipse, $17 - F > 0 \quad \Leftrightarrow \quad F < 17.$
 (b) For a single point, $17 - F = 0 \quad \Leftrightarrow \quad F = 17.$
 (c) For the empty set, $17 - F < 0 \quad \Leftrightarrow \quad F > 17.$

33. (a) $x^2 = 4p(y + p)$, for $p = -2, -\frac{3}{2}, -1, -\frac{1}{2}, \frac{1}{2}, 1, \frac{3}{2}, 2.$

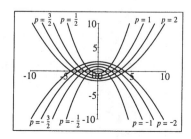

(b) The graph of $x^2 = 4p(y + p)$ is obtained by shifting the graph of $x^2 = 4py$ vertically $-p$ units so that the vertex is at $(0, -p)$. The focus of $x^2 = 4py$ is at $(0, p)$, so this point is also shifted $-p$ units vertically to the point $(0, p - p) = (0, 0)$. Thus, the focus is located at the origin.

(c) The parabolas become narrower as the vertex moves toward the origin.

Exercises 11.5

1. $(x, y) = (1, 1)$, $\phi = 45°$. Then, $X = x \cos \phi + y \sin \phi = 1 \cdot \frac{1}{\sqrt{2}} + 1 \cdot \frac{1}{\sqrt{2}} = \sqrt{2}$, and $Y = -x \sin \phi + y \cos \phi = -1 \cdot \frac{1}{\sqrt{2}} + 1 \cdot \frac{1}{\sqrt{2}} = 0$. Therefore, the XY-coordinates of the given point are $(X, Y) = (\sqrt{2}, 0)$.

3. $(x, y) = (3, -\sqrt{3})$, $\phi = 60°$. Then, $X = x \cos \phi + y \sin \phi = 3 \cdot \frac{1}{2} - \sqrt{3} \cdot \frac{\sqrt{3}}{2} = 0$, and $Y = -x \sin \phi + y \cos \phi = -3 \cdot \frac{\sqrt{3}}{2} - \sqrt{3} \cdot \frac{1}{2} = -2\sqrt{3}$. Therefore, the XY-coordinates of the given point are $(X, Y) = (0, -2\sqrt{3})$.

5. $(x, y) = (0, 2)$, $\phi = 55°$. Then, $X = x \cos \phi + y \sin \phi = 0 \cos 55° + 2 \sin 55° \approx 1.6383$ and $Y = -x \sin \phi + y \cos \phi = -0 \sin 55° + 2 \cos 55° \approx 1.1472$. Therefore, the XY-coordinates of the given point are approximately $(X, Y) = (1.6383, 1.1472)$.

7. $y = (x - 1)^2$, $\phi = 45°$. Then $x = X \cos 45° - Y \sin 45° = \frac{X}{\sqrt{2}} - \frac{Y}{\sqrt{2}}$ and $y = X \cos 45° + Y \cos 45° = \frac{X}{\sqrt{2}} + \frac{Y}{\sqrt{2}}$. Substituting these values into the equation, we get

$$\left(\frac{X}{\sqrt{2}} + \frac{Y}{\sqrt{2}}\right) = \left(\frac{X}{\sqrt{2}} - \frac{Y}{\sqrt{2}} - 1\right)^2 = \frac{X^2}{2} - 2\frac{X}{\sqrt{2}}\left(\frac{Y}{\sqrt{2}} + 1\right) + \left(\frac{Y}{\sqrt{2}} + 1\right)^2$$

$$= \frac{X^2}{2} - 2\frac{X}{\sqrt{2}}\left(\frac{Y}{\sqrt{2}} + 1\right) + \frac{Y^2}{2} + \sqrt{2}Y + 1 \quad \Leftrightarrow$$

$$\frac{X^2}{2} - \frac{X}{\sqrt{2}} - \sqrt{2}X\left(\frac{Y}{\sqrt{2}} + 1\right) + \frac{Y^2}{2} + \sqrt{2}Y + 1 - \frac{Y}{\sqrt{2}} = 0 \quad \Leftrightarrow$$

$$X^2 + Y^2 - 2XY - 3\sqrt{2}X + \sqrt{2}Y + 2 = 0.$$

9. $x^2 + 2\sqrt{3}\,xy - y^2 = 4$, $\phi = 30°$. Then $x = X \cos 30° - Y \sin 30° = \frac{\sqrt{3}}{2}X - \frac{1}{2}Y$ $= \frac{1}{2}\left(\sqrt{3}X - Y\right)$, and $y = X \sin 30° + Y \cos 30° = \frac{1}{2}X + \frac{\sqrt{3}}{2}Y = \frac{1}{2}(X + \sqrt{3}Y)$. Substituting these values into the equation, we get

$$\left[\frac{1}{2}\left(\sqrt{3}X - Y\right)\right]^2 + 2\sqrt{3}\left[\frac{1}{2}\left(\sqrt{3}X - Y\right)\right]\left[\frac{1}{2}\left(X + \sqrt{3}Y\right)\right] - \left[\frac{1}{2}\left(X + \sqrt{3}Y\right)\right]^2 = 4 \quad \Leftrightarrow$$

$$\left(\sqrt{3}X - Y\right)^2 + 2\sqrt{3}\left(\sqrt{3}X - Y\right)\left(X + \sqrt{3}Y\right) - \left(X + \sqrt{3}Y\right)^2 = 16 \quad \Leftrightarrow$$

$$\left(3X^2 - 2\sqrt{3}\,XY + Y^2\right) + \left(6X^2 + 4\sqrt{3}\,XY - 6Y^2\right) - \left(X^2 + 2\sqrt{3}\,XY + 3Y^2\right) = 16 \quad \Leftrightarrow$$

$$8X^2 - 8Y^2 = 16 \quad \Leftrightarrow \quad \frac{X^2}{2} - \frac{Y^2}{2} = 1. \text{ This is a hyperbola.}$$

11. (a) $xy = 8 \quad \Leftrightarrow \quad 0x^2 + xy + 0y^2 = 8$. So $A = 0, B = 1, C = 0$, and so the discriminant is $B^2 - 4AC = 1^2 - 4(0)(0) = 1$. Since the discriminant is positive the equation represents a hyperbola.

(b) $\cot 2\phi = \frac{A-C}{B} = 0 \quad \Rightarrow \quad 2\phi = 90° \quad \Leftrightarrow \quad \phi = 45°$. Therefore, $x = \frac{X}{\sqrt{2}} - \frac{Y}{\sqrt{2}}$ and $y = \frac{X}{\sqrt{2}} + \frac{Y}{\sqrt{2}}$. After substitution, the original equation becomes $\left(\frac{X}{\sqrt{2}} - \frac{Y}{\sqrt{2}}\right)\left(\frac{X}{\sqrt{2}} + \frac{Y}{\sqrt{2}}\right) = 8$

$\Leftrightarrow \quad \frac{(X - Y)(X + Y)}{2} = 8 \quad \Leftrightarrow \quad \frac{X^2}{16} - \frac{Y^2}{16} = 1$. This is a hyperbola with $a = 4, b = 4$, and $c = 4\sqrt{2}$. Hence, the vertices are at $V(\pm 4, 0)$ and the foci are at $F(\pm 4\sqrt{2}, 0)$.

(c)

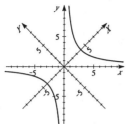

13. (a) $x^2 + 2xy + y^2 + x - y = 0$. So $A = 1$, $B = 2$, $C = 1$, and so the discriminant is $B^2 - 4AC = 2^2 - 4(1)(1) = 0$. Since the discriminant is zero the equation represents a parabola.

(b) $\cot 2\phi = \frac{A-C}{B} = 0 \Rightarrow 2\phi = 90° \Leftrightarrow \phi = 45°$. Therefore, $x = \frac{X}{\sqrt{2}} - \frac{Y}{\sqrt{2}}$ and $y = \frac{X}{\sqrt{2}} + \frac{Y}{\sqrt{2}}$. After substitution, the original equation becomes:

$$\left(\frac{X}{\sqrt{2}} - \frac{Y}{\sqrt{2}}\right)^2 + 2\left(\frac{X}{\sqrt{2}} - \frac{Y}{\sqrt{2}}\right)\left(\frac{X}{\sqrt{2}} + \frac{Y}{\sqrt{2}}\right) + \left(\frac{X}{\sqrt{2}} + \frac{Y}{\sqrt{2}}\right)^2 + \left(\frac{X}{\sqrt{2}} - \frac{Y}{\sqrt{2}}\right) - \left(\frac{X}{\sqrt{2}} + \frac{Y}{\sqrt{2}}\right) = 0$$

$$\Leftrightarrow \frac{X^2}{2} - XY + \frac{Y^2}{2} + X^2 - Y^2 + \frac{X^2}{2} + XY + \frac{Y^2}{2} - \sqrt{2}Y = 0 \Leftrightarrow 2X^2 - \sqrt{2}Y$$

$$\Leftrightarrow X^2 = \frac{Y}{\sqrt{2}}.$$ This is a parabola that has $4p = \frac{1}{\sqrt{2}}$ and hence the focus is at $F\left(0, \frac{1}{4\sqrt{2}}\right)$.

(c)

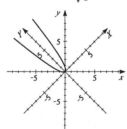

15. (a) $x^2 + 2\sqrt{3}xy - y^2 + 2 = 0$. So $A = 1$, $B = 2\sqrt{3}$, $C = -1$, and so the discriminant is $B^2 - 4AC = (2\sqrt{3})^2 - 4(1)(-1) > 0$. Since the discriminant is positive the equation represents a hyperbola.

(b) $\cot 2\phi = \frac{A-C}{B} = \frac{1+1}{2\sqrt{3}} = \frac{1}{\sqrt{3}} \Rightarrow 2\phi = 60° \Leftrightarrow \phi = 30°$. Therefore, $x = \frac{\sqrt{3}X}{2} - \frac{Y}{2}$ and $y = \frac{X}{2} + \frac{\sqrt{3}Y}{2}$. After substitution, the original equation becomes:

$$\left(\frac{\sqrt{3}X}{2} - \frac{Y}{2}\right)^2 + 2\sqrt{3}\left(\frac{\sqrt{3}X}{2} - \frac{Y}{2}\right)\left(\frac{X}{2} + \frac{\sqrt{3}Y}{2}\right) - \left(\frac{X}{2} + \frac{\sqrt{3}Y}{2}\right)^2 + 2 = 0 \Leftrightarrow$$

$$\frac{3X^2}{4} - \frac{2\sqrt{3}XY}{4} + \frac{Y^2}{4} + \frac{\sqrt{3}}{2}\left(\sqrt{3}X^2 + 2XY - \sqrt{3}Y^2\right) - \frac{X^2}{4} - \frac{2\sqrt{3}XY}{4} - \frac{3Y^2}{4} + 2 = 0 \Leftrightarrow$$

$$X^2\left(\frac{3}{4} + \frac{3}{2} - \frac{1}{4}\right) + XY\left(-\frac{\sqrt{3}}{2} + \sqrt{3} - \frac{\sqrt{3}}{2}\right) + Y^2\left(\frac{1}{4} - \frac{3}{2} - \frac{3}{4}\right) = -2 \Leftrightarrow$$

$$2X^2 - 2Y^2 = -2 \Leftrightarrow Y^2 - X^2 = 1.$$

(c)

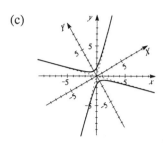

17. (a) $11x^2 - 24xy + 4y^2 + 20 = 0$. So $A = 11$, $B = -24$, $C = 4$, and so the discriminant is $B^2 - 4AC = (-24)^2 - 4(11)(4) > 0$. Since the discriminant is positive the equation represents a hyperbola.

(b) $\cot 2\phi = \frac{A-C}{B} = \frac{11-4}{-24} = \frac{-7}{24}$ $\Rightarrow$ $\cos 2\phi = \frac{-7}{25}$. Therefore, $\cos \phi = \sqrt{\frac{1+(-7/25)}{2}} = \frac{3}{5}$ and $\sin \phi = \sqrt{\frac{1-(-7/25)}{2}} = \frac{4}{5}$. Hence, $x = \frac{3X}{5} - \frac{4Y}{5}$ and $y = \frac{4X}{5} + \frac{3Y}{5}$. After substitution, the original equation becomes:

$11\left(\frac{3X}{5} - \frac{4Y}{5}\right)^2 - 24\left(\frac{3X}{5} - \frac{4Y}{5}\right)\left(\frac{4X}{5} + \frac{3Y}{5}\right) + 4\left(\frac{4X}{5} + \frac{3Y}{5}\right)^2 + 20 = 0$ $\Leftrightarrow$

$\frac{11}{25}(9X^2 - 24XY + 16Y^2) - \frac{24}{25}(12X^2 - 7XY - 12Y^2) + \frac{4}{25}(16X^2 + 24XY + 9Y^2) + 20 = 0$ $\Leftrightarrow$ $X^2(99 - 288 + 64) + XY(-264 + 168 + 96) + Y^2(176 + 288 + 36) = -500$ $\Leftrightarrow$ $-125X^2 + 500Y^2 = -500$ $\Leftrightarrow$ $\frac{X^2}{4} - Y^2 = 1$.

(c)

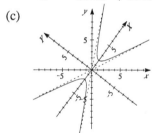

Since $\cos 2\phi = -\frac{7}{25}$, we have $2\phi = \cos^{-1}\left(-\frac{7}{25}\right) \approx 106.26°$ and so $\phi \approx 53°$.

19. (a) $\sqrt{3}x^2 + 3xy = 3$. So $A = \sqrt{3}$, $B = 3$, $C = 0$, and so the discriminant is $B^2 - 4AC = (3)^2 - 4(\sqrt{3})(0) = 9$. Since the discriminant is positive the equation represents a hyperbola.

(b) $\cot 2\phi = \frac{A-C}{B} = \frac{1}{\sqrt{3}}$ $\Rightarrow$ $2\phi = 60°$ $\Leftrightarrow$ $\phi = 30°$. Therefore, $x = \frac{\sqrt{3}X}{2} - \frac{Y}{2}$ and $y = \frac{X}{2} + \frac{\sqrt{3}Y}{2}$. After substitution, the equation becomes

$\sqrt{3}\left(\frac{\sqrt{3}X}{2} - \frac{Y}{2}\right)^2 + 3\left(\frac{\sqrt{3}X}{2} - \frac{Y}{2}\right)\left(\frac{X}{2} + \frac{\sqrt{3}Y}{2}\right) = 3$ $\Leftrightarrow$

$\frac{\sqrt{3}}{4}(3X^2 - 2\sqrt{3}XY + Y^2) + \frac{3}{4}(\sqrt{3}X^2 + 2XY - \sqrt{3}Y^2) = 3$ $\Leftrightarrow$

$X^2\left(\frac{3\sqrt{3}}{4} + \frac{3\sqrt{3}}{4}\right) + XY\left(\frac{-6}{4} + \frac{6}{4}\right) + Y^2\left(\frac{\sqrt{3}}{4} - \frac{3\sqrt{3}}{4}\right) = 3$ $\Leftrightarrow$ $\frac{3\sqrt{3}X^2}{2} - \frac{\sqrt{3}Y^2}{2} = 3$ $\Leftrightarrow$

$\frac{\sqrt{3}X^2}{2} - \frac{Y^2}{2\sqrt{3}} = 1$. This is a hyperbola that has $a = \sqrt{\frac{2}{\sqrt{3}}}$ and $b = \sqrt{2\sqrt{3}}$.

(c)

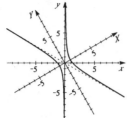

21. (a) $2\sqrt{3}x^2 - 6xy + \sqrt{3}x + 3y = 0$. So $A = 2\sqrt{3}$, $B = -6$, and $C = 0$, and so the discriminant is $B^2 - 4AC = (-6)^2 - 4(2\sqrt{3})(0) = 36$. Since the discriminant is positive the equation represents a hyperbola.

(b) $\cot 2\phi = \frac{A-C}{B} = \frac{2\sqrt{3}}{-6} = -\frac{1}{\sqrt{3}} \Rightarrow 2\phi = 120° \Leftrightarrow \phi = 60°$. Therefore,

$x = \frac{X}{2} - \frac{\sqrt{3}Y}{2}$ and $y = \frac{\sqrt{3}X}{2} + \frac{Y}{2}$, and substituting gives

$2\sqrt{3}\left(\frac{X}{2} - \frac{\sqrt{3}Y}{2}\right)^2 - 6\left(\frac{X}{2} - \frac{\sqrt{3}Y}{2}\right)\left(\frac{\sqrt{3}X}{2} + \frac{Y}{2}\right) + \sqrt{3}\left(\frac{X}{2} - \frac{\sqrt{3}Y}{2}\right) + 3\left(\frac{\sqrt{3}X}{2} + \frac{Y}{2}\right) = 0$

$\Leftrightarrow \frac{\sqrt{3}}{2}(X^2 - 2\sqrt{3}XY + 3Y^2) - \frac{3}{2}(\sqrt{3}X^2 - 2XY - \sqrt{3}Y^2) + \frac{\sqrt{3}}{2}(X - \sqrt{3}Y) +$

$\frac{3}{2}(\sqrt{3}X + Y) = 0 \Leftrightarrow$

$X^2\left(\frac{\sqrt{3}}{2} - \frac{3\sqrt{3}}{2}\right) + X\left(\frac{\sqrt{3}}{2} + \frac{3\sqrt{3}}{2}\right) + XY(-3 + 3) + Y^2\left(\frac{3\sqrt{3}}{2} + \frac{3\sqrt{3}}{2}\right) + Y\left(-\frac{3}{2} + \frac{3}{2}\right) = 0$

$\Leftrightarrow -\sqrt{3}X^2 + 2\sqrt{3}X + 3\sqrt{3}Y^2 = 0 \Leftrightarrow -X^2 + 2X + 3Y^2 = 0 \Leftrightarrow$

$3Y^2 - (X^2 - 2X + 1) = -1 \Leftrightarrow (X-1)^2 - 3Y^2 = 1$. This is a hyperbola that has

$a = 1, b = \frac{1}{\sqrt{3}}, c = \sqrt{1 + \frac{1}{3}} = \frac{2}{\sqrt{3}}$, and $C(1, 0)$.

(c)

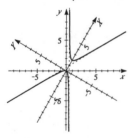

As in Exercise 17, we get $\phi \approx 53°$.

23. (a) $52x^2 + 72xy + 73y^2 = 40x - 30y + 75$. So $A = 52$, $B = 72$, $C = 73$, and so the discriminant is $B^2 - 4AC = (72)^2 - 4(52)(73) = -10000$. Since the discriminant is negative the equation represents an ellipse.

(b) $\cot 2\phi = \frac{A-C}{B} = \frac{52-73}{72} = \frac{-7}{24}$. Therefore, as in Exercise 17(b), we get $\cos \phi = \frac{3}{5}$, $\sin \phi = \frac{4}{5}$, and $x = \frac{3X}{5} - \frac{4Y}{5}$, $y = \frac{4X}{5} + \frac{3Y}{5}$. By substitution,

$52\left(\frac{3X}{5} - \frac{4Y}{5}\right)^2 + 72\left(\frac{3X}{5} - \frac{4Y}{5}\right)\left(\frac{4X}{5} + \frac{3Y}{5}\right) + 73\left(\frac{4X}{5} + \frac{3Y}{5}\right)^2$

$= 40\left(\frac{3X}{5} - \frac{4Y}{5}\right) - 30\left(\frac{4X}{5} + \frac{3Y}{5}\right) + 75 \Leftrightarrow$

$\frac{52}{25}(9X^2 - 24XY + 16Y^2) + \frac{72}{25}(12X^2 - 7XY - 12Y^2) + \frac{73}{25}(16X^2 + 24XY + 9Y^2)$

$= 24X - 32Y - 24X - 18Y + 75 \Leftrightarrow$

$468X^2 + 832Y^2 + 864X^2 - 864Y^2 + 1168X^2 + 657Y^2 = -1250Y + 1875 \Leftrightarrow$

$2500X^2 + 625Y^2 + 1250Y = 1875 \Leftrightarrow 100X^2 + 25Y^2 + 50Y = 75 \Leftrightarrow$

$X^2 + \frac{(Y+1)^2}{4} = 1$. This is an ellipse with $a = 2$, $b = 1$, $c = \sqrt{4-1} = \sqrt{3}$, and center $C(0, -1)$.

(c)

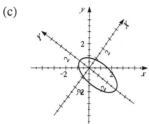

25. (a) $7x^2 + 48xy - 7y^2 - 200x - 150y + 600 = 0$. Then, $A = 7$, $B = 48$, and $C = -7$, and so the discriminant is $B^2 - 4AC = (48)^2 - 4(7)(7) > 0$. Since the discriminant is positive the equation represents a hyperbola.

We now find the equation in terms of XY-coordinates. We have $\cot 2\phi = \frac{A-C}{B} = \frac{7}{24}$ $\Rightarrow$ $\cos\phi = \frac{4}{5}$ and $\sin\phi = \frac{3}{5}$. Therefore, $x = \frac{4X}{5} - \frac{3Y}{5}$ and $y = \frac{3X}{5} + \frac{4Y}{5}$, and substitution gives

$7\left(\frac{4X}{5} - \frac{3Y}{5}\right)^2 + 48\left(\frac{4X}{5} - \frac{3Y}{5}\right)\left(\frac{3X}{5} + \frac{4Y}{5}\right) - 7\left(\frac{3X}{5} + \frac{4Y}{5}\right)^2 - 200\left(\frac{4X}{5} - \frac{3Y}{5}\right)$

$\qquad\qquad\qquad\qquad\qquad\qquad\qquad\qquad\qquad - 150\left(\frac{3X}{5} + \frac{4Y}{5}\right) + 600 = 0 \quad \Leftrightarrow$

$\frac{7}{25}(16X^2 - 24XY + 9Y^2) + \frac{48}{25}(12X^2 + 7XY - 12Y^2) - \frac{7}{25}(9X^2 + 24XY + 16Y^2)$

$\qquad\qquad\qquad\qquad\qquad\qquad -160X + 120Y - 90X - 120Y + 600 = 0 \quad \Leftrightarrow$

$112X^2 - 168XY + 63Y^2 + 576X^2 + 336XY - 576Y^2 - 63X^2$

$\qquad\qquad\qquad\qquad\qquad\qquad - 168XY - 112Y^2 - 6250X + 15{,}000 = 0 \quad \Leftrightarrow$

$25X^2 - 25Y^2 - 250X + 600 = 0 \quad \Leftrightarrow \quad 25(X^2 - 10X + 25) - 25Y^2 = -600 + 625$

$\Leftrightarrow \quad (X-5)^2 - Y^2 = 1$. This is a hyperbola that has $a = 1$, $b = 1$, $c = \sqrt{1+1} = \sqrt{2}$, and its center at $C(5, 0)$.

(b) In the XY-plane, the center is $C(5, 0)$, the vertices are $V(5 \pm 1, 0) = V_1(4, 0)$ and $V_2(6, 0)$, and the foci are $F(5 \pm \sqrt{2}, 0)$. In the xy-plane, the center is
$C\left(\frac{4}{5} \cdot 5 - \frac{3}{5} \cdot 0, \; \frac{3}{5} \cdot 5 + \frac{4}{5} \cdot 0\right) = C(4, 3)$, the vertices are
$V_1\left(\frac{4}{5} \cdot 4 - \frac{3}{5} \cdot 0, \; \frac{3}{5} \cdot 4 + \frac{4}{5} \cdot 0\right) = V_1\left(\frac{16}{5}, \frac{12}{5}\right)$ and $V_2\left(\frac{4}{5} \cdot 6 - \frac{3}{5} \cdot 0, \; \frac{3}{5} \cdot 6 + \frac{4}{5} \cdot 0\right)$
$= V_2\left(\frac{24}{5}, \frac{18}{5}\right)$, and the foci are $F_1\left(4 + \frac{4}{5}\sqrt{2}, \; 3 + \frac{3}{5}\sqrt{2}\right)$ and $F_2\left(4 - \frac{4}{5}\sqrt{2}, \; 3 - \frac{3}{5}\sqrt{2}\right)$.

(c) In the XY-plane the equations of the asymptotes are $Y = X - 5$ and $Y = -X + 5$. In the xy-plane, these equations become $-x \cdot \frac{3}{5} + y \cdot \frac{4}{5} = x \cdot \frac{4}{5} + y \cdot \frac{3}{5} - 5 \quad \Leftrightarrow \quad y = 7x - 25$ and
$-x \cdot \frac{3}{5} + y \cdot \frac{4}{5} = -\left(x \cdot \frac{4}{5} + y \cdot \frac{3}{5}\right) + 5 \quad \Leftrightarrow \quad y = -\frac{1}{7}x + \frac{25}{7}$.

27. We use the hint and eliminate the Y by adding:

$x = X \cos\phi - Y \sin\phi \qquad \Leftrightarrow \qquad\qquad x \cos\phi = X \cos^2\phi - Y \sin\phi \cos\phi$

$y = X \sin\phi + Y \cos\phi \qquad \Leftrightarrow \qquad \underline{\qquad y \sin\phi = X \sin^2\phi + Y \sin\phi \cos\phi}$

$\qquad\qquad\qquad\qquad\qquad\qquad x \cos\phi + y \sin\phi = X(\cos^2\phi + \sin^2\phi)$

$\Leftrightarrow \quad x \cos\phi + y \sin\phi = X$.

In a similar manner we eliminate the X by subtracting:

$$x = X \cos \phi - Y \sin \phi \qquad \Leftrightarrow \qquad -x \sin \phi = -X \cos \phi \sin \phi + Y \sin^2 \phi$$

$$y = X \sin \phi + Y \cos \phi \qquad \Leftrightarrow \qquad \underline{y \cos \phi = X \sin \phi \cos \phi + Y \cos^2 \phi}$$

$$-x \sin \phi + y \cos \phi = Y (\cos^2 \phi + \sin^2 \phi)$$

$$\Leftrightarrow \quad -x \sin \phi + y \cos \phi = Y. \text{ So } X = x \cos \phi + y \sin \phi \text{ and } Y = -x \sin \phi + y \cos \phi.$$

29. $Z = \begin{bmatrix} x \\ y \end{bmatrix}$, $Z' = \begin{bmatrix} X \\ Y \end{bmatrix}$, and $R = \begin{bmatrix} \cos \phi & -\sin \phi \\ \sin \phi & \cos \phi \end{bmatrix}$.

Thus $Z = RZ' \quad \Leftrightarrow \quad \begin{bmatrix} x \\ y \end{bmatrix} = \begin{bmatrix} \cos \phi & -\sin \phi \\ \sin \phi & \cos \phi \end{bmatrix} \begin{bmatrix} X \\ Y \end{bmatrix} = \begin{bmatrix} X \cos \phi - Y \sin \phi \\ X \sin \phi + Y \cos \phi \end{bmatrix}.$

Equating the entries in this matrix equation gives the first pair of rotation of axes formulas. Now

$$R^{-1} = \frac{1}{\cos^2 \phi + \sin^2 \phi} \begin{bmatrix} \cos \phi & \sin \phi \\ -\sin \phi & \cos \phi \end{bmatrix} = \begin{bmatrix} \cos \phi & \sin \phi \\ -\sin \phi & \cos \phi \end{bmatrix} \text{ and so } Z' = R^{-1} Z \quad \Leftrightarrow$$

$$\begin{bmatrix} X \\ Y \end{bmatrix} = \begin{bmatrix} \cos \phi & \sin \phi \\ -\sin \phi & \cos \phi \end{bmatrix} \begin{bmatrix} x \\ y \end{bmatrix} = \begin{bmatrix} x \cos \phi + y \sin \phi \\ -x \sin \phi + y \cos \phi \end{bmatrix}. \text{ Equating the entries in this matrix}$$

equation gives the second pair of rotation of axes formulas.

31. Let P be the point (x_1, y_1) and Q be the point (x_2, y_2) and let $P'(X_1, Y_1)$ and $Q'(X_2, Y_2)$ be the images of P and Q under the rotation of ϕ. So $X_1 = x_1 \cos \phi + y_1 \sin \phi$, $Y_1 = -x_1 \sin \phi + y_1 \cos \phi$, $X_2 = x_2 \cos \phi + y_2 \sin \phi$, and $Y_2 = -x_2 \sin \phi + y_2 \cos \phi$. Thus $d(P', Q') = \sqrt{(X_2 - X_1)^2 + (Y_2 - Y_1)^2}$ where

$$(X_2 - X_1)^2 = [(x_2 \cos \phi + y_2 \sin \phi) - (x_1 \cos \phi + y_1 \sin \phi)]^2 = [(x_2 - x_1) \cos \phi + (y_2 - y_1) \sin \phi]^2$$

$$= (x_2 - x_1)^2 \cos^2 \phi + (x_2 - x_1)(y_2 - y_1) \sin \phi \cos \phi + (y_2 - y_1)^2 \sin^2 \phi \text{ and}$$

$$(Y_2 - Y_1)^2 = [(-x_2 \sin \phi + y_2 \cos \phi) - (-x_1 \sin \phi + y_1 \cos \phi)]^2$$

$$= [-(x_2 - x_1) \sin \phi + (y_2 - y_1) \cos \phi]^2$$

$$= (x_2 - x_1)^2 \sin^2 \phi - (x_2 - x_1)(y_2 - y_1) \sin \phi \cos \phi + (y_2 - y_1)^2 \cos^2 \phi$$

So $(X_2 - X_1)^2 + (Y_2 - Y_1)^2 = (x_2 - x_1)^2 \cos^2 \phi + (x_2 - x_1)(y_2 - y_1) \sin \phi \cos \phi +$

$(y_2 - y_1)^2 \sin^2 \phi + (x_2 - x_1)^2 \sin^2 \phi - (x_2 - x_1)(y_2 - y_1) \sin \phi \cos \phi + (y_2 - y_1)^2 \cos^2 \phi$

$$= (x_2 - x_1)^2 \cos^2 \phi + (y_2 - y_1)^2 \sin^2 \phi + (x_2 - x_1)^2 \sin^2 \phi + (y_2 - y_1)^2 \cos^2 \phi$$

$$= (x_2 - x_1)^2 (\cos^2 \phi + \sin^2 \phi) + (y_2 - y_1)^2 (\sin^2 \phi + \cos^2 \phi) = (x_2 - x_1)^2 + (y_2 - y_1)^2. \text{ Putting}$$

these equations together gives

$$d(P', Q') = \sqrt{(X_2 - X_1)^2 + (Y_2 - Y_1)^2} = \sqrt{(x_2 - x_1)^2 + (y_2 - y_1)^2} = d(P, Q).$$

Exercises 11.6

Solutions to Exercises 1 − 5 will vary. Some possible solutions are given.

1. $(3, \frac{\pi}{2})$ also has polar coordinates $(3, \frac{5\pi}{2})$ or $(-3, \frac{3\pi}{2})$.

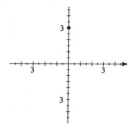

3. $(-1, \frac{7\pi}{6})$ also has polar coordinates $(1, \frac{\pi}{6})$ or $(-1, -\frac{5\pi}{6})$.

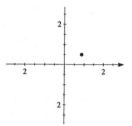

5. $(-5, 0)$ also has polar coordinates $(5, \pi)$ or $(-5, 2\pi)$.

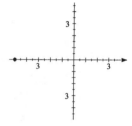

7. $(r, \theta) = (4, \frac{\pi}{6})$. So $x = r \cos\theta = 4\cos\frac{\pi}{6} = 4 \cdot \frac{\sqrt{3}}{2} = 2\sqrt{3}$ and $y = r\sin\theta = 4\sin\frac{\pi}{6} = 4 \cdot \frac{1}{2} = 2$. Thus, the rectangular coordinates are $(2\sqrt{3}, 2)$.

9. $(r, \theta) = (\sqrt{2}, -\frac{\pi}{4})$. So $x = r\cos\theta = \sqrt{2}\cos(-\frac{\pi}{4}) = \sqrt{2} \cdot \frac{1}{\sqrt{2}} = 1$, and $y = r\sin\theta = \sqrt{2}\sin(-\frac{\pi}{4})$
    $= \sqrt{2} \cdot \frac{-1}{\sqrt{2}} = -1$. Thus, the rectangular coordinates are $(1, -1)$.

11. $(r, \theta) = (5, 5\pi)$. So $x = r\cos\theta = 5\cos 5\pi = -5$, and $y = r\sin\theta = 5\sin 5\pi = 0$. Thus, the rectangular coordinates are $(-5, 0)$.

13. $(x, y) = (-1, 1)$. Since $r^2 = x^2 + y^2$, we have $r^2 = (-1)^2 + 1^2 = 2$, so $r = \sqrt{2}$. Now $\tan\theta = \frac{y}{x} = \frac{1}{-1} = -1$, so, since the point is in the second quadrant, $\theta = \frac{3\pi}{4}$. Thus, the polar coordinates are $(\sqrt{2}, \frac{3\pi}{4})$.

15. $(x, y) = (\sqrt{8}, \sqrt{8})$. Since $r^2 = x^2 + y^2$, we have $r^2 = (\sqrt{8})^2 + (\sqrt{8})^2 = 16$, so $r = 4$. Now $\tan\theta = \frac{y}{x} = \frac{\sqrt{8}}{\sqrt{8}} = 1$, so, since the point is in the first quadrant, $\theta = \frac{\pi}{4}$. Thus, the polar coordinates are $(4, \frac{\pi}{4})$.

17. $(x, y) = (3, 4)$. Since $r^2 = x^2 + y^2$, we have $r^2 = 3^2 + 4^2 = 25$, so $r = 5$. Now $\tan\theta = \dfrac{y}{x} = \dfrac{4}{3}$, so, since the point is in the first quadrant, $\theta = \tan^{-1}\frac{4}{3}$. Thus, the polar coordinates are $(5, \tan^{-1}\frac{4}{3})$.

19. $x = y \quad \Leftrightarrow \quad r\cos\theta = r\sin\theta \quad \Leftrightarrow \quad \tan\theta = 1$, and so $\theta = \frac{\pi}{4}$.

21. $y = x^2$. We substitute and then solve for r: $r\sin\theta = (r\cos\theta)^2 = r^2\cos^2\theta \quad \Leftrightarrow \quad \sin\theta = r\cos^2\theta$
$\Leftrightarrow \quad r = \dfrac{\sin\theta}{\cos^2\theta} = \tan\theta\sec\theta$.

23. $x = 4$. We substitute and then solve for r: $r\cos\theta = 4 \quad \Leftrightarrow \quad r = \dfrac{4}{\cos\theta} = 4\sec\theta$.

25. $r = 7$. But $r^2 = x^2 + y^2$, so $x^2 + y^2 = r^2 = 49$. Hence, the equivalent equation in rectangular coordinates is $x^2 + y^2 = 49$.

27. $r\cos\theta = 6$. But $x = r\cos\theta$, and so $x = 6$ is the equation.

29. $r^2 = \tan\theta$. Substituting $r^2 = x^2 + y^2$ and $\tan\theta = \dfrac{y}{x}$, we get $x^2 + y^2 = \dfrac{y}{x}$.

31. $r = \dfrac{1}{\sin\theta - \cos\theta} \quad \Rightarrow \quad r(\sin\theta - \cos\theta) = 1 \quad \Leftrightarrow \quad r\sin\theta - r\cos\theta = 1$, and since $r\cos\theta = x$ and $r\sin\theta = y$, we get $y - x = 1$.

33. $r = 1 + \cos\theta$. If we multiply both sides of this equation by r we get $r^2 = r + r\cos\theta$. Thus $r^2 - r\cos\theta = r$, and squaring both sides gives $(r^2 - r\cos\theta)^2 = r^2 \quad \Leftrightarrow$
$(x^2 + y^2 - x)^2 = x^2 + y^2$

35. $r = 2\sec\theta \quad \Leftrightarrow \quad r = 2 \cdot \dfrac{1}{\cos\theta} \quad \Leftrightarrow \quad r\cos\theta = 2 \quad \Leftrightarrow \quad x = 2$.

37. $\sec\theta = 2 \quad \Leftrightarrow \quad \cos\theta = \frac{1}{2} \quad \Leftrightarrow \quad \theta = \pm\frac{\pi}{3} \quad \Leftrightarrow \quad \tan\theta = \pm\sqrt{3} \quad \Leftrightarrow \quad \dfrac{y}{x} = \pm\sqrt{3} \quad \Leftrightarrow$
$y = \pm\sqrt{3}x$.

39. $r = 3$. Circle.

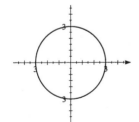

41. $\theta = -\frac{\pi}{2}$. Line.

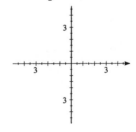

43. $r = 6\sin\theta$. Circle.

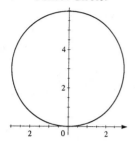

45. $r = -2\cos\theta$. Circle.

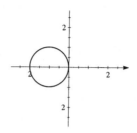

47. $r = 2 - 2\cos\theta$. Cardioid.

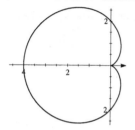

49. $r = -3\left(1 + \sin\theta\right)$. Cardioid.

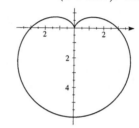

51. $r = \theta, \theta \geq 0$

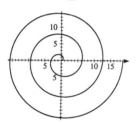

53. $r = \sin 2\theta$

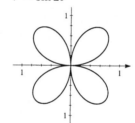

55. $r^2 = \cos 2\theta$

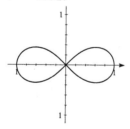

57. $r = 2 + \sin\theta$

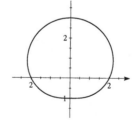

59. $r = 2 + \sec\theta$

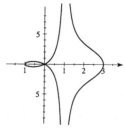

61. $r = \cos\left(\frac{\theta}{2}\right)$; Domain: $[0, 4\pi]$

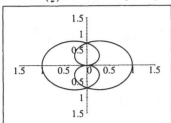

63. $r = 1 + 2\sin\left(\frac{\theta}{2}\right)$; Domain: $[0, 4\pi]$

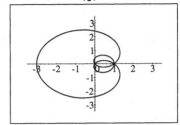

65. $r = 1 + \sin n\theta$. The number of loops is n.

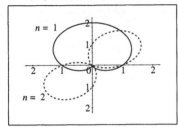
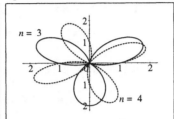
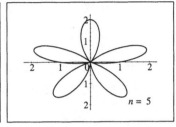

67. The graph of $r = \sin\left(\frac{\theta}{2}\right)$ is IV, since the graph must contain the points $(0, 0)$, $\left(\frac{1}{\sqrt{2}}, \frac{\pi}{2}\right)$, $(1, \pi)$, and so on.

69. The graph of $r = \theta\sin\theta$ is III, since for $\theta = \frac{\pi}{2}, \frac{5\pi}{2}, \frac{7\pi}{2}, \ldots$ the values of r are also $\frac{\pi}{2}, \frac{5\pi}{2}, \frac{7\pi}{2}, \ldots$. Thus the graph must cross the vertical axis at an infinite number of points.

71. (a) In rectangular coordinates, the points (r_1, θ_1) and (r_2, θ_2) are $(x_1, y_1) = (r_1\cos\theta_1, r_1\sin\theta_1)$ and $(x_2, y_2) = (r_2\cos\theta_2, r_2\sin\theta_2)$. Then, the distance between the points is
$$D = \sqrt{(x_1 - x_2)^2 + (y_1 - y_2)^2} = \sqrt{(r_1\cos\theta_1 - r_2\cos\theta_2)^2 + (r_1\sin\theta_1 - r_2\sin\theta_2)^2} =$$
$$\sqrt{r_1^2\left(\cos^2\theta_1 + \sin^2\theta_1\right) + r_2^2\left(\cos^2\theta_2 + \sin^2\theta_2\right) - 2r_1r_2\left(\cos\theta_1\cos\theta_2 + \sin\theta_1\sin\theta_2\right)} =$$
$$\sqrt{r_1^2 + r_2^2 - 2r_1r_2\cos(\theta_1 - \theta_2)}$$

(b) The distance between the points $\left(3, \frac{3\pi}{4}\right)$ and $\left(-1, \frac{7\pi}{6}\right)$ is
$$D = \sqrt{3^2 + (-1)^2 - 2(3)(-1)\cos\left(\frac{3\pi}{4} - \frac{7\pi}{6}\right)} = \sqrt{9 + 1 + 6\cos\left(-\frac{5\pi}{12}\right)} =$$
$$\sqrt{10 + 6\cos\frac{5\pi}{12}} \approx 3.40.$$

73. The graphs of $r = 1 + \sin\left(\theta - \frac{\pi}{6}\right)$ and $r = 1 + \sin\left(\theta - \frac{\pi}{3}\right)$ have the same shape as $r = 1 + \sin\theta$, rotated through angles of $\frac{\pi}{6}$ and $\frac{\pi}{3}$, respectively. Similarly, the graph of $r = f(\theta - \alpha)$ is the graph of $r = f(\theta)$ rotated by the angle α.

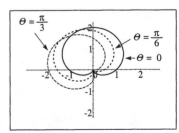

75. $y = 2 \quad \Leftrightarrow \quad r\sin\theta = 2 \quad \Leftrightarrow \quad r = 2\csc\theta$. The rectangular coordinate system gives the simpler equation here. It is easier to study lines in rectangular coordinates.

Exercises 11.7

1. Substituting $e = \frac{2}{3}$ and $d = 3$ into the general equation of a conic with vertical directrix, we get
$$r = \frac{\frac{2}{3} \cdot 3}{1 + \frac{2}{3}\cos\theta} \quad \Leftrightarrow \quad r = \frac{6}{3 + 2\cos\theta}.$$

3. Substituting $e = 1$ and $d = 2$ into the general equation of a conic with horizontal directrix, we get
$$r = \frac{1 \cdot 2}{1 + \sin\theta} \quad \Leftrightarrow \quad r = \frac{2}{1 + \sin\theta}.$$

5. $r = 5\sec\theta \quad \Leftrightarrow \quad r\cos\theta = 5 \quad \Leftrightarrow \quad x = 5$. So $d = 5$ and $e = 4$ gives $r = \dfrac{4 \cdot 5}{1 + 4\cos\theta} \quad \Leftrightarrow$
$$r = \frac{20}{1 + 4\cos\theta}.$$

7. Since this is a parabola whose focus is at the origin and vertex at $(5, \pi/2)$, the directrix must be $y = 10$. So $d = 10$ and $e = 1$ gives $r = \dfrac{1 \cdot 10}{1 + \sin\theta} = \dfrac{10}{1 + \sin\theta}$.

9. (a) $r = \dfrac{4}{1 + 3\cos\theta} \quad \Rightarrow \quad e = 3$, thus the conic is a hyperbola.

   (b) The vertices occur where $\theta = 0$ and $\theta = \pi$.

   Now $\theta = 0 \quad \Rightarrow \quad r = \dfrac{4}{1 + 3\cos 0} = 1$,

   and $\theta = \pi \quad \Rightarrow \quad r = \dfrac{4}{1 + 3\cos\pi} = \dfrac{4}{-2} = -2$.

   Thus the vertices are $(1, 0)$ and $(-2, \pi)$.

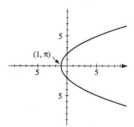

11. (a) $r = \dfrac{2}{1 - \cos\theta} \quad \Rightarrow \quad e = 1$, thus the conic is a parabola.

    (b) Substituting $\theta = \pi$, we have

    $r = \dfrac{2}{1 - \cos\pi} = \dfrac{2}{2} = 1$. Thus the vertex is $(1, \pi)$.

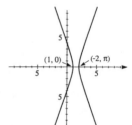

13. (a) $r = \dfrac{6}{2 + \sin\theta} \quad \Leftrightarrow \quad r = \dfrac{\frac{1}{2} \cdot 6}{1 + \frac{1}{2}\sin\theta} \quad \Rightarrow \quad e = \frac{1}{2}$, thus the conic is an ellipse.

(b)   The vertices occur where $\theta = \frac{\pi}{2}$ and $\theta = \frac{3\pi}{2}$.

Now $\theta = \frac{\pi}{2}$ $\Rightarrow$ $r = \dfrac{6}{2 + \sin\frac{\pi}{2}} = \frac{6}{3} = 2$,

and $\theta = \frac{3\pi}{2}$ $\Rightarrow$ $r = \dfrac{6}{2 + \sin\frac{3\pi}{2}} = \frac{6}{1} = 6$.

Thus the vertices are $\left(2,\frac{\pi}{2}\right)$ and $\left(6,\frac{3\pi}{2}\right)$.

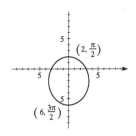

15.   (a) $r = \dfrac{7}{2 - 5\sin\theta}$ $\Leftrightarrow$ $r = \dfrac{\frac{7}{2}}{1 - \frac{5}{2}\sin\theta}$ $\Rightarrow$ $e = \frac{5}{2}$, thus this conic is a hyperbola.

(b)   The vertices occur where $\theta = \frac{\pi}{2}$ and $\theta = \frac{3\pi}{2}$.

$r = \dfrac{7}{2 - 5\sin\frac{\pi}{2}} = \frac{7}{-3} = -\frac{7}{3}$, and $\theta = \frac{3\pi}{2}$

$\Rightarrow$ $r = \dfrac{7}{2 - 5\sin\frac{3\pi}{2}} = \frac{7}{7} = 1$.

Thus the vertices are $\left(-\frac{7}{3}, \frac{\pi}{2}\right)$ and $\left(1, \frac{3\pi}{2}\right)$.

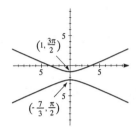

17.   (a) $r = \dfrac{1}{4 - 3\cos\theta}$ $\Leftrightarrow$ $r = \dfrac{\frac{1}{4}}{1 - \frac{3}{4}\cos\theta}$

$\Rightarrow$ $e = \frac{3}{4}$. Since $ed = \frac{1}{4}$ we have

$d = \dfrac{\frac{1}{4}}{e} = \frac{4}{3} \cdot \frac{1}{4} = \frac{1}{3}$. Therefore the

eccentricity is $\frac{3}{4}$ and the directrix is

$x = -\frac{1}{3}$ $\Leftrightarrow$ $r = -\frac{1}{3}\sec\theta$.

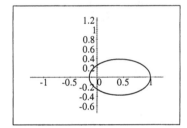

(b)   We replace θ by $\theta - \frac{\pi}{3}$ to get

$r = \dfrac{1}{4 - 3\cos(\theta - \frac{\pi}{3})}$.

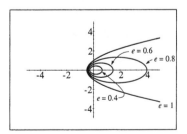

19.   The ellipse is nearly circular when e is close to
0 and becomes more elongated as $e \to 1^-$. At
$e = 1$, the curve becomes a parabola.

21.   (a)   Since the polar form of an ellipse with directrix $x = -d$ is $r = \dfrac{ed}{1 - e\cos\theta}$ we need to show
that $ed = a(1 - e^2)$. From the proof of the Equivalent Description of Conics we have

$a^2 = \dfrac{e^2 d^2}{(1-e^2)^2}$. Since the conic is an ellipse, $e < 1$ and so the variables a, d, and $(1-e^2)$ are

all positive. Thus we can take the square roots of both sides. Thus $a^2 = \dfrac{e^2 d^2}{(1-e^2)^2} \quad \Leftrightarrow$

$a = \dfrac{ed}{1-e^2} \quad \Leftrightarrow \quad ed = a(1-e^2)$. As a result $r = \dfrac{ed}{1-e\cos\theta} \quad \Leftrightarrow \quad r = \dfrac{a(1-e^2)}{1-e\cos\theta}$.

(b)   Since $2a = 2.99 \times 10^8$ we have $a = 1.495 \times 10^8$ so the polar equation for the earth's orbit

(using $e \approx 0.017$) is $r = \dfrac{1.495 \times 10^8 [1-(0.017)^2]}{1-0.017\cos\theta} \approx \dfrac{1.49 \times 10^8}{1-0.017\cos\theta}$.

23.   From Exercise 22, we have:
at perihelion, $r = 4.43 \times 10^9 = a(1-e)$;
at aphelion, $r = 7.37 \times 10^9 = a(1+e)$.

Dividing these equations gives $\dfrac{7.37 \times 10^9}{4.43 \times 10^9} = \dfrac{a(1+e)}{a(1-e)} \quad \Leftrightarrow \quad 1.664 = \dfrac{1+e}{1-e} \quad \Leftrightarrow$

$1.664(1-e) = 1+e \quad \Leftrightarrow \quad 1.664 - 1 = e + 1.664 \quad \Leftrightarrow \quad 0.664 = 2.664e \quad \Leftrightarrow$
$e = \frac{0.664}{2.664} \approx 0.25$.

25.   The r-coordinate of the satellite will be it's distance from the focus (the center of the earth). From
the r-coordinate we can easily calculate the height of the satellite

Exercises 11.8

1. (a) $x = 2t,\ y = t + 6$

    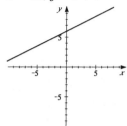

    (b) Since $x = 2t$, $t = \dfrac{x}{2}$ and so $y = \dfrac{x}{2} + 6$

    $\Leftrightarrow\ x - 2y + 12 = 0.$

3. (a) $x = t^2,\ y = t - 2,\ 2 \le t \le 4$

    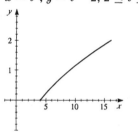

    (b) Since $x = t^2$, $t = \sqrt{x}$ and since $2 \le t \le 4$ we have $4 \le x \le 16$. Thus $y = \sqrt{x} - 2$ with $4 \le x \le 16$.

5. (a) $x = \sqrt{t},\ y = 1 - t\ \Rightarrow\ t \ge 0$

    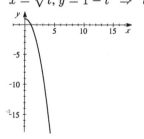

    (b) Since $x = \sqrt{t}$, we have $x^2 = t$, and so $y = 1 - x^2$ with $x \ge 0$.

7. (a) $x = \dfrac{1}{t},\ y = t + 1$

    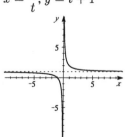

    (b) Since $x = \dfrac{1}{t}$ we have $t = \dfrac{1}{x}$ and so $y = \dfrac{1}{x} + 1.$

9. (a) $x = 4t^2,\ y = 8t^3$

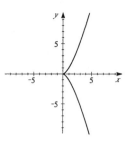

(b) Since $x = 4t^2$ we have
$x\sqrt{x} = 4t^2 \cdot 2t = 8t^3$, and so $y = x\sqrt{x}$.

11. (a) $x = 2\sin t, y = 2\cos t, 0 \le t \le \pi$

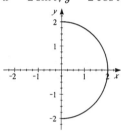

(b) $x^2 = (2\sin t)^2 = 4\sin^2 t$, and
$y^2 = (2\cos t)^2 = 4\cos^2 t$. Hence,
$x^2 + y^2 = 4\sin^2 t + 4\cos^2 t = 4 \quad \Leftrightarrow$
$x^2 + y^2 = 4$, where $x \ge 0$.

13. (a) $x = \sin^2 t, y = \sin^4 t$

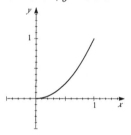

(b) Since $x = \sin^2 t$ we have $x^2 = \sin^4 t$ and
so $y = x^2$. But since $0 \le \sin^2 t \le 1$ we
only get the part of this parabola for
which $0 \le x \le 1$.

15. (a) $x = \cos t, y = \cos 2t$

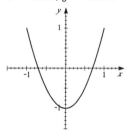

(b) Since $x = \cos t$ we have $x^2 = \cos^2 t$ and
so $2x^2 - 1 = 2\cos^2 t - 1 = \cos 2t = y$.
Hence, the rectangular coordinate
equation is $y = 2x^2 - 1$, with
$-1 \le x \le 1$.

17. (a) $x = \sec t, y = \tan t, 0 \le t < \frac{\pi}{2} \quad \Rightarrow$
$x \ge 1$ and $y \ge 0$.

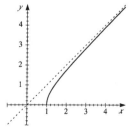

(b) $x^2 = \sec^2 t$, $y^2 = \tan^2 t$, and $y^2 + 1 = \tan^2 t + 1 = \sec^2 t = x^2$. Therefore, $y^2 + 1 = x^2$ $\Leftrightarrow$
     $x^2 - y^2 = 1$, $x \ge 1$, $y \ge 0$.

19. (a) $x = e^t$, $y = e^{-t}$ $\Rightarrow$ $x > 0$, $y > 0$.

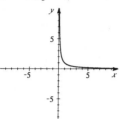

     (b) $xy = e^t \cdot e^{-t} = e^0 = 1$. Hence, the
          equation is $xy = 1$, with $x > 0$, $y > 0$.

21. (a) $x = \cos^2 t$, $y = \sin^2 t$

     (b) $x + y = \cos^2 t + \sin^2 t = 1$. Hence, the
          equation is $x + y = 1$ with
          $0 \le x, y \le 1$.

23. Since the line passes through the point $(4, -1)$ and has slope $\frac{1}{2}$, parametric equations for the line are
     $x = 4 + t$, $y = -1 + \frac{1}{2}t$.

25. Since the line passes through the points $(6, 7)$ and $(7, 8)$, its slope is $\frac{8-7}{7-6} = 1$. Thus, parametric
     equations for the line are $x = 6 + t$, $y = 7 + t$.

27. Since $\cos^2 t + \sin^2 t = 1$, we have $a^2 \cos^2 t + a^2 \sin^2 t = a^2$. If we let $x = a \cos t$ and $y = a \sin t$, then
     $x^2 + y^2 = a^2$. Hence, parametric equations for the circle are $x = a \cos t$, $y = a \sin t$.

29. $x = a \tan \theta$ $\Leftrightarrow$ $\tan \theta = \dfrac{x}{a}$ $\Rightarrow$ $\tan^2 \theta = \dfrac{x^2}{a^2}$. Also, $y = b \sec \theta$ $\Leftrightarrow$ $\sec \theta = \dfrac{y}{b}$ $\Rightarrow$
     $\sec^2 \theta = \dfrac{y^2}{b^2}$. Since $\tan^2 \theta = \sec^2 \theta - 1$, we have $\dfrac{x^2}{a^2} = \dfrac{y^2}{b^2} - 1$ $\Leftrightarrow$ $\dfrac{y^2}{b^2} - \dfrac{x^2}{a^2} = 1$, which is the
     equation of a hyperbola.

31. $x = t \cos t$, $y = t \sin t$, $t \ge 0$

t	x	y
0	0	0
$\frac{\pi}{4}$	$\frac{\pi\sqrt{2}}{8}$	$\frac{\pi\sqrt{2}}{8}$
$\frac{\pi}{2}$	0	$\frac{\pi}{2}$
$\frac{3\pi}{4}$	$\frac{-3\pi\sqrt{2}}{8}$	$\frac{3\pi\sqrt{2}}{8}$
π	$-\pi$	0

t	x	y
$\frac{5\pi}{4}$	$-\frac{5\pi\sqrt{2}}{8}$	$-\frac{5\pi\sqrt{2}}{8}$
$\frac{3\pi}{2}$	0	$-\frac{3\pi}{2}$
$\frac{7\pi}{4}$	$\frac{7\pi\sqrt{2}}{8}$	$-\frac{7\pi\sqrt{2}}{8}$
2π	2π	0

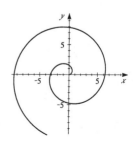

33. $x = \dfrac{3t}{1 + t^3}, y = \dfrac{3t^2}{1 + t^3}, t \neq -1$

t	x	y	t	x	y
−0.90	−9.96	8.97	−1.10	9.97	−10.97
−0.75	−3.89	2.92	−1.25	3.93	−4.92
−0.50	−1.71	0.86	−1.5	1.89	−2.84
0	0	0	−2	0.86	−1.71
0.5	1.33	0.67	−2.5	0.51	−1.28
1	1.5	1.5	−3	0.35	−1.04
1.5	1.03	1.54	−3.5	0.25	−0.88
2	0.67	1.33	−4	0.19	−0.76
2.5	0.45	1.13	−4.5	015	−0.67
3	0.32	0.96	−5	0.12	−0.60
4	0.18	0.74	−6	0.08	−0.50
5	0.12	0.60	−7	0.06	−0.43
6	0.08	0.50	−8	0.05	−0.38

As $t \to -1^-$ we have $x \to -\infty$ and $y \to \infty$.
As $t \to -1^+$ we have $x \to \infty$ and $y \to -\infty$.
As $t \to \infty$ we have $x \to 0^+$ and $y \to 0^+$.
As $t \to -\infty$ we have $x \to 0^+$ and $y \to 0^-$.

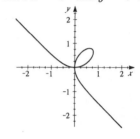

35. $x = (v_0 \cos \alpha)\, t, y = (v_0 \sin \alpha)\, t - 16t^2$. From the equation for x, $t = \dfrac{x}{v_0 \cos \alpha}$. Substituting into

the equation for y gives $y = (v_0 \sin \alpha)\dfrac{x}{v_0 \cos \alpha} - 16\left(\dfrac{x}{v_0 \cos \alpha}\right)^2 = x \tan \alpha - \dfrac{16x^2}{v_0^2 \cos^2\alpha}$. Thus the

equation is of the form $y = c_1 x - c_2 x^2$, where c_1 and c_2 are constants, so its graph is a parabola.

37. $x = \sin t, y = 2 \cos 3t$

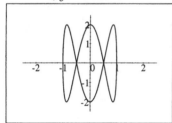

39. $x = 3 \sin 5t, y = 5 \cos 3t$

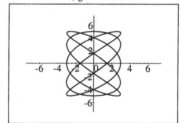

41. $x = \sin(\cos t), y = \cos t^{3/2}, 0 \leq t \leq 2\pi$

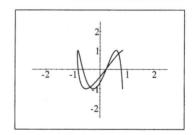

43. (a) $r = e^{\theta/12}$, $0 \le \theta \le 4\pi$ $\Rightarrow$
       $x = e^{t/12} \cdot \cos t$ and $y = e^{t/12} \cdot \sin t$

    (b)

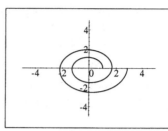

45. (a) $r = \frac{4}{2 - \cos\theta}$ $\Rightarrow$
       $x = \frac{4\cos t}{2 - \cos t}$, $y = \frac{4\sin t}{2 - \cos t}$

    (b)

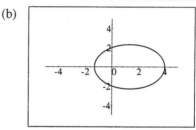

47. This is III, since $y = t^2 - t = (t^2 - t + \frac{1}{4}) - \frac{1}{4} = (t - \frac{1}{2})^2 - \frac{1}{4}$, and so $y \ge -\frac{1}{4}$ on this curve, while x is unbounded.

49. This is II, since the values of x and y should oscillate about their values on the line $x = t$, $y = t$
    $\Leftrightarrow$ $y = x$.

51. If we modify Figure 6 so that $|PC| = b$, then
    by the same reasoning as in Example 5, we see
    that $x = |OT| - |PQ| = a\theta - b\sin\theta$ and
    $y = |TC| - |CQ| = a - b\cos\theta$. We graph
    the case where $a = 3$ and $b = 2$.

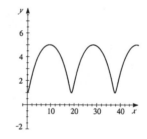

53. (a) We first note that the center of circle C (the small
        circle) has coordinates $([a - b]\cos\theta, [a - b]\sin\theta)$.
        Now the arc PQ has the same length as the arc
        $P'Q$ (these two arcs are shown as thicker lines in
        the figure on the right). Thus $b\phi = a\theta$ $\Leftrightarrow$
        $\phi = \frac{a}{b}\theta$, and so $\phi - \theta = \frac{a}{b}\theta - \theta = \frac{a - b}{b}\theta$. Thus
        the x-coordinate of P is the x-coordinate of the
        center of circle C plus
        $b \cdot \cos(\phi - \theta) = b\cos\left(\frac{a - b}{b}\theta\right)$, and the y-
        coordinate of P is the y-coordinate of the center of
        circle C minus $b \cdot \sin(\phi - \theta) = b\sin\left(\frac{a - b}{b}\theta\right)$.
        So $x = (a - b)\cos\theta + b\cos\left(\frac{a - b}{b}\theta\right)$ and $y = (a - b)\sin\theta - b\sin\left(\frac{a - b}{b}\theta\right)$.

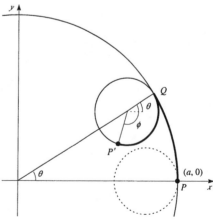

    (b) If $a = 4b$, $b = \frac{a}{4}$, and $x = \frac{3}{4}a\cos\theta + \frac{1}{4}a\cos 3\theta$,

$y = \frac{3}{4}a\sin\theta - \frac{1}{4}a\sin 3\theta$. From Example 2 in Section 7.3, we see that $\cos 3\theta = 4\cos^3\theta - 3\cos\theta$. Similarly, one can prove that $\sin 3\theta = 3\sin\theta - 4\sin^3\theta$. Substituting, we get

$x = \frac{3}{4}a\cos\theta + \frac{1}{4}a(4\cos^3\theta - 3\cos\theta) = a\cos^3\theta$, and

$y = \frac{3}{4}a\sin\theta - \frac{1}{4}a(3\sin\theta - 4\sin^3\theta) = a\sin^3\theta$. Thus,

$x^{2/3} + y^{2/3} = a^{2/3}\cos^2\theta + a^{2/3}\sin^2\theta = a^{2/3}$, so $x^{2/3} + y^{2/3} = a^{2/3}$.

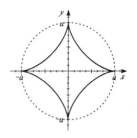

55. The polar coordinate equation for the circle is $r = 2a\sin\theta$. Thus the coordinates of Q are $x = r\cos\theta = 2a\sin\theta\cos\theta$ and $y = r\sin\theta = 2a\sin^2\theta$. The coordinates of R are $x = 2a\cot\theta$ and $y = 2a$. Since P is the midpoint of QR, we use the midpoint formula to get $x = a(\sin\theta\cos\theta + \cot\theta)$ and $y = a(1 + \sin^2\theta)$.

57. We use the equation for y from Example 5 and solve for θ. Thus for $0 \le \theta \le \pi$, $y = a(1 - \cos\theta)$
$\Leftrightarrow \quad \dfrac{a - y}{a} = \cos\theta \quad \Leftrightarrow \quad \theta = \cos^{-1}\left(\dfrac{a - y}{a}\right)$. Substituting into the equation for x, we get:

$x = a\left[\cos^{-1}\left(\dfrac{a - y}{a}\right) - \sin\left(\cos^{-1}\left(\dfrac{a - y}{a}\right)\right)\right]$. However,

$\sin\left(\cos^{-1}\left(\dfrac{a - y}{a}\right)\right) = \sqrt{1 - \left(\dfrac{a - y}{a}\right)^2} = \dfrac{\sqrt{2ay - y^2}}{a}$. Thus

$x = a\left[\cos^{-1}\left(\dfrac{a - y}{a}\right) - \dfrac{\sqrt{2ay - y^2}}{a}\right]$, and we have $\dfrac{\sqrt{2ay - y^2} + x}{a} = \cos^{-1}\left(\dfrac{a - y}{a}\right) \quad \Rightarrow$

$1 - \dfrac{y}{a} = \cos\left(\dfrac{\sqrt{2ay - y^2} + x}{a}\right) \quad \Rightarrow \quad y = a\left[1 - \cos\left(\dfrac{\sqrt{2ay - y^2} + x}{a}\right)\right]$.

Review Exercises for Chapter 11

1. $x^2 + 8y = 0 \iff x^2 = -8y$. This is a parabola that has
   $4p = -8 \iff p = -2$.
   Vertex: $(0,0)$; focus: $(0,-2)$; directrix: $y = 2$.

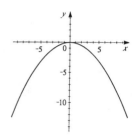

3. $x - y^2 + 4y - 2 = 0 \iff x - (y^2 - 4y + 4) - 2 = -4 \iff$
   $x - (y-2)^2 = -2 \iff (y-2)^2 = x + 2$. This is a parabola
   that has $4p = 1 \iff p = \frac{1}{4}$.
   Vertex: $(-2,2)$; focus: $\left(-2 + \frac{1}{4}, 2\right) = \left(-\frac{7}{4}, 2\right)$; directrix:
   $x = -2 - \frac{1}{4} = -\frac{9}{4}$.

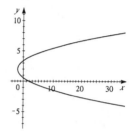

5. $x^2 + 4y^2 = 16 \iff \dfrac{x^2}{16} + \dfrac{y^2}{4} = 1$. This is an ellipse with
   $a = 4$, $b = 2$, and $c = \sqrt{16 - 4} = 2\sqrt{3}$.
   Center: $(0,0)$; vertices: $(\pm 4, 0)$; foci: $(\pm 2\sqrt{3}, 0)$; length of the
   major axis: $2a = 8$; length of the minor axis: $2b = 4$.

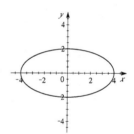

7. $4x^2 + 9y^2 = 36y \iff 4x^2 + 9(y^2 - 4y + 4) = 36 \iff$
   $4x^2 + 9(y-2)^2 = 36 \iff \dfrac{x^2}{9} + \dfrac{(y-2)^2}{4} = 1$. This is an
   ellipse with $a = 3$, $b = 2$, and $c = \sqrt{9 - 4} = \sqrt{5}$.
   Center: $(0,2)$; vertices: $(\pm 3, 2)$; foci: $(\pm\sqrt{5}, 2)$; length of the
   major axis: $2a = 6$; length of the minor axis: $2b = 4$.

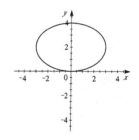

9. $x^2 - 2y^2 = 16 \iff \dfrac{x^2}{16} - \dfrac{y^2}{8} = 1$. This is a hyperbola that has
   $a = 4$, $b = 2\sqrt{2}$, and $c = \sqrt{16 + 8} = \sqrt{24} = 2\sqrt{6}$.
   Center: $(0,0)$; vertices: $(\pm 4, 0)$; foci: $(\pm 2\sqrt{6}, 0)$; asymptotes:
   $y = \pm\dfrac{2\sqrt{2}}{4}x \iff y = \pm\dfrac{1}{\sqrt{2}}x$.

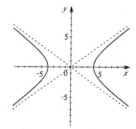

11. $9y^2 + 18y = x^2 + 6x + 18$ $\Leftrightarrow$

$9(y^2 + 2y + 1) = (x^2 + 6x + 9) + 9 - 9 + 18$ $\Leftrightarrow$.

$9(y+1)^2 - (x+3)^2 = 18$ $\Leftrightarrow$ $\dfrac{(y+1)^2}{2} - \dfrac{(x+3)^2}{18} = 1$. This is

a hyperbola that has $a = \sqrt{2}$, $b = 3\sqrt{2}$, and $c = \sqrt{2+18} = 2\sqrt{5}$.

Center: $(-3, -1)$; vertices: $(-3, -1 \pm \sqrt{2})$; foci:

$(-3, -1 \pm 2\sqrt{5})$; asymptotes: $y + 1 = \pm\frac{1}{3}(x+3)$ $\Leftrightarrow$ $y = \frac{1}{3}x$

and $y = -\frac{1}{3}x - 2$.

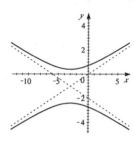

13. This is a parabola that opens to the right with its vertex at $(0,0)$ and the focus at $(2,0)$. So $p = 2$, and the equation is $y^2 = 4(2)x$ $\Leftrightarrow$ $y^2 = 8x$.

15. From the graph, the center is $(0,0)$, and the vertices are $(0,-4)$ and $(0,4)$. Since a is the distance from the center to a vertex, we have $a = 4$. Because one focus is $(0,5)$, we have $c = 5$, and since $c^2 = a^2 + b^2$, we have $25 = 16 + b^2$ $\Leftrightarrow$ $b^2 = 9$. Thus the equation of the hyperbola is $\dfrac{y^2}{16} - \dfrac{x^2}{9} = 1$.

17. From the graph, the center of the ellipse is $(4, 2)$, and so $a = 4$ and $b = 2$. The equation is $\dfrac{(x-4)^2}{4^2} + \dfrac{(y-2)^2}{2^2} = 1$ $\Leftrightarrow$ $\dfrac{(x-4)^2}{16} + \dfrac{(y-2)^2}{4} = 1$.

19. $\dfrac{x^2}{12} + y = 1$ $\Leftrightarrow$ $\dfrac{x^2}{12} = -(y-1)$ $\Leftrightarrow$ $x^2 = -12(y-1)$. This

is a parabola that has $4p = -12$ $\Leftrightarrow$ $p = -3$.

Vertex: $(0, 1)$; focus: $(0, 1-3) = (0, -2)$.

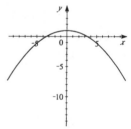

21. $x^2 - y^2 + 144 = 0$ $\Leftrightarrow$ $\dfrac{y^2}{144} - \dfrac{x^2}{144} = 1$. This is a hyperbola

that has $a = 12$, $b = 12$, and $c = \sqrt{144 + 144} = 12\sqrt{2}$.

Vertices: $(0, \pm12)$; foci: $(0, \pm12\sqrt{2})$.

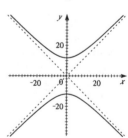

23. $4x^2 + y^2 = 8(x+y)$ $\Leftrightarrow$ $4(x^2 - 2x) + (y^2 - 8y) = 0$ $\Leftrightarrow$

$4(x^2 - 2x + 1) + (y^2 - 8y + 16) = 4 + 16$ $\Leftrightarrow$

$4(x-1)^2 + (y-4)^2 = 20$ $\Leftrightarrow$ $\dfrac{(x-1)^2}{5} + \dfrac{(y-4)^2}{20} = 1$. This

is an ellipse that has $a = 2\sqrt{5}$, $b = \sqrt{5}$, and $c = \sqrt{20-5} = \sqrt{15}$.

Vertices: $(1, 4 \pm 2\sqrt{5})$; foci: $(1, 4 \pm \sqrt{15})$.

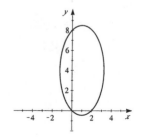

25. $x = y^2 - 16y$ $\Leftrightarrow$ $x + 64 = y^2 - 16y + 64$ $\Leftrightarrow$
$(y - 8)^2 = x + 64$. This is a parabola that has $4p = 1$ $\Leftrightarrow$ $p = \frac{1}{4}$.

Vertex: $(-64, 8)$; focus: $\left(-64 + \frac{1}{4}, 8\right) = \left(-\frac{255}{4}, 8\right)$.

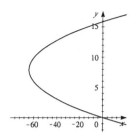

27. $2x^2 - 12x + y^2 + 6y + 26 = 0$ $\Leftrightarrow$
$2(x^2 - 6x) + (y^2 + 6y) = -26$ $\Leftrightarrow$
$2(x^2 - 6x + 9) + (y^2 + 6y + 9) = -26 + 18 + 9$ $\Leftrightarrow$
$2(x - 3)^2 + (y + 3)^2 = 1$ $\Leftrightarrow$ $\dfrac{(x - 3)^2}{\frac{1}{2}} + (y + 3)^2 = 1$. This is

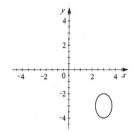

an ellipse that has $a = 1$, $b = \frac{1}{\sqrt{2}}$, and $c = \sqrt{1 - \frac{1}{2}} = \frac{1}{\sqrt{2}}$.

Vertices: $(3, -3 \pm 1)$, so the vertices are $(3, -4)$ and $(3, -2)$; foci:
$\left(3, -3 \pm \frac{1}{\sqrt{2}}\right)$.

29. $9x^2 + 8y^2 - 15x + 8y + 27 = 0$ $\Leftrightarrow$ $9\left(x^2 - \frac{5}{3}x + \frac{25}{36}\right) + 8\left(y^2 + y + \frac{1}{4}\right) = -27 + \frac{25}{4} + 2$ $\Leftrightarrow$
$9\left(x - \frac{5}{6}\right)^2 + 8\left(y + \frac{1}{2}\right)^2 = -\frac{75}{4}$. However, since the left-hand side of the equation is greater than or
equal to 0, there is no point that satisfies this equation. The graph is empty.

31. The parabola has focus $(0, 1)$ and directrix $y = -1$. Therefore, $p = 1$ and so $4p = 4$. Since the
focus is on the y-axis and the vertex is $(0, 0)$, the equation of the parabola is $x^2 = 4y$.

33. The hyperbola has vertices $(0, \pm 2)$ and asymptotes $y = \pm\frac{1}{2}x$. Therefore, $a = 2$, and the foci are on
the y-axis. Since the slopes of the asymptotes are $\pm\frac{1}{2} = \pm\frac{a}{b}$ $\Leftrightarrow$ $b = 2a = 4$, the equation of
the hyperbola is $\dfrac{y^2}{4} - \dfrac{x^2}{16} = 1$.

35. The ellipse has foci $F_1(1, 1)$ and $F_2(1, 3)$, and one vertex is on the x-axis. Thus, $2c = 3 - 1 = 2$
  $\Leftrightarrow$ $c = 1$, and so the center of the ellipse is $C(1, 2)$. Also, since one vertex is on the x-axis,
$a = 2 - 0 = 2$, and thus $b^2 = 4 - 1 = 3$. So the equation of the ellipse is
$\dfrac{(x - 1)^2}{3} + \dfrac{(y - 2)^2}{4} = 1$.

37. The ellipse has vertices $V_1(7, 12)$ and $V_2(7, -8)$ and passes through the point $P(1, 8)$. Thus,
$2a = 12 - (-8) = 20$ $\Leftrightarrow$ $a = 10$, and the center is $\left(7, \frac{-8+12}{2}\right) = (7, 2)$. Thus the equation of
the ellipse has the form $\dfrac{(x - 7)^2}{b^2} + \dfrac{(y - 2)^2}{100} = 1$. Since the point $P(1, 8)$ is on the ellipse,
$\dfrac{(1 - 7)^2}{b^2} + \dfrac{(8 - 2)^2}{100} = 1$ $\Leftrightarrow$ $3600 + 36b^2 = 100b^2$ $\Leftrightarrow$ $64b^2 = 3600$ $\Leftrightarrow$ $b^2 = \dfrac{225}{4}$.
Therefore, the equation of the ellipse is $\dfrac{(x - 7)^2}{225/4} + \dfrac{(y - 2)^2}{100} = 1$ $\Leftrightarrow$
$\dfrac{4(x - 7)^2}{225} + \dfrac{(y - 2)^2}{100} = 1$.

39. Since $(0,0)$ and $(1600,0)$ are both points on the parabola, the x-coordinate of the vertex is 800. And since the highest point it reaches is 3200, the y-coordinate of the vertex is 3200. Thus the vertex is $(800, 3200)$, and the equation is of the form $(x - 800)^2 = 4p(y - 3200)$. Substituting the point $(0,0)$, we get $(0 - 800)^2 = 4p(0 - 3200)$ $\Leftrightarrow$ $640000 = -12800p$ $\Leftrightarrow$ $p = -50$. So the equation is $(x - 800)^2 = 4(-50)(y - 3200)$ $\Leftrightarrow$ $(x - 800)^2 = -200(y - 3200)$.

41. The length of the major axis is $2a = 186{,}000{,}000$ $\Leftrightarrow$ $a = 93{,}000{,}000$. The eccentricity is $e = \frac{c}{a} = 0.017$, and so $c = 0.017(93{,}000{,}000) = 1{,}581{,}000$.

   (a) The earth is closest to the sun when the distance is $a - c = 93{,}000{,}000 - 1{,}581{,}000$
      $= 91{,}419{,}000$.

   (b) The earth is furthest from the sun when the distance is $a + c = 93{,}000{,}000 + 1{,}581{,}000$
      $= 94{,}581{,}000$.

43. (a) The graphs of $\dfrac{x^2}{16 + k^2} + \dfrac{y^2}{k^2} = 1$ for
      $k = 1, 2, 4,$ and 8 are shown in the figure.

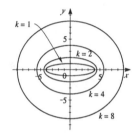

   (b) $c^2 = (16 + k^2) - k^2 = 16$ $\Rightarrow$ $c = \pm 4$.
      Since the center is $(0,0)$, the foci of each of
      the ellipses are $(\pm 4, 0)$.

45. (a) $x^2 + 4xy + y^2 = 1$. Then $A = 1$, $B = 4$, and $C = 1$, so the discriminant is
      $4^2 - 4(1)(1) = 12$. Since the discriminant is positive the equation represents a hyperbola.

   (b) $\cot 2\phi = \frac{A-C}{B} = \frac{1-1}{4} = 0$ $\Rightarrow$ $2\phi = 90°$ $\Leftrightarrow$ $\phi = 45°$. Therefore, $x = \dfrac{X}{\sqrt{2}} - \dfrac{Y}{\sqrt{2}}$ and
      $y = \frac{X}{\sqrt{2}} + \frac{Y}{\sqrt{2}}$. Substituting into the original equation gives:

$$\left(\frac{X}{\sqrt{2}} - \frac{Y}{\sqrt{2}}\right)^2 + 4\left(\frac{X}{\sqrt{2}} - \frac{Y}{\sqrt{2}}\right)\left(\frac{X}{\sqrt{2}} + \frac{Y}{\sqrt{2}}\right) + \left(\frac{X}{\sqrt{2}} + \frac{Y}{\sqrt{2}}\right)^2 = 1 \quad \Leftrightarrow$$

$$\tfrac{1}{2}(X^2 - 2XY + Y^2) + 2(X^2 + XY - XY - Y^2) + \tfrac{1}{2}(X^2 + 2XY + Y^2) = 1 \quad \Leftrightarrow$$

$$3X^2 - Y^2 = 1 \quad \Leftrightarrow \quad \frac{X^2}{\frac{1}{3}} - Y^2 = 1. \text{ This is a hyperbola that has } a = \frac{1}{\sqrt{3}}, b = 1, \text{ and}$$

      $c = \sqrt{\frac{1}{3} + 1} = \frac{2}{\sqrt{3}}$. Therefore, the hyperbola has vertices $V\left(\pm \frac{1}{\sqrt{3}}, 0\right)$ and foci $F\left(\pm \frac{2}{\sqrt{3}}, 0\right)$, in
      XY-coordinates.

   (c)

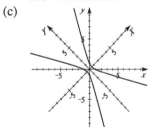

47. (a) $7x^2 - 6\sqrt{3}xy + 13y^2 - 4\sqrt{3}x - 4y = 0$. Then $A = 7$, $B = -6\sqrt{3}$, and $C = 13$, so the discriminant is $(-6\sqrt{3})^2 - 4(7)(13) = -256$. Since the discriminant is negative the equation represents an ellipse.

(b) $\cot 2\phi = \dfrac{A - C}{B} = \dfrac{7-13}{-6\sqrt{3}} = \dfrac{1}{\sqrt{3}} \;\Rightarrow\; 2\phi = 60° \;\Leftrightarrow\; \phi = 30°.$ Therefore,

$x = \dfrac{\sqrt{3}X}{2} - \dfrac{Y}{2}$ and $y = \dfrac{X}{2} + \dfrac{\sqrt{3}Y}{2}.$ Substituting into the original equation gives:

$$7\left(\dfrac{\sqrt{3}X}{2} - \dfrac{Y}{2}\right)^2 - 6\sqrt{3}\left(\dfrac{\sqrt{3}X}{2} - \dfrac{Y}{2}\right)\left(\dfrac{X}{2} + \dfrac{\sqrt{3}Y}{2}\right) + 13\left(\dfrac{X}{2} + \dfrac{\sqrt{3}Y}{2}\right)^2$$

$$- 4\sqrt{3}\left(\dfrac{\sqrt{3}X}{2} - \dfrac{Y}{2}\right) - 4\left(\dfrac{X}{2} + \dfrac{\sqrt{3}Y}{2}\right) = 0 \;\Leftrightarrow$$

$\tfrac{7}{4}(3X^2 - 2\sqrt{3}XY + Y^2) - \tfrac{3\sqrt{3}}{2}(\sqrt{3}X^2 + 3XY - XY - \sqrt{3}Y^2)$

$\qquad + \tfrac{13}{4}(X^2 + 2\sqrt{3}XY + 3Y^2) - 6X + 2\sqrt{3}Y - 2X - 2\sqrt{3}Y = 0 \;\Leftrightarrow$

$X^2\left(\tfrac{21}{4} - \tfrac{9}{2} + \tfrac{13}{4}\right) - 8X + Y^2\left(\tfrac{7}{4} + \tfrac{9}{2} + \tfrac{39}{4}\right) = 0 \;\Leftrightarrow\; 4X^2 - 8X + 16Y^2 = 0 \;\Leftrightarrow$

$4(X^2 - 2X + 1) + 16Y^2 = 4 \;\Leftrightarrow\; (X - 1)^2 + 4Y^2 = 1.$ This ellipse has $a = 1, b = \tfrac{1}{2},$

and $c = \sqrt{1 - \tfrac{1}{4}} = \tfrac{1}{2}\sqrt{3}.$ Therefore, the vertices are at $V(1 \pm 1, 0) = V_1(0, 0)$ and $V_2(2, 0)$

and the foci are at $F\left(1 \pm \tfrac{1}{2}\sqrt{3}, 0\right).$

(c)

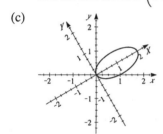

49. (a)

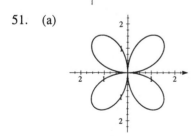

(b) $r = 3 + 3\cos\theta \;\Leftrightarrow$
$r^2 = 3r + 3r\cos\theta$ which gives
$x^2 + y^2 = 3\sqrt{x^2 + y^2} + 3x \;\Leftrightarrow$
$x^2 - 3x + y^2 = 3\sqrt{x^2 + y^2}.$ Squaring
both sides gives
$(x^2 - 3x + y^2)^2 = 9(x^2 + y^2).$

51. (a)

(b) $r = 2\sin 2\theta \;\Leftrightarrow\; r = 2 \cdot 2\sin\theta\cos\theta$
$\Leftrightarrow\; r^3 = 4r^2\sin\theta\cos\theta$
$= 4(r\sin\theta)(r\cos\theta)$ and so, since
$x = r\cos\theta$ and $y = r\sin\theta$ we get
$(x^2 + y^2)^{3/2} = 4xy.$

53. (a)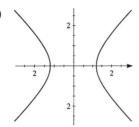

(b) $r^2 = \sec 2\theta = \dfrac{1}{\cos 2\theta} = \dfrac{1}{\cos^2\theta - \sin^2\theta}$

$\Leftrightarrow \quad r^2(\cos^2\theta - \sin^2\theta) = 1 \quad \Leftrightarrow$

$r^2\cos^2\theta - r^2\sin^2\theta = 1 \quad \Leftrightarrow$

$(r\cos\theta)^2 - (r\sin\theta)^2 = 1 \quad \Leftrightarrow$

$x^2 - y^2 = 1.$

55. (a)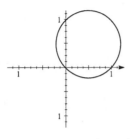

(b) $r = \sin\theta + \cos\theta \quad \Leftrightarrow$

$r^2 = r\sin\theta + r\cos\theta$ which gives

$x^2 + y^2 = y + x \quad \Leftrightarrow$

$(x^2 - x + \tfrac{1}{4}) + (y^2 - y + \tfrac{1}{4}) = \tfrac{1}{2} \quad \Leftrightarrow$

$\left(x - \tfrac{1}{2}\right)^2 + \left(y - \tfrac{1}{2}\right)^2 = \tfrac{1}{2}.$

57. Domain of θ shown is $[0,\, 6\pi]$.

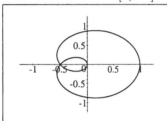

59. Domain of θ shown is $[0,\, 6\pi]$.

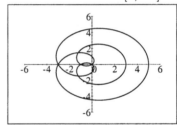

61. (a) $r = \dfrac{1}{1 - \cos\theta} \quad \Rightarrow \quad e = 1.$ Therefore, this is a parabola.

(b)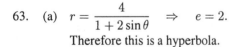

$\left(\tfrac{1}{2}, \pi\right)$

63. (a) $r = \dfrac{4}{1 + 2\sin\theta} \quad \Rightarrow \quad e = 2.$

Therefore this is a hyperbola.

(b)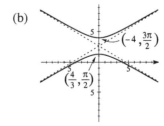

$\left(-4, \tfrac{3\pi}{2}\right)$

$\left(\tfrac{4}{3}, \tfrac{\pi}{2}\right)$

65. $x = 1 - t^2, y = 1 + t \quad \Leftrightarrow \quad t = y - 1.$
     Substituting for t gives $x = 1 - (y - 1)^2$
     $\Leftrightarrow \quad (y - 1)^2 = 1 - x$ which is the
     rectangular coordinate equation.

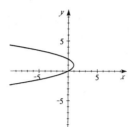

67. $x = 1 + \cos t \quad \Leftrightarrow \quad \cos t = x - 1$, and
     $y = 1 - \sin t \quad \Leftrightarrow \quad \sin t = 1 - y.$ Since
     $\cos^2 t + \sin^2 t = 1$, it follows that
     $(x - 1)^2 + (1 - y)^2 = 1 \quad \Leftrightarrow$
     $(x - 1)^2 + (y - 1)^2 = 1.$ Since t is restricted
     by $0 \le t \le \frac{\pi}{2}, 1 + \cos 0 \le x \le 1 + \cos \frac{\pi}{2}$
     $\Leftrightarrow \quad 1 \le x \le 2$, and similarly, $0 \le y \le 1.$
     (This is the lower right quarter of the circle.)

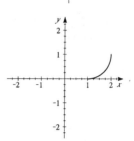

69. $x = \cos 2t, y = \sin 3t$

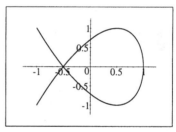

71. $C: x = t, y = t^2.$ $D: x = \sqrt{t}, y = t, t \ge 0.$ $E: x = \sin t, y = 1 - \cos^2 t.$ $F: x = e^t, y = e^{2t}.$
     (a) For C: $x = t, y = t^2 \quad \Rightarrow \quad y = x^2.$
          For D: $x = \sqrt{t}, y = t \quad \Rightarrow \quad y = x^2.$
          For E: $x = \sin t \quad \Rightarrow \quad x^2 = \sin^2 t = 1 - \cos^2 t = y$ and so $y = x^2.$
          For F: $x = e^t \quad \Rightarrow \quad x^2 = e^{2t} = y$ and so $y = x^2.$ Therefore, the points on all four curves
          satisfy the same rectangular coordinate equation.

     (b) The curve in C is the entire parabola
          $y = x^2.$

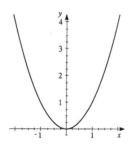

The curve in D is the right-half of the
parabola (because $t \geq 0$ and so $x \geq 0$).

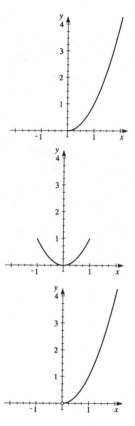

The curve in E is the portion of the
parabola where $-1 \leq x \leq 1$.

The curve in F is the portion of the
parabola where $x > 0$.

Chapter 11 Test

1. $x^2 = -12y$. This is a parabola that has $4p = -12$
   $\Leftrightarrow \quad p = -3$.

   Focus: $(0, -3)$; directrix: $y = 3$.

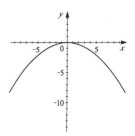

2. $\dfrac{x^2}{16} + \dfrac{y^2}{4} = 1$. This is an ellipse that has $a = 4$, $b = 2$, and
   $c = \sqrt{16 - 4} = 2\sqrt{3}$.

   Vertices: $(\pm 4, 0)$; foci: $(\pm 2\sqrt{3}, 0)$; length of the major axis: $2a = 8$;
   length of the minor axis: $2b = 4$.

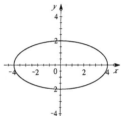

3. $\dfrac{y^2}{9} - \dfrac{x^2}{16} = 1$. This is a hyperbola that has $a = 3$, $b = 4$, and
   $c = \sqrt{9 + 16} = 5$.

   Vertices: $(0, \pm 3)$; foci: $(0, \pm 5)$; asymptotes: $y = \pm \frac{3}{4}x$.

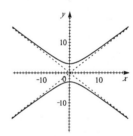

4. This is a parabola that opens to the left with its vertex at $(0, 0)$. So its equation is of the form
   $y^2 = 4px$ with $p < 0$. Substituting the point $(-4, 2)$, we have $2^2 = 4p(-4) \quad \Leftrightarrow \quad 4 = -16p$
   $\Leftrightarrow \quad p = -\frac{1}{4}$. So the equation is $y^2 = 4\left(-\frac{1}{4}\right)x \quad \Leftrightarrow \quad y^2 = -x$.

5. This is an ellipse tangent to the x-axis at $(0, 0)$ and with one vertex at the point $(4, 3)$. The center is
   $(0, 3)$, and $a = 4$ and $b = 3$. Thus the equation is $\dfrac{x^2}{16} + \dfrac{(y - 3)^2}{9} = 1$.

6. This a hyperbola with a horizontal transverse axis, vertices at $(1, 0)$ and $(3, 0)$, and foci at $(0, 0)$ and
   $(4, 0)$. Thus the center is $(2, 0)$, and $a = 3 - 2 = 1$ and $c = 4 - 2 = 2$. Thus $b^2 = 2^2 - 1^2 = 3$. So
   the equation is $\dfrac{(x - 2)^2}{1^2} - \dfrac{y^2}{3} = 1 \quad \Leftrightarrow \quad (x - 2)^2 - \dfrac{y^2}{3} = 1$.

7. $16x^2 + 36y^2 - 96x + 36y + 9 = 0 \quad \Leftrightarrow$
   $16(x^2 - 6x) + 36(y^2 + y) = -9 \quad \Leftrightarrow$
   $16(x^2 - 6x + 9) + 36\left(y^2 + y + \frac{1}{4}\right) = -9 + 144 + 9 \quad \Leftrightarrow$

   $16(x - 3)^2 + 36\left(y + \frac{1}{2}\right)^2 = 144 \quad \Leftrightarrow \quad \dfrac{(x - 3)^2}{9} + \dfrac{\left(y + \frac{1}{2}\right)^2}{4} = 1$.

   This is an ellipse that has $a = 3$, $b = 2$, and $c = \sqrt{9 - 4} = \sqrt{5}$.

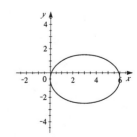

Center: $\left(3, -\frac{1}{2}\right)$; vertices: $\left(3 \pm 3, -\frac{1}{2}\right)$, so vertices are $\left(0, -\frac{1}{2}\right)$ and $\left(6, -\frac{1}{2}\right)$; foci: $(h \pm c, k)$ which are $\left(3 \pm \sqrt{5}, -\frac{1}{2}\right)$, or $\left(3 + \sqrt{5}, -\frac{1}{2}\right)$ and $\left(3 - \sqrt{5}, -\frac{1}{2}\right)$.

8. $9x^2 - 8y^2 + 36x + 64y = 92 \quad \Leftrightarrow \quad 9(x^2 + 4x) - 8(y^2 - 8y) = 92$

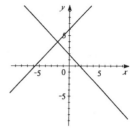

$\Leftrightarrow \quad 9(x^2 + 4x + 4) - 8(y^2 - 8y + 16) = 92 + 36 - 128 \quad \Leftrightarrow$

$9(x + 2)^2 - 8(y - 4)^2 = 0 \quad \Leftrightarrow \quad 9(x + 2)^2 = 8(y - 4)^2 \quad \Rightarrow$

$3(x + 2) = \pm 2\sqrt{2}(y - 4) \quad \Leftrightarrow \quad y = 4 \pm \frac{3}{2\sqrt{2}}(x + 2)$. This is a

degenerate hyperbola that consists of two intersecting lines.

9. $2x + y^2 + 8y + 8 = 0 \quad \Leftrightarrow \quad y^2 + 8y + 16 = -2x - 8 + 16 \quad \Leftrightarrow$

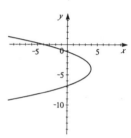

$(y + 4)^2 = -2(x - 4)$. This is a parabola that has $4p = -2 \quad \Leftrightarrow$

$p = -\frac{1}{2}$.

Vertex: $(4, -4)$; focus: $\left(4 - \frac{1}{2}, -4\right) = \left(\frac{7}{2}, -4\right)$. No asymptotes.

10.   The hyperbola has foci $(0, \pm 5)$ and asymptotes $y = \pm \frac{3}{4} x$. Since the foci are $(0, \pm 5)$, $c = 5$, the foci are on the y-axis, and the center is $(0, 0)$. Also, since $y = \pm \frac{3}{4} x = \pm \frac{a}{b} x$, it follows that $\frac{a}{b} = \frac{3}{4} \quad \Leftrightarrow$

$a = \frac{3}{4} b$. Then $c^2 = 5^2 = 25 = a^2 + b^2 = \left(\frac{3}{4} b\right)^2 + b^2 = \frac{25}{16} b^2 \quad \Leftrightarrow \quad b^2 = 16$, and by substitution, $a = \frac{3}{4}(4) = 3$. Therefore, the equation of the hyperbola is $\dfrac{y^2}{9} - \dfrac{x^2}{16} = 1$.

11.   The parabola has focus $(2, 4)$ and directrix the x-axis $(y = 0)$. Therefore, $2p = 4 - 0 = 4 \quad \Leftrightarrow$ $p = 2 \quad \Leftrightarrow \quad 4p = 8$, and the vertex is $(2, 4 - p) = (2, 2)$. Hence, the equation of the parabola is $(x - 2)^2 = 8(y - 2) \quad \Leftrightarrow \quad x^2 - 4x + 4 = 8y - 16 \quad \Leftrightarrow \quad x^2 - 4x - 8y + 20 = 0$.

12.   We place the vertex of the parabola at the origin, so the parabola contains the points $(3, \pm 3)$, and the equation is of the form $y^2 = 4px$. Substituting the point $(3, 3)$, we get $3^2 = 4p(3) \quad \Leftrightarrow \quad 9 = 12p$ $\Leftrightarrow \quad p = \frac{3}{4}$. So the focus is $\left(\frac{3}{4}, 0\right)$, and we should place the light bulb $\frac{3}{4}$ inches from the vertex.

13.   (a) $5x^2 + 4xy + 2y^2 = 18$. Then $A = 5$, $B = 4$, and $C = 2$, so the discriminant is $(4)^2 - 4(5)(2) = -24$. Since the discriminant is negative the equation represents an ellipse.

(b) $\cot 2\phi = \dfrac{A - C}{B} = \dfrac{5 - 2}{4} = \dfrac{3}{4}$. Thus, $\cos 2\phi = \frac{3}{5}$ and so $\cos \phi = \sqrt{\dfrac{1 + (3/5)}{2}} = \dfrac{2}{\sqrt{5}}$,

$\sin \phi = \sqrt{\dfrac{1 - (3/5)}{2}} = \dfrac{1}{\sqrt{5}}$. It follows that $x = \dfrac{2X}{\sqrt{5}} - \dfrac{Y}{\sqrt{5}}$ and $y = \dfrac{X}{\sqrt{5}} + \dfrac{2Y}{\sqrt{5}}$. By substitution,

$5\left(\dfrac{2X}{\sqrt{5}} - \dfrac{Y}{\sqrt{5}}\right)^2 + 4\left(\dfrac{2X}{\sqrt{5}} - \dfrac{Y}{\sqrt{5}}\right)\left(\dfrac{X}{\sqrt{5}} + \dfrac{2Y}{\sqrt{5}}\right) + 2\left(\dfrac{X}{\sqrt{5}} + \dfrac{2Y}{\sqrt{5}}\right)^2 = 18$

$\Leftrightarrow \quad 4X^2 - 4XY + Y^2 + \frac{4}{5}(2X^2 + 4XY - XY - 2Y^2) + \frac{2}{5}(X^2 + 4XY + 4Y^2) = 18$

$\Leftrightarrow \quad X^2\left(4 + \frac{8}{5} + \frac{2}{5}\right) + XY\left(-4 + \frac{12}{5} + \frac{8}{5}\right) + Y^2\left(1 - \frac{8}{5} + \frac{4}{5}\right) = 18 \quad \Leftrightarrow \quad 6X^2 + Y^2 = 18$

$\Leftrightarrow \quad \dfrac{X^2}{3} + \dfrac{Y^2}{18} = 1$. This is an ellipse that has $a = 3\sqrt{2}$ and $b = \sqrt{3}$.

(c)

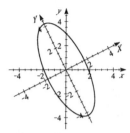

Since $\cos 2\phi = \frac{3}{5}$ we have $2\phi = \cos^{-1}\frac{3}{5} \approx 53.13°$ so $\phi \approx 27°$.

(d) In XY-coordinates, the vertices are $V(0, \pm 3\sqrt{2})$. Therefore, in xy-coordinates, the vertices are $x = \frac{-3\sqrt{2}}{\sqrt{5}}$ and $y = \frac{6\sqrt{2}}{\sqrt{5}} \Rightarrow V_1\left(\frac{-3\sqrt{2}}{\sqrt{5}}, \frac{6\sqrt{2}}{\sqrt{5}}\right)$, and $x = \frac{3\sqrt{2}}{\sqrt{5}}$ and $y = -\frac{6\sqrt{2}}{\sqrt{5}} \Rightarrow V_2\left(\frac{3\sqrt{2}}{\sqrt{5}}, \frac{-6\sqrt{2}}{\sqrt{5}}\right)$.

14. $r = 2 + \cos\theta$. This is a limacon.

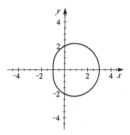

15. $r = 2\cos\theta - 4\sin\theta \quad \Leftrightarrow \quad r^2 = 2r\cos\theta - 4r\sin\theta$ which gives $x^2 + y^2 = 2x - 4y \quad \Leftrightarrow$
       $(x^2 - 2x + 1) + (y^2 + 4y + 4) = 5 \quad \Leftrightarrow \quad (x - 1)^2 + (y + 2)^2 = 5$. The graph of this equation is
       a circle with center $(1, -2)$ and radius $\sqrt{5}$.

16.   (a) $x = 3\sin\theta + 3, y = 2\cos\theta, 0 \le \theta \le \pi$.
            From the work of part (b), we see that
            this is the half-ellipse shown on the right.

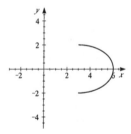

       (b) $x = 3\sin\theta + 3 \quad \Leftrightarrow \quad x - 3 = 3\sin\theta \quad \Leftrightarrow \quad \dfrac{x - 3}{3} = \sin\theta$. Squaring both sides gives
            $\dfrac{(x - 3)^2}{9} = \sin^2\theta$. Similarly, $y = 2\cos\theta \quad \Leftrightarrow \quad \dfrac{y}{2} = \cos\theta$, and squaring both sides gives
            $\dfrac{y^2}{4} = \cos^2\theta$. Since $\sin^2\theta + \cos^2\theta = 1$, it follows that $\dfrac{(x - 3)^2}{9} + \dfrac{y^2}{4} = 1$. Since $0 \le \theta \le \pi$,
            we get that $\sin\theta \ge 0$, so $3\sin\theta \ge 0 \quad \Rightarrow \quad 3\sin\theta + 3 \ge 3$, and so $x \ge 3$. Thus the curve
            consists only of the right half of the ellipse.

Focus on Modeling

1. From $x = (v_0\cos\theta)t$ we get $t = \dfrac{x}{v_0\cos\theta}$. Substituting this value for t into the equation for y, we get

$$y = (v_0\sin\theta)t - \tfrac{1}{2}gt^2 \quad\Leftrightarrow\quad y = (v_0\sin\theta)\left(\dfrac{x}{v_0\cos\theta}\right) - \dfrac{1}{2}g\left(\dfrac{x}{v_0\cos\theta}\right)^2 \quad\Leftrightarrow$$

$y = \tan\theta\, x - \dfrac{g}{2v_0^2\cos^2\theta}x^2$. This shows that y is a quadratic function of x, so its graph is a parabola as long as $\theta \neq 90°$. When $\theta = 90°$, the path of the projectile is a straight line up (then down).

3. (a) We use the equation $t = \dfrac{2v_0\sin\theta}{g}$ from page 682 of the text. Substituting $g \approx 32$ ft/s^2, $\theta = 5°$, and $v_0 = 1000$, we get $t = \dfrac{2\cdot 1000\cdot\sin 5°}{32} \approx 5.447$ seconds.

(b) Substituting the given values into $y = (v_0\sin\theta)t - \tfrac{1}{2}gt^2$, we get $y = 87.2t - 16t^2$. The maximum value of y is attained at the vertex of the parabola; thus
$y = 87.2t - 16t^2 = -16(t^2 - 5.45t) \quad\Leftrightarrow\quad y = -16[t^2 - 2(2.725)t + 7.425625] + 118.7$.
Thus the greatest height is 118.7 ft.

(c) The rocket hits the ground after 5.447 s, so substituting this into the expression for the horizontal distance gives $x = (1000\cos 5°)\,5.447 = 5426$ ft.

(d)

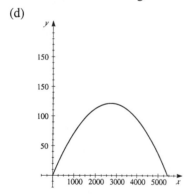

5. We use the equation of the parabola from Exercise 1 and find its vertex: $y = \tan\theta\, x - \dfrac{g}{2v_0^2\cos^2\theta}x^2$

$$\Leftrightarrow\quad y = -\dfrac{g}{2v_0^2\cos^2\theta}\left[x^2 - \dfrac{2v_0^2\sin\theta\cos\theta\, x}{g}\right] \quad\Leftrightarrow$$

$$y = -\dfrac{g}{2v_0^2\cos^2\theta}\left[x^2 - \dfrac{2v_0^2\sin\theta\cos\theta\, x}{g} + \left(\dfrac{v_0^2\sin\theta\cos\theta}{g}\right)^2\right] + \dfrac{g}{2v_0^2\cos^2\theta}\cdot\left(\dfrac{v_0^2\sin\theta\cos\theta}{g}\right)^2 \quad\Leftrightarrow$$

$y = -\dfrac{g}{2v_0^2\cos^2\theta}\left[x - \dfrac{v_0^2\sin\theta\cos\theta}{g}\right]^2 + \dfrac{v_0^2\sin^2\theta}{2g}$. Thus the vertex is at $\left(\dfrac{v_0^2\sin\theta\cos\theta}{g}, \dfrac{v_0^2\sin^2\theta}{2g}\right)$ so

the maximum height is $\dfrac{v_0^2\sin^2\theta}{2g}$.

7. In Problem 6 we derived the equations
$x = (v_0\cos\theta - w)t$,
$y = (v_0\sin\theta)t - \frac{1}{2}gt^2$. We plot the graphs for
the given values of v_0, w, and θ in the figure to
the right. The projectile will be blown
backwards if the horizontal component of its
velocity is less than the speed of the wind, that
is, $32\cos\theta < 24 \quad \Leftrightarrow \quad \cos\theta < \frac{3}{4} \quad \Rightarrow$
$\theta > 41.4°$.

The optimal firing angle appears to be
between 15° and 30°. We graph the trajectory
for $\theta = 20°$, $\theta = 23°$, and $\theta = 25°$. The
solution appears to be close to 23°.

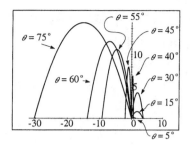

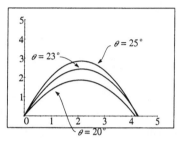

Chapter Twelve
Exercises 12.1

1. $a_n = n + 1$. Then $a_1 = 1 + 1 = 2$; $a_2 = 2 + 1 = 3$; $a_3 = 3 + 1 = 4$; $a_4 = 4 + 1 = 5$; and $a_{1000} = 1000 + 1 = 1001$.

3. $a_n = \dfrac{1}{n+1}$. Then $a_1 = \dfrac{1}{1+1} = \dfrac{1}{2}$; $a_2 = \dfrac{1}{2+1} = \dfrac{1}{3}$; $a_3 = \dfrac{1}{3+1} = \dfrac{1}{4}$; $a_4 = \dfrac{1}{4+1} = \dfrac{1}{5}$; and $a_{1000} = \dfrac{1}{1000+1} = \dfrac{1}{1001}$.

5. $a_n = \dfrac{(-1)^n}{n^2}$. Then $a_1 = \dfrac{(-1)^1}{1^2} = -1$; $a_2 = \dfrac{(-1)^2}{2^2} = \dfrac{1}{4}$; $a_3 = \dfrac{(-1)^3}{3^2} = -\dfrac{1}{9}$; $a_4 = \dfrac{(-1)^4}{4^2} = \dfrac{1}{16}$; and $a_{1000} = \dfrac{(-1)^{1000}}{1000^2} = \dfrac{1}{1,000,000}$.

7. $a_n = 1 + (-1)^n$. Then $a_1 = 1 + (-1)^1 = 0$; $a_2 = 1 + (-1)^2 = 2$; $a_3 = 1 + (-1)^3 = 0$; $a_4 = 1 + (-1)^4 = 2$; and $a_{1000} = 1 + (-1)^{1000} = 2$.

9. $a_n = n^n$. Then $a_1 = 1^1 = 1$; $a_2 = 2^2 = 4$; $a_3 = 3^3 = 27$; $a_4 = 4^4 = 256$; and $a_{1000} = 1000^{1000} = 10^{3000}$.

11. $a_n = 2(a_{n-1} - 2)$ and $a_1 = 3$. Then $a_2 = 2[(3) - 2] = 2$; $a_3 = 2[(2) - 2] = 0$; $a_4 = 2[(0) - 2] = -4$; and $a_5 = 2[(-4) - 2] = -12$.

13. $a_n = 2a_{n-1} + 1$ and $a_1 = 1$. Then $a_2 = 2(1) + 1 = 3$; $a_3 = 2(3) + 1 = 7$; $a_4 = 2(7) + 1 = 15$; and $a_5 = 2(15) + 1 = 31$.

15. $a_n = a_{n-1} + a_{n-2}$; $a_1 = 1$; and $a_2 = 2$. Then $a_3 = 2 + 1 = 3$; $a_4 = 3 + 2 = 5$; and $a_5 = 5 + 3 = 8$.

17. $2, 4, 8, 16, \ldots$. All are powers of 2, so $a_1 = 2$, $a_2 = 2^2$, $a_3 = 2^3$, $a_4 = 2^4, \ldots$ Thus $a_n = 2^n$.

19. $1, 4, 7, 10, \ldots$. The difference between any two consecutive terms is 3, so $a_1 = 3(1) - 2$, $a_2 = 3(2) - 2$, $a_3 = 3(3) - 2$, $a_4 = 3(4) - 2, \ldots$ Thus $a_n = 3n - 2$.

21. $1, \frac{3}{4}, \frac{5}{9}, \frac{7}{16}, \frac{9}{25}, \ldots$. We consider the numerator separately from the denominator. The numerators of the terms differ by 2, and the denominators are perfect squares. So $a_1 = \dfrac{2(1) - 1}{1^2}$, $a_2 = \frac{2(2)-1}{2^2}$, $a_3 = \dfrac{2(3) - 1}{3^2}$, $a_4 = \dfrac{2(4) - 1}{4^2}$, $a_5 = \dfrac{2(5) - 1}{5^2}, \ldots$. Thus $a_n = \dfrac{2n - 1}{n^2}$.

23. $0, 2, 0, 2, 0, 2, \ldots$. These terms alternate between 0 and 2. So $a_1 = 1 - 1$, $a_2 = 1 + 1$, $a_3 = 1 - 1$, $a_4 = 1 + 1$, $a_5 = 1 - 1$, $a_6 = 1 + 1, \ldots$ Thus $a_n = 1 + (-1)^n$.

25. $a_1 = 1, a_2 = 3, a_3 = 5, a_4 = 7, \ldots$. Therefore, $a_n = 2n - 1$. So $S_1 = 1$; $S_2 = 1 + 3 = 4$; $S_3 = 1 + 3 + 5 + \, = 9$; $S_4 = 1 + 3 + 5 + 7 = 16$; $S_5 = 1 + 3 + 5 + 7 + 9 = 25$; and $S_6 = 1 + 3 + 5 + 7 + 9 + 11 = 36$.

27. $a_1 = \frac{1}{3}, a_2 = \frac{1}{3^2}, a_3 = \frac{1}{3^3}, a_4 = \frac{1}{3^4}, \ldots$. Therefore, $a_n = \frac{1}{3^n}$. So $S_1 = \frac{1}{3}$; $S_2 = \frac{1}{3} + \frac{1}{3^2} = \frac{4}{9}$; $S_3 = \frac{1}{3} + \frac{1}{3^2} + \frac{1}{3^3} = \frac{13}{27}$; $S_4 = \frac{1}{3} + \frac{1}{3^2} + \frac{1}{3^3} + \frac{1}{3^4} = \frac{40}{81}$; and $S_5 = \frac{1}{3} + \frac{1}{3^2} + \frac{1}{3^3} + \frac{1}{3^4} + \frac{1}{3^5} = \frac{121}{243}$; $S_6 = \frac{1}{3} + \frac{1}{3^2} + \frac{1}{3^3} + \frac{1}{3^4} + \frac{1}{3^5} + \frac{1}{3^6} = \frac{364}{729}$.

29. $a_n = \frac{2}{3^n}$. So $S_1 = \frac{2}{3}$; $S_2 = \frac{2}{3} + \frac{2}{3^2} = \frac{8}{9}$; $S_3 = \frac{2}{3} + \frac{2}{3^2} + \frac{2}{3^3} = \frac{26}{27}$; and $S_4 = \frac{2}{3} + \frac{2}{3^2} + \frac{2}{3^3} + \frac{2}{3^4} = \frac{80}{81}$. Therefore, $S_n = \frac{3^n - 1}{3^n}$.

31. $a_n = \sqrt{n} - \sqrt{n+1}$. So $S_1 = \sqrt{1} - \sqrt{2} = \underline{1 - \sqrt{2}}$;

$S_2 = \left(\sqrt{1} - \sqrt{2}\right) + \left(\sqrt{2} - \sqrt{3}\right) = 1 + \left(-\sqrt{2} + \sqrt{2}\right) - \sqrt{3} = \underline{1 - \sqrt{3}}$;

$S_3 = \left(\sqrt{1} - \sqrt{2}\right) + \left(\sqrt{2} - \sqrt{3}\right) + \left(\sqrt{3} - \sqrt{4}\right)$

$= 1 + \left(-\sqrt{2} + \sqrt{2}\right) + \left(-\sqrt{3} + \sqrt{3}\right) - \sqrt{4} = \underline{1 - \sqrt{4}}$;

$S_4 = \left(\sqrt{1} - \sqrt{2}\right) + \left(\sqrt{2} - \sqrt{3}\right) + \left(\sqrt{3} - \sqrt{4}\right) + \left(\sqrt{4} - \sqrt{5}\right)$

$= 1 + \left(-\sqrt{2} + \sqrt{2}\right) + \left(-\sqrt{3} + \sqrt{3}\right) + \left(-\sqrt{4} + \sqrt{4}\right) - \sqrt{5} = \underline{1 - \sqrt{5}}$. Therefore,

$S_n = \left(\sqrt{1} - \sqrt{2}\right) + \left(\sqrt{2} - \sqrt{3}\right) + \cdots + \left(\sqrt{n} - \sqrt{n+1}\right)$

$= 1 + \left(-\sqrt{2} + \sqrt{2}\right) + \left(-\sqrt{3} + \sqrt{3}\right) + \cdots + \left(-\sqrt{n} + \sqrt{n}\right) - \sqrt{n+1} = \underline{1 - \sqrt{n+1}}$.

33. $\displaystyle\sum_{k=1}^{4} k = 1 + 2 + 3 + 4 = 10$

35. $\displaystyle\sum_{k=1}^{3} \frac{1}{k} = 1 + \frac{1}{2} + \frac{1}{3} = \frac{6}{6} + \frac{3}{6} + \frac{2}{6} = \frac{11}{6}$.

37. $\displaystyle\sum_{i=1}^{8} [1 + (-1)^i] = 0 + 2 + 0 + 2 + 0 + 2 + 0 + 2 = 8$

39. $\displaystyle\sum_{k=1}^{5} 2^{k-1} = 2^0 + 2^1 + 2^2 + 2^3 + 2^4 = 1 + 2 + 4 + 8 + 16 = 31$

41. $\displaystyle\sum_{k=1}^{5} \sqrt{k} = \sqrt{1} + \sqrt{2} + \sqrt{3} + \sqrt{4} + \sqrt{5}$

43. $\displaystyle\sum_{k=0}^{6} \sqrt{k+4} = \sqrt{4} + \sqrt{5} + \sqrt{6} + \sqrt{7} + \sqrt{8} + \sqrt{9} + \sqrt{10}$

45. $\displaystyle\sum_{k=3}^{100} x^k = x^3 + x^4 + x^5 + \cdots + x^{100}$

47. $1 + 2 + 3 + 4 + \cdots + 100 = \displaystyle\sum_{k=1}^{100} k$

49. $1^2 + 2^2 + 3^2 + \cdots + 10^2 = \displaystyle\sum_{k=1}^{10} k^2$

51. $\dfrac{1}{1 \cdot 2} + \dfrac{1}{2 \cdot 3} + \dfrac{1}{3 \cdot 4} + \cdots + \dfrac{1}{999 \cdot 1000} = \displaystyle\sum_{k=1}^{999} \dfrac{1}{k(k+1)}$

53. $1 + x + x^2 + x^3 + \cdots + x^{100} = \displaystyle\sum_{k=0}^{100} x^k$

55. $\sqrt{2}, \sqrt{2\sqrt{2}}, \sqrt{2\sqrt{2\sqrt{2}}}, \sqrt{2\sqrt{2\sqrt{2\sqrt{2}}}}, \ldots$. We simplify each term in an attempt to determine a

formula for a_n. So $a_1 = 2^{1/2}$; $a_2 = \sqrt{2 \cdot 2^{1/2}} = \sqrt{2^{3/2}} = 2^{3/4}$; $a_3 = \sqrt{2 \cdot 2^{3/4}} = \sqrt{2^{7/4}} = 2^{7/8}$;
$a_4 = \sqrt{2 \cdot 2^{7/8}} = \sqrt{2^{15/8}} = 2^{15/16}$; $\ldots$. Thus $a_n = 2^{(2^n - 1)/2^n}$.

57. (a) $a_n = n^2$. Then $a_1 = 1^2 = 1$, $a_2 = 2^2 = 4$, $a_3 = 3^2 = 9$, $a_4 = 4^2 = 16$.

   (b) $a_n = n^2 + (n-1)(n-2)(n-3)(n-4)$;
       $a_1 = 1^2 + (1-1)(1-2)(1-3)(1-4) = 1 + 0(-1)(-2)(-3) = 1$;
       $a_2 = 2^2 + (2-1)(2-2)(2-3)(2-4) = 4 + 1 \cdot 0(-1)(-2) = 4$;
       $a_3 = 3^2 + (3-1)(3-2)(3-3)(3-4) = 9 + 2 \cdot 1 \cdot 0(-1) = 9$;
       $a_4 = 4^2 + (4-1)(4-2)(4-3)(4-4) = 16 + 3 \cdot 2 \cdot 1 \cdot 0 = 16$.
       Hence, the sequences agree in the first four terms. However, for the second sequence,
       $a_5 = 5^2 + (5-1)(5-2)(5-3)(5-4) = 25 + 4 \cdot 3 \cdot 2 \cdot 1 = 49$, and for the first sequence,
       $a_5 = 5^2 = 25$, and thus the sequences disagree from the fifth term on.

   (c) $a_n = n^2 + (n-1)(n-2)(n-3)(n-4)(n-5)(n-6)$ agrees with $a_n = n^2$ in the first six
       terms only.

   (d) $a_n = 2^n$ and $b_n = 2^n + (n-1)(n-2)(n-3)(n-4)$.

59. $a_n = a_{n-a_{n-1}} + a_{n-a_{n-2}}$; $a_1 = 1$; and $a_2 = 1$. So
    $a_3 = a_{3-1} + a_{3-1} = a_2 + a_2 = 1 + 1 = 2$;
    $a_4 = a_{4-2} + a_{4-1} = a_2 + a_3 = 1 + 2 = 3$;
    $a_5 = a_{5-3} + a_{5-2} = a_2 + a_3 = 1 + 2 = 3$;
    $a_6 = a_{6-3} + a_{6-3} = a_3 + a_3 = 2 + 2 = 4$;
    $a_7 = a_{7-4} + a_{7-3} = a_3 + a_4 = 2 + 3 = 5$;
    $a_8 = a_{8-5} + a_{8-4} = a_3 + a_4 = 2 + 3 = 5$;
    $a_9 = a_{9-5} + a_{9-5} = a_4 + a_4 = 3 + 3 = 6$;
    $a_{10} = a_{10-6} + a_{10-5} = a_4 + a_5 = 3 + 3 = 6$.
    The definition of a_n depends on the <u>values</u> of certain preceding terms. So a_n is the sum of two
    preceding terms whose choice depends on the <u>values</u> of a_{n-1} and a_{n-2} (not on $n-1$ and $n-2$).

Exercises 12.2

1. $a_4 - a_3 = 14 - 11 = 3$; $a_3 - a_2 = 11 - 8 = 3$; $a_2 - a_1 = 8 - 5 = 3$. This sequence is arithmetic with the common difference 3.

3. Since $a_2 - a_1 = 4 - 2 = 2$ and $a_4 - a_3 = 16 - 8 = 8$, the terms of the sequence do not have a common difference. This sequence is not arithmetic.

5. $a_4 - a_3 = -\frac{3}{2} - 0 = -\frac{3}{2}$; $a_3 - a_2 = 0 - \frac{3}{2} = -\frac{3}{2}$; $a_2 - a_1 = \frac{3}{2} - 3 = -\frac{3}{2}$. This sequence is arithmetic with the common difference $-\frac{3}{2}$.

7. $2, 5, 8, 11, \ldots$. Then $d = a_2 - a_1 = 5 - 2 = 3$; $a_5 = a_4 + 3 = 11 + 3 = 14$; $a_n = 2 + 3(n-1)$; and $a_{100} = 2 + 3(99) = 299$.

9. $4, 9, 14, 19, \ldots$. Then $d = a_2 - a_1 = 9 - 4 = 5$; $a_5 = a_4 + 5 = 19 + 5 = 24$; $a_n = 4 + 5(n-1)$; and $a_{100} = 4 + 5(99) = 499$.

11. $-12, -8, -4, 0, \ldots$. Then $d = a_2 - a_1 = -8 - (-12) = 4$; $a_5 = a_4 + 4 = 0 + 4 = 4$; $a_n = -12 + 4(n-1)$; and $a_{100} = -12 + 4(99) = 384$.

13. $25, 26.5, 28, 29.5, \ldots$. Then $d = a_2 - a_1 = 26.5 - 25 = 1.5$; $a_5 = a_4 + 1.5 = 29.5 + 1.5 = 31$; $a_n = 25 + 1.5(n-1)$; $a_{100} = 25 + 1.5(99) = 173.5$.

15. $2, 2 + s, 2 + 2s, 2 + 3s, \ldots$. Then $d = a_2 - a_1 = 2 + s - 2 = s$; $a_5 = a_4 + s = 2 + 3s + s = 2 + 4s$; $a_n = 2 + (n-1)s$; and $a_{100} = 2 + 99s$.

17. $a_{10} = \frac{55}{2}$, $a_2 = \frac{7}{2}$, and $a_n = a + d(n-1)$. Then $a_2 = a + d = \frac{7}{2}$ $\Leftrightarrow$ $d = \frac{7}{2} - a$. Substituting into $a_{10} = a + 9d = \frac{55}{2}$ gives $a + 9\left(\frac{7}{2} - a\right) = \frac{55}{2}$ $\Leftrightarrow$ $a = \frac{1}{2}$. Thus, the first term is $a_1 = \frac{1}{2}$.

19. $a_{100} = 98$ and $d = 2$. Note that $a_{100} = a + 99d = a + 99(2) = a + 198$. Since $a_{100} = 98$, we have $a + 198 = a_{100} = 98$ $\Leftrightarrow$ $a = -100$. Hence, $a_1 = -100$, $a_2 = -100 + 2 = -98$, and $a_3 = -100 + 4 = -96$.

21. The arithmetic sequence is $1, 4, 7, \ldots$. So $d = 4 - 1 = 3$ and $a_n = 1 + 3(n-1)$. Then $a_n = 88$ $\Leftrightarrow$ $1 + 3(n-1) = 88$ $\Leftrightarrow$ $3(n-1) = 87$ $\Leftrightarrow$ $n - 1 = 29$ $\Leftrightarrow$ $n = 30$. So, 88 is the 30th term.

23. $a = 1$, $d = 2$, $n = 10$. Then $S_{10} = \frac{10}{2}[2a + (10-1)d] = \frac{10}{2}[2 \cdot 1 + 9 \cdot 2] = 100$.

25. $a = 4$, $d = 2$, $n = 20$. Then $S_{20} = \frac{20}{2}[2a + (20-1)d] = \frac{20}{2}[2 \cdot 4 + 19 \cdot 2] = 460$.

27. $a_1 = 55$, $d = 12$, $n = 10$. Then $S_{10} = \frac{10}{2}[2a + (10-1)d] = \frac{10}{2}[2 \cdot 55 + 9 \cdot 12] = 1090$.

29. $1 + 5 + 9 + \cdots + 401$ is a partial sum of an arithmetic series, where $a = 1$ and $d = 5 - 1 = 4$. The last term is $401 = a_n = 1 + 4(n-1)$, so $n - 1 = 100$ $\Leftrightarrow$ $n = 101$. So, the partial sum is $S_{101} = \frac{101}{2}(1 + 401) = 101 \cdot 201 = 20{,}301$.

31. $0.7 + 2.7 + 4.7 + \cdots + 56.7$ is a partial sum of an arithmetic series, where $a = 0.7$ and $d = 2.7 - 0.7 = 2$. The last term is $56.7 = a_n = 0.7 + 2(n-1)$ $\Leftrightarrow$ $28 = n - 1$ $\Leftrightarrow$ $n = 29$. So, the partial sum is $S_{29} = \frac{29}{2}(0.7 + 56.7) = 832.3$.

33. $\sum_{k=0}^{10}(3+0.25k)$ is a partial sum of an arithmetic series, where $a = 3 + 0.25 \cdot 0 = 3$ and $d = 0.25$.
The last term is $a_{11} = 3 + 0.25 \cdot 10 = 5.5$. So the partial sum is $S_{11} = \frac{11}{2}(3 + 5.5) = 46.75$.

35. The diminishing values of the computer form an arithmetic sequence with $a_1 = 12500$ and common difference $d = -1875$. Thus the value of the computer after 6 years is
$a_7 = 12500 + (7 - 1)(-1875) = \$1250$.

37. The increasing values of the man's salary form an arithmetic sequence with $a_1 = 30000$ and common difference $d = 2300$. Then his total earnings for a ten-year period are
$S_{10} = \frac{10}{2}[2(30000) + 9(2300)] = 403500$. Thus his total earnings for the 10 year period are $\$403,500$.

39. The number of seats in the nth row is given by the nth term of an arithmetic sequence with $a_1 = 15$ and common difference $d = 3$. We need to find n such that $S_n = 870$. So we solve
$870 = S_n = \frac{n}{2}[2(15) + (n - 1)3]$ for n. We have $870 = \frac{n}{2}(27 + 3n)$ $\Leftrightarrow$ $1740 = 3n^2 + 27n$
$\Leftrightarrow$ $3n^2 + 27n - 1740 = 0$ $\Leftrightarrow$ $n^2 + 9n - 580 = 0$ $\Leftrightarrow$ $(x - 20)(x + 29) = 0$ $\Rightarrow$
$n = 20$ or $n = -29$. Since the number of rows is positive, the theater must have 20 rows.

41. The sequence is $16, 48, 80, \ldots$. This is an arithmetic sequence with $a = 16$ and $d = 48 - 16 = 32$.
(a) The total distance after 6 seconds is $S_6 = \frac{6}{2}(32 + 5 \cdot 32) = 3 \cdot 192 = 576$ ft.
(b) The total distance after n seconds is $S_n = \frac{n}{2}[32 + 32(n - 1)] = 16n^2$ ft.

43. Let x denote the length of the side between the length of the other two sides. Then the lengths of the three sides of the triangle are $x - a$, x, and $x + a$, for some $a > 0$. Since $x + a$ is the longest side, it is the hypotenuse, and by the Pythagorean Theorem, we know that $(x - a)^2 + x^2 = (x + a)^2$ $\Leftrightarrow$
$x^2 - 2ax + a^2 + x^2 = x^2 + 2ax + a^2$ $\Leftrightarrow$ $x^2 - 4ax = 0$ $\Leftrightarrow$ $x(x - 4a) = 0$ $\Rightarrow$
$x = 4a$ ($x = 0$ is not a possible solution). Thus, the lengths of the three sides are
$x - a = 4a - a = 3a$, $x = 4a$, and $x + a = 4a + a = 5a$. The lengths $3a, 4a, 5a$ are proportional to $3, 4, 5$, and so the triangle is similar to a 3-4-5 triangle.

45. The sequence $1, \frac{3}{5}, \frac{3}{7}, \frac{1}{3}, \ldots$ is harmonic if $1, \frac{5}{3}, \frac{7}{3}, 3, \ldots$ forms an arithmetic sequence. Since
$\frac{5}{3} - 1 = \frac{7}{3} - \frac{5}{3} = 3 - \frac{7}{3} = \frac{2}{3}$, the sequence of reciprocals is arithmetic and thus the original sequence is harmonic.

47. (a) We want an arithmetic sequence with 4 terms, so let $a_1 = 10$ and $a_4 = 18$. Since the sequence is arithmetic, $a_4 - a_1 = 3d = 18 - 10 = 8$ $\Leftrightarrow$ $3d = 8$ $\Leftrightarrow$ $d = \frac{8}{3}$. Therefore,
$a_2 = 10 + \frac{8}{3} = \frac{38}{3}$ and $a_3 = 10 + 2(\frac{8}{3}) = \frac{46}{3}$ are the two arithmetic means between 10 and 18.

(b) We want an arithmetic sequence with 5 terms, so let $a_1 = 10$ and $a_5 = 18$. Since the sequence is arithmetic, $a_5 - a_1 = 4d = 18 - 10 = 8$ $\Leftrightarrow$ $4d = 8$ $\Leftrightarrow$ $d = 2$. Therefore,
$a_2 = 10 + 2 = 12$, $a_3 = 10 + 2(2) = 14$, and $a_4 = 10 + 3(2) = 16$ are the three arithmetic means between 10 and 18.

(c) We want an arithmetic sequence with 6 terms, with the starting dosage $a_1 = 100$ and the final dosage $a_6 = 300$. Since the sequence is arithmetic, $a_6 - a_1 = 5d = 300 - 100 = 200$ $\Leftrightarrow$
$5d = 200$ $\Leftrightarrow$ $d = 40$. Therefore, $a_2 = 140$, $a_3 = 180$, $a_4 = 220$, $a_5 = 260$, and
$a_6 = 300$. The patient should take 140 mg, then 180 mg, then 220 mg, then 260 mg, and finally arrive at 300 mg.

Exercises 12.3

1. $\dfrac{a_2}{a_1} = \dfrac{4}{2} = 2;\ \dfrac{a_3}{a_2} = \dfrac{8}{4} = 2;\ \dfrac{a_4}{a_3} = \dfrac{16}{8} = 2.$ Since these ratios are the same, the sequence is geometric with the common ratio 2.

3. $\dfrac{a_2}{a_1} = \dfrac{3/2}{3} = \dfrac{1}{2};\ \dfrac{a_3}{a_2} = \dfrac{3/4}{3/2} = \dfrac{1}{2};\ \dfrac{a_4}{a_3} = \dfrac{3/8}{3/4} = \dfrac{1}{2}.$ Since these ratios are the same, the sequence is geometric with the common ratio $\frac{1}{2}$.

5. $\dfrac{a_2}{a_1} = \dfrac{1/3}{1/2} = \dfrac{2}{3};\ \dfrac{a_4}{a_3} = \dfrac{1/5}{1/4} = \dfrac{4}{5}.$ Since these ratios are not the same, this is not a geometric sequence.

7. $2, 6, 18, 54, \ldots$. Then $r = \dfrac{a_2}{a_1} = \dfrac{6}{2} = 3;\ a_5 = a_4 \cdot 3 = 54(3) = 162;$ and $a_n = 2 \cdot 3^{n-1}.$

9. $0.3, -0.09, 0.027, -0.0081, \ldots$. Then $r = \dfrac{a_2}{a_1} = \dfrac{-0.09}{0.3} = -0.3;\ a_5 = a_4 \cdot (-0.3) = -0.0081(-0.3) = 0.00243;$ and $a_n = 0.3(-0.3)^{n-1}.$

11. $144, -12, 1, -\dfrac{1}{12}, \ldots$. Then $r = \dfrac{a_2}{a_1} = \dfrac{-12}{144} = -\dfrac{1}{12};\ a_5 = a_4 \cdot \left(-\dfrac{1}{12}\right) = -\dfrac{1}{12}\left(-\dfrac{1}{12}\right) = \dfrac{1}{144};$ $a_n = 144\left(-\dfrac{1}{12}\right)^{n-1}.$

13. $3, 3^{5/3}, 3^{7/3}, 27, \ldots$. Then $r = \dfrac{a_2}{a_1} = \dfrac{3^{5/3}}{3} = 3^{2/3};\ a_5 = a_4 \cdot \left(3^{2/3}\right) = 27 \cdot 3^{2/3} = 3^{11/3};$ and $a_n = 3\left(3^{2/3}\right)^{n-1} = 3 \cdot 3^{(2n-2)/3} = 3^{(2n+1)/3}.$

15. $1, s^{2/7}, s^{4/7}, s^{6/7}, \ldots$. Then $r = \dfrac{a_2}{a_1} = \dfrac{s^{2/7}}{1} = s^{2/7};\ a_5 = a_4 \cdot s^{2/7} = s^{6/7} \cdot s^{2/7} = s^{8/7};$ and $a_n = \left(s^{2/7}\right)^{n-1} = s^{(2n-2)/7}.$

17. $a_1 = 8,\ a_2 = 4.$ Thus $r = \dfrac{a_2}{a_1} = \dfrac{4}{8} = \dfrac{1}{2}$ and $a_5 = a_1 r^{5-1} = 8\left(\dfrac{1}{2}\right)^4 = \dfrac{8}{16} = \dfrac{1}{2}.$

19. $r = \dfrac{2}{5},\ a_4 = \dfrac{5}{2}.$ Since $r = \dfrac{a_4}{a_3}$, we have $a_3 = \dfrac{a_4}{r} = \dfrac{\frac{5}{2}}{\frac{2}{5}} = \dfrac{25}{4}.$

21. The geometric sequence is $2, 6, 18, \ldots$. Thus $r = \dfrac{a_2}{a_1} = \dfrac{6}{2} = 3.$ We need to find n so that

$a_n = 2 \cdot 3^{n-1} = 118{,}098 \quad \Leftrightarrow \quad 3^{n-1} = 59{,}049 \quad \Leftrightarrow \quad n - 1 = \log_3 59{,}049 = 10 \quad \Leftrightarrow \quad n = 11.$ Therefore, 118,098 is the 11th term of the geometric sequence.

23. $a = 5,\ r = 2,\ n = 6.$ Then $S_6 = 5\dfrac{1 - 2^6}{1 - 2} = (-5)(-63) = 315.$

25. $a_3 = 28,\ a_6 = 224,\ n = 6.$ So $\dfrac{a_6}{a_3} = \dfrac{ar^5}{ar^2} = r^3.$ So we have $r^3 = \dfrac{a_6}{a_3} = \dfrac{224}{28} = 8,$ and hence $r = 2.$

Since $a_3 = a \cdot r^2$, we get $a = \dfrac{a_3}{r^2} = \dfrac{28}{2^2} = 7.$ So $S_6 = 7\dfrac{1 - 2^6}{1 - 2} = (-7)(-63) = 441.$

27. $1 + 3 + 9 + \cdots + 2187$ is a partial sum of a geometric sequence, where $a = 1$ and $r = \dfrac{a_2}{a_1} = \dfrac{3}{1} = 3$.

Then the last term is $2187 = a_n = 1 \cdot 3^{n-1}$ $\Leftrightarrow$ $n - 1 = \log_3 2187 = 7$ $\Leftrightarrow$ $n = 8$. So the

partial sum is $S_8 = (1)\dfrac{1 - 3^8}{1 - 3} = 3280$.

29. $\displaystyle\sum_{k=0}^{10} 3(\tfrac{1}{2})^k$ is partial sum of a geometric sequence, where $a = 3$, $r = \tfrac{1}{2}$, and $n = 11$. So the partial sum

is $S_{11} = (3)\dfrac{1 - \left(\tfrac{1}{2}\right)^{11}}{1 - \left(\tfrac{1}{2}\right)} = 6\left[1 - \left(\tfrac{1}{2}\right)^{11}\right] = 5.997070313$.

31.   Since the ball is dropped from a height of 80 feet, $a = 80$. Also since the ball rebounds three-
fourths of the distance fallen, $r = \tfrac{3}{4}$. So on the nth bounce, the ball attains a height of
$a_n = 80\left(\tfrac{3}{4}\right)^n$. Hence, on the 5th bounce, the ball goes $a_5 = 80\left(\tfrac{3}{4}\right)^5 = \dfrac{80 \cdot 243}{1024} \approx 19$ ft high.

33.   Let a_n be the amount of water remaining at the nth stage. We start with 5 gallons, so $a = 5$. When
1 gallon, that is $\tfrac{1}{5}$ of the mixture, is removed, $\tfrac{4}{5}$ of the mixture (and hence $\tfrac{4}{5}$ of the water in the
mixture) remains. Thus, $a_1 = 5 \cdot \tfrac{4}{5}$, $a_2 = 5 \cdot \tfrac{4}{5} \cdot \tfrac{4}{5}, \ldots$, and in general, $a_n = 5\left(\tfrac{4}{5}\right)^n$. The amount of
water remaining after 3 repetitions is $a_3 = 5\left(\tfrac{4}{5}\right)^3 = \dfrac{64}{25}$, and after 5 repetitions is $a_5 = 5\left(\tfrac{4}{5}\right)^5 = \dfrac{1024}{625}$.

35.   Let a_n be the height the ball reaches on the nth bounce. From the given information, a_n is the
geometric sequence $a_n = 9 \cdot \left(\tfrac{1}{3}\right)^n$. (Notice that the ball hits the ground for the fifth time after the 4th
bounce.)

(a) $a_0 = 9$, $a_1 = 9 \cdot \tfrac{1}{3} = 3$, $a_2 = 9 \cdot \left(\tfrac{1}{3}\right)^2 = 1$, $a_3 = 9 \cdot \left(\tfrac{1}{3}\right)^3 = \tfrac{1}{3}$, and $a_4 = 9 \cdot \left(\tfrac{1}{3}\right)^4 = \tfrac{1}{9}$. The
total distance traveled is $a_0 + 2a_1 + 2a_2 + 2a_3 + 2a_4 =$
$9 + 2 \cdot 3 + 2 \cdot 1 + 2 \cdot \tfrac{1}{3} + 2 \cdot \tfrac{1}{9} = \dfrac{161}{9} = 17\tfrac{8}{9}$ ft.

(b)   The total distance traveled at the instant the ball hits the ground for the nth time is
$$D_n = 9 + 2 \cdot 9 \cdot \tfrac{1}{3} + 2 \cdot 9 \cdot \left(\tfrac{1}{3}\right)^2 + 2 \cdot 9 \cdot \left(\tfrac{1}{3}\right)^3 + 2 \cdot 9 \cdot \left(\tfrac{1}{3}\right)^4 + \cdots + 2 \cdot 9 \cdot \left(\tfrac{1}{3}\right)^{n-1}$$
$$= 2\left[9 + 9 \cdot \tfrac{1}{3} + 9 \cdot \left(\tfrac{1}{3}\right)^2 + 9 \cdot \left(\tfrac{1}{3}\right)^3 + 9 \cdot \left(\tfrac{1}{3}\right)^4 + \cdots + 9 \cdot \left(\tfrac{1}{3}\right)^{n-1}\right] - 9$$
$$= 2\left[9 \cdot \dfrac{1 - \left(\tfrac{1}{3}\right)^n}{1 - \tfrac{1}{3}}\right] - 9 = 27\left[1 - \left(\tfrac{1}{3}\right)^n\right] - 9 = 18 - \left(\tfrac{1}{3}\right)^{n-3}.$$

37. $1 + \tfrac{1}{3} + \tfrac{1}{9} + \tfrac{1}{27} + \cdots$ is an infinite geometric series with $a = 1$ and $r = \tfrac{1}{3}$. Therefore, the sum of the
series is $S = \dfrac{a}{1 - r} = \dfrac{1}{1 - \left(\tfrac{1}{3}\right)} = \dfrac{3}{2}$.

39. $1 - \tfrac{1}{3} + \tfrac{1}{9} - \tfrac{1}{27} + \cdots$ is an infinite geometric series with $a = 1$ and $r = -\tfrac{1}{3}$. Therefore, the sum of
the series is $S = \dfrac{a}{1 - r} = \dfrac{1}{1 - \left(-\tfrac{1}{3}\right)} = \dfrac{3}{4}$.

41. $\dfrac{1}{3^6} + \dfrac{1}{3^8} + \dfrac{1}{3^{10}} + \dfrac{1}{3^{12}} + \cdots$ is an infinite geometric series with $a = \dfrac{1}{3^6}$ and $r = \dfrac{1}{3^2} = \dfrac{1}{9}$. Therefore,

the sum of the series is $S = \dfrac{a}{1 - r} = \dfrac{\tfrac{1}{3^6}}{1 - \left(\tfrac{1}{9}\right)} = \dfrac{1}{3^6} \cdot \dfrac{9}{8} = \dfrac{1}{648}$.

43. $-\dfrac{100}{9} + \dfrac{10}{3} - 1 + \dfrac{3}{10} - \cdots$ is an infinite geometric series with $a = -\dfrac{100}{9}$ and $r = \dfrac{\frac{10}{3}}{-\frac{100}{9}} = \dfrac{3}{10}$.

Therefore, the sum of the series is $S = \dfrac{a}{1-r} = \dfrac{-\frac{100}{9}}{1 - \left(-\frac{3}{10}\right)} = \dfrac{-\frac{100}{9}}{\frac{13}{10}} = -\dfrac{100}{9} \cdot \dfrac{10}{13} = -\dfrac{1000}{117}$.

45. $0.777\ldots = \dfrac{7}{10} + \dfrac{7}{100} + \dfrac{7}{1000} + \cdots$ is an infinite geometric series with $a = \dfrac{7}{10}$ and $r = \dfrac{1}{10}$. Thus

$0.777\ldots = \dfrac{a}{1-r} = \dfrac{\frac{7}{10}}{1 - \frac{1}{10}} = \dfrac{7}{9}$.

47. $0.030303\ldots = \dfrac{3}{100} + \dfrac{3}{10{,}000} + \dfrac{3}{1{,}000{,}000} + \cdots$ is an infinite geometric series with $a = \dfrac{3}{100}$ and

$r = \dfrac{1}{100}$. Thus $0.030303\ldots = \dfrac{a}{1-r} = \dfrac{\frac{3}{100}}{1 - \frac{1}{100}} = \dfrac{3}{99} = \dfrac{1}{33}$.

49. $0.\overline{112} = 0.112112112\ldots = \dfrac{112}{1000} + \dfrac{112}{1{,}000{,}000} + \dfrac{112}{1{,}000{,}000{,}000} + \cdots$ is an infinite geometric

series with $a = \dfrac{112}{1000}$ and $r = \dfrac{1}{1000}$. Thus $0.112112112\ldots = \dfrac{a}{1-r} = \dfrac{\frac{112}{1000}}{1 - \frac{1}{1000}} = \dfrac{112}{999}$.

51.   Let a_n be the height the ball reaches on the nth bounce. We have $a_0 = 1$ and $a_n = \frac{1}{2}a_{n-1}$. Since
the total distance d traveled includes the bounce up as well and the distance down, we have
$d = a_0 + 2 \cdot a_1 + 2 \cdot a_2 + \ldots = 1 + 2\left(\frac{1}{2}\right) + 2\left(\frac{1}{2}\right)^2 + 2\left(\frac{1}{2}\right)^3 + 2\left(\frac{1}{2}\right)^4 + \cdots$

$= 1 + 1 + \frac{1}{2} + \left(\frac{1}{2}\right)^2 + \left(\frac{1}{2}\right)^3 + \cdots = 1 + \displaystyle\sum_{i=0}^{\infty}\left(\frac{1}{2}\right)^i = 1 + \dfrac{1}{1 - \frac{1}{2}} = 3$. Thus the total distance

traveled is about 3m.

53.   (a)   If a square has side x, then the side of the square formed by joining the midpoints is (by the

Pythagorean Theorem) $\sqrt{\left(\dfrac{x}{2}\right)^2 + \left(\dfrac{x}{2}\right)^2} = \sqrt{\dfrac{x^2}{4} + \dfrac{x^2}{4}} = \dfrac{x}{\sqrt{2}}$. In our case, $x = 1$ and the

side of the first inscribed square is $\dfrac{1}{\sqrt{2}}$, the side of the second inscribed square is

$\dfrac{1}{\sqrt{2}} \cdot \dfrac{1}{\sqrt{2}} = \left(\dfrac{1}{\sqrt{2}}\right)^2$, the side of the third inscribed square is $\left(\dfrac{1}{\sqrt{2}}\right)^3, \ldots$. Since this pattern

continues, the total area of all the squares is $A = 1^2 + \left(\dfrac{1}{\sqrt{2}}\right)^2 + \left(\dfrac{1}{\sqrt{2}}\right)^4 + \left(\dfrac{1}{\sqrt{2}}\right)^6 + \cdots$

$= 1 + \frac{1}{2} + \left(\frac{1}{2}\right)^2 + \left(\frac{1}{2}\right)^3 + \cdots = \dfrac{1}{1 - \frac{1}{2}} = 2$.

(b)   As in part (a), the sides of the squares are $1, \dfrac{1}{\sqrt{2}}, \left(\dfrac{1}{\sqrt{2}}\right)^2, \left(\dfrac{1}{\sqrt{2}}\right)^3, \ldots$. Thus the sum of the

perimeters is $S = 4 \cdot 1 + 4 \cdot \dfrac{1}{\sqrt{2}} + 4 \cdot \left(\dfrac{1}{\sqrt{2}}\right)^2 + 4 \cdot \left(\dfrac{1}{\sqrt{2}}\right)^3 + \cdots$, which is an infinite

geometric series with $a = 4$ and $r = \dfrac{1}{\sqrt{2}}$. Thus the sum of the perimeters is

$S = \dfrac{4}{1 - \frac{1}{\sqrt{2}}} = \dfrac{4\sqrt{2}}{\sqrt{2}-1} = \dfrac{4\sqrt{2}}{\sqrt{2}-1} \cdot \dfrac{\sqrt{2}+1}{\sqrt{2}+1} = \dfrac{4\cdot 2 + 4\sqrt{2}}{2-1} = 8 + 4\sqrt{2}$.

55.   Let a_n denote the area colored blue at nth stage. Since only the middle squares are colored blue,
$a_n = \frac{1}{9} \times$ (area remaining yellow at the $(n-1)^{\text{st}}$ stage). Also, the area remaining yellow at the nth

stage is $\frac{8}{9}$ of the area remaining yellow at the preceding stage. So $a_1 = \frac{1}{9}$, $a_2 = \frac{1}{9}\left(\frac{8}{9}\right)$, $a_3 = \frac{1}{9}\left(\frac{8}{9}\right)^2$, $a_4 = \frac{1}{9}\left(\frac{8}{9}\right)^3, \ldots$. Thus the total area colored blue $A = \frac{1}{9} + \frac{1}{9}\left(\frac{8}{9}\right) + \frac{1}{9}\left(\frac{8}{9}\right)^2 + \frac{1}{9}\left(\frac{8}{9}\right)^3 + \ldots$ is an

infinite geometric series with $a = \frac{1}{9}$ and $r = \frac{8}{9}$. So the total area is $A = \dfrac{\frac{1}{9}}{1 - \frac{8}{9}} = 1$.

57. The sum is given by $(a + b) + (a^2 + 2b) + (a^3 + 3b) + \cdots + (a^{10} + 10b) =$

$(a + a^2 + a^3 + \cdots + a^{10}) + (b + 2b + 3b + \cdots + 10b) = a\dfrac{1 - a^9}{1 - a} + \dfrac{10}{2}(b + 10b)$

$= a\dfrac{1 - a^9}{1 - a} + 55b.$

59. Let $a_1, a_2, a_3, \ldots$ be a geometric sequence with common ratio r. Thus $a_2 = a_1 r$, $a_3 = a_1 \cdot r^2, \ldots$,

$a_n = a_1 \cdot r^{n-1}$. Hence, $\dfrac{1}{a_2} = \dfrac{1}{a_1 \cdot r} = \dfrac{1}{a_1} \cdot \dfrac{1}{r}$, $\dfrac{1}{a_3} = \dfrac{1}{a_1 \cdot r^2} = \dfrac{1}{a_1} \cdot \dfrac{1}{r^2} = \dfrac{1}{a_1} \cdot \left(\dfrac{1}{r}\right)^2, \ldots$

$\dfrac{1}{a_n} = \dfrac{1}{a_1 \cdot r^{n-1}} = \dfrac{1}{a_1} \cdot \dfrac{1}{r^{n-1}} = \dfrac{1}{a_1}\left(\dfrac{1}{r}\right)^{n-1}$, and so $\dfrac{1}{a_1}, \dfrac{1}{a_2}, \dfrac{1}{a_3}, \ldots$ is a geometric sequence with common ratio $\frac{1}{r}$.

61. Since $a_1, a_2, a_3, \ldots$ is an arithmetic sequence with common difference d, the terms can be expressed as $a_2 = a_1 + d$, $a_3 = a_1 + 2d, \ldots$, $a_n = a_1 + (n - 1)d$. So $10^{a_2} = 10^{a_1 + d} = 10^{a_1} \cdot 10^d$, $10^{a_3} = 10^{a_1 + 2d} = 10^{a_1} \cdot (10^d)^2, \ldots$, $10^{a_n} = 10^{a_1 + (n-1)d} = 10^{a_1} \cdot (10^d)^{n-1}$, and so $10^{a_1}, 10^{a_2}, 10^{a_3}$, $\ldots$ is a geometric sequence with common ratio $r = 10^d$.

Exercises 12.4

1. $n = 10$, $R = \$1000$, $i = 0.06$. So, $A_f = R\dfrac{(1+i)^n - 1}{i} = 1000\dfrac{(1+0.06)^{10} - 1}{0.06} = \$13,180.79$.

3. $n = 20$, $R = \$5000$, $i = 0.12$. So, $A_f = R\dfrac{(1+i)^n - 1}{i} = 5000\dfrac{(1+0.12)^{20} - 1}{0.12} = \$360,262.21$.

5. $n = 16$, $R = \$300$, $i = \dfrac{0.08}{4} = 0.02$. So, $A_f = R\dfrac{(1+i)^n - 1}{i} = 300\dfrac{(1+0.02)^{16} - 1}{0.02}$
   $= \$5,591.79$.

7. $A_f = \$2000$, $i = \dfrac{0.06}{12} = 0.005$, $n = 8$. Then $R = \dfrac{iA_f}{(1+i)^n - 1} = \dfrac{(0.005)(2000)}{(1+0.005)^8 - 1} = \$245.66$.

9. $R = \$200$, $n = 20$, $i = \dfrac{0.09}{2} = 0.045$. So, $A_p = R\dfrac{1 - (1+i)^{-n}}{i} = (200)\dfrac{1 - (1+0.045)^{-20}}{0.045}$
   $= \$2601.59$.

11. $A_p = \$12,000$, $i = \dfrac{0.105}{12} = 0.00875$, $n = 48$. Then $R = \dfrac{iA_p}{1 - (1+i)^{-n}} = \dfrac{(0.00875)(12000)}{1 - (1+0.00875)^{-48}}$
    $= \$307.24$.

13. $A_p = \$100,000$, $i = \dfrac{0.08}{12} \approx 0.006667$, $n = 360$. Then $R = \dfrac{iA_p}{1 - (1+i)^{-n}}$
    $= \dfrac{(0.006667)(100,000)}{1 - (1+0.006667)^{-360}} = \$733.76$. Therefore, the total amount paid on this loan over the 30
    year period is $(360)(733.76) = \$264,153.60$.

15. $A_p = 100,000$, $n = 360$, $i = \frac{0.0975}{12} = 0.008125$.
    (a) $R = \dfrac{i A_p}{1 - (1+i)^{-n}} = \dfrac{(0.008125)(100,000)}{1 - (1+0.008125)^{-360}} = \$859.15$.
    (b) The total amount that will be paid over the 30 year period is $(360)(859.15) = \$309,294.00$.
    (c) $R = \$859.15$, $i = \dfrac{0.0975}{12} = 0.008125$, $n = 360$. So, $A_f = 859.15\dfrac{(1+0.008125)^{360} - 1}{0.008125}$
        $= \$1,841,519.29$.

17. $R = \$30$, $i = \dfrac{0.10}{12} \approx 0.008333$, $n = 12$. Then $A_p = R\dfrac{1 - (1+i)^{-n}}{i}$
    $= 30\dfrac{1 - \left(1 + 0.008333\right)^{-12}}{0.008333} = \$341.24$.

19. $A_p = \$640$, $R = \$32$, $n = 24$. We want to solve the equation $R = \dfrac{iA_p}{1 - (1+i)^n}$ for the interest
    rate i. Let x be the interest rate, then $i = \frac{x}{12}$. So we can express R as a function of x by
    $R(x) = \dfrac{\frac{x}{12} \cdot 640}{1 - \left(1 + \frac{x}{12}\right)^{-24}}$. We graph $R(x)$ and $y = 32$ in the rectangle $[0.12, 0.22] \times [30, 34]$. The
    x-coordinate of the intersection is about 0.1816, which corresponds to an interest rate of 18.16%.

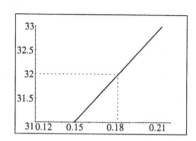

21. $A_p = \$189.99$, $R = \$10.50$, $n = 20$. We want to solve the

equation $R = \dfrac{iA_p}{1 - (1+i)^n}$ for the interest rate i. Let x be the

interest rate, then $i = \frac{x}{12}$. So we can express R as a function of

x by $R(x) = \dfrac{\frac{x}{12} \cdot 189.99}{1 - \left(1 + \frac{x}{12}\right)^{-20}}$. We graph $R(x)$ and $y = 10.50$

in the rectangle $[0.10, 0.18] \times [10, 11]$. The x-coordinate of the
intersection is about 0.1168, which corresponds to an interest
rate of 11.68%.

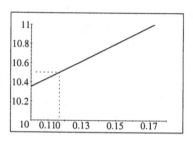

23. (a) The present value of the kth payment is $PV = \dfrac{R}{(1+i)^k}$. The amount of money to be invested

now (A_p) to ensure an annuity in perpetuity is the (infinite) sum of the present values of each of
the payments, as the following time line shows.

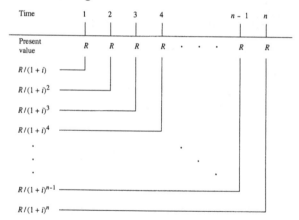

(b) $A_p = \dfrac{R}{1+i} + \dfrac{R}{(1+i)^2} + \dfrac{R}{(1+i)^3} + \cdots + \dfrac{R}{(1+i)^n} + \cdots$. This is an infinite geometric series

with $a = \dfrac{R}{1+i}$ and $r = \dfrac{1}{1+i}$. Therefore $A_p = \dfrac{\frac{R}{1+i}}{1 - \frac{1}{1+i}} = \dfrac{R}{1+i} \cdot \dfrac{1+i}{i} = \dfrac{R}{i}$

(c) Using the result from Part (b), we have $R = 5000$ and $i = 0.10$. Then $A_p = \dfrac{R}{i} = \dfrac{5000}{0.10}$

= \$50,000.

(d) We are given two different time periods: the interest is compounded quarterly, while the
annuity is paid yearly. In order to use the formula in Part (b), we need to find the effective

annual interest rate produced by 8% interest compounded quarterly, which is $\left(1 + \frac{i}{n}\right)^n - 1$. Now $i = 8\%$ and it is compounded quarterly, $n = 4$, so the effective annual yield is $(1 + 0.02)^4 - 1 \approx 0.08243216$. Thus by the formula in Part (b), the amount that must be invested is $A_p = \dfrac{3000}{0.08243216} = \$36{,}393.56$.

Exercises 12.5

1. Let $P(n)$ denote the statement $2 + 4 + 6 + \cdots + 2n = n(n+1)$.

    <u>Step 1</u> $P(1)$ is the statement that $2 = 1(1+1)$, which is true.

    <u>Step 2</u> Assume that $P(k)$ is true; that is, $2 + 4 + 6 + \cdots + 2k = k(k+1)$. We want to use this to show that $P(k+1)$ is true. Now,

    $$2 + 4 + 6 + \cdots + 2k + 2(k+1) =$$
    $$k(k+1) + 2(k+1) = \qquad\qquad\qquad \text{induction hypothesis}$$
    $$(k+1)(k+2) = (k+1)[(k+1)+1].$$

    Thus, $P(k+1)$ follows from $P(k)$. So by the Principle of Mathematical Induction, $P(n)$ is true for all n.

3. Let $P(n)$ denote the statement $5 + 8 + 11 + \cdots + (3n+2) = \dfrac{n(3n+7)}{2}$.

    <u>Step 1</u> We need to show that $P(1)$ is true. But $P(1)$ says that $5 = \frac{1 \cdot (3 \cdot 1 + 7)}{2}$, which is true.

    <u>Step 2</u> Assume that $P(k)$ is true; that is, $5 + 8 + 11 + \cdots + (3k+2) = \dfrac{k(3k+7)}{2}$. We want to use this to show that $P(k+1)$ is true. Now,

    $$5 + 8 + 11 + \cdots + (3k+2) + [3(k+1)+2] =$$
    $$\frac{k(3k+7)}{2} + 3k + 5 = \qquad\qquad \text{induction hypothesis}$$
    $$\frac{3k^2 + 7k}{2} + \frac{6k+10}{2} =$$
    $$\frac{3k^2 + 13k + 10}{2} = \frac{(3k+10)(k+1)}{2} = \frac{(k+1)[3(k+1)+7]}{2}.$$

    Thus, $P(k+1)$ follows from $P(k)$. So by the Principle of Mathematical Induction, $P(n)$ is true for all n

5. Let $P(n)$ denote the statement $1 \cdot 2 + 2 \cdot 3 + 3 \cdot 4 + \cdots + n(n+1) = \dfrac{n(n+1)(n+2)}{3}$.

    <u>Step 1</u> $P(1)$ is the statement that $1 \cdot 2 = \frac{1 \cdot (1+1) \cdot (1+2)}{3}$, which is true.

    <u>Step 2</u> Assume that $P(k)$ is true; that is, $1 \cdot 2 + 2 \cdot 3 + 3 \cdot 4 + \cdots + k(k+1) = \dfrac{k(k+1)(k+2)}{3}$. We want to use this to show that $P(k+1)$ is true. Now,

    $$1 \cdot 2 + 2 \cdot 3 + 3 \cdot 4 + \cdots + k(k+1) + (k+1)[(k+1)+1] =$$
    $$\frac{k(k+1)(k+2)}{3} + (k+1)(k+2) = \qquad\qquad \text{induction hypothesis}$$
    $$\frac{k(k+1)(k+2)}{3} + \frac{3(k+1)(k+2)}{3} = \frac{(k+1)(k+2)(k+3)}{3}.$$

    Thus, $P(k+1)$ follows from $P(k)$. So by the Principle of Mathematical Induction, $P(n)$ is true for all n.

7. Let $P(n)$ denote the statement $1^3 + 2^3 + 3^3 + \cdots + n^3 = \dfrac{n^2(n+1)^2}{4}$.

    <u>Step 1</u> $P(1)$ is the statement that $1^3 = \frac{1^2 \cdot (1+1)^2}{4}$, which is clearly true.

Step 2 Assume that $P(k)$ is true; that is, $1^3 + 2^3 + 3^3 + \cdots + k^3 = \dfrac{k^2(k+1)^2}{4}$. We want to use this

to show that $P(k+1)$ is true. Now,

$1^3 + 2^3 + 3^3 + \cdots + k^3 + (k+1)^3 =$

$\dfrac{k^2(k+1)^2}{4} + (k+1)^3 =$ induction hypothesis

$\dfrac{(k+1)^2[k^2 + 4(k+1)]}{4} =$

$\dfrac{(k+1)^2[k^2 + 4k + 4]}{4} = \dfrac{(k+1)^2(k+2)^2}{4} = \dfrac{(k+1)^2[(k+1) + 1]^2}{4}.$

Thus, $P(k+1)$ follows from $P(k)$. So by the Principle of Mathematical Induction, $P(n)$ is true for all n.

9. Let $P(n)$ denote the statement $2^3 + 4^3 + 6^3 + \cdots + (2n)^3 = 2n^2(n+1)^2$.

Step 1 $P(1)$ is true since $2^3 = 2(1)^2(1+1)^2 = 2 \cdot 4 = 8$.

Step 2 Assume that $P(k)$ is true; that is, $2^3 + 4^3 + 6^3 + \cdots + (2k)^3 = 2k^2(k+1)^2$. We want to use this to show that $P(k+1)$ is true. Now,

$2^3 + 4^3 + 6^3 + \cdots + (2k)^3 + [2(k+1)]^3 =$

$2k^2(k+1)^2 + [2(k+1)]^3 =$ induction hypothesis

$2k^2(k+1)^2 + 8(k+1)(k+1)^2 =$

$(k+1)^2(2k^2 + 8k + 8) = 2(k+1)^2(k+2)^2 = 2(k+1)^2[(k+1) + 1]^2.$

Thus, $P(k+1)$ follows from $P(k)$. So by the Principle of Mathematical Induction, $P(n)$ is true for all n.

11. Let $P(n)$ denote the statement $1 \cdot 2 + 2 \cdot 2^2 + 3 \cdot 2^3 + 4 \cdot 2^4 + \cdots + n \cdot 2^n = 2[1 + (n-1)2^n]$.

Step 1 $P(1)$ is the statement that $1 \cdot 2 = 2[1 + 0]$, which is clearly true.

Step 2 Assume that $P(k)$ is true; that is,

$1 \cdot 2 + 2 \cdot 2^2 + 3 \cdot 2^3 + 4 \cdot 2^4 + \cdots + k \cdot 2^k = 2[1 + (k-1)2^k]$. We want to use this to show that $P(k+1)$ is true. Now,

$1 \cdot 2 + 2 \cdot 2^2 + 3 \cdot 2^3 + 4 \cdot 2^4 + \cdots + k \cdot 2^k + (k+1) \cdot 2^{(k+1)} =$

$2[1 + (k-1)2^k] + (k+1) \cdot 2^{(k+1)} =$ induction hypothesis

$2[1 + (k-1) \cdot 2^k + (k+1) \cdot 2^k] =$

$2[1 + 2k \cdot 2^k] = 2[1 + k \cdot 2^{k+1}] = 2\{1 + [(k+1) - 1]2^{k+1}\}.$

Thus $P(k+1)$ follows from $P(k)$. So by the Principle of Mathematical Induction, $P(n)$ is true for all n.

13. Let $P(n)$ denote the statement $n^2 + n$ is divisible by 2.

Step 1 $P(1)$ is the statement that $1^2 + 1 = 2$ is divisible by 2, which is clearly true.

Step 2 Assume that $P(k)$ is true; that is, $k^2 + k$ is divisible by 2. Now,

$(k+1)^2 + (k+1) = k^2 + 2k + 1 + k + 1 = (k^2 + k) + 2k + 2 = (k^2 + k) + 2(k+1).$

By the induction hypothesis, $k^2 + k$ is divisible by 2, and clearly $2(k+1)$ is divisible by 2. Thus, the sum is divisible by 2, so $P(k+1)$ true. Therefore, $P(k+1)$ follows from $P(k)$. So by the Principle of Mathematical Induction, $P(n)$ is true for all n.

15. Let $P(n)$ denote the statement that $n^2 - n + 41$ is odd.

Step 1 $P(1)$ is the statement that $1^2 - 1 + 41 = 41$ is odd, which is clearly true.

Step 2 Assume that $P(k)$ is true; that is, $k^2 - k + 41$ is odd. We want to use this to show that $P(k+1)$ is true. Now,
$(k+1)^2 - (k+1) + 41 = k^2 + 2k + 1 - k - 1 + 41 = (k^2 - k + 41) + 2k$, which is also odd because $k^2 - k + 41$ is odd by the induction hypothesis, $2k$ is always even, and an odd number plus an even number is always odd. Therefore, $P(k+1)$ follows from $P(k)$. So by the Principle of Mathematical Induction, $P(n)$ is true for all n.

17. Let $P(n)$ denote the statement that $8^n - 3^n$ is divisible by 5.

Step 1 $P(1)$ is the statement that $8^1 - 3^1 = 5$ is divisible by 5, which is clearly true.

Step 2 Assume that $P(k)$ is true; that is, $8^k - 3^k$ is divisible by 5. We want to use this to show that $P(k+1)$ is true. Now, $8^{k+1} - 3^{k+1} = 8 \cdot 8^k - 3 \cdot 3^k = 8 \cdot 8^k - (8-5) \cdot 3^k$
$= 8 \cdot (8^k - 3^k) + 5 \cdot 3^k$, which is divisible by 5 because $8^k - 3^k$ is divisible by 5 by our induction hypothesis, and $5 \cdot 3^k$ is divisible by 5. Thus $P(k+1)$ follows from $P(k)$. So by the Principle of Mathematical Induction, $P(n)$ is true for all n.

19. Let $P(n)$ denote the statement $n < 2^n$.

Step 1 $P(1)$ is the statement that $1 < 2^1 = 2$, which is clearly true.

Step 2 Assume that $P(k)$ is true; that is, $k < 2^k$. We want to use this to show that $P(k+1)$ is true. Adding 1 to both sides of $P(k)$ we have $k + 1 < 2^k + 1$. Since $1 < 2^k$ for $k \geq 1$, we have $2^k + 1 < 2^k + 2^k = 2 \cdot 2^k = 2^{k+1}$. Thus $k + 1 < 2^{k+1}$, which is exactly $P(k+1)$. Therefore, $P(k+1)$ follows from $P(k)$. So by the Principle of Mathematical Induction, $P(n)$ is true for all n.

21. Let $P(n)$ denote the statement $(1+x)^n \geq 1 + nx$, if $x > -1$.

Step 1 $P(1)$ is the statement that $(1+x)^1 \geq 1 + 1x$, which is clearly true.

Step 2 Assume that $P(k)$ is true; that is, $(1+x)^k \geq 1 + kx$. Now,
$(1+x)^{k+1} = (1+x)(1+x)^k \geq (1+x)(1+kx)$, by the induction hypothesis. Since $(1+x)(1+kx) = 1 + (k+1)x + kx^2 \geq 1 + (k+1)x$ (since $kx^2 \geq 0$), we have $(1+x)^{k+1} \geq 1 + (k+1)x$, which is $P(k+1)$. Thus $P(k+1)$ follows from $P(k)$. So the Principle of Mathematical Induction, $P(n)$ is true for all n.

23. Let $P(n)$ be the statement that $a_n = 5 \cdot 3^{n-1}$.

Step 1 $P(1)$ is the statement that $a_1 = 5 \cdot 3^0 = 5$, which is true.

Step 2 Assume that $P(k)$ is true; that is, $a_k = 5 \cdot 3^{k-1}$. We want to use this to show that $P(k+1)$ is true. Now, $a_{k+1} = 3a_k = 3 \cdot (5 \cdot 3^{k-1})$, by the induction hypothesis. Therefore, $a_{k+1} = 3 \cdot (5 \cdot 3^{k-1}) = 5 \cdot 3^k$, which is exactly $P(k+1)$. Thus, $P(k+1)$ follows from $P(k)$. So by the Principle of Mathematical Induction, $P(n)$ is true for all n.

25. Let $P(n)$ be the statement that $x - y$ is a factor of $x^n - y^n$ for all natural numbers n.

Step 1 $P(1)$ is the statement that $x - y$ is a factor of $x^1 - y^1$, which is clearly true.

Step 2 Assume that $P(k)$ is true; that is, $x - y$ is a factor of $x^k - y^k$. We want to use this to show that $P(k+1)$ is true. Now, $x^{k+1} - y^{k+1} = x^{k+1} - x^k y + x^k y - y^{k+1}$
$= x^k(x - y) + (x^k - y^k)y$, for which $x - y$ is a factor because $x - y$ is a factor of $x^k(x - y)$, and $x - y$ is a factor of $(x^k - y^k)y$, by the induction hypothesis. Thus $P(k+1)$ follows from $P(k)$. So by the Principle of Mathematical Induction, $P(n)$ is true for all n.

27. Let $P(n)$ denote the statement that F_{3n} is even for all natural numbers n.

    Step 1 $P(1)$ is the statement that F_3 is even. Since $F_3 = F_2 + F_1 = 1 + 1 = 2$, this statement is true.

    Step 2 Assume that $P(k)$ is true; that is, F_{3k} is even. We want to use this to show that $P(k+1)$ is true. Now, $F_{3(k+1)} = F_{3k+3} = F_{3k+2} + F_{3k+1} = F_{3k+1} + F_{3k} + F_{3k+1} = F_{3k} + 2 \cdot F_{3k+1}$, which is even because F_{3k} is even by the induction hypothesis, and $2 \cdot F_{3k+1}$ is even. Thus $P(k+1)$ follows from $P(k)$. So by the Principle of Mathematical Induction, $P(n)$ is true for all n.

29. Let $P(n)$ denote the statement that $F_1^2 + F_2^2 + F_3^2 + \cdots + F_n^2 = F_n \cdot F_{n+1}$.

    Step 1 $P(1)$ is the statement that $F_1^2 = F_1 \cdot F_2$ or $1^2 = 1 \cdot 1$, which is true.

    Step 2 Assume that $P(k)$ is true, that is, $F_1^2 + F_2^2 + F_3^2 + \cdots + F_k^2 = F_k \cdot F_{k+1}$. We want to use this to show that $P(k+1)$ is true. Now,

    $$F_1^2 + F_2^2 + F_3^2 + \cdots + F_k^2 + F_{k+1}^2 =$$
    $$F_k \cdot F_{k+1} + F_{k+1}^2 = \qquad\qquad \text{induction hypothesis}$$
    $$F_{k+1}(F_k + F_{k+1}) = F_{k+1} \cdot F_{k+2}. \qquad \text{by definition of the Fibonacci sequence}$$

    Thus $P(k+1)$ follows from $P(k)$. So by the Principle of Mathematical Induction, $P(n)$ is true for all n.

31. Let $P(n)$ denote the statement $\begin{bmatrix} 1 & 1 \\ 1 & 0 \end{bmatrix}^n = \begin{bmatrix} F_{n+1} & F_n \\ F_n & F_{n-1} \end{bmatrix}$.

    Step 1 Since $\begin{bmatrix} 1 & 1 \\ 1 & 0 \end{bmatrix}^2 = \begin{bmatrix} 1 & 1 \\ 1 & 0 \end{bmatrix}\begin{bmatrix} 1 & 1 \\ 1 & 0 \end{bmatrix} = \begin{bmatrix} 2 & 1 \\ 1 & 1 \end{bmatrix} = \begin{bmatrix} F_3 & F_2 \\ F_2 & F_1 \end{bmatrix}$, it follows that $P(2)$ is true.

    Step 2 Assume that $P(k)$ is true; that is, $\begin{bmatrix} 1 & 1 \\ 1 & 0 \end{bmatrix}^k = \begin{bmatrix} F_{k+1} & F_k \\ F_k & F_{k-1} \end{bmatrix}$. We show that $P(k+1)$ follows from this. Now,

    $$\begin{bmatrix} 1 & 1 \\ 1 & 0 \end{bmatrix}^{k+1} = \begin{bmatrix} 1 & 1 \\ 1 & 0 \end{bmatrix}^k \begin{bmatrix} 1 & 1 \\ 1 & 0 \end{bmatrix} = \begin{bmatrix} F_{k+1} & F_k \\ F_k & F_{k-1} \end{bmatrix}\begin{bmatrix} 1 & 1 \\ 1 & 0 \end{bmatrix} \quad \text{(by the induction hypothesis)}$$
    $$= \begin{bmatrix} F_{k+1} + F_k & F_{k+1} \\ F_k + F_{k-1} & F_k \end{bmatrix} = \begin{bmatrix} F_{k+2} & F_{k+1} \\ F_{k+1} & F_k \end{bmatrix}. \quad \text{(by definition of the Fibonacci sequence)}$$

    Thus $P(k+1)$ follows from $P(k)$. So by the Principle of Mathematical Induction, $P(n)$ is true for all $n \geq 2$.

33. Since $F_1 = 1$, $F_2 = 1$, $F_3 = 2$, $F_4 = 3$, $F_5 = 5$, $F_6 = 8$, $F_7 = 13$, $\ldots$ our conjecture is that $F_n \geq n$, for all $n \geq 5$. Let $P(n)$ denote the statement that $F_n \geq n$.

    Step 1 $P(5)$ is the statement that $F_5 = 5 \geq 5$, which is clearly true.

    Step 2 Assume that $P(k)$ is true; that is, $F_k \geq k$, for some $k \geq 5$. We want to use this to show that $P(k+1)$ is true. Now, $F_{k+1} = F_k + F_{k-1} \geq k + F_{k-1}$ (by the induction hypothesis) $\geq k + 1$ (because $F_{k-1} \geq 1$). Thus $P(k+1)$ follows from $P(k)$. So by the Principle of Mathematical Induction, $P(n)$ is true for all $n \geq 5$.

35. (a) $P(n) = n^2 - n + 11$ is prime for all n. This is false as the case for $n = 11$ demonstrates: $P(11) = 11^2 - 11 + 11 = 121$, which is not prime since $11^2 = 121$.

    (b) $n^2 > n$, for all $n \geq 2$. This is true. Let $P(n)$ denote the statement that $n^2 > n$.

Step 1 $P(2)$ is the statement that $2^2 = 4 > 2$, which is clearly true.

Step 2 Assume that $P(k)$ is true; that is, $k^2 > k$. We want to use this to show that $P(k+1)$ is true. Now $(k+1)^2 = k^2 + 2k + 1$. Using the induction hypothesis (to replace k^2), we have $k^2 + 2k + 1 > k + 2k + 1 = 3k + 1 > k + 1$, since $k \geq 2$. Therefore, $(k+1)^2 > k + 1$, which is exactly $P(k+1)$. Thus $P(k+1)$ follows from $P(k)$. So by the Principle of Mathematical Induction, $P(n)$ is true for all n.

(c) $2^{2n+1} + 1$ is divisible by 3, for all $n \geq 1$. This is true. Let $P(n)$ denote the statement that $2^{2n+1} + 1$ is divisible by 3.

Step 1 $P(1)$ is the statement that $2^3 + 1 = 9$ is divisible by 3, which is clearly true.

Step 2 Assume that $P(k)$ is true; that is, $2^{2k+1} + 1$ is divisible by 3. We want to use this to show that $P(k+1)$ is true. Now, $2^{2(k+1)+1} + 1 = 2^{2k+3} + 1 = 4 \cdot 2^{2k+1} + 1$ $= (3+1)2^{2k+1} + 1 = 3 \cdot 2^{2k+1} + (2^{2k+1} + 1)$, which is divisible by 3 since $2^{2k+1} + 1$ is divisible by 3 by the induction hypothesis, and $3 \cdot 2^{2k+1}$ is clearly divisible by 3. Thus $P(k+1)$ follows from $P(k)$. So by the Principle of Mathematical Induction, $P(n)$ is true for all n.

(d) The statement $n^3 \geq (n+1)^2$ for all $n \geq 2$ is false. The statement fails when $n = 2$: $2^3 = 8 < (2+1)^2 = 9$.

(e) $n^3 - n$ is divisible by 3, for all $n \geq 2$. This is true. Let $P(n)$ denote the statement that $n^3 - n$ is divisible by 3.

Step 1 $P(2)$ is the statement that $2^3 - 2 = 6$ is divisible by 3, which is clearly true.

Step 2 Assume that $P(k)$ is true; that is, $k^3 - k$ is divisible by 3. We want to use this to show that $P(k+1)$ is true. Now, $(k+1)^3 - (k+1) = k^3 + 3k^2 + 3k + 1 - (k+1) = k^3 + 3k^2 + 2k = k^3 - k + 3k^2 + 2k + k = (k^3 - k) + 3(k^2 + k)$. The term $k^3 - k$ is divisible by 3 by our induction hypothesis, and the term $3(k^2 + k)$ is clearly divisible by 3. Thus $(k+1)^3 - (k+1)$ is divisible by 3, which is exactly $P(k+1)$. So by the Principle of Mathematical Induction, $P(n)$ is true for all n.

(f) $n^3 - 6n^2 + 11n$ is divisible by 6, for all $n \geq 1$. This is true. Let $P(n)$ denote the statement that $n^3 - 6n^2 + 11n$ is divisible by 6.

Step 1 $P(1)$ is the statement that $(1)^3 - 6(1)^2 + 11(1) = 6$ is divisible by 6, which is clearly true.

Step 2 Assume that $P(k)$ is true; that is, $k^3 - 6k^2 + 11k$ is divisible by 6. We show that $P(k+1)$ is then also true. Now, $(k+1)^3 - 6(k+1)^2 + 11(k+1)$ $= k^3 + 3k^2 + 3k + 1 - 6k^2 - 12k - 6 + 11k + 11 = k^3 - 3k^2 + 2k + 6$ $= k^3 - 6k^2 + 11k + (3k^2 - 9k + 6) = (k^3 - 6k^2 + 11k) + 3(k^2 - 3k + 2)$ $= (k^3 - 6k^2 + 11k) + 3(k-1)(k-2)$. In this last expression, the first term is divisible by 6 by our induction hypothesis. The second term is also divisible by 6. To see this, notice that $(k-1)$ and $(k-2)$ are consecutive natural numbers, and so one of them must be even (divisible by 2). Since 3 also appears in this second term, it follows that this term is divisible by 2 and 3 and so is divisible by 6. Thus $P(k+1)$ follows from $P(k)$. So by the Principle of Mathematical Induction, $P(n)$ is true for all n.

Exercises 12.6

1. $(x + y)^6 = x^6 + 6x^5y + 15x^4y^2 + 20x^3y^3 + 15x^2y^4 + 6xy^5 + y^6$

3. $\left(x + \dfrac{1}{x}\right)^4 = x^4 + 4x^3 \cdot \dfrac{1}{x} + 6x^2\left(\dfrac{1}{x}\right)^2 + 4x\left(\dfrac{1}{x}\right)^3 + \left(\dfrac{1}{x}\right)^4 = x^4 + 4x^2 + 6 + \dfrac{4}{x^2} + \dfrac{1}{x^4}$

5. $(x - 1)^5 = x^5 - 5x^4 + 10x^3 - 10x^2 + 5x - 1$

7. $(x^2y - 1)^5 = (x^2y)^5 - 5(x^2y)^4 + 10(x^2y)^3 - 10(x^2y)^2 + 5x^2y - 1$
$= x^{10}y^5 - 5x^8y^4 + 10x^6y^3 - 10x^4y^2 + 5x^2y - 1$

9. $(2x - 3y)^3 = (2x)^3 - 3(2x)^23y + 3 \cdot 2x(3y)^2 - (3y)^3 = 8x^3 - 36x^2y + 54xy^2 - 27y^3$

11. $\left(\dfrac{1}{x} - \sqrt{x}\right)^5 = \left(\dfrac{1}{x}\right)^5 - 5\left(\dfrac{1}{x}\right)^4\sqrt{x} + 10\left(\dfrac{1}{x}\right)^3 x - 10\left(\dfrac{1}{x}\right)^2 x\sqrt{x} + 5\left(\dfrac{1}{x}\right)x^2 - x^2\sqrt{x}$

$= \dfrac{1}{x^5} - \dfrac{5}{x^{7/2}} + \dfrac{10}{x^2} - \dfrac{10}{x^{1/2}} + 5x - x^{5/2}$

13. $\dbinom{6}{4} = \dfrac{6!}{4!\,2!} = \dfrac{6 \cdot 5 \cdot 4!}{2 \cdot 1 \cdot 4!} = 15$

15. $\dbinom{100}{98} = \dfrac{100!}{98!\,2!} = \dfrac{100 \cdot 99 \cdot 98!}{98! \cdot 2 \cdot 1} = 4950$

17. $\dbinom{3}{1}\dbinom{4}{2} = \dfrac{3!}{1!\,2!}\dfrac{4!}{2!\,2!} = \dfrac{3 \cdot 2! \cdot 4 \cdot 3 \cdot 2!}{1 \cdot 2! \cdot 2 \cdot 1 \cdot 2!} = 18$

19. $\dbinom{5}{0} + \dbinom{5}{1} + \dbinom{5}{2} + \dbinom{5}{3} + \dbinom{5}{4} + \dbinom{5}{5} = (1 + 1)^5 = 2^5 = 32$

21. $(x + 2y)^4 = \dbinom{4}{0}x^4 + \dbinom{4}{1}x^3 \cdot 2y + \dbinom{4}{2}x^2 \cdot 4y^2 + \dbinom{4}{3}x \cdot 8y^3 + \dbinom{4}{4}16y^4$
$= x^4 + 8x^3y + 24x^2y^2 + 32xy^3 + 16y^4$

23. $\left(1 + \dfrac{1}{x}\right)^6 = \dbinom{6}{0}1^6 + \dbinom{6}{1}1^5\left(\dfrac{1}{x}\right) + \dbinom{6}{2}1^4\left(\dfrac{1}{x}\right)^2 + \dbinom{6}{3}1^3\left(\dfrac{1}{x}\right)^3 + \dbinom{6}{4}1^2\left(\dfrac{1}{x}\right)^4$

$+ \dbinom{6}{5}1\left(\dfrac{1}{x}\right)^5 + \dbinom{6}{6}\left(\dfrac{1}{x}\right)^6 = 1 + \dfrac{6}{x} + \dfrac{15}{x^2} + \dfrac{20}{x^3} + \dfrac{15}{x^4} + \dfrac{6}{x^5} + \dfrac{1}{x^6}$

25. The first three terms in the expansion of $(x + 2y)^{20}$ are $\dbinom{20}{0}x^{20} = x^{20}$, $\dbinom{20}{1}x^{19} \cdot 2y = 40x^{19}y$,

and $\dbinom{20}{2}x^{18} \cdot (2y)^2 = 760x^{18}y^2$.

27. The last two terms in the expansion of $(a^{2/3} + a^{1/3})^{25}$ are $\dbinom{25}{24}a^{2/3} \cdot (a^{1/3})^{24} = 25a^{26/3}$, and

$\dbinom{25}{25}a^{25/3} = a^{25/3}$.

29. The middle term in the expansion of $(x^2+1)^{18}$ occurs when both terms are raised to the 9th power. So this term is $\binom{18}{9}(x^2)^9 1^9 = 48{,}620x^{18}$.

31. The 24th term in the expansion of $(a+b)^{25}$ is $\binom{25}{23}a^2 b^{23} = 300a^2 b^{23}$.

33. The 100th term in the expansion of $(1+y)^{100}$ is $\binom{100}{99}1^1 \cdot y^{99} = 100y^{99}$.

35. The term that contains x^4 in the expansion of $(x+2y)^{10}$ has exponent $r=4$. So this term is $\binom{10}{4}x^4 \cdot (2y)^{10-4} = 13{,}440x^4 y^6$.

37. The rth term is $\binom{12}{r}a^r(b^2)^{12-r} = \binom{12}{r}a^r b^{24-2r}$. Thus the term that contains b^8 occurs where
$24 - 2r = 8 \quad \Leftrightarrow \quad r = 8$. So the term is $\binom{12}{8}a^8 b^8 = 495a^8 b^8$.

39. $x^4 + 4x^3 y + 6x^2 y^2 + 4xy^3 + y^4 = (x+y)^4$.

41. $8a^3 + 12a^2 b + 6ab^2 + b^3 = \binom{3}{0}(2a)^3 + \binom{3}{1}(2a)^2 b + \binom{3}{2}2ab^2 + \binom{3}{3}b^3 = (2a+b)^3$.

43. $\dfrac{(x+h)^3 - x^3}{h} = \dfrac{x^3 + 3x^2 h + 3xh^2 + h^3 - x^3}{h} = \dfrac{3x^2 h + 3xh^2 + h^3}{h} = \dfrac{h(3x^2 + 3xh + h^2)}{h}$
$= 3x^2 + 3xh + h^2$.

45. $(1.01)^{100} = (1+0.01)^{100}$. Now the 1st term in the expansion is $\binom{100}{0}1^{100} = 1$, and the 2nd term is $\binom{100}{1}1^{99}(0.01) = 1$, and the 3rd term is $\binom{100}{2}1^{98}(0.01)^2 = 0.495$. Now each term is non-negative, so $(1.01)^{100} = (1+0.01)^{100} > 1 + 1 + .0.495 > 2$. Thus $(1.01)^{100} > 2$.

47. $\binom{n}{1} = \dfrac{n!}{1!\,(n-1)!} = \dfrac{n(n-1)!}{1(n-1)!} = \dfrac{n}{1} = n$. $\binom{n}{n-1} = \dfrac{n!}{(n-1)!\,1!} = \dfrac{n(n-1)!}{(n-1)!\,1} = n$.
Therefore, $\binom{n}{1} = \binom{n}{n-1} = n$.

49. (a) $\binom{n}{r-1} + \binom{n}{r} = \dfrac{n!}{(r-1)!\,[n-(r-1)]!} + \dfrac{n!}{r!\,(n-r)!}$.

(b) $\dfrac{n!}{(r-1)!\,[n-(r-1)]!} + \dfrac{n!}{r!\,(n-r)!} = \dfrac{r \cdot n!}{r \cdot (r-1)!\,(n-r+1)!} + \dfrac{(n-r+1)\,n!}{r!\,(n-r+1)(n-r)!}$
$= \dfrac{r \cdot n!}{r!\,(n-r+1)!} + \dfrac{(n-r+1)\,n!}{r!\,(n-r+1)!}$. Thus a common denominator is $r!\,(n-r+1)!$.

(c) Therefore, using the results of parts (a) and (b),
$\binom{n}{r-1} + \binom{n}{r} = \dfrac{n!}{(r-1)!\,[n-(r-1)]!} + \dfrac{n!}{r!\,(n-r)!} = \dfrac{r \cdot n!}{r!\,(n-r+1)!} + \dfrac{(n-r+1)\,n!}{r!\,(n-r+1)!}$
$= \dfrac{r \cdot n! + (n-r+1)n!}{r!\,(n-r+1)!} = \dfrac{n! \cdot (r+n-r+1)}{r!\,(n-r+1)!} = \dfrac{n! \cdot (n+1)}{r!\,(n+1-r)!}$

$$= \frac{(n+1)!}{r\,!(n+1-r)!} = \binom{n+1}{r}.$$

51. Notice that $(100!)^{101} = (100!)^{100} \cdot 100!$ and $(101!)^{100} = (101 \cdot 100!)^{100} = 101^{100} \cdot (100!)^{100}$. Now, $100! = 1 \cdot 2 \cdot 3 \cdot 4 \cdots 99 \cdot 100$ and $101^{100} = 101 \cdot 101 \cdot 101 \cdots 101$. Thus each of these last two expressions consists of 100 factors multiplied together, and since each factor in the product for 101^{100} is larger than each factor in the product for $100!$, it follows that $100! < 101^{100}$. Thus $(100!)^{100} \cdot 100! < (100!)^{100} \cdot 101^{100}$. So, $(100!)^{101} < (101!)^{100}$.

53. $0 = 0^n = (-1+1)^n$

$$= \binom{n}{0}(-1)^0(1)^n + \binom{n}{1}(-1)^1(1)^{n-1} + \binom{n}{2}(-1)^2(1)^{n-2} + \cdots + \binom{n}{n}(-1)^n(1)^0$$

$$= \binom{n}{0} - \binom{n}{1} + \binom{n}{2} - \cdots + (-1)^k\binom{n}{k} + \cdots + (-1)^n\binom{n}{n}.$$

Review Exercises for Chapter 12

1. $a_n = \dfrac{n^2}{n+1}$. Then $a_1 = \dfrac{1^2}{1+1} = \dfrac{1}{2}$; $a_2 = \dfrac{2^2}{2+1} = \dfrac{4}{3}$; $a_3 = \dfrac{3^2}{3+1} = \dfrac{9}{4}$; $a_4 = \dfrac{4^2}{4+1} = \dfrac{16}{5}$; and $a_{10} = \dfrac{10^2}{10+1} = \dfrac{100}{11}$.

3. $a_n = \dfrac{(-1)^n + 1}{n^3}$. Then $a_1 = \dfrac{(-1)^1 + 1}{1^3} = 0$; $a_2 = \dfrac{(-1)^2 + 1}{2^3} = \dfrac{2}{8} = \dfrac{1}{4}$; $a_3 = \dfrac{(-1)^3 + 1}{3^3} = 0$; $a_4 = \dfrac{(-1)^4 + 1}{4^3} = \dfrac{2}{64} = \dfrac{1}{32}$; and $a_{10} = \dfrac{(-1)^{10} + 1}{10^3} = \dfrac{1}{500}$.

5. $a_n = \dfrac{(2n)!}{2^n n!}$. Then $a_1 = \dfrac{(2 \cdot 1)!}{2^1 \cdot 1!} = 1$; $a_2 = \dfrac{(2 \cdot 2)!}{2^2 \cdot 2!} = 3$; $a_3 = \dfrac{(2 \cdot 3)!}{2^3 \cdot 3!} = \dfrac{6 \cdot 5 \cdot 4}{8} = 15$; $a_4 = \dfrac{(2 \cdot 4)!}{2^4 \cdot 4!} = \dfrac{8 \cdot 7 \cdot 6 \cdot 5}{16} = 105$; and $a_{10} = \dfrac{(2 \cdot 10)!}{2^{10} \cdot 10!} = 654{,}729{,}075$.

7. $a_n = a_{n-1} + 2n - 1$ and $a_1 = 1$. Then $a_2 = a_1 + 4 - 1 = 4$; $a_3 = a_2 + 6 - 1 = 9$; $a_4 = a_3 + 8 - 1 = 16$; $a_5 = a_4 + 10 - 1 = 25$; $a_6 = a_5 + 12 - 1 = 36$; and $a_7 = a_6 + 14 - 1 = 49$.

9. $a_n = a_{n-1} + 2a_{n-2}$, $a_1 = 1$ and $a_2 = 3$. Then $a_3 = a_2 + 2a_1 = 5$; $a_4 = a_3 + 2a_2 = 11$; $a_5 = a_4 + 2a_3 = 21$; $a_6 = a_5 + 2a_4 = 43$; and $a_7 = a_6 + 2a_5 = 85$.

11. $5, 5.5, 6, 6.5, \ldots$. Since $5.5 - 5 = 6 - 5.5 = 6.5 - 6 = 0.5$, this is an arithmetic sequence with $a_1 = 5$ and $d = 0.5$. Then $a_5 = a_4 + 0.5 = 7$.

13. $\sqrt{2}, 2\sqrt{2}, 3\sqrt{2}, 4\sqrt{2}, \ldots$. Since $2\sqrt{2} - \sqrt{2} = 3\sqrt{2} - 2\sqrt{2} = 4\sqrt{2} - 3\sqrt{2} = \sqrt{2}$, this is an arithmetic sequence with $a_1 = \sqrt{2}$ and $d = \sqrt{2}$. Then $a_5 = a_4 + \sqrt{2} = 4\sqrt{2} + \sqrt{2} = 5\sqrt{2}$.

15. $t - 3, t - 2, t - 1, t, \ldots$. Since $(t - 2) - (t - 3) = (t - 1) - (t - 2) = t - (t - 1) = 1$, this is an arithmetic sequence with $a_1 = t - 3$ and $d = 1$. Then $a_5 = a_4 + 1 = t + 1$.

17. $\frac{3}{4}, \frac{1}{2}, \frac{1}{3}, \frac{2}{9}, \ldots$. Since $\dfrac{\frac{1}{2}}{\frac{3}{4}} = \dfrac{\frac{1}{3}}{\frac{1}{2}} = \dfrac{\frac{2}{9}}{\frac{1}{3}} = \frac{2}{3}$, this is a geometric sequence with $a_1 = \frac{3}{4}$ and $r = \frac{2}{3}$. Then $a_5 = a_4 \cdot r = \frac{2}{9} \cdot \frac{2}{3} = \frac{4}{27}$.

19. $3, 6i, -12, -24i, \ldots$. Since $\dfrac{6i}{3} = 2i$, $\dfrac{-12}{6i} = \dfrac{-2}{i} = \dfrac{-2i}{i^2} = 2i$, $\dfrac{-24i}{-12} = 2i$, this is a geometric sequence with common ratio $r = 2i$.

21. $a_6 = 17 = a + 5d$ and $a_4 = 11 = a + 3d$. Then, $a_6 - a_4 = 17 - 11 \iff (a + 5d) - (a + 3d) = 6 \iff 6 = 2d \iff d = 3$. Substituting into $11 = a + 3d$ gives $11 = a + 3 \cdot 3$, and so $a = 2$. Thus $a_2 = a + (2 - 1)d = 2 + 3 = 5$.

23. $a_3 = 9$ and $r = \frac{3}{2}$. Then $a_5 = a_3 \cdot r^2 = 9 \cdot \left(\frac{3}{2}\right)^2 = \frac{81}{4}$.

25. Let a_{mc} denote the term of the geometric series that is the frequency of middle C. Then $a_{mc} = 256$ and $a_{mc+1} = 512$. Since this is a geometric sequence, $r = \frac{512}{256} = 2$, and so $a_{mc-2} = \dfrac{a_{mc}}{r^2} = \dfrac{256}{2^2} = 64$.

27. Let a_n be the number of bacteria in the dish at the end of $5n$ seconds. So, $a_0 = 3$, $a_1 = 3 \cdot 2$, $a_2 = 3 \cdot 2^2$, $a_3 = 3 \cdot 2^3$, ... Then, clearly, a_n is a geometric sequence with $r = 2$ and $a = 3$. Thus at the end of $60 = 5(12)$ seconds, the number of bacteria is $a_{12} = 3 \cdot 2^{12} = 12{,}288$.

29. Suppose that the common ratio in the sequence $a_1, a_2, a_3, \ldots$ is r. Also, suppose that the common ratio in the sequence $b_1, b_2, b_3, \ldots$ is s. Then $a_n = a_1 r^{n-1}$ and $b_n = b_1 s^{n-1}$, $n = 1, 2, 3, \ldots$. Thus $a_n b_n = a_1 r^{n-1} \cdot b_1 s^{n-1} = (a_1 b_1)(rs)^{n-1}$. So the sequence $a_1 b_1, a_2 b_2, a_3 b_3, \ldots$ is geometric with first term $a_1 b_1$ and common ratio rs.

31. (a) $6, x, 12, \ldots$ is arithmetic if $x - 6 = 12 - x$ $\Leftrightarrow$ $2x = 18$ $\Leftrightarrow$ $x = 9$.

    (b) $6, x, 12, \ldots$ is geometric if $\dfrac{x}{6} = \dfrac{12}{x}$ $\Leftrightarrow$ $x^2 = 72$ $\Leftrightarrow$ $x = \pm 6\sqrt{2}$.

33. $\displaystyle\sum_{k=3}^{6} (k+1)^2 = (3+1)^2 + (4+1)^2 + (5+1)^2 + (6+1)^2 = 16 + 25 + 36 + 49 = 126$

35. $\displaystyle\sum_{k=1}^{6} (k+1)2^{k-1} = 2 \cdot 2^0 + 3 \cdot 2^1 + 4 \cdot 2^2 + 5 \cdot 2^3 + 6 \cdot 2^4 + 7 \cdot 2^5 =$
    $2 + 6 + 16 + 40 + 96 + 224 = 384$

37. $\displaystyle\sum_{k=1}^{10} (k-1)^2 = 0^2 + 1^2 + 2^2 + 3^2 + 4^2 + 5^2 + 6^2 + 7^2 + 8^2 + 9^2$

39. $\displaystyle\sum_{k=1}^{50} \frac{3^k}{2^{k+1}} = \frac{3}{2^2} + \frac{3^2}{2^3} + \frac{3^3}{2^4} + \frac{3^4}{2^5} + \cdots + \frac{3^{49}}{2^{50}} + \frac{3^{50}}{2^{51}}$

41. $3 + 6 + 9 + 12 + \cdots + 99 = 3(1) + 3(2) + 3(3) + \cdots + 3(33) = \displaystyle\sum_{k=1}^{33} 3k$

43. $1 \cdot 2^3 + 2 \cdot 2^4 + 3 \cdot 2^5 + 4 \cdot 2^6 + \cdots + 100 \cdot 2^{102} = (1)2^{(1)+2} + (2)2^{(2)+2} + (3)2^{(3)+2} + (4)2^{(4)+2}$
    $+ \cdots + (100)2^{(100)+2} = \displaystyle\sum_{k=1}^{100} k \cdot 2^{k+2}$

45. $1 + 0.9 + (0.9)^2 + \cdots + (0.9)^5$ is a geometric series with $a = 1$ and $r = \frac{0.9}{1} = 0.9$. Thus, the sum of the series is $S_6 = \dfrac{1 - (0.9)^6}{1 - 0.9} = \dfrac{1 - 0.531441}{0.1} = 4.68559$.

47. $\sqrt{5} + 2\sqrt{5} + 3\sqrt{5} + \cdots + 100\sqrt{5}$ is an arithmetic series with $a = \sqrt{5}$ and $d = \sqrt{5}$. Then $100\sqrt{5} = a_n = \sqrt{5} + \sqrt{5}(n-1)$ $\Leftrightarrow$ $n = 100$. So the sum is $S_{100} = \frac{100}{2}\left(\sqrt{5} + 100\sqrt{5}\right) = 50(101\sqrt{5}) = 5050\sqrt{5}$.

49. $\displaystyle\sum_{n=0}^{6} 3 \cdot (-4)^n$ is a geometric series with $a = 3$, $r = -4$, and $n = 7$. Therefore, the sum of the series is $S_7 = 3 \cdot \dfrac{1 - (-4)^7}{1 - (-4)} = \dfrac{3}{5}(1 + 4^7) = 9831$.

51. We have an arithmetic sequence with $a = 7$ and $d = 3$. Then $S_n = 325 = \dfrac{n}{2}[2a + (n-1)d]$
    $= \dfrac{n}{2}[14 + 3(n-1)] = \dfrac{n}{2}(11 + 3n)$ $\Leftrightarrow$ $650 = 3n^2 + 11n$ $\Leftrightarrow$ $(3n + 50)(n - 13) = 0$
    $\Leftrightarrow$ $n = 13$ (because $n = -\frac{50}{3}$ is inadmissible). Thus, 13 terms must be added.

53. This is a geometric sequence with $a = 2$ and $r = 2$. Then

$$S_{15} = 2 \cdot \frac{1 - 2^{15}}{1 - 2} = 2(2^{15} - 1) = 65{,}534, \text{ and so the total number of ancestors is 65,534.}$$

55. $A = 10{,}000$, $i = 0.03$, and $n = 4$. Thus, $10{,}000 = R\dfrac{(1.03)^4 - 1}{0.03} \quad \Leftrightarrow \quad R = \dfrac{10{,}000 \cdot 0.03}{(1.03)^4 - 1}$

   $= \$2390.27$.

57. $1 - \frac{2}{5} + \frac{4}{25} - \frac{8}{125} + \cdots$ is a geometric series with $a = 1$ and $r = -\frac{2}{5}$. Therefore, the sum is

   $$S = \frac{a}{1 - r} = \frac{1}{1 - \left(-\frac{2}{5}\right)} = \frac{5}{7}.$$

59. $1 + \dfrac{1}{3^{1/2}} + \dfrac{1}{3} + \dfrac{1}{3^{3/2}} + \cdots$ is an infinite geometric series with $a = 1$ and $r = \dfrac{1}{\sqrt{3}}$. Thus, the sum is

   $$S = \frac{1}{1 - \left(\frac{1}{\sqrt{3}}\right)} = \frac{\sqrt{3}}{\sqrt{3} - 1} = \frac{1}{2}(3 + \sqrt{3}).$$

61. Let $P(n)$ denote the statement that $1 + 4 + 7 + \cdots + (3n - 2) = \dfrac{n(3n - 1)}{2}$.

   <u>Step 1</u> $P(1)$ is the statement that $1 = \dfrac{1[3(1) - 1]}{2} = \dfrac{1 \cdot 2}{2}$, which is true.

   <u>Step 2</u> Assume that $P(k)$ is true; that is, $1 + 4 + 7 + \cdots + (3k - 2) = \dfrac{k(3k - 1)}{2}$. We want to use

   this to show that $P(k + 1)$ is true. Now,

   $1 + 4 + 7 + 10 + \cdots + (3k - 2) + [3(k + 1) - 2] =$

   $\dfrac{k(3k - 1)}{2} + 3k + 1 =$ induction hypothesis

   $\dfrac{k(3k - 1)}{2} + \dfrac{6k + 2}{2} = \dfrac{3k^2 - k + 6k + 2}{2} =$

   $\dfrac{3k^2 + 5k + 2}{2} = \dfrac{(k + 1)(3k + 2)}{2} = \dfrac{(k + 1)[3(k + 1) - 1]}{2}.$

   Thus, $P(k + 1)$ follows from $P(k)$. So by the Principle of Mathematical Induction, $P(n)$ is
   true for all n.

63. Let $P(n)$ denote the statement that $\left(1 + \dfrac{1}{1}\right)\left(1 + \dfrac{1}{2}\right)\left(1 + \dfrac{1}{3}\right) \cdots \cdot \left(1 + \dfrac{1}{n}\right) = n + 1$.

   <u>Step 1</u> $P(1)$ is the statement that $\left(1 + \dfrac{1}{1}\right) = 1 + 1$, which is clearly true.

   <u>Step 2</u> Assume that $P(k)$ is true; that is, $\left(1 + \dfrac{1}{1}\right)\left(1 + \dfrac{1}{2}\right)\left(1 + \dfrac{1}{3}\right) \cdots \cdot \left(1 + \dfrac{1}{k}\right) = k + 1$. We

   want to use this to show that $P(k + 1)$ is true. Now,

   $\left(1 + \dfrac{1}{1}\right)\left(1 + \dfrac{1}{2}\right)\left(1 + \dfrac{1}{3}\right) \cdots \cdot \left(1 + \dfrac{1}{k}\right)\left(1 + \dfrac{1}{k+1}\right) =$

   $\left[\left(1 + \dfrac{1}{1}\right)\left(1 + \dfrac{1}{2}\right)\left(1 + \dfrac{1}{3}\right) \cdots \cdot \left(1 + \dfrac{1}{k}\right)\right]\left(1 + \dfrac{1}{k+1}\right) =$

   $(k + 1)\left(1 + \dfrac{1}{k+1}\right) = (k + 1) + 1.$ induction hypothesis

   Thus, $P(k + 1)$ follows from $P(k)$. So by the Principle of Mathematical Induction, $P(n)$ is
   true for all n.

65. $a_{n+1} = 3a_n + 4$ and $a_1 = 4$. Let $P(n)$ denote the statement that $a_n = 2 \cdot 3^n - 2$.

Step 1 $P(1)$ is the statement that $a_1 = 2 \cdot 3^1 - 2 = 4$, which is clearly true.

Step 2 Assume that $P(k)$ is true; that is, $a_k = 2 \cdot 3^k - 2$. We want to use this to show that $P(k+1)$ is true. Now,

$$\begin{aligned}
a_{k+1} &= 3a_k + 4 & \text{definition of } a_{k+1} \\
&= 3(2 \cdot 3^k - 2) + 4 & \text{induction hypothesis} \\
&= 2 \cdot 3^{k+1} - 6 + 4 \\
&= 2 \cdot 3^{k+1} - 2.
\end{aligned}$$

Thus $P(k+1)$ follows from $P(k)$. So by the Principle of Mathematical Induction, $P(n)$ is true for all n.

67. Let $P(n)$ denote the statement that $n! > 2^n$, for all natural numbers $n \geq 4$.

Step 1 $P(4)$ is the statement that $4! = 24 > 2^4 = 16$, which is true.

Step 2 Assume that $P(k)$ is true; that is, $k! > 2^k$. We want to use this to show that $P(k+1)$ is true. Now,

$$\begin{aligned}
(k+1)! &= (k+1)k! \\
&> (k+1) \cdot 2^k & \text{induction hypothesis} \\
&> 2 \cdot 2^k & \text{because } k+1 > 2 \text{ (for } k \geq 4\text{)} \\
&= 2^{k+1}.
\end{aligned}$$

Thus $P(k+1)$ follows from $P(k)$. So by the Principle of Mathematical Induction, $P(n)$ is true for all $n \geq 4$.

69. $\binom{10}{2} + \binom{10}{6} = \dfrac{10!}{2!\,8!} + \dfrac{10!}{6!\,4!} = \dfrac{10 \cdot 9}{2} + \dfrac{10 \cdot 9 \cdot 8 \cdot 7}{4 \cdot 3 \cdot 2} = 45 + 210 = 255.$

71. $\displaystyle\sum_{k=0}^{8} \binom{8}{k}\binom{8}{8-k} = 2\binom{8}{0}\binom{8}{8} + 2\binom{8}{1}\binom{8}{7} + 2\binom{8}{2}\binom{8}{6} + 2\binom{8}{3}\binom{8}{5} + \binom{8}{4}\binom{8}{4}$

$= 2 + 2 \cdot 8^2 + 2 \cdot (28)^2 + 2 \cdot (56)^2 + (70)^2 = 12{,}870.$

73. $(2x + y)^4 = \binom{4}{0}(2x)^4 + \binom{4}{1}(2x)^3 y + \binom{4}{2}(2x)^2 y^2 + \binom{4}{3} \cdot 2xy^3 + \binom{4}{4}y^4$

$= 16x^4 + 32x^3 y + 24x^2 y^2 + 8xy^3 + y^4.$

75. The first three terms in the expansion of $(b^{-2/3} + b^{1/3})^{20}$ are $\binom{20}{0}(b^{-2/3})^{20} = b^{-40/3}$;

$\binom{20}{1}(b^{-2/3})^{19}(b^{1/3}) = 20b^{-37/3}$; and $\binom{20}{2}(b^{-2/3})^{18}(b^{1/3})^2 = 190b^{-34/3}.$

Chapter 12 Test

1. $a_1 = 1^2 - 1 = 0$; $a_2 = 2^2 - 1 = 3$; $a_3 = 3^2 - 1 = 8$; $a_4 = 4^2 - 1 = 15$; and $a_{10} = 10^2 - 1 = 99$.

2. $d = 76 - 80 = -4$. So $a_4 = a_3 + (-4) = 72 + (-4) = 68$ and
    $a_n = a_1 + (n-1)d = 80 + (n-1)(-4) = 84 - 4n$.

3. Since $a_n = a_1 r^{n-1}$, we substitute $a_1 = 25$, $a_4 = \frac{1}{5}$, and $n = 4$ and solve for r. This gives $\frac{1}{5} = 25 \, r^3$
    $\Leftrightarrow$ $r^3 = \frac{1}{125} = \frac{1}{5^3}$ $\Leftrightarrow$ $r = \frac{1}{5}$. So the common ratio is $\frac{1}{5}$ and $a_5 = a_1 r^4 = 25\left(\frac{1}{5}\right)^4 = \frac{1}{25}$.

4. (a) If $a_1, a_2, a_3, \ldots$ is an arithmetic sequence, then the sequence $a_1^2, a_2^2, a_3^2, \ldots$ is also arithmetic.
        This statement is FALSE. For a counterexample, consider $a_1 = 1$, $a_2 = 2$, $a_3 = 3 \ldots$. Then
        $a_1, a_2, a_3, \ldots$ is an arithmetic sequence with $a = 1$ and $d = 1$. However, the sequence a_1^2, a_2^2,
        $a_3^2, \ldots$ becomes $1, 4, 9, \ldots$, which is NOT arithmetic because there is no common difference.

    (b) If $a_1, a_2, a_3, \ldots$ is a geometric sequence, then the sequence $a_1^2, a_2^2, a_3^2, \ldots$ is also geometric.
        This statement is TRUE. Let the common ratio for the geometric series $a_1, a_2, a_3, \ldots$ be r so
        that $a_n = a_1 r^{n-1}$, $n = 1, 2, 3, \ldots$. Then $a_n^2 = (a_1 r^{n-1})^2 = (a_1^2)(r^2)^{n-1}$. Therefore, the
        sequence $a_1^2, a_2^2, a_3^2, \ldots$ is geometric with common ratio r^2.

5. (a) The nth partial sum an arithmetic sequence is $S_n = \dfrac{n}{2}(a + a_n)$ or $S_n = \dfrac{n}{2}[2a + (n-1)d]$.

    (b) $a_1 = 10$ and $a_{10} = 2$. Then $S_{10} = \frac{10}{2}(10 + 2) = 60$.

    (c) Since $S_{10} = \frac{10}{2}(2 \cdot 10 + 9d) = 60$, then $5(20 + 9d) = 60$ $\Leftrightarrow$ $d = -\frac{8}{9}$, and so the common
        difference is $-\frac{8}{9}$. Then $a_{100} = a + 99d = 10 - \frac{8}{9} \cdot 99 = -78$.

6. (a) The nth partial sum a geometric sequence is $S_n = a\dfrac{1 - r^n}{1 - r}$, where $r \neq 1$.

    (b) The geometric series $\dfrac{1}{3} + \dfrac{2}{3^2} + \dfrac{2^2}{3^3} + \dfrac{2^3}{3^4} + \cdots + \dfrac{2^9}{3^{10}}$ has $a = \frac{1}{3}$, $r = \frac{2}{3}$, and $n = 10$. So

    $$S_{10} = \frac{1}{3} \cdot \frac{1 - \left(\frac{2}{3}\right)^{10}}{1 - \left(\frac{2}{3}\right)} = \frac{1}{3} \cdot 3\left(1 - \frac{1024}{59{,}049}\right) = \frac{58{,}025}{59{,}049}.$$

7. The infinite geometric series $1 + \dfrac{1}{2^{1/2}} + \dfrac{1}{2} + \dfrac{1}{2^{3/2}} + \cdots$ has $a = 1$ and $r = 2^{-1/2} = \dfrac{1}{\sqrt{2}}$. Thus,
    $S = \dfrac{1}{1 - \left(\frac{1}{\sqrt{2}}\right)} = \dfrac{\sqrt{2}}{\sqrt{2} - 1} = \dfrac{\sqrt{2}}{\sqrt{2} - 1} \cdot \dfrac{\sqrt{2} + 1}{\sqrt{2} + 1} = 2 + \sqrt{2}$.

8. Let $P(n)$ denote the statement that $1^2 + 2^2 + 3^2 + \cdots + n^2 = \dfrac{n(n+1)(2n+1)}{6}$.

    <u>Step 1</u> Show that $P(1)$ is true. But $P(1)$ says that $1^2 = \frac{1 \cdot 2 \cdot 3}{6}$, which is true.

    <u>Step 2</u> Assume that $P(k)$ is true; that is, $1^2 + 2^2 + 3^2 + \cdots + k^2 = \dfrac{k(k+1)(2k+1)}{6}$. We want to
    use this to show that $P(k+1)$ is true. Now,

$$1^2 + 2^2 + 3^2 + \cdots + k^2 + (k+1)^2 = \frac{k(k+1)(2k+1)}{6} + (k+1)^2 \qquad \text{induction hypothesis}$$

$$= \frac{k(k+1)(2k+1) + 6(k+1)^2}{6}$$

$$= \frac{(k+1)(2k^2 + k) + (6k+6)(k+1)}{6}$$

$$= \frac{(k+1)(2k^2 + k + 6k + 6)}{6}$$

$$= \frac{(k+1)(2k^2 + 7k + 6)}{6}$$

$$= \frac{(k+1)(k+2)(2k+3)}{6}$$

$$= \frac{(k+1)[(k+1) + 1][2(k+1) + 1]}{6}.$$

Thus $P(k+1)$ follows from $P(k)$. So by the Principle of Mathematical Induction, $P(n)$ is true for all n.

9. (a) $\displaystyle\sum_{n=1}^{5}(1 - n^2) = (1 - 1^2) + (1 - 2^2) + (1 - 3^2) + (1 - 4^2) + (1 - 5^2)$

$$= 0 - 3 - 8 - 15 - 24 = -50$$

(b) $\displaystyle\sum_{n=3}^{6}(-1)^n 2^{n-2} = (-1)^3 2^{3-2} + (-1)^4 2^{4-2} + (-1)^5 2^{5-2} + (-1)^6 2^{6-2}$

$$= -2 + 4 - 8 + 16 = 10$$

10. $\displaystyle (2x + y^2)^5 = \binom{5}{0}(2x)^5 + \binom{5}{1}(2x)^4 y^2 + \binom{5}{2}(2x)^3 (y^2)^2 + \binom{5}{3}(2x)^2 (y^2)^3 + \binom{5}{4}(2x)(y^2)^4$

$\displaystyle + \binom{5}{5}(y^2)^5 = 32x^5 + 80x^4 y^2 + 80x^3 y^4 + 40x^2 y^6 + 10xy^8 + y^{10}.$

11. $a_{n+2} = (a_n)^2 - a_{n+1}$, $a_1 = 1$ and $a_2 = 1$. Then $a_3 = a_1^2 - a_2 = 1^2 - 1 = 0$,
$a_4 = a_2^2 - a_3 = 1^2 - 0 = 1$, and $a_5 = a_3^2 - a_4 = 0^2 - 1 = -1$.

Focus on Problem Solving

1. By dividing the successive gnomons into squares with side 1, we see that the sum of their areas is $1 + 3 + 5 + \cdots + (2n - 1) = n^2$, because the total area is the area of an $n \times n$ square.

3. Using the formula for the sum of cubes on page 635, we have
$$1^3 + 2^3 + \cdots + (n - 1)^3 + n^3 = (1 + 2 + \cdots + n)^2$$
$$1^3 + 2^3 + \cdots + (n - 1)^3 = (1 + 2 + \cdots + (n - 1))^2.$$

    Subtracting gives $n^3 = (1 + 2 + \cdots + n)^2 - (1 + 2 + \cdots + (n - 1))^2$.

5. Since all rows have the same sum, S, and there are n rows, the total sum is nS. But this total sum must be $1 + 2 + 3 + \cdots + n^2 = \dfrac{n^2(n^2 + 1)}{2}$ (using the formula for the sum of an arithmetic series or the formula in Example 2 in Section 9.5). Therefore, $nS = \dfrac{n^2(n^2 + 1)}{2} \Leftrightarrow S = \dfrac{n(n^2 + 1)}{2}$.

7. (a) At each stage, each side is replaced by 4 shorter sides, each of length $\frac{1}{3}$ having the length of a side at the previous stage. Writing S_0 and L_0 for the number of sides and the length of the side of the initial figure, we have

$$
\begin{array}{ll}
S_0 = 3 & L_0 = 1 \\
S_1 = 3 \cdot 4 & L_1 = \frac{1}{3} \\
S_2 = 3 \cdot 4^2 & L_2 = \frac{1}{3^2} \\
S_3 = 3 \cdot 4^3 & L_3 = \frac{1}{3^3} \\
\vdots & \vdots
\end{array}
$$

In general, we have $S_n = 3 \cdot 4^n$ and $L_n = \dfrac{1}{3^n}$. The area of each of the small triangles that are added at a given stage is $\frac{1}{9}$th of the area of the triangle added at the preceding stage. Let a be the area of the original triangle. Then the area a_n of each of the small triangles added at stage n is $a_n = a \cdot \dfrac{1}{9^n} = \dfrac{a}{9^n}$. Since a small triangle is added to each side at every stage, it follows that the total area A_n added to the figure at the nth stage is
$$A_n = S_{n-1} \cdot a_n = 3 \cdot 4^{n-1} \cdot \frac{a}{9^n} = a \cdot \frac{4^{n-1}}{3^{2n-1}}.$$
Writing A_0 for the area of the original triangle, we see that the total area enclosed by the snowflake curve is
$$A = A_0 + A_1 + A_2 + A_3 + \cdots = a + a \cdot \tfrac{1}{3} + a \cdot \tfrac{4}{3^3} + a \cdot \tfrac{4^2}{3^5} + a \cdot \tfrac{4^3}{3^7} + \cdots$$
$$= a + \frac{a}{3}\left(1 + \frac{4}{3^2} + \frac{4^2}{3^4} + \frac{4^3}{3^6} + \cdots\right).$$
The sum in the brackets is an infinite geometric series with the first term 1 and the common ratio $\frac{4}{3^2}$. Thus $1 + \frac{4}{3^2} + \frac{4^2}{3^4} + \frac{4^3}{3^6} + \cdots = \frac{1}{1-(4/3^2)} = \frac{9}{5}$. So the area enclosed by the snowflake curve is $A = a + \frac{a}{3} \cdot \frac{9}{5} = \frac{8}{5}a$. It remains to find the area, a, of the original triangle. This area can be easily found using $Area = \frac{1}{2}(base)(height)$ $= \frac{1}{2} \cdot 1 \cdot \cos 60° = \frac{1}{2} \cdot \frac{\sqrt{3}}{2} = \frac{\sqrt{3}}{4}$. Thus $a = \frac{\sqrt{3}}{4}$. Finally, $A = \frac{8}{5}a = \frac{8}{5} \cdot \frac{\sqrt{3}}{4} = \frac{2\sqrt{3}}{5}$.

(b)   At each stage, each side is replaced by 4 shorter sides, each having length $\frac{1}{3}$ of the length of a side at the previous stage. In general, we have $S_n = 3 \cdot 4^n$ and $L_n = \dfrac{1}{3^n}$. Thus the length of the

perimeter at the nth stage is $P_n = S_n \cdot L_n = 3 \cdot 4^n \cdot \dfrac{1}{3^n} = \dfrac{4^n}{3^{n-1}}$. But P_n is a geometric sequence with $r = \frac{4}{3} > 1$, so $P_n \to \infty$ as $n \to \infty$.

9.   (a)   The number of 1×1 squares is 6^2, the number of 2×2 squares is 5^2, the number of 3×3 squares is 4^2, the number of 4×4 squares is 3^2, the number of 5×5 squares is 2^2, and there is just one 6×6 square. So the total number of squares is $1^2 + 2^2 + 3^2 + 4^2 + 5^2 + 6^2 = 91$. For an $n \times n$ grid, the number is $1^2 + 2^2 + 3^2 + \cdots + n^2 = \dfrac{n(n+1)(2n+1)}{6}$. (See Exercise 4 in Section 9.5.)

(b)   A rectangle is determined by 2 vertical lines and 2 horizontal lines. For an $n \times n$ grid, the 2 vertical lines can be chosen in $C(n+1, 2)$ ways and the horizontal lines can also be chosen in $C(n+1, 2)$ ways (see Section 10.2 for the definition of $C(n, r)$). So the number of rectangles is $C(n+1, 2) \cdot C(n+1, 2) = \left[\dfrac{n(n+1)}{2} \right]^2$. For a 6×6 grid, the number is $\left(\dfrac{6 \cdot 7}{2} \right)^2 = 441$.

Chapter Thirteen
Exercises 13.1

1. By the Fundamental Counting Principle, the number of possible single-scoop ice cream cones is $\left(\begin{smallmatrix} \text{number of ways to} \\ \text{choose the flavor} \end{smallmatrix}\right) \cdot \left(\begin{smallmatrix} \text{number of ways to} \\ \text{choose the type of cone} \end{smallmatrix}\right) = 4 \cdot 3 = 12.$

3. By the Fundamental Counting Principle, the possible number of 3-letter words is $\left(\begin{smallmatrix} \text{number of ways to} \\ \text{choose the 1}^{\text{st}} \text{ letter} \end{smallmatrix}\right) \cdot \left(\begin{smallmatrix} \text{number of ways to} \\ \text{choose the 2}^{\text{nd}} \text{ letter} \end{smallmatrix}\right) \cdot \left(\begin{smallmatrix} \text{number of ways to} \\ \text{choose the 3}^{\text{rd}} \text{ letter} \end{smallmatrix}\right).$

   (a) Since repetitions are allowed, we have 4 choices for each letter. Thus there are $4 \cdot 4 \cdot 4 = 64$ words.

   (b) Since repetitions are *not* allowed, we have 4 choices for the 1st letter, 3 choices for the 2nd letter, and 2 choices for the 3rd letter. Thus there are $4 \cdot 3 \cdot 2 = 24$ words.

5. Since there are four choices for each of the five questions, by the Fundamental Counting Principle there are $4 \cdot 4 \cdot 4 \cdot 4 \cdot 4 = 1024$ different ways the test can be completed.

7. Since a runner can only finish once, there are no repetitions. And since we are assuming that there is no tie, the number of different finishes is $\left(\begin{smallmatrix} \text{number of ways to} \\ \text{choose the 1}^{\text{st}} \text{ runner} \end{smallmatrix}\right) \cdot \left(\begin{smallmatrix} \text{number of ways to} \\ \text{choose the 2}^{\text{nd}} \text{ runner} \end{smallmatrix}\right) \cdot \left(\begin{smallmatrix} \text{number of ways to} \\ \text{choose the 3}^{\text{rd}} \text{ runner} \end{smallmatrix}\right) \cdot$ $\left(\begin{smallmatrix} \text{number of ways to} \\ \text{choose the 4}^{\text{th}} \text{ runner} \end{smallmatrix}\right) \cdot \left(\begin{smallmatrix} \text{number of ways to} \\ \text{choose the 5}^{\text{th}} \text{ runner} \end{smallmatrix}\right) = 5 \cdot 4 \cdot 3 \cdot 2 \cdot 1 = 120.$

9. Since there are 6 main courses, there are 6 ways to choose a main course. Likewise, there are 8 drinks and 3 desserts so there are 8 ways to choose a drink and 3 ways to choose a dessert. So the number of different meals consisting of a main course, a drink, and a dessert is $\left(\begin{smallmatrix} \text{number of ways to} \\ \text{choose the main course} \end{smallmatrix}\right) \cdot \left(\begin{smallmatrix} \text{number of ways to} \\ \text{choose a drink} \end{smallmatrix}\right) \cdot \left(\begin{smallmatrix} \text{number of ways to} \\ \text{choose a dessert} \end{smallmatrix}\right) = (6)(8)(3) = 144.$

11. By the Fundamental Counting Principle, the number of different routes from town A to town D via towns B and C is $\left(\begin{smallmatrix} \text{number of routes} \\ \text{from A to B} \end{smallmatrix}\right) \cdot \left(\begin{smallmatrix} \text{number of routes} \\ \text{from B to C} \end{smallmatrix}\right) \cdot \left(\begin{smallmatrix} \text{number of routes} \\ \text{from C to D} \end{smallmatrix}\right) = (4)(5)(6) = 120.$

13. The number of possible sequences of heads and tails when a coin is flipped 5 times is $\left(\begin{smallmatrix} \text{number of possible} \\ \text{outcomes on the 1}^{\text{st}} \text{ flip} \end{smallmatrix}\right) \cdot \left(\begin{smallmatrix} \text{number of possible} \\ \text{outcomes on the 2}^{\text{nd}} \text{ flip} \end{smallmatrix}\right) \cdot \left(\begin{smallmatrix} \text{number of possible} \\ \text{outcomes on the 3}^{\text{rd}} \text{ flip} \end{smallmatrix}\right) \cdot \left(\begin{smallmatrix} \text{number of possible} \\ \text{outcomes on the 4}^{\text{th}} \text{ flip} \end{smallmatrix}\right) \cdot$ $\left(\begin{smallmatrix} \text{number of possible} \\ \text{outcomes on the 5}^{\text{th}} \text{ flip} \end{smallmatrix}\right) = (2)(2)(2)(2)(2) = 2^5 = 32.$ (Here there are only two choices, heads or tails, for each flip.)

15. Since there are six different faces on each die, the number of possible outcomes when a red die and a blue die and a white die are rolled is $\left(\begin{smallmatrix} \text{number of possible} \\ \text{outcomes on the red die} \end{smallmatrix}\right) \cdot \left(\begin{smallmatrix} \text{number of possible} \\ \text{outcomes on the blue die} \end{smallmatrix}\right) \cdot \left(\begin{smallmatrix} \text{number of possible} \\ \text{outcomes on the white die} \end{smallmatrix}\right) = (6)(6)(6) = 6^3 = 216.$

17. The number of possible skirt-blouse-shoe outfits is $\left(\begin{smallmatrix} \text{number of ways} \\ \text{to choose a skirt} \end{smallmatrix}\right) \cdot \left(\begin{smallmatrix} \text{number of ways} \\ \text{to choose a blouse} \end{smallmatrix}\right) \cdot \left(\begin{smallmatrix} \text{number of ways} \\ \text{to choose shoes} \end{smallmatrix}\right) = (5)(8)(12) = 480.$

19. The number of possible ID numbers of one letter followed by two digits is $\left(\begin{smallmatrix} \text{number of ways} \\ \text{to choose a letter} \end{smallmatrix}\right) \cdot \left(\begin{smallmatrix} \text{number of ways} \\ \text{to choose a digit} \end{smallmatrix}\right) \cdot \left(\begin{smallmatrix} \text{number of ways} \\ \text{to choose a digit} \end{smallmatrix}\right) = (26)(10)(10) = 2600.$ Since $2600 < 2844$, it is not possible to give each of the company's employees a different ID number using this scheme.

21. The number of different California license plates possible is $\left(\begin{smallmatrix} \text{number of ways to} \\ \text{choose a nonzero digit} \end{smallmatrix}\right) \cdot \left(\begin{smallmatrix} \text{number of ways} \\ \text{to choose 3 letters} \end{smallmatrix}\right) \cdot \left(\begin{smallmatrix} \text{number of ways} \\ \text{to choose 3 digits} \end{smallmatrix}\right) = (9)(26^3)(10^3) = 158{,}184{,}000.$

23. There are two possible ways to answer each question on a true-false test. Thus the possible number of different ways to complete a true-false test containing 10 questions is

$$\binom{\text{number of ways}}{\text{to answer question 1}} \cdot \binom{\text{number of ways}}{\text{to answer question 2}} \cdot \ldots \cdot \binom{\text{number of ways}}{\text{to answer question 10}} = 2^{10} = 1024.$$

25. The number of different classifications possible is

$$\binom{\text{number of ways}}{\text{to choose a major}} \cdot \binom{\text{number of ways to}}{\text{choose a minor}} \cdot \binom{\text{number of ways}}{\text{to choose a year}} \cdot \binom{\text{number of ways}}{\text{to choose a sex}} = (32)(32)(4)(2) = 8192. \text{ To}$$

see that there are 32 ways to choose a minor, we start with 32 fields, subtract the major (you can not major and minor in the same subject), and then add one for the possibility of *NO MINOR*.

27. The number of possible license plates of two letters followed by three digits is

$$\binom{\text{number of ways to}}{\text{choose 2 letters}} \cdot \binom{\text{number of ways}}{\text{to choose 3 digits}} = (26^2)(10^3) = 676{,}000. \text{ Since } 676{,}000 < 8{,}000{,}000, \text{ there will}$$

not be enough different license plates for the state's 8 million registered cars.

29. Since a student can hold only one office, the number of ways that a president, a vice-president and a secretary can be chosen from a class of 30 students is

$$\binom{\text{number of ways}}{\text{to choose a president}} \cdot \binom{\text{number of ways to}}{\text{choose a vice-president}} \cdot \binom{\text{number of ways}}{\text{to choose a secretary}} = (30)(29)(28) = 24{,}360.$$

31. The number of ways a chairman, vice-chairman, and a secretary can be chosen if the chairman must be a Democrat and the vice-chairman must be a Republican is

$$\binom{\text{number of ways to}}{\substack{\text{choose a Dem. chairman} \\ \text{from the 10 Dem.}}} \cdot \binom{\text{number of ways to choose}}{\substack{\text{a Rep. vice-chairman} \\ \text{from the 7 Rep}}} \cdot \binom{\text{number of ways to choose}}{\substack{\text{a secretary from the} \\ \text{remaining 15 members}}} = (10)(7)(15) = 1050.$$

33. The possible number of 5-letter words formed using the letters A, B, C, D, E, F, and G is calculated as follows:

   (a) If repetition of letters is allowed, then there are 7 ways to pick each letter in the word. Thus the number of 5-letter words is

$$\binom{\text{number of ways to}}{\text{choose the 1st letter}} \cdot \binom{\text{number of ways to}}{\text{choose the 2nd letter}} \cdot \ldots \cdot \binom{\text{number of ways to}}{\text{choose the 5th letter}} = 7^5 = 16{,}807.$$

   (b) If no letter can be repeated in a word, then the number of 5-letter words is

$$\binom{\text{number of ways to}}{\text{choose the 1st letter}} \cdot \binom{\text{number of ways to}}{\text{choose the 2nd letter}} \cdot \ldots \cdot \binom{\text{number of ways to}}{\text{choose the 5th letter}} = (7)(6)(5)(4)(3) = 2520.$$

   (c) If each word must begin with the letter A, then the number of 5-letter words (letters can be repeated) is

$$\binom{\text{number of ways to}}{\text{choose the letter A}} \cdot \binom{\text{number of ways to}}{\text{choose the 2nd letter}} \cdot \ldots \cdot \binom{\text{number of ways to}}{\text{choose the 5th letter}} = (1)(7)(7)(7)(7) = 7^4 = 2401.$$

   (d) If the letter C must be in the middle, then the number of 5-letter words (letters can be repeated) is

$$\binom{\text{number of ways to}}{\text{choose the 1st letter}} \cdot \binom{\text{number of ways to}}{\text{choose the 2nd letter}} \cdot \binom{\text{number of ways to}}{\text{choose the letter C}} \cdot \binom{\text{number of ways to}}{\text{choose the 4th letter}} \cdot \binom{\text{number of ways to}}{\text{choose the 5th letter}}$$
$$= (7)(7)(1)(7)(7) = 7^4 = 2401.$$

   (e) If the middle letter must be a vowel, then the number of 5-letter words is $(7)(7)(2)(7)(7)$
$$= 2 \cdot 7^4 = 4802.$$

35. The number of possible variable names is

$$\binom{\text{number of ways to}}{\text{choose a letter}} \cdot \binom{\text{number of ways to choose}}{\text{a letter or a digit}} = (26)(36) = 936.$$

37. The number of ways 4 men and 4 women may be seated in a row of 8 seats is calculated as follows:

(a) If the women are to be seated together and the men together, then the number of ways they can be seated is $\left(\begin{smallmatrix}\text{number of ways to seat the}\\\text{women first and the men second}\end{smallmatrix}\right) + \left(\begin{smallmatrix}\text{number of ways to seat the}\\\text{the men first and the women second}\end{smallmatrix}\right)$

$= (4)(3)(2)(1)(4)(3)(2)(1) + (4)(3)(2)(1)(4)(3)(2)(1) = (4!)(4!) + (4!)(4!) = 576 + 576$
$= 1152.$

(b) If they are to be seated alternately by gender , then the number of ways they can be seated is $\left(\begin{smallmatrix}\text{number of ways to seat them}\\\text{alternately if a woman is first}\end{smallmatrix}\right) + \left(\begin{smallmatrix}\text{number of ways to seat them}\\\text{alternately if a man is first}\end{smallmatrix}\right)$

$= (4)(4)(3)(3)(2)(2)(1)(1) + (4)(4)(3)(3)(2)(2)(1)(1) = 1152.$

39. The number of ways that 8 mathematics books and 3 chemistry books may be placed on a shelf if the math books are to be next to each other and the chemistry books next to each other is

$\left(\begin{smallmatrix}\text{number of ways to place them}\\\text{if the math books are first}\end{smallmatrix}\right) + \left(\begin{smallmatrix}\text{number of ways to place them}\\\text{if the chemistry books are first}\end{smallmatrix}\right) = (8!)(3!) + (3!)(8!) = 483,840.$

41. In order for a number to be odd, the last digit must be odd. In this case, it must be a 1. Thus, the other two digits must be chosen from the three digits 2, 4, and 6. So, the number of 3-digit odd numbers that can be formed using the digits 1, 2, 4, and 6, if no digit may be used more than once is

$\left(\begin{smallmatrix}\text{number of ways}\\\text{to choose a digit}\\\text{from the 3 digits}\end{smallmatrix}\right) \cdot \left(\begin{smallmatrix}\text{number of ways}\\\text{to choose from the}\\\text{2 remaining digits}\end{smallmatrix}\right) \cdot \left(\begin{smallmatrix}\text{number of ways}\\\text{to choose}\\\text{an odd digit}\end{smallmatrix}\right) = (3)(2)(1) = 6.$

43. (a) From Exercise 6 the number of seven-digit phone numbers is 8,000,000. Under the old rules, the possible number of telephone area codes consisting of 3 digits with the given conditions is $(8)(2)(10) = 160$. So the possible number of area code + telephone number combinations under the old rules is $160 \cdot 8,000,000 = 1,280,000,000$.

(b) Under the new rules, where the second digit can be any number, the number of area codes possible is $(8)(10)(10) = 800$. Thus the possible number of area code + telephone number combinations under the new rules is $800 \cdot 8,000,000 = 6,400,000,000$.

(c) Although over $1\frac{1}{4}$ billion telephone numbers may seem to be enough for a population of 300,000,000, so many individuals and businesses now have multiple lines for fax, Internet access, and cellular phones that this number has proven insufficient.

(d) There are now more "new" area codes than "old" area codes. Numerous answers are possible.

Exercises 13.2

1. $P(8,3) = \dfrac{8!}{(8-3)!} = \dfrac{8!}{5!} = 8 \cdot 7 \cdot 6 = 336$

3. $P(11,4) = \dfrac{11!}{(11-4)!} = \dfrac{11!}{7!} = 11 \cdot 10 \cdot 9 \cdot 8 = 7920$

5. $P(100,1) = \dfrac{100!}{(100-1)!} = \dfrac{100!}{99!} = 100$

7. We need the number of ways of selecting three students *in order* for the positions of president, vice president, and secretary from the 15 students in the class. Since a student cannot hold more than one position, this number is $P(15,3) = \dfrac{15!}{(15-3)!} = \dfrac{15!}{12!} = 15 \cdot 14 \cdot 13 = 2730.$

9. Since a person cannot occupy more than one chair at a time, the number of ways of seating 6 people from 10 people in a row of 6 chairs is $P(10,6) = 10 \cdot 9 \cdot 8 \cdot 7 \cdot 6 \cdot 5 = 151{,}200.$

11. The number of ways of selecting 3 objects in order (a 3-letter word) from 6 distinct objects (the 6 letters) assuming that the letters cannot be repeated is $P(6,3) = 6 \cdot 5 \cdot 4 = 120.$

13. The number of ways of selecting 3 objects in order (a 3-digit number) from 4 distinct objects (the 4 digits) with no repetition of the digits is $P(4,3) = 4 \cdot 3 \cdot 2 = 24.$

15. The number of ways of ordering 9 distinct objects (the contestants) is $P(9,9) = 9! = 362{,}880.$ Here a runner cannot finish more than once, so no repetitions are allowed, and order is important.

17. The number of ways of ordering 1000 distinct objects (the contestants) taking 3 at a time is $P(1000,3) = 1000 \cdot 999 \cdot 998 = 997{,}002{,}000.$ We are assuming that a person cannot win more than once, that is, there are no repetitions.

19. We first place Jack in the first seat, and then seat the remaining 4 students. Thus the number of these arrangements is $\left(\substack{\text{number of ways to} \\ \text{seat Jack in the 1}^{\text{st}} \text{ seat}} \right) \cdot \left(\substack{\text{number of ways to seat} \\ \text{the remaining 4 students}} \right) = P(1,1) \cdot P(4,4) = 1! \, 4! = 24.$

21. Here we have 6 letters, of which 3 are A's, 2 are B's, and 1 is a C. Thus the number of distinguishable permutations is $\dfrac{6!}{3! \, 2! \, 1!} = \dfrac{(6)(5)(4)(3!)}{(2)(3!)} = 60.$

23. Here we have 5 letters, of which 2 are A's, 1 is a B, 1 is a C, and 1 is a D. Thus the number of distinguishable permutations is $\dfrac{5!}{2! \, 1! \, 1! \, 1!} = \dfrac{(5)(4)(3)(2!)}{2!} = 60.$

25. Here we have 6 objects, of which 2 are blue marbles and 4 are red marbles. Thus the number of distinguishable permutations is $\dfrac{6!}{2! \, 4!} = \dfrac{(6)(5)(4!)}{2 \cdot 4!} = 15.$

27. The number of distinguishable permutations of 12 objects (the 12 coins), from like groups of size 4 (the pennies), of size 3 (the nickels), of size 2 (the dimes) and of size 3 (the quarters) is $\dfrac{12!}{4! \, 3! \, 2! \, 3!} = 277{,}200.$

29. The number of distinguishable permutations of 12 objects (the 12 ice cream cones) from like groups of size 3 (the vanilla cones), of size 2 (the chocolate cones), of size 4 (the strawberry cones), and of size 5 (the butterscotch cones) is $\dfrac{14!}{3!\,2!\,4!\,5!} = 2{,}522{,}520$.

31. The number of distinguishable permutations of 8 objects (the 8 cleaning tasks) from like groups of size 5, 2, and 1 workers, respectively is $\dfrac{8!}{5!\,2!\,1!} = 168$.

33. $C(8,3) = \dfrac{8!}{3!\,(8-3)!} = \dfrac{8!}{3!\,5!} = \dfrac{8\cdot 7\cdot 6}{3\cdot 2\cdot 1} = 56$

35. $C(11,4) = \dfrac{11!}{4!\,7!} = \dfrac{11\cdot 10\cdot 9\cdot 8}{4\cdot 3\cdot 2\cdot 1} = 330$

37. $C(100,1) = \dfrac{100!}{1!\,99!} = \dfrac{100}{1} = 100$

39. We want the number of ways of choosing a group of three from a group of six. This number is $C(6,3) = \dfrac{6!}{3!\,3!} = 20$.

41. We want the number of ways of choosing a group of six people from a group of ten people. The number of combinations of 10 objects (10 people) taken 6 at a time is $C(10,6) = \dfrac{10!}{6!\,4!} = 210$.

43. We want the number of ways of choosing a group (the 5-card hand) where order of selection is not important. The number of combinations of 52 objects (the 52 cards) taken 5 at a time is $C(52,5) = \dfrac{52!}{5!\,47!} = 2{,}598{,}960$.

45. The order of selection is not important, hence we must calculate the number of combinations of 10 objects (the 10 questions) taken 7 at a time, this gives $C(10,7) = \dfrac{10!}{7!\,3!} = 120$.

47. We assume that the order in which he plays the pieces in the recital is *not* important, so the number of combinations of 12 objects (the 12 pieces) taken 8 at a time is $C(12,8) = \dfrac{12!}{8!\,4!} = 495$.

49. We are just interested in the group of seven students taken from the class of 30 students, not the order in which they are picked. Thus the number is $C(30,7) = \dfrac{30!}{7!\,23!} = 2{,}035{,}800$.

51. We first take Jack out of the class of 30 students and select the 7 students from the remaining 29 students. Thus there are $C(29,7) = \dfrac{29!}{7!\,22!} = 1{,}560{,}780$ ways to pick the 7 students for the field trip.

53. We must count the number of different ways to choose the group of 6 numbers from the 53 numbers. There are $C(53,6) = \dfrac{53!}{6!\,47!} = 22{,}957{,}480$ possible tickets, so it would cost \$22,957,480.

55. (a) The number of ways to select 5 of the 8 objects is $C(8,5) = \dfrac{8!}{5!\,3!} = 56$.

    (b) A set with 8 elements has $2^8 = 256$ subsets.

57. Each subset of toppings constitutes a different way a hamburger can be ordered. Since a set with 10 elements has $2^{10} = 1024$ subsets, there are 1024 different ways to order a hamburger.

59. We pick the two men from the group of ten men, and we pick the two women from the group of ten women. So the number of ways 2 men and 2 women can be chosen is
$$\left(\begin{smallmatrix}\text{number of ways to pick}\\ \text{2 of the 10 men}\end{smallmatrix}\right) \cdot \left(\begin{smallmatrix}\text{number of ways to pick}\\ \text{2 of the 10 women}\end{smallmatrix}\right) = C(10,2) \cdot C(10,2) = (45)(45) = 2025$$

61. The leading and supporting roles are different (order counts), while the extra roles are not (order doesn't count). Also, the male roles must be filled by the male actors and the female roles filled by the female actresses. Thus, number of ways the actors and actresses can be chosen is
$$\left(\begin{smallmatrix}\text{number of ways the}\\ \text{leading and the supporting}\\ \text{actors can be chosen}\end{smallmatrix}\right) \cdot \left(\begin{smallmatrix}\text{number of ways the}\\ \text{leading and supporting}\\ \text{actresses can be chosen}\end{smallmatrix}\right) \cdot \left(\begin{smallmatrix}\text{number of ways to choose}\\ \text{5 of 8 male extras}\end{smallmatrix}\right) \cdot \left(\begin{smallmatrix}\text{number of ways to choose}\\ \text{3 of 10 female extras}\end{smallmatrix}\right)$$
$$= P(10,2) \cdot P(12,2) \cdot C(8,5) \cdot C(10,3) = (90)(132)(56)(120) = 79{,}833{,}600.$$

63. To order a pizza, we must make several choices. First the size (4 choices), the type of crust (2 choices), and then the toppings. Since there are 14 toppings, the number of possible choices is the number of subsets of the 14 toppings, that is 2^{14} choices. So by the Fundamental Counting Principle, the number of possible pizzas is $(4)(2) \cdot 2^{14} = 131{,}072$.

65. We are only interested in selecting a set of three marbles to give to Luke and a set of two marbles to give to Mark, not the order in which we hand out the marbles. Since both $C(10,3) \cdot C(7,2)$ and $C(10,2) \cdot C(8,3)$ count the number of ways this can be done, these numbers must be equal. (Calculating these values shows that they are indeed equal.) In general, if we wish to find two distinct sets of k and r objects selected from n objects ($k + r \leq n$), then we can either first select the k objects from the n objects and then select the r objects from the $n - k$ remaining objects, or we can first select the r objects from the n objects and then the k objects from the $n - r$ remaining objects. Thus $\binom{n}{r} \cdot \binom{n-r}{k} = \binom{n}{k} \cdot \binom{n-k}{r}$.

Exercises 13.3

1. Let H stand for head and T for tails.

   (a) The sample space is $S = \{HH, HT, TH, TT\}$.

   (b) Let E be the event of getting exactly two heads, so $E = \{HH\}$. Then $P(E) = \dfrac{n(E)}{n(S)} = \frac{1}{4}$.

   (c) Let F be the event of getting at least one head. Then $F = \{HH, HT, TH\}$, and
   $$P(F) = \frac{n(F)}{n(S)} = \tfrac{3}{4}.$$

   (d) Let G be the event of getting exactly one head, that is, $G = \{HT, TH\}$. Then
   $$P(G) = \frac{n(G)}{n(S)} = \tfrac{2}{4} = \tfrac{1}{2}.$$

3. (a) Let E be the event of rolling a six. Then $P(E) = \dfrac{n(E)}{n(S)} = \frac{1}{6}$.

   (b) Let F be the event of rolling an even number. Then $F = \{2, 4, 6\}$. So
   $$P(F) = \frac{n(F)}{n(S)} = \tfrac{3}{6} = \tfrac{1}{2}.$$

   (c) Let G be the event of rolling a number greater than 5. Since 6 is the only face greater than 5,
   $$P(G) = \frac{n(G)}{n(S)} = \tfrac{1}{6}.$$

5. (a) Let E be the event of choosing a king. Since a deck has 4 kings, $P(E) = \dfrac{n(E)}{n(S)} = \frac{4}{52} = \frac{1}{13}$.

   (b) Let F be the event of choosing a face card. Since there are 3 face cards per suit and 4 suits,
   $$P(F) = \frac{n(F)}{n(S)} = \tfrac{12}{52} = \tfrac{3}{13}.$$

   (c) Let F be the event of choosing a face card. Then $P(F') = 1 - P(F) = 1 - \frac{3}{13} = \frac{10}{13}$.

7. (a) Let E be the event of selecting a red ball. Since the jar contains 5 red balls,
   $$P(E) = \frac{n(E)}{n(S)} = \tfrac{5}{8}.$$

   (b) Let F be the event of selecting a yellow ball. Since there is only one yellow ball,
   $$P(F') = 1 - P(F) = 1 - \frac{n(F)}{n(S)} = 1 - \tfrac{1}{8} = \tfrac{7}{8}.$$

   (c) Let G be the event of selecting a black ball. Since there are no black balls in the jar,
   $$P(G) = \frac{n(G)}{n(S)} = \tfrac{0}{8} = 0.$$

9. (a) Let E be the event of drawing a red sock. Since 3 pairs are red, the drawer contains 6 red socks, and so $P(E) = \dfrac{n(E)}{n(S)} = \frac{6}{18} = \frac{1}{3}$.

   (b) Let F be the event of drawing another red sock. Since there are 17 socks left of which 5 are red, $P(F) = \dfrac{n(F)}{n(S)} = \frac{5}{17}$.

11. (a) Let E be the event of choosing a "T". Since 3 of the 16 letters are T's, $P(E) = \frac{3}{16}$.

(b) Let F be the event of choosing a vowel. Since there are 6 vowels, $P(F) = \frac{6}{16} = \frac{3}{8}$.

(c) Let F be the event of choosing a vowel. Then $P(F') = 1 - \frac{3}{8} = \frac{5}{8}$.

13. Let E be the event of choosing 5 cards of the same suit. Since there are 4 suits and 13 cards in each suit, $n(E) = 4 \cdot C(13, 5)$. Also by Exercise 43, Section 10.2, $n(S) = C(52, 5)$. Therefore,
$$P(E) = \frac{4 \cdot C(13, 5)}{C(52, 5)} = \frac{5,148}{2,598,960} \approx .00198.$$

15. Let E be the event of dealing a royal flush (ace, king, queen, jack, and 10 of the same suit). Since there is only one such sequence for each suit, there are only 4 royal flushes, so
$$P(E) = \frac{4}{C(52, 5)} = \frac{4}{2,598,960} \approx 1.53908 \times 10^{-6}.$$

17. (a) Let B stand for "boy" and G stand for "girl". Then $S = \{$BBBB, GBBB, BGBB, BBGB, BBBG, GGBB, GBGB, GBBG, BGGB, BGBG, BBGG, BGGG, GBGG, GGBG, GGGB, GGGG$\}$.

(b) Let E be the event that the couple has only boys. Then $E = \{$BBBB$\}$ and $P(E) = \frac{1}{16}$.

(c) Let F be the event that the couple has 2 boys and 2 girls. Then $F = \{$GGBB, GBGB, GBBG, BGGB, BGBG, BBGG$\}$, so $P(F) = \frac{6}{16} = \frac{3}{8}$.

(d) Let G be the event that the couple has 4 children of the same sex. Then $G = \{$BBBB, GGGG$\}$, and $P(G) = \frac{2}{16} = \frac{1}{8}$.

(e) Let H be the event that the couple has at least 2 girls. Then H' is the event that the couple has fewer than two girls. Thus, $H' = \{$BBBB, GBBB, BGBB, BBGB, BBBG$\}$, so $n(H') = 5$, and $P(H) = 1 - P(H') = 1 - \frac{5}{16} = \frac{11}{16}$.

19. Let E be the event that the ball lands in an odd numbered slot. Since there are 18 odd numbers between 1 and 36, $P(E) = \frac{18}{38} = \frac{9}{19}$.

21. Let E be the event of picking the 6 winning numbers. Since there is only one way to pick these,
$$P(E) = \frac{1}{C(49, 6)} = \frac{1}{13,983,816} \approx 7.15 \times 10^{-8}.$$

23. The sample space consist of all possible True-False combinations, so $n(S) = 2^{10}$.

(a) Let E be the event that the student answers all 10 questions correctly. Since there is only one way to answer all 10 questions correctly, $P(E) = \frac{1}{2^{10}} = \frac{1}{1024}$.

(b) Let F be the event that the student answers exactly 7 questions correctly. The number of ways to answer exactly 7 of the 10 questions correctly is the number of ways to choose 7 of the 10 questions, so $n(F) = C(10, 7)$. Therefore, $P(E) = \frac{C(10, 7)}{2^{10}} = \frac{120}{1024} = \frac{15}{128}$.

25. (a) Let E be the event that the monkey types "Hamlet" as his first word. Since "Hamlet" contains 6 letters and there are 48 typewriter keys, $P(E) = \frac{1}{48^6} \approx 8.18 \times 10^{-11}$.

(b) Let F be the event that the monkey types "to be or not to be" as his first words. Since this phrase has 18 characters (including the blanks), $P(F) = \frac{1}{48^{18}} \approx 5.47 \times 10^{-31}$.

27. Let E be the event that the monkey will arrange the 11 blocks to spell "PROBABILITY" as his first word. The number of ways of arranging these blocks is the number of distinguishable permutations

of 11 blocks. Since there are two blocks labeled 'B' and two blocks labeled 'I', the number of distinguishable permutations is $\frac{11!}{2!\,2!}$. Only one of these arrangements spells the word "PROBABILITY". Thus $P(E) = \dfrac{1}{\frac{11!}{2!\,2!}} = \frac{2!\,2!}{11!} \approx 1.00 \times 10^{-7}$.

29. (a) Let E be the event that the pea is tall. Since tall is dominant, $E = \{TT, Tt, tT\}$. So $P(E) = \frac{3}{4}$.

  (b) E' is the event that the pea is short. So $P(E') = 1 - P(E) = 1 - \frac{3}{4} = \frac{1}{4}$.

31. (a) YES, the events are mutually exclusive since a person cannot be both male and female.

  (b) NO, the events are not mutually exclusive since a person can be both tall and blond.

33. (a) YES, the events are mutually exclusive since the number cannot be both even and odd. So $P(E \cup F) = P(E) + P(F) = \frac{3}{6} + \frac{3}{6} = 1$.

  (b) NO, the events are not mutually exclusive since 6 is both even and greater than 4. So $P(E \cup F) = P(E) + P(F) - P(E \cap F) = \frac{3}{6} + \frac{2}{6} - \frac{1}{6} = \frac{2}{3}$.

35. (a) NO, the events E and F are not mutually exclusive since the Jack, Queen, and King of spades are both face cards and spades. So $P(E \cup F) = P(E) + P(F) - P(E \cap F) = \frac{13}{52} + \frac{12}{52} - \frac{3}{52} = \frac{11}{26}$.

  (b) YES, the events E and F are mutually exclusive since the card cannot be both a heart and a spade. So $P(E \cup F) = P(E) + P(F) = \frac{13}{52} + \frac{13}{52} = \frac{1}{2}$.

37. (a) Let E be the event that the spinner stops on red. Since 12 of the regions are red, $P(E) = \frac{12}{16} = \frac{3}{4}$.

  (b) Let F be the event that the spinner stops on an even number. Since 8 of the regions are even-numbered, $P(F) = \frac{8}{16} = \frac{1}{2}$.

  (c) Since 4 of the even-numbered regions are red, $P(E \cup F) = P(E) + P(F) - P(E \cap F)$ $= \frac{3}{4} + \frac{1}{2} - \frac{4}{16} = 1$.

39. Let E be the event that the ball lands in an odd numbered slot and F be the event that it lands in a slot with a number higher than 31. Since there are two odd-numbered slots with numbers greater than 31, $P(E \cup F) = P(E) + P(F) - P(E \cap F) = \frac{18}{38} + \frac{5}{38} - \frac{2}{38} = \frac{21}{38}$.

41. Let E be the event that the committee is all male and F the event it is all female. The sample space is the set of all ways that 5 people can be chosen from the group of 14. These events are mutually exclusive, so $P(E \cup F) = P(E) + P(F) = \dfrac{C(6,5)}{C(14,5)} + \dfrac{C(8,5)}{C(14,5)} = \frac{6+56}{2002} = \frac{31}{1001}$.

43. Let E be the event that the marble is red and F be the event that the number is odd-numbered. Then E' is the event that the marble is blue, and F' is the event that the marble is even-numbered.

  (a) $P(E) = \frac{6}{16} = \frac{3}{8}$

  (b) $P(F) = \frac{8}{16} = \frac{1}{2}$

  (c) $P(E \cup F) = P(E) + P(F) - P(E \cap F) = \frac{6}{16} + \frac{8}{16} - \frac{3}{16} = \frac{11}{16}$

  (d) $P(E' \cup F') = P(E') + P(F') - P(E' \cap F') = \frac{10}{16} + \frac{8}{16} - \frac{5}{16} = \frac{13}{16}$.

45. (a) YES, the first roll does not influence the outcome of the second roll.

(b)   The probability of getting a six on both rolls is $P(E \cap F) = P(E) \cdot P(F) = \left(\frac{1}{6}\right)\left(\frac{1}{6}\right) = \frac{1}{36}$.

47.   (a)   Let E_A and E_B be the event that the respective spinners stop on a purple region. Since these events are independent, $P(E_A \cap E_B) = P(E_A) \cdot P(E_B) = \frac{1}{4} \cdot \frac{2}{8} = \frac{1}{16}$.

(b)   Let F_A and F_B be the event that the respective spinners stop on a blue region. Since these events are independent, $P(F_A \cap F_B) = P(F_A) \cdot P(F_B) = \frac{1}{4} \cdot \frac{1}{8} = \frac{1}{32}$.

49.   Let E be the event of getting a 1 on the first roll, and let F be the event of getting an even number on the second roll. Since these events are independent, $P(E \cap F) = P(E) \cdot P(F) = \frac{1}{6} \cdot \frac{3}{6} = \frac{1}{12}$.

51.   Let E be the event that the player wins on spin 1, and let F be the event that the player wins on spin 2. What happens on the first spin does not influence what happens on the second spin, so the events are independent. Thus, $P(E \cap F) = P(E) \cdot P(F) = \frac{1}{38} \cdot \frac{1}{38} = \frac{1}{1444}$.

53.   Let E, F and G denote the events of rolling two ones on the first, second, and third rolls, respectively, of a pair of dice. The events are independent, so
$P(E \cap F \cap G) = P(E) \cdot P(F) \cdot P(G) = \frac{1}{36} \cdot \frac{1}{36} \cdot \frac{1}{36} = \frac{1}{36^3} \approx 2.14 \times 10^{-5}$.

55.   The probability of getting 2 red balls by picking from jar B is $\left(\frac{5}{7}\right)\left(\frac{4}{6}\right) = \frac{10}{21}$. The probability of getting 2 red balls by picking one ball from each jar is $\left(\frac{3}{7}\right)\left(\frac{5}{7}\right) = \frac{15}{49}$. The probability of getting 2 red balls after putting all balls in one jar is $\left(\frac{8}{14}\right)\left(\frac{7}{13}\right) = \frac{4}{13}$. Hence, picking both balls from jar B gives the greatest probability.

57.   Let E be the event that two of the students have the same birthday. It is easier to consider the complementary event E': no two students have the same birthday. Then
$$P(E') = \frac{\text{number of ways to assign 8 different birthdays}}{\text{number of ways to assign 8 birthdays}} = \frac{P(365, 8)}{365^8}$$
$$= \frac{365 \cdot 364 \cdot 363 \cdot 362 \cdot 361 \cdot 360 \cdot 359 \cdot 358}{365 \cdot 365 \cdot 365 \cdot 365 \cdot 365 \cdot 365 \cdot 365 \cdot 365} \approx 0.92566. \text{ So } P(E) = 1 - P(E') \approx .07434.$$

59.   The sample space for the genders of Mrs. Smith's children is $S_S = \{(b, g), (b, b)\}$, where we list the first born first. The sample space for the genders of Mrs. Jones' children is $S_J = \{(b, g), (g, b), (b, b)\}$. Thus, the probability that Mrs. Smith's other child is a boy is $\frac{1}{2}$, while the probability that Mrs. Jones' other child is a boy is $\frac{1}{3}$. The reason there is a difference is that we *know* that Mrs. Smith's first born is a boy.

Exercises 13.4

1. Mike gets $2 with probability $\frac{1}{2}$ and $1 with probability $\frac{1}{2}$. Thus, $E = (2)\left(\frac{1}{2}\right) + (1)\left(\frac{1}{2}\right) = 1.5$, and so his expected winnings are $1.50 per game.

3. Since the probability of drawing the ace of spades is $\frac{1}{52}$, the expected value of this game is $E = (100)\left(\frac{1}{52}\right) + (-1)\left(\frac{51}{52}\right) = \frac{49}{52} \approx 0.94$. So your expected winnings are $0.94 per game.

5. Since the probability that Carol rolls a six is $\frac{1}{6}$, the expected value of this game is $E = (3)\left(\frac{1}{6}\right) + (0.50)\left(\frac{5}{6}\right) = \frac{5.5}{6} \approx 0.9167$. So Carol expects to win $0.92 per game.

7. Since the probability that the die shows an even number equals the probability that that die shows an odd number, the expected value of this game is $E = (2)\left(\frac{1}{2}\right) + (-2)\left(\frac{1}{2}\right) = 0$. So Tom should expect to break even after playing this game many times.

9. Since it cost 50¢ to play, if you get a silver dollar, you only win $1 - 0.50 = \$.50$. Thus the expected value of this game is $E = (0.50)\left(\frac{2}{10}\right) + (-0.50)\left(\frac{8}{10}\right) = -0.30$. So your expected winnings are $-$0.30 per game. In other words, you should expect to lose $0.30 per game.

11. You can either win $35 or lose $1, so the expected value of this game is $E = (35)\left(\frac{1}{38}\right) + (-1)\left(\frac{37}{38}\right) = -\frac{2}{38} = -0.0526$. Thus the expected value is $-$0.0526 per game.

13. By the rules of the game, a player can win $10, $5, $0 or lose $100. Thus the expected value of this game is $E = (10)\left(\frac{10}{100}\right) + (5)\left(\frac{10}{100}\right) + (-100)\left(\frac{2}{100}\right) + (0)\left(\frac{78}{100}\right) = -0.50$. So the expected winnings per game are $-$0.50.

15. If the stock goes up to $20, she expects to make $20 - \$5 = \$15$. And if the stock falls to $1, then she has lost $5 - \$1 = \$4$. So the expected value of her profit is $E = (15)(0.1) - (4)(0.9) = -2.1$. Thus, her expected profit per share is $-$2.10, that is, she should expect to lose $2.10 per share. She did not make a wise investment.

17. There are $C(49, 6)$ ways to select a group of six numbers from the group of 49 numbers, of which only one is a winning set. Thus the expected value of this game is
$$E = (10^6 - 1)\left(\frac{1}{C(49, 6)}\right) + (-1)\left(1 - \frac{1}{C(49, 6)}\right) \approx -\$0.93.$$

19. Let x be the fair price to pay to play this game. Then the game is fair whenever $E = 0 \Leftrightarrow (13 - x)\left(\frac{4}{52}\right) + (-x)\left(\frac{48}{52}\right) = 0 \Leftrightarrow 52 - 52x = 0 \Leftrightarrow x = 1$. Thus, a fair price to pay to play this game is $1.

Review Exercises for Chapter 13

1. The number of possible outcomes is $\left(\substack{\text{number of outcomes}\\\text{when a coin is tossed}}\right) \cdot \left(\substack{\text{number of outcomes}\\\text{a die is rolled}}\right) \cdot \left(\substack{\text{number of ways}\\\text{to draw a card}}\right)$
    $= (2)(6)(52) = 624.$

3. (a) Order is not important, and there are no repetitions, so the number of different two-element
        subsets is $C(5,2) = \dfrac{5!}{2!\,3!} = \dfrac{5 \cdot 4}{2} = 10.$

    (b) Order is important, and there are no repetitions, so the number of different two-letter words is
        $P(5,2) = \frac{5!}{3!} = 20.$

5. You earn a score of 70% by answering exactly 7 of the 10 questions correctly. The number of
    different ways to answer the questions correctly is $C(10,7) = \dfrac{10!}{7!\,3!} = 120.$

7. You must choose two of the ten questions to omit, and the number of ways of choosing these two
    questions is $C(10,2) = \dfrac{10!}{2!\,8!} = 45.$

9. The maximum number of employees using this security system is
    $\left(\substack{\text{number of choices}\\\text{for the first letter}}\right) \cdot \left(\substack{\text{number of choices}\\\text{for the second letter}}\right) \cdot \left(\substack{\text{number of choices}\\\text{for the third letter}}\right) = (26)(26)(26) = 17{,}576.$

11. We could count the number of ways of choosing 7 of the flips to be HEADS; equivalently we could
    count the number of ways of choosing 3 of the flips to be TAILS. Thus, the number of different
    ways this can occur is $C(10,7) = C(10,3) = \dfrac{10!}{3!\,7!} = 120.$

13. Let x be the number of people in the group. Then $C(x,2) = 10 \quad \Leftrightarrow \quad \dfrac{x!}{2!(x-2)!} = 10 \quad \Leftrightarrow$

    $\dfrac{x!}{(x-2)!} = 20 \quad \Leftrightarrow \quad x(x-1) = 20 \quad \Leftrightarrow \quad x^2 - x - 20 = 0 \quad \Leftrightarrow \quad (x-5)(x+4) = 0 \quad \Leftrightarrow$
    $x = 5$ or $x = -4$. So there are 5 people in this group.

15. A letter can be represented by a sequence of length 1, a sequence of length 2, or a sequence of length
    3. Since each symbol is either a dot or a dash, the possible number of letters is
    $\left(\substack{\text{number of letters}\\\text{using 3 symbols}}\right) + \left(\substack{\text{number of letters}\\\text{using 2 symbols}}\right) + \left(\substack{\text{number of letters}\\\text{using 1 symbol}}\right) = 2^3 + 2^2 + 2 = 14.$

17. (a) Since we cannot choose a major and a minor in the same subject, the number of ways a student
        can select a major and a minor is $P(16,2) = 16 \cdot 15 = 240.$

    (b) Again, since we cannot have repetitions and the order of selection is important, the number of
        ways to select a major, a first minor, and a second minor is $P(16,3) = 16 \cdot 15 \cdot 14 = 3360.$

    (c) When we select a major and 2 minors, the order in which we choose the minors is not
        important. Thus the number of ways to select a major and 2 minors is
        $\left(\substack{\text{number of ways}\\\text{to select a major}}\right) \cdot \left(\substack{\text{number of ways to}\\\text{select two minors}}\right) = 16 \cdot C(15,2) = 16 \cdot 105 = 1680.$

19. Since the letters are distinct, the number of anagrams of the word TRIANGLE is $8! = 40{,}320.$

21. The probability that you select a mathematics book is $\dfrac{\text{number of ways to select}}{\text{number of ways to select a book}} = \dfrac{4}{10}$
$= \dfrac{2}{5} = .4$.

23. (a) $S = \{HHH, HHT, HTH, HTT, THH, THT, TTH, TTT\}$.

    (b) $P(\text{HHH}) = \frac{1}{8}$.

    (c) $P(\text{2 or more heads}) = P(\text{exactly 2 heads}) + P(\text{3 heads}) = \frac{3}{8} + \frac{1}{8} = \frac{4}{8} = \frac{1}{2}$.

    (d) $P(\text{tails of the first toss}) = \frac{4}{8} = \frac{1}{2}$.

25. (a) $P(\text{ace}) = \frac{4}{52} = \frac{1}{13}$.

    (b) Let E be the event the card chosen is an ace, and let F be the event the card chosen is a jack. Then $P(E \cup F) = P(E) + P(F) = \frac{4}{52} + \frac{4}{52} = \frac{2}{13}$.

    (c) Let E be the event the card chosen is an ace, and let F be the event the card chosen is a spade. Then $P(E \cup F) = P(E) + P(F) - P(E \cap F) = \frac{4}{52} + \frac{13}{52} - \frac{1}{52} = \frac{4}{13}$.

    (d) Let E be the event the card chosen is an ace, and let F be the event the card chosen is a red card. Then $P(E \cap F) = \dfrac{n(E \cap F)}{n(S)} = \frac{2}{52} = \frac{1}{26}$.

27. (a) The probability the first die shows some number is 1, and the probability the second die shows the same number is $\frac{1}{6}$. So the probability each die shows the same number is $1 \cdot \frac{1}{6} = \frac{1}{6}$.

    (b) By part (a), the event of showing the same number has probability of $\frac{1}{6}$, and the complement of this event is that the dice show different numbers. Thus the probability that the dice show different numbers is $1 - \frac{1}{6} = \frac{5}{6}$.

29. In the numbers game lottery, there are 1000 possible "winning" numbers.

    (a) The probability that John wins \$500 is $\frac{1}{1000}$.

    (b) There are $P(3,3) = 6$ ways to arrange the digits "1", "5", "9". However, if John wins only \$50, it means that his number 159 was not the winning number. Thus the probability is $\frac{5}{1000} = \frac{1}{200}$.

31. There are 36 possible outcomes in rolling two dice and 6 ways in which both dice show the same numbers, namely, $(1,1), (2,2), (3,3), (4,4), (5,5),$ and $(6,6)$. So the expected value of this game is $E = (5)\left(\frac{6}{36}\right) + (-1)\left(\frac{30}{36}\right) = 0$.

33. Since Mary makes a guess as to the order of ratification of the 13 original states, the number of such guesses is $P(13,13) = 13!$, while the probability that she guesses the correct order is $\frac{1}{13!}$. Thus the expected value is $E = (1{,}000{,}000)\left(\frac{1}{13!}\right) + (0)\left(\frac{13!-1}{13!}\right) = 0.00016$. So Mary's expected winnings are \$0.00016.

35. (a) Since there are only two colors of socks, any 3 socks must contain a matching pair.

    (b) <u>Solution 1:</u> If the two socks drawn form a matching pair then they are either both red or both blue. So $P(\text{choosing a matching pair}) = P(\text{both red or both blue})$

$$= P(\text{both red}) + P(\text{both blue}) = \dfrac{C(20,2)}{C(50,2)} + \dfrac{C(30,2)}{C(50,2)} \approx .51.$$

Solution 2: The complement of choosing a matching pair is choosing one sock of each color.
So $P(\text{choosing a matching pair}) = 1 - P(\text{different colors}) = 1 - \dfrac{C(20, 1) \cdot C(30, 1)}{C(50, 2)}$
$\approx 1 - .49 = .51.$

37. (a) $\left(\begin{array}{c}\text{number of different} \\ \text{zip codes}\end{array}\right) = \left(\begin{array}{c}\text{number of ways to} \\ \text{choose the 1}^{\text{st}}\text{ digit}\end{array}\right) \cdot \left(\begin{array}{c}\text{number of ways to} \\ \text{choose the 2}^{\text{nd}}\text{ digit}\end{array}\right) \cdot \ \cdots \ \cdot \left(\begin{array}{c}\text{number of ways to} \\ \text{choose the 5}^{\text{th}}\text{ digit}\end{array}\right)$
$= 10 \cdot 10 \cdot 10 \cdot 10 \cdot 10 = 10^5 = 100,000.$

(b) Since there are five numbers (0, 1, 6, 8 and 9) that can be read upside down, we have
$\left(\begin{array}{c}\text{number of different} \\ \text{zip codes}\end{array}\right) = \left(\begin{array}{c}\text{number of ways to} \\ \text{choose the 1}^{\text{st}}\text{ digit}\end{array}\right) \cdot \left(\begin{array}{c}\text{number of ways to} \\ \text{choose the 2}^{\text{nd}}\text{ digit}\end{array}\right) \cdot \ \cdots \ \cdot \left(\begin{array}{c}\text{number of ways to} \\ \text{choose the 5}^{\text{th}}\text{ digit}\end{array}\right) = 5^5 = 3125.$

(c) Let E be the event that a zip code can be read upside down. Then by parts (a) and (b),
$P(E) = \dfrac{n(E)}{n(S)} = \dfrac{5^5}{10^5} = \dfrac{1}{32}.$

(d) Suppose a zip code is turned upside down. Then the middle digit remains the middle digit, so it must be a digit that reads the same when turned upside down, that is, a 0, 1 or 8. Also, the last digit becomes the first digit, and the next to last digit becomes the second digit. Thus, once the first two digits are chosen, the last two are determined. Therefore, the number of zip codes that read the same upside down as right side up is
$\left(\begin{array}{c}\text{number of ways to} \\ \text{choose the 1}^{\text{st}}\text{ digit}\end{array}\right) \cdot \left(\begin{array}{c}\text{number of ways to} \\ \text{choose the 2}^{\text{nd}}\text{ digit}\end{array}\right) \cdot \left(\begin{array}{c}\text{number of ways to} \\ \text{choose the 3}^{\text{rd}}\text{ digit}\end{array}\right) \cdot \left(\begin{array}{c}\text{number of ways to} \\ \text{choose the 4}^{\text{th}}\text{ digit}\end{array}\right) \cdot \left(\begin{array}{c}\text{number of ways to} \\ \text{choose the 5}^{\text{th}}\text{ digit}\end{array}\right)$
$= 5 \cdot 5 \cdot 3 \cdot 1 \cdot 1 = 75.$

39. (a) Using the rule for the number of distinguishable combinations, the number of divisors of N is $(7 + 1)(2 + 1)(5 + 1) = 144.$

(b) An even divisor of N must contain 2 as a factor. Thus we place a 2 as one of the factors and count the number of distinguishable combinations of $M = 2^6 3^2 5^5$. So using the rule for the number of distinguishable combinations, the number of even divisors of N is $(6 + 1)(2 + 1)(5 + 1) = 126.$

(c) A divisor is a multiple of 6 if 2 is a factor and 3 is a factor. Thus we place a 2 as one of the factors and a 3 as one of the factors. Then we count the number of distinguishable combinations of $K = 2^6 3^1 5^5$. So using the rule for the number of distinguishable combinations, the number of even divisors of N is $(6 + 1)(1 + 1)(5 + 1) = 84.$

(d) Let E be the event that the divisor is even. Then using parts (a) and (b), $P(E) = \dfrac{n(E)}{n(S)}$
$= \dfrac{126}{144} = \dfrac{7}{8}.$

Chapter 13 Test

1. (a) If repetition is allowed, then each letter of the word can be chosen in 10 ways since there are 10 letters. Thus the number of possible 5 letter words is $10 \cdot 10 \cdot 10 \cdot 10 \cdot 10 = 10^5 = 100{,}000$.

    (b) If repetition is not allowed, then since order is important, we need the number of permutations of 10 objects (the 10 letters) taken 5 at a time. Therefore, the number of possible 5 letter words is $P(10,5) = 10 \cdot 9 \cdot 8 \cdot 7 \cdot 6 = 30{,}240$.

2. There are three choices to be made: one choice each of a main course, a dessert, and a drink. Since a main course can be chosen in one of five ways, a dessert in one of three ways, and a drink in one of four ways, there are $5 \cdot 3 \cdot 4 = 60$ possible ways that a customer could order a meal.

3. We choose the office first. Here order is important, because the offices are different. Thus there are $P(30,3)$ ways to do this. Next we choose the other 5 members from the remaining 27 members. Here order is not important, so there are $C(27,5)$ ways to do this. Therefore the number of ways that the board of directors can be chosen is
    $P(30,3) \cdot C(27,5) = 30 \cdot 29 \cdot 28 \cdot \frac{27!}{5!\,22!} = 1{,}966{,}582{,}800$.

4. There are two choices to be made: choose a road to travel from Ajax to Barrie, and then choose a different road from Barrie to Ajax. Since there are 4 roads joining the two cities, we need the number of permutations of 4 objects (the roads) taken 2 at a time (the road there and the road back). This number is $P(4,2) = 4 \cdot 3 = 12$.

5. A customer must choose a size of pizza and must make a choice of toppings. There are 4 sizes of pizza, and each choice of toppings from the 14 available corresponds to a subset of the 14 objects. Since a set with 14 objects has 2^{14} subsets, the number of different pizzas this parlor offers is $4 \cdot 2^{14} = 65{,}536$.

6. (a) We want the number of ways of arranging 4 distinct objects (the letters L, O, V, E). This is the number of permutations of 4 objects taken 4 at a time. Therefore, the number of anagrams of the word LOVE is $P(4,4) = 4! = 24$.

    (b) We want the number of distinguishable permutations of 6 objects (the letters K, I, S, S, E, S) consisting of three like groups of size 1 and a like group of size 3 (the S's). Therefore, the number of different anagrams of the word KISSES is $\dfrac{6!}{1!\,1!\,1!\,3!} = \dfrac{6!}{3!} = 120$.

7. Let E be the event of choosing 3 men. Then
    $$P(E) = \frac{n(E)}{n(S)} = \frac{\text{number of ways to choose 3 men}}{\text{number of ways to choose 3 people}} = \frac{C(5,3)}{C(15,3)} \approx .022.$$

8. Two dice are rolled. Let E be the event of getting doubles. Since a double may occur in 6 ways,
    $$P(E) = \frac{n(E)}{n(S)} = \frac{6}{36} = \frac{1}{6}.$$

9. One card is drawn from a deck.

    (a) Since there are 26 red cards, the probability that the card is red is $\frac{26}{52} = \frac{1}{2}$.

    (b) Since there are 4 kings, the probability that the card is a king is $\frac{4}{52} = \frac{1}{13}$.

    (c) Since there are 2 red kings, the probability that the card is a red king is $\frac{2}{52} = \frac{1}{26}$.

10. Let R be the event that the ball chosen is red. Let E be the event that the ball chosen is even-numbered.

    (a) Since 5 of the 13 balls are red, $P(R) = \frac{5}{13} \approx .3846$.

    (b) Since 6 of the 13 balls are even-numbered, $P(E) = \frac{6}{13} \approx .4615$.

    (c) $P(R \text{ or } E) = P(R) + P(E) - P(R \cap E) = \frac{5}{13} + \frac{6}{13} - \frac{2}{13} = \frac{9}{13} \approx .6923$.

11. A deck of cards contains 4 aces, 12 face cards, and 36 other cards. So the probability of an ace is $\frac{4}{52} = \frac{1}{13}$, the probability of a face card is $\frac{12}{52} = \frac{3}{13}$, and the probability of a non-ace, non-face card is $\frac{36}{52} = \frac{9}{13}$. Thus the expected value of this game is $E = (10)\left(\frac{1}{13}\right) + (1)\left(\frac{3}{13}\right) + (-.5)\left(\frac{9}{13}\right) = \frac{8.5}{13} \approx 0.654$, that is, about \$0.65.

12. There are 4 students and 12 astrological signs. Let E be the event that at least 2 have the same astrological sign. Then E' is the event that no 2 have the same astrological sign. It is easier to find E'. So $P(E') = \dfrac{\text{number of ways to assign 4 different astrological signs}}{\text{number of ways to assign 4 astrological signs}} = \dfrac{P(12,4)}{12^4} = \dfrac{12 \cdot 11 \cdot 10 \cdot 9}{12 \cdot 12 \cdot 12 \cdot 12} = \dfrac{55}{96}$. Therefore, $P(E) = 1 - P(E') = 1 - \frac{55}{96} = \frac{41}{96} \approx .427$.

Focus on Modeling

1. (a) You should find that with the switching strategy, you win about 90% of the time. The more games you play, the closer to 90% your winning ratio will be.

   (b) The probability that the contestant has selected the winning door to begin with is $\frac{1}{10}$, since there are ten doors and only one is a winner. So the probability that he has selected a losing door is $\frac{9}{10}$. If the contestant switches, he exchanges a losing door for a winning door (and vice versa), so the probability that he loses is now $\frac{1}{10}$, and the probability that he wins is now $\frac{9}{10}$.

3. (a) You should find that A wins about $\frac{7}{8}$ of the time. That is, if you play this game 80 times, A should win approximately 70 times.

   (b) The game will end when either A gets one more "head" or B gets three more "tails". Each toss is independent, and both "heads" and "tails" have probability $\frac{1}{2}$, so we obtain the following probabilities.

Outcome	Probability
H	$\frac{1}{2}$
TH	$\frac{1}{2} \cdot \frac{1}{2} = \frac{1}{4}$
TTH	$\frac{1}{2} \cdot \frac{1}{2} \cdot \frac{1}{2} = \frac{1}{8}$
TTT	$\frac{1}{2} \cdot \frac{1}{2} \cdot \frac{1}{2} = \frac{1}{8}$

   Since A wins for any outcome that ends in "heads", the probability that he wins is $\frac{1}{2} + \frac{1}{4} + \frac{1}{8} = \frac{7}{8}$.

5. With 1000 trials, you should obtain an estimate for π that is reasonably close to 3.1 or 3.2.